Siegmund Brandt

Data Analysis

Statistical and Computational Methods for Scientists and Engineers

Third Edition

Translated by Glen Cowan

With 273 Illustrations

Includes CD-ROM

Springer

Siegmund Brandt
Physics Department
University of Siegen
Siegen, D-57068, Germany

Cover illustration: Determination of mean foot length. Woodcut from Jacob Köbel's "Geometrei" published 1575 in Frankfurt.

Library of Congress Cataloging-in-Publication Data
Brandt, Siegmund.
[Datenanalyse, English]
Data analysis: statistical and computational methods for scientists and engineers / Siegmund Brandt. — 3rd ed.
p. cm.
Includes bibliographical references and index.
ISBN 0-387-98498-4 (alk. paper)
1. Probabilities. 2. Mathematical statistics. I. Title.
QA273.B86213 1998
519.2—dc21 98-11969

Printed on acid-free paper.

Production managed by Terry Kornak; manufacturing supervised by Nancy Wu.
Photocomposed copy prepared by the author.
Printed and bound by Hamilton Printing Co., Rensselaer, NY.
Printed in the United States of America.

9 8 7 6 5 4 3 2 1

ISBN 0-387-98498-4 Springer-Verlag New York Berlin Heidelberg SPIN 10670849

Preface to the Third English Edition

For the present edition large parts of this book have undergone major revision. In this way newer developments could be taken into account, for example, in solving systems of linear equations, minimization, and generation of random numbers. In particular, consideration has been taken of the widespread use of personal computers by giving a greatly increased number of computer programs for all of the problems treated in this book. The programs themselves are printed in the text and in addition are contained on a CD-ROM included with this book.

The basic concept of this book – concise but sufficiently rigorous mathematical treatment and a particular emphasis on applications – has been retained. It is assumed that the reader possesses a sound basic knowledge of differential and integral calculus. Other necessary mathematical tools are briefly introduced, in particular from the calculus of probabilities. The use of vector and matrix notation is essential for the present treatment. With this notation many problems can be formulated in a particularly clear way. An introduction to calculation with vectors and matrices and a collection of related computer routines is given in Appendix A.

The development of the computer routines and corresponding example programs and problems was supported in the framework of a study grant from IBM Germany. These programs have been developed and tested on an IBM RISC/6000 computer and subsequently transported to the IBM PS/2 and IBM PC (or compatible devices). I thank IBM Germany for their generous support.

In the completion of the present edition I was supported by many friends and collaborators to whom I express my heartfelt thanks: Dr. Karin Dahmen, Dr. Martin S. Brandt, Mr. Tilo Stroh, and Mr. Clemens Stupperich helped to develop several of the programs, in particular those for computer graphics. Mr. Tilo Stroh performed the conversion of all FORTRAN programs into C and tested the translated programs. Together with Mr. Ingo Struck he also established an interface with WINDOWS and Linux/XWindow for those compilers which support such an interface. Dr. Glen Cowan, Dr. Helge Meinhard, and Mr. Tilo Stroh critically read the manuscript and provided many corrections and changes. My special thanks go to Dr. Glen Cowan for his careful translation into English and to Mrs. Ute Bender and Mr. Tilo Stroh for their expert work in the computer preparation of the manuscript. Finally I would like to thank readers from earlier editions of this book and its translations for pointing out necessary corrections and desirable additions.

Siegen, April 1997 Siegmund Brandt

Contents

List of Examples

List of Programs

Frequently Used Symbols and Notation

$x, y, \xi, \eta, \ldots$	(ordinary) variable
$\mathbf{x}, \mathbf{y}, \boldsymbol{\xi}, \boldsymbol{\eta}, \ldots$	vector variable
$\mathsf{x}, \mathsf{y}, \ldots$	random variable
$\boldsymbol{\mathsf{x}}, \boldsymbol{\mathsf{y}}, \ldots$	vector of random variables
$A, B, C, \ldots$	matrices
B	bias
$\mathrm{cov}(\mathsf{x}, \mathsf{y})$	covariance
F	variance ratio
$f(x)$	probability density
$F(x)$	distribution function
$E(\mathsf{x}) = \hat{x}$	mean value, expectation value
H	hypothesis
H_0	null hypothesis
L, ℓ	likelihood functions
$L(S_c, \lambda)$	operating characteristic function
$M(S_c, \lambda)$	test function, power (of a test)
M	minimum function, target function
$P(A)$	probability of the event A
Q	sum of squares

s^2, s_x^2	sample variance
S	estimator
S_c	critical region
t	variable of Student's distribution
x_m	most probable value (mode)
$x_{0.5}$	median
x_q	quantile
$\bar{\mathsf{x}}$	sample mean
$\widetilde{x}$	estimator from maximum likelihood or least squares
λ	parameter of a distribution
$\varphi(t)$	characteristic function
$\phi(x)$, $\psi(x)$	probability density and distribution function of the normal distribution
$\phi_0(x)$, $\psi_0(x)$	probability density and distribution function of the standard normal distribution
$\sigma(\mathsf{x}) = \varDelta(\mathsf{x})$	standard deviation
$\sigma^2(\mathsf{x})$	variance
$\varOmega(P)$	inverse function of the normal distribution

1. Introduction

1.1 Typical Problems of Data Analysis

Every branch of experimental science, after passing through an early stage of qualitative description, concerns itself with quantitative studies of the phenomena of interest, i.e., *measurements*. In addition to designing and carrying out the experiment, an important task is the accurate evaluation and *complete* exploitation of the data obtained. Let us list a few typical problems.

1. A study is made of the weight of laboratory animals under the influence of various drugs. After the application of drug A to 25 animals, an average increase of 5 % is observed. Drug B, used on 10 animals, yields a 3 % increase. Is drug A more effective? The *averages* 5 % and 3 % give practically no answer to this question, since the lower value may have been caused by a single animal that lost weight for some unrelated reason. One must therefore study the *distribution* of individual weights and their spread around the average value. Moreover, one has to decide whether the number of test animals used will enable one to differentiate with a certain accuracy between the effects of the two drugs.

2. In experiments on crystal growth it is essential to maintain exactly the ratios of the different components. From a total of 500 crystals, a *sample* of 20 is selected and analyzed. What conclusions can be drawn about the composition of the remaining 480? This problem of sampling comes up, for example, in quality control, reliability tests of automatic measuring devices, and opinion polls.

3. A certain experimental result has been obtained. It must be decided whether it is in contradiction with some predicted theoretical value or with previous experiments. The experiment is used for *hypothesis testing*.

4. A general law is known to describe the dependence of measured variables, but parameters of this law must be obtained from experiment. In

radioactive decay, for example, the number N of atoms that decay per second decreases exponentially with time: $N(t) = \text{const} \cdot \exp(-\lambda t)$. One wishes to determine the decay constant λ and its measurement error by making maximal use of a series of measured values $N_1(t_1)$, $N_2(t_2), \ldots$. One is concerned here with the problem of *fitting* a function containing unknown parameters to the data and the determination of the numerical values of the parameters and their errors.

5. From these examples some of the aspects of data analysis become apparent. We see in particular that the outcome of an experiment is not uniquely determined by the experimental procedure but is also subject to chance: it is a *random variable*. This stochastic tendency is either rooted in the nature of the experiment (test animals are necessarily different, radioactivity is a stochastic phenomenon), or it is a consequence of the inevitable uncertainties of the experimental equipment, i.e., measurement errors. It is often useful to simulate with a computer the variable or stochastic characteristics of the experiment in order to get an idea of the expected uncertainties of the results before carrying out the experiment itself. This *simulation* of random quantities on a computer is called the *Monte Carlo method*, so named in reference to games of chance.

1.2 On the Structure of this Book

The basis for using random quantities is the *calculus of probabilities*. The most important concepts and rules for this are collected in Chapter 2. *Random variables* are introduced in Chapter 3. Here one considers distributions of random variables, and parameters are defined to characterize the distributions, such as the expectation value and variance. Special attention is given to the interdependence of several random variables. In addition, *transformations* between different sets of variables are considered; this forms the basis of *error propagation*.

Generating random numbers on a computer and the *Monte Carlo method* are the topics of Chapter 4. In addition to methods for generating random numbers, a well-tested program and also examples for generating arbitrarily distributed random numbers are given. Use of the Monte Carlo method for problems of integration and simulation is introduced by means of examples. The method is also used to generate simulated data with measurement errors, with which the data analysis routines of later chapters can be demonstrated.

In Chapter 5 we introduce a number of distributions which are of particular interest in applications. This applies especially to the Gaussian or normal distribution, whose properties are studied in detail.

In practice a distribution must be determined from a finite number of observations, i.e., from a *sample*. Various cases of sampling are considered in Chapter 6. Computer programs are presented for a first rough numerical treatment and graphical display of empirical data. Functions of the sample, i.e., of the individual observations, can be used to estimate the parameters characterizing the distribution. The requirements that a good estimate should satisfy are derived. At this stage the quantity χ^2 is introduced. This is the sum of the squares of the deviations between observed and expected values and is therefore a suitable indicator of the goodness-of-fit.

The *maximum-likelihood method*, discussed in Chapter 7, forms the core of modern statistical analysis. It allows one to construct estimators with optimum properties. The method is discussed for the single and multiparameter cases and illustrated in a number of examples. Chapter 8 is devoted to *hypothesis testing*. It contains the most commonly used F, t, and χ^2 tests and in addition outlines the general points of test theory.

The *method of least squares*, which is perhaps the most widely used statistical procedure, is the subject of Chapter 9. The special cases of direct, indirect, and constrained measurements, often encountered in applications, are developed in detail before the general case is discussed. Programs and examples are given for all cases. Every least-squares problem, and in general every problem of maximum likelihood, involves determining the minimum of a function of several variables. In Chapter 10 various methods are discussed in detail, by which such a minimization can be carried out. The relative efficiency of the procedures is shown by means of programs and examples.

The analysis of variance (Chapter 11) can be considered as an extension of the F-test. It is widely used in biological and medical research to study the dependence, or rather to test the independence, of a measured quantity from various experimental conditions expressed by other variables. For several variables rather complex situations can arise. Some simple numerical examples are calculated using a computer program.

Linear and polynomial *regression*, the subject of Chapter 12, is a special case of the least-squares method and has therefore already been treated in Chapter 9. Before the advent of computers, usually only linear least-squares problems were tractable. A special terminology, still used, was developed for this case. It seemed therefore justified to devote a special chapter to this subject. At the same time it extends the treatment of Chapter 9. For example the determination of confidence intervals for a solution and the relation between regression and analysis of variance are studied. A general program for polynomial regression is given and its use is shown in examples.

In the last chapter the elements of time series analysis are introduced. This method is used if data are given as a function of a controlled variable (usually time) and no theoretical prediction for the behavior of the data as a function of the controlled variable is known. It is used to try to reduce the statistical fluctuation of the data without destroying the genuine dependence on the controlled variable. Since the computational work in time series analysis is rather involved, a computer program is also given.

The field of data analysis, which forms the main part of this book, can be called *applied mathematical statistics*. In addition, wide use is made of other branches of mathematics and of specialized computer techniques. This material is contained in the appendices.

In Appendix A titled "Matrix Calculations", the most important concepts and methods from *linear algebra* are summarized. Of central importance are procedures for solving systems of linear equations, in particular the singular value decomposition, which provides the best numerical properties. Appendix A contains an extensive collection of computer routines for vector and matrix calculations.

Necessary concepts and relations of *combinatorics* are compiled in Appendix B. The numerical value of functions of mathematical statistics must often be computed. The necessary formulas and programs are contained in Appendix C. Many of these functions are related to the Euler *gamma function* and like it can only be computed with approximation techniques. In Appendix D formulas and programs for gamma and related functions are given. Appendix E contains further utility routines for numerical differentiation and determination of zeros.

The *graphical representation* of measured data and their errors and in many cases also of a fitted function is of special importance in data analysis. In Appendix F a complete graphics programming package is presented. The most important concepts of computer graphics are introduced and all of the necessary explanations for using the programming package are given.

The set of appendices is concluded with a *collection of formulas* in Appendix H, which should facilitate reference to the most important equations, and with a short collection of statistical tables in Appendix I. Although all of the tabulated values can be computed (and in fact were computed) with the programs of Appendix C, it is easier to look up one or two values from the tables than to use a computer.

Near the end of many chapters there is a *Problems* section. These problems can be solved with paper and pencil. They should help the reader to understand the basic concepts and theorems. In some cases also simple numerical calculations must be carried out. Under the heading *Hints and Solutions* at the end of the chapter either the solution of problems is sketched or the result is simply given.

1.3 About the Computer Programs

All of the programs of this book are written in the programming language FORTRAN77, as this continues to be the most widely used computer language in the natural sciences and engineering. But since the programming language C has become increasingly popular, all programs have also been converted into C. The conversion was done using an automatic translator. Therefore, the resulting C programs look somewhat different from what they would be had they been coded in C in the first place. Users are free to choose whether they want to install FORTRAN or C programs (cf. installation procedure in Appendix G). The programming of the routines adheres strictly to the FORTRAN77 or ANSI C standards. The programs should therefore be usable on any computer that has a correctly working FORTRAN77 or C compiler. The only exceptions are the graphics routines, which can only be used on IBM PC or PS/2 compatible computers.

There are four groups of computer programs discussed in this book and stored on the accompanying CD-ROM. These are

- the data analysis library,
- the graphics library,
- a collection of example programs,
- solutions to the programming problems.

The *data analysis library* is a collection of subprograms. This allows a format-independent input and output of values as a list of arguments or function values. Each program processes a precisely delimited task discussed in detail in the text. Its use is explained in a schematic functional description. After the description, the program listing (source code) is given. In order not to use up too much space only the FORTRAN source code is reproduced; users interested in details of the code in C can always display or print it. Certain aids help users to find their way within the collection. For example, all routines of a certain chapter or appendix, i.e., of a certain topic, begin with the same letter. A list of all routines is given after the table of contents at the beginning of this book, and an alphabetic index of the routine names is given before the subject index at the book's end. It has been attempted to structure the programs as clearly and understandably as possible. In cases of doubt, clarity took precedence over computing speed. Floating point operations are all carried out with *double precision*. One must note that all floating point quantities and function subprograms in the program collection are of the type `DOUBLE PRECISION`. After the installation procedure described in

Appendix G both the source code and a library file ready for linking with the user's own routines are available.

For the *graphics library* only a functional description of the individual graphics routines and a library file are provided. The source code is not reproduced since computer graphics is not a topic but rather only a tool of data analysis.

In order to solve a specific problem the user has to write a short main program, which essentially consists of calls to routines from the collection, and which in certain cases organizes the input of the user's data and output of the results. The *example programs* are a collection of such main programs. The application of each program from the data analysis and graphics libraries is demonstrated in at least one example program. Such example programs are described in a special section near the end of most chapters under subsection headings such as "*Main Program* `E3RN` *to demonstrate subprograms* `RNMLCG`, `RNECUY`, *and* `RNSTNR`". Other example programs are presented for specific tasks of data analysis or for the simulation of measured data and their graphical representation. The source code is not reproduced in the book but all source files are contained on the CD-ROM.

The *programming problems* like the example programs are designed to help the reader in using computer methods. Working through these problems should enable readers to formulate their own specific tasks in data analysis to be solved on a computer. For the programming problems, programs exist on the CD-ROM which represent a possible solution.

In data analysis, of course, *data* play a special role. The type of data and the format in which they are given to the computer cannot be defined in a general textbook since it depends very much on the particular problem at hand. In order to have somewhat realistic data for our examples and problems we have decided to produce them in most cases within the program using the Monte Carlo method. It is particularly instructive to simulate data with known properties and a given error distribution and to subsequently analyze these data. In the analysis one must in general make an assumption about the distribution of the errors. If this assumption is not correct, then the results of the analysis are not optimal. Effects that are often decisively important in practice can be "experienced" with exercises combining simulation and analysis.

2. Probabilities

2.1 Experiments, Events, Sample Space

Since in this book we are concerned with the analysis of data originating from experiments, we will have to state first what we mean by an experiment and its result. Just as in the laboratory, we define an experiment to be a strictly followed procedure, as a consequence of which a quantity or a set of quantities is obtained that constitutes the result. These quantities are continuous (temperature, length, current) or discrete (number of particles, birthday of a person, one of three possible colors). No matter how accurately all conditions of the procedure are maintained, the results of repetitions of an experiment will in general differ. This is caused either by the intrinsic statistical nature of the phenomenon under investigation or by the finite accuracy of the measurement. The possible results will therefore always be spread over a finite region for each quantity. All of these regions for all quantities that make up the result of an experiment constitute the *sample space* of that experiment. Since it is difficult and often impossible to determine exactly the accessible regions for the quantities measured in a particular experiment, the sample space actually used may be larger and may contain the true sample space as a subspace. We shall use this somewhat looser concept of a sample space.

Example 2.1: Sample space for continuous variables

In the manufacture of resistors it is important to maintain the values R (electrical resistance measured in ohms) and N (maximum heat dissipation measured in watts) at given values. The sample space for R and N is a plane spanned by axes labeled R and N. Since both quantities are always positive, the first quadrant of this plane is itself a sample space. ■

Example 2.2: Sample space for discrete variables

In practice the exact values of R and N are unimportant as long as they are contained within a certain interval about the nominal value (e.g., $99\,\mathrm{k}\Omega < R < 101\,\mathrm{k}\Omega$, $0.49\,\mathrm{W} < N < 0.60\,\mathrm{W}$). If this is the case, we shall say that the resistor has the properties R_n, N_n. If the value falls below (above) the lower (upper) limit, then we shall substitute the index n by $-(+)$. The possible values of resistance and heat dissipation are therefore R_-, R_n, R_+, N_-, N_n, N_+. The sample space now consists of 9 points:

$$\begin{array}{lll} R_- N_-, & R_- N_n, & R_- N_+, \\ R_n N_-, & R_n N_n, & R_n N_+, \\ R_+ N_-, & R_+ N_n, & R_+ N_+. \end{array} \quad \blacksquare$$

Often one or more particular subspaces of the sample space are of special interest. In Example 2.2, for instance, the point R_n, N_n represents the case where the resistors meet the production specifications. We can give such subspaces names, e.g., $A, B, \ldots$ and say that if the result of an experiment falls into one such subspace, then the *event* A (or $B, C, \ldots$) has occurred. If A has not occurred, we speak of the complementary event $\bar{A}$ (i.e., not A). The whole sample space corresponds to an event that will occur in every experiment, which we call E. In the rest of this chapter we shall define what we mean by the probability of the occurrence of an event and present rules for computations with probabilities.

2.2 The Concept of Probability

Let us consider the simplest experiment, namely, the tossing of a coin. Like the throwing of dice or certain problems with playing cards it is of no practical interest but is useful for didactic purposes. What is the probability that a "fair" coin shows "heads" when tossed once? Our intuition suggests that this probability is equal to $1/2$. It is based on the assumption that all points in sample space (there are only two points: "heads" and "tails") are equally probable and on the convention that we give the event E (here: "heads" or "tails") a probability of unity. This way of determining probabilities can be applied only to symmetric experiments and is therefore of little practical use. (It is, however, of great importance in statistical physics and quantum statistics, where the equal probabilities of all allowed states is an essential postulate of very successful theories.) If no such perfect symmetry exists — which will even be the case with normal "physical" coins — the following procedure seems reasonable. In a large number N of experiments the event A is observed to occur n times. We define

$$P(A) = \lim_{N \to \infty} \frac{n}{N} \tag{2.2.1}$$

as the probability of the occurrence of the event A. This somewhat loose *frequency definition* of probability is sufficient for practical purposes, although it is mathematically unsatisfactory. One of the difficulties with this definition is the need for an infinity of experiments, which are of course impossible to perform and even difficult to imagine. Although we shall in fact use the frequency definition in this book, we will indicate the basic concepts of an axiomatic theory of probability due to KOLMOGOROV [1]. The minimal set of axioms generally used is the following:

(a) To each event A there corresponds a non-negative number, its probability,

$$P(A) \geq 0 \quad . \tag{2.2.2}$$

(b) The event E has unit probability,

$$P(E) = 1 \quad . \tag{2.2.3}$$

(c) If A and B are *mutually exclusive* events, then the probability of A *or* B (written $A + B$) is

$$P(A + B) = P(A) + P(B) \quad . \tag{2.2.4}$$

From these axioms* one obtains immediately the following useful results. From (b) and (c):

$$P(\bar{A} + A) = P(A) + P(\bar{A}) = 1 \quad , \tag{2.2.5}$$

and furthermore with (a):

$$0 \leq P(A) \leq 1 \quad . \tag{2.2.6}$$

From (c) one can easily obtain the more general theorem for mutually exclusive events $A, B, C, \ldots,$

$$P(A + B + C + \cdots) = P(A) + P(B) + P(C) + \cdots \quad . \tag{2.2.7}$$

It should be noted that summing the probabilities of events combined with "or" here refers only to mutually exclusive events. If one must deal with events that are not of this type, then they must first be decomposed into mutually exclusive ones. In throwing a die, A may signify even, B odd, C less than 4 dots, D 4 or more dots. Suppose one is interested in the probability for the event A *or* C, which are obviously not exclusive. One forms A *and* C (written AC) as well as AD, BC, and BD, which are mutually exclusive, and finds for A or C (sometimes written $A \dot{+} C$) the expression $AC + AD + BC$. Note that the axioms do not prescribe a method for assigning the value of a particular probability $P(A)$.

*Sometimes the definition (2.3.1) is included as a fourth axiom.

Finally it should be pointed out that the word probability is often used in common language in a sense that is different or even opposed to that considered by us. This is subjective probability, where the probability of an event is given by the measure of our belief in its occurrence. An example of this is: "The probability that the party A will win the next election is $1/3$." As another example consider the case of a certain track in nuclear emulsion which could have been left by a proton or pion. One often says: "The track was caused by a pion with probability $1/2$." But since the event had already taken place and only one of the two kinds of particle could have caused that particular track, the probability in question is either 0 or 1, but we do not know which.

2.3 Rules of Probability Calculus. Conditional Probability

Suppose the result of an experiment has the property A. We now ask for the probability that it also has the property B, i.e., the probability of B under the condition A. We define this *conditional probability* as

$$P(B|A) = \frac{P(A\,B)}{P(A)} \quad . \tag{2.3.1}$$

It follows that

$$P(A\,B) = P(A)\,P(B|A) \quad . \tag{2.3.2}$$

One can also use (2.3.2) directly for the definition, since here the requirement $P(A) \neq 0$ is not necessary. From Fig. 2.1 it can be seen that this definition is reasonable. Consider the event A to occur if a point is in the region labeled A, and correspondingly for the event (and region) B. For the overlap region both A and B occur, i.e., the event (AB) occurs. Let the area of the different regions be proportional to the probabilities of the corresponding events. Then the probability of B under the condition A is the ratio of the area AB to that of A. In particular this is equal to unity if A is contained in B and zero if the overlapping area vanishes.

Using conditional probability we can now formulate the *rule of total probability*. Consider an experiment that can lead to one of n possible mutually exclusive events,

$$E = A_1 + A_2 + \cdots + A_n \quad . \tag{2.3.3}$$

The probability for the occurrence of any event with the property B is

$$P(B) = \sum_{i=1}^{n} P(A_i)P(B|A_i) \quad , \tag{2.3.4}$$

as can be seen easily from (2.3.2) and (2.2.7).

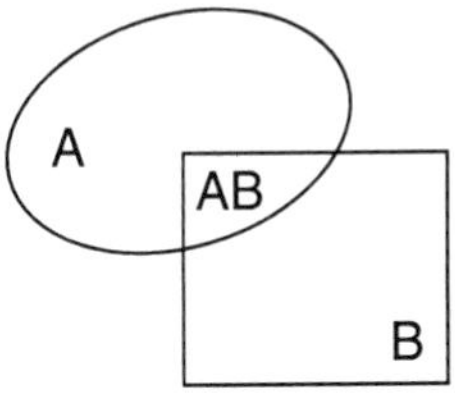

Fig. 2.1. Illustration of conditional probability.

We can now also define the *independence* of events. Two events A and B are said to be independent if the knowledge that A has occurred does not change the probability for B and vice versa, i.e., if

$$P(B|A) = P(B) \quad , \tag{2.3.5}$$

or, by use of (2.3.2),

$$P(A\,B) = P(A)P(B) \quad . \tag{2.3.6}$$

In general several decompositions of the type (2.3.3),

$$\begin{aligned} E &= A_1 + A_2 + \cdots + A_n \quad , \\ E &= B_1 + B_2 + \cdots + B_m \quad , \\ &\vdots \\ E &= Z_1 + Z_2 + \cdots + Z_\ell \quad , \end{aligned} \tag{2.3.7}$$

are said to be independent, if for all possible combinations $\alpha, \beta, \ldots, \omega$ the condition

$$P(A_\alpha B_\beta \cdots Z_\omega) = P(A_\alpha)P(B_\beta) \cdots P(Z_\omega) \tag{2.3.8}$$

is fulfilled.

2.4 Examples

2.4.1 Probability for n Dots in the Throwing of Two Dice

If n_1 and n_2 are the number of dots on the individual dice and if $n = n_1 + n_2$, then one has $P(n_i) = 1/6$; $i = 1, 2$; $n_i = 1, 2, \ldots, 6$. Because the two dice are independent of each other one has $P(n_1, n_2) = P(n_1)P(n_2) = 1/36$. By considering in how many different ways the sum $n = n_i + n_j$ can be formed one obtains

$$\begin{aligned} P_2(2) &= P(1,1) = 1/36 \quad , \\ P_2(3) &= P(1,2) + P(2,1) = 2/36 \quad , \\ P_2(4) &= P(1,3) + P(2,2) + P(3,1) = 3/36 \quad , \end{aligned}$$

$$
\begin{aligned}
P_2(5) &= P(1,4)+P(2,3)+P(3,2)+P(4,1)=4/36 \quad , \\
P_2(6) &= P(1,5)+P(2,4)+P(3,3)+P(4,2) \\
&\quad + P(5,1)=5/36 \quad , \\
P_2(7) &= P(1,6)+P(2,5)+P(3,4)+P(4,3) \\
&\quad + P(5,2)+P(6,1)=6/36 \quad , \\
P_2(8) &= P_2(6)=5/36 \quad , \\
P_2(9) &= P_2(5)=4/36 \quad , \\
P_2(10) &= P_2(4)=3/36 \quad , \\
P_2(11) &= P_2(3)=2/36 \quad , \\
P_2(12) &= P_2(2)=1/36 \quad .
\end{aligned}
$$

Of course, the normalization condition $\sum_{k=2}^{12} P_2(k) = 1$ is fulfilled.

2.4.2 Lottery 6 out of 49

A container holds 49 balls numbered 1 through 49. During the drawing 6 balls are taken out of the container consecutively and none are put back in. We compute the probabilities $P(1), P(2), \ldots, P(6)$ that a player, who before the drawing has chosen six of the numbers 1, 2, …, 49, has predicted exactly 1, 2, …, or 6 of the drawn numbers.

First we compute $P(6)$. The probability to choose as the first number the one which will also be drawn first is obviously 1/49. If that step was successful, then the probability to choose as the second number the one which is also drawn second is 1/48. We conclude that the probability for choosing six numbers correctly in the order in which they are drawn is

$$
\frac{1}{49 \cdot 48 \cdot 47 \cdot 46 \cdot 45 \cdot 44} = \frac{43!}{49!} \quad .
$$

The order, however, is irrelevant. Since there are $6!$ possible ways to arrange 6 numbers in different orders we have

$$
P(6) = \frac{6!43!}{49!} = \frac{1}{\binom{49}{6}} = \frac{1}{C_6^{49}} \quad .
$$

That is exactly the inverse of the number of combinations C_6^{49} of 6 elements out of 49 (see Appendix B), since all of these combinations are equally probable but only one of them contains only the drawn numbers.

We may now argue that the container holds two kinds of balls, namely 6 balls in which the player is interested since they carry the numbers which he selected, and 43 balls whose numbers the player did not select. The result of the drawing is a sample from a set of 49 elements of which 6 are of one kind and 43 are of the other. The sample itself contains 6 elements which are drawn without putting elements back into the container. This method of sampling is described by the hypergeometric distribution (see Sect. 5.3). The probability for predicting correctly ℓ out of the 6 drawn numbers is

$$P(\ell) = \frac{\binom{6}{\ell}\binom{43}{6-\ell}}{\binom{49}{6}} \quad , \qquad \ell = 0, \dots, 6 \quad .$$

2.4.3 Three-Door Game

In a TV game show a candidate is given the following problem. Three rooms are closed by three identical doors. One room contains a luxury car, the other two each contain a goat. The candidate is asked to guess behind which of the doors the car is. He chooses a door which we will call A. The door A, however, remains closed for the moment. Of course, behind at least one of the other doors there is a goat. The quiz master now opens one door which we will call B to reveal a goat. He now gives the candidate the chance to either stay with the original choice A or to choose remaining closed door C. Can the candidate increase his or her chances by choosing C instead of A?

The answer (astonishing for many) is yes. The probability to find the car behind the door A obviously is $P(A) = 1/3$. Then the probability that the car is behind one of the other doors is $P(\bar{A}) = 2/3$. The candidate exhausts this probability fully if he chooses the door C since through the opening of B it is shown to be a door without the car, so that $P(C) = P(\bar{A})$.

2.5 Problems

2.5.1 Determination of Probabilities through Symmetry Considerations

There are n students in a classroom. What is the probability for the fact that at least two of them have their birthday on the same day? Solve the problem by working through the following questions:

(a) What is the number N of possibilities to distribute the n birthdays over the year (365 days)?

(b) How large is the number N' of possibilities for which all n birthdays are different?

(c) How large then is the probability P_{diff} that the birthdays are different?

(d) How large finally is the probability P that at least two birthdays are not different?

2.5.2 Probability for Non-exclusive Events

The probabilities $P(A)$, $P(B)$, and $P(AB) \neq 0$ for non-exclusive events A and B are given. How large is the probability $P(A+B)$ for the observation of A or B? As an example compute the probability that a playing card which was drawn at random out of a deck of 52 cards is either an ace or a diamond.

2.5.3 Dependent and Independent Events

Are the events A and B that a playing card out of a deck is an ace or a diamond independent

(a) if an ordinary deck of 52 cards is used,

(b) if a joker is added to the deck?

2.5.4 Complementary Events

Show that $\bar{A}$ and $\bar{B}$ are independent if A and B are independent. Use the result of Problem 2.5.2 to express $P(\overline{AB})$ by $P(A)$, $P(B)$, and $P(AB)$.

2.5.5 Probabilities Drawn from Large and Small Populations

A container holds a large number (>1000) of coins. They are divided into three types A, B, and C, which make up 20 %, 30 %, and 50 % of the total.

(a) What are the probabilities $P(A)$, $P(B)$, $P(C)$ of picking a coin of type A, B, or C if one coin is taken at random? What are the probabilities $P(AB)$, $P(AC)$, $P(BC)$, $P(AA)$, $P(BB)$, $P(CC)$, P(2 identical coins), P(2 different coins) for picking 2 coins?

(b) What are the probabilities if 10 coins (2 of type A, 3 of type B, and 5 of type C) are in the container?

2.6 Hints and Solutions

Problem 2.5.1

(a) $N = 365^n$.

(b) $N' = 365 \cdot 364 \cdot \ldots \cdot (365 - n + 1)$.

(c) $$P_{\text{diff}} = N'/N = \frac{364}{365} \cdot \ldots \cdot \frac{365-n+1}{365}$$
$$= \left(1 - \frac{1}{365}\right)\left(1 - \frac{2}{365}\right) \cdots \left(1 - \frac{n-1}{365}\right) \quad .$$

(d) $P = 1 - P_{\text{diff}}$.

Putting in numbers one obtains $P \approx 0.5$ for $n = 23$ and $P \approx 0.99$ for $n = 57$.

Problem 2.5.2

$$P(A+B) = P(A) + P(B) - P(AB) \quad ,$$

$$\begin{aligned} P(\text{ace or diamond}) &= P(\text{ace}) + P(\text{diamond}) - P(\text{ace} + \text{diamond}) \\ &= \frac{4}{52} + \frac{13}{52} - \frac{1}{52} = \frac{4}{13} \quad . \end{aligned}$$

Problem 2.5.3

(a) $P(A) = \frac{4}{52}, P(B) = \frac{13}{52}, P(AB) = \frac{1}{52}$,

i.e., $P(AB) = P(A)P(B)$.

(b) $P(A) = \frac{4}{53}, P(B) = \frac{13}{53}, P(AB) = \frac{1}{53}$,

i.e., $P(AB) \neq P(A)P(B)$.

Problem 2.5.4

$$P(\bar{A}\bar{B}) = 1 - P(A+B) = 1 - P(A) - P(B) + P(AB) \quad .$$

For A and B independent one has $P(AB) = P(A)P(B)$. Therefore,

$$\begin{aligned} P(\bar{A}\bar{B}) &= 1 - P(A) - P(B) + P(A)P(B) = (1 - P(A))(1 - P(B)) \\ &= P(\bar{A})P(\bar{B}) \quad . \end{aligned}$$

Problem 2.5.5

(a) $P(A) = 0.2\,,\ P(B) = 0.3\,,\ P(C) = 0.5\,,$
$P(AB) = 2 \cdot 0.2 \cdot 0.3\,,\ P(AC) = 2 \cdot 0.2 \cdot 0.5\,,$
$P(BC) = 2 \cdot 0.3 \cdot 0.5\,,\ P(AA) = 0.2^2\,,\ P(BB) = 0.3^2\,,$
$P(CC) = 0.5^2\,.$

(b) $P(A) = 2/10 = 0.2\,,\ P(B) = 3/10 = 0.3\,,\ P(C) = 5/10 = 0.5\,,$

$$P(AB) = \frac{2}{10} \cdot \frac{3}{9} + \frac{3}{10} \cdot \frac{2}{9}\,,\ P(AC) = \frac{2}{10} \cdot \frac{5}{9} + \frac{5}{10} \cdot \frac{2}{9}\,,$$

$$P(BC) = \frac{3}{10} \cdot \frac{5}{9} + \frac{5}{10} \cdot \frac{3}{9}\,,$$

$$P(AA) = \frac{2}{10} \cdot \frac{1}{9}\,,\ P(BB) = \frac{3}{10} \cdot \frac{2}{9}\,,\ P(CC) = \frac{5}{10} \cdot \frac{4}{9}\,.$$

For (a) and (b) it holds that

$$\begin{aligned} P(\text{2 identical coins}) &= P(AA) + P(BB) + P(CC)\,, \\ P(\text{2 different coins}) &= P(AB) + P(AC) + P(BC) \\ &= 1 - P(\text{2 identical coins})\,. \end{aligned}$$

3. Random Variables. Distributions

3.1 Random Variables

We will now consider not the probability of observing particular events but rather the events themselves and try to find a particularly simple way of classifying them. We can, for instance, associate the event "heads" with the number 0 and the event "tails" with the number 1. Generally we can classify the events of the decomposition (2.3.3) by associating each event A_i with the real number i. In this way each event can be characterized by one of the possible values of a *random variable*. Random variables can be discrete or continuous. We denote them by symbols like x, y, ...

Example 3.1: Discrete random variable

It may be of interest to study the number of coins still in circulation as a function of their age. It is obviously most convenient to use the year of issue stamped on each coin directly as the (discrete) random variable, e.g., $\mathsf{x} = \ldots$, 1949, 1950, 1951, ... ■

Example 3.2: Continuous random variable

All processes of measurement or production are subject to smaller or larger imperfections or fluctuations that lead to variations in the result, which is therefore described by one or several random variables. Thus the values of electrical resistance and maximum heat dissipation characterizing a resistor in Example 2.1 are continuous random variables. ■

3.2 Distributions of a Single Random Variable

From the classification of events we return to probability considerations. We consider the random variable x and a real number x, which can assume any value between $-\infty$ and $-\infty$, and study the probability for the event $\mathsf{x} < x$.

This probability is a function of x and is called the *(cumulative) distribution function* of x:

$$F(x) = P(\mathsf{x} < x) \quad . \tag{3.2.1}$$

If x can assume only a finite number of discrete values, e.g., the number of dots on the faces of a die, then the distribution function is a step function. It is shown in Fig. 3.1 for the example mentioned above. Obviously distribution functions are always monotonic and non-decreasing.

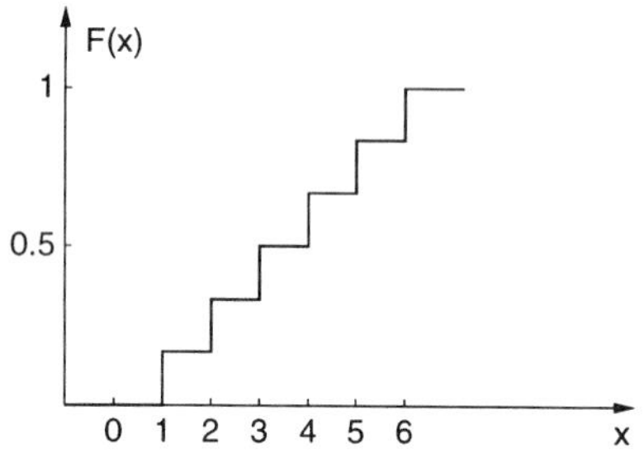

Fig. 3.1. Distribution function for throwing of a symmetric die.

Because of (2.2.3) one has the limiting case

$$\lim_{x\to\infty} F(x) = \lim_{x\to\infty} P(\mathsf{x} < x) = P(E) = 1 \quad . \tag{3.2.2}$$

Applying Eq. (2.2.5) to (3.2.1) we obtain

$$P(\mathsf{x} \geq x) = 1 - F(x) = 1 - P(\mathsf{x} < x) \tag{3.2.3}$$

and therefore

$$\lim_{x\to-\infty} F(x) = \lim_{x\to-\infty} P(\mathsf{x} < x) = 1 - \lim_{x\to-\infty} P(\mathsf{x} \geq x) = 0 \quad . \tag{3.2.4}$$

Of special interest are distribution functions $F(x)$ that are continuous and differentiable. The first derivative

$$f(x) = \frac{\mathrm{d}F(x)}{\mathrm{d}x} = F'(x) \tag{3.2.5}$$

is called the *probability density (function)* of x. It is a measure of the probability of the event $(x \leq \mathsf{x} < x + \mathrm{d}x)$. From (3.2.1) and (3.2.5) it immediately follows that

$$P(\mathsf{x} < a) = F(a) = \int_{-\infty}^{a} f(x)\,\mathrm{d}x \quad , \tag{3.2.6}$$

$$P(a \leq \mathsf{x} < b) = \int_{a}^{b} f(x)\,\mathrm{d}x = F(b) - F(a) \quad , \tag{3.2.7}$$

and in particular

$$\int_{-\infty}^{\infty} f(x)\,\mathrm{d}x = 1 \quad . \tag{3.2.8}$$

A trivial example of a continuous distribution is given by the angular position of the hand of a watch read at random intervals. We obtain a constant probability density (Fig. 3.2).

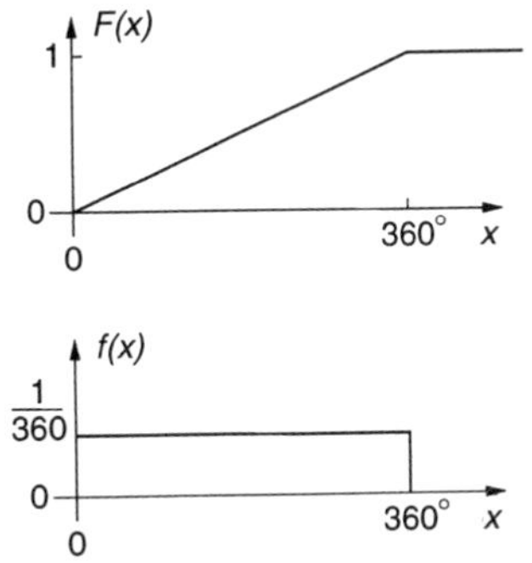

Fig. 3.2. Distribution function and probability density for the angular position of a watch hand.

3.3 Functions of a Single Random Variable, Expectation Value, Variance, Moments

In addition to the distribution of a random variable x, we are often interested in the distribution of a function of x. Such a function of a random variable is also a random variable:

$$\mathsf{y} = H(\mathsf{x}) \quad . \tag{3.3.1}$$

The variable y then possesses a distribution function and probability density in the same way as x.

In the two simple examples of the last section we were able to give the distribution function immediately because of the symmetric nature of the problems. Usually this is not possible. Instead, we have to obtain it from experiment. Often we are limited to determining a few characteristic parameters instead of the complete distribution.

The *mean* or *expectation value* of a random variable is the sum of all possible values x_i of x multiplied by their corresponding probabilities

$$E(\mathsf{x}) = \widehat{x} = \sum_{i=1}^{n} x_i P(\mathsf{x} = x_i) \quad . \tag{3.3.2}$$

Note that $\widehat{x}$ is not a random variable but rather has a fixed value. Correspondingly the expectation value of a function (3.3.1) is defined to be

$$E\{H(\mathsf{x})\} = \sum_{i=1}^{n} H(x_i) P(\mathsf{x} = x_i) \quad . \tag{3.3.3}$$

In the case of a continuous random variable (with a differentiable distribution function), we define by analogy

$$E(\mathsf{x}) = \widehat{x} = \int_{-\infty}^{\infty} x f(x)\,\mathrm{d}x \tag{3.3.4}$$

and

$$E\{H(\mathsf{x})\} = \int_{-\infty}^{\infty} H(x) f(x)\,\mathrm{d}x \quad . \tag{3.3.5}$$

If we choose in particular

$$H(\mathsf{x}) = (\mathsf{x} - c)^\ell \quad , \tag{3.3.6}$$

we obtain the expectation values

$$\alpha_\ell = E\{(\mathsf{x} - c)^\ell\} \quad , \tag{3.3.7}$$

which are called the *ℓ–th moments* of the variable about the point c. Of special interest are the *moments about the mean*,

$$\mu_\ell = E\{(\mathsf{x} - \widehat{x})^\ell\} \quad . \tag{3.3.8}$$

The lowest moments are obviously

$$\mu_0 = 1 \quad , \qquad \mu_1 = 0 \quad . \tag{3.3.9}$$

The quantity

$$\mu_2 = \sigma^2(\mathsf{x}) = \mathrm{var}(\mathsf{x}) = E\{(\mathsf{x} - \widehat{x})^2\} \tag{3.3.10}$$

is the lowest moment containing information about the average deviation of the variable x from its mean. It is called the *variance* of x.

We will now try to visualize the practical meaning of the expectation value and variance of a random variable x. Let us consider the measurement of some quantity, for example, the length x_0 of a small crystal using a microscope. Because of the influence of different factors, such as the imperfections of the different components of the microscope and observational errors, repetitions of the measurement will yield slightly different results for x. The individual measurements will, however, tend to group themselves in the neighborhood of the true value of the length to be measured, i.e., it will be more probable to find a value of x near to x_0 than far from it, providing no systematic biases exist. The probability density of x will therefore have a bell-shaped form as sketched in Fig. 3.3, although it need not be symmetric. It seems reasonable – especially in the case of a symmetric probability density – to interpret the expectation value (3.3.4) as the best estimate of the true value. It is interesting to note that (3.3.4) has the mathematical form of a center of gravity, i.e., $\widehat{x}$ can be visualized as the x-coordinate of the center of gravity of the surface under the curve describing the probability density.

The variance (3.3.10),

$$\sigma^2(\mathsf{x}) = \int_{-\infty}^{\infty} (x - \widehat{x})^2 f(x)\,\mathrm{d}x \quad , \tag{3.3.11}$$

which has the form of a moment of inertia, is a measure of the width or dispersion of the probability density about the mean. If it is small, the individual

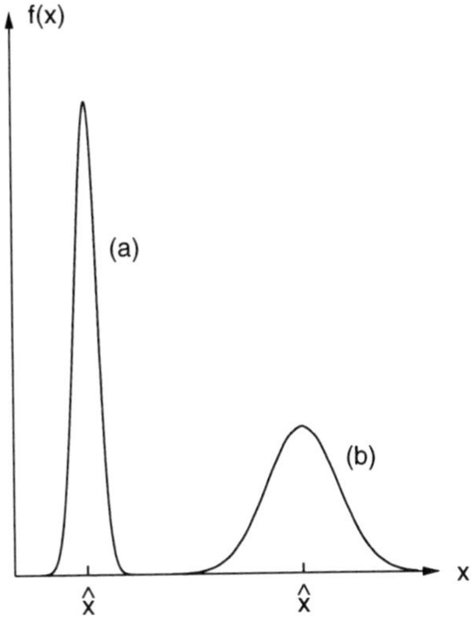

Fig. 3.3. Distribution with small variance (a) and large variance (b).

measurements lie close to $\widehat{x}$ (Fig. 3.3a); if it is large, they will in general be further from the mean (Fig. 3.3b). The positive square root of the variance

$$\sigma = \sqrt{\sigma^2(\mathsf{x})} \tag{3.3.12}$$

is called the *standard deviation* (or sometimes the *dispersion*) of x. Like the variance itself it is a measure of the average deviation of the measurements x from the expectation value.

Since the standard deviation has the same dimension as x (in our example both have the dimension of length), it is identified with the error of the measurement,

$$\sigma(\mathsf{x}) = \Delta x \quad .$$

This definition of measurement error is discussed in more detail in Sects. 5.6 – 5.10. It should be noted that the definitions (3.3.4) and (3.3.10) do not provide completely a way of calculating the mean or the measurement error, since the probability density describing a measurement is in general unknown.

The third moment about the mean is sometimes called *skewness*. We prefer to define the dimensionless quantity

$$\gamma = \mu_3/\sigma^3 \tag{3.3.13}$$

to be the skewness of x. It is positive (negative) if the distribution is skew to the right (left) of the mean. For symmetric distributions the skewness vanishes. It contains information about a possible difference between positive and negative deviation from the mean.

We will now obtain a few important rules about means and variances. In the case where

$$H(\mathsf{x}) = c\mathsf{x} \quad , \qquad c = \text{const} \quad , \tag{3.3.14}$$

it follows immediately that

$$\begin{aligned} E(c\mathsf{x}) &= cE(\mathsf{x}) \quad , \\ \sigma^2(c\mathsf{x}) &= c^2\sigma^2(\mathsf{x}) \quad , \end{aligned} \tag{3.3.15}$$

and therefore

$$\sigma^2(\mathsf{x}) = E\{(\mathsf{x} - \widehat{x})^2\} = E\{\mathsf{x}^2 - 2\mathsf{x}\widehat{x} + \widehat{x}^2\} = E(\mathsf{x}^2) - \widehat{x}^2 \quad . \tag{3.3.16}$$

We now consider the function

$$\mathsf{u} = \frac{\mathsf{x} - \widehat{x}}{\sigma(\mathsf{x})} \quad . \tag{3.3.17}$$

It has the expectation value

$$E(\mathsf{u}) = \frac{1}{\sigma(\mathsf{x})} E(\mathsf{x} - \widehat{x}) = \frac{1}{\sigma(\mathsf{x})}(\widehat{x} - \widehat{x}) = 0 \tag{3.3.18}$$

and variance

$$\sigma^2(\mathsf{u}) = \frac{1}{\sigma^2(\mathsf{x})} E\{(\mathsf{x} - \widehat{x})^2\} = \frac{\sigma^2(\mathsf{x})}{\sigma^2(\mathsf{x})} = 1 \quad . \tag{3.3.19}$$

The function u – which is also a random variable – has particularly simple properties, which makes its use in more involved calculations preferable. We will call such a variable (having zero mean and unit variance) a *reduced variable*. It is also called a standardized, normalized, or dimensionless variable.

Although a distribution is mathematically most easily described by its expectation value, variance, and higher moments (in fact any distribution can be completely specified by these quantities, cf. Sect. 5.5), it is often convenient to introduce further definitions so as to better visualize the form of a distribution.

The *mode* x_m (or *most probable value*) of a distribution is defined as that value of the random variable that corresponds to the highest probability:

$$P(\mathsf{x} = x_m) = \max \quad . \tag{3.3.20}$$

If the distribution has a differentiable probability density, the mode, which corresponds to its maximum, is easily determined by the conditions

$$\frac{\mathrm{d}}{\mathrm{d}x} f(x) = 0 \quad , \qquad \frac{\mathrm{d}^2}{\mathrm{d}x^2} f(x) < 0 \quad . \tag{3.3.21}$$

In many cases only one maximum exists; the distribution is said to be *unimodal*. The *median* $x_{0.5}$ of a distribution is defined as that value of the random variable for which the distribution function equals 1/2:

$$F(x_{0.5}) = P(\mathsf{x} < x_{0.5}) = 0.5 \quad . \tag{3.3.22}$$

In the case of a continuous probability density Eq. (3.3.22) takes the form

$$\int_{-\infty}^{x_{0.5}} f(x)\,\mathrm{d}x = 0.5 \quad , \tag{3.3.23}$$

i.e., the median divides the total range of the random variable into two regions each containing equal probability.

It is clear from these definitions that in the case of a unimodal distribution with continuous probability density that is symmetric about its maximum, the values of mean, mode, and median coincide. This is not, however, the case for asymmetric distributions (Fig. 3.4).

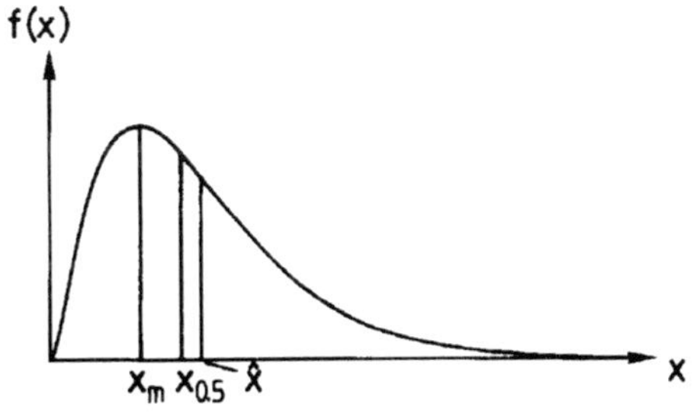

Fig. 3.4. Most probable value (mode) x_m, mean $\widehat{x}$, and median $x_{0.5}$ of an asymmetric distribution.

The definition (3.3.22) can easily be generalized. The quantities $x_{0.25}$ and $x_{0.75}$ defined by

$$F(x_{0.25}) = 0.25 \quad , \qquad F(x_{0.75}) = 0.75 \tag{3.3.24}$$

are called lower and upper *quartiles*. Similarly we can define *deciles* $x_{0.1}$, $x_{0.2}$, ..., $x_{0.9}$, or in general *quantiles* x_q, by

$$F(x_q) = \int_{-\infty}^{x_q} f(x)\,\mathrm{d}x = q \tag{3.3.25}$$

with $0 \le q \le 1$.

The definition of quantiles is most easily visualized from Fig. 3.5. In a plot of the distribution function $F(x)$, the quantile x_q can be read off as the abscissa corresponding to the value q on the ordinate. The quantile $x_q(q)$, regarded as a function of the probability q, is simply the inverse of the distribution function.

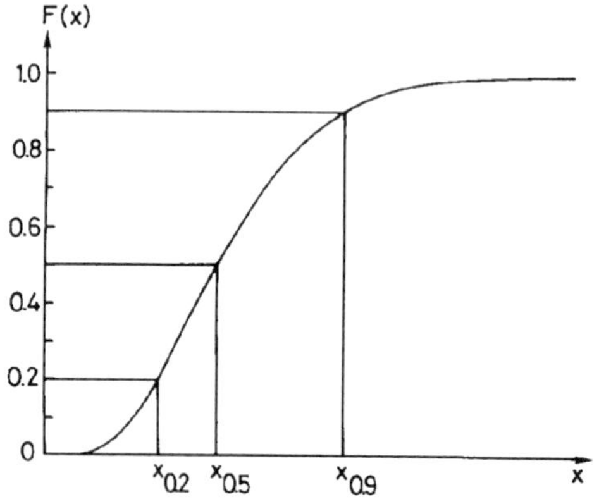

Fig. 3.5. Median and quantiles of a continuous distribution.

Example 3.3: Uniform distribution

We will now discuss the simplest case of a distribution function of a continuous variable. Suppose that in the interval $a \le x < b$ the probability density of x is constant, and it is zero outside of this interval:

$$\begin{aligned} f(x) &= c \quad , \quad a \le x < b \quad , \\ f(x) &= 0 \quad , \quad x < a \quad , \quad x \ge b \quad . \end{aligned} \tag{3.3.26}$$

Because of (3.2.8) one has

$$\int_{-\infty}^{\infty} f(x)\,\mathrm{d}x = c\int_a^b \mathrm{d}x = c(b-a) = 1$$

or

$$\begin{aligned} f(x) &= \frac{1}{b-a} \quad , \quad a \le x < b \quad , \\ f(x) &= 0 \quad , \quad x < a \quad , \quad x \ge b \quad . \end{aligned} \tag{3.3.27}$$

The distribution function is

$$\begin{aligned} F(x) &= \int_a^x \frac{\mathrm{d}x}{b-a} = \frac{x-a}{b-a} \quad , \quad a \le x < b \quad , \\ F(x) &= 0 \quad , \quad x < a \quad , \\ F(x) &= 1 \quad , \quad x \ge b \quad . \end{aligned} \tag{3.3.28}$$

By symmetry arguments the expectation value of x must be the arithmetic mean of the boundaries a and b. In fact, (3.3.4) immediately gives

$$E(\mathsf{x}) = \widehat{x} = \frac{1}{b-a}\int_a^b x\,\mathrm{d}x = \frac{1}{2}\frac{1}{(b-a)}(b^2-a^2) = \frac{b+a}{2} \quad . \tag{3.3.29}$$

Correspondingly, one obtains from (3.3.10)

$$\sigma^2(\mathsf{x}) = \frac{1}{12}(b-a)^2 \quad . \tag{3.3.30}$$

The uniform distribution is not of great practical interest. It is, however, particularly easy to handle, being the simplest distribution of a continuous variable. It is often advantageous to transform a distribution function by means of a transformation of variables into a uniform distribution or the reverse, to express the given distribution in terms of a uniform distribution. This method is used particularly in the "Monte Carlo method" discussed in Chapter 4. ■

Example 3.4: Cauchy distribution
In the (x, y) plane a gun is mounted at the point $(x, y) = (0, -1)$ such that its barrel lies in the (x, y) plane and can rotate around an axis parallel to the z axis (Fig. 3.6).

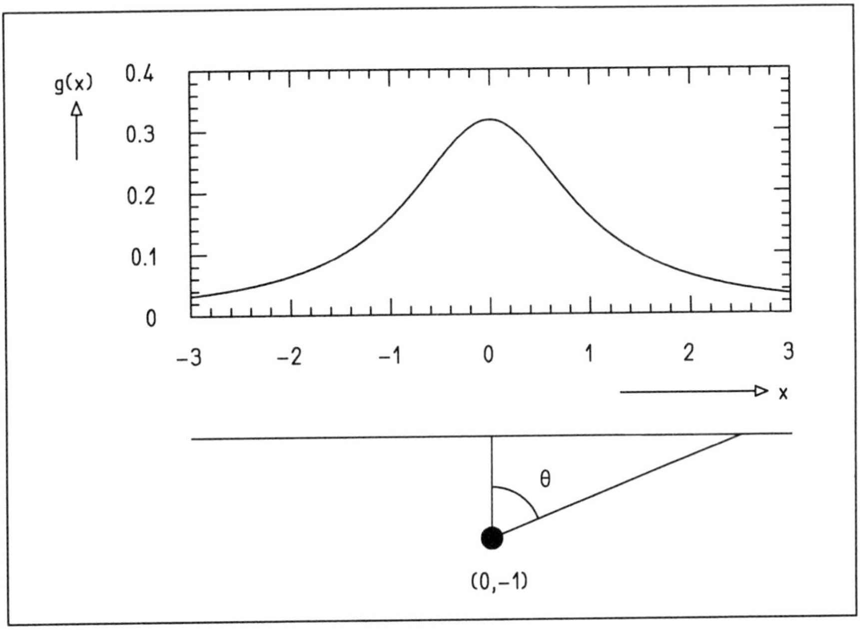

Fig. 3.6. Model for producing a Cauchy distribution (below) and probability density of the Cauchy distribution (above).

The gun is fired such that the angle θ between the barrel and the y axis is chosen at random from uniform distribution in the range $-\pi/2 \leq \theta < \pi/2$, i.e., the probability density of θ is

$$f(\theta) = \frac{1}{\pi} \quad .$$

Since

$$\theta = \arctan x \quad , \qquad \frac{\mathrm{d}\theta}{\mathrm{d}x} = \frac{1}{1+x^2} \quad ,$$

we find by the transformation (cf. Sect. 3.7) $\theta \to x$ of the variable for the probability density in x

$$g(x) = \left|\frac{\mathrm{d}\theta}{\mathrm{d}x}\right| f(\theta) = \frac{1}{\pi}\frac{1}{1+x^2} \quad . \tag{3.3.31}$$

A distribution with this probability density (in our example of the position of hits on the x axis) is called the *Cauchy distribution*.

The expectation value of x is (taking the principal value for the integral)

$$\widehat{x} = \frac{1}{\pi}\int_{-\infty}^{\infty}\frac{x\,\mathrm{d}x}{1+x^2} = 0 \quad .$$

The expression for the variance,

$$\begin{aligned}\int_{-\infty}^{\infty} x^2 g(x)\,\mathrm{d}x &= \frac{1}{\pi}\int_{-\infty}^{\infty}\frac{x^2\,\mathrm{d}x}{1+x^2} = \frac{1}{\pi}(x - \arctan x)\Big|_{x=-\infty}^{x=\infty} \\ &= \frac{2}{\pi}\lim_{x\to\infty}(x - \arctan x) \quad ,\end{aligned}$$

yields an infinite result. One says that the variance of the Cauchy distribution does not exist.

One can, however, construct another measure for the width of the distribution, the *full width at half maximum 'FWHM'* (cf. Example 6.3). The function $g(x)$ has its maximum at $x = \widehat{x} = 0$ and reaches half its maximum value at the points $x_a = -1$ and $x_\ell = 1$. Therefore,

$$\Gamma = 2$$

is the full width at half maximum of the Cauchy distribution. ∎

Example 3.5: Lorentz (Breit–Wigner) distribution

With $\widehat{x} = a = 0$ and $\Gamma = 2$ we can write the probability density (3.3.31) of the Cauchy distribution in the form

$$g(x) = \frac{2}{\pi\Gamma}\frac{\Gamma^2}{4(x-a)^2+\Gamma^2} \quad . \tag{3.3.32}$$

This function is a normalized probability density for all values of a and full width at half maximum $\Gamma > 0$. It is called the probability density of the *Lorentz* or also *Breit–Wigner distribution* and plays an important role in the physics of resonance phenomena. ∎

3.4 Distribution Function and Probability Density of Two Variables. Conditional Probability

We now consider two random variables x and y and ask for the probability that both $\mathsf{x} < x$ and $\mathsf{y} < y$. As in the case of a single variable we expect there to exist of a *distribution function* (see Fig. 3.7)

$$F(x, y) = P(\mathsf{x} < x,\ \mathsf{y} < y) \quad . \tag{3.4.1}$$

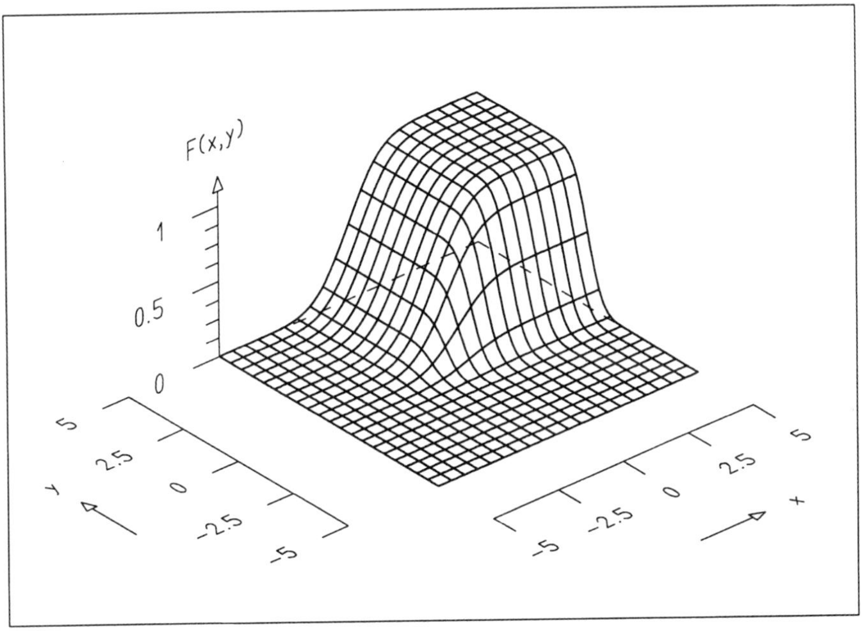

Fig. 3.7. Distribution function of two variables.

We will not enter here into axiomatic details and into the conditions for the existence of F, since these are always fulfilled in cases of practical interest. If F is a differentiable function of x and y, then the *joint probability density* of x and y is

$$f(x, y) = \frac{\partial}{\partial x}\frac{\partial}{\partial y}F(x, y) \quad . \tag{3.4.2}$$

One then has

$$P(a \le \mathsf{x} < b, c \le \mathsf{y} < d) = \int_a^b \left[\int_c^d f(x, y)\,\mathrm{d}y \right] \mathrm{d}x \quad . \tag{3.4.3}$$

Often we are faced with the following experimental problem. One determines approximately with many measurements the joint distribution function

$F(x,y)$. One wishes to find the probability for x without consideration of y. (For example, the probability density for the appearance of a certain infectious disease might be given as a function of date and geographic location. For some investigations the dependence on the time of year might be of no interest.)

We integrate Eq. (3.4.3) over the whole range of y and obtain

$$P(a \le x < b, -\infty < y < \infty) = \int_a^b \left[\int_{-\infty}^{\infty} f(x,y)\,\mathrm{d}y\right] \mathrm{d}x = \int_a^b g(x)\,\mathrm{d}x \quad ,$$

where

$$g(x) = \int_{-\infty}^{\infty} f(x,y)\,\mathrm{d}y \tag{3.4.4}$$

is the probability density for x. It is called the *marginal probability density* of x. The corresponding distribution for y is

$$h(y) = \int_{-\infty}^{\infty} f(x,y)\,\mathrm{d}x \quad . \tag{3.4.5}$$

In analogy to the independence of events [Eq. (2.3.6)] we can now define the *independence of random variables*. The variables x and y are said to be independent if

$$f(x,y) = g(x)h(y) \quad . \tag{3.4.6}$$

Using the marginal distributions we can also define conditional probability for y under the condition that x is known,

$$P(y \le \mathsf{y} < y + \mathrm{d}y \mid x \le \mathsf{x} \le x + \mathrm{d}x) \quad . \tag{3.4.7}$$

We define the *conditional probability density* as

$$f(y|x) = \frac{f(x,y)}{g(x)} \quad , \tag{3.4.8}$$

so that the probability of Eq. (3.4.7) is given by

$$f(y|x)\,\mathrm{d}y \quad .$$

The rule of total probability can now also be expressed for distributions:

$$h(y) = \int_{-\infty}^{\infty} f(x,y)\,\mathrm{d}x = \int_{-\infty}^{\infty} f(y|x)g(x)\,\mathrm{d}x \quad . \tag{3.4.9}$$

In the case of independent variables as defined by Eq. (3.4.6) one obtains directly from Eq. (3.4.8)

$$f(y|x) = \frac{f(x,y)}{g(x)} = \frac{g(x)h(y)}{g(x)} = h(y) \quad . \tag{3.4.10}$$

This was expected since, in the case of independent variables, any constraint on one variable cannot contribute information about the probability distribution of the other.

3.5 Expectation Values, Variance, Covariance, and Correlation

In analogy to Eq. (3.3.5) we define the expectation value of a function $H(\mathsf{x},\mathsf{y})$ to be

$$E\{H(\mathsf{x},\mathsf{y})\} = \int_{-\infty}^{\infty}\int_{-\infty}^{\infty} H(x,y)f(x,y)\,\mathrm{d}x\,\mathrm{d}y \quad . \tag{3.5.1}$$

Similarly, the variance of $H(\mathsf{x},\mathsf{y})$ is defined to be

$$\sigma^2\{H(\mathsf{x},\mathsf{y})\} = E\{[H(\mathsf{x},\mathsf{y}) - E(H(\mathsf{x},\mathsf{y}))]^2\} \quad . \tag{3.5.2}$$

For the simple case $H(\mathsf{x},\mathsf{y}) = a\mathsf{x} + b\mathsf{y}$, Eq. (3.5.1) clearly gives

$$E(a\mathsf{x} + b\mathsf{y}) = aE(\mathsf{x}) + bE(\mathsf{y}) \quad . \tag{3.5.3}$$

We now choose

$$H(\mathsf{x},\mathsf{y}) = x^\ell y^m \qquad (\ell, m \text{ non-negative integers}) \quad . \tag{3.5.4}$$

The expectation values of such functions are the ℓmth *moments* of x,y about the origin,

$$\lambda_{\ell m} = E(x^\ell y^m) \quad . \tag{3.5.5}$$

If we choose more generally

$$H(\mathsf{x},\mathsf{y}) = (\mathsf{x} - a)^\ell (\mathsf{y} - b)^m \quad , \tag{3.5.6}$$

the expectation values

$$\alpha_{\ell m} = E\{(\mathsf{x} - a)^\ell (\mathsf{y} - b)^m\} \tag{3.5.7}$$

are the ℓm-th moments about the point a, b. Of special interest are the moments about the point $\lambda_{10}, \lambda_{01}$,

$$\mu_{\ell m} = E\{(\mathsf{x} - \lambda_{10})^\ell (\mathsf{y} - \lambda_{01})^m\} \quad . \tag{3.5.8}$$

As in the case of a single variable, the lower moments have a special significance, in particular,

$$\begin{aligned}
\mu_{00} &= \lambda_{00} = 1 \quad , \\
\mu_{10} &= \mu_{01} = 0 \quad ; \\
\\
\lambda_{10} &= E(\mathsf{x}) = \widehat{x} \quad , \\
\lambda_{01} &= E(\mathsf{y}) = \widehat{y} \quad ; \\
\\
\mu_{11} &= E\{(\mathsf{x} - \widehat{x})(\mathsf{y} - \widehat{y})\} = \mathrm{cov}(\mathsf{x},\mathsf{y}) \quad , \\
\mu_{20} &= E\{(\mathsf{x} - \widehat{x})^2\} = \sigma^2(\mathsf{x}) \quad , \\
\mu_{02} &= E\{(\mathsf{y} - \widehat{y})^2\} = \sigma^2(\mathsf{y}) \quad .
\end{aligned} \tag{3.5.9}$$

We can now express the variance of $ax + by$ in terms of these quantities:

$$\begin{aligned} \sigma^2(ax+by) &= E\{[(ax+by) - E(ax+by)]^2\} \\ &= E\{[a(x-\widehat{x}) + b(y-\widehat{y})]^2\} \\ &= E\{a^2(x-\widehat{x})^2 + b^2(y-\widehat{y})^2 + 2ab(x-\widehat{x})(y-\widehat{y})\} \quad , \\ \sigma^2(ax+by) &= a^2\sigma^2(x) + b^2\sigma^2(y) + 2ab\,\mathrm{cov}(x,y) \quad . \end{aligned} \tag{3.5.10}$$

In deriving (3.5.10) we have made use of (3.3.14). As another example we consider

$$H(x,y) = xy \quad . \tag{3.5.11}$$

In this case we have to assume the independence of x and y in the sense of (3.4.6) in order to obtain the expectation value. Then according to (3.5.1) one has

$$\begin{aligned} E(xy) &= \int_{-\infty}^{\infty}\int_{-\infty}^{\infty} x\,y\,g(x)h(y)\,dx\,dy \\ &= \left(\int_{-\infty}^{\infty} x\,g(x)\,dx\right)\left(\int_{-\infty}^{\infty} y\,h(y)\,dy\right) \end{aligned} \tag{3.5.12}$$

or

$$E(xy) = E(x)E(y) \quad . \tag{3.5.13}$$

While the quantities $E(x)$, $E(y)$, $\sigma^2(x)$, $\sigma^2(y)$ are very similar to those obtained in the case of a single variable, we still have to explain the meaning of $\mathrm{cov}(x,y)$. The concept of *covariance* is of considerable importance for the understanding of many of our subsequent problems. From its definition we see that $\mathrm{cov}(x,y)$ is positive if values $x > \widehat{x}$ appear preferentially together with values $y > \widehat{y}$. On the other hand, $\mathrm{cov}(x,y)$ is negative if in general $x > \widehat{x}$ implies $y < \widehat{y}$. If, finally, the knowledge of the value of x does not give us additional information about the probable position of y, the covariance vanishes. These cases are illustrated in Fig. 3.8.

It is often convenient to use the *correlation coefficient*

$$\rho(x,y) = \frac{\mathrm{cov}(x,y)}{\sigma(x)\sigma(y)} \tag{3.5.14}$$

rather than the covariance.

Both the covariance and the correlation coefficient offer a (necessarily crude) measure of the mutual dependence of x and y. To investigate this further we now consider two reduced variables u and v in the sense of Eq. (3.3.17) and determine the variance of their sum by using (3.5.9),

$$\sigma^2(u+v) = \sigma^2(u) + \sigma^2(v) + 2\rho(u,v)\sigma(u)\sigma(v) \quad . \tag{3.5.15}$$

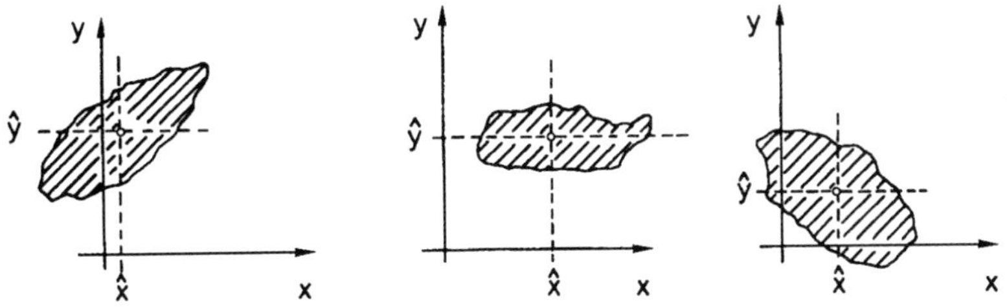

Fig. 3.8. Illustration of the covariance between the variables x and y. (a) $\mathrm{cov}(x, y) > 0$; (b) $\mathrm{cov}(x, y) \approx 0$; (c) $\mathrm{cov}(x, y) < 0$.

From Eq. (3.3.19) we know that $\sigma^2(u) = \sigma^2(v) = 1$. Therefore we have

$$\sigma^2(u + v) = 2(1 + \rho(u, v)) \tag{3.5.16}$$

and correspondingly

$$\sigma^2(u - v) = 2(1 - \rho(u, v)) \quad . \tag{3.5.17}$$

Since the variance always fulfills

$$\sigma^2 \geq 0 \quad , \tag{3.5.18}$$

it follows that

$$-1 \leq \rho(u, v) \leq 1 \quad . \tag{3.5.19}$$

If one now returns to the original variables x, y, then it is easy to show that

$$\rho(u, v) = \rho(x, y) \quad . \tag{3.5.20}$$

Thus we have finally shown that

$$-1 \leq \rho(x, y) \leq 1 \quad . \tag{3.5.21}$$

We now investigate the limiting cases ± 1. For $\rho(u, v) = 1$ the variance is $\sigma(u - v) = 0$, i.e., the random variable $(u - v)$ is a constant. Expressed in terms of x, y one has therefore

$$u - v = \frac{x - \widehat{x}}{\sigma(x)} - \frac{y - \widehat{y}}{\sigma(y)} = \text{const} \quad . \tag{3.5.22}$$

The equation is always fulfilled if

$$y = a + bx \quad , \tag{3.5.23}$$

where b is positive. Therefore in the case of a linear dependence (b positive) between x and y the correlation coefficient takes the value $\rho(\mathsf{x},\mathsf{y}) = +1$. Correspondingly one finds $\rho(\mathsf{x},\mathsf{y}) = -1$ for a negative linear dependence (b negative). We would expect the covariance to vanish for two independent variables x and y, i.e., for which the probability density obeys Eq. (3.4.6). Indeed with (3.5.9) and (3.5.1) we find

$$\begin{aligned}
\operatorname{cov}(x,y) &= \int_{-\infty}^{\infty}\int_{-\infty}^{\infty}(x-\widehat{x})(y-\widehat{y})g(x)h(y)\,\mathrm{d}x\,\mathrm{d}y \\
&= \left(\int_{-\infty}^{\infty}(x-\widehat{x})g(x)\,\mathrm{d}x\right)\left(\int_{-\infty}^{\infty}(y-\widehat{y})h(y)\,\mathrm{d}y\right) \\
&= 0 \quad .
\end{aligned}$$

3.6 More than Two Variables. Vector and Matrix Notation

In analogy to (3.4.1) we now define a *distribution function of n variables* $\mathsf{x}_1, \mathsf{x}_2, \ldots, \mathsf{x}_n$:

$$F(x_1,x_2,\ldots,x_n) = P(\mathsf{x}_1 < x_1,\ \mathsf{x}_2 < x_2,\ \ldots, \mathsf{x}_n < x_n) \quad . \tag{3.6.1}$$

If the function F is differentiable with respect to the x_i, then the *joint probability density* is given by

$$f(x_1,x_2,\ldots,x_n) = \frac{\partial^n}{\partial x_1\,\partial x_2\cdots\partial x_n}F(x_1,x_2,\ldots,x_n) \quad . \tag{3.6.2}$$

The probability density of one of the variables x_r, the *marginal probability density*, is given by

$$g_r(x_r) = \int_{-\infty}^{\infty}\cdots\int_{-\infty}^{\infty} f(x_1,x_2,\ldots,x_n)\,\mathrm{d}x_1\cdots\mathrm{d}x_{r-1}\,\mathrm{d}x_{r+1}\cdots\mathrm{d}x_n \quad . \tag{3.6.3}$$

If $H(\mathsf{x}_1,\mathsf{x}_2,\ldots,\mathsf{x}_n)$ is a function of n variables, then the *expectation value* of H is

$$\begin{aligned}
&E\{H(\mathsf{x}_1,\mathsf{x}_2,\ldots,\mathsf{x}_n)\} \\
&= \int_{-\infty}^{\infty}\cdots\int_{-\infty}^{\infty} H(x_1,x_2,\ldots,x_n)f(x_1,x_2,\ldots,x_n)\,\mathrm{d}x_1\cdots\mathrm{d}x_n \quad .
\end{aligned} \tag{3.6.4}$$

With $H(\mathsf{x}) = \mathsf{x}_r$ one obtains

$$\begin{aligned}
E(\mathsf{x}_r) &= \int_{-\infty}^{\infty}\cdots\int_{-\infty}^{\infty} x_r f(x_1,x_2,\ldots,x_n)\,\mathrm{d}x_1\cdots\mathrm{d}x_n \quad , \\
E(\mathsf{x}_r) &= \int_{-\infty}^{\infty} x_r g_r(x_r)\,\mathrm{d}x_r \quad .
\end{aligned} \tag{3.6.5}$$

The variables are *independent* if

$$f(x_1, x_2, \ldots, x_n) = g_1(x_1) g_2(x_2) \cdots g_n(x_n) \quad . \tag{3.6.6}$$

In analogy to Eq. (3.6.3) one can define the joint marginal probability density of ℓ out of the n variables,* by integrating (3.6.3) over only the $n-\ell$ remaining variables,

$$g(x_1, x_2, \ldots, x_\ell) = \int_{-\infty}^{\infty} \cdots \int_{-\infty}^{\infty} f(x_1, x_2, \ldots, x_n)\, \mathrm{d}x_{\ell+1} \cdots \mathrm{d}x_n \quad . \tag{3.6.7}$$

These ℓ variables are independent if

$$g(x_1, x_2, \ldots, x_\ell) = g_1(x_1) g_2(x_2) \cdots g_\ell(x_\ell) \quad . \tag{3.6.8}$$

The *moments of order* $\ell_1, \ell_2, \ldots, \ell_n$ *about the origin* are the expectation values of the functions

$$H = x_1^{\ell_1} x_2^{\ell_2} \cdots x_n^{\ell_n} \quad ,$$

that is,

$$\lambda_{\ell_1 \ell_2 \ldots \ell_n} = E(\mathrm{x}_1^{\ell_1} \mathrm{x}_2^{\ell_2} \cdots \mathrm{x}_n^{\ell_n}) \quad .$$

In particular one has

$$\begin{aligned} \lambda_{100\ldots0} &= E(\mathrm{x}_1) = \widehat{x}_1 \quad , \\ \lambda_{010\ldots0} &= E(\mathrm{x}_2) = \widehat{x}_2 \quad , \\ &\vdots \\ \lambda_{000\ldots1} &= E(\mathrm{x}_n) = \widehat{x}_n \quad . \end{aligned} \tag{3.6.9}$$

The moments about $(\widehat{x}_1, \widehat{x}_2, \ldots, \widehat{x}_n)$ are correspondingly

$$\mu_{\ell_1 \ell_2 \ldots \ell_n} = E\{(\mathrm{x}_1 - \widehat{x}_1)^{\ell_1} (\mathrm{x}_2 - \widehat{x}_2)^{\ell_2} \cdots (\mathrm{x}_n - \widehat{x}_n)^{\ell_n}\} \quad . \tag{3.6.10}$$

The *variances* of the x_i are then

$$\begin{aligned} \mu_{200\ldots0} &= E\{(\mathrm{x}_1 - \widehat{x}_1)^2\} = \sigma^2(\mathrm{x}_1) \quad , \\ \mu_{020\ldots0} &= E\{(\mathrm{x}_2 - \widehat{x}_2)^2\} = \sigma^2(\mathrm{x}_2) \quad , \\ &\vdots \\ \mu_{000\ldots2} &= E\{(\mathrm{x}_n - \widehat{x}_n)^2\} = \sigma^2(\mathrm{x}_n) \quad . \end{aligned} \tag{3.6.11}$$

The moment with $\ell_i = \ell_j = 1$, $\ell_k = 0$ $(i \neq k \neq j)$ is called the *covariance* between the variables x_i and x_j,

$$c_{ij} = \mathrm{cov}(\mathrm{x}_i, \mathrm{x}_j) = E\{(\mathrm{x}_i - \widehat{x}_i)(\mathrm{x}_j - \widehat{x}_j)\} \quad . \tag{3.6.12}$$

---

*Without loss of generality we can take these to be the variables $\mathrm{x}_1, \mathrm{x}_2, \ldots, \mathrm{x}_\ell$.

It proves useful to represent the n variables $\mathrm{x}_1, \mathrm{x}_2, \ldots, \mathrm{x}_n$ as components of a vector $\mathbf{x}$ in an n-dimensional space. We can then write the distribution function (3.6.1) as

$$F = F(\mathbf{x}) \quad . \tag{3.6.13}$$

Correspondingly, the probability density (3.6.2) is then

$$f(\mathbf{x}) = \frac{\partial^n}{\partial x_1 \partial x_2 \cdots \partial x_n} F(\mathbf{x}) \quad . \tag{3.6.14}$$

The expectation value of a function $H(\mathbf{x})$ is then simply

$$E\{H(\mathbf{x})\} = \int H(\mathbf{x}) f(\mathbf{x})\, \mathrm{d}\mathbf{x} \quad . \tag{3.6.15}$$

We would now like to express the variances and covariances by means of a matrix.† This is called the *covariance matrix*

$$C = \begin{pmatrix} c_{11} & c_{12} & \cdots & c_{1n} \\ c_{21} & c_{22} & \cdots & c_{2n} \\ \vdots & & & \\ c_{n1} & c_{n2} & \cdots & c_{nn} \end{pmatrix} \quad . \tag{3.6.16}$$

The elements c_{ij} are given by (3.6.12); the diagonal elements are the variances $c_{ii} = \sigma^2(\mathrm{x}_i)$. The covariance matrix is clearly symmetric, since

$$c_{ij} = c_{ji} \quad . \tag{3.6.17}$$

If we now also write the expectation values of the x_i as a vector,

$$E(\mathbf{x}) = \widehat{\mathbf{x}} \quad , \tag{3.6.18}$$

we see that each element of the covariance matrix

$$c_{ij} = E\{(\mathrm{x}_i - \widehat{x}_i)(\mathrm{x}_j - \widehat{x}_j)^{\mathrm{T}}\}$$

is given by the expectation value of the product of the row vector $(\mathbf{x} - \widehat{\mathbf{x}})^{\mathrm{T}}$ and the column vector $(\mathbf{x} - \widehat{\mathbf{x}})$, where

$$\mathbf{x}^{\mathrm{T}} = (x_1, x_2, \ldots, x_n) \quad , \qquad \mathbf{x} = \begin{pmatrix} x_1 \\ x_2 \\ \vdots \\ x_n \end{pmatrix} \quad .$$

The covariance matrix can therefore be written simply as

$$C = E\{(\mathbf{x} - \widehat{\mathbf{x}})(\mathbf{x} - \widehat{\mathbf{x}})^{\mathrm{T}}\} \quad . \tag{3.6.19}$$

† For details on matrix notation see Appendix A.

3.7 Transformation of Variables

As already mentioned in Sect. 3.3, a function of a random variable is itself a random variable, e.g.,

$$\mathsf{y} = \mathsf{y}(\mathsf{x}) \quad .$$

We now ask for the probability density $g(y)$ for the case where the probability density $f(x)$ is known.

Clearly the probability

$$g(y)\,\mathrm{d}y$$

that y falls into a small interval $\mathrm{d}y$ must be equal to the probability $f(x)\,\mathrm{d}x$ that x falls into the "corresponding interval" $\mathrm{d}x$, $f(x)\,\mathrm{d}x = g(y)\,\mathrm{d}y$. This is illustrated in Fig. 3.9. The intervals $\mathrm{d}x$ and $\mathrm{d}y$ are related by

$$\mathrm{d}y = \left|\frac{\mathrm{d}y}{\mathrm{d}x}\right| \mathrm{d}x \quad , \qquad \text{i.e.,} \qquad \mathrm{d}x = \left|\frac{\mathrm{d}x}{\mathrm{d}y}\right| \mathrm{d}y \quad .$$

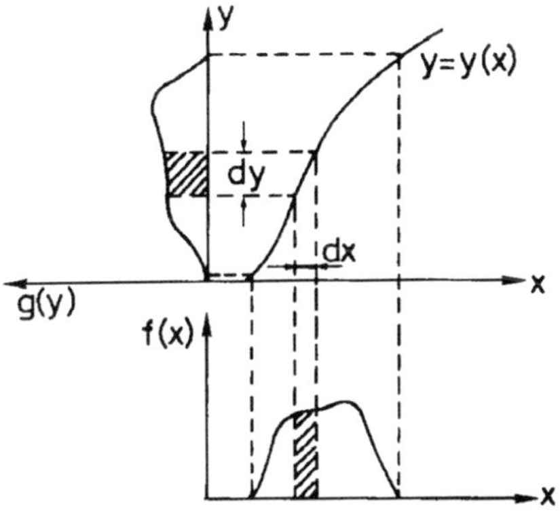

Fig. 3.9. Transformation of variables for a probability density of x to y.

The absolute value ensures that we consider the values $\mathrm{d}x$, $\mathrm{d}y$ as intervals without a given direction. Only in this way are the probabilities $f(x)\,\mathrm{d}x$ and $g(x)\,\mathrm{d}y$ always positive. The probability density is then given by

$$g(y) = \left|\frac{\mathrm{d}x}{\mathrm{d}y}\right| f(x) \quad . \tag{3.7.1}$$

We see immediately that $g(y)$ is defined only in the case of a single-valued function $y(x)$ since only then is the derivative in (3.7.1) uniquely defined. For functions where this is not the case, e.g., $y = \sqrt{x}$, one must consider the individual single-valued parts separately, i.e., $y = +\sqrt{x}$. Equation (3.7.1) also guarantees that the probability distribution of y is normalized to unity:

$$\int_{-\infty}^{\infty} g(y)\,\mathrm{d}y = \int_{-\infty}^{\infty} f(x)\,\mathrm{d}x = 1 \quad .$$

In the case of two independent variables x, y the transformation to the new variables

$$\mathsf{u} = \mathsf{u}(\mathsf{x}, \mathsf{y}) \quad , \qquad \mathsf{v} = \mathsf{v}(\mathsf{x}, \mathsf{y}) \tag{3.7.2}$$

can be illustrated in a similar way. One must find the quantity J that relates the probabilities $f(x, y)$ and $g(u, v)$:

$$g(u,v) = f(x,y) \left| J\left(\frac{x,y}{u,v}\right) \right| \quad . \tag{3.7.3}$$

Figure 3.10 shows in the (x, y) plane two lines each for $u =$ const and $v =$ const. They bound the surface element $\mathrm{d}A$ of the transformed variables u, v corresponding to the element $\mathrm{d}x\,\mathrm{d}y$ of the original variables.

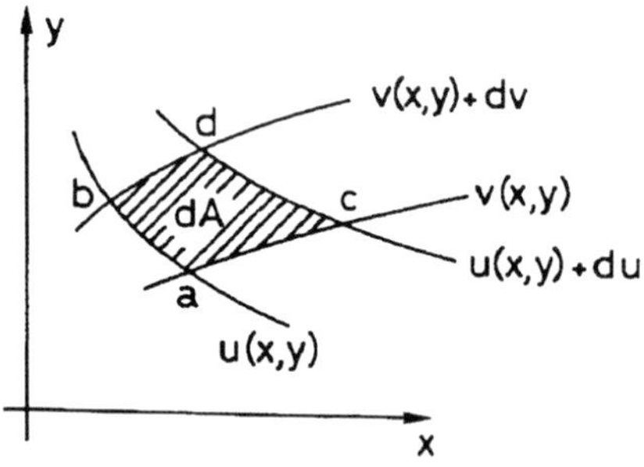

Fig. 3.10. Variable transformation from x, y to u, v.

These curves of course need not be straight lines. Since, however, $\mathrm{d}A$ is an "infinitesimal" surface element, it can be treated as a parallelogram, whose area we will now compute. The coordinates of the corner points a, b, c are

$$\begin{aligned} x_a &= x(u,v) \quad , & y_a &= y(u,v) \quad , \\ x_b &= x(u, v+\mathrm{d}v) \quad , & y_b &= y(u, v+\mathrm{d}v) \quad , \\ x_c &= x(u+\mathrm{d}u, v) \quad , & y_c &= y(u+\mathrm{d}u, v) \quad . \end{aligned}$$

We can expand the last two lines in a series and obtain

$$\begin{aligned} x_b &= x(u,v) + \frac{\partial x}{\partial v}\,\mathrm{d}v \quad , & y_b &= y(u,v) + \frac{\partial y}{\partial v}\,\mathrm{d}v \quad , \\ x_c &= x(u,v) + \frac{\partial x}{\partial u}\,\mathrm{d}u \quad , & y_c &= y(u,v) + \frac{\partial y}{\partial u}\,\mathrm{d}u \quad . \end{aligned}$$

The area of the parallelogram is equal to the absolute value of the determinant

$$\mathrm{d}A = \begin{vmatrix} 1 & x_a & y_a \\ 1 & x_b & y_b \\ 1 & x_c & y_c \end{vmatrix} = \begin{vmatrix} \frac{\partial x}{\partial u} & \frac{\partial y}{\partial u} \\ \frac{\partial x}{\partial v} & \frac{\partial y}{\partial v} \end{vmatrix} \mathrm{d}u\,\mathrm{d}v = J\left(\frac{x,y}{u,v}\right)\mathrm{d}u\,\mathrm{d}v \quad , \tag{3.7.4}$$

where the sign is of no consequence because of the absolute value in Eq. (3.7.3). The expression

$$J\left(\frac{x,y}{u,v}\right) = \begin{vmatrix} \frac{\partial x}{\partial u} & \frac{\partial y}{\partial u} \\ \frac{\partial x}{\partial v} & \frac{\partial y}{\partial v} \end{vmatrix} \tag{3.7.5}$$

is called the *Jacobian (determinant)* of the transformation (3.7.2).

For the general case of n variables $\mathbf{x} = (\mathsf{x}_1, \mathsf{x}_2, \ldots, \mathsf{x}_n)$ and the transformation

$$\begin{aligned} \mathsf{y}_1 &= \mathsf{y}_1(\mathbf{x}) \quad , \\ \mathsf{y}_2 &= \mathsf{y}_2(\mathbf{x}) \quad , \\ &\vdots \\ \mathsf{y}_n &= \mathsf{y}_n(\mathbf{x}) \quad , \end{aligned} \tag{3.7.6}$$

the probability density of the transformed variables is given by

$$g(\mathbf{y}) = \left| J\left(\frac{\mathbf{x}}{\mathbf{y}}\right) \right| f(\mathbf{x}) \quad , \tag{3.7.7}$$

where the Jacobian is

$$J\left(\frac{\mathbf{x}}{\mathbf{y}}\right) = J\left(\frac{x_1, x_2, \ldots, x_n}{y_1, y_2, \ldots, y_n}\right) = \begin{vmatrix} \frac{\partial x_1}{\partial y_1} & \frac{\partial x_2}{\partial y_1} & \cdots & \frac{\partial x_n}{\partial y_1} \\ \frac{\partial x_1}{\partial y_2} & \frac{\partial x_2}{\partial y_2} & \cdots & \frac{\partial x_n}{\partial y_2} \\ \vdots & & & \\ \frac{\partial x_1}{\partial y_n} & \frac{\partial x_2}{\partial y_n} & \cdots & \frac{\partial x_n}{\partial y_n} \end{vmatrix} \quad . \tag{3.7.8}$$

A requirement for the existence of $g(\mathbf{y})$ is of course again the uniqueness of all derivatives occurring in J.

3.8 Linear and Orthogonal Transformations. Error Propagation

In practice we deal frequently with linear transformations of variables. The main reason is that they are particularly easy to handle, and we try therefore to approximate other transformations by linear ones using Taylor series techniques.

Consider r linear functions of the n variables $\mathbf{x} = (\mathsf{x}_1, \mathsf{x}_2, \ldots, \mathsf{x}_n)$:

$$\begin{aligned} \mathsf{y}_1 &= a_1 + t_{11}\mathsf{x}_1 + t_{12}\mathsf{x}_2 + \cdots + t_{1n}\mathsf{x}_n \quad , \\ \mathsf{y}_2 &= a_2 + t_{21}\mathsf{x}_1 + t_{22}\mathsf{x}_2 + \cdots + t_{2n}\mathsf{x}_n \quad , \\ &\vdots \\ \mathsf{y}_r &= a_r + t_{r1}\mathsf{x}_1 + t_{r2}\mathsf{x}_2 + \cdots + t_{rn}\mathsf{x}_n \quad , \end{aligned} \tag{3.8.1}$$

or in matrix notation,

$$\mathbf{y} = T\mathbf{x} + \mathbf{a} \quad . \tag{3.8.2}$$

The expectation value of $\mathbf{y}$ follows from the generalization of (3.5.3)

$$E(\mathbf{y}) = \widehat{\mathbf{y}} = T\widehat{\mathbf{x}} + \mathbf{a} \quad . \tag{3.8.3}$$

Together with (3.6.19) one obtains the covariance matrix for $\mathbf{y}$,

$$\begin{aligned} C_y &= E\{(\mathbf{y} - \widehat{\mathbf{y}})(\mathbf{y} - \widehat{\mathbf{y}})^{\mathrm{T}}\} \\ &= E\{(T\mathbf{x} + \mathbf{a} - T\widehat{\mathbf{x}} - \mathbf{a})(T\mathbf{x} + \mathbf{a} - T\widehat{\mathbf{x}} - \mathbf{a})^{\mathrm{T}}\} \\ &= E\{T(\mathbf{x} - \widehat{\mathbf{x}})(\mathbf{x} - \widehat{\mathbf{x}})^{\mathrm{T}}T^{\mathrm{T}}\} \\ &= TE\{(\mathbf{x} - \widehat{\mathbf{x}})(\mathbf{x} - \widehat{\mathbf{x}})^{\mathrm{T}}\}T^{\mathrm{T}} \quad , \\ C_y &= TC_xT^{\mathrm{T}} \quad . \end{aligned} \tag{3.8.4}$$

Equation (3.8.4) expresses the well-known law of *error propagation*. Suppose the expectation values $\widehat{x}_i$ have been measured. Suppose as well that the errors (i.e., the standard deviations or variances) and the covariances of $\mathbf{x}$ are known. One would like to know the errors of an arbitrary function $\mathbf{y}(\mathbf{x})$. If the errors are relatively small, then the probability density is only significantly large in a small region (on the order of the standard deviation) around the point $\widehat{\mathbf{x}}$. One then performs a Taylor expansion of the functions,

$$\begin{aligned} \mathsf{y}_i &= \mathsf{y}_i(\widehat{x}) + \left(\frac{\partial y_i}{\partial x_1}\right)_{\mathbf{x}=\widehat{\mathbf{x}}} (\mathsf{x}_1 - \widehat{x}_1) + \cdots + \left(\frac{\partial y_i}{\partial x_n}\right)_{\mathbf{x}=\widehat{\mathbf{x}}} (\mathsf{x}_n - \widehat{x}_n) \\ &\quad + \text{higher-order terms} \quad , \end{aligned}$$

or in matrix notation,

$$\mathbf{y} = \mathbf{y}(\widehat{\mathbf{x}}) + T(\mathbf{x} - \widehat{\mathbf{x}}) + \text{ higher-order terms} \tag{3.8.5}$$

with

$$T = \begin{pmatrix} \dfrac{\partial y_1}{\partial x_1} & \dfrac{\partial y_1}{\partial x_2} & \cdots & \dfrac{\partial y_1}{\partial x_n} \\ \dfrac{\partial y_2}{\partial x_1} & \dfrac{\partial y_2}{\partial x_2} & \cdots & \dfrac{\partial y_2}{\partial x_n} \\ \vdots & & & \\ \dfrac{\partial y_r}{\partial x_1} & \dfrac{\partial y_r}{\partial x_2} & \cdots & \dfrac{\partial y_r}{\partial x_n} \end{pmatrix}_{\mathbf{x}=\widehat{\mathbf{x}}} \quad . \tag{3.8.6}$$

If one neglects the higher-order terms and substitutes the first partial derivatives of the matrix T into Eq. (3.8.4), then one obtains the law of error propagation. We see in particular that not only the errors (i.e., the variances) of $\mathbf{x}$ but *also the covariances* make a significant contribution to the errors of $\mathbf{y}$, that is, to the diagonal elements of C_y. If these are not taken into account in the error propagation, then the result cannot be trusted.

The covariances can only be neglected when they vanish anyway, i.e., in the case of independent original variables $\mathbf{x}$. In this case C_x simplifies to a diagonal matrix. The diagonal elements of C_y then have the simple form

$$\sigma^2(\mathsf{y}_i) = \sum_{j=1}^{n} \left(\frac{\partial y_i}{\partial x_j} \right)^2_{\mathbf{x}=\widehat{\mathbf{x}}} \sigma^2(\mathsf{x}_j) \quad . \tag{3.8.7}$$

If we now call the standard deviation, i.e., the positive square root of the variance, the error of the corresponding quantity and use for this the symbol Δ, Eq. (3.8.7) leads immediately to the formula

$$\Delta y_i = \sqrt{\sum_{j=1}^{n} \left(\frac{\partial y_i}{\partial x_j} \right)^2 (\Delta x_j)^2} \quad , \tag{3.8.8}$$

known commonly as the law of the propagation of errors. It must be emphasized that this expression is incorrect in cases of non-vanishing covariances. This is illustrated in the following example.

Example 3.6: Error propagation and covariance

In a Cartesian coordinate system a point (x, y) is measured. The measurement is performed with a coordinate measuring device whose error in y is three times larger than that in x. The measurements of x and y are independent. We therefore can take the covariance matrix to be (up to a factor common to all elements)

$$C_{x,y} = \begin{pmatrix} 1 & 0 \\ 0 & 9 \end{pmatrix} \quad .$$

We now evaluate the errors (i.e., the covariance matrix) in polar coordinates

$$r = \sqrt{(x^2 + y^2)} \quad , \qquad \varphi = \arctan \frac{y}{x} \quad .$$

The transformation matrix (3.8.6) is

$$T = \begin{pmatrix} \frac{x}{r} & \frac{y}{r} \\ -\frac{y}{r^2} & \frac{x}{r^2} \end{pmatrix} \quad .$$

To simplify the numerical calculations we consider only the point $(1, 1)$. Then

$$T = \begin{pmatrix} \frac{1}{\sqrt{2}} & \frac{1}{\sqrt{2}} \\ -\frac{1}{2} & \frac{1}{2} \end{pmatrix}$$

and therefore

$$C_{r\varphi} = \begin{pmatrix} \frac{1}{\sqrt{2}} & \frac{1}{\sqrt{2}} \\ -\frac{1}{2} & \frac{1}{2} \end{pmatrix} \begin{pmatrix} 1 & 0 \\ 0 & 9 \end{pmatrix} \begin{pmatrix} \frac{1}{\sqrt{2}} & -\frac{1}{2} \\ \frac{1}{\sqrt{2}} & \frac{1}{2} \end{pmatrix} = \begin{pmatrix} 5 & \frac{4}{\sqrt{2}} \\ \frac{4}{\sqrt{2}} & \frac{5}{2} \end{pmatrix} \quad .$$

We can now return to the original Cartesian coordinate system

$$x = r\cos\varphi \quad , \qquad y = r\sin\varphi$$

by use of the transformation

$$T' = \begin{pmatrix} \cos\varphi & -r\sin\varphi \\ \sin\varphi & r\cos\varphi \end{pmatrix} = \begin{pmatrix} \frac{1}{\sqrt{2}} & -1 \\ \frac{1}{\sqrt{2}} & 1 \end{pmatrix} \quad .$$

As expected we obtain

$$C_{xy} = \begin{pmatrix} \frac{1}{\sqrt{2}} & -1 \\ \frac{1}{\sqrt{2}} & 1 \end{pmatrix} \begin{pmatrix} 5 & \frac{4}{\sqrt{2}} \\ \frac{4}{\sqrt{2}} & \frac{5}{2} \end{pmatrix} \begin{pmatrix} \frac{1}{\sqrt{2}} & \frac{1}{\sqrt{2}} \\ -1 & 1 \end{pmatrix} = \begin{pmatrix} 1 & 0 \\ 0 & 9 \end{pmatrix} \quad .$$

If instead we had used the formula (3.8.8), i.e., if we had neglected the covariances in the transformation of r, φ to x, y, then we would have obtained

$$C'_{xy} = \begin{pmatrix} \frac{1}{\sqrt{2}} & -1 \\ \frac{1}{\sqrt{2}} & 1 \end{pmatrix} \begin{pmatrix} 5 & 0 \\ 0 & \frac{5}{2} \end{pmatrix} \begin{pmatrix} \frac{1}{\sqrt{2}} & \frac{1}{\sqrt{2}} \\ -1 & 1 \end{pmatrix} = \begin{pmatrix} 5 & 0 \\ 0 & 5 \end{pmatrix} \quad ,$$

which is different from the original covariance matrix. This example stresses the importance of covariances, since it is obviously not possible to change errors of measurements by simply transforming back and forth between coordinate systems. ■

Finally we discuss a special type of linear transformation. We consider the case of exactly n functions y of the n variables x. In particular we take $\mathbf{a} = 0$ in (3.8.2). One then has

$$\mathbf{y} = R\mathbf{x} \quad , \tag{3.8.9}$$

where R is a square matrix. We now require that the transformation (3.8.9) leaves the modulus of a vector invariant

$$\mathbf{y}^2 = \sum_{i=1}^{n} \mathrm{y}_i^2 = \mathbf{x}^2 = \sum_{i=1}^{n} \mathrm{x}_i^2 \quad . \tag{3.8.10}$$

Using Eq. (A.1.9) we can write

$$\mathbf{y}^{\mathrm{T}}\mathbf{y} = (R\mathbf{x})^{\mathrm{T}}(R\mathbf{x}) = \mathbf{x}^{\mathrm{T}}R^{\mathrm{T}}R\mathbf{x} = \mathbf{x}^{\mathrm{T}}\mathbf{x} \quad .$$

This means

$$R^{\mathrm{T}}R = I \quad ,$$

or written in terms of components,

$$\sum_{i=1}^{n} r_{ik}r_{i\ell} = \delta_{k\ell} = \begin{cases} 0\,, & \ell \neq k \quad , \\ 1\,, & \ell = k \quad . \end{cases} \tag{3.8.11}$$

A transformation of the type (3.8.9) that fulfills condition (3.8.11) is said to be *orthogonal*. We now consider the determinant of the transformation matrix

$$D = \begin{vmatrix} r_{11} & r_{12} & \cdots & r_{1n} \\ r_{21} & r_{22} & \cdots & r_{2n} \\ \vdots & & & \\ r_{n1} & r_{n2} & \cdots & r_{nn} \end{vmatrix}$$

and form its square. According to rules for computing determinants we obtain from Eq. (3.8.11)

$$D^2 = \begin{vmatrix} 1 & 0 & \cdots & 0 \\ 0 & 1 & \cdots & 0 \\ \cdots & & & \\ 0 & 0 & \cdots & 1 \end{vmatrix} \quad ,$$

i.e., $D = \pm 1$. The determinant D, however, is the Jacobian determinant of the transformation (3.8.9),

$$J\left(\frac{\mathbf{y}}{\mathbf{x}}\right) = \pm 1 \quad . \tag{3.8.12}$$

We multiply the system of equations (3.8.9) on the left with R^{T} and obtain

$$R^{\mathrm{T}}\mathbf{y} = R^{\mathrm{T}}R\mathbf{x} \quad .$$

Because of Eq. (3.8.11) this expression reduces to

$$\mathbf{x} = R^{\mathrm{T}}\mathbf{y} \quad . \tag{3.8.13}$$

The inverse transformation of an orthogonal transformation is described simply by the transposed transformation matrix. It is itself orthogonal.

An important property of any linear transformation of the type

$$\mathrm{y}_1 = r_{11}\mathrm{x}_1 + r_{12}\mathrm{x}_2 + \cdots + r_{1n}\mathrm{x}_n$$

is the following. By constructing additional functions $\mathrm{y}_2, \mathrm{y}_3, \ldots, \mathrm{y}_n$ of equivalent form it can be extended to yield an orthogonal transformation as long as the condition

$$\sum_{i=1}^{n} r_{1i}^2 = 1$$

is fulfilled.

3.9 Problems

3.9.1 Mean, Variance, and Skewness of a Discrete Distribution

The throwing of a die yields as possible results $x_i = 1, 2, \ldots, 6$. For an ideally symmetric die one has $p_i = P(x_i) = 1/6$, $i = 1, 2, \ldots, 6$. Determine the expectation value $\widehat{x}$, the variance $\sigma^2(x) = \mu_2$, and the skewness γ of the distribution,

(a) for an ideally symmetric die,

(b) for a die with

$$\begin{array}{lll} p_1 = \frac{1}{6} \quad , & p_2 = \frac{1}{12} \quad , & p_3 = \frac{1}{12} \quad , \\ p_4 = \frac{1}{6} \quad , & p_5 = \frac{3}{12} \quad , & p_6 = \frac{3}{12} \quad . \end{array}$$

3.9.2 Mean, Mode, Median, and Variance of a Continuous Distribution

Consider the probability density $f(x)$ of a *triangular distribution* of the form shown in Fig. 3.11, given by

$$\begin{aligned} f(x) &= 0 \quad , \qquad x < a \quad , \qquad x \geq b \quad , \\ f(x) &= \frac{2}{(b-a)(c-a)}(x-a) \quad , \qquad a \leq x < c \quad , \\ f(x) &= \frac{2}{(b-a)(b-c)}(b-x) \quad , \qquad c \leq x < b \quad . \end{aligned}$$

Determine the mean $\widehat{x}$, the mode x_m, the median $x_{0.5}$, and the variance σ^2 of the distribution. For simplicity choose $c = 0$ (which corresponds to the substitution of x by $x' = x - c$). Give explicit results for the symmetric case $a = -b$ and for the case $a = -2b$.

3.9.3 Transformation of a Single Variable

In Appendix D it is shown that

$$\int_{-\infty}^{\infty} \exp(-x^2/2)\,\mathrm{d}x = \sqrt{2\pi} \quad .$$

Use the transformation $y = x/\sigma$ to show that

$$\int_{-\infty}^{\infty} \exp(-x^2/2\sigma^2)\,\mathrm{d}x = \sigma\sqrt{2\pi} \quad .$$

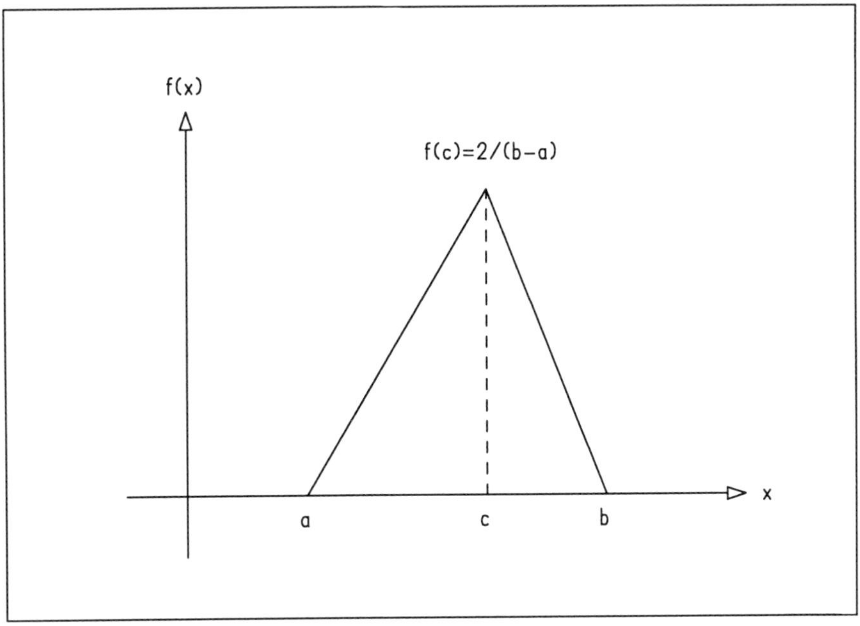

Fig. 3.11. Triangular probability density.

3.9.4 Transformation of Several Variables

A "normal distribution of two variables" (see Sect. 5.10) can take the form

$$f(x,y) = \frac{1}{2\pi\sigma_x\sigma_y} \exp\left(-\frac{1}{2}\frac{x^2}{\sigma_x^2} - \frac{1}{2}\frac{y^2}{\sigma_y^2}\right) \quad .$$

(a) Determine the marginal probability densities $f(x)$, $f(y)$ by using the results of Problem 3.9.3.

(b) Are x and y independent?

(c) Transform the distribution $f(x,y)$ to the variables

$$u = x\cos\phi + y\sin\phi \quad , \qquad v = y\cos\phi - x\sin\phi \quad .$$

(The u, v coordinate system has the same origin as the x, y coordinate system, but it is rotated with respect to the latter by an angle ϕ.)
Hint: Show that the transformation is orthogonal and use (3.8.12).

(d) Show that u and v are independent variables only if $\phi = 0°, 90°, 180°, 270°$ or if $\sigma_x = \sigma_y = \sigma$.

(e) Consider the case $\sigma_x = \sigma_y = \sigma$, i.e.,

$$f(x) = \frac{1}{2\pi\sigma^2} \exp\left[-\frac{1}{2\sigma^2}(x^2 + y^2)\right] \quad .$$

Transform the distribution to polar coordinates r, ϕ, determine the marginal probability densities $g(r)$ and $g(\phi)$ and show that r and ϕ are independent.

3.9.5 Error Propagation

The period T of a pendulum is given by $T = 2\pi\sqrt{\ell/g}$. Here ℓ is the length of the pendulum and g is the gravitational acceleration. Compute g and Δg using the measured values $\ell = 99.8\,\mathrm{cm}$, $\Delta\ell = 0.3\,\mathrm{cm}$, $T = 2.03\,\mathrm{s}$, $\Delta T = 0.05\,\mathrm{s}$ and assuming that the measurements of ℓ and T are uncorrelated.

3.9.6 Covariance and Correlation

We denote the mass and the velocity of an object by m and v and their measurement errors by $\Delta m = \sqrt{\sigma^2(m)}$ and $\Delta v = \sqrt{\sigma^2(v)}$. The measurements are assumed to be independent, i.e., $\mathrm{cov}(m, v) = 0$. Furthermore, the relative errors of measurement are known, i.e.,

$$\Delta m/m = a \quad , \qquad \Delta v/v = b \quad .$$

(a) Consider the momentum $p = mv$ and the kinetic energy $E = \frac{1}{2}mv^2$ of the object and compute $\sigma^2(p)$, $\sigma^2(E)$, $\mathrm{cov}(p, E)$, and the correlation $\rho(p, E)$. Discuss $\rho(p, E)$ for the special cases $a = 0$ and $b = 0$. Hint: Form the vectors $\mathbf{x} = (m, v)$ and $\mathbf{y} = (p, E)$. Then approximate $\mathbf{y} = \mathbf{y}(\mathbf{x})$ by a linear transformation and finally compute the covariance matrix.

(b) For the case where the measured values of E, p and the covariance matrix are known, compute the mass m and its error by error propagation. Use the results from (a) to verify your result. Note that you will obtain the correct result only if $\mathrm{cov}(p, E)$ is taken into account in the error propagation.

3.10 Hints and Solutions

Problem 3.9.1

(a)

$$\begin{aligned}\widehat{x} &= \sum_{i=1}^{6} x_i p_i = \frac{1}{6}\sum_{i=1}^{6} i = \frac{21}{6} = 3.5 \quad , \\ \sigma^2(x) &= \sum_{i=1}^{6} (x_i - \widehat{x})^2 p_i \\ &= \frac{1}{6}(2.5^2 + 1.5^2 + 0.5^2 + 0.5^2 + 1.5^2 + 2.5^2) \\ &= \frac{2}{6}(6.25 + 2.25 + 0.25) = 2.92 \quad , \\ \mu_3 &= \sum_{i=1}^{6} (x_i - \widehat{x})^3 p_i = \frac{1}{6}\sum_{i=1}^{6} (i - 3.5)^3 = 0 \quad , \\ \gamma &= \mu_3/\sigma^3 = 0 \quad .\end{aligned}$$

(b)

$$\begin{aligned}\widehat{x} &= \frac{1}{12}(2 + 2 + 3 + 8 + 15 + 18) = 4 \quad , \\ \sigma^2(x) &= \frac{1}{12}(2 \cdot 3^2 + 1 \cdot 2^2 + 1 \cdot 1^2 + 3 \cdot 1^2 + 3 \cdot 2^2) \\ &= \frac{38}{12} = 3.167 \quad , \\ \mu_3 &= \frac{1}{12}(-2 \cdot 3^3 - 1 \cdot 2^3 - 1 \cdot 1^3 + 3 \cdot 1^3 + 3 \cdot 2^3) = -3 \quad , \\ \gamma &= \mu_3/\sigma^3 = -3/3.167^{3/2} = -0.533 \quad .\end{aligned}$$

Problem 3.9.2

For $a = -b$:

$$\widehat{x} = 0 \quad , \qquad x_{0.5} = 0 \quad , \qquad \sigma^2(x) = \frac{b^2}{6} \quad .$$

For $a = -2b$:

$$\widehat{x} = -\frac{b}{3} = -0.33b \ , \quad x_{0.5} = b(\sqrt{3} - 2) = -0.27b \ , \quad \sigma^2(x) = \frac{7}{18}b^2 \quad .$$

Problem 3.9.3

$$\exp(-y^2/2) = \frac{\mathrm{d}x}{\mathrm{d}y}\exp(-x^2/2) = \sigma\exp(-x^2/2) \quad .$$

Problem 3.9.4

(a) $f_x(x) = \dfrac{1}{\sqrt{2\pi}\sigma_x} \exp\left(-\dfrac{1}{2}\dfrac{x^2}{\sigma_x^2}\right) \quad , \quad f_y(y) = \dfrac{1}{\sqrt{2\pi}\sigma_y} \exp\left(-\dfrac{1}{2}\dfrac{y^2}{\sigma_y^2}\right) \quad .$

(b) Yes, since (3.4.6) is fulfilled.

(c) The transformation can be written in the form

$$\begin{pmatrix} u \\ v \end{pmatrix} = R\begin{pmatrix} x \\ y \end{pmatrix} \quad , \qquad R = \begin{pmatrix} \cos\phi & \sin\phi \\ -\sin\phi & \cos\phi \end{pmatrix} \quad .$$

It is orthogonal, since $R^{\mathrm{T}}R = I$. Therefore,

$$\begin{aligned} g(u,v) &= f(x,y) = \frac{1}{2\pi\sigma_x\sigma_y} \exp\left(-\frac{x^2}{2\sigma_x^2} - \frac{y^2}{2\sigma_y^2}\right) \\ &= \frac{1}{2\pi\sigma_x\sigma_y} \exp\left(-\frac{(u\cos\phi - v\sin\phi)^2}{2\sigma_x^2}\right. \\ &\qquad \left. - \frac{(u\sin\phi + v\cos\phi)^2}{2\sigma_y^2}\right) \\ &= \frac{1}{2\pi\sigma_x\sigma_y} \exp\left(-\frac{u^2\cos^2\phi + v^2\sin^2\phi - 2uv\cos\phi\sin\phi}{2\sigma_x^2}\right. \\ &\qquad \left. - \frac{u^2\sin^2\phi + v^2\cos^2\phi + 2uv\cos\phi\sin\phi}{2\sigma_y^2}\right) \quad . \end{aligned}$$

(d) For $\phi = 90°$: $\cos\phi = 0$, $\sin\phi = 1$, etc. The expression $g(u,v)$ factorizes, i.e., $g(u,v) = g_u(u)g_v(v)$.

For $\sigma_x = \sigma_y = \sigma$:

$$g(u,v) = \frac{1}{2\pi\sigma^2} \exp\left(-\frac{u^2+v^2}{2\sigma^2}\right) = g(u)g(v) \quad .$$

(e)

$$J = \begin{vmatrix} \dfrac{\partial x}{\partial r} & \dfrac{\partial y}{\partial r} \\[2ex] \dfrac{\partial x}{\partial \phi} & \dfrac{\partial y}{\partial \varphi} \end{vmatrix} = \begin{vmatrix} \cos\phi & \sin\phi \\ -r\sin\phi & r\cos\phi \end{vmatrix} = r \quad ,$$

$$\begin{aligned} g(r,\phi) &= rf(x,y) = \frac{r}{2\pi\sigma^2} \exp\left(-\frac{x^2+y^2}{2\sigma^2}\right) \\ &= \frac{r}{2\pi\sigma^2} \exp\left(-\frac{r^2}{2\sigma^2}\right) \quad ; \end{aligned}$$

$$
\begin{aligned}
g_r(r) &= \int_0^{2x} g(r,\phi)\,\mathrm{d}\phi = \frac{r}{\sigma^2}\exp\left(-\frac{r^2}{2\sigma^2}\right) \quad , \\
g_\phi(\phi) &= \frac{1}{2\pi\sigma^2}\int_0^\infty r\exp\left(-\frac{r^2}{2\sigma^2}\right)\mathrm{d}r \\
&= \frac{1}{4\pi\sigma^2}\int_0^\infty \exp\left(-\frac{u}{2\sigma^2}\right)\mathrm{d}u \\
&= -\frac{1}{2\pi}\left[\exp\left(-\frac{u}{2\sigma^2}\right)\right]_0^\infty = \frac{1}{2\pi} \quad ; \\
g(r,\phi) &= g_r(r)g_\phi(\phi) \quad ,
\end{aligned}
$$

therefore, r and ϕ are independent.

Problem 3.9.5

$$
\begin{aligned}
g &= 4\pi^2\frac{\ell}{T^2} = 4\pi^2\frac{99.8}{2.03^2}\,\mathrm{cm\,s^{-2}} = 956.09\,\mathrm{cm\,s^{-2}} \quad , \\
\frac{\partial g}{\partial \ell} &= \frac{4\pi^2}{T^2} = 9.58\,\mathrm{s^{-2}} \quad , \qquad \frac{\partial g}{\partial T} = -\frac{8\pi^2\ell}{T^3} = -942\,\mathrm{cm\,s^{-3}} \quad , \\
(\Delta g)^2 &= \left(\frac{\partial g}{\partial \ell}\right)^2(\Delta\ell)^2 + \left(\frac{\partial g}{\partial T}\right)^2\Delta T^2 = 2226\,\mathrm{cm^2 s^{-4}} \quad , \\
\Delta g &= 47.19\,\mathrm{cm\,s^{-2}} \quad .
\end{aligned}
$$

Problem 3.9.6

(a) $y = Tx$,

$$
\begin{aligned}
T &= \begin{pmatrix} \dfrac{\partial y_1}{\partial x_1} & \dfrac{\partial y_1}{\partial x_2} \\ \dfrac{\partial y_2}{\partial x_1} & \dfrac{\partial y_2}{\partial x_2} \end{pmatrix} = \begin{pmatrix} \dfrac{\partial p}{\partial m} & \dfrac{\partial p}{\partial v} \\ \dfrac{\partial E}{\partial m} & \dfrac{\partial E}{\partial v} \end{pmatrix} \\
&= \begin{pmatrix} v & m \\ \frac{1}{2}v^2 & mv \end{pmatrix} \quad , \\
C_y &= TC_xT^{\mathrm{T}} \\
&= \begin{pmatrix} v & m \\ \frac{1}{2}v^2 & mv \end{pmatrix}\begin{pmatrix} a^2m^2 & 0 \\ 0 & b^2v^2 \end{pmatrix}\begin{pmatrix} v & \frac{1}{2}v^2 \\ m & mv \end{pmatrix} \\
&= \begin{pmatrix} (a^2+b^2)m^2v^2 & (\frac{1}{2}a^2+b^2)m^2v^3 \\ (\frac{1}{2}a^2+b^2)m^2v^3 & (\frac{1}{4}a^2+b^2)m^2v^4 \end{pmatrix} \quad ,
\end{aligned}
$$

$$\begin{aligned}\rho(p,E) &= \frac{\mathrm{cov}(p,E)}{\sigma(p)\sigma(E)} = \frac{(\frac{1}{2}a^2+b^2)m^2v^3}{\sqrt{(a^2+b^2)m^2v^2}\sqrt{(\frac{1}{4}a^2+b^2)m^2v^4}} \\ &= \frac{\frac{1}{2}a^2+b^2}{\sqrt{(a^2+b^2)(\frac{1}{4}a^2+b^2)}} \quad .\end{aligned}$$

For $a = 0$ or $b = 0$: $\rho = 1$. (In this case either m or v are completely determined. Therefore, there is a strict relation between E and p.) If, however, $a, b \neq 0$, one has $\rho \neq 1$, e.g., for $a = b$ one obtains $\rho = 3/\sqrt{10}$.

(b) $m = \frac{1}{2}p^2/E \quad , \qquad v = E/2p \quad ,$

$$\begin{aligned}C_y &= \begin{pmatrix} (a^2+b^2)p^2 & (a^2+2b^2)Ep \\ (a^2+2b^2)Ep & (a^2+4b^2)E^2 \end{pmatrix} \quad , \\ m &= Ty \quad , \\ T &= \left(\frac{\partial m}{\partial y_1}, \frac{\partial m}{\partial y_2}\right) = \left(\frac{\partial m}{\partial p}, \frac{\partial m}{\partial E}\right) = \left(\frac{p}{E}, -\frac{p^2}{2E^2}\right) \quad , \\ C_m &= \sigma^2(m) = TC_yT^{\mathrm{T}} \\ &= \left(\frac{p}{E}, -\frac{p^2}{2E^2}\right) \begin{pmatrix} (a^2+b^2)p^2 & (a^2+2b^2)Ep \\ (a^2+2b^2)Ep & (a^2+4b^2)E^2 \end{pmatrix} \\ &\quad \times \begin{pmatrix} \frac{p}{E} \\ -\frac{p^2}{2E^2} \end{pmatrix} \\ &= a^2\frac{p^4}{4E^2} = a^2m^2 \quad .\end{aligned}$$

4. Computer Generated Random Numbers. The Monte Carlo Method

4.1 Random Numbers

Up to now in this book we have considered the observation of random variables, but not prescriptions for producing them. It is in many applications useful, however, to have a sequence of values of a randomly distributed variable x. Since operations must often be carried out with a large number of such *random numbers*, it is particularly convenient to have them directly available on a computer. The correct procedure to create such random numbers would be to use a statistical process, e.g., the measurement of the time between two decays from a radioactive source, and to transfer the measured results into the computer. In practical applications, however, the random numbers are almost always calculated directly by the computer. Since this works in a strictly deterministic way, the resulting values are not really random, but rather can be exactly predicted. They are therefore called *pseudorandom*.

Computations with random numbers currently make up a large part of all computer calculations in the planning and evaluation of experiments. The statistical behavior which stems either from the nature of the experiment or from the presence of measurement errors can be simulated on the computer. The use of random numbers in computer programs is often called *the Monte Carlo method.*

We begin this chapter with a discussion of the representation of numbers in a computer (Sect. 4.2), which is indispensable for an understanding of what follows. The best studied method for the creation of uniformly distributed random numbers is the subject of Sects. 4.3 to 4.7. In Sect. 4.8 we present a FORTRAN program which implements this procedure. Sections 4.9 and 4.10 cover the creation of random numbers that follow an arbitrary distribution and the especially common case of normally distributed numbers. In the last two sections one finds discussion and examples of the Monte Carlo method in applications of numerical integration and simulation.

In many examples and exercises we will simulate measurements with the Monte Carlo method and then analyze them. We possess in this way a *computer laboratory*, which allows us to study individually the influence of simulated measurement errors on the results of an analysis.

4.2 Representation of Numbers in a Computer

For most applications the representation of numbers used in a computation is unimportant. It can be of decisive significance, however, for the properties of computer-generated random numbers. We will restrict ourselves to the binary representation, which is used today in practically all computers. The elementary unit of information is the *bit**, which can assume the values of 0 or 1. This is realized physically by two distinguishably different electric or magnetic states of a component in the computer.

If one has k bits available for the representation of an integer, then one bit is sufficient to encode the sign. The remaining $k-1$ bits are used for the binary representation of the absolute value in the form

$$a = a^{(k-2)}2^{k-2} + a^{(k-3)}2^{k-3} + \cdots + a^{(1)}2^{1} + a^{(0)}2^{0} \quad . \tag{4.2.1}$$

Here each of the coefficients $a^{(j)}$ can assume only the values 0 or 1, and thus can be represented by a single bit.

The binary representation for non-negative integers is

$$\begin{aligned} 00\cdots000 &= 0 \\ 00\cdots001 &= 1 \\ 00\cdots010 &= 2 \\ 00\cdots011 &= 3 \\ &\vdots \end{aligned}$$

One could simply use the first bit to encode the sign and represent the corresponding negative numbers such that in the first bit the 0 is replaced by a 1. That would give, however, two different representations for the number zero, or rather $+0$ and -0. In fact, one uses for negative numbers the "complementary representation"

$$\begin{aligned} 11\cdots111 &= -1 \\ 11\cdots110 &= -2 \\ 11\cdots101 &= -3 \\ &\vdots \end{aligned}$$

*Abbreviation of *binary digit*

Then using k bits, integers in the interval

$$-2^{k-1} \le x \le 2^{k-1} - 1 \qquad (4.2.2)$$

can be represented.

In most computers eight bits are grouped together into one *byte*. Four bytes are generally used for the representation of integers, i.e., $k = 32$, $2^{k-1} - 1 = 2\,147\,483\,647$. In many small computers only two bytes are available, $k = 16$, $2^{k-1} - 1 = 32\,767$. This constraint (4.2.2) must be taken into consideration when designing a program to generate random numbers.

Before turning to the representation of fractional numbers in a computer, let us consider a finite decimal fraction, which we can write in various ways, e.g.,

$$x = 17.23 = 0.1723 \cdot 10^2$$

or in general

$$x = M \cdot 10^e \quad .$$

The quantities M and e are called the *mantissa* and *exponent*, respectively. One chooses the exponent such that the mantissa's nonzero digits are all to the right of the decimal point, and the first place after the decimal point is not zero. If one has available n decimal places for the representation of the value M, then

$$m = M \cdot 10^n$$

is an integer. In our example, $n = 4$ and $m = 1723$. In this way the decimal fraction d is represented by the two integers m and e.

The representation of fractions in the binary system is done in a completely analogous way. One decomposes a number of the form

$$x = M \cdot 2^e \qquad (4.2.3)$$

into a mantissa M and exponent e. If n_m bits are available for the representation of the mantissa (including sign), it can be expressed by the integer

$$m = M \cdot 2^{n_m - 1} \quad , \qquad -2^{n_m - 1} \le m \le 2^{n_m - 1} - 1 \quad . \qquad (4.2.4)$$

If the exponent with its sign is represented by n_e bits, then it can cover the interval

$$-2^{n_e} \le e \le 2^{n_e} - 1 \quad . \qquad (4.2.5)$$

Normally 32 bits are used for the representation of floating point numbers, from which usually $n_m = 24$ bits are used to represent the mantissa and $n_e = 8$ for the exponent. For computations of higher accuracy 64 bits are used for floating point numbers.

For the interval of values in which a floating point number can be represented in a computer, the constraint (4.2.2) no longer applies but one has rather the weaker condition

$$2^{e_{\min}} < |x| < 2^{e_{\max}} \quad . \tag{4.2.6}$$

Here $e_{\min}$ and $e_{\max}$ are given by (4.2.5). If eight bits are available for representing the exponent (including sign), then one has $e_{\max} = 2^7 - 1 = 127$. Therefore, one has the constraint $|x| < 2^{127} \approx 1.7 \cdot 10^{38}$.

When computing with floating point numbers, the concept of the *relative precision* of the representation is of considerable significance. There are a fixed number of binary digits corresponding to a fixed number of decimal places available for the representation of the mantissa M. If we designate by α the smallest possible mantissa, then two numbers x_1 and x_2 can still be represented as being distinct if

$$x_1 = x = M \cdot 2^e \quad , \qquad x_2 = (M + \alpha) \cdot 2^e \quad .$$

The *absolute precision* in the representation of x is thus

$$\Delta x = x_1 - x_2 = \alpha \cdot 2^e \quad ,$$

which depends on the exponent of x. The relative precision

$$\frac{\Delta x}{x} = \frac{\alpha}{M}$$

is in contrast independent of x. If n binary digits are available for the representation of the mantissa, then one has $M \approx 2^n$, since the exponent is chosen such that all n places for the mantissa are completely used. The smallest possible mantissa is $\alpha = 2^0$, so that the *relative precision* in the representation of x is

$$\frac{\Delta x}{x} = 2^{-n} \quad . \tag{4.2.7}$$

4.3 Linear Congruential Generators

Since, as mentioned, computers work in a strictly deterministic way, all (pseudo)random numbers generated in a computer are in the most general case a function of all of the preceding (pseudo)random numbers†

$$x_{j+1} = f(x_j, x_{j-1}, \ldots, x_1) \quad . \tag{4.3.1}$$

†Since the numbers are pseudorandom and not strictly random, we use the notation x in place of x

Programs for creating random numbers are called *random number generators*.

The best studied algorithm is based on the following rule,

$$x_{j+1} = (a\,x_j + c) \bmod m \quad . \tag{4.3.2}$$

All of the quantities in (4.3.2) are integer valued. Generators using this rule are called linear congruential generators (LCG). The symbol mod m or modulo m in (4.3.2) means that the expression before the symbol is divided by m and only the remainder of the result is taken, e.g., $6 \bmod 5 = 1$. Each random number made by an LCG according to the rule (4.3.2) depends only on the number immediately preceding it and on the constant a (the *multiplier*), on c (the *increment*), and on m (the *modulus*). When these three constants and one *initial value* x_0 are given, the infinite sequence of random numbers $x_0, x_1, \ldots$ is determined.

The sequence is clearly periodic. The maximum period length is m. Only partial sequences that are short compared to the period length are useful for computations.

Theorem on the maximum period of an LCG with $c \neq 0$:
An LCG defined by the values m, a, c, and x_0 has the period m if and only if

(a) c and m have no common factors;

(b) $b = a - 1$ is a multiple of p for every prime number p that is a factor of m;

(c) b is a multiple of 4 if m is a multiple of 4.

The proof of this theorem as well as the theorems of Sect. 4.4 can be found in, e.g., KNUTH [2].

A simple example is $c = 3$, $a = 5$, $m = 16$. One can easily compute that $x_0 = 0$ results in the sequence

$$0, 3, 2, 13, 4, 7, 6, 1, 8, 11, 10, 5, 12, 15, 14, 9, 0, \ldots \quad .$$

Since the period m can only be attained when all m possible values are actually assumed, the choice of the initial value x_0 is unimportant.

4.4 Multiplicative Linear Congruential Generators

If one chooses $c = 0$ in (4.3.2), then the algorithm simplifies to

$$x_{j+1} = (a x_j) \bmod m \quad . \tag{4.4.1}$$

Generators based on this rule are called multiplicative linear congruential generators (MLCG). The computation becomes somewhat shorter and thus faster. The exact value zero, however, can no longer be produced (except for the unusable sequence $0, 0, \ldots$). In addition the period becomes shorter. Before giving the theorem on the maximum period length for this case, we introduce the concept of the primitive element modulo m.

Let a be an integer having no common factors (except unity) with m. We consider all a for which $a^\lambda \bmod m = 1$ for integer λ. The smallest value of λ for which this relation is valid is called the *order* of a modulo m. All values a having the same largest possible order $\lambda(m)$ are called *primitive elements* modulo m.

Theorem on the order $\lambda(m)$ of a primitive element modulo m: For every integer e and prime number p

$$\begin{aligned} &\lambda(2) = 1 \quad ; \\ &\lambda(4) = 2 \quad ; \\ &\lambda(2^e) = 2^{e-2} \quad , \qquad e > 2 \quad ; \\ &\lambda(p^e) = p^{e-1}(p-1) \quad , \qquad p > 2 \quad . \end{aligned} \tag{4.4.2}$$

Theorem on primitive elements modulo p^e: The number a is a primitive element modulo p^e if and only if

$$\begin{aligned} &a \text{ odd} \;,\quad p^e = 2 \;; \\ &a \bmod 4 = 3 \;,\quad p^e = 4 \;; \\ &a \bmod 8 = 3, 5, 7 \;,\quad p^e = 8 \;; \\ &a \bmod 8 = 3, 5 \;,\quad p = 2 \;,\quad e > 3 \;; \\ &a \bmod p \neq 0 \;,\quad a^{(p-1)/q} \bmod p \neq 1 \;,\quad p > 2 \;,\quad e = 1 \;, \\ &\qquad q \text{ every prime factor of } p-1 \;; \\ &a \bmod p \neq 0 \;,\quad a^{p-1} \bmod p^2 \neq 1 \;,\quad a^{(p-1)/q} \bmod p \neq 1 \;, \\ &\qquad p > 2 \;,\quad e > 1 \;,\quad q \text{ every prime factor of } p-1 \;. \end{aligned} \tag{4.4.3}$$

For large values of p the primitive elements must be determined with computer programs with the aid of this theorem.

Theorem on the maximum period of an MLCG: The maximum period of an MLCG defined by the quantities m, a, $c = 0$, x_0 is equal to the order $\lambda(m)$. This is attained if the multiplier a is a primitive element modulo m and when the initial value x_0 and the multiplier m have no common factors (except unity).

In fact, MLC generators with $c = 0$ are frequently used in practice. There are two cases of practical significance in choosing the multiplier m.

(i) $m = 2^e$: Here $m - 1$ can be the largest integer that can be represented on the computer. According to (4.4.2) the maximum attainable period length is $m/4$.

(ii) $m = p$: If m is a prime number, the period of $m - 1$ can be attained according to (4.4.2).

4.5 Quality of an MLCG. Spectral Test

When producing random numbers, the main goal is naturally not just to attain the longest possible period. This could be achieved very simply with the sequence 0, 1, 2, ..., $m - 1$, 0, 1, ... Much more importantly, the individual elements within a period should follow each other "randomly". First the modulus m is chosen, and then one chooses various multipliers a corresponding to (4.4.3) that guaranty a maximum period. One then constructs generators with the constants a, m, and $c = 0$ in the form of computer programs and checks with statistical tests the randomness of the resulting numbers. General tests, also applicable to this particular question, will be discussed in Sect. 8. The spectral test was especially developed for investigating random numbers, in particular for detecting non-random dependencies between neighboring elements in a sequence.

In a simple example we first consider the case $a = 3$, $m = 7$, $c = 0$, $x_0 = 1$ and obtain the sequence

$$1, 3, 2, 6, 4, 5, 1, \ldots \quad .$$

We now form pairs of neighboring numbers

$$(x_j, x_{j+1}) \quad , \qquad j = 0, 1, \ldots, n-1 \quad . \tag{4.5.1}$$

Here n is the period, which in our example is $n = m - 1 = 6$. In Fig. 4.1 the number pairs (4.5.1) are represented as points in a two-dimensional Cartesian coordinate system. We note – possibly with surprise – that they form a regular lattice. The surprise is somewhat less, however, when we consider two features of the algorithm (4.3.2):

(i) All coordinate values x_j are integers. In the accessible range of values $1 \le x_j \le n$ there are, however, only n^2 number pairs (4.5.1) for which both elements are integer. They lie on a lattice of horizontal and vertical lines. Two neighboring lines have a separation of one.

(ii) There are only n different pairs (4.5.1), so that only a fraction of the n^2 points mentioned in (i) are actually occupied.

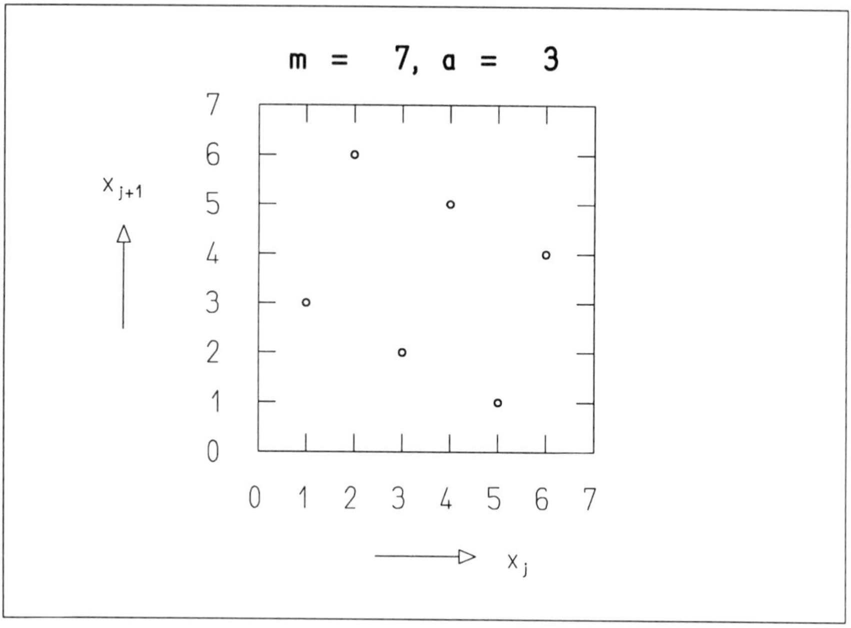

Fig. 4.1. Diagram of number pairs (4.5.1) for $a = 3$, $m = 7$.

We now go from integer numbers x_j to transformed numbers

$$u_j = x_j/m \tag{4.5.2}$$

with the property

$$0 < u_j < 1 \quad . \tag{4.5.3}$$

For simplicity we assume that the sequence $x_0, x_1, \ldots$ has the maximum possible period m for an MLC generator. The pairs

$$(u_j,\ u_{j+1}) \quad , \qquad j = 0, 1, \ldots, m-1 \quad , \tag{4.5.4}$$

lie in a square whose side has unit length. Because the x_j are integers, the spacing between the horizontal or vertical lattice lines on which the points (4.5.4) must lie is $1/m$. By far not all of these points, however, are occupied. A finite family of lines can be constructed which pass through those points that are actually occupied. We consider now the spacing of neighboring lines within a family, look for the family for which this distance is a maximum, and call this d_2.

If the distances between neighboring lattice lines for all families are approximately equal, we can then be certain of having a maximally uniform distribution of the occupied lattice points on the unit square. Since this distance is $1/m$ for a completely occupied lattice (m^2 points), we obtain for a uniformly occupied lattice with m points a distance of $d_2 \approx m^{-1/2}$. With a very nonuniform lattice one obtains the considerably larger value $d_2 \gg m^{-1/2}$.

If one now considers not only pairs (4.5.4), but t-tuples of numbers

$$(u_j\,, u_{j+1}\,, \ldots\,, u_{j+t-1}) \quad , \tag{4.5.5}$$

one sees that the corresponding points lie on families of $(t-1)$-dimensional hyperplanes in a t-dimensional cube whose side has unit length. Let us investigate as before the distance between neighboring hyperplanes of a family. We determine the family with the largest spacing and designate this by d_t. One expects for a uniform distribution of points (4.5.5) a distance

$$d_t \approx m^{-1/t} \quad . \tag{4.5.6}$$

If the lattice is nonuniform, however, we expect

$$d_t \gg m^{-1/t} \quad . \tag{4.5.7}$$

The situations (4.5.6) and (4.5.7) are shown in Fig. 4.2. Naturally one tries to achieve as uniform a lattice as possible. One should note that the smallest decimal places of random numbers are not random, but rather reflect the structure of the lattice.

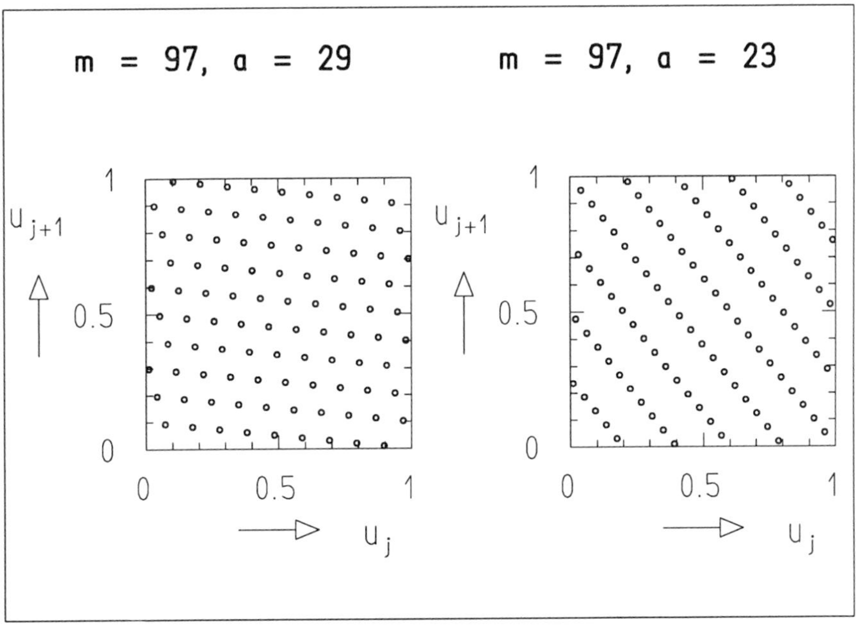

Fig. 4.2. Diagram of number pairs (4.5.4) for various small values of a and m.

Table 4.1. Suitable moduli m and multipliers a for portable MLC generators for computers with 32-bit (16-bit) integer arithmetic.

32 Bit		16 Bit	
m	a	m	a
2 147 483 647	39 373	32 749	162
2 147 483 563	40 014	32 363	157
2 147 483 399	40 692	32 143	160
2 147 482 811	41 546	32 119	172
2 147 482 801	42 024	31 727	146
2 147 482 739	45 742	31 657	142

Theoretical considerations give an upper limit on the smallest possible lattice spacing,

$$d_t \geq d_t^* = c_t m^{-1/t} \quad . \tag{4.5.8}$$

The constants c_t are of order unity. They have the numerical values [2]

$$\begin{aligned} c_2 &= (4/3)^{-1/4} \;, & c_3 &= 2^{-1/6} \;, & c_4 &= 2^{-1/4} \;, & c_5 &= 2^{-3/10} \;, \\ c_6 &= (64/3)^{-1/12} \;, & c_7 &= 2^{-3/7} \;, & c_8 &= 2^{-1/2} \;. \end{aligned} \tag{4.5.9}$$

The *spectral test* can now be carried out as follows. For given values (m, a) of the modulus and multiplier of an MLCG one determines with a computer algorithm [2] the values $d_t(m, a)$ for small t, e.g., $t = 2, 3, \ldots, 6$. One constructs the test quantities

$$S_t(m, a) = \frac{d_t^*(m)}{d_t(m, a)} \tag{4.5.10}$$

and accepts the generator as usable if the $S_t(m, a)$ do not exceed a given limit. Table 4.1 gives the results of extensive investigations by L'ECUYER [3]. The moduli m are prime numbers close to the maximum integer values representable by 16 or 32 bits. The multipliers are primitive elements modulo m. They fulfill the requirement $a < \sqrt{m}$ (see Sect. 4.6). The prime numbers were chosen such that a does not have to be much smaller than $\sqrt{m}$, but the condition (m, a) in Table 4.1 $S_t(m, a) > 0.65$, $t = 2, 3, \ldots, 6$, still applies.

4.6 Implementation and Portability of an MLCG

By *implementation* of an algorithm one means its realization as a computer program for a specific type of computer. If the program can be easily transferred to other computer types and gives there (essentially) the same results, then the program is said to be *portable*. In this section we will give a portable implementation of an MLCG, as realized by WICHMANN and HILL [4] and L'ECUYER [3].

A program that implements the rule (4.4.1) is certain to be portable if the computations are carried out exclusively with integers. If the computer has k bits for the representation of an integer, then all numbers between $-m-1$ and m for $m < 2^{k-1}$ are available.

We now choose a multiplier a with

$$a^2 < m \tag{4.6.1}$$

and define

$$q = m \text{ div } a \quad , \qquad r = m \bmod a \quad , \tag{4.6.2}$$

so that

$$m = aq + r \quad . \tag{4.6.3}$$

The expression m div a defined by (4.6.2) and (4.6.3) is the integer part of the quotient m/a. We now compute the right-hand side of (4.4.1), where we leave off the index j and note that $[(x\text{div } q)m] \bmod m = 0$, since x div q is an integer:

$$\begin{aligned}
[ax] \bmod m &= [ax - (x \text{ div } q)m] \bmod m \\
&= [ax - (x \text{ div } q)(aq + r)] \bmod m \\
&= [a\{x - (x \text{ div } q)q\} - (x \text{ div } q)r] \bmod m \\
&= [a(x \bmod q) - (x \text{ div } q)r] \bmod m \quad .
\end{aligned} \tag{4.6.4}$$

Since one always has $0 < x < m$, it follows that

$$a(x \bmod q) < aq \le m \quad , \tag{4.6.5}$$

$$(x \text{ div } q)r < [(aq + r) \text{ div } q]r = ar < a^2 < m \quad . \tag{4.6.6}$$

In this way both terms in square brackets in the last line of (4.6.4) are less than m, so that the bracketed expression remains in the interval between $-m$ and m.

The expression (4.6.4) can be implemented as a FORTRAN program in the following three lines, in which all variables are integer:

```
K = IX/IQ
IX = IA*(IX-K*IQ)-K*IR
IF(IX.LT.0) IX=IX+M
```

One should note that division of two integer FORTRAN variables results directly in the integer part of the quotient. The first line therefore yields x div q and the last line ax mod m, respectively.

Program 4.1 gives as output a partial sequence of random numbers of length N. Each time the subroutine is called, an additional partial sequence is produced. The period of the entire sequence is $m - 1 = 2\,147\,483\,562$. The computation is carried out entirely with integer arithmetic, ensuring portability. The output values are, however, floating point valued because of the division by m, and therefore correspond to a uniform distribution between 0 and 1.

Often one would like to interrupt a computation requiring many random numbers and continue it later starting from the same place. In this case one can read out and store the last computed (integer) random number directly before the interruption, and use it later for producing the next random number. In the technical terminology one calls such a number the *seed* of the generator.

It is sometimes desirable to be able to produce non-overlapping partial sequences of random numbers not one after the other but rather independently. In this way one can, for example, carry out parts of larger simulation problems simultaneously on several computers. As seeds for such partial sequences one uses elements of the total sequence separated by an amount greater than the length of each partial sequence. Such seeds can be determined without having to run through the entire sequence. From (4.4.1) it follows that

$$x_{j+n} = (a^n x_j) \bmod m = [(a^n \bmod m)x_j] \bmod m \quad . \tag{4.6.7}$$

L'ECUYER [3] suggests setting $n = 2^d$ and choosing some seed x_0. The expression a^{2^d} mod m can be computed by beginning with a and squaring it d times modulo m. Then one computes x_n using (4.6.7) and obtains correspondingly $x_{2n}, x_{3n}, \ldots$. One should note that 32-bit integer arithmetic is not sufficient for this procedure; 64-bit precision must be used instead.

Program 4.1: Subroutine `RNMLCG`. MLC generator for producing uniformly distributed random numbers

CALLING SEQUENCE: `CALL RNMLCG(U,N)`

INPUT ARGUMENT: `N` — number of random numbers

OUTPUT ARGUMENT: `U` — array containing the random numbers in the first `N` positions

REMARKS: The call `CALL RNMSIN(ISEED)` initializes sequence of random numbers with the seed `ISEED`. The call `CALL RNMSOT(ISEED)` delivers as output to the calling program the last produced random number in the argument `ISEED` for later use as a seed.

```
      SUBROUTINE RNMLCG(U,N)
      IMPLICIT DOUBLE PRECISION (A-H,O-Z)
      DIMENSION U(N)
      PARAMETER (M=2147483563, IA=40014, IQ=53668, IR=12211)
      PARAMETER (XMINV=4.656613D-10)
      DATA IX /123456/
C Compute N random numbers
      DO 10 I=1,N
        K=IX/IQ
        IX=IA*(IX-K*IQ)-K*IR
        IF (IX.LT.0) IX=IX+M
        U(I)=DBLE(IX)*XMINV
   10 CONTINUE
      RETURN
C Input of seed
      ENTRY RNMSIN(ISEED)
      IX=ISEED
      RETURN
C Output of seed
      ENTRY RNMSOT(ISEED)
      ISEED=IX
      RETURN
      END
```

We recommend, whenever programming circumstances allow, to produce a relatively large number of random numbers ($N \gg 1$) with each call of the routine and to process them later, since the computing time spent on a single subroutine call can easily be greater than that taken by the few arithmetic operations necessary to produce a single random number.

4.7 Combination of Several MLCGs

Since the period of an MLCG is at most $m - 1$, and since m is restricted to $m < 2^{k-1} - 1$ where k is the number of bits available in the computer for the representation of an integer, only a relatively short period can be attained with a single MLCG. WICHMANN and HILL [4] and L'ECUYER [3] have given a procedure for combining several MLCGs, which allows for very long periods. We will use this as the basis for the program given in Sect. 4.8. The technique is based on the following two theorems.

Theorem on the sum of discrete random variables, one of which comes from a discrete uniform distribution: If $\mathsf{x}_1, \ldots, \mathsf{x}_\ell$ are independent random variables that can only assume integer values, and if x_1 follows a discrete uniform distribution, so that

$$P(\mathsf{x}_1 = n) = \frac{1}{d} \quad , \qquad n = 0, 1, \ldots, d-1 \quad ,$$

then

$$\mathsf{x} = \left(\sum_{j=1}^{\ell} \mathsf{x}_j\right) \bmod d \tag{4.7.1}$$

also follows this distribution.

We first demonstrate the proof for $\ell = 2$, using the abbreviations min $(\mathsf{x}_2) = a$, $\max(\mathsf{x}_2) = b$. One has

$$\begin{aligned} P(\mathsf{x} = n) &= \sum_{k=0}^{\infty} P(\mathsf{x}_1 + \mathsf{x}_2 = n + kd) \\ &= \sum_{i=a}^{b} P(\mathsf{x}_2 = i) P(\mathsf{x}_1 = (n-i) \bmod d) \\ &= \frac{1}{d} \sum_{i=a}^{b} P(\mathsf{x}_2 = i) = \frac{1}{d} \quad . \end{aligned}$$

For $\ell = 3$ we first construct the variable $\mathsf{x}'_1 = \mathsf{x}_1 + \mathsf{x}_2$, which follows a discrete uniform distribution between 0 and $d - 1$, and then the sum $\mathsf{x}'_1 + \mathsf{x}_3$, which has only two terms and therefore possesses the same property. The generalization for $\ell > 3$ is obvious.

Theorem on the period of a family of generators: Consider the random variables $\mathsf{x}_{j,i}$ coming from a generator j with a period p_j, so that the generator gives a sequence $\mathsf{x}_{j,0}, \mathsf{x}_{j,1}, \ldots, \mathsf{x}_{j,p_j-1}$. We consider now ℓ generators $j = 1, 2, \ldots, \ell$ and the sequence of ℓ-tuples

$$\mathbf{x}_i = \{\mathsf{x}_{1,i}, \mathsf{x}_{2,i}, \ldots, \mathsf{x}_{\ell,i}\} \quad , \qquad i = 0, 1, \ldots \quad . \tag{4.7.2}$$

Its period p is the smallest common multiple of the periods $p_1, p_2, \ldots, p_\ell$ of the individual generators. The proof is obtained directly from the fact that p is clearly a multiple of each p_j.

We now determine the maximum value of the period p. If the ℓ individual MLCGs have prime numbers m_j as moduli, then their periods are $p_j = m_j - 1$ and are therefore even. Therefore one has

$$p \le \frac{\prod_{j=1}^{\ell}(m_j - 1)}{2^{\ell-1}} \quad . \tag{4.7.3}$$

Equality results if the quantities $(m_j - 1)/2$ possess no common factors.

The first theorem of this section can now be used to construct a sequence with period given by (4.7.3). One forms first the integer quantity

$$z_i = \left(\sum_{j=1}^{\ell} (-1)^{j-1} x_{j,i}\right) \bmod (m_1 - 1) \quad . \tag{4.7.4}$$

The alternating sign in (4.7.4), which simplifies the construction of the modulus function, does not contradict the prescription of (4.7.1), since one could also use in place of $x_2, x_4, \ldots$, the variables $x'_2 = -x_2$, $x'_4 = -x_4$, ... The quantity z_i can take on the values

$$z_i \in \{0, 1, \ldots, m_1 - 2\} \quad . \tag{4.7.5}$$

The transformation to floating point numbers

$$u_i = \begin{cases} z_i/m_1 \, , & z_i > 0 \\ (m_1 - 1)/m_1 \, , & z_i = 0 \end{cases} \tag{4.7.6}$$

gives values in the range $0 < u_i < 1$.

4.8 Program for Generation of Uniformly Distributed Random Numbers

We now use the methods developed in Sects. 4.6 and 4.7 to give a program for producing uniformly distributed random numbers with a long period. Here we require that the computer carry out integer arithmetic with 32 bits. We combine two MLCGs with $m_1 = 2\,147\,483\,563$, $a_1 = 40\,014$, $m_2 = 2\,147\,483\,399$, $a_2 = 40\,692$. The numbers $(m_1 - 1)/2$ and $(m_2 - 1)/2$ have no common factor. Therefore the period of the combined generator is, according to (4.7.3),

$$p = (m_1 - 1)(m_2 - 1)/2 \approx 2.3 \cdot 10^{18} \quad .$$

The absolute values of all integers occurring during the computation remain in the range $\leq 2^{31} - 85$. The resulting floating point values u are in the range $0 < u < 1$. One does not obtain the values 0 or 1, at least if 23 or more bits are available for the mantissa, which is almost always the case when representing floating point numbers with 32 bits. The program with the given values of m_1, m_2, a_1, a_2 has been subjected to the spectral test and to many other tests by L'ECUYER [3], who has provided a PASCAL version. He determined that it satisfied all of the requirements of the tests.

Program 4.2: Subroutine RNECUY. Combined random number generator with very long period by L'Ecuyer

CALLING SEQUENCE: CALL RNECUY(U,N)

INPUT ARGUMENT: N — number of random numbers

OUTPUT ARGUMENT: U — array containing the random numbers in the first N positions

REMARKS: The call `CALL RNESIN(ISEED1,ISEED2)` initializes the two component generators with the seeds `ISEED1, ISEED2`. The call `CALL RNESOT(ISEED1,ISEED2)` returns the last generated values of the two generators for later use as seeds.

```
      SUBROUTINE RNECUY(U,N)
      IMPLICIT DOUBLE PRECISION (A-H,O-Z)
      DIMENSION U(N)
      PARAMETER (M1=2147483563,M1MIN1=2147483562,IA1=40014,
     +           IQ1=53668,IR1=12211)
      PARAMETER (M2=2147483399,IA2=40692,IQ2=52774,IR2=2791)
      PARAMETER (XM1INV=4.656613D-10)
      DATA IX1,IX2 /123456, 654321/
      DO 10 I=1,N
C Produce integer random number X1 from first MLCG
        K=IX1/IQ1
        IX1=IA1*(IX1-K*IQ1)-K*IR1
        IF (IX1.LT.0) IX1=IX1+M1
C Produce integer random number X2 from second MLCG
        K=IX2/IQ2
        IX2=IA2*(IX2-K*IQ2)-K*IR2
        IF (IX2.LT.0) IX2=IX2+M2
C Combine
        IZ=IX1-IX2
        IF(IZ.LT.1) IZ=IZ+M1MIN1
C Normalize and transform to floating point
        U(I)=DBLE(IZ)*XM1INV
   10 CONTINUE
      RETURN
C Input of seed
      ENTRY RNE2IN(ISEED1,ISEED2)
      IX1=ISEED1
      IX2=ISEED2
      RETURN
C Output of seed
      ENTRY RNE2OT(ISEED1,ISEED2)
      ISEED1=IX1
      ISEED2=IX2
      RETURN
      END
```

Figure 4.3 illustrates the difference between the simple MLCG and the combined generator. For the simple MLCG one can still recognize a structure in a scatter plot of the number pairs (4.5.4), although with an expansion of the abscissa by a factor of 1000. The corresponding diagram for the combined generator appears, in contrast, to be completely without structure. For each diagram one million pairs of random numbers were generated. The plots correspond only to a narrow strip on the left-hand edge of the unit square.

In order to initialize non-overlapping partial sequences one can use two methods:

(i) One applies the procedure discussed in connection with (4.6.7) to both MLCGs, naturally with the same value n, in order to construct pairs of seeds for each partial sequence.

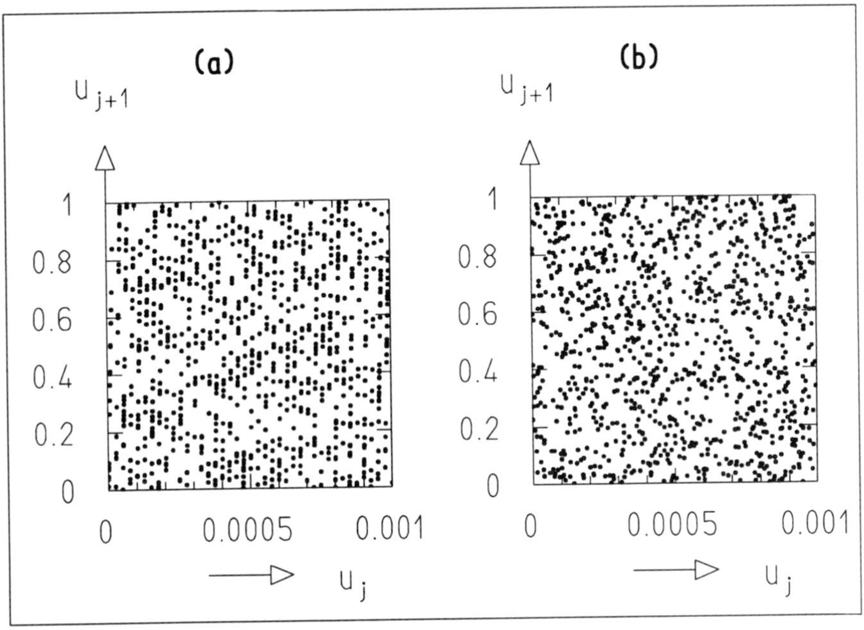

Fig. 4.3. Scatter plot of number pairs (4.5.4) from (a) the MLC generator RNMLCG and (b) the combined generator RNECUY.

(ii) It is considerably easier to use the same seed for the first MLCG for every partial sequence. For the second MLCG one uses an arbitrary seed for the first partial sequence, the following random number from the second MLCG for the second partial sequence, etc. In this way one obtains partial sequences that can reach a length of $(m_1 - 1)$ without overlapping.

4.9 Generation of Arbitrarily Distributed Random Numbers

4.9.1 Generation by Transformation of the Uniform Distribution

If x is a random variable following the uniform distribution,

$$f(x) = 1 \quad , \quad 0 \le x < 1 \quad ; \qquad f(x) = 0 \quad , \quad x < 0 \quad , \quad x \ge 1 \quad , \tag{4.9.1}$$

and y is a random variable described by the probability density $g(y)$, the transformation (3.7.1) simplifies to

$$g(y)\,\mathrm{d}y = \mathrm{d}x \quad . \tag{4.9.2}$$

We use the distribution function $G(y)$, which is related to $g(y)$ through $\mathrm{d}G(y)/\mathrm{d}y = g(y)$, and write (4.9.2) in the form

$$\mathrm{d}x = g(y)\,\mathrm{d}y = \mathrm{d}G(y) \quad , \tag{4.9.3}$$

or after integration,

$$x = G(y) = \int_{-\infty}^{y} g(t)\,\mathrm{d}t \quad . \tag{4.9.4}$$

This equation has the following meaning. If a random number x is taken from a uniform distribution between 0 and 1 and the function $\mathsf{x} = G(\mathsf{y})$ is inverted,

$$\mathsf{y} = G^{-1}(\mathsf{x}) \quad , \tag{4.9.5}$$

then one obtains a random number y described by the probability density $g(y)$. The relationship is depicted in Fig. 4.4a. The probability to obtain a random number x between x and $x + \mathrm{d}x$ is equal to the probability to have a value $\mathsf{y}(\mathsf{x})$ between y and $y + \mathrm{d}y$.

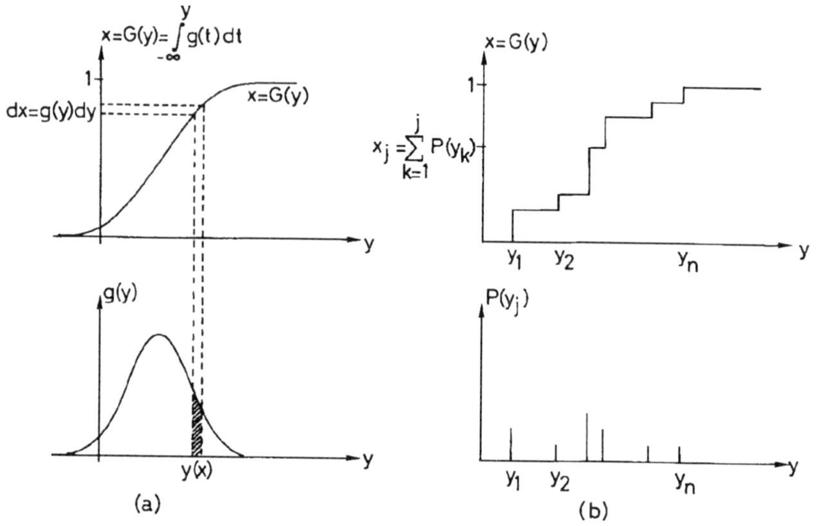

Fig. 4.4. Transformation from a uniformly distributed variable x to a variable y with the distribution function $G(y)$. The variable y can be continuous (a) or discrete (b).

The relationship (4.9.4) can be also be used to produce discrete probability distributions. An example is shown in Fig. 4.4b. The random variable y can take on the values $y_1, y_2, \ldots, y_n$ with the probabilities $P(y_1), P(y_2), \ldots, P(y_n)$. The distribution function as given by (3.2.1) is $G(y) = P(\mathsf{y} < y)$. The construction of a step function $x = G(y)$ according to this equation gives the values

$$x_j = G(y_j) = \sum_{k=1}^{i} P(y_k) \quad , \tag{4.9.6}$$

which lie in the range between 0 and 1. From this one can produce random numbers according to a discrete distribution $G(y)$ by first producing random numbers x uniformly distributed between 0 and 1. Depending on the interval j in which x falls, $x_{j-1} < \mathsf{x} < x_j$, the number y_j is then produced.

Example 4.1: Exponentially distributed random numbers

We would like to generate random numbers according to the probability density

$$g(t) = \begin{cases} \dfrac{1}{\tau}\mathrm{e}^{-t/\tau} \,, & t \geq 0 \quad , \\ 0 \,, & t < 0 \quad . \end{cases} \tag{4.9.7}$$

This is the probability density describing the time t of the decay of a radioactive nucleus that exists at time $t = 0$ and has a mean lifetime τ. The distribution function is

$$x = G(t) = \frac{1}{\tau} \int_{t'=0}^{t} g(t')\,\mathrm{d}t' = 1 - \mathrm{e}^{-t/\tau} \quad . \tag{4.9.8}$$

According to (4.9.4) and (4.9.5) we can obtain exponentially distributed random numbers t by first generating random numbers uniformly distributed between 0 and 1 and then finding the inverse function $\mathsf{t} = G^{-1}(\mathsf{x})$, i.e.,

$$\mathsf{t} = -\tau \ln(1 - \mathsf{x}) \quad .$$

Since $1 - \mathsf{x}$ is also uniformly distributed between 0 and 1, it is sufficient to compute

$$\mathsf{t} = -\tau \ln \mathsf{x} \quad . \blacksquare \tag{4.9.9}$$

Example 4.2: Generation of random numbers following a Breit–Wigner distribution

To generate random numbers y which follow a Breit–Wigner distribution (3.3.32),

$$g(y) = \frac{2}{\pi\Gamma} \frac{\Gamma^2}{4(y-a)^2 + \Gamma^2} \quad ,$$

we proceed as discussed in Sect. 4.9.1. We form the distribution function

$$x = G(y) = \int_{-\infty}^{y} g(y)\,\mathrm{d}y = \frac{2}{\pi\Gamma} \int_{-\infty}^{y} \frac{\Gamma^2}{4(y-a)^2 + \Gamma^2}\,\mathrm{d}y$$

and perform the integration using the substitution

$$u = \frac{2(y-a)}{\Gamma} \quad , \qquad du = \frac{2}{\Gamma}\,dy \quad .$$

Thus we obtain

$$\begin{aligned} x &= G(y) = \frac{1}{\pi}\int_{\theta=-\infty}^{\theta=2(y-a)/\Gamma} \frac{1}{1+u^2}\,du = \frac{1}{\pi}\,[\arctan u]_{-\infty}^{2(y-a)/\Gamma} \\ &= \frac{\arctan 2(y-a)/\Gamma}{\pi} + \frac{1}{2} \quad . \end{aligned}$$

By inversion we obtain

$$\begin{aligned} 2(y-a)/\Gamma &= \tan\left\{\pi\left(x-\frac{1}{2}\right)\right\} \quad , \\ y &= a + \frac{\Gamma}{2}\tan\left\{\pi\left(x-\frac{1}{2}\right)\right\} \quad . \end{aligned} \tag{4.9.10}$$

If x are random numbers uniformly distributed in the interval $0 < \mathsf{x} < 1$, then y follows a Breit–Wigner distribution. ■

Example 4.3: Generation of random numbers with a triangular distribution

In order to generate random numbers y following a triangular distribution as in Problem 3.9.2 we form the distribution function

$$F(y) = \begin{cases} 0\,, & y < a \quad , \\ \dfrac{(y-a)^2}{(b-a)(c-a)}\,, & a \le y < c \quad , \\ 1 - \dfrac{(y-b)^2}{(b-a)(b-c)}\,, & c \le y < b \quad , \\ 1\,, & b \le y \quad . \end{cases}$$

In particular we have

$$F(c) = \frac{c-a}{b-a} \quad .$$

Inverting $x = F(y)$ gives

$$\begin{aligned} y &= a + \sqrt{(b-a)(c-a)x} \quad , & x &< (c-a)/(b-a) \quad , \\ y &= b - \sqrt{(b-a)(b-c)(1-x)} \quad , & x &\ge (c-a)/(b-a) \quad . \end{aligned}$$

If x is uniformly distributed with $0 < \mathsf{x} < 1$, then y follows a triangular distribution. ■

4.9.2 Generation with the von Neumann Acceptance–Rejection Technique

The elegant technique of the previous section requires that the distribution function $x = G(y)$ be known and that the inverse function $y = G^{-1}(x)$ exists and be known as well.

Often one only knows the probability density $g(y)$. One can then use the VON NEUMANN acceptance–rejection technique, which we introduce with a simple example before discussing it in its general form.

Example 4.4: Semicircle distribution with the simple acceptance–rejection method

As a simple example we generate random numbers following a semicircular probability density,

$$g(y) = \begin{cases} (2/\pi R^2)\sqrt{R^2 - y^2}, & |y| \leq R \quad , \\ 0, & |y| > R \quad . \end{cases} \tag{4.9.11}$$

Instead of trying to find and invert the distribution function $G(y)$, we generate pairs of random numbers $(\mathsf{y}_i, \mathsf{u}_i)$. Here y_i is uniformly distributed in the interval available to y, $-R \leq y \leq R$, and u_i is uniformly distributed in the range of values assumed by the function $g(y)$, $0 \leq u \leq R$. For each pair we test if

$$\mathsf{u}_i \geq g(\mathsf{y}_i) \quad . \tag{4.9.12}$$

If this inequality is fulfilled, we reject the random number y_i. The set of random numbers y_i that are not rejected then follow a probability density $g(y)$, since each was accepted with a probability proportional to $g(\mathsf{y}_i)$. ■

The technique of Example 4.4 can easily be described geometrically. To generate random numbers in the interval $a \leq y \leq b$ according to the probability density $g(y)$, one must consider in the region $a \leq y \leq b$ the curve

$$u = g(y) \tag{4.9.13}$$

and a constant

$$u = d \quad , \qquad d \geq g_{\max} \quad , \tag{4.9.14}$$

which is greater than or equal to the maximum value of $g(y)$ in that region. In the (y, u) plane this constant is described by the line $u = d$. Pairs of random numbers $(\mathsf{y}_i, \mathsf{u}_i)$ uniformly distributed in the interval $a \leq \mathsf{y}_i \leq b$, $0 \leq \mathsf{u}_i \leq d$ correspond to a uniform distribution of points in the corresponding rectangle of the (y, u)-plane. If all of the points for which (4.9.12) holds are rejected, then only points under the curve $u = g(y)$ remain. Figure 4.5 shows this situation

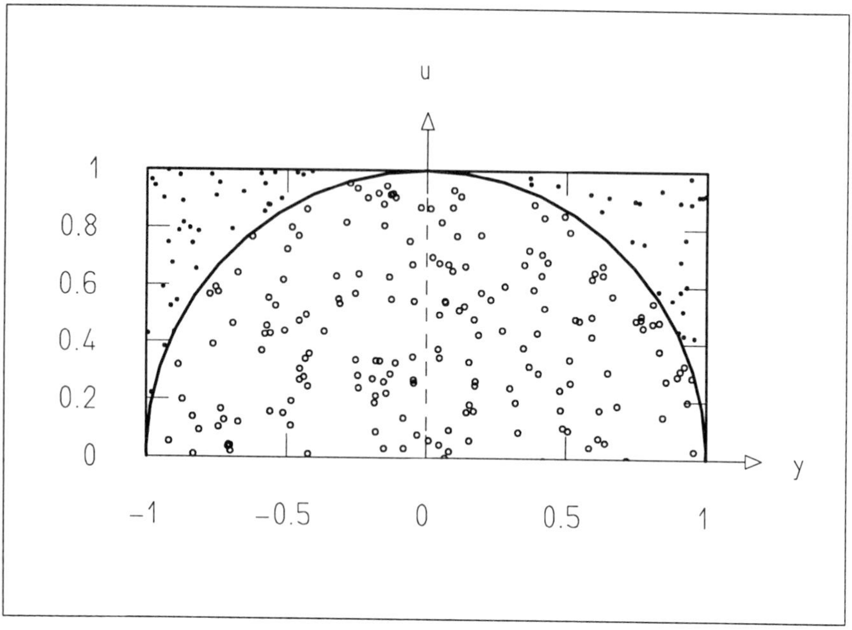

Fig. 4.5. All the pairs (y_i, u_i) produced are marked as points in the (y, u)-plane. Points above the curve $u = g(y)$ (small points) are rejected.

for the Example 4.4. [It is clear that the technique also gives meaningful results if the function is not normalized to one. In Fig. 4.5 we have simply set $g(y) = \sqrt{R^2 - y^2}$ and $R = 1$.]

For the transformation technique of Sect. 4.9.1, each random number y_i required only that exactly one random number x_i be generated from a uniform distribution and that it be transformed according to (4.9.5). In the acceptance–rejection technique, pairs y_i, u_i must always be generated, and a considerable fraction of the numbers y_i – depending on the value of u_i according to (4.9.12) – are rejected. The probability for y_i to be accepted is

$$E = \frac{\int_a^b g(y)\,dy}{(b-a)d} \quad . \tag{4.9.15}$$

We can call E the *efficiency* of the procedure. If the interval $a \le y \le b$ includes the entire allowed range of y, then the numerator of (4.9.15) is equal to unity, and one obtains

$$E = \frac{1}{(b-a)d} \quad . \tag{4.9.16}$$

The numerator and denominator of (4.9.15) are simply the areas contained in the region $a \le y \le b$ under the curves (4.9.13) and (4.9.14), respectively.

One distributes points $(\mathsf{y}_i, \mathsf{u}_i)$ uniformly under the curve (4.9.14) and rejects the random numbers y_i if the inequality (4.9.12) holds. The efficiency of the procedure is certainly higher if one uses as the upper curve not the constant (4.9.14) but rather a curve that is closer to $g(y)$.

With this in mind the acceptance–rejection technique can be stated in its general form:

(i) One finds a probability density $s(y)$ that is sufficiently simple that random numbers can be generated from it using the transformation method, and a constant c such that

$$g(y) \leq c \cdot s(y) \quad , \qquad a < y < b \quad , \tag{4.9.17}$$

holds.

(ii) One generates one random number y uniformly distributed in the interval $a < \mathsf{y} < b$ and a second random number u uniformly distributed in the interval $0 < \mathsf{u} < 1$.

(iii) One rejects y if

$$\mathsf{u} \geq \frac{g(\mathsf{y})}{c \cdot s(\mathsf{y})} \quad . \tag{4.9.18}$$

After the points (ii) and (iii) have been repeated enough times, the resulting set of accepted random numbers y follows the probability density $g(y)$, since

$$P(\mathsf{y} < y) = \int_a^y s(t) \frac{g(t)}{c \cdot s(t)} \,\mathrm{d}t = \frac{1}{c} \int_a^y g(t) \,\mathrm{d}t = \frac{1}{c}[G(y) - G(a)] \quad .$$

If the interval $a \leq y \leq b$ includes the entire range of y for both $g(y)$ as well as for $s(y)$, then one obtains an efficiency

$$E = \frac{1}{c} \quad . \tag{4.9.19}$$

Example 4.5: Semicircle distribution with the general acceptance–rejection method

One chooses for $c \cdot s(y)$ the polygon

$$c \cdot s(y) = \begin{cases} 0\,, & y < -R \quad , \\ 3R/2 + y\,, & -R \leq y < -R/2 \quad , \\ R\,, & -R/2 \leq y < R/2 \quad , \\ 3R/2 - y\,, & R/2 \leq y < R \quad , \\ 0\,, & R \leq y \quad . \end{cases}$$

The efficiency is clearly

$$E = \frac{\pi R^2}{2} \cdot \frac{1}{2R^2 - R^2/4} = \frac{2\pi}{7}$$

in comparison to

$$E = \frac{\pi R^2}{2} \cdot \frac{1}{2R^2} = \frac{\pi}{4}$$

as in Example 4.4. ■

4.10 Generation of Normally Distributed Random Numbers

By far the most important distribution for data analysis is the normal distribution, which we will discuss in Sect. 5.7. We present here a program that can produce random numbers x_i following the standard normal distribution with the probability density

$$f(x) = \frac{1}{\sqrt{2\pi}} e^{-x^2/2} \quad . \tag{4.10.1}$$

The corresponding distribution function $F(x)$ can only be computed and inverted numerically (Appendix C). Therefore the simple transformation method of Sect. 4.9.1 cannot be used. The *polar method* by BOX and MULLER [5] described here combines in an elegant way acceptance–rejection with transformation. The algorithm consists of the following steps:

(i) Generate two independent random numbers u_1, u_2 from a uniform distribution between 0 and 1. Transform $v_1 = 2u_1 - 1$, $v_2 = 2u_2 - 1$.

(ii) Compute $s = v_1^2 + v_2^2$.

(iii) If $s \geq 1$, return to step (i).

(iv) $x_1 = v_1\sqrt{-(2/s)\ln s}$ and $x_2 = v_2\sqrt{-(2/s)\ln s}$ are two independent random numbers following the standard normal distribution.

The number pairs (v_1, v_2) obtained from step (i) are the coordinates of a set of points uniformly distributed inside the unit circle. The point (v_1, v_2) has the polar coordinates $v_1 = r\cos\theta$, $v_2 = r\sin\theta$ with $r = \sqrt{s}$, $\theta = \arctan(v_2/v_1)$. The point (x_1, x_2) has the polar coordinates

$$x_1 = \cos\theta\,\sqrt{-2\ln s} \quad , \qquad x_2 = \sin\theta\,\sqrt{-2\ln s} \quad .$$

We now ask for the probability

$$\begin{aligned} F(r) &= P(\sqrt{-2\ln \mathsf{s}} \le r) = P(-2\ln \mathsf{s} \le r^2) \\ &= P(\mathsf{s} > \mathrm{e}^{-r^2/2}) \quad . \end{aligned}$$

Since $\mathsf{s} = \mathsf{r}^2$ is by construction uniformly distributed between 0 and 1, one has

$$F(r) = P(\mathsf{s} > \mathrm{e}^{-r^2/2}) = 1 - \mathrm{e}^{-r^2/2} \quad .$$

The probability density of r is

$$f(r) = \frac{\mathrm{d}F(r)}{\mathrm{d}r} = r\mathrm{e}^{-r^2/2} \quad .$$

The joint distribution function of x_1 and x_2,

$$\begin{aligned} F(x_1, x_2) &= P(\mathsf{x}_1 \le x_1,\ \mathsf{x}_2 \le x_2) = P(\mathsf{r}\cos\theta \le x_1,\ \mathsf{r}\sin\theta \le x_2) \\ &= \frac{1}{2\pi}\int\int_{(\mathsf{x}_1<x_1,\mathsf{x}_2<x_2)} r\mathrm{e}^{-r^2/2}\,\mathrm{d}r\,\mathrm{d}\varphi \\ &= \frac{1}{2\pi}\int\int_{(\mathsf{x}_1<x_1,\mathsf{x}_2<x_2)} \mathrm{e}^{-(x_1^2+x_2^2)/2}\,\mathrm{d}x\,\mathrm{d}y \\ &= \left(\frac{1}{\sqrt{2\pi}}\int_{-\infty}^{x_1} \mathrm{e}^{-x_1^2/2}\,\mathrm{d}x_1\right)\left(\frac{1}{\sqrt{2\pi}}\int_{-\infty}^{x_2} \mathrm{e}^{-x_2^2/2}\,\mathrm{d}x_2\right) \quad , \end{aligned}$$

is the product of two distribution functions of the standard normal distribution. The procedure is illustrated in Fig. 4.6.

Program 4.3: Subroutine RNSTNR for generating random numbers according to the standard normal distribution

CALLING SEQUENCE: `CALL RNSTNR(R,N)`

INPUT ARGUMENT: N — number of random numbers

OUTPUT ARGUMENT: R — array containing the random numbers in the first N positions

REMARKS: Subroutine 4.2 RNECUY is used. In order to produce non-overlapping partial sequences, RNECUY must be appropriately initialized.

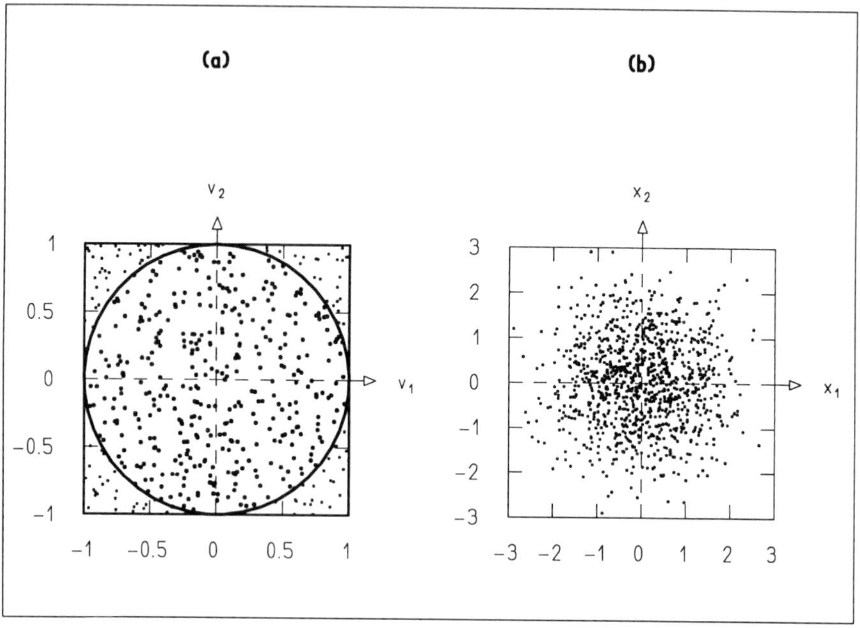

Fig. 4.6. Illustration of the Box–Muller procedure. (a) Number pairs (v_1, v_2) are generated that uniformly populate the square. Those pairs are then rejected that do not lie inside the unit circle (marked by small points). (b) This is followed by the transformation $(v_1, v_2) \rightarrow (x_1, x_2)$.

```
      SUBROUTINE RNSTNR(R,N)
      IMPLICIT DOUBLE PRECISION (A-H,O-Z)
      DIMENSION R(N), U(2)
      I=1
   10 CONTINUE
      CALL RNECUY(U,2)
      V1=2.*U(1)-1.
      V2=2.*U(2)-1.
      S=V1*V1+V2*V2
      IF(S.GE.1.) GO TO 10
      ROOT=SQRT(-2.*LOG(S)/S)
      R(I)=V1*ROOT
      IF(I.LT.N) R(I+1)=V2*ROOT
      I=I+2
      IF(I.LE.N) GO TO 10
      END
```

Many other procedures are described in the literature for the generation of normally distributed random numbers. They are to a certain extent more efficient, but are generally more difficult to program than the BOX–MULLER procedure.

4.11 Generation of Random Numbers According to a Multivariate Normal Distribution

The probability density of a multivariate normal distribution of n variables $\mathbf{x} = (x_1, x_2, \ldots, x_n)$ is according to (5.10.1)

$$\phi(\mathbf{x}) = k \exp\left\{-\frac{1}{2}(\mathbf{x}-\mathbf{a})^{\mathrm{T}} B(\mathbf{x}-\mathbf{a})\right\} \quad .$$

Here $\mathbf{a}$ is the vector of expectation values and $B = C^{-1}$ is the inverse of the positive-definite symmetric covariance matrix. With the Cholesky decomposition $B = D^{\mathrm{T}} D$ and the substitution $\mathbf{u} = D(\mathbf{x}-\mathbf{a})$ the exponent takes on the simple form

$$-\frac{1}{2}\mathbf{u}^{\mathrm{T}}\mathbf{u} = -\frac{1}{2}(u_1^2 + u_2^2 + \cdots + u_n^2) \quad .$$

Thus the elements u_i of the vectors $\mathbf{u}$ follow independent standard normal distributions [cf. (5.10.9)]. One obtains vectors $\mathbf{x}$ of random numbers by first forming a vector $\mathbf{u}$ of elements u_i which follow the standard normal distribution and then performing the transformation

$$\mathbf{x} = D^{-1}\mathbf{u} + \mathbf{a} \quad .$$

Program 4.4: Subroutine RNMNPR to initialize the generation of random numbers according to an n-dimensional normal distribution

CALLING SEQUENCE: CALL RNMNPR(C,DPLUS,N)

INPUT ARGUMENTS: C — $n \times n$ covariance matrix C of the n-dimensional normal distribution
N — number of variables n

OUTPUT ARGUMENT: DPLUS — $n \times n$ matrix D^{-1}, which is needed as an input argument for program RNMNGN.

```
      SUBROUTINE RNMNPR(C,DPLUS,N)
C prepares for generation of random numbers from multivaraiate normal
      IMPLICIT DOUBLE PRECISION(A-H,O-Z)
      PARAMETER(NMAX=10)
      DIMENSION B(NMAX,NMAX),D(NMAX,NMAX),R(NMAX)
      DIMENSION C(N,N),DPLUS(N,N)
      LOGICAL OK
      CALL MTXTRA(C,B,N,N)
      CALL MTXCHI(B,D,N)
      CALL MTXCHL(B,D,N)
      CALL MTXUNT(B,N)
      CALL MTXSVD(D,B,DPLUS,R,N,N,N,0.D0,OK)
      END
```

Program 4.5: Subroutine RNMNGN to generate n-tuples of random numbers according to an n-dimensional normal distribution

CALLING SEQUENCE: CALL RNMNGN(DPLUS,A,X,N)

INPUT ARGUMENTS: DPLUS — $n \times n$ matrix D^{-1}, which was created by calling RNMNPR
A — n-vector **a** of the expectation values of the distribution
N — number of variables n

OUTPUT ARGUMENT: X — n-vector **x**, whose components make up an n-tuple of random numbers

REMARKS: After the matrix DPLUS has been prepared by calling RNMNPR, each call of RNMNGN gives a new random vector **x**.

```
      SUBROUTINE RNMNGN(DPLUS,A,X,N)
C generation of random numbers from multivaraiate normal
      IMPLICIT DOUBLE PRECISION(A-H,O-Z)
      PARAMETER(NMAX=10)
      DIMENSION R(NMAX)
      DIMENSION DPLUS(N,N),A(N),X(N)
      CALL RNSTNR(R,N)
      CALL MTXMLT(DPLUS,R,X,N,N,1)
      CALL MTXADD(X,A,X,N,1)
      END
```

4.12 The Monte Carlo Method for Integration

It follows directly from its construction that the acceptance–rejection technique, Sect. 4.9.2, provides a very simple method for numerical integration. If N pairs of random numbers $(\mathrm{y}_1, \mathrm{u}_i)$, $i = 1, 2, \ldots, N$ are generated according to the prescription of the general acceptance–rejection technique, and if $N - n$ of them are rejected because they fulfill condition (4.9.18), then the numbers N (or n) are proportional to the areas under the curves $c \cdot s(y)$ (or $g(y)$), at least in the limit of large N, i.e.,

$$\frac{\int_a^b g(y)\,\mathrm{d}y}{c\int_a^b s(y)\,\mathrm{d}y} = \lim_{N\to\infty}\frac{n}{N} \quad . \tag{4.12.1}$$

Since the function $s(y)$ is chosen to be particularly simple [in the simplest case one has $s(y) = 1/(b-a)$], the ratio n/N is a direct measure of the value of the integral

$$I = \int_a^b g(y)\,\mathrm{d}y = \left(\lim_{N\to\infty}\frac{n}{N}\right) c\int_a^b s(y)\,\mathrm{d}y \quad . \tag{4.12.2}$$

Here the integrand $g(y)$ does not necessarily have to be normalized, i.e., one does not need to require

$$\int_{-\infty}^{\infty} g(y)\,\mathrm{d}y = 1$$

as long as c is chosen such that (4.9.17) is fulfilled.

Example 4.6: Computation of π

Referring to Example 4.4 we compute the integral using (4.9.11) with $R = 1$:

$$I = \int_0^1 g(y)\,\mathrm{d}y = \pi/4 \quad .$$

Choosing $s(y) = 1$ and $c = 1$ we obtain

$$I = \lim_{N\to\infty} \frac{n}{N} \quad .$$

We expect that when N points are distributed according to a uniform distribution in the square $0 \le y \le 1, 0 \le u \le 1$, and when n of them lie inside the unit circle, then the ratio n/N approaches the value $I = \pi/4$ in the limit $N \to \infty$. Table 4.2 shows the results for various values of n and for various sequences of random numbers. The exact value of n/N clearly depends on the particular sequence. In Sect. 6.8 we will determine that the typical fluctuations of the number n are approximately $\Delta n = \sqrt{n}$. Therefore one has for the relative precision for the determination of the integral (4.12.2)

$$\frac{\Delta I}{I} = \frac{\Delta n}{n} = \frac{1}{\sqrt{n}} \quad . \tag{4.12.3}$$

We expect therefore in the columns of Table 4.2 to find the value of π with precisions of 10 %, 1 %, and 0.1 %. We find in fact in the three columns fluctuations in the first, second, and third places after the decimal point. ■

Table 4.2. Numerical values of $4\,n/N$ for various values of n. The entries in the columns correspond to various sequences of random numbers.

$4\,n/N$		
$n = 10^2$	$n = 10^4$	$n = 10^6$
3.419	3.122	3.141
3.150	3.145	3.143
3.279	3.159	3.144
3.419	3.130	3.143

The Monte Carlo method of integration can now be implemented by a very simple program. For integration of single variable functions it is usually better to use other numerical techniques for reasons of computing time. For integrals with many variables, however, the Monte Carlo method is more straightforward and often faster as well.

4.13 The Monte Carlo Method for Simulation

Many real situations that are determined by statistical processes can be simulated in a computer with the aid of random numbers. Examples are automobile traffic in a given system of streets or the behavior of neutrons in a nuclear reactor. The Monte Carlo method was originally developed for the latter problem by VON NEUMANN and ULAM. A change of the parameters of the distributions corresponds then to a change in the actual situation. In this way the effect of additional streets or changes in the reactor can be investigated without having to undertake costly and time consuming changes in the real system. Not only processes of interest following statistical laws can be simulated with the Monte Carlo method, but also the measurement errors which occur in every measurement.

Example 4.7: Simulation of measurement errors of points on a line
We consider a line in the (t, y)-plane. It is described by the equation

$$y = at + b \quad . \tag{4.13.1}$$

If we choose discrete values of t

$$t_0 \quad , \qquad t_1 = t_0 + \Delta t \quad , \qquad t_2 = t_0 + 2\Delta t \quad , \qquad \ldots \quad , \tag{4.13.2}$$

then they correspond to values of y

$$y_i = at_i + b \quad , \qquad i = 0, 1, \ldots, n-1 \quad . \tag{4.13.3}$$

We assume that the values t_0, t_1, ... of the "controlled variable" t can be set without error. Because of measurement errors, however, instead of the values y_i, one obtains different values

$$y_i' = y_i + \varepsilon_i \quad . \tag{4.13.4}$$

Here ε_i are the measurement errors, which follow a normal distribution with mean of zero and standard deviation σ_y (cf. Sect. 5.7). Program 4.6 generates pairs of points according to Eq. (4.13.4).

Figure 4.7 shows the result of the program call
`CALL RNLINE(0.5D0,1.D0,0.D0,1.D0,10,0.5D0,T,Y)` ■

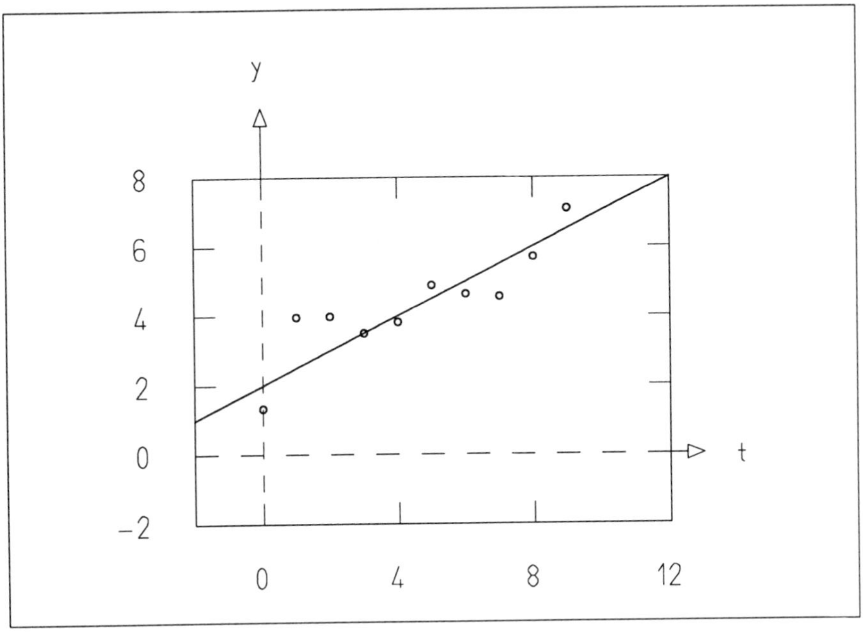

Fig. 4.7. Line in the (t, y)-plane and simulated measured values with errors in y.

Program 4.6: Subroutine `RNLINE` for generating values on a line with measurement errors

CALLING SEQUENCE: `CALL RNLINE(A,B,T0,DT,N,SIGMAY,T,Y)`

INPUT ARGUMENTS: `A,B` — parameters a, b of the line; cf. (4.13.1).
`T0,DT,N` — t_0, Δt, n determine the number and position of the points without error
`SIGMAY` — σ_y standard deviation of the normal distribution for the measurement errors

OUTPUT ARGUMENTS: `T,Y` — arrays containing the coordinates of the n measured points (t_i, y_i')

```
      SUBROUTINE RNLINE(A,B,T0,DT,N,SIGMAY,T,Y)
      IMPLICIT DOUBLE PRECISION (A-H,O-Z)
      DIMENSION T(N), Y(N)
      DO 10 I =  1,N
        T(I)=T0+(I-1)*DT
        Y(I)=A*T(I)+B
        CALL RNSTNR(R,1)
        Y(I)=Y(I)+R*SIGMAY
   10 CONTINUE
      END
```

Example 4.8: Generation of decay times for a mixture of two different radioactive substances

At time $t = 0$ a source consists of N radioactive nuclei of which aN decay with a lifetime τ_1 and $(a - 1)N$ with a mean lifetime τ_2, with $0 \le a \le 1$. Random numbers for two different problems must be used in the simulation the decay times occurring: for the choice of the type of nucleus and for the determination of the decay time of the nucleus chosen, cf. (4.9.9). ■

Program 4.7: Subroutine `RNRADI` for generating decay times of radioactive nuclei

CALLING SEQUENCE: `CALL RNRADI(A,TAU1,TAU2,T)`

INPUT ARGUMENTS: `A` — fraction a of the nucleus type with lifetime τ_1
`TAU1,TAU2` — mean lifetimes τ_1, τ_2 of the two types of nuclei

OUTPUT ARGUMENT: `T` — decay time t of a nucleus

REMARKS: A decay time is returned for each call.

```
      SUBROUTINE RNRADI(A,TAU1,TAU2,T)
      IMPLICIT DOUBLE PRECISION (A-H,O-Z)
      DIMENSION R(3)
      CALL RNECUY (R,3)
      IF(R(1).LT.A) THEN
C Mean life is TAU1
        T=-TAU1*LOG(R(2))
      ELSE
C Mean life is TAU2
        T=-TAU2*LOG(R(3))
      END IF
      END
```

4.14 Example Programs

The source files of the example programs are available in the subdirectory `\DATAN\EXASRC`.

4.14.1 Main Program E1RN to Demonstrate Subprograms RNMLCG, RNECUY, and RNSTNR

The following main program demonstrates the calls of the subroutines `RNMLCG` and `RNECUY`, which generate uniformly distributed random numbers, and of the program `RNSTNR`, which generates normally distributed random numbers. The user chooses a random number generator, is shown the current seeds, and can – if wanted – input new seeds. Finally a list of 100 random numbers is shown. The user may then exit the program or start it again.

4.14.2 Main Program E2RN to Demonstrate Subprogram RNLINE

The main program generates N pairs of numbers (t, y) with $t = t_0 + i\Delta t$, $i = 0, 1, \ldots, N-1$, and $y = at + b + \Delta y$. Here the measurement error Δy is sampled from a normal distribution of width σ. The pairs of numbers are either shown numerically or graphically as chosen by the user. To produce the graphical output it is sufficient to call the subroutine SMSDGR.

4.14.3 Main Program E3RN to Demonstrate Subprogram RNRADI

The program generates N decay times t_i. Each individual time corresponds with probability A to an individual decay with mean life τ_1, and with the probability $1 - A$ to a decay with mean life τ_2. At the users choice the decay times are shown numerically or graphically (as a histogram). To produce the graphics SMHSGR is called.

4.14.4 Main Program E4RN to Simulate Molecular Movement of a Gas

The behavior of an ideal gas in a closed box can be simulated as follows. Gas molecules are idealized as spheres of radius r, i.e., of diameter $d = 2r$. Movement is limited to the (x_1, x_2)-plane. The box is thus the rectangle $x_{1A} \le x_1 \le x_{1B}$, $x_{2A} \le x_2 \le x_{2B}$. In a first step the spheres are distributed randomly within the box, but in such a way that they do not overlap with each other or with the boundary. The position of each sphere is determined by the vector $\mathbf{x}_i = (x_{1i}, x_{2i})$ of its center. In a second step the velocity vectors $\mathbf{v}$ are determined. We choose the totally unphysical initial condition in which only the vector $\mathbf{v}_1$ is different from zero. It has the modulus v_0 and a random direction.

We now consider n_t consecutive moments in time which follow each other with the small interval $\mathrm{d}t$. With each step every sphere i is shifted along a straight line by the vector $\mathbf{v}_i\,\mathrm{d}t$,

$$\mathbf{x}_i \to \mathbf{x}_i + \mathbf{v}_i\,\mathrm{d}t \quad .$$

After each step a test is performed to find out whether a collision has taken place with the wall or with another sphere. If a sphere has collided with the wall, the velocity component perpendicular to the wall is reversed. For the collision of the sphere i with a sphere j the laws of elastic collisions are used. Suppose $\mathbf{e}_\perp = (\mathbf{x}_i - \mathbf{x}_j)/|\mathbf{x}_i - \mathbf{x}_j|$ is the unit vector in the direction connecting the centers of the two spheres, and $\mathbf{e}_\parallel$ is a unit vector perpendicular to that direction, $\mathbf{w} = (\mathbf{v}_i + \mathbf{v}_j)/2$ is the velocity of the center of mass of the two spheres and finally, $\mathbf{u}_i = \mathbf{v}_i - \mathbf{w}$ and $\mathbf{u}_j = \mathbf{v}_j - \mathbf{w}$ are the velocities in a reference system in which the center of mass is at rest, so that

$$\mathbf{u}_i = -\mathbf{u}_j \quad .$$

Then with the decompositions

$$\begin{aligned}
\mathbf{u}_i &= \mathbf{u}_{i\perp} + \mathbf{u}_{i\parallel} \quad , \\
\mathbf{u}_{i\perp} &= (\mathbf{u}_i \cdot \mathbf{e}_\perp)\mathbf{e}_\perp = u_{i\perp}\mathbf{e}_\perp \quad , \\
\mathbf{u}_{i\parallel} &= (\mathbf{u}_i \cdot \mathbf{e}_\parallel)\mathbf{e}_\parallel = u_{i\parallel}\mathbf{e}_\parallel \quad ,
\end{aligned}$$

one obtains for the velocities $\mathbf{u}'_i$, $\mathbf{u}'_j$ after the collision

$$\begin{aligned}
u'_{i\parallel} &= u_{i\parallel} \quad , & u'_{j\parallel} &= u_{j\parallel} = -u_{i\parallel} \quad , \\
u'_{i\perp} &= -u_{i\perp} \quad , & u'_{j\perp} &= -u_{j\perp} = u_{i\perp} \quad .
\end{aligned}$$

With the transformation $\mathbf{v}'_i = \mathbf{u}'_i + \mathbf{w}$, $\mathbf{v}'_j = \mathbf{u}'_j + \mathbf{w}$ one finally obtains the velocities after the collision in the original reference system.

In this way we can simulate a rather complex physical system and study its time development in detail. In our example program we study in particular the time evolution of the velocity component v_1. There in time intervals $m_t\,dt$ we construct histograms of v_1. The histograms contain bins in the interval $-v_0 < x < v_0$.

4.14.5 Main Program E5RN to Demonstrate Subprograms RNMNPR and RNMNGN

The program generates pairs of random numbers following a two-dimensional normal distribution with means $\mathbf{a} = (a_1, a_2)$ and the 2×2 covariance matrix C. The largest part of the program is a dialogue with the user. First there is the possibility to input new seeds. Next, the program asks for the elements of C and the components of $\mathbf{a}$. Calling RNMNPR initializes the procedure. Finally, RNMNGN is called 20 times within a loop and the generated pairs of random numbers are displayed.

4.15 Programming Problems

The source files of the example solutions are available in the subdirectory \DATAN\SOLSRC.

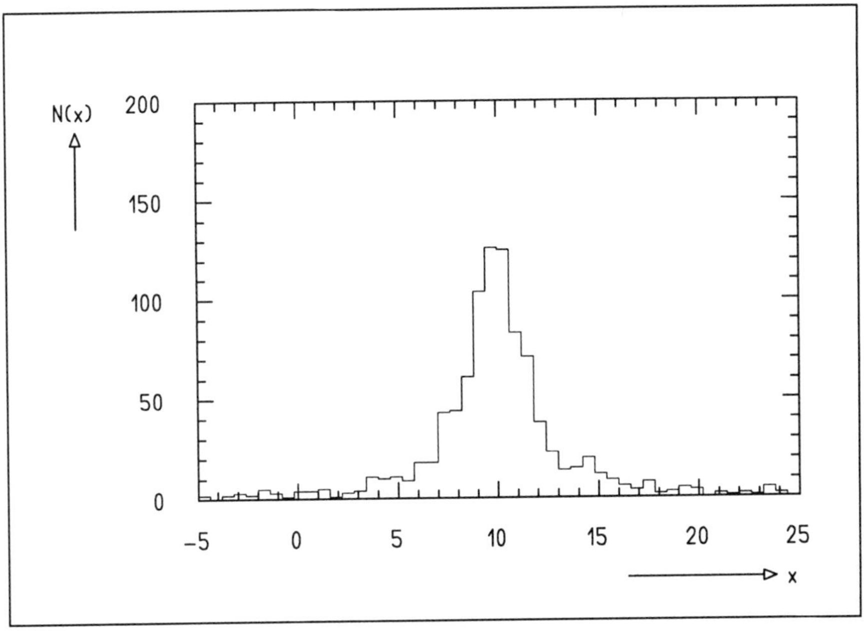

Fig. 4.8. Histogram of 1000 random numbers following a Breit–Wigner distribution with $a = 10$ and $\Gamma = 3$.

4.15.1 Program to Generate Breit–Wigner-Distributed Random Numbers

Write a subprogram with the following specifications:

CALL: `CALL RNBW(A,GAMMA,R,N)`

INPUT ARGUMENTS: `A` — expectation value a

`GAMMA` — full width at half maximum (FWHM) Γ

`N` — number n of generated random numbers

OUTPUT ARGUMENT: `R` — n-dimensional vector of random numbers following a Breit–Wigner distribution characterized by a and Γ

Write a main program which calls this subprogram and displays the generated numbers numerically or, optionally, graphically as a histogram (Fig. 4.8).

4.15.2 Program to Generate Random Numbers from a Triangular Distribution

Write the subprogram specified as follows:

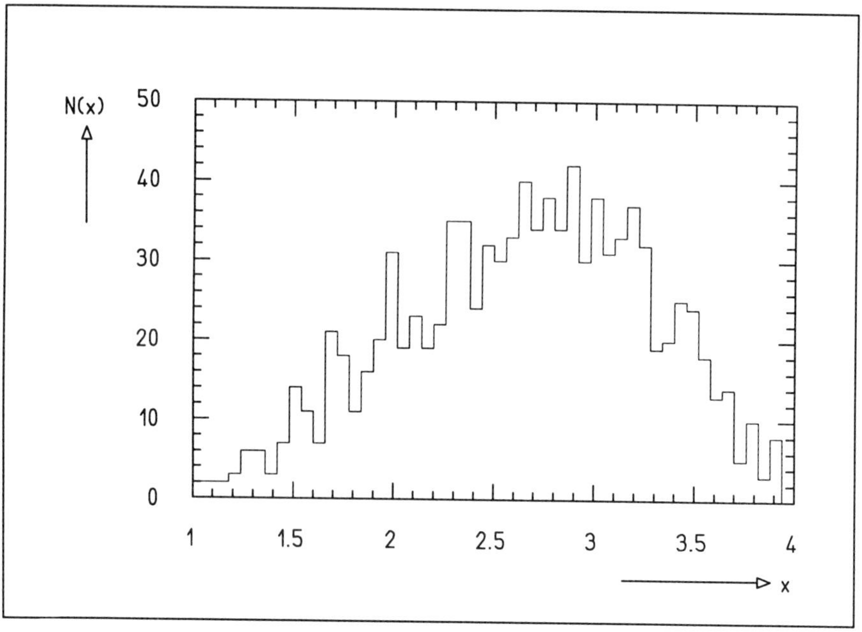

Fig. 4.9. Histogram of 1000 random numbers following a triangular distribution with $a = 1$, $b = 4$, $c = 3$.

CALL: `CALL RNTRIT(A,B,C,R,N)`

INPUT ARGUMENTS: `A,B,C` — parameters a, b, c of a triangular distribution as in Example 4.3

`N` — number n of random numbers to be generated

OUTPUT ARGUMENT: `R` — n-dimensional vector of random variables following a triangular distribution with the parameters a, b, c, generated using the transformation method as in Example 4.3.

Write a second subprogram with the calling sequence

```
CALL RNTRIR(A,B,C,R,N)
```

which solves the same problem, but uses von Neumann's acceptance–rejection method. Which of the two programs is faster?

Write a main program which allows the user to call either of the two subprograms and which (again at the user's choice) displays the generated numbers either numerically or graphically in the form of a histogram (Fig. 4.9).

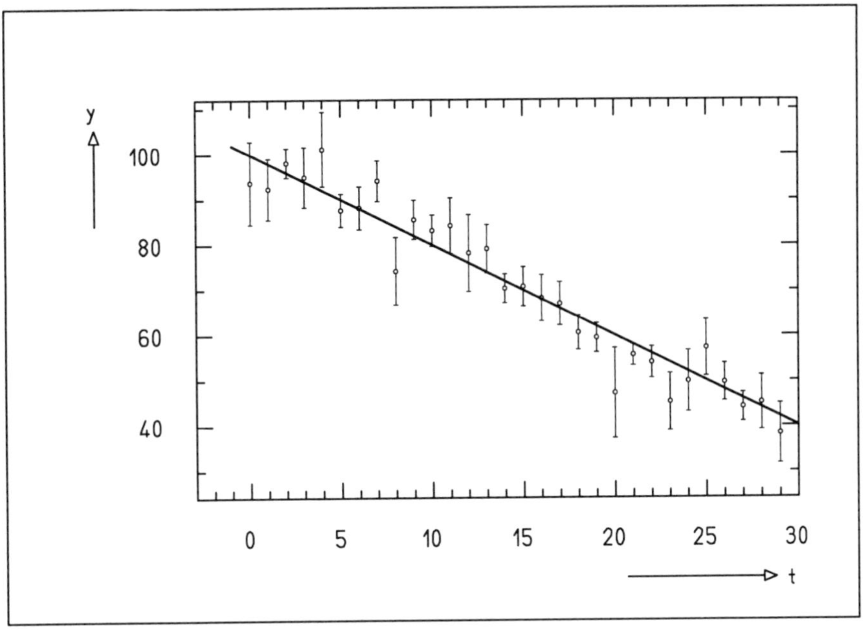

Fig. 4.10. Data points with errors of different size.

4.15.3 Program to Generate Data Points with Errors of Different Size

Write a subprogram similar to RNLINE which generates data points $y = at + b + \Delta y$ (cf. Example Program 4.14.1). The errors Δy, however, are not to be taken for all data points y_i from the same uniform distribution with the width σ, but Δy_i is to be sampled from a normal distribution with the width σ_i. The widths σ_i are to be taken from a uniform distribution within the region $\sigma_{\text{min}} < \sigma_i < \sigma_{\text{max}}$. The specifications of the program are:

CALL: `CALL RNLNDF(A,B,T0,DT,N,SIGMIN,SIGMAX,T,Y, SIGMAY)`

INPUT ARGUMENTS: `A,B,T0,DT,N` — as in RNLINE
`SIGMIN,SIGMAX` — σ_{min}, σ_{max}

OUTPUT ARGUMENTS: `T,Y,SIGMAY` — n-dimensional vectors with the elements t_i, y_i, σ_i

Write a main program which calls the subprogram and which displays graphically the straight line $y = at + b$ as well as the simulated data points with error bars $y_i \pm \Delta y_i$ (Fig. 4.10). Use the program GRDATP.

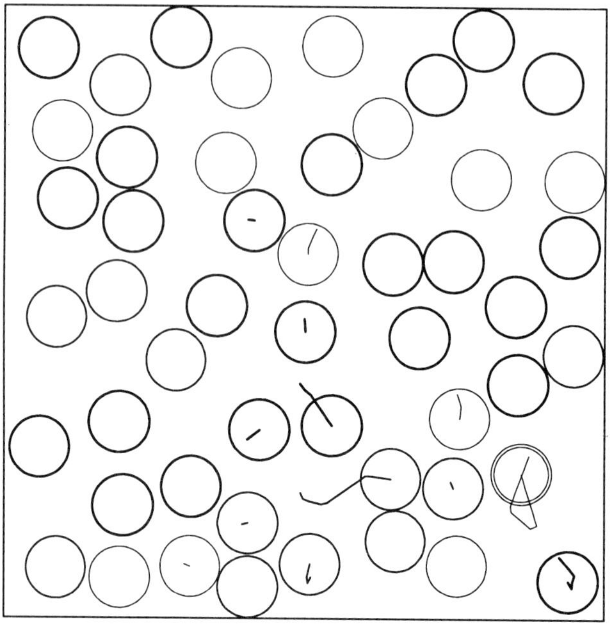

Fig. 4.11. Simulation of molecular motion in an ideal gas in two dimensions. Shown are the initial positions of the molecules (circles) and the trajectories of their centers during the first steps in time (continuous lines). Initially only the molecule shown as a double circle has a non-vanishing velocity. It moves towards the lower right of the picture. It collides with the molecule at the lower right which in turn is reflected by both walls. Through successive collisions the motion spreads towards the upper left of the picture.

4.15.4 Programs to Simulate Molecular Movement

Modify the program E4RN (Example Program 4.14.4) so that the user can choose (a) to draw the trajectories of the centers of the spheres in the (x_1, x_2) plane, Fig 4.11, or (b) to display the histograms made at regular time intervals. Allow the choice of histograms of the quantities v_1, v_2, $|\mathbf{v}|$, $|\mathbf{v}|^2$.

Analyze in particular the histograms that result for large times for v_1 and v_2 and use these results to formulate a more realistic distribution of the velocities at time $t = 0$.

5. Some Important Distributions and Theorems

We shall now discuss in detail some specific distributions. This chapter could therefore be regarded as a collection of examples. These distributions, however, are of great practical importance and are often encountered in many applications. Moreover, their study will lead us to a number of important theorems.

5.1 The Binomial and Multinomial Distributions

Consider an experiment having only two possible outcomes. The sample space can therefore be expressed as

$$E = A + \bar{A} \tag{5.1.1}$$

with the probabilities

$$P(A) = p \quad , \qquad P(\bar{A}) = 1 - p = q \quad . \tag{5.1.2}$$

One now performs n independent trials of the experiment defined by (5.1.1). One wishes to find the probability distribution for the quantity $\mathsf{x} = \sum_{i=1}^{n} \mathsf{x}_i$, where one has $\mathsf{x}_i = 1$ (or 0) when A (or $\bar{A}$) occurs as the result of the ith experiment.

The probability that the first k trials result in A and all of the rest in $\bar{A}$ is, using Eq. (2.3.8),

$$p^k \, q^{n-k} \quad .$$

Using the rules of combinatorics, the event "outcome A k times in n trials" occurs in $\binom{n}{k} = \frac{n!}{k!(n-k)!}$ different ways, according to the order of the occurrences of A and $\bar{A}$ (see Appendix B). The probability of this event is therefore

$$P(k) = W_k^n = \binom{n}{k} p^k \, q^{n-k} \quad . \tag{5.1.3}$$

We are interested in the mean value and variance of x. We first find these quantities for the variable x_i of an individual event. According to (3.3.2) one has

$$E(\mathsf{x}_i) = 1 \cdot p + 0 \cdot q \qquad (5.1.4)$$

and

$$\begin{aligned} \sigma^2(\mathsf{x}_i) &= E\{(x_i - p)^2\} = (1-p)^2 p + (0-p)^2 q \quad , \\ \sigma^2(\mathsf{x}_i) &= pq \quad . \end{aligned} \qquad (5.1.5)$$

From the generalization of (3.5.3) for $\mathsf{x} = \sum \mathsf{x}_i$ it follows that

$$E(\mathsf{x}) = \sum_{i=1}^{n} p = np \quad , \qquad (5.1.6)$$

and from (3.5.10), since all of the covariances vanish because the x_i are independent, one has

$$\sigma^2(\mathsf{x}) = npq \quad . \qquad (5.1.7)$$

Figure 5.1 shows the distribution W_k^n for various n and for fixed p, and Fig. 5.2 shows it for fixed n and various values of p. Finally in Fig. 5.3 n and p are both varied but the product np is held constant. The figures will help us to see relationships between the *binomial distribution* (5.1.3) and other distributions.

A logical extension of the binomial distribution deals with experiments where more than two different outcomes are possible. Equation (5.1.1) is then replaced by

$$E = A_1 + A_2 + \cdots + A_\ell \quad . \qquad (5.1.8)$$

Let the probability for the outcome A_j be

$$P(A_j) = p_j \quad , \qquad \sum_{j=1}^{\ell} p_j = 1 \quad . \qquad (5.1.9)$$

We consider again n trials and ask for the probability that the outcome A_j occurs k_j times. This is given by

$$W^n_{(k_1,k_2,\ldots,k_\ell)} = \frac{n!}{\prod_{j=1}^{\ell} k_j!} \prod_{j=1}^{\ell} p_j^{kj} \quad , \qquad \sum_{j=1}^{\ell} k_j = n \quad . \qquad (5.1.10)$$

The proof is left to the reader. The probability distribution (5.1.10) is called the *multinomial distribution*.

We can define a random variable x_{ij} that takes on the value 1 when the ith trial leads to the outcome A_j, and is zero otherwise. In addition define $\mathsf{x}_j = \sum_{i=1}^{n} \mathsf{x}_{ij}$. The expectation value of x_j is then

$$E(\mathsf{x}_j) = \widehat{x}_j = np_j \quad . \qquad (5.1.11)$$

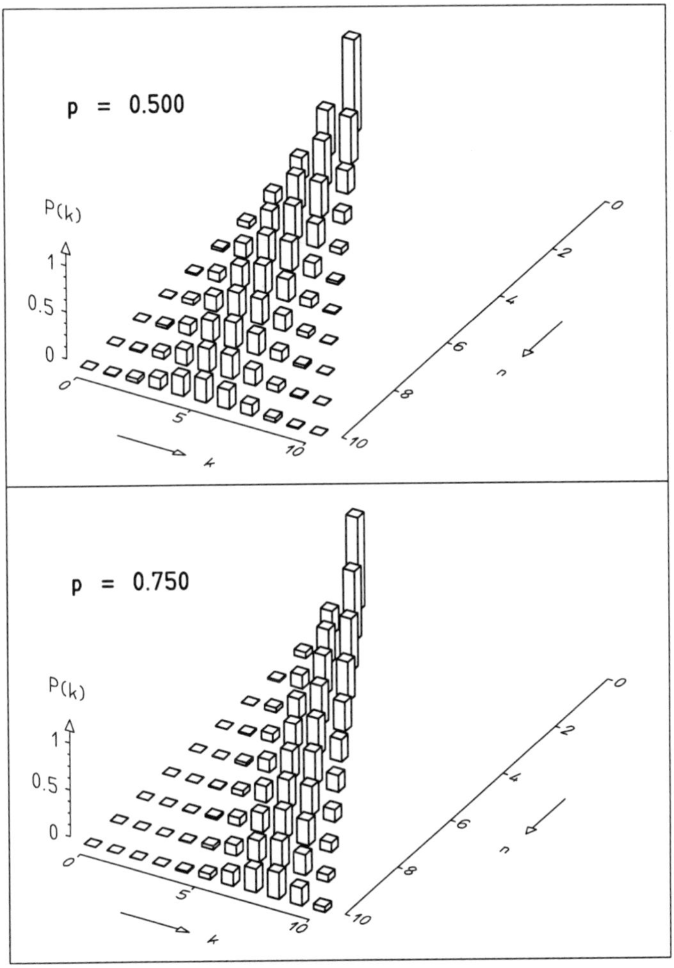

Fig. 5.1. Binomial distribution for various values of n, with p fixed.

The elements of the covariance matrix of the x_j are

$$c_{ij} = np_i(\delta_{ij} - p_j) \quad . \tag{5.1.12}$$

The off-diagonal elements are clearly not zero. This was to be expected, since from Eq. (5.1.9) the variables x_j are not independent.

5.2 Frequency. The Law of Large Numbers

Usually the probabilities for the different types of events, e.g., p_j in the case of the multinomial distribution, are not known but have to be obtained from

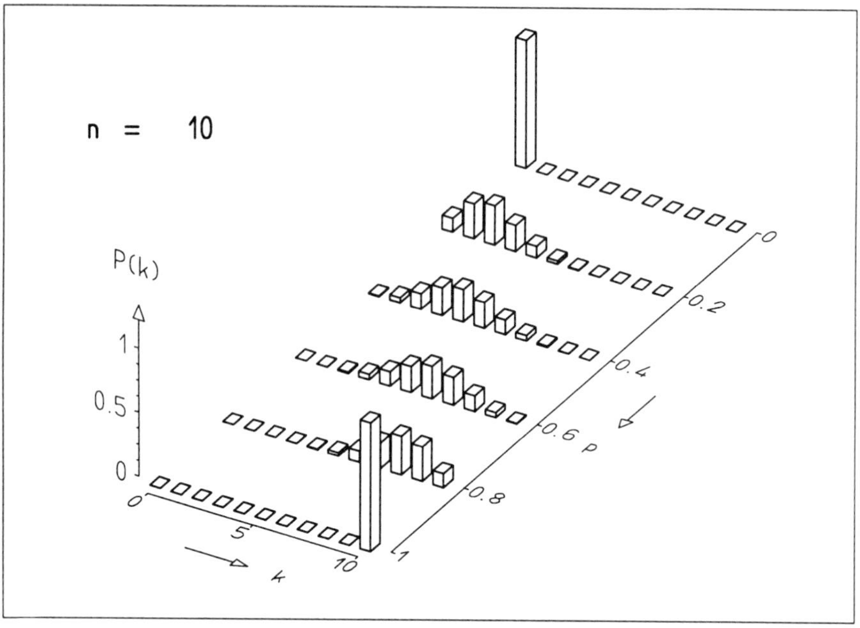

Fig. 5.2. Binomial distribution for various values of p, with n fixed.

experiment. One first measures the *frequency* of the events in n experiments,

$$\mathsf{h}_j = \frac{1}{n}\sum_{i=1}^{n}\mathsf{x}_{ij} = \frac{1}{n}\mathsf{x}_j \quad . \tag{5.2.1}$$

Unlike the probability, the frequency is a random quantity, since it depends on the outcomes of the n individual experiments. By use of (5.1.11), (5.1.12), and (3.3.15) we obtain

$$E(\mathsf{h}_j) = \widehat{h_j} = E\left(\frac{\mathsf{x}_j}{n}\right) = p_j \tag{5.2.2}$$

and

$$\sigma^2(\mathsf{h}_j) = \sigma^2\left(\frac{\mathsf{x}_j}{n}\right) = \frac{1}{n^2}\sigma^2(\mathsf{x}_j) = \frac{1}{n}p_j(1-p_j) \quad . \tag{5.2.3}$$

The product $p_j(1-p_j)$ in Eq. (5.2.3) is at most $1/4$. One sees that the expectation value of the frequency of an event is exactly equal to the probability that the event will occur, and that the variance of frequency about this expectation value can be made arbitrarily small as the number of trials increases. Since pq is at most $1/4$, one can always say that the standard deviation of h_j is at most $1/\sqrt{n}$. This property of the frequency is known as the *law of large numbers*. It is clearly the reason for the frequency definition of probability given by Eq. (2.2.1).

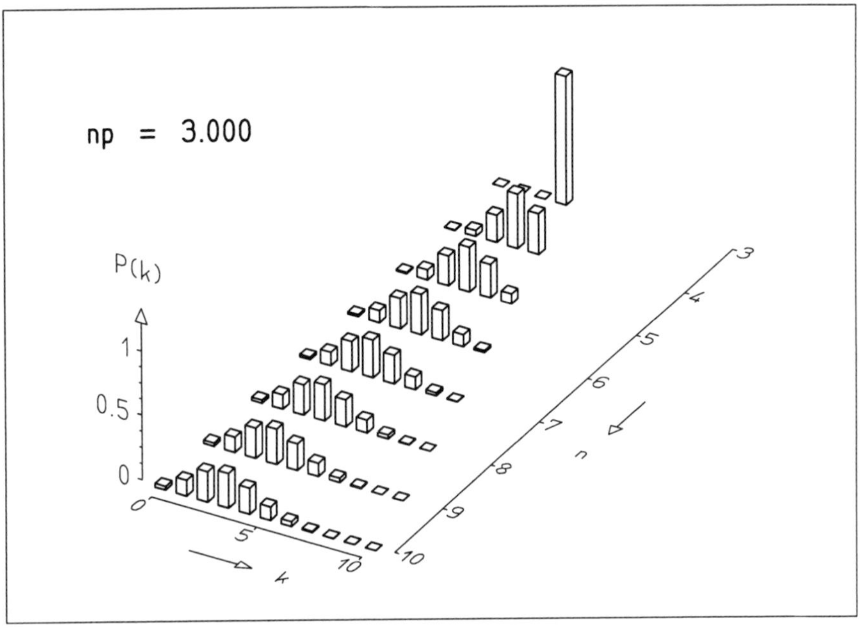

Fig. 5.3. Binomial distribution for various values of n but fixed product np. For higher values of n the distribution changes very little.

Frequently the purpose of an experimental investigation is to determine the probability for the occurrence of a certain type of event. According to (5.2.2) we can use the frequency as an approximation of the probability. The square of the error of this approximation is then inversely proportional to the number of individual experiments. This kind of error, which originates from the fact that only a finite number of experiments can be performed, is called the *statistical error*. It is of prime importance for applications that are concerned with the counting of individual events, e.g., nuclear particles passing through a counter, animals with certain traits in heredity experiments, defective items in quality control, and so forth.

Example 5.1: Statistical error

Suppose it is known from earlier experiments that a fraction $R \approx 1/200$ of a sample of fruit flies *(Drosophila)* develop a certain property A if exposed to a given dose of X-rays. An experiment is planned to determine the fraction R with an accuracy of 1 %. How large must the original sample be in order to achieve this accuracy?

We use Eq. (5.2.3) and find $p_j = 0.005$, $(1-p_j) \approx 1$. We must now choose n such that $\sigma(h_j)/h_j = 200\sigma(h_j) = 0.01$. This gives $\sigma(h_j) = 0.000\,05$ and $\sigma^2(h_j) = 0.25 \times 10^{-8}$. Equation (5.2.3) gives

$$0.25 \times 10^{-8} = \frac{1}{n} \times 0.005$$

and therefore

$$n = 2 \times 10^6 \quad .$$

A total of two million fruit flies would have to be used. This is practically impossible. To determine the fraction R with an accuracy of 10 % would require 20 000 flies. ■

5.3 The Hypergeometric Distribution

Although we shall rigorously introduce the concept of random sampling at a later point, we will now discuss a typical problem of sampling. We consider a container – we shall not break with the habit of mathematicians of calling such a container an urn – with K white and $L = N - K$ black balls. We want to determine the probability that in drawing n balls (without replacing them) we will find exactly k white and $l = n - k$ black ones. The problem is rendered difficult by the fact that the drawing of a ball of a particular color changes the ratio of white and black balls and therefore influences the outcome of the next draw. One clearly has $\binom{N}{n}$ equally likely ways to choose n out of N balls. The probability that one of these possibilities will occur is therefore $1/\binom{N}{n}$. There are $\binom{K}{k}$ ways to choose k of the K white balls, and $\binom{L}{\ell}$ ways to choose ℓ of the L black ones. The required probability is therefore

$$W_k = \frac{\binom{K}{k}\binom{L}{\ell}}{\binom{N}{n}} \quad . \tag{5.3.1}$$

As in Sect. 5.1 we define the random variable $\mathsf{x} = \sum_{i=1}^{n} \mathsf{x}_i$ with $\mathsf{x}_i = 1$ when the ith draw results in a black ball, and $\mathsf{x}_i = 0$ otherwise. (In other words, we define k as the random variable x.)

To compute the expectation values of x we cannot simply add the expectation values of the x_i, since these are no longer independent. Instead we must return to the definition (3.3.2),

$$\begin{aligned}
E(\mathsf{x}) &= \frac{1}{\binom{N}{n}} \sum_{i=1}^{n} i \binom{K}{i}\binom{N-K}{n-i} \\
&= \frac{(N-n)!n!}{N!} \sum_{i=1}^{n} \frac{iK!(N-K)!}{i!(K-i)!(n-i)!(N-K-n+i)!} \\
&= \frac{n(n-1)!(N-n)!}{N(N-1)!} \sum_{i=1}^{n} \frac{K!}{(i-1)!(K-1-(i-1))!} \\
&\quad \times \frac{(N-K)!}{(n-1-(i-1))!(N-K-(n-1)+(i-1))!}
\end{aligned}$$

If we substitute $i - 1 = j$, this gives

$$\begin{aligned} E(\mathsf{x}) &= n\frac{K}{N}\frac{(n-1)!(N-n)!}{(N-1)!} \\ &\quad \times \sum_{j=0}^{n-1} \frac{(K-1)!(N-K)!}{j!(K-1-j)!(n-1-j)!(N-K-(n-1)+j)!} \\ &= n\frac{K}{N}\frac{1}{\binom{N-1}{n-1}} \sum_{j=0}^{n-1} \binom{K-1}{j}\binom{N-K}{n-1-j} \quad . \end{aligned}$$

With Eq. (B.5) we obtain

$$E(\mathsf{x}) = n\frac{K}{N} \quad . \tag{5.3.2}$$

The calculation of the variance follows along the same lines but is rather lengthy. The result is

$$\sigma^2(\mathsf{x}) = \frac{n\,K(N-K)(N-n)}{N^2(N-1)} \quad . \tag{5.3.3}$$

Figures 5.4 and 5.5 depict several examples of the distribution. If $n \ll N$, then drawing a white ball has little influence on the probabilities for the next draw. We therefore expect that in this case W_k behaves in a manner similar to

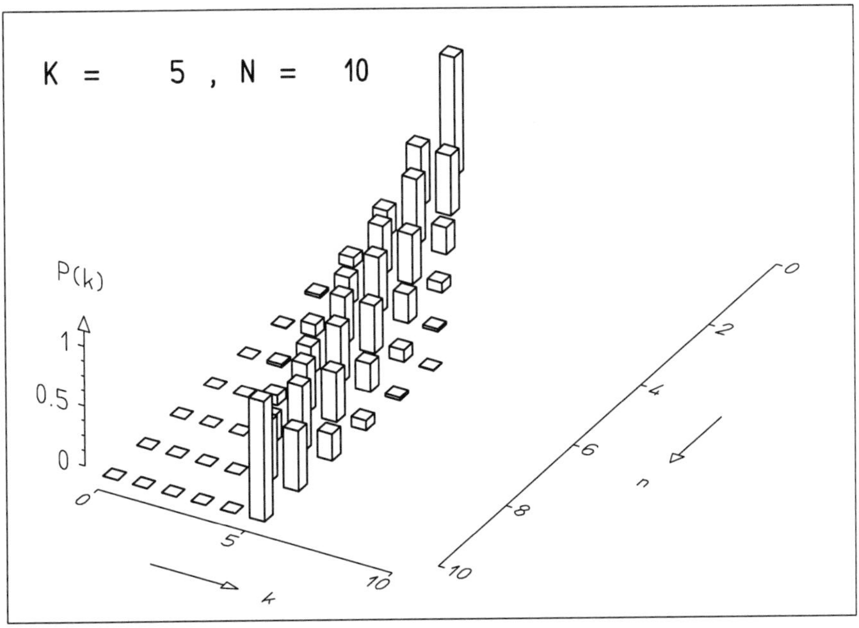

Fig. 5.4. Hypergeometric distribution for various values of n and small values of K and N.

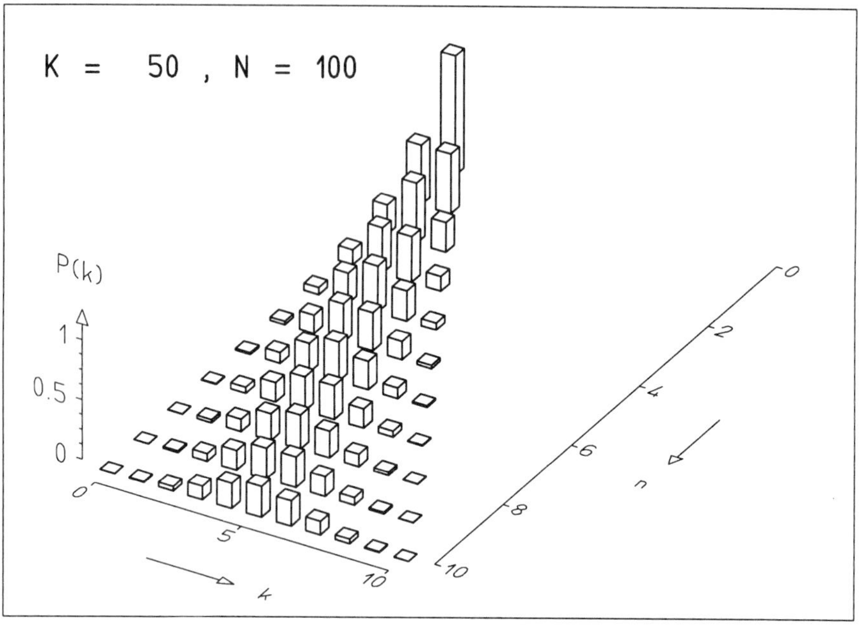

Fig. 5.5. Hypergeometric distribution for various values of n and for large values of K and N.

a binomial distribution with $p = \frac{K}{N}$ and $q = \frac{N-K}{N}$. This is also made clear by the similarity of Figs. 5.5 and 5.1. One obtains in fact the same expectation value,

$$E(\mathrm{x}) = n\frac{K}{N} = np \quad ,$$

as for the binomial distribution. The variance is then

$$\sigma^2(\mathrm{x}) = \frac{npq(N-n)}{N-1} \quad ,$$

which for the case $n \ll N$ becomes

$$\sigma^2 = npq \quad .$$

There are many applications of the hypergeometric distribution. Opinion polls, quality controls, and so forth are all based on the experimental scheme of taking (polling) an object without replacement back into the original sample or *population*. The distribution can be generalized in two ways. First we can of course consider more properties instead of just two (white and black balls). This leads us to a similar transition as the one from the binomial to the multinomial distribution. The original sample (population) contains N elements each of which possesses one of l properties,

$$N = N_1 + N_2 + \cdots + N_\ell \quad .$$

The probability that n draws (without replacement) will be composed as

$$n = n_1 + n_2 + \cdots + n_\ell$$

is, in analogy to Eq. (5.3.1),

$$W_{n_1,n_2,\ldots,n_\ell} = \frac{\binom{N_1}{n_1}\binom{N_2}{n_2}\cdots\binom{N_\ell}{n_\ell}}{\binom{N}{n}} \quad . \tag{5.3.4}$$

Another extension of the hypergeometric distribution is obtained in the following way. We saw earlier that consecutive drawings ceased to be independent because the balls were not replaced. If now each time we draw a ball of one type we place more balls of that type back in the urn, this dependence can be enhanced. One then obtains the *Polya distribution*. It is of importance in the study of epidemic diseases, where the appearance of a case of the disease enhances the probability of future cases.

Example 5.2: Application of the hypergeometric distribution for determination of zoological populations

From a pond K fish are taken and marked. They are then returned to the pond. After a short while n fish are caught, k of which are found to be marked. Before the second time that the fish are taken, the pond contains a total of N fish, of which K are marked. The probability of finding k marked out of n removed fish is given by Eq. (5.3.1). We will return to this problem in Example 7.3. ■

5.4 The Poisson Distribution

Looking at Fig. 5.3 it appears that if n tends to infinity, but at the same time $np = \lambda$ is kept constant, the binomial distribution approaches a certain fixed distribution. We rewrite Eq. (5.1.3) as

$$\begin{aligned} W_k^n &= \binom{n}{k} p^k q^{n-k} \\ &= \frac{n!}{k!(n-k)!} \left(\frac{\lambda}{n}\right)^k \frac{\left(1-\frac{\lambda}{n}\right)^n}{\left(1-\frac{\lambda}{n}\right)^k} \end{aligned}$$

$$= \frac{\lambda^k}{k!} \frac{n(n-1)(n-2)\cdots(n-k+1)}{n^k} \frac{\left(1-\frac{\lambda}{n}\right)^n}{\left(1-\frac{\lambda}{n}\right)^k}$$

$$= \frac{\lambda^k}{k!}\left(1-\frac{\lambda}{n}\right)^n \frac{\left(1-\frac{1}{n}\right)\left(1-\frac{2}{n}\right)\cdots\left(1-\frac{k-1}{n}\right)}{\left(1-\frac{\lambda}{n}\right)^k} \quad .$$

In the limiting case all of the many individual factors of the term on the right approach unity. In addition one has

$$\lim_{n\to\infty}\left(1-\frac{\lambda}{n}\right)^n = e^{-\lambda} \quad ,$$

so that in the limit one has

$$\lim_{n\to\infty} W_k^n = f(k) = \frac{\lambda^k}{k!}e^{-\lambda} \quad . \tag{5.4.1}$$

The quantity $f(k)$ is the probability of the *Poisson distribution*. It is plotted in Fig. 5.6 for various values of λ. As is the case for the other distributions we have encountered so far, the Poisson distribution is only defined for integer values of k.

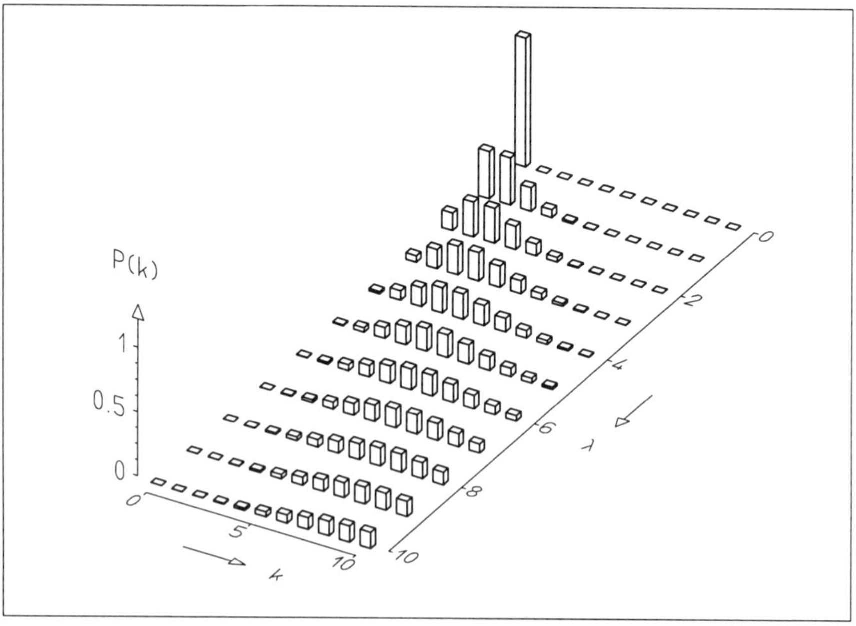

Fig. 5.6. Poisson distribution for various values of λ.

The distribution satisfies the requirement that the total probability is equal to unity,

$$\begin{aligned}\sum_{k=0}^{\infty} f(k) &= \sum_{k=0}^{\infty} \frac{e^{-\lambda}\lambda^k}{k!} \\ &= e^{-\lambda}\left(1+\lambda+\frac{\lambda^2}{2!}+\frac{\lambda^3}{3!}+\cdots\right) \\ &= e^{-\lambda}e^{\lambda} \quad , \\ \sum_{k=0}^{\infty} f(k) &= 1 \quad . \end{aligned} \tag{5.4.2}$$

The expression in parentheses is in fact the Taylor expansion of e^{λ}.

We now want to determine the mean, variance, and skewness of the Poisson distribution. The definition (3.3.2) gives

$$\begin{aligned}E(\mathsf{k}) &= \sum_{k=0}^{\infty} k\frac{\lambda^k}{k!}e^{-\lambda} = \sum_{k=1}^{\infty} k\frac{\lambda^k}{k!}e^{-\lambda} \\ &= \sum_{k=1}^{\infty} \frac{\lambda\,\lambda^{k-1}}{(k-1)!}e^{-\lambda} = \lambda\sum_{j=0}^{\infty}\frac{\lambda^j}{j!}e^{-\lambda}\end{aligned}$$

and using this with (5.4.2),

$$E(\mathsf{k}) = \lambda \quad . \tag{5.4.3}$$

We would now like to find $E(\mathsf{k}^2)$. One obtains in a corresponding way

$$\begin{aligned}E(\mathsf{k}^2) &= \sum_{k=1}^{\infty} k^2\frac{\lambda^k}{k!}e^{-\lambda} = \lambda\sum_{k=1}^{\infty} k\frac{\lambda^{k-1}}{(k-1)!}e^{-\lambda} \\ &= \lambda\sum_{j=0}^{\infty}(j+1)\frac{\lambda^j}{j!}e^{-\lambda} = \lambda\left(\sum_{j=0}^{\infty} j\frac{\lambda^j}{j!}e^{-\lambda}+1\right)\end{aligned}$$

and therefore

$$E(\mathsf{k}^2) = \lambda(\lambda+1) \quad . \tag{5.4.4}$$

We will use Eqs. (5.4.3) and (5.4.4) to compute the variance. According to Eq. (3.3.16) one has

$$\sigma^2(\mathsf{k}) = E(\mathsf{k}^2) - \{E(\mathsf{k})\}^2 = \lambda(\lambda+1) - \lambda^2 \tag{5.4.5}$$

or

$$\sigma^2(\mathsf{k}) = \lambda \quad . \tag{5.4.6}$$

We now consider the skewness (3.3.13) of the Poisson distribution. Following Sect. 3.3 we easily find that

$$\mu_3 = E\{(\mathsf{k}-\widehat{k})^3\} = \lambda \quad .$$

The skewness (3.3.13) is then

$$\gamma = \frac{\mu_3}{\sigma^3} = \frac{\lambda}{\lambda^{\frac{3}{2}}} = \lambda^{-\frac{1}{2}} \quad , \tag{5.4.7}$$

that is, the Poisson distribution becomes increasingly symmetric as λ increases. Figure 5.6 shows the distribution for various values of λ. In particular the distribution with $\lambda = 3$ should be compared with Fig. 5.3.

We have obtained the Poisson distribution from the binomial distribution with large n but constant $\lambda = np$, i.e., small p. We therefore expect it to apply to processes in which a large number of events occur but of which only very few have a certain property of interest to us (i.e., a large number of "trials" but few "successes").

Example 5.3: Poisson distribution and independence of radioactive decays

We consider a radioactive nucleus with mean lifetime τ and observe it for a time $T \ll \tau$. The probability that it decays within this time interval is $W \ll 1$. We break the observation time T into n smaller time intervals of length t, so that $T = nt$. The probability for the nucleus to decay in a particular time interval is $p \approx W/n$. We now observe a radioactive source containing N nuclei which decay independently from each other for a total time T, and detect a_1 decays in time interval 1, a_2 decays in interval 2, etc. Let $h(k)$ be the frequency of decays observed in the interval k, with $(k = 0, 1, \ldots)$. That is, if n_k is the number of intervals with k decays, then $h(k) = n_k/n$. In the limit $N \to \infty$ and for large n the frequency distribution $h(k)$ becomes the probability distribution (5.4.1). The statistical nature of radioactive decay was established in this way in a famous experiment by RUTHERFORD and GEIGER.
■

Similarly, the frequency of finding k stars per element of the celestial sphere or k raisins per volume element of a fruit cake is distributed according to the Poisson law, but not, however, the frequency of finding k animals of a given species per element of area, at least if these animals live in herds, since in this case the assumption of independence is not fulfilled.

As a quantitative example of the Poisson distribution many textbooks discuss the number of Prussian cavalrymen killed during a period of 20 years by horse kicks, an example originally due to VON BORTKIEWICZ (1898) [6]. We prefer to turn our attention to a somewhat less macabre example taken from a lecture of DE SOLLA PRICE [7].

Example 5.4: Poisson distribution and the independence of scientific discoveries

The author first constructs the model of an apple tree with 1000 apples and 1000 pickers with blindfolded eyes who each try at the same time to pick an apple. Since we are dealing with a model, they do not hinder each other but it can happen that two or several of them will attempt to pick the same apple at the same time. The number of apples grabbed simultaneously by k people ($k = 0, 1, 2, \ldots$) follows a Poisson distribution. It was determined by DE SOLLA PRICE that the number of scientific discoveries made independently twice, three times, etc. is also distributed according to the Poisson law, in a way similar to the principle of the blindfolded apple pickers (Table 5.1). One gets the impression that scientists are not concerned with the activities of their colleagues. DE SOLLA PRICE believes that this can be explained by the assumption that scientists have a strong urge write papers, but feel only a relatively mild need to read them. ■

Table 5.1. Simultaneous Discovery and the Poisson Distribution.

Number of simultaneous discoveries	Cases of simultaneous discovery	Prediction of Poisson distribution
0	not defined	368
1	not known	368
2	179	184
3	51	61
4	17	15
5	6	3
≥ 6	8	1

5.5 The Characteristic Function of a Distribution

So far we have only considered real random variables. In fact, in Sect. 3.1 we have introduced the concept of a random quantity as a real number associated with an event. Without changing this concept we can formally construct a *complex random variable* from two real ones by writing

$$\mathsf{z} = \mathsf{x} + \mathrm{i}\,\mathsf{y} \quad . \tag{5.5.1}$$

As its expectation value we define

$$E(\mathsf{z}) = E(\mathsf{x}) + \mathrm{i}\,E(\mathsf{y}) \quad . \tag{5.5.2}$$

By analogy with real variables, complex random variables are independent if the real and imaginary parts are independent among themselves.

If x is a real random variable with distribution function $F(x) = P(\mathsf{x} < x)$ and probability density $f(x)$, we define its characteristic function to be the expectation value of the quantity $\exp(it\mathsf{x})$:

$$\varphi(t) = E\{\exp(it\mathsf{x})\} \quad . \tag{5.5.3}$$

That is, in the case of a continuous variable the characteristic function is a Fourier integral with its known transformation properties:

$$\varphi(t) = \int_{-\infty}^{\infty} \exp(itx) f(x)\,\mathrm{d}x \quad . \tag{5.5.4}$$

For a discrete variable we obtain instead from (3.3.2)

$$\varphi(t) = \sum_i \exp(itx_i) P(\mathsf{x} = x_i) \quad . \tag{5.5.5}$$

We now consider the moments of x about the origin,

$$\lambda_n = E(\mathsf{x}^n) = \int_{-\infty}^{\infty} x^n f(x)\,\mathrm{d}x \quad , \tag{5.5.6}$$

and find that λ_n can be obtained simply by differentiating the characteristic function n times at the point $t = 0$:

$$\varphi^{(n)}(t) = \frac{\mathrm{d}^n\varphi(t)}{\mathrm{d}t^n} = \mathrm{i}^n \int_{-\infty}^{\infty} x^n \exp(itx) f(x)\,\mathrm{d}x$$

and therefore

$$\varphi^{(n)}(0) = \mathrm{i}^n \lambda_n \quad . \tag{5.5.7}$$

If we now introduce the simple coordinate translation

$$\mathsf{y} = \mathsf{x} - \widehat{x} \tag{5.5.8}$$

and construct the characteristic function

$$\varphi_y(t) = \int_{-\infty}^{\infty} \exp\{\mathrm{i}\,t(x - \widehat{x})\} f(x)\,\mathrm{d}x = \varphi(t)\exp(-\mathrm{i}\,t\widehat{x}) \quad , \tag{5.5.9}$$

then its nth derivative is (up to a power of i) equal to the nth moment of x about the expectation value [cf. (3.3.8)]:

$$\varphi_y^{(n)}(0) = \mathrm{i}^n \mu_n = \mathrm{i}^n E\{(\mathsf{x} - \widehat{x})^n\} \quad , \tag{5.5.10}$$

and in particular

$$\sigma^2(x) = -\varphi_y''(0) \quad . \tag{5.5.11}$$

Inverting the Fourier transform (5.5.4) we see that it is possible to obtain the probability density from the characteristic function,

$$f(x) = \frac{1}{2\pi}\int_{-\infty}^{\infty} \exp(-itx)\varphi(t)\,\mathrm{d}t \quad . \tag{5.5.12}$$

It is possible to show that a distribution is determined *uniquely* by its characteristic function. This is the case even for discrete variables where one has

$$F(b) - F(a) = \frac{\mathrm{i}}{2\pi}\int_{-\infty}^{\infty} \frac{\exp(itb) - \exp(ita)}{t}\varphi(t)\,\mathrm{d}t \quad , \tag{5.5.13}$$

since in this case the probability density is not defined. Often it is more convenient to use the characteristic function rather than the original distribution. Because of the unique relation between the two it is possible to switch back and forth at any place in the course of a calculation.

We now consider the sum of two independent random variables

$$\mathsf{w} = \mathsf{x} + \mathsf{y} \quad .$$

Its characteristic function is

$$\varphi_w(t) = E[\exp\{it(\mathsf{x}+\mathsf{y})\}] = E\{\exp(it\mathsf{x})\exp(it\mathsf{y})\} \quad .$$

Generalizing relation (3.5.13) to complex variables we obtain

$$\varphi_w(t) = E\{\exp(itx)\}E\{\exp(ity)\} = \varphi_x(t)\varphi_y(t) \quad , \tag{5.5.14}$$

i.e., the characteristic function of a sum of independent random variables is equal to the product of their respective characteristic functions.

Example 5.5: Addition of two Poisson distributed variables with use of the characteristic function

From Eqs. (5.5.5) and (5.4.1) one obtains for the characteristic function of the Poisson distribution

$$\begin{aligned} \varphi(t) &= \sum_{k=0}^{\infty} \exp(itk)\frac{\lambda^k}{k!}\exp(-\lambda) = \exp(-\lambda)\sum_{k=0}^{\infty}\frac{(\lambda\exp(it))^k}{k!} \\ &= \exp(-\lambda)\exp(\lambda\,\mathrm{e}^{\mathrm{i}t}) = \exp\{\lambda(\mathrm{e}^{\mathrm{i}t} - 1)\} \quad . \end{aligned} \tag{5.5.15}$$

We now form the characteristic function of the sum of two independent Poisson distributed variables with mean values λ_1 and λ_2,

$$\begin{aligned} \varphi_{\mathrm{sum}}(t) &= \exp\{\lambda_1(\mathrm{e}^{\mathrm{i}t} - 1)\}\exp\{\lambda_2(\mathrm{e}^{\mathrm{i}t} - 1)\} \\ &= \exp\{(\lambda_1 + \lambda_2)(\mathrm{e}^{\mathrm{i}t} - 1)\} \quad . \end{aligned} \tag{5.5.16}$$

This is again of the form of Eq. (5.5.15). Therefore the distribution of the sum of two independent Poisson distributed variables is itself a Poisson variable. Its mean is the sum of the means of the individual distributions. ■

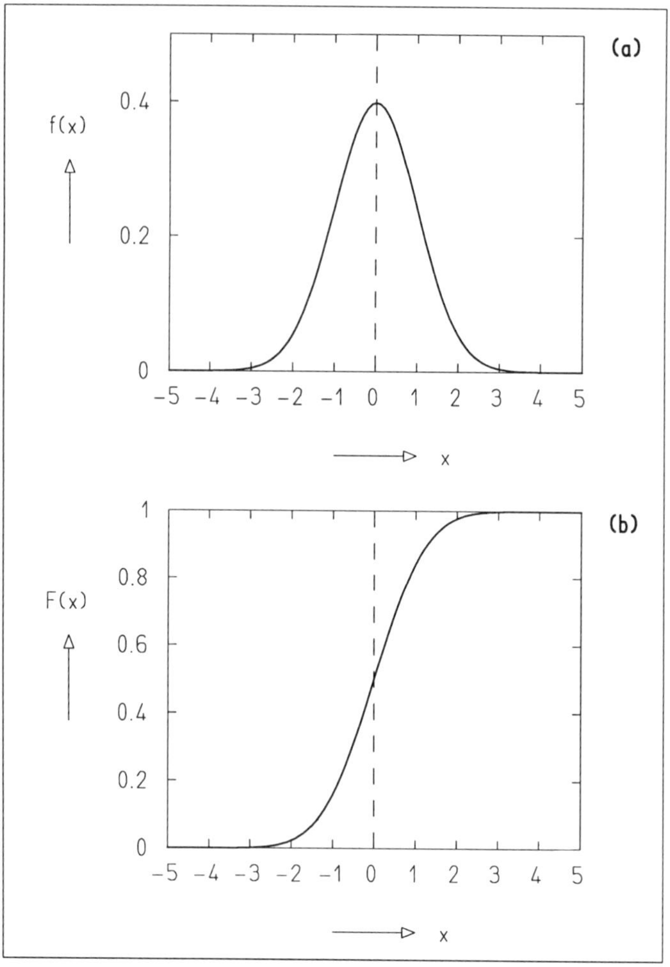

Fig. 5.7. Probability density (a) and distribution function (b) of the standard normal distribution.

5.6 The Standard Normal Distribution

The probability density of the *standard normal distribution* is defined as

$$f(x) = \phi_0(x) = \frac{1}{\sqrt{2\pi}} e^{-x^2/2} \quad . \tag{5.6.1}$$

This function is depicted in Fig. 5.7a. It has a bell shape with the maximum at $x = 0$. From Appendix D.1 we have

$$\int_{-\infty}^{\infty} e^{-x^2/2}\, dx = \sqrt{2\pi} \quad , \tag{5.6.2}$$

so that $\phi_0(x)$ is normalized to one as required. Using the symmetry of Fig. 5.7a, or alternatively, using the antisymmetry of the integrand we conclude that the expectation value is

$$\widehat{x} = \frac{1}{\sqrt{2\pi}} \int_{-\infty}^{\infty} x\, \mathrm{e}^{-x^2/2}\, \mathrm{d}x = 0 \quad . \tag{5.6.3}$$

By integrating by parts we can compute the variance to be

$$\begin{aligned} \sigma^2 &= \frac{1}{\sqrt{2\pi}} \int_{-\infty}^{\infty} x^2\, \mathrm{e}^{-x^2/2}\, \mathrm{d}x \\ &= \frac{1}{\sqrt{2\pi}} \left\{ \left[-x\, \mathrm{e}^{-x^2/2} \right]_{-\infty}^{\infty} + \int_{-\infty}^{\infty} \mathrm{e}^{-x^2/2}\, \mathrm{d}x \right\} = 1 \quad , \end{aligned} \tag{5.6.4}$$

since the expression in the square brackets vanishes at the integral's boundaries and the integral in curly brackets is given by Eq. (5.6.2).

The distribution function of the standard normal distribution

$$F(x) = \psi_0(x) = \frac{1}{\sqrt{2\pi}} \int_{-\infty}^{x} \mathrm{e}^{-t^2/2}\, \mathrm{d}t \tag{5.6.5}$$

is shown in Fig. 5.7b. It cannot be expressed in analytic form. It is tabulated numerically in Appendix C.4.

5.7 The Normal or Gaussian Distribution

The standardized distribution of the last section had the properties $\widehat{x} = E(\mathsf{x}) = 0$, $\sigma^2(\mathsf{x}) = 1$, i.e., the variable x had the properties of the standardized variable u in Eq. (3.3.17). If we now replace x by $(\mathsf{x} - a)/b$ in (5.6.1), we obtain the probability density of the *normal* or *Gaussian distribution*,

$$f(x) = \phi(x) = \frac{1}{\sqrt{2\pi}b} \exp\left\{ -\frac{(x-a)^2}{2b^2} \right\} \quad , \tag{5.7.1}$$

with

$$\widehat{x} = a \quad , \qquad \sigma^2(\mathsf{x}) = b^2 \quad . \tag{5.7.2}$$

The characteristic function of the normal distribution (5.7.1) is, using Eq. (5.5.4),

$$\varphi(t) = \frac{1}{\sqrt{2\pi}b} \int_{-\infty}^{\infty} \exp(\mathrm{i}tx) \exp\left(-\frac{(x-a)^2}{2b^2} \right) \mathrm{d}x \quad . \tag{5.7.3}$$

With $u = (x - a)/b$ one obtains

$$\begin{aligned}\varphi(t) &= \frac{1}{\sqrt{2\pi}} \int_{-\infty}^{\infty} \exp\{-\frac{1}{2}u^2 + it(bu+a)\}\,du \\ &= \frac{1}{\sqrt{2\pi}} \exp(ita) \int_{-\infty}^{\infty} \exp\{-\frac{1}{2}u^2 + itbu\}\,du \quad . \end{aligned} \qquad (5.7.4)$$

By completing the square the integral can be rewritten as

$$\begin{aligned}&\int_{-\infty}^{\infty} \exp\{-\frac{1}{2}u^2 + itbu\}\,du \\ &\qquad = \int_{-\infty}^{\infty} \exp\{-\frac{1}{2}(u-itb)^2 - \frac{1}{2}t^2b^2\}\,du \\ &\qquad = \exp\{-\frac{1}{2}t^2b^2\} \int_{-\infty}^{\infty} \exp\{-\frac{1}{2}(u-itb)^2\}\,du \quad . \end{aligned} \qquad (5.7.5)$$

With $r = u - itb$ the last integral takes on the form

$$\int_{-\infty-itb}^{\infty-itb} \exp\{-\frac{1}{2}r^2\}\,dr \quad .$$

The integrand does not have any singularities in the complex r plane. According to the residue theorem, therefore, the contour integral around any closed path vanishes. Consider a path that runs along the real axis from $r = -L$ to $r = L$, and then parallel to the imaginary axis from $r = L$ to $r = L - itb$ and from there antiparallel to the real axis to $r = -L - itb$, and finally back to the starting point $r = L$. In the limit $L \to \infty$ the integrand vanishes along the parts of the path that run parallel to the imaginary axis. One then has

$$\int_{-\infty-itb}^{\infty-itb} \exp\{-\frac{1}{2}r^2\}\,dr = \int_{-\infty}^{\infty} \exp\{-\frac{1}{2}r^2\}\,dr \quad ,$$

i.e., we can extend the integral to cover the entire real axis. The integral is computed in Appendix D.1 and has the value

$$\int_{-\infty}^{\infty} \exp\{-\frac{1}{2}r^2\}\,dr = \sqrt{2\pi} \quad . \qquad (5.7.6)$$

Substituting this into Eqs. (5.7.5) and (5.7.4) we obtain finally the characteristic function of the normal distribution

$$\varphi(t) = \exp(ita)\exp(-\frac{1}{2}b^2t^2) \quad . \qquad (5.7.7)$$

For the case $a = 0$ one obtains from this the following interesting **theorem**:

> A normal distribution with mean value zero has a characteristic function that has itself (up to normalization) the form of a normal distribution. The product of the variances of both functions is one.

If we now consider the sum of two independent normal distributions, then by applying Eq. (5.5.14) one immediately sees that the characteristic function of the sum is again of the form of Eq. (5.7.7). The sum of independent normally distributed quantities is therefore itself normally distributed. The Poisson distribution behaves in a similar way (cf. Example 5.5).

5.8 Quantitative Properties of the Normal Distribution

Figure 5.7a shows the probability density of the standard Gaussian distribution $\phi_0(x)$ and the corresponding distribution function. By simple computation one can determine that the points of inflection of (5.6.1) are at $x = \pm 1$. [In the case of a general Gaussian distribution (5.7.1) they are at $x = a \pm b$.] The distribution function $\psi_0(x)$ gives the probability for the random variable to take on a value smaller than x:

$$\psi_0(x) = P(\mathsf{x} < x) \quad . \tag{5.8.1}$$

By symmetry one has

$$P(|\mathsf{x}| > x) = 2\psi_0(-|x|) = 2\{1 - \psi_0(|x|)\} \tag{5.8.2}$$

or conversely, the probability to obtain a random value within an interval of width $2x$ about zero (the expectation value) is

$$P(|\mathsf{x}| \le x) = 2\psi_0(|x|) - 1 \quad . \tag{5.8.3}$$

Since the integral (5.6.5) is not easy to evaluate, one typically finds the values of (5.8.1) and (5.8.3) from statistical tables, e.g., in Tables I.2 and I.3 of the appendix.

One can now extend this relation to the general Gaussian distribution given by Eq. (5.7.1). Its distribution function is

$$\psi(x) = \psi_0\left(\frac{x-a}{b}\right) \quad . \tag{5.8.4}$$

We are interested in finding the probability to obtain a random value inside (or outside) of a given multiple of $\sigma = b$ about the mean value:

$$P(|\mathsf{x} - a| \le n\sigma) = 2\psi_0\left(\frac{nb}{b}\right) - 1 = 2\psi_0(n) - 1 \quad . \tag{5.8.5}$$

From Table I.3 we find

$$\begin{aligned} P(|\mathsf{x}-a| \le \sigma) &= 68.3\,\% \quad , & P(|\mathsf{x}-a| > \sigma) &= 31.7\,\% \quad , \\ P(|\mathsf{x}-a| \le 2\sigma) &= 95.4\,\% \quad , & P(|\mathsf{x}-a| > 2\sigma) &= 4.6\,\% \quad , \\ P(|\mathsf{x}-a| \le 3\sigma) &= 99.8\,\% \quad , & P(|\mathsf{x}-a| > 3\sigma) &= 0.2\,\% \quad . \end{aligned} \tag{5.8.6}$$

As we will see later in more detail, one can often assume that the measurement errors of a quantity are distributed according to a Gaussian distribution about zero. This means that the probability to obtain a value between x and $x + \mathrm{d}x$ is given by

$$P(x \le \mathsf{x} < x + \mathrm{d}x) = \phi(x)\,\mathrm{d}x \quad .$$

The dispersion σ of the distribution $\phi(x)$ is called the *standard deviation* or *standard error*. If the standard error of an instrument is known and one carries out a single measurement, then Eq. (5.8.6) tells us that the probability that the true value is within an interval given by plus or minus the standard error about the measured value is 68.3 %. It is therefore a common practice to multiply the standard error with a more or less arbitrary factor in order to improve this percentage. (One obtains around 99.8 % for the factor 3.) This procedure is, however, misleading and often harmful. If this factor is not explicitly stated, a comparison of different measurements of the same quantity and especially the calculation of a weighted average (cf. Example 9.1) is rendered impossible or is liable to be erroneous.

The quantiles [see Eq. (3.3.25)] of the standard normal distribution are of considerable interest. For the distribution function (5.6.5) one obtains by definition

$$P(x_p) = P(\mathsf{x} < x_p) = \psi_0(x_p) \quad . \tag{5.8.7}$$

The quantile x_p is therefore given by the inverse function

$$x_p = \Omega(P) \tag{5.8.8}$$

of the distribution function $\psi_0(x_p)$. This is computed numerically in Appendix C.4 and is given in Table I.4. Figure 5.8 shows a graphical representation.

We now consider the probability

$$P'(x) = P(|\mathsf{x}| < x) \quad , \qquad x > 0 \quad , \tag{5.8.9}$$

for a quantity distributed according to the standard normal distribution to differ from zero in absolute value by less than x. Since

$$\begin{aligned} P(x) &= P(\mathsf{x} < x) = \psi_0(x) = \int_{-\infty}^{x} \phi_0(x)\,\mathrm{d}x \\ &= \frac{1}{2} + \int_0^x \phi_0(x)\,\mathrm{d}x = \frac{1}{2} + \frac{1}{2}P'(x) = \frac{1}{2}(P'(x) + 1) \quad , \end{aligned}$$

it is the inverse function and with it the quantiles of the distribution function $P'(x)$ are obtained by substituting $(P'(x) + 1)/2$ for the argument of the inverse function of P:

$$x_p = \Omega'(P') = \Omega((P' + 1)/2) \quad . \tag{5.8.10}$$

This function is tabulated in Table I.5.

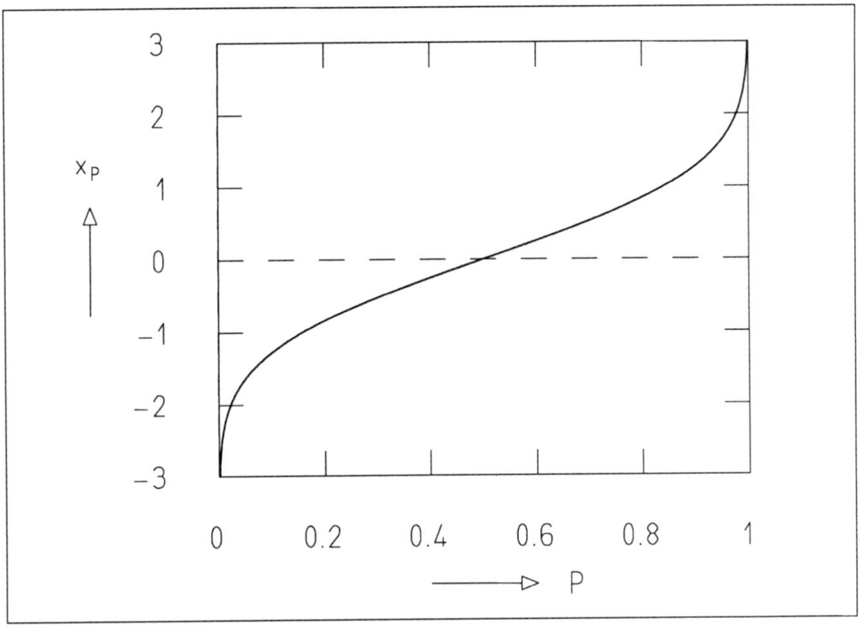

Fig. 5.8. Quantile of the standard normal distribution.

5.9 The Central Limit Theorem

We will now prove the following important theorem. If x_i are independent random variables with mean values a and variances b^2, then the variable

$$\mathsf{x} = \lim_{n\to\infty} \sum_{i=1}^{n} \mathsf{x}_i \tag{5.9.1}$$

is normally distributed with

$$E(\mathsf{x}) = na \quad , \qquad \sigma^2(\mathsf{x}) = nb^2 \quad . \tag{5.9.2}$$

From Eq. (3.3.15) one then has that the variable

$$\xi = \frac{1}{n}\mathsf{x} = \lim_{n\to\infty} \frac{1}{n} \sum_{i=1}^{n} \mathsf{x}_i \tag{5.9.3}$$

is normally distributed with

$$E(\xi) = a \quad , \qquad \sigma^2(\xi) = b^2/n \quad . \tag{5.9.4}$$

To prove this we assume for simplicity that all of the x_i have the same distribution. If we denote the characteristic function of the x_i by $\varphi(t)$, then the sum of n variables has the characteristic function $\{\varphi(t)\}^n$. We now assume that $a = 0$. (The general case can be related to this by a simple coordinate translation $\mathsf{x}_i' = \mathsf{x}_i - a$.) From (5.5.10) we have the first two derivatives of $\varphi(t)$ at $t = 0$,

$$\varphi'(0) = 0 \quad , \qquad \varphi''(0) = -\sigma^2 \quad .$$

We can therefore expand,

$$\varphi_{x'}(t) = 1 - \frac{1}{2}\sigma^2 t^2 + \cdots \quad .$$

Instead of x_i let us now choose

$$\mathsf{u}_i = \frac{\mathsf{x}_i'}{b\sqrt{n}} = \frac{\mathsf{x}_i - a}{b\sqrt{n}}$$

as the variable. If we consider n to be fixed for the moment, then this implies a simple translation and a change of scale. The corresponding characteristic function is

$$\varphi_{u_i}(t) = E\{\exp(itu_i)\} = E\left\{\exp\left(it\frac{x_i - a}{b\sqrt{n}}\right)\right\} = \varphi_{x_i'}\left(\frac{t}{b\sqrt{n}}\right)$$

or

$$\varphi_{u_i}(t) = 1 - \frac{t^2}{2n} + \cdots \quad .$$

The higher-order terms are at most of the order n^{-2}. If we now consider the limiting case and use

$$\mathsf{u} = \lim_{n\to\infty} \sum_{i=1}^{n} \mathsf{u}_i = \lim_{n\to\infty} \sum_{i=1}^{n} \frac{\mathsf{x}_i - a}{b\sqrt{n}} = \lim_{n\to\infty} \frac{(\mathsf{x} - na)}{b\sqrt{n}} \tag{5.9.5}$$

we obtain

$$\varphi_u(t) = \lim_{n\to\infty} \{\varphi_{u_i}(t)\}^n = \lim_{n\to\infty} \left(1 - \frac{t^2}{2n} + \cdots\right)^n$$

or

$$\varphi_u(t) = \exp\left(-\frac{1}{2}t^2\right) \quad . \tag{5.9.6}$$

This, however, is exactly the characteristic function of the standard normal distribution $\phi_0(u)$. One therefore has $E(\mathsf{u}) = 0$, $\sigma^2(\mathsf{u}) = 1$. Using Eqs. (5.9.5) and (3.3.15) leads directly to the theorem.

Example 5.6: Normal distribution as the limiting case of the binomial distribution

Suppose that the individual variables x_i in (5.9.1) are described by the simple distribution given by (5.1.1) and (5.1.2), i.e., they can only take on the value 1 (with probability p) or 0 (with probability $1-p$). One then has $E(\mathsf{x}_i) = p$, $\sigma^2(\mathsf{x}_i) = p(1-p)$. The variable

$$\mathsf{x}^{(n)} = \sum_{i=1}^{n} \mathsf{x}_i \tag{5.9.7}$$

then follows the binomial distribution, $P(\mathsf{x}^{(n)} = k) = W_k^n$ [see Eqs. (5.1.3), (5.1.6), (5.1.7)]. As done for (5.9.5) let us consider the distribution of

$$\mathsf{u}^{(n)} = \sum_{i=1}^{n} \frac{\mathsf{x}_i - p}{\sqrt{np(1-p)}} = \frac{1}{\sqrt{np(1-p)}} \left(\sum_{i=1}^{n} \mathsf{x}_i - np \right) \quad . \tag{5.9.8}$$

One clearly has $P(\mathsf{x} = k) = P\left(\mathsf{u}^{(n)} = (k - np)/\sqrt{np(1-p)}\right) = W_k^n$. These values, however, lie increasingly closer to each other on the $\mathsf{u}^{(n)}$ axis as n increases. Let us denote the distance between two neighboring values of $\mathsf{u}^{(n)}$ by $\Delta\mathsf{u}^{(n)}$. Then the distribution of a discrete variable $P(\mathsf{u}^{(n)})/\Delta\mathsf{u}^{(n)}$ finally becomes the probability density of a continuous variable. According to the Central Limit Theorem this must be a standard normal distribution. This is illustrated in Fig. 5.9, where $P(\mathsf{u}^{(n)})/\Delta\mathsf{u}^{(n)}$ is shown for various possible values of $\mathsf{u}^{(n)}$. ■

Example 5.7: Error model of Laplace

In 1783 LAPLACE made the following remarks concerning the origin of errors of an observation. Suppose the true value of a quantity to be measured is m_0. Now let the measurement be disturbed by a large number n of independent causes, each resulting in a disturbance of magnitude ε. For each disturbance there exists an equal probability for a variation of the measured value in either direction, i.e., $+\varepsilon$ or $-\varepsilon$. The measurement error is then composed of the sum of the individual disturbances. It is clear that in this model the probability distribution of measurement errors will be given by the binomial distribution. It is interesting nevertheless to follow the model somewhat further, since it leads directly to the famous Pascal triangle.

Figure 5.10 shows how the probability distribution is derived from the model. The starting point is with no disturbance where the probability of measuring m_0 is equal to one. With one disturbance this probability is split equally between the neighboring possibilities $m_0 + \varepsilon$ and $m_0 - \varepsilon$. The same happens with every further disturbance. Of course the individual probabilities leading to the same measured value must be added.

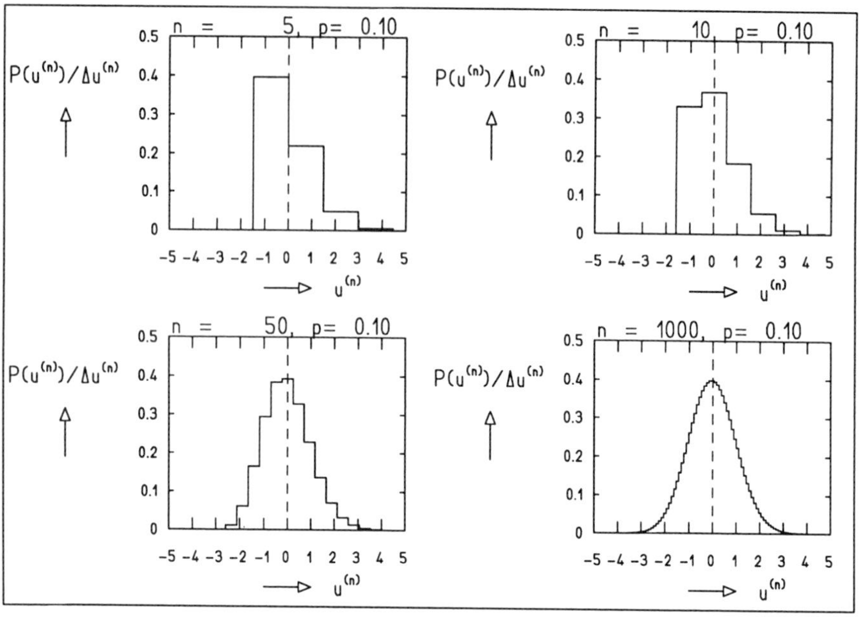

Fig. 5.9. The quantity $P(u^{(n)})/\Delta u^{(n)}$ for various values of the discrete variable $u^{(n)}$ for increasing n.

Number of disturbances	Deviation from true value						
n	-3ε	-2ε	$-\varepsilon$	0	$+\varepsilon$	$+2\varepsilon$	$+3\varepsilon$
0				1			
1			$\frac{1}{2}$		$\frac{1}{2}$		
2		$\frac{1}{4}$	$\frac{1}{4}$, $\frac{1}{4}$ $\frac{1}{2}$			$\frac{1}{4}$	
3	$\frac{1}{8}$		$\frac{1}{8}$, $\frac{1}{4}$ $\underbrace{\quad}_{\frac{3}{8}}$		$\frac{1}{4}$, $\frac{1}{8}$ $\underbrace{\quad}_{\frac{3}{8}}$		$\frac{1}{8}$

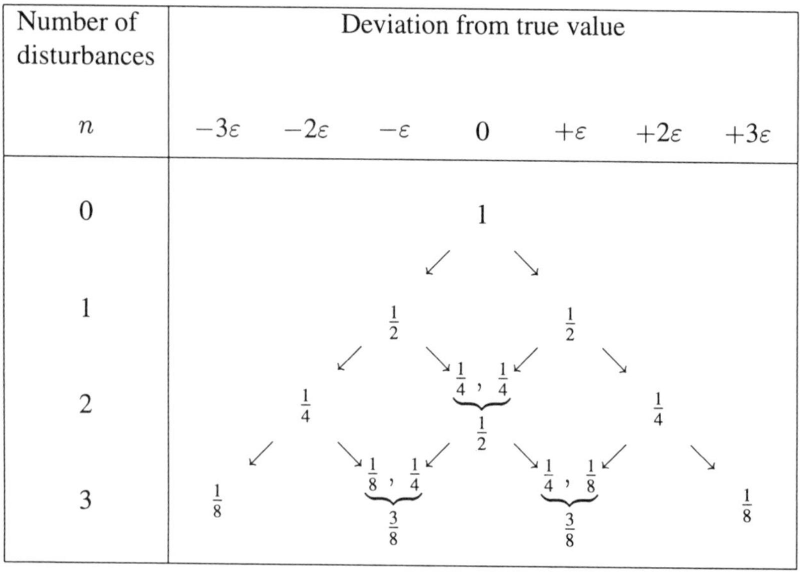

Fig. 5.10. Connection between the Laplacian error model and the binomial distribution.

Each line of the resulting triangle contains the distribution $W_k^n(k = 0, 1, \ldots, n)$ of Eq. (5.1.3) for the case $p = q = 1/2$. Multiplied by $1/(p^k q^{n-k}) = 2^n$ it becomes a line of binomial coefficients of Pascal's triangle (cf. Appendix B).

It is easy to relate this to Example 5.6 by extending Eq. (5.9.8) and substituting $p = 1/2$. For $n \to \infty$ the quantity

$$\mathsf{u}^{(n)} = \frac{2\left(\sum_{i=1}^{n} \varepsilon \mathsf{x}_i - n\varepsilon/2\right)}{\sqrt{n\varepsilon}}$$

follows a normal distribution with expectation value zero and standard deviation $\sqrt{n\varepsilon}/2$. Thus Gaussian measurement errors can result from a large number of small independent disturbances. ■

The identification of the measurement error distribution as Gaussian is of great significance in many computations, particularly for the method of least squares. The normal distribution for measurement errors is, however, not a law of nature. The causes of experimental errors can be individually very complicated. One cannot, therefore, find a distribution function that describes the behavior of measurement errors in all possible experiments. In particular, it is not always possible to guaranty symmetry and independence. One must ask in each individual case whether the measurement errors can be modeled by a Gaussian distribution. This can be done, for example, by means of a χ^2-test, applied to the distribution of a measured quantity (see Sect. 8.7). It is always necessary to check the distribution of experimental errors before more lengthy computations can be used whose results are only meaningful for the case of a Gaussian error distribution.

5.10 The Multivariate Normal Distribution

Consider a vector **x** of n variables,

$$\mathbf{x} = (\mathsf{x}_1, \mathsf{x}_2, \ldots, \mathsf{x}_n) \quad .$$

We define the probability density of the joint normal distribution of the x_i to be

$$\phi(\mathbf{x}) = k \exp\{-\frac{1}{2}(\mathbf{x} - \mathbf{a})^{\mathrm{T}} B (\mathbf{x} - \mathbf{a})\} = k \exp\{-\frac{1}{2} g(\mathbf{x})\} \qquad (5.10.1)$$

with

$$g(\mathbf{x}) = (\mathbf{x} - \mathbf{a})^{\mathrm{T}} B (\mathbf{x} - \mathbf{a}) \quad . \qquad (5.10.2)$$

Here **a** is an n-component vector and B is an $n \times n$ matrix, which is symmetric and positive definite. Since $\phi(\mathbf{x})$ is clearly symmetric about the point $\mathbf{x} = \mathbf{a}$, one has

$$\int_{-\infty}^{\infty} \cdots \int_{-\infty}^{\infty} (\mathbf{x} - \mathbf{a})\phi(\mathbf{x})\, \mathrm{d}x_1\, \mathrm{d}x_2 \ldots \mathrm{d}x_n = 0 \quad , \tag{5.10.3}$$

that is,

$$E(\mathbf{x} - \mathbf{a}) = 0$$

or

$$E(\mathbf{x}) = \mathbf{a} \quad . \tag{5.10.4}$$

The vector of expectation values is therefore given directly by **a**.

We now differentiate Eq. (5.10.3) with respect to **a**,

$$\int_{-\infty}^{\infty} \cdots \int_{-\infty}^{\infty} [I - (\mathbf{x} - \mathbf{a})(\mathbf{x} - \mathbf{a})^{\mathrm{T}} B]\phi(\mathbf{x})\, \mathrm{d}x_1\, \mathrm{d}x_2 \ldots \mathrm{d}x_n = 0 \quad .$$

This means that the expectation value of the quantity in square brackets vanishes,

$$E\{(\mathbf{x} - \mathbf{a})(\mathbf{x} - \mathbf{a})^{\mathrm{T}}\}\, B = I$$

or

$$C = E\{(\mathbf{x} - \mathbf{a})(\mathbf{x} - \mathbf{a})^{\mathrm{T}}\} = B^{-1} \quad . \tag{5.10.5}$$

Comparing with Eq. (3.6.19) one sees that C is the covariance matrix of the variables $\mathbf{x} = (\mathrm{x}_1, \mathrm{x}_2, \ldots, \mathrm{x}_n)$.

Because of the practical importance of the normal distribution, we would like to investigate the case of two variables in somewhat more detail. In particular we are interested in the correlation of the variables. One has

$$C = B^{-1} = \begin{pmatrix} \sigma_1^2 & \mathrm{cov}(\mathrm{x}_1, \mathrm{x}_2) \\ \mathrm{cov}(\mathrm{x}_1, \mathrm{x}_2) & \sigma_2^2 \end{pmatrix} \quad . \tag{5.10.6}$$

By inversion one obtains for B

$$B = \frac{1}{\sigma_1^2\sigma_2^2 - \mathrm{cov}(\mathrm{x}_1, \mathrm{x}_2)^2} \begin{pmatrix} \sigma_2^2 & -\mathrm{cov}(\mathrm{x}_1, \mathrm{x}_2) \\ -\mathrm{cov}(\mathrm{x}_1, \mathrm{x}_2) & \sigma_1^2 \end{pmatrix} \quad . \tag{5.10.7}$$

One sees that B is a diagonal matrix if the covariances vanish. One then has

$$B_0 = \begin{pmatrix} 1/\sigma_1^2 & 0 \\ 0 & 1/\sigma_2^2 \end{pmatrix} \quad . \tag{5.10.8}$$

If we substitute B_0 into Eq. (5.10.1), we obtain – as expected – the joint probability density of two *independently* normally distributed variables as the product of two normal distributions:

$$\phi = k \exp\left(-\frac{1}{2}\frac{(x_1 - a_1)^2}{\sigma_1^2}\right) \exp\left(-\frac{1}{2}\frac{(x_2 - a_2)^2}{\sigma_2^2}\right) \quad . \tag{5.10.9}$$

In this simple case the constant k takes on the value

$$k_0 = \frac{1}{2\pi\sigma_1\sigma_2} \quad ,$$

as can be determined by integration of (5.10.9) or simply by comparison with Eq. (5.7.1). In the general case of n variables with non-vanishing covariances, one has

$$k = \left(\frac{\det B}{(2\pi)^n}\right)^{\frac{1}{2}} \quad . \tag{5.10.10}$$

Here $\det B$ is the determinant of the matrix B. If the variables are not independent, i.e., if the covariance does not vanish, then the expression for the normal distribution of two variables is somewhat more complicated.

Let us consider the reduced variables

$$\mathsf{u}_i = \frac{\mathsf{x}_i - a_i}{\sigma_i} \quad , \qquad i = 1, 2 \quad ,$$

and make use of the correlation coefficient

$$\rho = \frac{\operatorname{cov}(\mathsf{x}_1, \mathsf{x}_2)}{\sigma_1\sigma_2} = \operatorname{cov}(\mathsf{u}_1, \mathsf{u}_2) \quad .$$

Equation (5.10.1) then takes on the simple form

$$\phi(u_1, u_2) = k \exp(-\frac{1}{2}\mathbf{u}^{\mathrm{T}} B \mathbf{u}) = k \exp\left(-\frac{1}{2} g(\mathbf{u})\right) \tag{5.10.11}$$

with

$$B = \frac{1}{1-\rho^2}\begin{pmatrix} 1 & -\rho \\ -\rho & 1 \end{pmatrix} \quad . \tag{5.10.12}$$

Contours of equal probability density are characterized by a constant exponent in (5.10.11):

$$-\frac{1}{2} \cdot \frac{1}{(1-\rho^2)}(u_1^2 + u_2^2 - 2u_1u_2\rho) = -\frac{1}{2} g(\mathbf{u}) = \text{const} \quad . \tag{5.10.13}$$

Let us take for the moment $g(\mathbf{u}) = 1$.

In the original variables Eq. (5.10.13) becomes

$$\frac{(x_1 - a_1)^2}{\sigma_1^2} - 2\rho \frac{x_1 - a_1}{\sigma_1}\frac{x_2 - a_2}{\sigma_2} + \frac{(x_2 - a_2)^2}{\sigma_2^2} = 1 - \rho^2 \quad . \tag{5.10.14}$$

This is the equation of an ellipse centered around the point (a_1, a_2). The principal axes of the ellipse make an angle α with respect to the axes x_1 and x_2. This angle and the half-diameters p_1 and p_2 can be determined from Eq. (5.10.14) by using the known properties of conic sections:

$$\tan 2\alpha = \frac{2\rho\sigma_1\sigma_2}{\sigma_1^2 - \sigma_2^2} \,, \tag{5.10.15}$$

$$p_1^2 = \frac{\sigma_1^2\sigma_2^2(1-\rho^2)}{\sigma_2^2\cos^2\alpha - 2\rho\sigma_1\sigma_2\sin\alpha\cos\alpha + \sigma_1^2\sin^2\alpha} \,, \tag{5.10.16}$$

$$p_2^2 = \frac{\sigma_1^2\sigma_2^2(1-\rho^2)}{\sigma_2^2\sin^2\alpha + 2\rho\sigma_1\sigma_2\sin\alpha\cos\alpha + \sigma_1^2\cos^2\alpha} \,. \tag{5.10.17}$$

The ellipse with these properties is called the *covariance ellipse* of the bivariate normal distribution. Several such ellipses are depicted in Fig. 5.11. The covariance ellipse always lies inside a rectangle determined by the point (a_1, a_2) and the standard deviations σ_1 and σ_2. It touches the rectangle at four points. For the extreme cases $\rho = \pm 1$ the ellipse becomes one of the two diagonals of this rectangle.

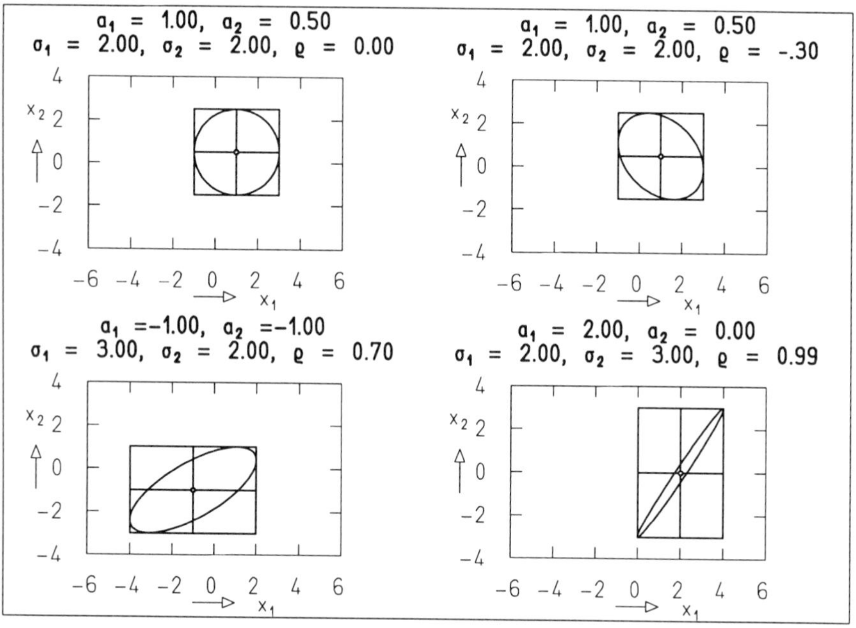

Fig. 5.11. Covariance ellipses.

From (5.10.14) it is clear that other lines of constant probability (for $g \neq 1$) are also ellipses, concentric and similar to the covariance ellipse and situated inside (outside) of it for larger (smaller) probability. The bivariate normal distribution therefore corresponds to a surface in the three-dimensional space (x_1, x_2, ϕ) (Fig. 5.12), whose horizontal sections are concentric ellipses. For the largest probability this ellipse collapses to the point (a_1, a_2). The vertical sections through the center have the form of a Gaussian distribution whose width is directly proportional to the diameter of the covariance ellipse along

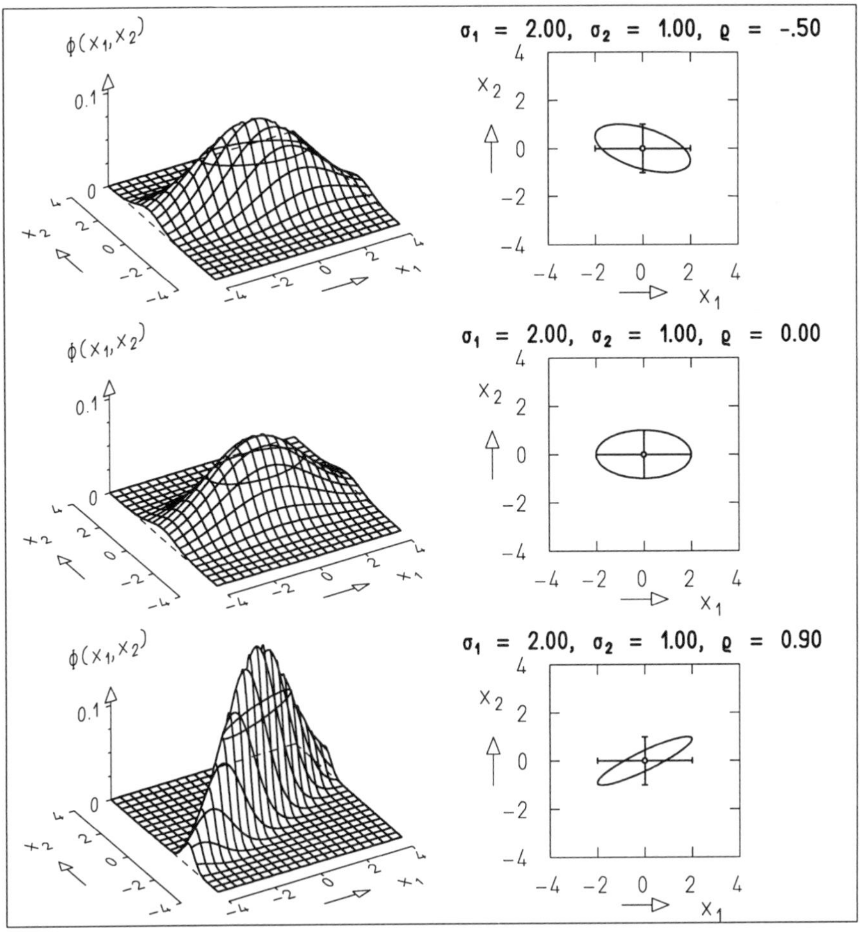

Fig. 5.12. Probability density of a bivariate Gaussian distribution (left) and the corresponding covariance ellipse (right). The three rows of the figure differ only in the numerical value of the correlation coefficient ρ.

which the section extends. The probability of observing a pair $\mathrm{x}_1, \mathrm{x}_2$ of random variables inside the covariance ellipse is equal to the integral

$$\int_A k\phi(\mathbf{x})\,\mathrm{d}\mathbf{x} = 1 - \mathrm{e}^{-\frac{1}{2}} = \text{const} \quad , \tag{5.10.18}$$

where the region of integration A is given by the area within the covariance ellipse (5.10.14). The relation (5.10.18) is obtained by application of the transformation of variables $\mathbf{y} = T\,\mathbf{x}$ with $T = B^{-1}$ to the distribution $\phi(\mathbf{x})$. The resulting distribution has the properties $\sigma(y_1) = \sigma(y_2) = 1$ and $\mathrm{cov}(y_1, y_2) = 0$, i.e., it is of the form of (5.10.9). In this way the region of integration is transformed to a unit circle centered about (a_1, a_2).

In our consideration of the normal distribution of measurement errors of a variable we found the interval $a - \sigma \leq x \leq a + \sigma$ to be the region in which the probability density $f(x)$ exceeded a given fraction, namely $e^{-1/2}$ of its maximum value. The integral over this region was independent of σ. In the case of two variables, the role of this region is taken over by the covariance ellipse determined by σ_1, σ_2, and ρ, and not – as is sometimes incorrectly assumed – by the rectangle that circumscribes the ellipse in Fig. 5.11. The meaning of the covariance ellipse can also be seen from Fig. 5.13. Points 1 and 2, which lie on the covariance ellipse, correspond to equal probabilities ($P(1) = P(2) = P_e$), although the distance of point 1 from the middle is less in both coordinate directions. In addition, point 3 is more probable, and point 4 less probable ($P(4) < P_e$, $P(3) > P_e$), even though point 4 even closer is to (a_1, a_2) than point 3.

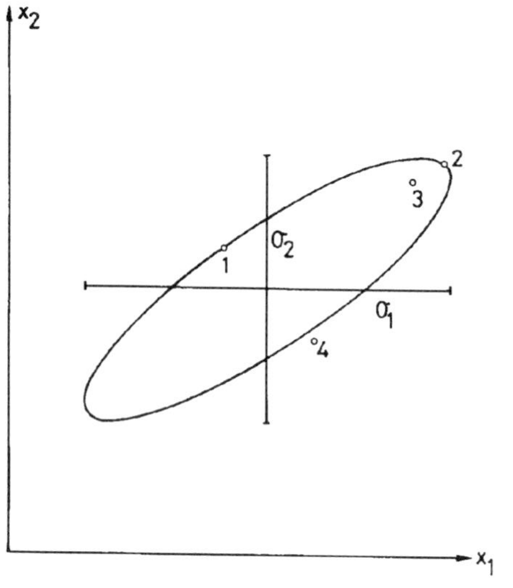

Fig. 5.13. Relative probability for various points from a bivariate Gaussian distribution ($P_1 = P_2 = P_e$, $P_3 > P_e$, $P_4 < P_e$).

For three variables one obtains instead of the covariance ellipse a *covariance ellipsoid*, for n variables a hyperellipsoid in an n-dimensional space (see also Sect. A.12). According to our construction, the covariance ellipsoid is the hypersurface in the n-dimensional space on which the function $g(\mathbf{x})$ in the exponent of the normal distribution (5.10.1) has the constant value $g(\mathbf{x}) = 1$. For other values $g(\mathbf{x}) = \text{const}$ one obtains similar ellipsoids which lie inside ($g < 1$) or outside ($g > 1$) of the covariance ellipsoid. In Sect. 6.6 it will be shown that the function $g(\mathbf{x})$ follows a χ^2-distribution with n degrees of freedom if $\mathbf{x}$ follows the normal distribution (5.10.1). The probability to find $\mathbf{x}$ inside the ellipsoid $g = \text{const}$ is therefore

$$W = \int_0^g f(\chi^2; n)\, d\chi^2 = P\left(\frac{n}{2}, \frac{g}{2}\right) \quad . \tag{5.10.19}$$

Here P is the incomplete gamma function given in Sect. D.5. For $g = 1$, that is, for the covariance ellipsoid in n dimensions, this probability is

$$W_n = P\left(\frac{n}{2}, \frac{1}{2}\right) \quad . \tag{5.10.20}$$

Numerical values for small n are

$$\begin{aligned} W_1 &= 0.682\,69 \quad , & W_2 &= 0.393\,47 \quad , & W_3 &= 0.198\,75 \quad , \\ W_4 &= 0.090\,20 \quad , & W_5 &= 0.037\,34 \quad , & W_6 &= 0.014\,39 \quad . \end{aligned}$$

The probability decreases rapidly as n increases. In order to be able to give regions for various n which correspond to equal probability content, one specifies a value W on the left-hand side of (5.10.19) and determines the corresponding value of g. Then g is the quantile with probability W of the χ^2-distribution with n degrees of freedom (see also Appendix C.5),

$$g = \chi^2_W(n) \quad . \tag{5.10.21}$$

The ellipsoid that corresponds to the value of g that contains $\mathbf{x}$ with the probability W is called the *confidence ellipsoid* of probability W. This expression can be understood to mean that, e.g., for $W = 0.9$ one should have 90 % confidence that $\mathbf{x}$ lies within the confidence ellipsoid.

The variances σ_i^2 or the standard deviations $\Delta_i = \sigma_i$ also have a certain meaning for n variables. The probability to observe the variable x_i in the region $a_i - \sigma_i < x_i < a_i + \sigma_i$ is, as before, 68.3 %, independent of the number n of the variables. This only holds, however, when one places no requirements on the positions of any of the other variables x_j, $j \neq i$.

5.11 Convolutions of Distributions

5.11.1 Folding Integrals

On various occasions we have already discussed sums of random variables, and in the derivation of the Central Limit Theorem, for example, we found the characteristic function to be a useful tool in such considerations. We would now like to discuss the distribution of the sum of two quantities, but for greater clarity we will not make use of the characteristic function.

A sum of two distributions is often observed in experiments. One could be interested, for example, in the angular distribution of secondary particles from the decay of an elementary particle. This can often be used to determine the spin of the particle. The observed angle is the distribution of a sum of random quantities, namely the decay angle and its measurement error. One speaks of the *convolution* of two distributions.

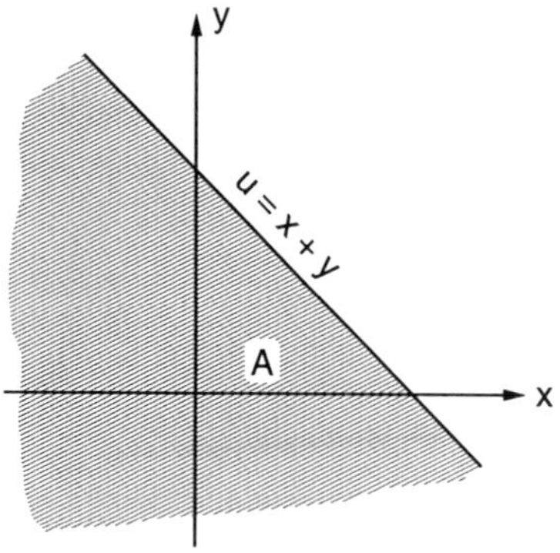

Fig. 5.14. Integration region for (5.11.4).

Let the original quantities be x and y and the sum

$$\mathsf{u} = \mathsf{x} + \mathsf{y} \quad . \tag{5.11.1}$$

A requirement for further treatment is that the original variables must be independent. In this case, the joint probability density is the product of simple densities,

$$f(x, y) = f_x(x) f_y(y) \quad . \tag{5.11.2}$$

If we now ask for the distribution function of u, i.e., for

$$F(u) = P(\mathsf{u} < u) = P(\mathsf{x} + \mathsf{y} < u) \quad , \tag{5.11.3}$$

then this is obtained by integration of (5.11.2) over the hatched region A in Fig. 5.14;

$$\begin{aligned} F(u) &= \iint_A f_x(x) f_y(y)\,\mathrm{d}x\,\mathrm{d}y = \int_{-\infty}^{\infty} f_x(x)\,\mathrm{d}x \int_{-\infty}^{u-x} f_y(y)\,\mathrm{d}y \\ &= \int_{-\infty}^{\infty} f_y(y)\,\mathrm{d}y \int_{-\infty}^{u-y} f_x(x)\,\mathrm{d}x \quad . \end{aligned} \tag{5.11.4}$$

By differentiation one obtains the probability density for u,

$$f(u) = \frac{\mathrm{d}F(u)}{\mathrm{d}u} = \int_{-\infty}^{\infty} f_x(x) f_y(u - x)\,\mathrm{d}x = \int_{-\infty}^{\infty} f_y(y) f_x(u - y)\,\mathrm{d}y \quad . \tag{5.11.5}$$

If x or y or both are only defined in a restricted region, then (5.11.5) is still true. The limits of integration, however, may be limited. We will consider various cases:

(a) $0 \le \mathsf{x} < \infty \,, \quad -\infty < \mathsf{y} < \infty$:

$$f(u) = \int_{-\infty}^{u} f_x(u - y) f_y(y)\,\mathrm{d}y \quad . \tag{5.11.6}$$

(Because $y = u - x$ and since for $x_{\mathrm{min}} = 0$ one has $y_{\mathrm{max}} = u$.)

(b) $0 \le \mathsf{x} < \infty\,,\quad 0 \le \mathsf{y} < \infty$:

$$f(u) = \int_0^u f_x(u-y) f_y(y)\,\mathrm{d}y \quad . \tag{5.11.7}$$

(c) $a \le \mathsf{x} < b\,,\quad -\infty < \mathsf{y} < \infty$:

$$f(u) = \int_a^b f_x(x) f_y(u-x)\,\mathrm{d}x \quad . \tag{5.11.8}$$

We will demonstrate case (d) in the following example, in which both x and y are bounded from below and from above.

Example 5.8: Convolution of uniform distributions

With

$$f_x(x) = \begin{cases} 1,\ 0 \le x < 1 \\ 0 \text{ otherwise} \end{cases} \quad \text{and} \quad f_y(y) = \begin{cases} 1,\ 0 \le y < 1 \\ 0 \text{ otherwise} \end{cases}$$

and Eq. (5.11.8) we obtain

$$f(u) = \int_0^1 f_y(u-x)\,\mathrm{d}x \quad .$$

We substitute $v = u - x$, $\mathrm{d}v = -\mathrm{d}x$ and obtain

$$f(u) = -\int_u^{u-1} f_y(v)\,\mathrm{d}v = \int_{u-1}^{u} f_y(v)\,\mathrm{d}v \quad . \tag{5.11.9}$$

Clearly one has $0 < u < 2$. We now consider separately the two cases

$$\begin{aligned} &\text{(a)}\quad 0 \le \mathsf{u} < 1 \quad : \qquad f_1(u) = \int_0^u f_y(v)\,\mathrm{d}v = \int_0^u \mathrm{d}v = u \quad , \\ &\text{(b)}\quad 1 \le \mathsf{u} < 2 \quad : \qquad f_2(u) = \int_{u-1}^1 f_y(v)\,\mathrm{d}v = \int_{u-1}^1 \mathrm{d}v = 2 - u \quad . \end{aligned} \tag{5.11.10}$$

Note that the lower (upper) limit of integration is not lower (higher) than the value 0 (1). The result is a triangular distribution (Fig. 5.15).

If this result is folded again with a uniform distribution, i.e., if u is the sum of three independent uniformly distributed variables, then one obtains

$$f(u) = \begin{cases} \frac{1}{2}u^2\,, & 0 \le \mathsf{u} < 1 \quad , \\ \frac{1}{2}(-2u^2 + 6u - 3)\,, & 1 \le \mathsf{u} < 2 \quad , \\ \frac{1}{2}(u-3)^2\,, & 2 \le \mathsf{u} < 3 \quad . \end{cases} \tag{5.11.11}$$

The proof is left to the reader. The distribution consists of three parabolic sections (Fig. 5.15) and is similar already to the Gaussian distribution predicted by the Central Limit Theorem. ■

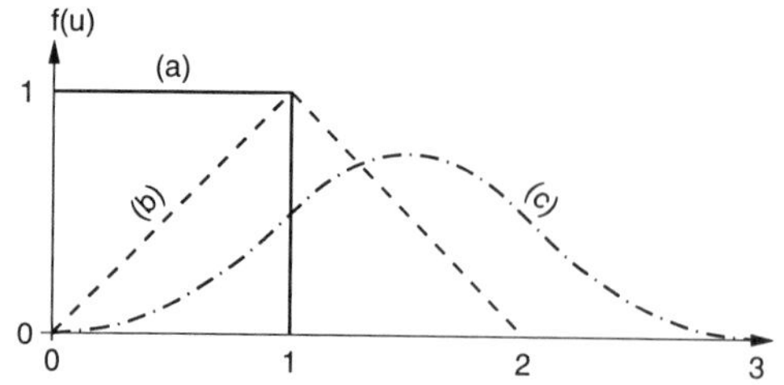

Fig. 5.15. Convolution of uniform distributions. Probability density of the sum u of uniformly distributed random variables x (a) $u = x$, (b) $u = x + x$, (c) $u = x + x + x$.

5.11.2 Convolutions with the Normal Distribution

Suppose a quantity of experimental interest x can be considered to be a random variable with probability density $f_x(x)$. It is measured with a measurement error y, which follows a normal distribution with a mean of zero and a variance of σ^2. The result of the measurement is then the sum

$$u = x + y \quad . \tag{5.11.12}$$

Its probability density is [see also (5.11.4)]

$$f(u) = \frac{1}{\sqrt{2\pi}\sigma} \int_{-\infty}^{\infty} f_x(x) \exp[-(u-x)^2/2\sigma^2] \, dx \quad . \tag{5.11.13}$$

By carrying out many measurements, $f(u)$ can be experimentally determined. The experimenter is interested, however, in the function $f_x(x)$. Unfortunately, Eq. (5.11.13) cannot in general be solved for $f_x(x)$. This is only possible for a restricted class of functions $f(u)$. Therefore one usually approaches the problem in a different way. From earlier measurements or theoretical considerations one possesses knowledge about the form of $f_x(x)$, e.g., one might assume that $f_x(x)$ is described by a uniform distribution, without, however, knowing its boundaries a and b. One then carries out the convolution (5.11.13), compares the resulting function $f(u)$ with the experiment and in this way determines the unknown parameters (in our example a and b).

In many cases it is not even possible to perform the integration (5.11.13) analytically. Numerical procedures, e.g., the Monte Carlo method, then have to be used. Sometimes approximations (cf. Example 5.11) give useful results. Because of the importance in many experiments of convolution with the normal distribution we will study some examples.

Example 5.9: Convolution of uniform and normal distributions

Using Eqs. (3.3.26) and (5.11.8) and substituting $v = (x-u)/\sigma$ we obtain

$$
\begin{aligned}
f(u) &= \frac{1}{b-a}\frac{1}{\sqrt{2\pi}\sigma}\int_a^b \exp[-(u-x)^2/2\sigma^2]\,\mathrm{d}x \\
&= \frac{1}{b-a}\frac{1}{\sqrt{2\pi}}\int_{(a-u)/\sigma}^{(b-u)/\sigma} \exp(-\frac{1}{2}v^2)\,\mathrm{d}v \quad , \\
f(u) &= \frac{1}{b-a}\left\{\psi_0\left(\frac{b-u}{\sigma}\right) - \psi_0\left(\frac{a-u}{\sigma}\right)\right\} \quad . \qquad (5.11.14)
\end{aligned}
$$

The function ψ has already been defined in (5.6.5). Figure 5.16 shows the result for $a = 0$, $b = 6$, $\sigma = 1$. If one has $|b-a| \gg \sigma$ (as is the case in Fig. 5.16), one of the terms in parentheses in (5.11.14) is either 0 or 1. The rising edge of the uniform distribution at $u = a$ is replaced by the distribution function of the normal distribution with standard deviation σ (see also Fig. 5.7). The falling edge at $u = b$ is its "mirror image". ∎

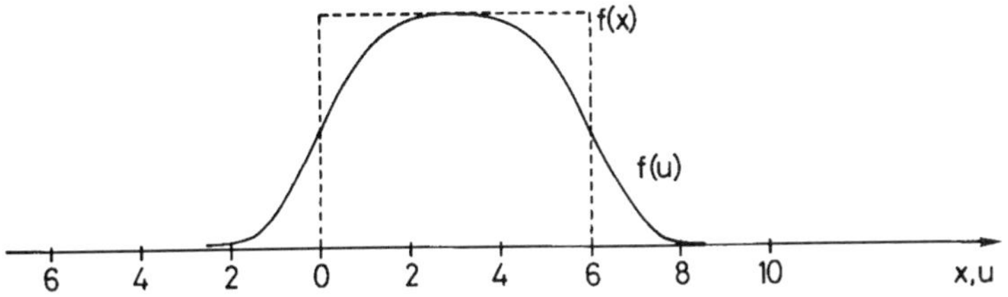

Fig. 5.16. Convolution of a uniform and Gaussian distribution.

Example 5.10: Convolution of two normal distributions. "Quadratic addition of errors"

If one convolutes two normal distributions with mean values 0 and variances σ_x^2 and σ_y^2, one obtains

$$f(u) = \frac{1}{\sqrt{2\pi}\sigma}\exp(-u^2/2\sigma^2) \quad , \qquad \sigma^2 = \sigma_x^2 + \sigma_y^2 \quad . \qquad (5.11.15)$$

The proof has already been shown in Sect. 5.7 with the help of the characteristic function. It can also be obtained by computation of the folding integral (5.11.5). If the distributions $f_x(x)$ and $f_y(y)$ describe two independent sources of measurement errors, the result (5.11.15) is known as the *"quadratic addition of errors"*. ∎

Example 5.11: Convolution of exponential and normal distributions
With

$$\begin{aligned} f_x(x) &= \frac{1}{\tau}\exp(-x/\tau) \quad , \qquad x > 0 \quad , \\ f_y(y) &= \frac{1}{\sqrt{2\pi}\sigma}\exp(-y^2/2\sigma^2) \quad , \end{aligned}$$

Eq. (5.11.6) takes on the following form:

$$f(u) = \frac{1}{\sqrt{2\pi}\sigma\tau}\int_{-\infty}^{u}\exp[-(u-y)/\tau]\exp(-y^2/2\sigma^2)\,dy \quad .$$

We can rewrite the exponent

$$\begin{aligned} -\frac{1}{2\sigma^2\tau}&[2\sigma^2(u-y)+\tau y^2] \\ &= -\frac{1}{2\sigma^2\tau}\left[2\sigma^2 u - 2\sigma^2 y + \tau y^2 + \frac{\sigma^4}{\tau} - \frac{\sigma^4}{\tau}\right] \\ &= -\frac{u}{\tau} + \frac{\sigma^2}{2\tau^2} - \frac{1}{2\sigma^2}\left(y - \frac{\sigma^2}{\tau}\right)^2 \end{aligned}$$

and obtain

$$f(u) = \frac{1}{\sqrt{2\pi}\sigma\tau}\exp\left\{\frac{\sigma^2}{2\tau^2} - \frac{u}{\tau}\right\}\int_{-\infty}^{u-\sigma^2/\tau}\exp\left(\frac{-v^2}{2\sigma^2}\right)dv \quad .$$

We now require that $\sigma \ll \tau$, i.e., that the measurement error is much smaller than the typical value (width) of the exponential distribution. In addition, we only consider values of u for which $u - \sigma^2/\tau \gg \sigma$, i.e., $u \gg \sigma$. The integral is then approximately equal to $\sqrt{2\pi}\sigma$ or

$$f(u) \approx \frac{1}{\tau}\exp\left\{-\frac{u}{\tau} + \frac{\sigma^2}{2\tau^2}\right\} \quad .$$

In a semi-logarithmic representation, i.e., in a plot of $\ln f(u)$ versus u, the curve $f(u)$ lies above the curve $f_x(x)$, by an amount $\sigma^2/2\tau^2$, since

$$\ln f(u) = \ln\frac{1}{\tau} + \frac{\sigma^2}{2\tau^2} - \frac{u}{\tau} = \ln f_x(x) + \frac{\sigma^2}{2\tau^2} \quad .$$

This is plotted in Fig. 5.17. The result can be qualitatively understood in the following way. For each small x interval of the exponential distribution, the convolution leads with equal probability to a shift to the left or to the right. Since, however, the exponential distribution for a given u is greater for small values of x, contributions to the convolution $f(u)$ originate with greater probability from the left than from the right. This leads to an overall shift to the right of $f(u)$ with respect to $f_x(x)$. ■

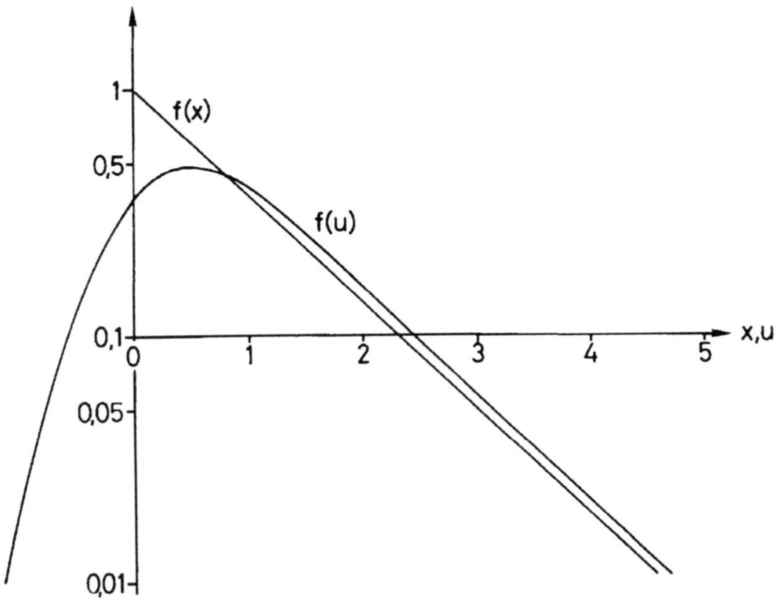

Fig. 5.17. Convolution of exponential and normal distributions.

5.12 Example Programs

The source files of the example programs are available in the subdirectory `\DATAN\EXASRC`.

5.12.1 Main Program `E1DS` to Simulate Empirical Frequency and Demonstrate Statistical Fluctuations

The program simulates the problem of Example 5.1. It first asks for three input values $n_{\rm exp}$, $n_{\rm fly}$, and $P(A)$, and then consecutively performs $n_{\rm exp}$ simulated experiments. In each experiment $n_{\rm fly}$ objects are analyzed. Each object has a probability $P(A)$ to have the property A. For each experiment one line of output is produced containing the current number $i_{\rm exp}$ of the experiment, the number N_A of objects with the property A and the frequency $h_A = N_A/n_{\rm fly}$ with which the property A was found. The fluctuation of h_A around the known input value $P(A)$ in the individual experiments gives a good impression of the statistical error of an experiment.

5.12.2 Main Program E2DS to Simulate the Experiment of Rutherford and Geiger

The principle of the experiment of Rutherford and Geiger is described in Example 5.3. It is simulated as follows. The program first asks for the number N of decays observed and the number $n_{\rm int}$ of partial intervals ΔT of the total observation time T. For simplicity the length of each partial interval is set equal to one. A total of N random events are simulated by simply generating N random numbers uniformly distributed between 0 and T. They are entered into a histogram with $n_{\rm int}$ intervals. The histogram is first displayed graphically and then analyzed numerically. For each number $k = 0, 1, \ldots, N_{\rm int}$ the program determines how many intervals $N(k)$ of the histogram have k entries. The numbers $N(k)$ themselves are presented in the form of another histogram.

Show that for the process simulated in this example program one obtains in the limit $N \to \infty$

$$N(k) = n_{\rm int} W_k^N(p = 1/n_{\rm int}) \quad .$$

If N is increased step by step and at the same time $\lambda = N_p = N/n_{\rm int}$ is kept constant, then for large N one has

$$W_k^N(p = \lambda/N) \to \frac{\lambda^k}{k!} \mathrm{e}^{-\lambda}$$

and, in the limit $N \to \infty$,

$$N(k) = n_{\rm int} \frac{\lambda^k}{k!} \mathrm{e}^{-\lambda} \quad .$$

Check the above statements by running the program with suitable pairs of numbers, e.g., $(N, n_{\rm int}) = (4, 2), (40, 20), \ldots, (2000, 1000)$, by reading the numbers $N(k)$ from the graphics display and by comparing them with the statements above.

5.12.3 Main Program E3DS to Simulate Galton's Board

Galton's board is a simple implementation of Laplace's model described in Example 5.7. The vertical board contains rows of horizontally oriented nails as shown in Fig. 5.18. The rows of nails are labeled $j = 1, 2, \ldots, n$, and row j has j nails. One by one a total of $N_{\rm exp}$ balls fall onto the nail in row 1. There each ball is deflected with probability p to the right and with probability $(1-p)$ to the left. (In a realistic board one has $p = 1/2$.) The distance between the nails is chosen in such a way that in each case the ball hits one of the two nails in row 2 and there again it is deflected with the probability p to the right.

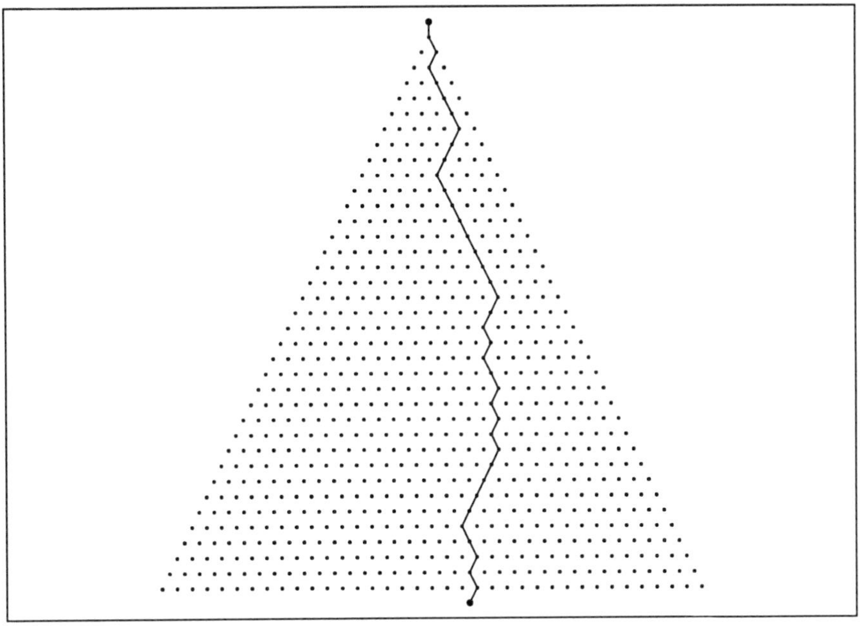

Fig. 5.18. Arrangement of nails in Galton's board and one possible trajectory of a ball.

After falling through n rows each ball assumes one of $n + 1$ places, which we denote by $k = 0$ (on the left), $k = 1, \ldots, k = n$ (on the right). After a total of N_{exp} experiments (i.e., balls) one finds $N(k)$ balls for each value k.

The program first asks for the numerical values of N_{exp}, n, and p. For each experiment the number k is first set to zero and n random numbers r_j are generated from a uniform distribution and analyzed. For each $r_j < p$ (corresponding to a deflection to the right in row j) the number k is increased by 1. For each experiment the value of k is entered into a histogram. After all experiments are simulated the histogram is displayed.

Show that in the limit $N \to \infty$ one has

$$N(k) = N\,W_k^n(p) \quad .$$

By choosing and entering suitable pairs of numbers (n, p), e.g., $(n, p) = (1, 0.5)$, $(2, 0.25)$, $(10, 0.05)$, $(100, 0.005)$ approximate the Poisson limit

$$N(k) = N\frac{\lambda^k}{k!}\mathrm{e}^{-\lambda} \quad , \qquad \lambda = np \quad ,$$

and compare these predictions with the results of your simulations.

5.13 Problems

5.13.1 Binomial Distribution

(a) Prove the recursion formula

$$W_{k+1}^n = \frac{n-k}{k+1}\frac{p}{q}W_k^n \quad .$$

(b) It may be known for a certain production process that a fraction $q = 0.2$ of all pieces produced are defective. This means that in 5 pieces produced the expected number of non-defective pieces is $np = n(1-q) = 5 \cdot 0.8 = 4$. What is the probability P_2 and the probability P_3 that at most 2 or at most 3 pieces are free from defects? Use relation (a) to simplify the calculation.

(c) Determine the value k_m for which the binomial distribution is maximum, i.e., k_m is the most probable value of the distribution. Hint: Since W_k^n is not a function of a continuous variable k, the maximum cannot be found by looking for a zero in the derivative. Therefore, one has to study finite differences $W_k^n - W_{k-1}^n$.

(d) In Sect. 5.1 the binomial distribution was constructed by considering the random variable $\mathsf{x} = \sum_{i=1}^n \mathsf{x}_i$. Here x_i was a random variable that took only the values 0 and 1 with the probabilities $P(\mathsf{x}_i = 1) = p$ and $P(\mathsf{x}_i = 0) = q$.

The binomial distribution

$$f(x) = f(k) = W_k^n = \binom{n}{k} p^k q^{n-k}$$

was then obtained by considering in detail the probability to have k cases of $\mathsf{x}_i = 1$ in a total of n observations of the variable x_i. Obtain the binomial distribution in a more formal way by constructing the characteristic function φ_{x_i} of the variable x_i. From the nth power of φ_{x_i} you obtain the characteristic function of x. Hint: Use the binomial theorem (B.6).

5.13.2 Poisson Distribution

In a certain hospital the doctor on duty is called on the average three times per night. The number of calls may be considered to be Poisson distributed. What is the probability for the doctor to have a completely quiet night?

5.13.3 Normal Distribution

The resistance R of electrical resistors produced by a particular machine may be described by a normal distribution with mean R_m and standard deviation σ.

The production cost for one resistor is C, and the price is $5C$ if $R = R_0 \pm \Delta_1$, and $2C$ if $R_0 - \Delta_2 < R < R_0 - \Delta_1$ or $R_0 + \Delta_1 < R < R_0 + \Delta_2$. Resistors outside these limits cannot be sold.

(a) Determine the profit P per resistor produced for $R_m = R_0$, $\Delta_1 = a_1 R_0$, $\Delta_2 = a_2 R_0$, $\sigma = bR_0$. Use the distribution function ψ_0.

(b) Use Table I.2 to compute the numerical values of P for $a_1 = 0.01$, $a_2 = 0.05$, $b = 0.05$.

(c) Show that the probability density (5.7.1) has points of inflection (i.e., second derivative equal to zero) at $x = a \pm b$.

5.13.4 Multivariate Normal Distribution

A planar xy coordinate system is used as target. The probability to observe a hit in the plane may be given by the normal distribution of Problem 3.9.4 (e). Use the result of that problem to determine

(a) the probability $P(R)$, to observe a hit within a given radius R around the origin,

(b) the radius R, within which a hit is observed with a given probability. Compute as a numerical example the value of R for $P = 90\,\%$ and $\sigma = 1$.

5.13.5 Convolution

(a) Prove the relation (5.11.11). Begin with (5.11.9) and use the expression (5.11.10) for $f_y(y)$. In the intervals $0 \le u < 1$ and $2 \le u < 3$ relations (5.11.10a) and (5.11.10b) hold, since in these intervals one always has $y < 1$ and $y > 1$, respectively. In the interval $1 \le u < 2$ the resulting distribution $f(u)$ must be constructed as sum of two integrals of the type (5.11.9) of which each contains one of the possible expression for $f_y(y)$. In this case particular care is necessary in the determination of the limits of integration. They are given by the limits of the intervals in which u and $f_y(y)$ are defined.

(b) Prove the relation (5.11.15) by performing the integration (5.11.5) for the case that f_x and f_y are normal distributions with means 0 and standard deviations σ_x and σ_y.

5.14 Programming Problems

The source files of the example solutions are available in the subdirectory `\DATAN\SOLSRC`.

5.14.1 Convolution of Uniform Distributions

Because of the Central Limit Theorem the quantity $\mathsf{x} = \sum_{i=1}^{N} \mathsf{x}_i$ follows in the limit $N \to \infty$ the standard normal distribution if the x_i come from an arbitrary distribution with mean zero and standard deviation $1/\sqrt{N}$. Choose for the x_i the uniform distribution with the limits

$$a = -\sqrt{3/N} \quad , \qquad b = -a \quad .$$

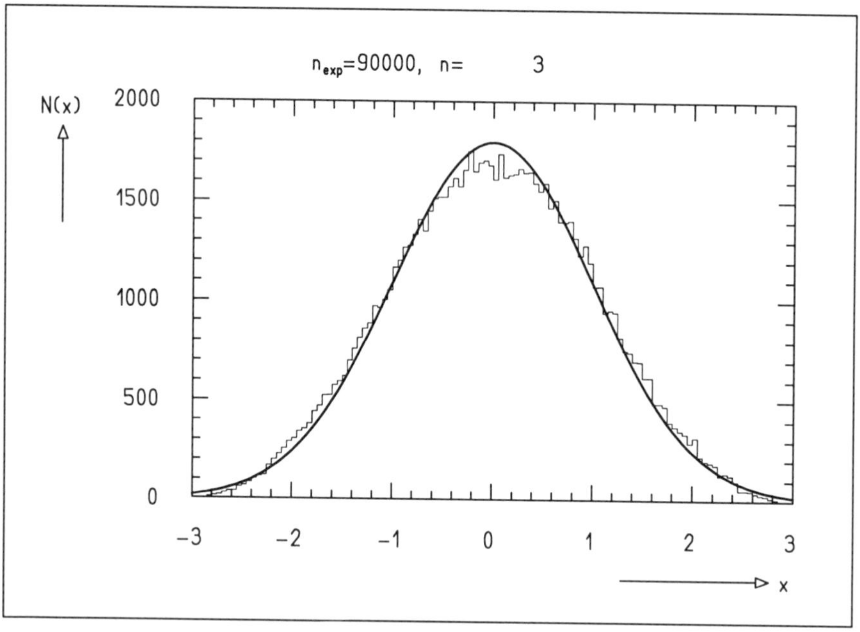

Fig. 5.19. Histogram of 90 000 random numbers x, each of which is a sum of three uniformly distributed random numbers. The curve corresponds to the standard normal distribution. Significant differences between curve and histogram are visible only because of the very large number of random numbers used.

Perform a large number n_{exp} of Monte Carlo experiments, each giving a random value x. Produce a histogram of the quantity x and show in addition the distribution of x as a continuous curve which you would expect from the standard normal distribution (Fig. 5.19). (Use the program GRHSCV for the simultaneous representation of histograms and curves.) Allow for interactive input of the quantities n_{exp} and N.

5.14.2 Convolution of Uniform Distribution and Normal Distribution

If x is taken from a uniform distribution between a and b and if y is taken from a normal distribution with mean zero and width σ, then the quantity $\mathsf{u} = \mathsf{x} + \mathsf{y}$ follows the distribution (5.11.14). Perform a large number n_{exp} of Monte Carlo experiments, each resulting in a random number u. Display a histogram of the quantity u and show in addition a curve of the distribution you would expect from (5.11.14), Fig. 5.20. Allow for the interactive input of the quantities n_{exp}, a, b, and σ.

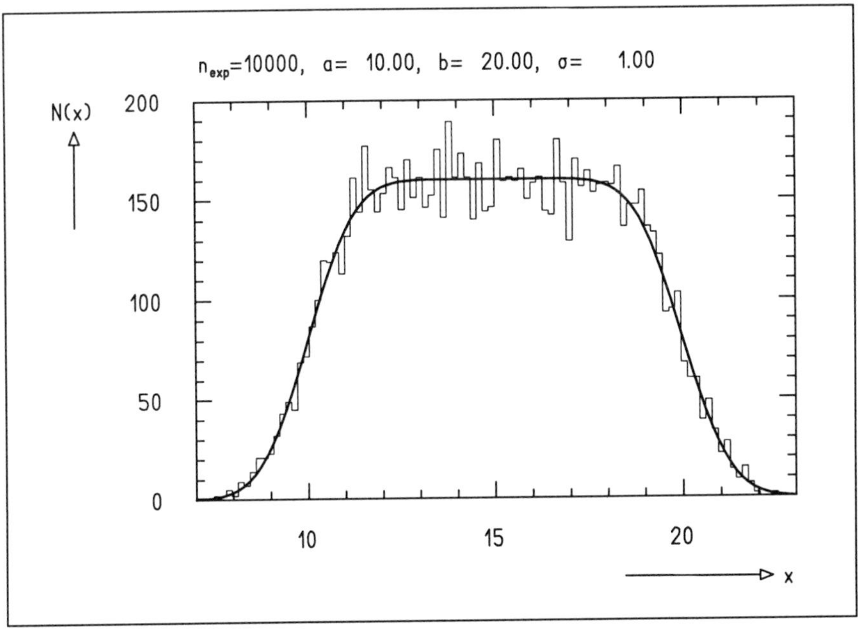

Fig. 5.20. A histogram of 10 000 random numbers, each of which is the sum of a uniformly distributed random number and a normally distributed random number. The curve corresponds to the convolution of a uniform and a normal distribution.

5.15 Hints and Solutions

Problem 5.13.1

(a)
$$W_{k+1}^n = \binom{n}{k+1} p^{k+1} q^{n-k-1} = \frac{n!}{(k+1)!(n-k-1)!} p^k q^{n-k} \frac{p}{q}$$
$$= \frac{n!}{k!(n-k)!} \frac{n-k}{k+1} p^k q^{n-k} \frac{p}{q} = W_k^n \frac{n-k}{k+1} \frac{p}{q} \quad .$$

(b)
$$W_3^5 = \binom{5}{3} 0.8^3 \cdot 0.2^2 = 10 \cdot 0.512 \cdot 0.04 = 0.2048 \quad ,$$
$$W_4^5 = W_3^3 \cdot \frac{2}{4} \cdot \frac{0.8}{0.2} = 0.2048 \cdot 2 = 0.4096 \quad ,$$
$$W_5^5 = W_4^5 \cdot \frac{1}{5} \cdot \frac{0.8}{0.2} = 0.4096 \cdot 0.8 = 0.327\,68 \quad ,$$
$$P_3 = 1 - W_4^5 - W_5^5 = 0.262\,72 \quad ,$$
$$P_2 = P_3 - W_3^5 = 0.057\,92 \quad .$$

(c) Using the result of (a) we obtain

$$W_k^n - W_{k-1}^n = W_{k-1}^n \left(\frac{n-k+1}{k} \frac{p}{q} - 1 \right) \quad .$$

Thus the probability W_k^n increases as long as the expression in brackets is positive, i.e.,

$$\frac{(n-k+1)p}{kq} - 1 > 0 \quad .$$

Since k and q are positive we have

$$(n-k+1)p > kq = k(1-p) \quad , \qquad k < (n+1)p \quad .$$

The most probable value k_m is the largest value of k for which this inequality holds.

(d) $\varphi_{\mathrm{x}_i}(t) = E\{\mathrm{e}^{\mathrm{i}t\mathrm{x}_i}\} = q\mathrm{e}^{\mathrm{i}t\cdot 0} + p\mathrm{e}^{\mathrm{i}t} = q + p\mathrm{e}^{\mathrm{i}t} \quad ;$

$$\varphi_{\mathrm{x}} = (q + p\mathrm{e}^{\mathrm{i}t})^n = \sum_{k=0}^{m} \binom{n}{k} q^{n-k} p^k \mathrm{e}^{\mathrm{i}tk} = \sum_{k=0}^{m} f(k) \mathrm{e}^{\mathrm{i}tk}$$
$$= E\{\mathrm{e}^{\mathrm{i}tk}\} \quad ,$$

$$f(k) = W_k^n \quad .$$

Problem 5.13.2

For $\lambda = 3$ one has $f(0) = \dfrac{\lambda^0}{0!}\mathrm{e}^{-\lambda} = \mathrm{e}^{-\lambda} = 0.0498 \approx 0.05 = 5\,\%$.

Problem 5.13.3

(a) The fraction

$$f_r = 2\psi_0\left(\frac{(R_0 - a_2 R_0) - R_0}{bR_0}\right) = 2\psi_0\left(-\frac{a_2}{b}\right)$$

is rejected, since $R < R_0 - \Delta_2$ or $R > R_0 + \Delta_2$. Correspondingly the fractions

$$f_2 = 2\psi_0\left(-\frac{a_1}{b}\right) - f_r \qquad \text{and} \qquad f_5 = 1 - f_2 - f_r$$

give prices $2C$ and $5C$, respectively. Therefore,

$$\begin{aligned} P &= 2f_2C + 5f_5C - C = C\{2f_2 + 5 - 5f_2 - 5f_r - 1\} \\ &= C\{4 - 3f_2 - 5f_r\} \\ &= C\left\{4 - 6\psi_0\left(-\frac{a_1}{b}\right) + 3f_r - 5f_r\right\} \\ &= C\left\{4 - 6\psi_0\left(-\frac{a_1}{b}\right) - 4\psi_0\left(-\frac{a_2}{b}\right)\right\} \quad . \end{aligned}$$

(b) $\psi_0(-0.2) = 0.421 \quad ; \qquad \psi_0(-1) = 0.159 \quad ,$

i.e., $P = C\{4 - 2.526 - 0.636\} = 0.838C \quad .$

(c) $\dfrac{\mathrm{d}^2 f}{\mathrm{d}x^2} = \dfrac{1}{\sqrt{2\pi}b^3}\exp(-(x-a)^2/2b^2)\{(x-a)^2/b^2 - 1\} = 0$

is fulfilled if the expression in the last set of brackets vanishes.

Problem 5.13.4

(a)
$$\begin{aligned} P(R) &= \int_0^R g(r)\,\mathrm{d}r = \frac{1}{\sigma^2}\int_0^R r\mathrm{e}^{-r^2/2\sigma^2}\,\mathrm{d}r \\ &= \frac{1}{2\sigma^2}\int_0^{R^2} \mathrm{e}^{-u/2\sigma^2}\,\mathrm{d}u \\ &= [\mathrm{e}^{-u/2\sigma^2}]^0_{R^2} = 1 - \mathrm{e}^{-R^2/2\sigma^2} \quad . \end{aligned}$$

(b) $1 - P = \exp(-R^2/2\sigma^2) \quad , \qquad R^2/2\sigma^2 = -\ln(1-P) \quad .$

For $\sigma = 1$, $P = 0.9$ one obtains

$$R = \sqrt{-2\ln 0.1} = \sqrt{4.61} = 2.15 \quad .$$

Problem 5.13.5

(a) $0 \le u < 1$:

$$f(u) = \int_{u-1}^{u} f_1(y)\,dy = \int_0^u y\,dy = \frac{1}{2}u^2 \quad .$$

$1 \le u < 2$:

$$\begin{aligned} f(u) &= \int_{u-1}^{u} f_1(y)\,dy + \int_{u-1}^{u} f_2(y)\,dy \\ &= \int_{u-1}^{1} y\,dy + \int_1^u (2-y)\,dy \\ &= \frac{1}{2}(1-(u-1)^2) + \frac{1}{2}(1-(2-u)^2) \\ &= \frac{1}{2}(-3+6u-2u^2) \quad . \end{aligned}$$

$2 \le u < 3$:

$$f(u) = \int_{u-1}^{u} f_2(y)\,dy = \int_{u-1}^{2} (2-y)\,dy = -\int_{3-u}^{0} z\,dz = \frac{1}{2}(3-u)^2 \quad .$$

(b) $f(u) = \dfrac{1}{2\pi\sigma_x\sigma_y}\displaystyle\int_{-\infty}^{\infty} \exp\left(-\frac{x^2}{2\sigma_x^2} - \frac{(u-x)^2}{2\sigma_y^2}\right) dx \quad .$

Completing the square yields for the exponent

$$-\frac{\sigma^2}{2\sigma_x^2\sigma_y^2}\left\{\left(x - \frac{\sigma_x^2}{\sigma^2}u\right)^2 - \frac{\sigma_x^4}{\sigma^4}u^2 + \frac{\sigma_x^2}{\sigma^2}u^2\right\} \quad .$$

With the change of variables $v = (\sigma/\sigma_x\sigma_y)(x - \sigma_x^2 u/\sigma^2)$ we obtain

$$\begin{aligned} f(u) &= \frac{1}{2\pi\sigma_x\sigma_y}\exp\left(\frac{\sigma_x^4 - \sigma^2\sigma_x^2}{2\sigma^2\sigma_x^2\sigma_y^2}u^2\right)\frac{\sigma_x\sigma_y}{\sigma} \\ &\quad \times \int_{-\infty}^{\infty} \exp\left(-\frac{1}{2}v^2\right) dv \\ &= \frac{1}{\sqrt{2\pi}\sigma}\exp\left(-\frac{u^2}{2\sigma^2}\right) \quad , \end{aligned}$$

since $\sigma_x^4 = \sigma_x^2(\sigma^2 - \sigma_y^2)$.

6. Samples

In the last chapter we discussed a number of distributions, but we have not specified how they are realized in a particular case. We have only given the probability that a random variable x will lie within an interval with boundaries x and $x + \mathrm{d}x$. This probability depends on certain parameters describing its distribution (like λ in the case of the Poisson distribution) which are usually unknown. We therefore have no direct knowledge of the probability distribution and have to approximate it by a *frequency distribution* obtained experimentally. The number of measurements performed for this purpose, called a *sample*, is necessarily finite. To discuss the elements of sampling theory we first have to introduce a number of new definitions.

6.1 Random Samples. Distribution of a Sample. Estimators

Every sample is taken from a set of elements which correspond to the possible results of an individual observation. Such a set, which usually has infinitely many elements, is called a *population*. If a sample of n elements is taken from it, then we say that the sample has the *size* n. Let the distribution of the random variable x in the population be given by the probability density $f(x)$. We are interested in the values of x assumed by the individual elements of the sample. Suppose that we take ℓ samples of size n and find the following values for x:

$$\begin{array}{ll} 1\text{st sample:} & \mathsf{x}_1^{(1)}, \mathsf{x}_2^{(1)}, \ldots, \mathsf{x}_n^{(1)}, \\ \vdots & \\ j\text{th sample:} & \mathsf{x}_1^{(j)}, \mathsf{x}_2^{(j)}, \ldots, \mathsf{x}_n^{(j)}, \\ \vdots & \\ \ell\text{th sample:} & \mathsf{x}_1^{(\ell)}, \mathsf{x}_2^{(\ell)}, \ldots, \mathsf{x}_n^{(\ell)}. \end{array}$$

We group the result of one sample into an n-dimensional vector

$$\mathbf{x}^{(j)} = (\mathsf{x}_1^{(j)}, \mathsf{x}_2^{(j)}, \ldots, \mathsf{x}_n^{(j)}) \quad , \tag{6.1.1}$$

which can be considered the position vector in an n-dimensional sample space (Sect. 2.1). Its probability density is

$$g(\mathbf{x}) = g(x_1, x_2, \ldots, x_n) \quad . \tag{6.1.2}$$

This function must fulfill two conditions in order for the sample to be *random*:

(a) The individual x_i must be independent, e.g., one must have

$$g(\mathbf{x}) = g_1(x_1)g_2(x_2)\ldots g_n(x_n) \quad . \tag{6.1.3}$$

(b) The individual marginal distributions must be identical and equal to the probability density $f(x)$ of the population,

$$g_1(x) = g_2(x) = \ldots = g_n(x) = f(x) \quad . \tag{6.1.4}$$

Comparing with (6.1.2) it is clear that there is a simple relation between a population and a sample only if these conditions are fulfilled. In the following we will mean by the word sample a random sample unless otherwise stated.

It should be emphasized that in the actual process of sampling it is often quite difficult to ensure randomness. Because of the large variety of applications, a general prescription cannot be given. In order to obtain reliable results from sampling, we have to take the utmost precautions to meet the requirements (6.1.3) and (6.1.4). Independence (6.1.3) can be checked to a certain extent by comparing the frequency distributions of the first, second, ... elements of a large number of samples. It is very difficult, however, to ensure that the samples in fact come from a population with the probability density $f(x)$. If the elements of the population can be numbered, it is often useful to use random numbers to select the elements for the sample.

We now suppose that the n elements of a sample are ordered according to the value of the variable, e.g., marked on the x axis, and we ask for the number of elements of the sample n_x for which $\mathsf{x} < x$, for arbitrary x. The function

$$W_n(x) = n_x/n \tag{6.1.5}$$

takes on the role of an empirical distribution function. It is a step function that increases by $1/n$ as soon as x is equal to one of the values x of an element of the sample. It is called the *sample distribution function*. It is clearly an approximation for $F(x)$, the distribution function of the population, which it approaches in the limit $n \to \infty$.

A function of the elements of a sample (6.1.1) is called a *statistic*. Since x is a random variable, a statistic is itself a random variable. The most important example is the *sample mean*,

$$\bar{\mathsf{x}} = \frac{1}{n}(\mathsf{x}_1 + \mathsf{x}_2 + \ldots + \mathsf{x}_n) \quad . \tag{6.1.6}$$

A typical problem of data analysis is the following. The general mathematical form of the probability density of the population is known. In radioactive decay, for example, the number of nuclei which decay before the time $t = \tau$ is $N_\tau = N_0(1 - \exp(-\lambda\tau))$, if N_0 nuclei existed at time $t = 0$. Here, however, the decay constant λ is, in general, not known. By taking a finite sample (measuring a finite number of decay times of individual nuclei) we want to determine the parameter λ as accurately as possible. Since such a task cannot be exactly solved, because the sample is finite, one speaks of the *estimation of parameters*. To estimate a parameter λ of a distribution function one uses an estimator

$$\mathsf{S} = \mathsf{S}(\mathsf{x}_1, \mathsf{x}_2, \ldots, \mathsf{x}_n) \quad . \tag{6.1.7}$$

An estimator is said to be *unbiased* if for arbitrary sample size the expectation value of the (random) quantity S is equal to the parameter to be estimated:

$$E\{\mathsf{S}(\mathsf{x}_1, \mathsf{x}_2, \ldots, \mathsf{x}_n)\} = \lambda \qquad \text{for all } n. \tag{6.1.8}$$

An estimator is said to be *consistent* if its variance vanishes for arbitrarily large sample size, i.e., if

$$\lim_{n\to\infty} \sigma(\mathsf{S}) = 0 \quad . \tag{6.1.9}$$

Often one can give a lower limit for the variance of an estimator of a parameter. If one finds an estimator S_0 whose variance is equal to this limit, then one apparently has the "best possible" estimator. S_0 is then said to be an *efficient estimator* for λ.

6.2 Samples from Continuous Populations. Mean and Variance of a Sample

The case of greatest interest in applications concerns a sample from an infinitely large continuous population described by the probability density $f(x)$. The sample mean (6.1.6) is a random variable, as are all statistics. Let us consider its expectation value

$$E(\bar{\mathsf{x}}) = \frac{1}{n}\{E(\mathsf{x}_1) + E(\mathsf{x}_2) + \ldots + E(\mathsf{x}_n)\} = \widehat{x} \quad . \tag{6.2.1}$$

This expectation value is equal to the expectation value of x. Since Eq. (6.2.1) holds for all values of n, the arithmetic mean of a sample is, as one would expect, an unbiased estimator for the mean of the population. The characteristic function of the random variable $\bar{\mathsf{x}}$ is

$$\varphi_{\bar{\mathsf{x}}}(t) = \left\{\varphi_{\frac{\mathsf{x}}{n}}(t)\right\}^n = \left\{\varphi_{\mathsf{x}}\left(\frac{t}{n}\right)\right\}^n \quad . \tag{6.2.2}$$

Next we are interested in the variance of $\bar{\mathsf{x}}$,

$$\begin{aligned}\sigma^2(\bar{\mathsf{x}}) &= E\{(\mathsf{x}-E(\bar{\mathsf{x}})\}^2 = E\left\{\left(\frac{\mathsf{x}_1+\mathsf{x}_2+\ldots+\mathsf{x}_n}{n}-\widehat{x}\right)^2\right\} \\ &= \frac{1}{n^2}E\{[(\mathsf{x}_1-\widehat{x})+(\mathsf{x}_2-\widehat{x})+\ldots+(\mathsf{x}_n-\widehat{x})]^2\} \quad .\end{aligned}$$

Since all of the x_i are independent, all of the cross terms of the type $E\{(\mathsf{x}_i - \widehat{x})(\mathsf{x}_j - \widehat{x})\}$, $i \neq j$ (i.e., all of the covariances) vanish, and we obtain

$$\sigma^2(\bar{\mathsf{x}}) = \frac{1}{n}\sigma^2(\mathsf{x}) \quad . \tag{6.2.3}$$

One thus shows that $\bar{\mathsf{x}}$ is a consistent estimator for $\widehat{x}$. The variance (6.2.3) is itself, however, not a random variable, and is therefore not directly obtainable by experiment. As a definition for the *sample variance* we could try using the arithmetic mean of squared differences

$$\mathsf{s}'^2 = \frac{1}{n}\{(\mathsf{x}_1-\bar{\mathsf{x}})^2+(\mathsf{x}_2-\bar{\mathsf{x}})^2+\ldots+(\mathsf{x}_n-\bar{\mathsf{x}})^2\} \quad . \tag{6.2.4}$$

Its expectation value is

$$\begin{aligned}E(\mathsf{s}'^2) &= \frac{1}{n}E\left\{\sum_{i=1}^{n}(\mathsf{x}_i-\bar{\mathsf{x}})^2\right\} \\ &= \frac{1}{n}E\left\{\sum_{i=1}^{n}(\mathsf{x}_i-\widehat{x}+\widehat{x}-\bar{\mathsf{x}})^2\right\} \\ &= \frac{1}{n}E\left\{\sum_{i=1}^{n}(\mathsf{x}_i-\widehat{x})^2+\sum_{i=1}^{n}(\widehat{x}-\bar{\mathsf{x}})^2+2\sum_{i=1}^{n}(\mathsf{x}_i-\widehat{x})(\widehat{x}-\bar{\mathsf{x}})\right\} \\ &= \frac{1}{n}\sum_{i=1}^{n}\{E((\mathsf{x}_i-\widehat{x})^2)-E((\bar{\mathsf{x}}-\widehat{x})^2)\} \\ &= \frac{1}{n}\left\{n\sigma^2(\mathsf{x})-n\left(\frac{1}{n}\sigma^2(\mathsf{x})\right)\right\} \quad , \\ E(\mathsf{s}'^2) &= \frac{n-1}{n}\sigma^2(\mathsf{x}) \quad .\end{aligned} \tag{6.2.5}$$

Hence one sees that the sample variance defined in this way is a biased estimator for the population variance, having an expectation value smaller than $\sigma^2(\mathsf{x})$. We can see directly from (6.2.5), however, the size of the bias. We therefore change our definition (6.2.4) and write for the sample variance

$$\mathsf{s}^2 = \frac{1}{n-1}\{(\mathsf{x}_1-\bar{\mathsf{x}})^2+(\mathsf{x}_2-\bar{\mathsf{x}})^2+\ldots+(\mathsf{x}_n-\bar{\mathsf{x}})^2\} \quad . \tag{6.2.6}$$

This is now an unbiased estimator for $\sigma^2(\mathsf{x})$. The value $(n-1)$ in the denominator appears at first to be somewhat strange. One must consider, however, that for $n = 1$ the sample mean is equal to the value x of the sole element of the sample ($\mathsf{x} = \bar{\mathsf{x}}$) and that therefore the quantity (6.2.4) would vanish. That is related to the fact that in (6.2.4) – and also in (6.2.6) – the sample mean $\bar{\mathsf{x}}$ was used instead of the population mean $\widehat{x}$, since the latter was not known. Part of the information contained in the sample first had to be used and was not available for the calculation of the variance. The effective number of elements available for calculating the variance is therefore reduced. This is taken into consideration by reducing the denominator of the arithmetic mean (6.2.4). The same line of reasoning is repeated quantitatively in Sect. 6.5.

If we substitute the estimator for the population variance (6.2.6) into (6.2.3), we obtain an estimator for the *variance of the mean*

$$\mathsf{s}^2(\bar{\mathsf{x}}) = \frac{1}{n}\mathsf{s}^2(\mathsf{x}) = \frac{1}{n(n-1)}\sum_{i=1}^{n}(\mathsf{x}_i - \bar{\mathsf{x}})^2 \quad . \tag{6.2.7}$$

The corresponding standard deviation can be considered to be the *error of the mean*

$$\Delta\bar{\mathsf{x}} = \sqrt{\mathsf{s}^2(\bar{\mathsf{x}})} = \mathsf{s}(\bar{\mathsf{x}}) = \frac{1}{\sqrt{n}}\mathsf{s}(\mathsf{x}) \quad . \tag{6.2.8}$$

Of course we are also interested in the error of the sample variance (6.2.6). In Sect. 6.6 we will show that this quantity can be determined under the assumption that the population follows a normal distribution. We will use the result here ahead of time. The variance of s^2 is

$$\mathrm{var}(\mathsf{s}^2) = \left(\frac{\sigma^2}{n-1}\right)^2 2(n-1) \quad . \tag{6.2.9}$$

If we substitute the estimator (6.2.6) into the right-hand side for σ^2 and take the square root, we obtain for the error of the sample variance

$$\Delta\mathsf{s}^2 = \mathsf{s}^2\sqrt{\frac{2}{(n-1)}} \quad . \tag{6.2.10}$$

Finally we give explicit expressions for estimators of the *sample standard deviation* and its error. The first is simply the square root of the sample variance

$$\mathsf{s} = \sqrt{\mathsf{s}^2} = \frac{1}{\sqrt{n-1}}\sqrt{\sum(\mathsf{x}_i - \bar{\mathsf{x}})^2} \quad . \tag{6.2.11}$$

The error of the sample standard deviation is obtained from (6.2.10) by error propagation, which gives

$$\Delta\mathsf{s} = \frac{\mathsf{s}}{\sqrt{2(n-1)}} \quad . \tag{6.2.12}$$

Example 6.1: Computation of the sample mean and variance from data

Suppose one has $n = 7$ measurements of a certain quantity (e.g., the length of an object). Their values are 10.5, 10.9, 9.2, 9.8, 9.0, 10.4, 10.7. The computation is made easier if one uses the fact that all of the measured values are near $a = 10$, i.e., they are of the form $\mathsf{x}_i = a + \delta_i$. The relation (6.1.6) then gives

$$\bar{\mathsf{x}} = \frac{1}{n}\sum_{i=1}^{n} \mathsf{x}_i = \frac{1}{n}\sum_{i=1}^{n}(a + \delta_i) = a + \frac{1}{n}\sum_{i=1}^{n}\delta_i = a + \Delta$$

with

$$\begin{aligned} \Delta &= \frac{1}{n}\sum_{i=1}^{n}\delta_i = \frac{1}{7}(0.5 + 0.9 - 0.8 - 0.2 - 1.0 + 0.4 + 0.7) \\ &= 0.5/7 = 0.07 \quad . \end{aligned}$$

We thus have $\bar{\mathsf{x}} = 10 + \Delta = 10.07$.

The sample variance is computed according to (6.2.6) to be

$$\begin{aligned} \mathsf{s}^2 &= \frac{1}{n-1}\sum_{i=1}^{n}(\mathsf{x}_i - \bar{\mathsf{x}})^2 \\ &= \frac{1}{n-1}\sum_{i=1}^{n}(\mathsf{x}_i^2 - 2\mathsf{x}_i\bar{\mathsf{x}} + \bar{\mathsf{x}}^2) \\ &= \frac{1}{n-1}\left\{\sum_{i=1}^{n}\mathsf{x}_i^2 - n\bar{\mathsf{x}}^2\right\} \quad . \end{aligned}$$

The result can be obtained either by the first or last line of the relation above. In the last line only one difference is computed, not n. The numbers to be squared, however, are usually considerably larger, and one must consider the problem of rounding errors. We therefore use the original expression

$$\begin{aligned} \mathsf{s}^2 &= \frac{1}{6}\{0.43^2 + 0.83^2 + 0.87^2 + 0.27^2 + 1.07^2 + 0.33^2 + 0.63^2\} \\ &= \frac{1}{6}\{0.1849 + 0.6889 + 0.7569 + 0.0729 + 1.1449 + 0.1089 \\ &\quad + 0.3969\} \\ &= 3.3543/6 \approx 0.56 \quad . \end{aligned}$$

The sample standard deviation is $\mathsf{s} \approx 0.75$. From (6.2.8), (6.2.10), and (6.2.12) we obtain finally $\Delta\bar{\mathsf{x}} = 0.28$, $\Delta\mathsf{s}^2 = 0.32$, and $\Delta\mathsf{s} = 0.21$. ■

Naturally one does not usually compute the sample mean and variance by hand, but rather with a simple computer program, such as the following.

Program 6.1: Subroutine SMMNVR for computing sample mean, variance, and standard deviation, and their errors

CALLING SEQUENCE: CALL SMMNVR(DATA,N,XMEAN,DELXM,S2, DELS2,S,DELS)

INPUT ARGUMENTS: DATA — array with elements of the sample
N — number of elements of the sample

OUTPUT ARGUMENTS: XMEAN — sample mean $\bar{x}$
DELXM — error of the mean $\Delta\bar{x}$
S2 — sample variance s^2
DELS2 — error of the variance Δs^2
S — sample standard deviation s
DELS — error of the standard deviation Δs

```
      SUBROUTINE SMMNVR(DATA,N,XMEAN,DELXM,S2,DELS2,S,DELS)
      IMPLICIT DOUBLE PRECISION (A-H,O-Z)
      DIMENSION DATA(*)
      PARAMETER(ZERO=0.D0, ONE=1.D0, TWO=2.D0)
      XMEAN=ZERO
      XN=DBLE(N)
      XN1=XN-ONE
      DO 10 I =  1, N
        XMEAN=XMEAN+DATA(I)
   10 CONTINUE
      XMEAN=XMEAN/XN
      IF(N.GT.1) THEN
        Q=ZERO
        DO 20 I =  1, N
          Q=Q+(DATA(I)-XMEAN)**2
   20   CONTINUE
        S2=Q/XN1
        S=SQRT(S2)
        DELXM=S/SQRT(XN)
        DELS2=S2*SQRT(TWO/XN1)
        DELS=S/SQRT(TWO*XN1)
      ELSE
        DELXM=ZERO
        S2=ZERO
        S=ZERO
        DELS2=ZERO
        DELS=ZERO
      END IF
      END
```

6.3 Graphical Representation of Samples. Histograms and Scatter Plots

After the theoretical considerations of the last sections we now turn to some simple practical aspects of the analysis of sample data. An important tool for this is the representation of the data in graphical form.

A sample

$$\mathrm{x}_1, \mathrm{x}_2, \ldots, \mathrm{x}_n \quad ,$$

which depends on a single variable x can be represented simply by means of tick marks on an x axis. We will call such a representation a *one-dimensional scatter plot*. It contains all the information about the sample. Table 6.1 contains the values $\mathrm{x}_1, \mathrm{x}_2, \ldots, \mathrm{x}_n$ of a sample of size 100, obtained from measuring the resistance R of 100 individual resistors of nominal value 200 kΩ. After obtaining the sample the measurements were ordered.

Table 6.1. Values of resistance R of 100 individual resistors of nominal value 200 kΩ. The data are graphically represented in Fig. 6.1.

193.199	195.673	195.757	196.051	196.092
196.596	196.679	196.763	196.847	197.267
197.392	197.477	198.189	198.650	198.944
199.070	199.111	199.153	199.237	199.698
199.572	199.614	199.824	199.908	200.118
200.160	200.243	200.285	200.453	200.704
200.746	200.830	200.872	200.914	200.956
200.998	200.998	201.123	201.208	201.333
201.375	201.543	201.543	201.584	201.711
201.878	201.919	202.004	202.004	202.088
202.172	202.172	202.297	202.339	202.381
202.507	202.591	202.633	202.716	202.884
203.051	203.052	203.094	203.094	203.177
203.178	203.219	203.764	203.765	203.848
203.890	203.974	204.184	204.267	204.352
204.352	204.729	205.106	205.148	205.231
205.357	205.400	205.483	206.070	206.112
206.154	206.155	206.615	206.657	206.993
207.243	207.621	208.124	208.375	208.502
208.628	208.670	208.711	210.012	211.394

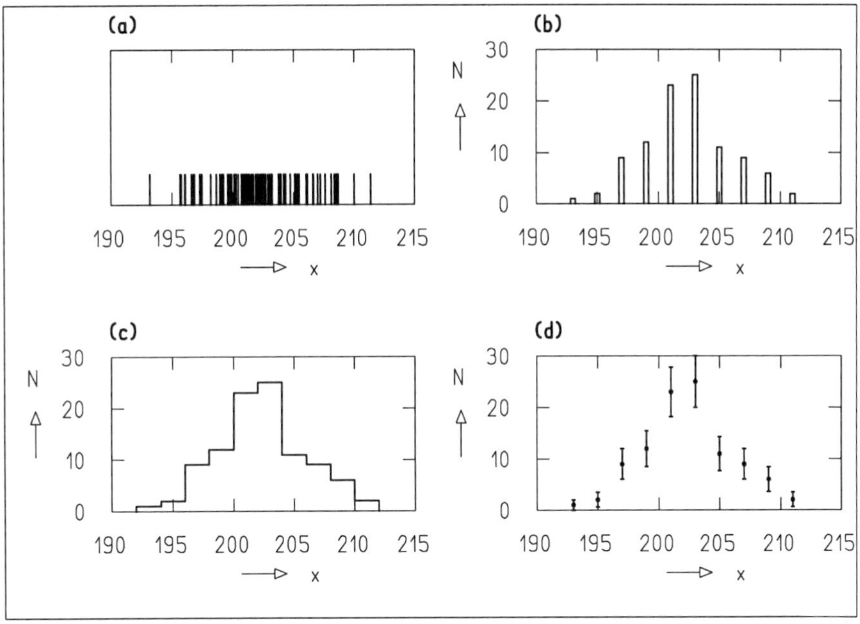

Fig. 6.1. Representation of the data from Table 6.1 as (a) a one-dimensional scatter plot, (b) a bar diagram, (c) a step diagram, and (d) a diagram of measured points with error bars.

Figure 6.1a shows the corresponding scatter plot. Qualitatively one can estimate the mean and variance from the position and width of the clustering of tick marks.

Another graphical representation is usually better suited to visualize the sample by using the second dimension available on the paper. The x axis is used as abscissa and divided into r intervals

$$\xi_1, \xi_2, \ldots, \xi_r$$

of equal width Δx. These intervals are called *bins*. The centers of the bins have the x-values

$$x_1, x_2, \ldots, x_r \quad .$$

On the vertical axis one plots the corresponding numbers of sample elements

$$n_1, n_2, \ldots, n_r$$

that fall into bins $\xi_1, \xi_2, \ldots, \xi_r$. The diagram obtained in this way is called a *histogram* of the sample. This can be interpreted as a frequency distribution, since $h_k = n_k/n$ is a frequency, i.e., a measure of the probability p_k to observe a sample element in the interval ξ_k. For the graphical form of histograms, various methods are used. In a *bar diagram* the values n_k are represented as bars perpendicular to the x axis on the x_k values (Fig. 6.1b). In a *step diagram* the n_k are represented as horizontal lines that cover the entire width ξ_k of the interval. Neighboring horizontal lines are connected by perpendicular lines (Fig. 6.1c). The fraction of the area covering each interval ξ_k of the x axis is then proportional to the number n_k of the sample elements in the interval. (If one uses the area in the interval for the graphical representation of n_k, then the bins can also have different widths.) In economics bar diagrams are most commonly used. (Sometimes one also sees diagrams in which, instead of bars, line segments are used to connect the tips of the bars. In contrast to the step diagram, the resulting figure does not have an area proportional to the sample size n.) In the natural sciences, step diagrams are more common.

In Sect. 6.8 we will determine that as long as the values n_k are not too small, their statistical errors are given by $\Delta n_k = \sqrt{n_k}$. In order to plot them on a graph, the observed values n_k can be drawn as points with vertical *error bars* ending at the points $n_k \pm \sqrt{n_k}$ (Fig. 6.1d).

It is clear that the relative errors $\Delta n_k / n_k = 1/\sqrt{n_k}$ decrease for increasing n_k, i.e., for a sample of fixed size n they decrease for increasing bin width of the histogram. On the other hand, by choosing a larger interval width, one loses any finer structure of the data with respect to the variable x. The ability of a histogram to convey information therefore depends crucially on the appropriate choice of the bin width, usually found only after several attempts.

Example 6.2: Histograms of the same sample with various choices of bin width

In Fig. 6.2 four histograms of the same sample are shown. The population is a Gaussian distribution, which is also shown as a continuous curve. This was scaled in such a way that the area under the histogram is equal to the area under the Gaussian curve. Although the information contained in the plot is greater for a smaller bin width – for vanishing bin width the histogram becomes a one dimensional scatter plot – one notices the similarity between the histogram and Gaussian curve much more easily for the larger bin width. This is because for the larger bin width the relative statistical fluctuations of the contents of individual bins are smaller. The individual steps of the histogram differ less from the curve. ■

Constructing a histogram from a sample is a simple programming task. Suppose the histogram has n_x bins of width Δx, with the first interval extending from $x = x_0$ to $x = x_0 + \Delta x$. The contents of the histogram is put into an array `HIST`, with the first bin in `HIST(1)`, the second bin in `HIST(2)`, etc. The histogram is then specified in the computer by the array `HIST` and the three values x_0, Δx, and n_x.

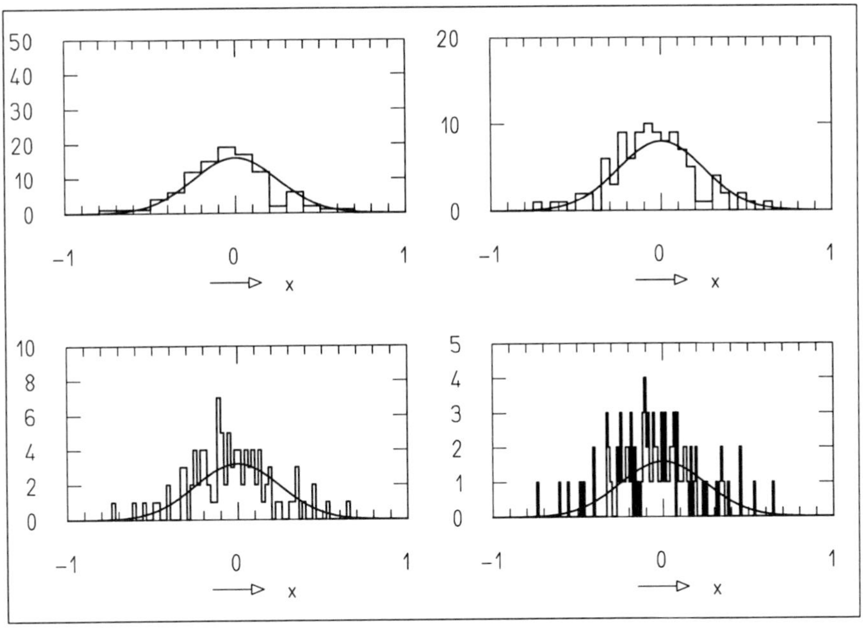

Fig. 6.2. Histogram of the same sample from a Gaussian distribution represented with four different bin widths.

Program 6.2: Subroutine `SMHSIN` initializes a histogram

CALLING SEQUENCE: `CALL SMHSIN(HIST,X0,DELX,NX)`

INPUT ARGUMENTS: `HIST` — array for the contents of the histogram bins, minimum length n_x
`X0` — lower limit of the first bin x_0
`DELX` — bin width Δx
`NX` — number of bins n_x

REMARKS: All histogram bins are initialized with the value zero.

```
      SUBROUTINE SMHSIN(HIST,XO,DELX,NX)
      IMPLICIT DOUBLE PRECISION (A-H,O-Z)
      DIMENSION HIST(*)
      DO 10 I=1,NX
        HIST(I)=0.
   10 CONTINUE
      END
```

Program 6.3: Subroutine SMHSFL enters an element of a sample into a histogram

CALLING SEQUENCE: CALL SMHSFL(HIST,X0,DELX,NX,XIN,WEIGHT)

INPUT ARGUMENTS: HIST,X0,DELX,NX — as in Program 6.2
XIN — element of the sample
WEIGHT — weight with which the element is entered into the histogram (usually WEIGHT = 1.0)

REMARKS: The routine must be called for each element of the sample. One must set WEIGHT=1., except for special cases where the elements of a sample should explicitly be given different weights.

```
SUBROUTINE SMHSFL(HIST,X0,DELX,NX,XIN,WEIGHT)
IMPLICIT DOUBLE PRECISION (A-H,O-Z)
DIMENSION HIST(*)
N=INT((XIN-X0)/DELX)+1
IF (N.GT.0 .AND. N.LE.NX) THEN
  HIST(N)=HIST(N)+WEIGHT
END IF
END
```

Graphical display of histograms can be accomplished with the graphics routines from Appendix F. With these routines the user can freely adjust all of the parameters that determine the appearance of the plot, such as the page format, scale factors, colors, line thickness, etc. We give here a short program that does not allow this freedom, but which accomplishes the graphical output of a histogram produced in the computer with a single subroutine call.

Program 6.4: Subroutine SMHSGR plots a histogram

CALLING SEQUENCE: CALL SMHSGR(HIST,X0,DELX,NX,TX,LTX,TY,LTY,CAPT,LCAPT,NWS)

INPUT ARGUMENTS: HIST,X0,DELX,NX — as in Program 6.2
TX — character string for labeling the abscissa of the histogram
TY — character string for labeling the ordinate of the histogram
CAPT — character string for labeling the histogram
LTX,LTY,LCAPT — lengths of the character strings
NWS — workstation number on which the plot is to be produced

REMARKS: The paper format is A5 landscape. The actual histogram is adjusted in a reasonable way so as to fit this format. The scaling of the abscissa is given by X0,DELX,NX. The ordinate is scaled in such a way that the bin with the greatest number of entries is not quite as high as the length of the ordinate.

```
      SUBROUTINE SMHSGR(HIST,X0,DELX,NX,TX,LTX,TY,LTY,CAPT,
     +                  LCAPT,NWS)
      IMPLICIT DOUBLE PRECISION (A-H,O-Z)
      DIMENSION HIST(*)
      CHARACTER*(*) TX,TY,CAPT
      CHARACTER*8 NUMSTR
C determine scale on ordinate
      HM=HIST(1)
      DO 10 I =  2, NX
        HM=MAX(HIST(I),HM)
   10 CONTINUE
      WRITE(NUMSTR,'(E8.2)')HM
      READ(NUMSTR(3:3),'(I1)')M1
      READ(NUMSTR(4:4),'(I1)')M2
      N0=0
      N1=M1+1
      N2=0
      IF(N1.GT.9) THEN
        N0=1
        N1=0
      END IF
      WRITE(NUMSTR(1:1),'(I1)')N0
      WRITE(NUMSTR(3:3),'(I1)')N1
      WRITE(NUMSTR(4:4),'(I1)')N2
      READ(NUMSTR,'(E8.2)')SCALE
C graphics
      CALL GROPEN
      CALL GROPWS(NWS)
      X1=X0+DBLE(NX)*DELX
      CALL GRWNCC(X0,X1,0.D0,SCALE)
      CALL GRVWWC(0.1D0,1.1D0,0.1D0,0.75D0)
      CALL GRWNWC(-0.214D0,1.2D0,-0.1D0,0.9D0)
      CALL GRSTFR(NWS,5.D0,0.D0)
      CALL GRFRAM
      CALL GRSCLX(TX,LTX)
      CALL GRSCLY(TY,LTY)
      CALL GRTXTC(1.D0,CAPT,LCAPT)
      CALL GRBOUN
      CALL GRSTCL(2)
      CALL GRHSDR(HIST,X0,DELX,NX)
      CALL GRCLWS(NWS)
      CALL GRCLSE
      END
```

A histogram allows a first direct look at the nature of the data. It answers questions such as "Are the data more or less distributed according to a Gaussian?" or "Do there exist points exceptionally far away from the average value?". If the histogram leads one to conclude that the population is distributed according to a Gaussian, then the mean and standard deviation can be estimated directly from the plot. The mean is the center of gravity of the histogram. The standard deviation is obtained as in the following example.

Example 6.3: Full width at half maximum (FWHM)

If the form of histogram allows one to assume that the represented sample originates from a Gaussian distribution, then one can draw by hand a Gaussian bell curve that follows the histogram as closely as possible. The position of the maximum is a good estimator for the mean of the sample. One then draws a horizontal line at half of the height of the maximum. This crosses the bell curve at the points x_a and x_b. The quantity

$$f = x_b - x_a$$

is called the *full width at half maximum* (FWHM). One can easily determine for a Gaussian distribution the simple relation

$$\sigma = \frac{f}{\sqrt{-8 \ln \frac{1}{2}}} \approx 0.4247\, f \tag{6.3.1}$$

between the standard deviation and FWHM. This expression can be used to estimate the standard deviation of a sample when f is obtained from a histogram. ■

We now use the Monte Carlo method (Chapter 4) together with histograms in order to illustrate the concepts of mean, standard deviation, and variance of a sample, and their errors, as introduced in Sect. 6.2.

Example 6.4: Investigation of characteristic quantities of samples from a Gaussian distribution with the Monte Carlo method

We generate successively 1000 samples of size $N = 100$ from the standard normal distribution, e.g., by calling

```
CALL RNSTNR(R,100) ,
```

and we compute the mean $\bar{\mathrm{x}}$, variance s^2, and standard deviation s for each sample as well as the errors $\Delta\bar{\mathrm{x}}$, $\Delta\mathrm{s}^2$, and $\Delta\mathrm{s}$, with the Program 6.1. We then produce for each of the six quantities a histogram (Fig. 6.3), containing 1000 entries. Since each of the quantities is defined as the sum of many random quantities, we expect in all cases that the histograms should resemble Gaussian distributions. From the histogram for $\bar{\mathrm{x}}$, we obtain a full width at half maximum of about 0.25, and hence a standard deviation of approximately 0.1. Indeed the histogram for $\Delta\bar{\mathrm{x}}$ shows an approximately Gaussian distribution with mean value $\Delta\bar{\mathrm{x}} = 0.1$. (From the width of this histogram one could determine the error of the error $\Delta\bar{\mathrm{x}}$ of the mean $\bar{\mathrm{x}}$!) From both histograms one obtains a very clear impression of the meaning of the error $\Delta\bar{\mathrm{x}}$ of the mean $\bar{\mathrm{x}}$ for a single sample, as computed according to (6.2.8). It gives (within its error) the standard deviation of the population, from which the sample mean $\bar{\mathrm{x}}$ comes. If many samples are successively taken (i.e., if the experiment is repeated many times) then the frequency distribution of the values $\bar{\mathrm{x}}$ follows a Gaussian distribution about the population mean with standard deviation $\Delta\bar{\mathrm{x}}$. The corresponding considerations also hold for the quantities s^2, s, and their errors $\Delta\mathrm{s}$ and $\Delta\mathrm{s}^2$. ■

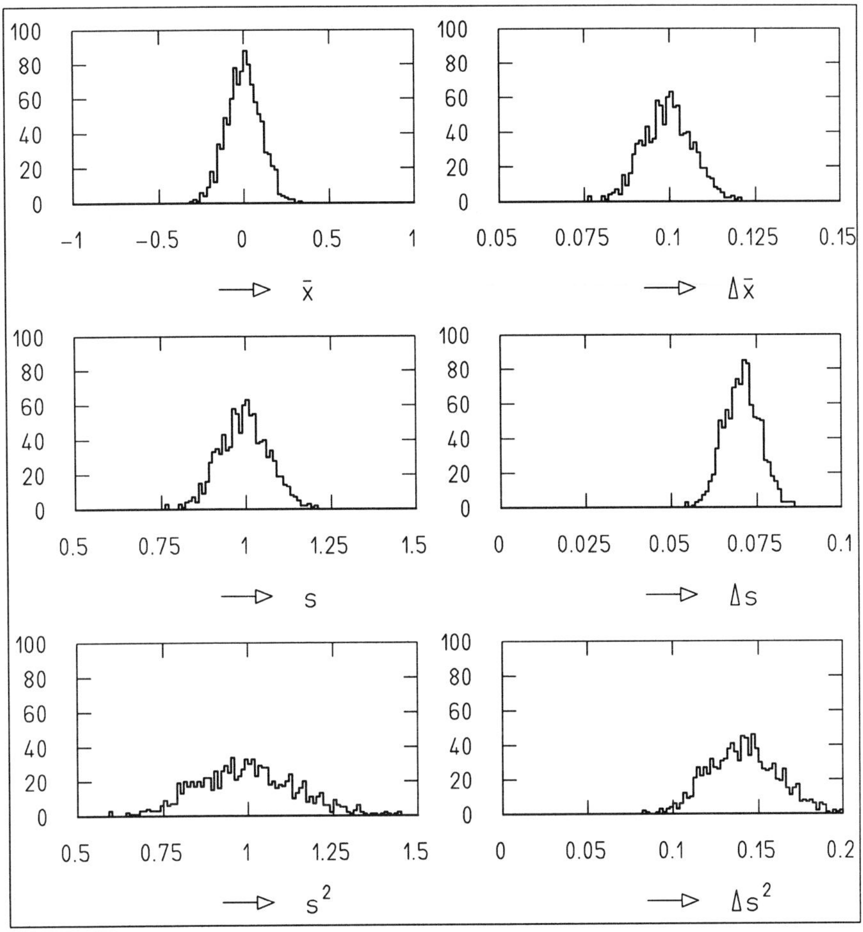

Fig. 6.3. Histograms of the quantities $\bar{x}$, $\Delta\bar{x}$, s, Δs, s^2, and Δs^2 from 1000 samples of size 100 from the standard normal distribution.

If the elements of the sample depend on two random variables x and y, then one can construct a scatter plot, where each element is represented as a point in a Cartesian coordinate system for the variables x and y. Such a *two-dimensional scatter plot* provides useful qualitative information about the relationship between the two variables.

Program 6.5: Subroutine SMSDGR produces a two-dimensional scatter plot

CALLING SEQUENCE: CALL SMSDGR(X,Y,N,XA,XB,YA,YB,TX,
LTX,TY,LTY,CAPT,LCAPT,NWS)

INPUT ARGUMENTS: X — array of x values

Y — array of y values

N — number of pairs (x, y)

XA, XB, YA, YB — the plot covers the region

$x_a \leq x \leq x_b$, $y_a \leq y \leq y_b$

TX — character string for labeling the horizontal axis

TY — character string for labeling the vertical axis

CAPT — character string for title

LTX, LTY, LCAPT — lengths of the character strings

NWS — workstation number

REMARKS: A picture in A5 landscape format is produced, into which the square scatter plot is placed. The program must be modified appropriately if another paper format or aspect ratio is to be used.

```
      SUBROUTINE SMSDGR(X,Y,N,XA,XB,YA,YB,TX,LTX,TY,LTY,CAPT,
     +                  LCAPT,NWS)
      IMPLICIT DOUBLE PRECISION (A-H,O-Z)
      DIMENSION X(*),Y(*)
      CHARACTER*(*) TX,TY,CAPT
      CALL GROPEN
      CALL GROPWS(NWS)
      CALL GRWNCC(XA,XB,YA,YB)
      CALL GRVWWC(0.2D0,.85D0,0.1D0,0.75D0)
      CALL GRWNWC(-0.214D0,1.2D0,-0.1D0,0.9D0)
      CALL GRSTFR(NWS,5.D0,0.D0)
      CALL GRFRAM
      CALL GRSCLX(TX,LTX)
      CALL GRSCLY(TY,LTY)
      CALL GRTXTC(1.D0,CAPT,LCAPT)
      CALL GRBOUN
      CALL GRSTCL(2)
      DO 10 I=1,N
        CALL GRDATP(1,0.1D0,X(I),Y(I),0.D0,0.D0,0.D0)
   10 CONTINUE
      CALL GRCLWS(NWS)
      CALL GRCLSE
      END
```

Example 6.5: Two-dimensional scatter plot: Dividend versus price for industrial stocks

Table 6.2 contains a list of the first 10 of 226 data sets, in which the dividend in 1967 (first column) and share price on December 31, 1967 (second column) are given for a number of industrial stocks. The third column shows the company name for all German corporations worth more than 10 million marks. The scatter plot of the number pairs (share price, dividend) is shown in Fig. 6.4.

As expected we see a strong correlation between dividend and share price. One can see, however, that the dividend does not grow linearly with the price.

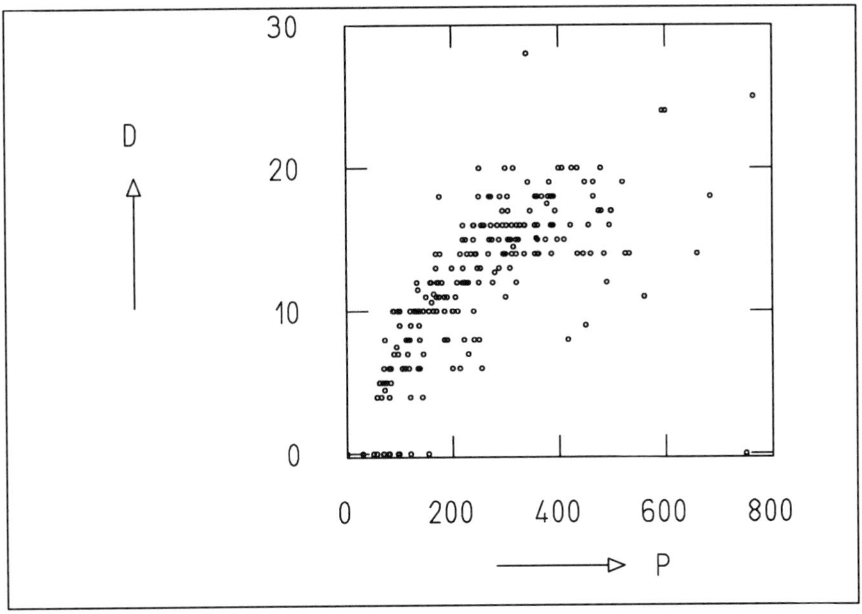

Fig. 6.4. Scatter plot of dividend D versus share price P for industrial stocks.

It appears that factors other than immediate profit determine the price of a stock.

Also shown are histograms for the share price (Fig. 6.5) and dividend (Fig. 6.6) which can be obtained as projections of the scatter plot onto the abscissa and ordinate. One clearly observes a non-statistical behavior for the dividends. It is given as a percent of the nominal value and is therefore almost always integer. One sees that even numbers are considerably more frequent than odd numbers. ■

6.4 Samples from Partitioned Populations

It is often advantageous to divide a population G (e.g., all of the students in Europe) into various *subpopulations* G_1, G_2, ..., G_t (students at university $1, 2, \ldots, t$). Suppose a quantity of interest x follows in the various subpopulations the probability densities $f_1(x), f_2(x), \ldots, f_t(x)$. The distribution function corresponding to $f_i(x)$ is then

$$F_i(x) = \int_{-\infty}^{x} f_i(x)\,\mathrm{d}x = P(\mathrm{x} < x | \mathrm{x} \in G_i) \qquad (6.4.1)$$

Table 6.2. Dividend, share price of a stock, and company name.

12.	133.	ACKERMANN-GOEGGINGEN
08.	417.	ADLERWERKE KLEYER
17.	346.	AGROB AG FUER GROB U. FEINKERAMIK
25.	765.	AG.F.ENERGIEWIRTSCHAFT
16.	355.	AG F. LICHT- U. KRAFTVERS.,MCHN.
20.	315.	AG.F. IND.U.VERKEHRSW.
08.	138.	AG. WESER
16.	295.	AEG ALLG.ELEKTR.-GES.
20.	479.	ANDREAE-NORIS ZAHN
10.	201.	ANKERWERKE

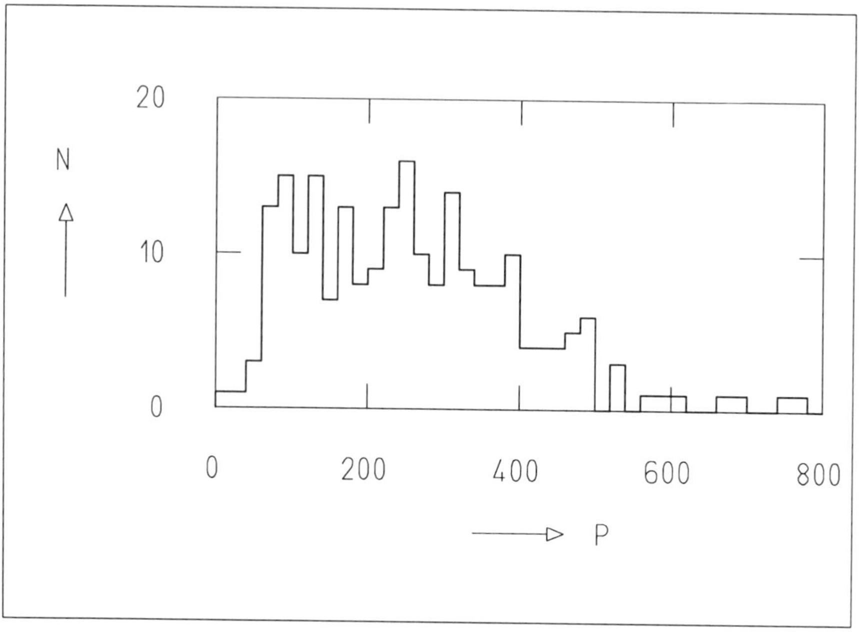

Fig. 6.5. Histogram of the price of industrial stocks.

This is equal to the conditional probability for $\mathsf{x} < x$ given that x is contained in the subpopulation G_i. The rule of total probability (2.3.4) provides the relationship between the various $F_i(x)$ and the distribution function $F(x)$ for G,

$$F(x) = P(\mathsf{x} < x | \mathsf{x} \in G) = \sum_{i=1}^{t} P(\mathsf{x} < x | \mathsf{x} \in G_i) P(\mathsf{x} \in G_i) \quad ,$$

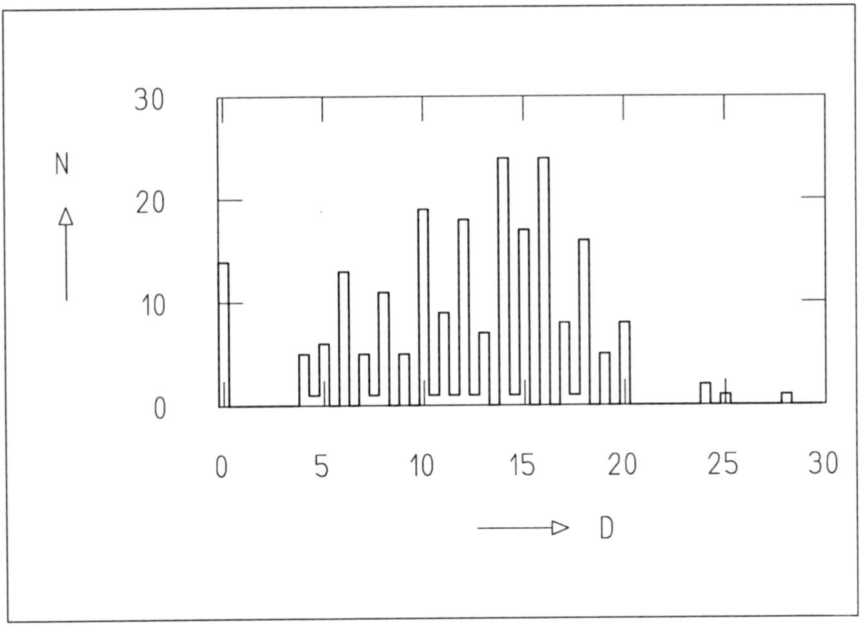

Fig. 6.6. Histogram of the dividend of industrial stocks.

i.e.,

$$F(x) = \sum_{i=1}^{t} P(\mathsf{x} \in G_i) F_i(x) \quad . \tag{6.4.2}$$

Correspondingly one has for the probability density

$$f(x) = \sum_{i=1}^{t} P(\mathsf{x} \in G_i) f_i(x) \quad . \tag{6.4.3}$$

If we now abbreviate $P(\mathsf{x} \in G_i)$ by p_i, then one has

$$\widehat{x} = E(\mathsf{x}) = \int_{-\infty}^{\infty} x f(x)\,\mathrm{d}x = \sum_{i=1}^{t} p_i \int_{-\infty}^{\infty} x f_i(x)\,\mathrm{d}x \quad ,$$

$$\widehat{x} = \sum_{i=1}^{t} p_i \widehat{x}_i \quad . \tag{6.4.4}$$

The population mean is thus the mean of the individual means of the subpopulations, each weighted by probability of its subpopulation. For the population variance one obtains

$$\sigma^2(\mathsf{x}) \;=\; \int_{-\infty}^{\infty} (x - \widehat{x})^2 f(x)\,\mathrm{d}x$$

$$= \int_{-\infty}^{\infty} (x - \widehat{x})^2 \sum_{i=1}^{t} p_i f_i(x)\,\mathrm{d}x$$

$$= \sum_{i=1}^{t} p_i \int_{-\infty}^{\infty} \{(x - \widehat{x}_i) + (\widehat{x}_i - \widehat{x})\}^2 f_i(x)\,\mathrm{d}x \quad .$$

All cross terms vanish since the x_i are independent, leading to

$$\sigma^2(\mathsf{x}) = \sum_{i=1}^{t} p_i \left\{ \int_{-\infty}^{\infty} (x - \widehat{x}_i)^2 f_i(x)\,\mathrm{d}x + (\widehat{x}_i - \widehat{x})^2 \int_{-\infty}^{\infty} f_i(x)\,\mathrm{d}x \right\}$$

or

$$\sigma^2(\mathsf{x}) = \sum_{i=1}^{t} p_i \{\sigma_i^2 + (\widehat{x}_i - \widehat{x})^2\} \quad . \tag{6.4.5}$$

One thus obtains the weighted mean of a sum of two terms. The first gives the dispersion of a subpopulation, the second gives the quadratic deviation of the mean of this subpopulation from the mean of the whole population.

Having discussed separating a population into parts, we now take from each subpopulation G_i a sample of size n_i (with $\sum_{i=1}^{t} n_i = n$) and examine the arithmetic mean of the total partitioned sample

$$\bar{\mathsf{x}}_p = \frac{1}{n} \sum_{i=1}^{t} \sum_{j=1}^{n_i} \mathsf{x}_{ij} = \frac{1}{n} \sum_{i=1}^{t} n_i \bar{\mathsf{x}}_i \tag{6.4.6}$$

with the expectation value and variance

$$E(\bar{\mathsf{x}}_p) = \frac{1}{n} \sum_{i=1}^{n} n_i \widehat{x}_i \quad , \tag{6.4.7}$$

$$\begin{aligned} \sigma^2(\bar{\mathsf{x}}_p) &= E\{(\bar{\mathsf{x}}_p - E(\bar{\mathsf{x}}_p))^2\} \\ &= E\left\{ \left(\sum_{i=1}^{t} \frac{n_i}{n} (\bar{\mathsf{x}}_i - \widehat{x}_i) \right)^2 \right\} \\ &= \frac{1}{n^2} \sum_{i=1}^{t} n_i^2 E\{(\bar{\mathsf{x}}_i - \widehat{x}_i)^2\} \quad , \end{aligned}$$

$$\sigma^2(\bar{\mathsf{x}}_p) = \frac{1}{n^2} \sum_{i=1}^{t} n_i^2 \sigma^2(\bar{\mathsf{x}}_i) \quad . \tag{6.4.8}$$

Using (6.2.3) this is finally

$$\sigma^2(\bar{\mathsf{x}}_p) = \frac{1}{n} \sum_{i=1}^{t} \frac{n_i}{n} \sigma_i^2 \quad . \tag{6.4.9}$$

One would obtain the same result by application of the law of error propagation (3.8.7) to Eq. (6.4.6).

It is clear that the arithmetic mean $\bar{\mathrm{x}}_p$ cannot in general be an estimator for the sample mean $\widehat{x}$, since it depends on the arbitrary choice of the size n_i of the samples from the subpopulations. A comparison of (6.4.7) with (6.4.4) shows that this is only true for the special case $p_i = n_i/n$.

The population mean $\widehat{x}$ can be estimated in the following way. One first determines the means $\bar{\mathrm{x}}_i$ for the subpopulations, and constructs then the expression

$$\widetilde{\mathrm{x}} = \sum_{i=1}^{t} p_i \bar{\mathrm{x}}_i \quad , \tag{6.4.10}$$

in analogy to Eq. (6.4.4). By error propagation one obtains for the variance of $\widetilde{\mathrm{x}}$

$$\sigma^2(\widetilde{\mathrm{x}}) = \sum_{i=1}^{t} p_i^2 \sigma^2(\bar{\mathrm{x}}_i) = \sum_{i=1}^{t} \frac{p_i^2}{n_i} \sigma_i^2 \quad . \tag{6.4.11}$$

Example 6.6: Optimal choice of the sample size for subpopulations

In order to minimize the variance $\sigma^2(\widetilde{\mathrm{x}})$, we cannot simply differentiate the relation (6.4.11) with respect to all n_i, since the n_i must satisfy a constraint, namely

$$\sum_{i=1}^{t} n_i - n = 0 \quad . \tag{6.4.12}$$

We must therefore use the method of *Lagrange multipliers*, by multiplying Eq. (6.4.12) with a factor μ, adding this to Eq. (6.4.11), and finally setting the partial derivatives of the n_i with respect to μ equal to zero:

$$L = \sigma^2(\widetilde{x}) + \mu\left(\sum n_i - n\right) = \sum (p_i^2/n_i)\sigma_i^2 + \mu\left(\sum n_i - n\right) \quad ,$$

$$\frac{\partial L}{\partial n_i} = -\frac{p_i^2 \sigma_i^2}{n_i^2} + \mu = 0 \quad , \tag{6.4.13}$$

$$\frac{\partial L}{\partial \mu} = \sum n_i - n = 0 \quad . \tag{6.4.14}$$

From (6.4.13) we obtain

$$n_i = p_i \sigma_i / \sqrt{\mu} \quad .$$

Together with (6.4.14) this gives

$$1/\sqrt{\mu} = n / \sum p_i \sigma_i$$

and therefore

$$n_i = n p_i \sigma_i / \sum p_i \sigma_i \quad . \tag{6.4.15}$$

The result (6.4.15) states that the sizes n_i of the samples from the subpopulations i should be chosen in such a way that they are proportional to the probability p_i of subpopulation i, weighted with the corresponding standard deviation.

As an example assume that a scientific publishing company wants to estimate the total amount spent for scientific books by two subpopulations: (1) students and (2) scientific libraries. Further, we will assume that there are one thousand libraries and 10^6 students in the population and that the standard deviation of the money spent by students is \$ 100, and for libraries (which are of greatly differing sizes) \$ $3 \cdot 10^5$. We then have

$$p_1 \approx 1 \quad , \qquad p_2 \approx 10^{-3} \quad , \qquad \sigma_1 = 100 \quad , \qquad \sigma_2 = 3 \times 10^5$$

and from (6.4.15)

$$n_1 = \text{const} \cdot 100 \quad , \qquad n_2 = \text{const} \cdot 300 \quad , \qquad n_2 = 3n_1 \quad .$$

Note that the result does not depend on the means of the partial populations. The quantities p_i, x_i, and σ_i are in general unknown. They must first be estimated from preliminary samples. ∎

The discussion of subpopulations will be taken up again in Chapter 11.

6.5 Samples without Replacement from Finite Discrete Populations. Mean Square Deviation. Degrees of Freedom

We first encountered the concept a sample in connection with the hypergeometric distribution (Sect. 5.3). There we determined that the independence of the individual sample elements was lost by the process of taking elements without replacing them from a finite (and hence discrete) population. We are therefore no longer dealing with genuine random sampling, even if no particular choice among the remaining elements is made.

To discuss this further let us introduce the following notation. Suppose the population consists of N elements $y_1, y_2, \ldots, y_N$. From it we take a sample of size n with the elements $\mathsf{x}_1, \mathsf{x}_2, \ldots, \mathsf{x}_n$. (In the hypergeometric distribution, the y_j and hence the x_i could only take on the values 0 and 1.)

Since it is equally probable for each of the remaining elements y_j to be chosen, we obtain for the expectation value

$$E(y) = \widehat{y} = \bar{y} = \frac{1}{N} \sum_{j=1}^{N} y_j \quad . \tag{6.5.1}$$

Although $\widehat{y}$ is not a random variable, this expression is the arithmetic mean of a finite number of elements of the population. A definition of the population variance encounters the difficulties discussed at the end of Sect. 6.2. We define it in analogy to (6.2.6) as

$$\begin{aligned} \sigma^2(y) &= \frac{1}{N-1}\sum_{j=1}^{N}(y_j-\bar{y})^2 \\ &= \frac{1}{N-1}\left\{\sum_{j=1}^{N} y_j^2 - \frac{1}{N}\left(\sum_{j=1}^{N} y_j\right)^2\right\} \quad . \end{aligned} \tag{6.5.2}$$

Let us now consider the *sum of squares*,

$$\sum_{j=1}^{N}(y_j-\bar{y})^2 \quad . \tag{6.5.3}$$

Since we have not constrained the population in any way, the y_j can take on all possible values. Therefore the first element in the sum in (6.5.2) can also take on any of the possible values. The same holds for the 2nd, 3rd, ..., $(N-1)$th terms summed. The Nth term in the sum is then, however, fixed, since

$$\sum_{j=1}^{N}(y_j-\bar{y}) = 0 \quad . \tag{6.5.4}$$

We say that the *number of degrees of freedom* of the sum of squares (6.5.3) is $N-1$. One can illustrate this connection geometrically. We consider the case $\bar{y}=0$ and construct an N-dimensional vector space with the y_j. The quadratic sum (6.5.3) is then the square of the absolute value of the position vector in this space. Because of the equation of constraint (6.5.4) the tip of the position vector can only move in a space of dimension $(N-1)$. In mechanics the dimension of such a constrained space is called the number of degrees of freedom. This is sketched in Fig. 6.7 for the case $N=2$. Here the position vector is constrained to lie on the line $y_2=-y_1$.

A sum of squares divided by the number of degrees of freedom, i.e., an expression of the form (6.5.2) is called a *mean square* or to be more complete, since we are dealing with the differences of the individual values from the expectation or mean value, *mean square deviation*. The square root of this expression, which is then a measure of the dispersion, is called the root mean square (RMS) deviation.

We now return to the sample $\mathrm{x}_1, \mathrm{x}_2, \ldots, \mathrm{x}_n$. For simplicity of notation we will introduce the *Kronecker symbol*, which describes the selection procedure for the sample. It is defined as

$$\delta_i^j = \begin{cases} 1, & \text{if } \mathrm{x}_i \text{ is the element } y_j, \\ 0 & \text{otherwise.} \end{cases} \tag{6.5.5}$$

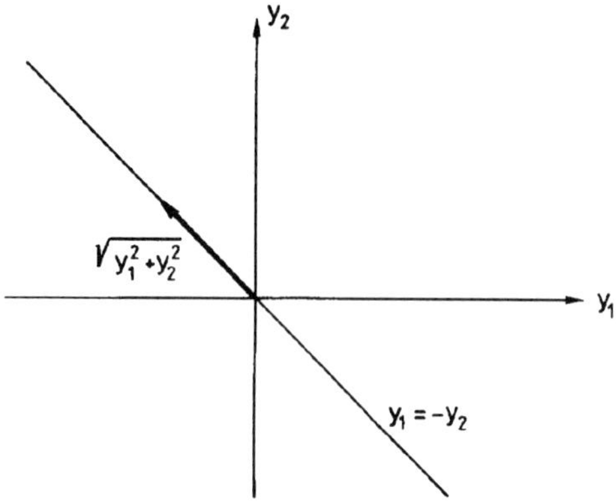

Fig. 6.7. A sample of size two gives a sum of squares with one degree of freedom.

In particular one has

$$\mathsf{x}_i = \sum_{j=1}^{N} \delta_i^j y_j \quad . \tag{6.5.6}$$

Since the selection of any of the y_j as the ith element is equally probable, one has

$$P(\delta_i^j = 1) = 1/N \quad . \tag{6.5.7}$$

Since δ_i^j describes a random procedure, it is clearly a random variable itself. Its expectation value is found from Eq. (3.3.2) (where $n = 2$, $x_1 = 0$, $x_2 = 1$) to be

$$E(\delta_i^j) = P(\delta_i^j = 1) = 1/N \quad . \tag{6.5.8}$$

If now one element x_i of the sample is determined, one then has only $(N-1)$ selection possibilities out of the population for a further element, e.g., x_k. That is,

$$P(\delta_i^j \delta_k^\ell = 1) = \frac{1}{N}\frac{1}{N-1} = E(\delta_i^j \delta_k^\ell) \quad . \tag{6.5.9}$$

Since the sample is taken without replacement, one has $j \neq \ell$, i.e.,

$$\delta_i^j \delta_k^j = 0 \quad . \tag{6.5.10}$$

Similarly one has

$$\delta_i^j \delta_i^\ell = 0 \quad , \tag{6.5.11}$$

since two different elements of the population cannot simultaneously occur as the ith element of the sample.

We consider now the expectation value of x_1,

$$E(\mathsf{x}_1) = E\left\{\sum_{j=1}^{N} \delta_1^j y_j\right\} = \sum_{j=1}^{N} y_j E(\delta_1^j) = \frac{1}{N}\sum_{j=1}^{N} y_j = \bar{y} \quad . \tag{6.5.12}$$

Since x_1 is in not in any way special, the expectation values of all elements of the sample, and thus also of their arithmetic mean, have the same value

$$E(\bar{\mathsf{x}}) = \frac{1}{n}\sum_{i=1}^{n} E(\mathsf{x}_i) = \bar{y} \quad . \tag{6.5.13}$$

The arithmetic mean of the sample is thus an unbiased estimator for the population mean.

Next we consider the *sample variance*

$$\mathsf{s}_x^2 = \frac{1}{n-1}\sum_{i=1}^{n}(\mathsf{x}_i - \bar{\mathsf{x}})^2 \quad . \tag{6.5.14}$$

By means of a somewhat longer calculation it can be shown that the expectation value is

$$E(\mathsf{s}_x^2) = \sigma^2(y) \quad . \tag{6.5.15}$$

The sample variance is thus an unbiased estimator for the population variance.

The *variance of the mean* is also of interest:

$$\sigma^2(\bar{\mathsf{x}}) = E\{(\bar{\mathsf{x}} - E(\bar{\mathsf{x}}))^2\} \quad .$$

$E(\bar{\mathsf{x}}) = \bar{y}$ is, however, a fixed, *not* a random quantity, whereas $\bar{\mathsf{x}}$ depends on the individual sample, and is hence a random variable. One therefore has

$$\begin{aligned} \sigma^2(\bar{\mathsf{x}}) &= E(\bar{\mathsf{x}}^2) - \bar{y}^2 = \frac{1}{n}\left\{\left(1 - \frac{n}{N}\right)\sigma^2(y) + n\bar{y}^2\right\} - \bar{y}^2 \quad , \\ \sigma^2(\bar{\mathsf{x}}) &= \frac{\sigma^2(y)}{n}\left(1 - \frac{n}{N}\right) \quad . \end{aligned} \tag{6.5.16}$$

Comparing with the case of an infinite continuous sample (6.2.3) one sees the additional factor $(1 - n/N)$. This corresponds to the fact that the variance of $\bar{\mathsf{x}}$ vanishes in the case $n = N$, where the "sample" contains the entire population and where one has exactly $\bar{\mathsf{x}} = \bar{y}$.

6.6 Samples from Gaussian Distributions. χ^2-Distribution

We return now to continuously distributed populations and consider in particular a Gaussian distribution with mean a and variance σ^2. According to (5.7.7), the characteristic function of such a Gaussian distribution is

$$\varphi_{\mathsf{x}}(t) = \exp(ita)\exp(-\frac{1}{2}\sigma^2 t^2) \quad . \tag{6.6.1}$$

We now take a sample of size n from the population. The characteristic function of the sample mean was given in (6.2.2) in terms of the characteristic function of the population. From this we have

$$\varphi_{\bar{x}}(t) = \left\{ \exp\left(\mathrm{i}\frac{t}{n}a - \frac{\sigma^2}{2}\left(\frac{t}{n}\right)^2 \right) \right\}^n \quad . \tag{6.6.2}$$

If we consider $(\bar{x} - a) = (\bar{x} - \widehat{x})$ in place of x, then one obtains

$$\varphi_{\bar{x}-a}(t) = \exp\left(-\frac{\sigma^2 t^2}{2n} \right) \quad . \tag{6.6.3}$$

This is again the characteristic function of a normal distribution, but with a different variance,

$$\sigma^2(\bar{x}) = \sigma^2(x)/n \quad . \tag{6.6.4}$$

For the simple case of a standard Gaussian distribution ($a = 0$, $\sigma^2 = 1$) we have

$$\varphi_{\bar{x}}(t) = \exp(-t^2/2n) \quad . \tag{6.6.5}$$

We take a sample from this distribution,

$$x_1, x_2, \ldots, x_n \quad ,$$

but we are interested in particular in the sum of the squares of the sample elements,

$$x^2 = x_1^2 + x_2^2 + \ldots + x_n^2 \quad . \tag{6.6.6}$$

We want to show that the quantity x^2 follows the distribution function*

$$F(\chi^2) = \frac{1}{\Gamma(\lambda)2^\lambda} \int_0^{\chi^2} u^{\lambda-1} \mathrm{e}^{-\frac{1}{2}u} \, \mathrm{d}u \quad , \tag{6.6.7}$$

where

$$\lambda = \frac{1}{2}n \quad . \tag{6.6.8}$$

The quantity n is called the *number of degrees of freedom.*

We first introduce the abbreviation

$$\frac{1}{\Gamma(\lambda)2^\lambda} = k \tag{6.6.9}$$

and determine the probability density to be

$$f(\chi^2) = k(\chi^2)^{\lambda-1} \mathrm{e}^{-\frac{1}{2}\chi^2} \quad . \tag{6.6.10}$$

---

*The symbol χ^2 (chi squared) was introduced by K. Pearson. Although it is written as something squared, which reminds one of its origin as a sum of squares, it is treated as a usual random variable.

For two degrees of freedom, the probability density is clearly an exponential function. First we want to prove what was claimed by (6.6.7) for one degree of freedom ($\lambda = 1/2$). Thus we ask for the probability that $\mathsf{x}^2 < \chi^2$, or rather, that $-\sqrt{\chi^2} < \mathsf{x} < +\sqrt{\chi^2}$. This is

$$\begin{aligned} F(\chi^2) &= P(\mathsf{x}^2 < \chi^2) = P(-\sqrt{\chi^2} < \mathsf{x} < +\sqrt{\chi^2}) \\ &= \frac{1}{\sqrt{2\pi}} \int_{-\sqrt{\chi^2}}^{\sqrt{\chi^2}} \mathrm{e}^{-\frac{1}{2}x^2}\,\mathrm{d}x = \frac{2}{\sqrt{2\pi}} \int_0^{\sqrt{\chi^2}} \mathrm{e}^{-\frac{1}{2}x^2}\,\mathrm{d}x \quad . \end{aligned}$$

Setting $x^2 = u$, $\mathrm{d}u = 2x\,\mathrm{d}x$, we obtain directly

$$F(\chi^2) = \frac{1}{\sqrt{2\pi}} \int_0^{\chi^2} u^{-\frac{1}{2}} \mathrm{e}^{-\frac{1}{2}u}\,\mathrm{d}u \quad . \tag{6.6.11}$$

To prove the general case we first find the characteristic function of the χ^2-distribution to be

$$\varphi_{\chi^2}(t) = \int_0^\infty k(\chi^2)^{\lambda-1} \exp(-\frac{1}{2}\chi^2 + \mathrm{i}t\chi^2)\,\mathrm{d}\chi^2 \tag{6.6.12}$$

or with $(1/2 - \mathrm{i}t)\chi^2 = \nu$,

$$\varphi_{\chi^2}(t) = 2^\lambda (1 - 2\mathrm{i}t)^{-\lambda} k \int_0^\infty \nu^{\lambda-1} \mathrm{e}^{-\nu}\,\mathrm{d}\nu \quad .$$

The integral on the right side is, according to (D.1.1), equal to $\Gamma(\lambda)$. One therefore has

$$\varphi_{\chi^2}(t) = (1 - 2\mathrm{i}t)^{-\lambda} \quad . \tag{6.6.13}$$

If we now consider the case of a second distribution with λ', then one has

$$\varphi'_{\chi^2}(t) = (1 - 2\mathrm{i}t)^{-\lambda'} \quad .$$

Since the characteristic function of a sum is equal to the product of the characteristic functions, one has the following important theorem:

> The sum of two independent χ^2 variables with n_1, n_2 degrees of freedom follows itself a χ^2-distribution with $n = n_1 + n_2$ degrees of freedom.

This theorem can now be used to easily generalize the claim (6.6.7), proven up to now only for $n = 1$. The proof follows from the fact that the individual terms of the sum of squares are independent and therefore (6.6.6) can be treated as the sum of n different χ^2 variables, each with one degree of freedom.

In order to obtain the expectation value and variance of the χ^2-distribution, we use the characteristic function, whose derivatives (5.5.7) give the central moments. We obtain

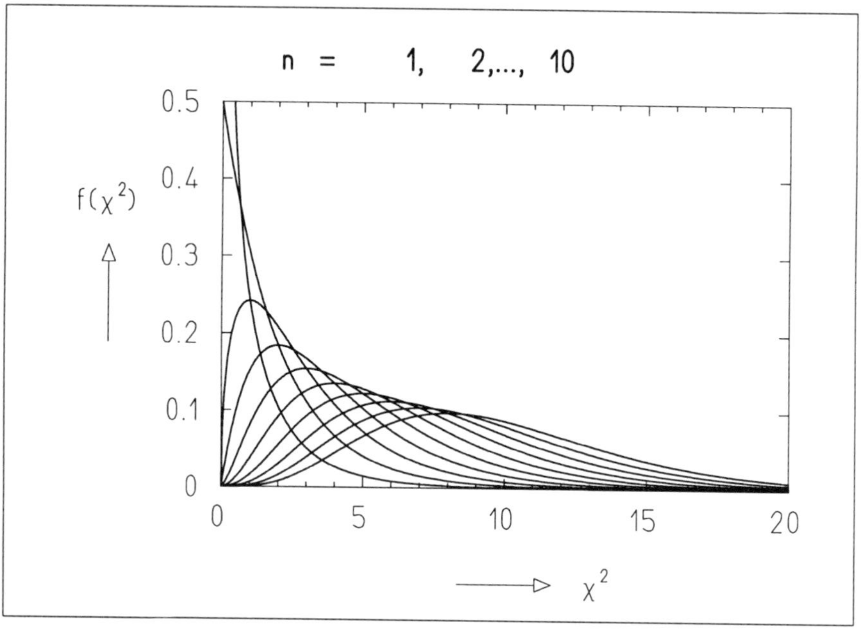

Fig. 6.8. Probability density of χ^2 for the number of degrees of freedom $n = 1, 2, \ldots, 10$. The expectation value $E(\chi^2) = n$ moves to the right as n increases.

$$
\begin{aligned}
E(\mathrm{x}^2) &= -\mathrm{i}\varphi'(0) = 2\lambda \quad , \\
E(\mathrm{x}^2) &= n
\end{aligned}
\tag{6.6.14}
$$

and

$$
\begin{aligned}
E\{(\mathrm{x}^2)^2\} &= -\varphi''(0) = 4\lambda^2 + 4\lambda \quad , \\
\sigma^2(\mathrm{x}^2) &= E\{(\mathrm{x}^2)^2\} - \{E(\mathrm{x}^2)\}^2 = 4\lambda \quad , \\
\sigma^2(\mathrm{x}^2) &= 2n \quad .
\end{aligned}
\tag{6.6.15}
$$

The expectation value of the χ^2-distribution is thus equal to the number of degrees of freedom, and the variance is two times larger. Figure 6.8 shows the probability density of the χ^2 distribution for various values of n. One sees [as can be directly seen also from (6.6.10)] that for $\chi^2 = 0$, the function diverges for $n = 1$, is equal to $1/2$ for $n = 2$, and vanishes for $n \geq 3$. A table of the χ^2-distribution is provided in the appendix (Table I.6).

The χ^2-distribution is of great significance in many applications, where the quantity χ^2 is used as a measure of confidence in a certain result. The smaller the value of χ^2, the greater is the confidence in the result. (After all, χ^2 was defined as the sum of squares of deviations of elements of a sample from the population mean. See Sect. 8.7.) The distribution function

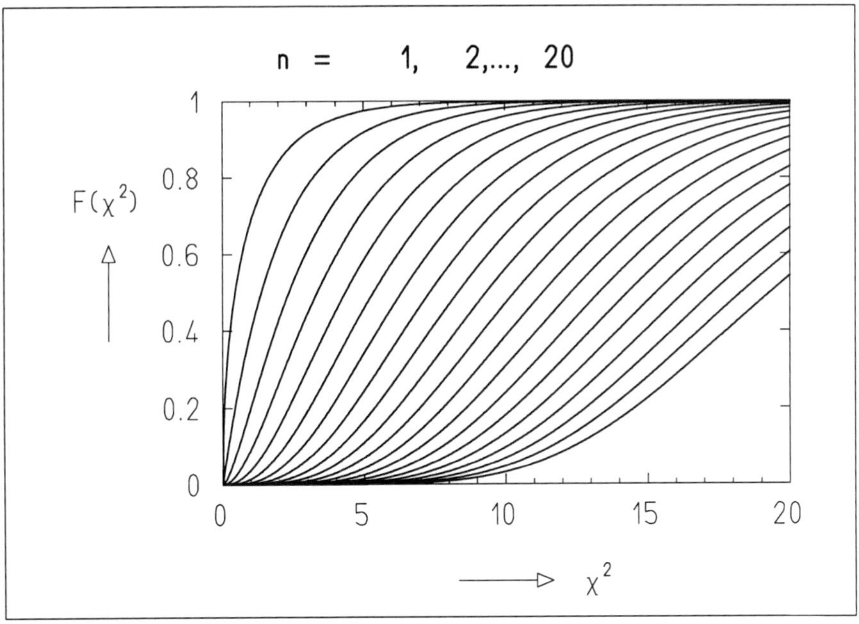

Fig. 6.9. Distribution function for χ^2 for the number of degrees of freedom $n = 1, 2, \ldots, 20$. The function for $n = 1$ is the curve at the far left, and the function for $n = 20$ is at the far right.

$$F(\chi^2) = P(\mathrm{x}^2 < \chi^2) \tag{6.6.16}$$

gives the probability that the random variable x^2 is not larger than χ^2. In practice, one frequently uses the quantity

$$W(\chi^2) = 1 - F(\chi^2) \tag{6.6.17}$$

as a measure of confidence in a result. $W(\chi^2)$ is often called the *confidence level*. $W(\chi^2)$ is large for small values of χ^2 and falls with increasing χ^2. The distribution function (6.6.16) is shown in Fig. 6.9 for various numbers of degrees of freedom n. The inverse function, which gives the quantiles of the χ^2 distribution,

$$\chi^2_F = \chi^2(F) = \chi^2(1 - W) \tag{6.6.18}$$

is used especially often in "hypothesis testing" (see Sect. 8.7). It is tabulated in the appendix (Table I.7).

Up to now we have restricted ourselves to the case where the population is described by the standard normal distribution. Usually, however, one has a normal distribution in general form with mean a and variance σ^2. Then the sum of squares (6.6.6) is clearly no longer distributed according to the

χ^2-distribution. One immediately obtains, however, a χ^2-distribution by considering the quantity

$$\mathrm{x}^2 = \frac{(\mathrm{x}_1 - a)^2 + (\mathrm{x}_2 - a)^2 + \cdots + (\mathrm{x}_n - a)^2}{\sigma^2} \quad . \tag{6.6.19}$$

This result follows directly from Eq. (5.8.4).

If the expectation values a_i and variances σ_i of the individual variables are different, then one has

$$\mathrm{x}^2 = \frac{(\mathrm{x}_1 - a_1)^2}{\sigma_1^2} + \frac{(\mathrm{x}_2 - a_2)^2}{\sigma_2^2} + \cdots + \frac{(\mathrm{x}_n - a_n)^2}{\sigma_n^2} \quad . \tag{6.6.20}$$

Finally if the n variables are not independent, but are described by a joint normal distribution (5.10.1) with the expectation values given by the vector **a** and the covariance matrix $C = B^{-1}$, then one has

$$\mathrm{x}^2 = (\mathbf{x} - \mathbf{a})^{\mathrm{T}} B (\mathbf{x} - \mathbf{a}) \quad . \tag{6.6.21}$$

6.7 χ^2 and Empirical Variance

In Eq. (6.2.6) we found that

$$\mathrm{s}^2 = \frac{1}{n-1} \sum_{i=1}^{n} (\mathrm{x}_i - \bar{\mathrm{x}})^2 \tag{6.7.1}$$

is a consistent, unbiased estimator for the variance σ^2 of a population. Let the x_i be independent and normally distributed with standard deviation σ. We want to show that the quantity

$$\frac{n-1}{\sigma^2} \mathrm{s}^2 \tag{6.7.2}$$

follows the χ^2-distribution with $f = n - 1$ degrees of freedom. We first carry out an orthogonal transformation of the n variables x_i (see Sect. 3.8):

$$\begin{aligned}
\mathrm{y}_1 &= \frac{1}{\sqrt{1 \cdot 2}} (\mathrm{x}_1 - \mathrm{x}_2) \quad , \\
\mathrm{y}_2 &= \frac{1}{\sqrt{2 \cdot 3}} (\mathrm{x}_1 + \mathrm{x}_2 - 2\mathrm{x}_3) \quad , \\
\mathrm{y}_3 &= \frac{1}{\sqrt{3 \cdot 4}} (\mathrm{x}_1 + \mathrm{x}_2 + \mathrm{x}_3 - 3\mathrm{x}_4) \quad , \\
&\vdots \\
\mathrm{y}_{n-1} &= \frac{1}{\sqrt{(n-1)n}} (\mathrm{x}_1 + \mathrm{x}_2 + \cdots + \mathrm{x}_{n-1} - (n-1)\mathrm{x}_n) \quad , \\
\mathrm{y}_n &= \frac{1}{\sqrt{n}} (\mathrm{x}_1 + \mathrm{x}_2 + \cdots + \mathrm{x}_n) = \sqrt{n}\bar{\mathrm{x}} \quad .
\end{aligned} \tag{6.7.3}$$

One can verify that this transformation is in fact orthogonal, i.e., that

$$\sum_{i=1}^{n} x_i^2 = \sum_{i=1}^{n} y_i^2 \quad . \tag{6.7.4}$$

Since a sum or difference of independent normally distributed quantities is again itself normally distributed, all of the y_i are normally distributed. The factors in (6.7.3) ensure that the y_i have mean values of zero and standard deviations σ.

From (6.7.1) and (6.7.2) one then has

$$\begin{aligned}(n-1)s^2 &= \sum_{i=1}^{n}(x_i - \bar{x})^2 = \sum_{i=1}^{n} x_i^2 - 2\bar{x}\sum_{i=1}^{n} x_i + n\bar{x}^2 \\ &= \sum_{i=1}^{n} x_i^2 - n\bar{x}^2 = \sum_{i=1}^{n} y_i^2 - y_n^2 = \sum_{i=1}^{n-1} y_i^2 \quad .\end{aligned}$$

This expression is a sum of only $(n-1)$ independent squared terms. A comparison with (6.6.19) shows that the quantity (6.7.2) in fact follows a χ^2-distribution with $(n-1)$ degrees of freedom.

The squared terms $(x_i - \bar{x})^2$ are not linearly independent. One has the following relation between them:

$$\sum_{i=1}^{n}(x_i - \bar{x}) = 0 \quad .$$

One can show that every additional relation between the squared terms reduces the number of degrees of freedom by one. Later we will make frequent use of this result, which we only state here without proof.

6.8 Sampling by Counting. Small Samples

Samples are often obtained in the following way. One draws n elements from a population, checks if they possess a given characteristic and accepts only those k elements into the sample that have the characteristic. The remaining $n - k$ elements are rejected, i.e., their properties are not recorded. This approach thus becomes the counting of k out of n elements drawn.

This approach corresponds exactly to selecting a sample according to a binomial distribution. The parameters p and q of this distribution correspond then to the occurrence or non-occurrence of the property in question. As will be shown in Example 7.5,

$$S(p) = \frac{k}{n} \tag{6.8.1}$$

is the maximum likelihood estimator of the parameter p. The variance of S is

$$\sigma^2(\mathsf{S}(p)) = \frac{p(1-p)}{n} \quad . \tag{6.8.2}$$

By using (6.8.1) it can be estimated from the sample by

$$\mathsf{s}^2(\mathsf{S}(p)) = \frac{1}{n}\frac{\mathsf{k}}{n}\left(1 - \frac{\mathsf{k}}{n}\right) \quad . \tag{6.8.3}$$

We define the error $\Delta\mathsf{k}$ as

$$\Delta\mathsf{k} = \sqrt{[\mathsf{s}^2(\mathsf{S}(np))]} \quad . \tag{6.8.4}$$

By using (6.8.3) we obtain

$$\Delta\mathsf{k} = \sqrt{\left[\mathsf{k}\left(1 - \frac{\mathsf{k}}{n}\right)\right]} \quad . \tag{6.8.5}$$

The error $\Delta\mathsf{k}$ only depends on the number of elements counted and on the size of the sample. It is called the *statistical error*. A particularly important case is that of small k, or more precisely, the case $\mathsf{k} \ll n$. In this limit we can define $\lambda = np$ and following Sect. 5.4 consider the counted number k as a single element of a sample taken from a Poisson distributed population with parameter λ. From (6.8.1) and (6.8.5) we obtain

$$\mathsf{S}(\lambda) = \mathsf{S}(np) = \mathsf{k} \quad , \tag{6.8.6}$$

$$\Delta\lambda = \sqrt{\mathsf{k}} \quad . \tag{6.8.7}$$

(This can be derived by using the result of Example 7.4 with $N = 1$.) The result (6.8.7) is often written in an actually incorrect but easy to remember form,

$$\Delta\mathsf{k} = \sqrt{\mathsf{k}} \quad ,$$

which is read: The *statistical error* of the *counted number* k is $\sqrt{\mathsf{k}}$.

In order to interpret the statistical error $\Delta\lambda = \sqrt{\mathsf{k}}$ we must examine the Poisson distribution somewhat more closely. Let us begin with the case where k is not too small (say, $\mathsf{k} > 20$). For large values of λ the Poisson distribution becomes a Gaussian distribution with mean λ and variance $\sigma^2 = \lambda$. This can be seen qualitatively from Fig. 5.6. As long as k is not too small, i.e., $\mathsf{k} \gg 1$, we can then treat the Poisson distribution in k with parameter λ as a normal distribution in x with mean λ and variance $\sigma^2 = \lambda$. The discrete variable k is then replaced by the continuous variable x. The probability density of x is

$$f(x;\lambda) = \frac{1}{\sigma\sqrt{2\pi}}\exp\left\{-\frac{(x-\lambda)^2}{2\sigma^2}\right\} = \frac{1}{\sqrt{2\pi\lambda}}\exp\left\{-\frac{(x-\lambda)^2}{2\lambda}\right\} \quad . \tag{6.8.8}$$

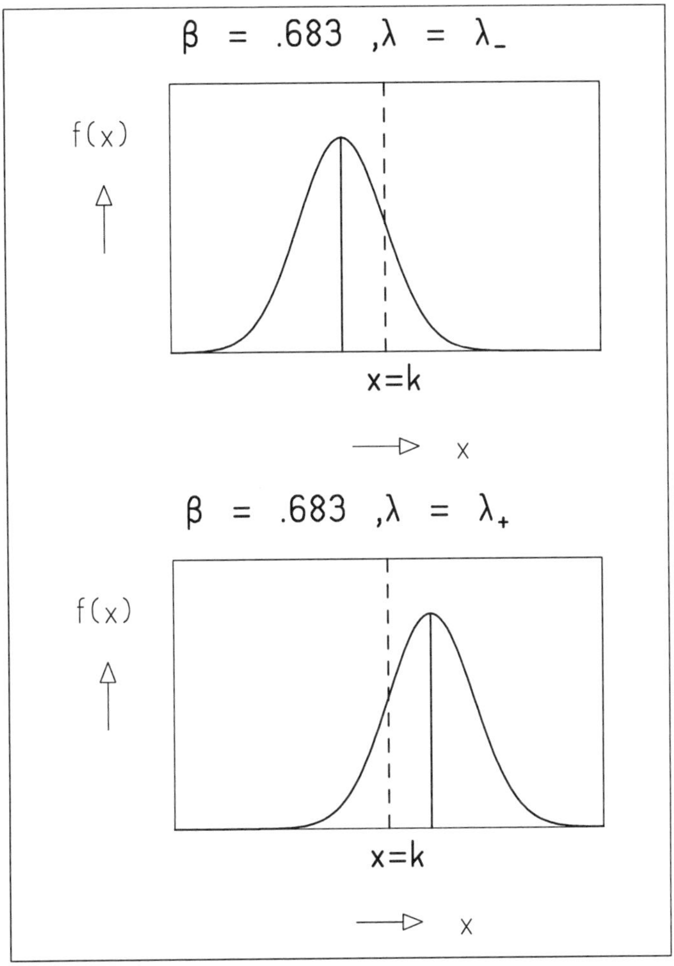

Fig. 6.10. Normal distribution with mean λ and standard deviation σ for $\lambda = \lambda_-$ and $\lambda = \lambda_+$.

The observation of k events corresponds to observing once the value of the random variable $\mathsf{x} = \mathsf{k}$.

With the help of the probability density (6.8.8) we now want to determine the *confidence limits* at a given *confidence level* $\beta = 1 - \alpha$ in such a way that

$$P(\lambda_- \leq \lambda \leq \lambda_+) = 1 - \alpha \quad . \tag{6.8.9}$$

That is, one requires that the probability that the true value of λ is contained within the confidence limits λ_- and λ_+ be equal to the confidence level $1 - \alpha$. The limiting cases $\lambda = \lambda_-$ and $\lambda = \lambda_+$ are depicted in Fig. 6.10. They are determined such that

$$P(x > \mathsf{k}|\lambda = \lambda_+) = 1 - \alpha/2 \ , \qquad P(x < \mathsf{k}|\lambda = \lambda_-) = 1 - \alpha/2 \ . \tag{6.8.10}$$

One clearly has

$$\begin{aligned}\alpha/2 &= \int_{x=-\infty}^{x=\mathsf{k}} f(x;\lambda_+)\,\mathrm{d}x = \frac{1}{\sigma\sqrt{2\pi}}\int_{-\infty}^{x=\mathsf{k}} \exp\left\{-\frac{(x-\lambda_+)^2}{2\sigma^2}\right\} \\ &= \int_{u=-\infty}^{u=(\mathsf{k}-\lambda_+)/\sigma} \phi_0(u)\,\mathrm{d}u = \psi_0\left(\frac{\mathsf{k}-\lambda_+}{\sigma}\right) \quad , \end{aligned} \qquad (6.8.11)$$

and correspondingly

$$1-\alpha/2 = \int_{-\infty}^{x=\mathsf{k}} f(x;\lambda_-)\,\mathrm{d}x = \psi_0\left(\frac{\mathsf{k}-\lambda_-}{\sigma}\right) \quad . \qquad (6.8.12)$$

Here ϕ_0 and ψ_0 are the probability density and distribution function of the standard normal distribution introduced in Sect. 5.8. By using the inverse function Ω of the distribution function ψ_0 [see Eq. (5.8.8)], one obtains

$$\frac{\mathsf{k}-\lambda_-}{\sigma} = \Omega(1-\alpha/2) \quad , \qquad \frac{\mathsf{k}-\lambda_+}{\sigma} = \Omega(\alpha/2) \quad . \qquad (6.8.13)$$

Because of (5.8.10) one has $\Omega(1-\alpha/2) = \Omega'(1-\alpha)$ and because of the symmetry of the function Ω, $\Omega(1-\alpha/2) = -\Omega(\alpha/2)$. Further, since $\alpha < 1$, one has $\Omega(1-\alpha/2) > 0$, $\Omega(\alpha/2) < 0$. From this we finally obtain

$$\lambda_- = \mathsf{k} - \sigma\Omega'(1-\alpha) \quad , \qquad \lambda_+ = \mathsf{k} + \sigma\Omega'(1-\alpha) \quad . \qquad (6.8.14)$$

According to (6.8.6), k is the best estimator for λ. Since $\sigma^2 = \lambda$, the best estimator for σ is given by $\mathsf{s} = \sqrt{\mathsf{k}}$. Since we have assumed that $\mathsf{k} \gg 1$, the uncertainty in s is significantly smaller than the uncertainty in k. We can therefore substitute $x = k$ and $\mathsf{s} = \sqrt{\mathsf{k}}$ in (6.8.9) and obtain for the confidence interval with confidence level $1-\alpha$

$$\lambda_- = \mathsf{k} - \sqrt{\mathsf{k}}\Omega'(1-\alpha) \le \lambda \le \mathsf{k} + \sqrt{\mathsf{k}}\Omega'(1-\alpha) = \lambda_+ \quad . \qquad (6.8.15)$$

For $1-\alpha = 68.3\,\%$ we find from Sect. 5.8 or Table I.5 that $\Omega'(\alpha) = 1$. What is usually reported,

$$\lambda = \mathsf{k} \pm \sqrt{\mathsf{k}} \quad ,$$

which was already the result of (6.8.6) and (6.8.7), thus gives the confidence limits at the confidence level of 68.3 %, but only for the case $\mathsf{k} \gg 1$. For the confidence level of 90 %, i.e., for $\alpha = 0.1$, we find $\Omega'(0.1) = 1.65$ and for the confidence level 99 % one has $\Omega'(0.01) = 2.57$.

For very small values of k, one can no longer replace the Poisson distribution with the normal distribution. We follow therefore reference [25]. We start again from Eq. (6.8.10), but use, instead of the probability density (6.8.8) for the continuous random variable x with fixed parameter λ, the Poisson probability for observing the discrete random variable n for a given λ,

$$f(n;\lambda) = \frac{\lambda^n}{n!}\mathrm{e}^{-\lambda} \quad . \qquad (6.8.16)$$

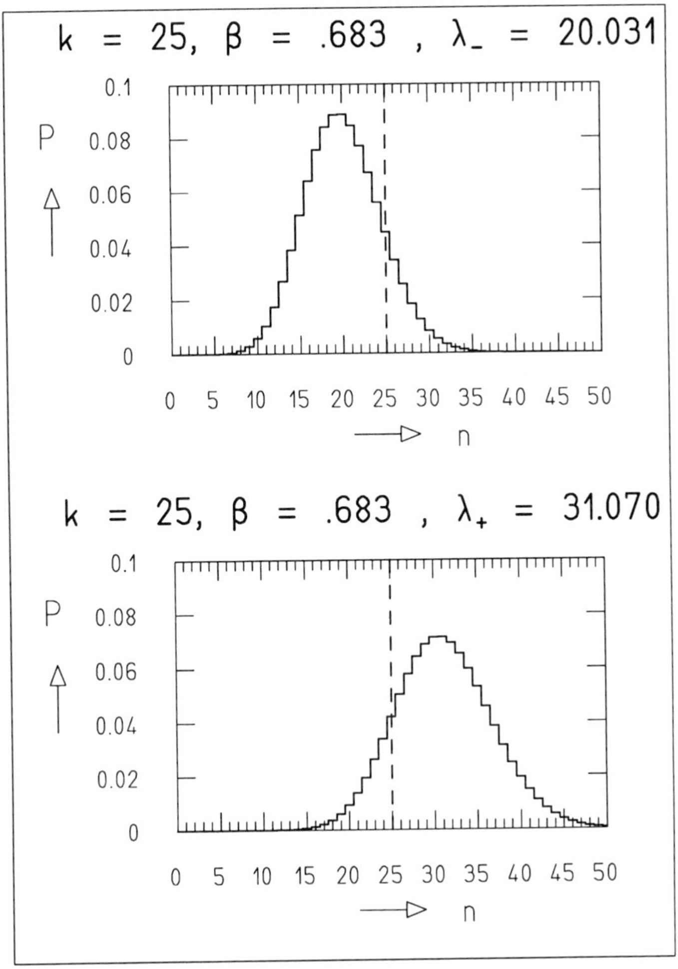

Fig. 6.11. Poisson distribution with parameter λ for $\lambda = \lambda_-$ and $\lambda = \lambda_+$.

For the observation k we now determine the confidence limits λ_- and λ_+, which fulfill (6.8.10) with $x = n$ (Fig. 6.11) and obtain in analogy to (6.8.11) and (6.8.12)

$$1 - \alpha/2 = \sum_{n=\mathsf{k}+1}^{\infty} f(n;\lambda_+) = 1 - \sum_{n=0}^{\mathsf{k}} f(n;\lambda_+) = 1 - F(\mathsf{k}+1;\lambda_+) \quad ,$$

$$1 - \alpha/2 = \sum_{n=0}^{\mathsf{k}-1} f(n;\lambda_-) = F(\mathsf{k};\lambda_-)$$

or

$$\alpha/2 = F(\mathsf{k}+1;\lambda_+) \quad , \tag{6.8.17}$$
$$1-\alpha/2 = F(\mathsf{k};\lambda_-) \quad . \tag{6.8.18}$$

Here,

$$F(k;\lambda) = \sum_{n=0}^{k-1} f(n;\lambda) = P(\mathsf{k} < k)$$

is the distribution function of the Poisson distribution.

In order to obtain numerical values for the confidence limits λ_+ and λ_-, we solve Eqs. (6.8.17) and (6.8.18). That is, we must construct the inverse function of the Poisson distribution for fixed k and given probability P (in our case $\alpha/2$ and $1-\alpha/2$),

$$\lambda = \lambda_P(k) \quad . \tag{6.8.19}$$

This is done numerically with the Program C.7. The function (6.8.19) is given in Table I.1 for frequently occurring values of P.

For extremely small samples one is often only interested in an *upper confidence limit* at confidence level $\beta = 1-\alpha$. This is obtained by requiring

$$P(n > \mathsf{k}|\lambda = \lambda^{(\mathrm{up})}) = \beta = 1-\alpha \tag{6.8.20}$$

instead of (6.8.10). Thus one has

$$\alpha = \sum_{n=0}^{\mathsf{k}} f(n;\lambda^{(\mathrm{up})}) = F(\mathsf{k}+1;\lambda^{(\mathrm{up})}) \quad . \tag{6.8.21}$$

For the extreme case $\mathsf{k} = 0$, i.e., for a sample in which no event was observed, one obtains the upper limit $\lambda^{(\mathrm{up})}$ by inverting $\alpha = F(1;\lambda^{(\mathrm{up})})$. The upper limit then has the following meaning. If the true value of the parameter were in fact $\lambda = \lambda^{(\mathrm{up})}$ and if one were to repeat the experiment many times, then the probability of observing at least one event is β. The observation $\mathsf{k} = 0$ is then expressed in the following way: One has $\lambda < \lambda^{(\mathrm{up})}$ with a confidence level of $1-\alpha$. From Table I.1 one finds that $k = 0$ corresponds to $\lambda < 2.996 \approx 3$ at a confidence level of 95 %.

Example 6.7: Determination of a lower limit for the lifetime of the proton from the observation of no decays

As already mentioned, the probability for the decay of a radioactive nucleus with the time t is

$$P(t) = \frac{1}{\tau}\int_0^t \mathrm{e}^{-x/\tau}\,\mathrm{d}x \quad .$$

Here τ is the mean lifetime of the nucleus. For $t \ll \tau$ the expression simplifies to

$$P(t) = t/\tau \quad .$$

For a total of N nuclei one expects that

$$k = NP(t) = N \cdot t/\tau$$

nuclei will decay within the time t. The mean lifetime τ is obtained by counting the number of such decays. If one observes k decays from a total of N nuclei in a time t, then one obtains as the measured value of τ

$$\tilde{\tau} = \frac{N}{\mathsf{k}} t \quad .$$

Of particular interest is the mean lifetime of the proton, one of the primary building blocks of matter. In experiments recently carried out with great effort, one observes large numbers of protons with detectors capable of detecting each individual decay. Up to now, not a single decay has been seen. According to Table I, the true expected number of decays λ does not exceed three (at a confidence level of 95 %). One has therefore

$$\tau > \frac{N}{3} t$$

at this confidence level. Typical experimental values are $t = 0.3$ years, $N = 10^{33}$, i.e.,

$$\tau > 10^{32} \text{ years} \quad .$$

The proton can therefore be considered as stable even over cosmological time scales, if one considers that the age of the universe is estimated to be only around 10^{10} years. ■

6.9 Small Samples with Background

In many experiments one is faced with the following situation. For the detected events one cannot determine whether they belong to the type that is actually of interest (*signal events*) or to another type (*background events*). For the expected number of events in the experiment one then has a Poisson distribution with the parameter $\lambda = \lambda_S + \lambda_B$. Here λ_S is the sought after parameter of the number of signal events, and λ_B is the parameter for the background events, which must of course be known if one wants to obtain information about λ_S. (In an experiment as in Example 6.7, one might have, for example, an admixture of radioactive nuclei whose decays cannot be distinguished from those of the proton. If the number of such nuclei and their lifetime is known, then λ_B can be computed.)

We are now tempted to simply take the results of the last section, to determine the confidence limits $\lambda_\pm$ and the upper limit $\lambda^{(\mathrm{up})}$ and to set $\lambda_{S\pm} =$

$\lambda_\pm - \lambda_B$, $\lambda_S^{(\mathrm{up})} = \lambda^{(\mathrm{up})} - \lambda_B$. This procedure can, however, lead to nonsensical results. (As seen in Example 6.7, one has $\lambda^{(\mathrm{up})} = 3$ at a confidence level of 95 %, for $k = 0$. For $\lambda_B = 4$, $k = 0$ we would obtain $\lambda_S^{(\mathrm{up})} = -1$, although a value $\lambda_S < 0$ has no meaning.)

The considerations up to now are based on the following. The probability for observing n events, $n = n_S + n_B$, is

$$f(n; \lambda_S + \lambda_B) = \frac{1}{n!} \mathrm{e}^{-(\lambda_S+\lambda_B)} (\lambda_S + \lambda_B)^n \quad , \tag{6.9.1}$$

and the probabilities to observe n_S signal events, and n_B background events are

$$f(n_S; \lambda_S) = \frac{1}{n_S!} \mathrm{e}^{-\lambda_S} \lambda_S^{n_S} \quad , \tag{6.9.2}$$

$$f(n_B; \lambda_B) = \frac{1}{n_B!} \mathrm{e}^{-\lambda_B} \lambda_B^{n_B} \quad . \tag{6.9.3}$$

The validity of (6.9.1) was shown in Example 5.5 with the help of the characteristic function, starting from the independence of the two Poisson distributions (6.9.2) and (6.9.3). One can also obtain them directly by summation of all products of the probabilities (6.9.2) and (6.9.3) that lead to $n = n_S + n_B$, by application of (B.4) and (B.6),

$$\begin{aligned}
&\sum_{n_S=0}^{n} f(n_S; \lambda_S) f(n - n_S; \lambda_B) \\
&= \mathrm{e}^{-(\lambda_S+\lambda_B)} \sum_{n_S=0}^{n} \frac{1}{n_S!(n-n_S)!} \lambda_S^{n_S} \lambda_B^{n-n_S} \\
&= \frac{1}{n!} \mathrm{e}^{-(\lambda_S+\lambda_B)} \sum_{n_S=0}^{n} \binom{n}{n_S} \lambda_S^{n_S} \lambda_B^{n-n_S} \\
&= \frac{1}{n!} \mathrm{e}^{-(\lambda_S+\lambda_B)} (\lambda_S + \lambda_B)^n \\
&= f(n; \lambda_S + \lambda_B) \quad .
\end{aligned}$$

The difficulties explained above are overcome with a method developed by ZECH [26]. In an experiment in which k events are recorded, the number of background events cannot simply be given by (6.9.3), since from the result of the experiment it is known that $n_B \le \mathsf{k}$. One must therefore replace (6.9.3) by

$$f'(n_B; \lambda) = f(n_B; \lambda_B) \Big/ \sum_{n_B=0}^{\mathsf{k}} f(n_B; \lambda_B) \quad , \qquad n_B \le \mathsf{k} \quad . \tag{6.9.4}$$

This distribution is normalized to unity in the region $0 \le n_B \le k$. In a corresponding way the distribution

$$f'(n;\lambda_S+\lambda_B) = f(n;\lambda_S+\lambda_B) \Big/ \sum_{n_B=0}^{\mathsf{k}} f(n_B;\lambda_B) \tag{6.9.5}$$

takes the place of (6.9.1).

In this way one obtains in analogy to (6.8.17) and (6.8.18) for the limits of the confidence interval $\lambda_{S-} \le \lambda_S \le \lambda_{S+}$ at a confidence level of $1-\alpha$

$$\alpha/2 = F'(\mathsf{k}+1, \lambda_{S+}+\lambda_B) \quad , \tag{6.9.6}$$

$$1-\alpha/2 = F'(\mathsf{k}, \lambda_{S-}+\lambda_B) \quad . \tag{6.9.7}$$

Here,

$$F'(\mathsf{k};\lambda_S+\lambda_B) = \sum_{n=0}^{\mathsf{k}-1} f'(n;\lambda_S+\lambda_B) = P(\mathsf{k} < k) \tag{6.9.8}$$

is the distribution function of the renormalized distribution (6.9.4). If only an upper limit at confidence level $1-\alpha$ is desired, then one clearly has in analogy to (6.8.21)

$$\alpha = F'(\mathsf{k}+1, \lambda_S^{(\mathrm{up})}+\lambda_B) \quad . \tag{6.9.9}$$

Table 6.3 gives some numerical values computed with the program SMERSS given below. Note that for $\mathsf{k}=0$, Eq. (6.9.7) has no meaning, so that λ_{S-} cannot be defined. In this case Eq. (6.9.6) and also λ_{S+} are not meaningful. (In the table, however, the values for λ_{S-} and λ_{S+} are shown as computed by the program. This sets $\lambda_{S-}=0$ and computes λ_{S+} according to (6.9.6).)

Program 6.6: Subroutine SMERSS computes the confidence limits of a Poisson parameter determined from a small sample, including consideration of possible background

CALLING SEQUENCE: CALL SMERSS(K,CONFID,ALBACK,ALSMIN, ALSPLS,ALSUPR)

INPUT ARGUMENTS: K — Size of the sample k
CONFID — Confidence level $\beta = 1-\alpha$
ALBACK — Poisson parameter of the background λ_B

OUTPUT ARGUMENTS: ALSMIN, ALSPLS — Confidence limits λ_{S-}, λ_{S+}
ALSUPR — Upper limit $\lambda_S^{(\mathrm{up})}$

Table 6.3. Limits λ_{S-} and λ_{S+} of the confidence interval and upper confidence limits $\lambda_S^{(\mathrm{up})}$, for various values of λ_B and various very small sample sizes k for a fixed confidence level of 90 %.

$\beta = 0.90$				
k	λ_B	λ_{S-}	λ_{S+}	$\lambda_S^{(\mathrm{up})}$
0	0.0	0.000	2.996	2.303
0	1.0	0.000	2.996	2.303
0	2.0	0.000	2.996	2.303
1	0.0	0.051	4.744	3.890
1	1.0	0.051	4.113	3.272
1	2.0	0.051	3.816	2.995
2	0.0	0.355	6.296	5.322
2	1.0	0.100	5.410	4.443
2	2.0	0.076	4.824	3.877
3	0.0	0.818	7.754	6.681
3	1.0	0.226	6.782	5.711
3	2.0	0.125	5.983	4.926
4	0.0	1.366	9.154	7.994
4	1.0	0.519	8.159	7.000
4	2.0	0.226	7.241	6.087
5	0.0	1.970	10.513	9.275
5	1.0	1.009	9.514	8.276
5	2.0	0.433	8.542	7.306

```
      SUBROUTINE SMERSS(K,CONFID,ALBACK,ALSMIN,ALSPLS,ALSUPR)
      IMPLICIT DOUBLE PRECISION (A-H,O-Z)
      PARAMETER(ZERO=0.D0,ONE=1.D0,HALF=0.5D0,EPSILN=1.D-6)
      COMMON /CMSMER/ CALPHA,CADNOM
      EXTERNAL SMZRSS
      ALPHA=ONE-CONFID
      IF(ALBACK.LE.ZERO) CADNOM=ONE
C compute lower limit ALSMIN of confidence region
      IF(K.EQ.0) THEN
        ALSMIN=ZERO
      ELSE
        KK=K
        IF(ALBACK.GT.ZERO) CADNOM=SCPOIS(KK,ALBACK)
        CALPHA=ONE-HALF*ALPHA
        ALS0=ZERO
        ALS1=ONE
        IF(K.GT.0)ALS1=DBLE(K)
        CALL AUXZBR(ALS0,ALS1,SMZRSS,ALBACK,KK,NDUM)
        CALL AUXZFN(ALS0,ALS1,ALSMIN,SMZRSS,ALBACK,KK,NDUM,EPSILN)
      END IF
      CALPHA=HALF*ALPHA
```

```
C compute upper limit ALSPLS of confidence region
      KK=K+1
      IF(ALBACK.GT.ZERO) CADNOM=SCPOIS(KK,ALBACK)
      ALS0=ZERO
      ALS1=DBLE(K)
      CALL AUXZBR(ALS0,ALS1,SMZRSS,ALBACK,KK,NDUM)
      CALL AUXZFN(ALS0,ALS1,ALSPLS,SMZRSS,ALBACK,KK,NDUM,EPSILN)
C compute upper limit ALSUPR of parameter
      CALPHA=ALPHA
      KK=K+1
      ALS0=ZERO
      ALS1=DBLE(K)
      CALL AUXZBR(ALS0,ALS1,SMZRSS,ALBACK,KK,NDUM)
      CALL AUXZFN(ALS0,ALS1,ALSUPR,SMZRSS,ALBACK,KK,NDUM,EPSILN)
      END
C------------------------------------------------------------------
      DOUBLE PRECISION FUNCTION SMZRSS(ALS,ALBACK,K,NDUM)
      IMPLICIT DOUBLE PRECISION (A-H,O-Z)
      COMMON /CMSMER/ CALPHA,CADNOM
      SMZRSS=CALPHA-SCPOIS(K,ALS+ALBACK)/CADNOM
      END
```

6.10 Determining a Ratio of Small Numbers of Events

Often a number of *signal* events k is measured and compared to a number of *reference* events d. One is interested in the true value r of the ratio of the number of signal events to the number of reference events, or more precisely, the ratio of the probability to observe a signal event to that of a reference event. As an estimator for this ratio one clearly uses

$$\tilde{r} = \mathsf{k}/\mathsf{d} \quad .$$

We now ask for the confidence limits of r. If k and d are sufficiently large, then they may be approximated as Gaussian variables with standard deviations $\sigma_k = \sqrt{\mathsf{k}}$, $\sigma_d = \sqrt{\mathsf{d}}$. Then according to the law of error propagation one has

$$\Delta r = \sqrt{\left(\frac{\partial r}{\partial \mathsf{k}}\right)^2 \mathsf{k} + \left(\frac{\partial r}{\partial \mathsf{d}}\right)^2 \mathsf{d}} = r\sqrt{\frac{1}{\mathsf{k}} + \frac{1}{\mathsf{d}}} \quad . \tag{6.10.1}$$

If in addition one has $\mathsf{d} \gg \mathsf{k}$, then Δr is simply

$$\Delta r = \frac{r}{\sqrt{\mathsf{k}}} \quad . \tag{6.10.2}$$

If the requirements for the validity of (6.10.1) or even of (6.10.2) are not fulfilled, i.e., if k and d are small numbers, then one must use considerations developed by James and Roos [23]. Clearly when one observes an individual event, the probability that it is a signal event is given by

$$p = \frac{r}{1+r} \quad , \tag{6.10.3}$$

and the probability that it is a reference event is

$$q = 1 - p = \frac{1}{1+r} \quad . \tag{6.10.4}$$

In an experiment in which a total of $\mathsf{N} = \mathsf{k} + \mathsf{d}$ are observed, the probability that exactly n signal events are present is given by a binomial distribution (5.1.3). This is

$$f(n;r) = \binom{\mathsf{N}}{n} p^n q^{N-n} = \binom{\mathsf{N}}{n} \left(\frac{r}{1+r}\right)^n \left(\frac{1}{1+r}\right)^{\mathsf{N}-n} \quad . \tag{6.10.5}$$

The probability to have $n < \mathsf{k}$ is then

$$P(n < \mathsf{k}) = \sum_{n=0}^{\mathsf{k}-1} f(n;r) = F(\mathsf{k};r) \tag{6.10.6}$$

with

$$F(\mathsf{k};r) = \sum_{n=0}^{\mathsf{k}-1} \binom{\mathsf{N}}{n} \left(\frac{r}{1+r}\right)^n \left(\frac{1}{1+r}\right)^{\mathsf{N}-n} \quad , \tag{6.10.7}$$

i.e., the distribution function of the binomial distribution. To determine the limits r_- and r_+ of the confidence interval at the confidence level $\beta = 1 - \alpha$, we use in analogy to (6.8.17) and (6.8.18)

$$\alpha/2 = F(\mathsf{k}+1; r_+) \quad , \tag{6.10.8}$$
$$1 - \alpha/2 = F(\mathsf{k}; r_-) \quad . \tag{6.10.9}$$

If one only seeks an upper limit at the confidence level $\beta = 1 - \alpha$, it can be obtained from [see Eq. (6.8.21)]

$$\alpha = F(\mathsf{k}+1; r^{(\mathsf{up})}) \quad . \tag{6.10.10}$$

The quantities r_+, r_-, and $r^{(\mathsf{up})}$ can be computed for given values of k, d, and β with the program SMERQS given at the end of the following section.

6.11 Ratio of Small Numbers of Events with Background

By combining as done by SWARTZ [24] the ideas of Sects. 6.9 and 6.10, one can deal with the following situation. In an experiment one has three types of events: *signal*, *background*, and *reference*. Signal and background events cannot be distinguished from each other. Suppose in the experiment one has detected a total of k signal and background events and d reference events. Let us label with r_S and r_B the true values (in the sense of the definition at the beginning of the previous section) of the ratios of the numbers of signal to reference and background to reference events. Then the probabilities that a randomly selected event is signal or background, p_S and p_B, are

$$p_S = \frac{r_S}{1 + r_S + r_B} \quad , \qquad p_B = \frac{r_B}{1 + r_S + r_B} \quad . \tag{6.11.1}$$

The probability that it is a reference event is then

$$p_R = 1 - p_S - p_B = \frac{1}{1 + r_S + r_B} \quad . \tag{6.11.2}$$

If one has a total of $\mathsf{N} = \mathsf{k} + \mathsf{d}$ events in the experiment, then the individual probabilities that one has exactly n_S signal events, n_B background events, and $n_R = \mathsf{N} - n_S - n_B$ reference events are

$$f_S(n_S; p_S) = \binom{\mathsf{N}}{n_S} p_S^{n_S} (1 - p_S)^{\mathsf{N} - n_S} \quad , \tag{6.11.3}$$

$$f_B(n_B; p_B) = \binom{\mathsf{N}}{n_B} p_B^{n_B} (1 - p_B)^{\mathsf{N} - n_B} \quad , \tag{6.11.4}$$

$$f_R(n_R; p_R) = \binom{\mathsf{N}}{n_R} p_R^{n_R} (1 - p_R)^{\mathsf{N} - n_R} \quad . \tag{6.11.5}$$

Since there are now three mutually exclusive types of events, one has instead of a binomial distribution (6.10.5) a trinomial distribution, i.e., a multinomial distribution (5.1.10) with $\ell = 3$. The probability that in an experiment with a total of N events one has exactly n_S signal, n_B background, and $\mathsf{N} - n_S - n_B$ reference events is therefore

$$f(n_S, n_B; r_S, r_B) = \frac{\mathsf{N}!}{n_S! n_B! (\mathsf{N} - n_S - n_B)!} p_S^{n_S} p_B^{n_B} (1 - p_S - p_B)^{\mathsf{N} - n_S - n_B} \quad . \tag{6.11.6}$$

Here, however, we have not taken into consideration that the number of background events cannot be greater than k. In a manner similar to (6.9.4) one uses

$$f'_B(n_B; p_B) = f_B(n_B; p_B) \Big/ \sum_{n_B=0}^{\mathsf{k}} f(n_B; p_B) \quad , \qquad n_B \leq \mathsf{k} \quad , \tag{6.11.7}$$

in place of f_B. This replacement must also be made in (6.11.6), which gives

$$f'(n_S, n_B; r_S, r_B) = f(n_S, n_B; r_S, r_B) \Big/ \sum_{n_B=0}^{\mathsf{k}} f(n_B; p_B) \quad . \qquad (6.11.8)$$

The probability to have n_S signal events regardless of the number of background events is

$$f'(n_S; r_S, r_B) = \sum_{n_B=0}^{\mathsf{k}} f'(n_S, n_B; r_S, r_B)$$

and finally the probability to have $n_S \leq \mathsf{k}$ is

$$\begin{aligned} F'(\mathsf{k}; r_S, r_B) &= \sum_{n_S=0}^{\mathsf{k}-n_B-1} f'(n_S; r_S, r_B) \\ &= \frac{\sum\limits_{n_S=0}^{\mathsf{k}-n_B-1} \sum\limits_{n_B=0}^{\mathsf{k}-1} \frac{\mathsf{N}!}{n_S! n_B! (\mathsf{N} - n_S - n_B)!} p_S^{n_S} p_B^{n_B} (1 - p_S - p_B)^{\mathsf{N}-n_S-n_B}}{\sum\limits_{n_B=0}^{\mathsf{k}} \frac{\mathsf{N}!}{n_B!(N - n_B)!} p_B^{n_B} (1 - p_B)^{\mathsf{N}-n_B}} \quad . \end{aligned}$$

Since r_B was assumed to be known, the quantity F' for a given k depends only on r_S. Similar to (6.9.6) and (6.9.7) one can determine the limits r_{S+} and r_{S-} of the confidence region for r_S with confidence level $\beta = 1 - \alpha$ from the following requirement:

$$\begin{aligned} \alpha/2 &= F'(\mathsf{k} + 1; r_{S+}, r_B) \quad , & (6.11.9) \\ 1 - \alpha/2 &= F'(\mathsf{k}; r_{S-}, r_B) \quad . & (6.11.10) \end{aligned}$$

If one only wants an upper limit with confidence level $\beta = 1 - \alpha$, this can be found according to (6.9.9) from

$$\alpha = F'(\mathsf{k} + 1; r_S^{(\mathsf{up})}, r_B) \quad . \qquad (6.11.11)$$

Table 6.4 contains some numerical values computed with the program SMERQS given below. For $\mathsf{k} = 0$, however, (6.11.10) and hence also r_{S-} have no meaning. Similarly for (6.11.9), r_{S+} for $\mathsf{k} = 0$ is not meaningful. (In the table, however, values for r_{S-} and r_{S+} are given for $\mathsf{k} = 0$ as computed by the program. For $\mathsf{k} = 0$ this sets $r_{S-} = 0$ and determines r_{S+} according to (6.11.9).)

Program 6.7: Subroutine `SMERQS` computes the confidence limits of a ratio of signal to reference events from a small sample with the possible presence of background

CALLING SEQUENCE: `CALL SMERQS(K,ID,CONFID,RBACK,RSMIN, RSPLS,RSUPR)`

Table 6.4. Limits r_{S-} and r_{S+} of the confidence interval and upper confidence limit $r_S^{(\mathrm{up})}$ for various values of r_B and various very small values of k for a fixed number of reference events d and fixed confidence level of 90 %.

$\beta = 0.90$, d = 10				
k	r_B	r_{S-}	r_{S+}	$r_S^{(up)}$
0	0.0	0.000	0.349	0.259
0	0.1	0.000	0.349	0.259
0	0.2	0.000	0.349	0.259
1	0.0	0.005	0.573	0.450
1	0.1	0.005	0.502	0.382
1	0.2	0.005	0.464	0.348
2	0.0	0.034	0.780	0.627
2	0.1	0.010	0.686	0.535
2	0.2	0.007	0.613	0.467
3	0.0	0.077	0.979	0.799
3	0.1	0.020	0.880	0.701
3	0.2	0.012	0.788	0.612
4	0.0	0.127	1.174	0.968
4	0.1	0.044	1.074	0.869
4	0.2	0.019	0.976	0.771
5	0.0	0.180	1.367	1.135
5	0.1	0.085	1.267	1.035
5	0.2	0.034	1.167	0.936

INPUT ARGUMENTS: `K` — number k of observed signal and background events
`ID` — number of observed reference events d
`CONFID` — confidence level $\beta = 1 - \alpha$
`RBACK` — expected ratio of background to reference events r_B

OUTPUT ARGUMENTS: `RSMIN, RSPLS` — confidence limits r_{S-}, r_{S+}
`RSUPR` — upper limit $r_S^{(\mathrm{up})}$

```
      SUBROUTINE SMERQS(K,ID,CONFID,RBACK,RSMIN,RSPLS,RSUPR)
      IMPLICIT DOUBLE PRECISION (A-H,O-Z)
      PARAMETER(ZERO=0.D0,ONE=1.D0,HALF=0.5D0,EPSILN=1.D-6)
      COMMON /CMSMER/ CALPHA
      EXTERNAL SMZRQS
      ALPHA=ONE-CONFID
C compute lower limit RSMIN of confidence region
      IF(K.EQ.0) THEN
```

```
            RSMIN=ZERO
          ELSE
            KK=K
            CALPHA=ONE-HALF*ALPHA
            RS0=ZERO
            RS1=ONE
            IF(K.GT.0)RS1=DBLE(K)
            CALL AUXZBR(RS0,RS1,SMZRQS,RBACK,KK,ID)
            CALL AUXZFN(RS0,RS1,RSMIN,SMZRQS,RBACK,KK,ID,EPSILN)
          END IF
          CALPHA=HALF*ALPHA
C compute upper limit RSPLS of confidence region
          KK=K+1
          RS0=ZERO
          RS1=DBLE(K)
          CALL AUXZBR(RS0,RS1,SMZRQS,RBACK,KK,ID)
          CALL AUXZFN(RS0,RS1,RSPLS,SMZRQS,RBACK,KK,ID,EPSILN)
C compute upper limit RSUPR of of parameter
          CALPHA=ALPHA
          KK=K+1
          RS0=ZERO
          RS1=DBLE(K)
          CALL AUXZBR(RS0,RS1,SMZRQS,RBACK,KK,ID)
          CALL AUXZFN(RS0,RS1,RSUPR,SMZRQS,RBACK,KK,ID,EPSILN)
          END
C-----------------------------------------------------------------------
          DOUBLE PRECISION FUNCTION SMZRQS(RS,RB,K,ID)
          IMPLICIT DOUBLE PRECISION (A-H,O-Z)
          COMMON /CMSMER/ CALPHA
          PARAMETER(ZERO=0.D0,ONE=1.D0,EPSILN=1.D-6)
          N=K-1+ID
          DEN=ONE+RS+RB
          PS=RS/DEN
          PB=RB/DEN
          IF(PB.LT.EPSILN) THEN
            DEN=ONE
          ELSE
            DEN=SCBINM(K,N,PB)
          END IF
          G=GLNGAM(DBLE(N+1))
          F=ZERO
          DO 20 NB=0,K-1
            DO 10 NS=0,K-NB-1
              IF(NS.EQ.0) THEN
                ES=ONE
              ELSE
                ES=PS**NS
              END IF
              IF(NB.EQ.0) THEN
                EB=ONE
              ELSE
                IF(PB.LT.EPSILN) THEN
                  EB=ZERO
                ELSE
                  EB=PB**NB
                END IF
              END IF
              IF(N-NS-NB.EQ.0) THEN
                E=ONE
              ELSE
                E=(ONE-PS-PB)**(N-NS-NB)
```

```
      END IF
      F=F+EXP(G-GLNGAM(DBLE(NS+1))-GLNGAM(DBLE(NB+1))-
+           GLNGAM(DBLE(N-NS-NB+1)))*ES*EB*E
10    CONTINUE
20  CONTINUE
    SMZRQS=CALPHA-F/DEN
    END
```

6.12 Example Programs

The source files of the example programs are available in the subdirectory `\DATAN\EXASRC`.

6.12.1 Main Program E1SM to Demonstrate Subprogram SMMNVR

This short program generates 10 samples of size $N = 100$ taken from the standard normal distribution. It computes for each sample six quantities: mean, error of the mean, variance, error of the variance, standard deviation, and error of the standard deviation, and outputs these quantities in a single line.

6.12.2 Main Program E2SM to Demonstrate Subprograms SMHSIN, SMHSFL, and SMHSGR

The initialization, filling, and graphical display of a histogram are demonstrated. The histogram of a sample with N elements from a standard normal distribution is shown. First, numerical values for N, the lower limit x_0, the bin width Δx, and the number of bins n_x are asked for. Next, the histogram is initialized. Then in a loop the elements of the sample are generated and filled into the histogram.

Finally, the number of the graphics workstation is requested and the histogram is displayed.

6.12.3 Main Program E3SM to Demonstrate Subprogram SMSDGR

A scatter plot is created and later displayed graphically. The coordinates of the points making up the scatter plot are given as pairs of random numbers from a bivariate normal distribution (cf. Sect. 4.11). The program asks the user for the parameters of the normal distribution (means a_1, a_2, standard deviations σ_1, σ_2, correlation coefficient ρ) and for the number of random number pairs to be generated. It generates the pairs and prepares the caption and the labeling of axes and scales. It asks for the number of the graphics workstation and displays the plot (Fig. 6.12).

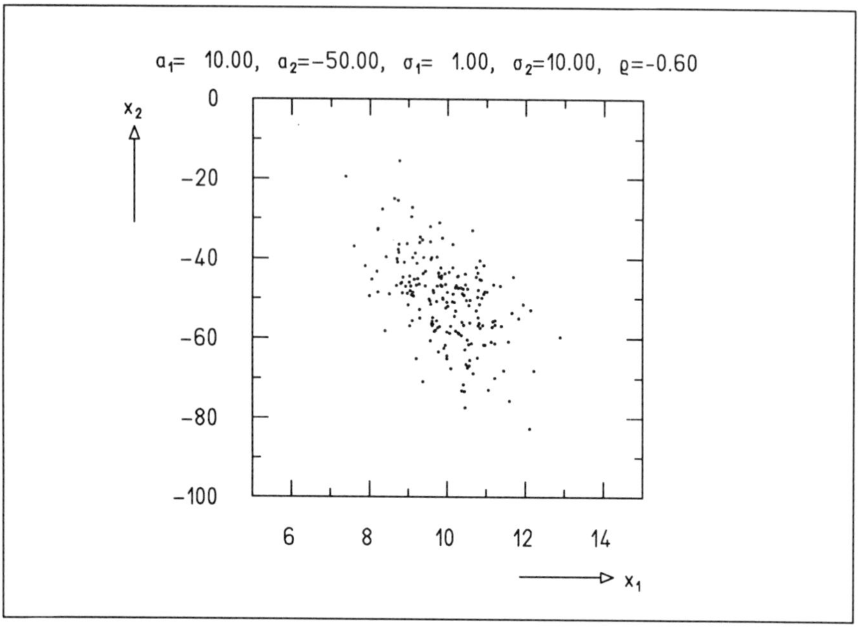

Fig. 6.12. Pairs of random numbers taken from a bivariate normal distribution.

6.12.4 Main Program `E4SM` to Demonstrate Subprogram `SMERSS`

The program computes the limits λ_{S-}, λ_{S+}, and $\lambda_S^{(\mathrm{up})}$ for the Poisson parameter of a signal. The user enters interactively the number of observed events k, the confidence level $\beta = 1 - \alpha$, and the Poisson parameter λ_B of the background.

Suggestions: (a) Verify a few lines from Table 6.3.
(b) Choose $\beta = 0.683$, $\lambda_B = 0$ and compare for different values k the values λ_{S-} and λ_{S+} with the naive statement $\lambda = \mathsf{k} \pm \sqrt{\mathsf{k}}$.

6.12.5 Main Program `E5SM` to Demonstrate Subprogram `SMERQS`

The program computes the limits r_{S-}, r_{S+}, and $r_S^{(\mathrm{up})}$ for the ratio r of the number of signal to reference events in the limit of a large number of events. More precisely, r is the ratio of the Poisson parameters λ_S to λ_R of the signal to reference events. The program first asks for the number k of observed (signal plus reference) events, for the number d of reference events, for the confidence level $\beta = 1-\alpha$, and for the expected ratio $r_B = \lambda_B/\lambda_R$ for background events.

Suggestion: Verify a few lines from Table 6.4.

6.12.6 Main Program E6SM to Simulate Experiments with Few Events and Background

A total of n_{exp} experiments are simulated. In each experiment N objects are analyzed. Each object yields with probability $p_S = \lambda_S/N$ a signal event and with probability $p_B = \lambda_B/N$ a background event. The numbers of events found in the simulated experiment are k_S, k_B, and $\mathsf{k} = \mathsf{k}_S + \mathsf{k}_B$. In the real experiment only k is known. The limits λ_{S-}, λ_{S+}, and $\lambda_S^{(\text{up})}$ for k are computed for a given confidence level $\beta = 1 - \alpha$ and a given value of λ_B and are displayed for each experiment.

Suggestion: Choose, e.g., $n_{\text{exp}} = 20$, $N = 1000$, $\lambda_S = 5$, $\lambda_B = 2$, $\beta = 0.9$ and find out whether, as expected, this simulation yields for 10 % of the experiments an interval $(\lambda_{S-}, \lambda_{S+})$, which does not contain the value λ_S. Keep in mind the meaning of the statistical error of your observation when using only 20 experiments.

6.12.7 Main Program E7SM to Simulate Experiments with Few Signal Events and with Reference Events

The program first asks for the quantities n_{exp}, N, λ_S, λ_B, λ_R, and the confidence level $\beta = 1 - \alpha$. It computes the probabilities $p_S = \lambda_S/N$, $p_B = \lambda_B/N$, $p_R = \lambda_R/N$, as well as the ratios $r_S = p_S/p_R$ and $r_B = p_B/p_R$ of a total of n_{exp} simulated experiments, in each of which N objects are analyzed. Each object is taken to be a signal event with probability p_S, a background event with probability p_B, and a reference event with probability p_R. (Here $p_S + p_B + p_R \ll 1$ is assumed.) The simulation yields the numbers k_S, k_B, and d for signal, background, and reference events, respectively. In a real experiment only the numbers $\mathsf{k} = \mathsf{k}_S + \mathsf{k}_B$ and d are known, since signal and background events cannot be distinguished. For the given values of β and r_B and for the quantities k and d which were found in the simulated experiments, the limits r_{S-}, r_{S+}, and $r_S^{(\text{up})}$ are computed and displayed.

Suggestion: Modify the suggestion accompanying Example Program 6.12.6 by choosing an additional input parameter $\lambda_S = 20$. Find out in how many cases the true value r_S used in the simulation is not contained in the interval (r_{S-}, r_{S+}).

6.13 Problems

6.13.1 Efficiency of Estimators

Let $\mathsf{x}_1, \mathsf{x}_2, \mathsf{x}_3$ be the elements of a sample from a continuous population with unknown mean $\widehat{x}$, but known variance σ^2.

(a) Show that the following quantities are unbiased estimators of $\widehat{x}$,

$$\begin{aligned} \mathsf{S}_1 &= \tfrac{1}{4}\mathsf{x}_1 + \tfrac{1}{4}\mathsf{x}_2 + \tfrac{1}{2}\mathsf{x}_3 \quad , \\ \mathsf{S}_2 &= \tfrac{1}{5}\mathsf{x}_1 + \tfrac{2}{5}\mathsf{x}_2 + \tfrac{2}{5}\mathsf{x}_3 \quad , \\ \mathsf{S}_3 &= \tfrac{1}{6}\mathsf{x}_1 + \tfrac{1}{3}\mathsf{x}_2 + \tfrac{1}{2}\mathsf{x}_3 \quad . \end{aligned}$$

Hint: It is simple to show that $\mathsf{S} = \sum_{i=1}^{n} a_i \mathsf{x}_i$ is unbiased if $\sum_{i=1}^{n} a_i = 1$ holds.

(b) Determine the variances $\sigma^2(\mathsf{S}_1)$, $\sigma^2(\mathsf{S}_2)$, $\sigma^2(\mathsf{S}_3)$ using (3.8.7) and the assumption that the elements $\mathsf{x}_1, \mathsf{x}_2, \mathsf{x}_3$ are independent.

(c) Show that the arithmetic mean $\bar{\mathsf{x}} = \frac{1}{3}\mathsf{x}_1 + \frac{1}{3}\mathsf{x}_2 + \frac{1}{3}\mathsf{x}_3$ has the smallest variance of all estimators of the type $\mathsf{S} = \sum_{i=1}^{3} a_i \mathsf{x}_i$ fulfilling the requirement $\sum_{i=1}^{3} a_i = 1$.
Hint: Minimize the variance of $\mathsf{S} = a_1\mathsf{x}_1 + a_2\mathsf{x}_2 + (1 - a_1 - a_2)\mathsf{x}_3$ with respect to a_1 and a_2. Compute this variance and compare it with the variances which you found (b).

6.13.2 Sample Mean and Sample Variance

Compute the sample mean $\bar{\mathsf{x}}$, the sample variance s^2, and an estimate for the variance of the sample mean $\mathsf{s}_{\bar{\mathsf{x}}}^2 = (1/n)\mathsf{s}^2$ for the following sample:

18, 21, 23, 19, 20, 21, 20, 19, 20, 17.

Use the method of Example 6.1.

6.13.3 Samples from a Partitioned Population

An opinion poll is performed on an upcoming election. In our (artificially constructed) example the population is partitioned into three subpopulations, and from each subpopulation a preliminary sample of size 10 is drawn. Each element of the sample can have the values 0 (vote for party A) or 1 (vote for party B). The samples are

$$\begin{aligned} i=1 \quad (p_i = 0.1) &: x_{ij} = 0,0,0,0,1,0,1,0,0,0, \\ i=2 \quad (p_i = 0.7) &: x_{ij} = 0,0,1,1,0,1,0,1,1,0, \\ i=3 \quad (p_i = 0.2) &: x_{ij} = 0,1,1,1,1,0,1,1,1,0, \end{aligned}$$

(a) Use these samples to form an estimator for the result of the election and its variance $s^2_{\tilde{x}}$. Does this result show a clear advantage for one party?

(b) Use these samples to determine for a much larger sample of size n the sizes n_i of the subsamples in such a way that $\tilde{x}$ has the smallest variance (cf. Example 6.6).

6.13.4 χ^2-distribution

(a) Determine the skewness $\gamma = \mu_3/\sigma^3$ of the χ^2-distribution using (5.5.7). Begin by expressing μ_3 by λ_3, $\widehat{x}$, and $E(\mathsf{x}^2)$.

(b) Show that $\gamma \to 0$ for $n \to \infty$.

6.13.5 Histogram

Construct a histogram from the following measured values:

26.02, 27.13, 24.78, 26.19, 22.57, 25.16, 24.39, 22.73, 25.35, 26.13,
23.15, 26.29, 24.89, 23.76, 26.04, 22.03, 27.48, 25.42, 24.27, 26.58,
27.17, 24.22, 21.86, 26.53, 27.09, 24.73, 26.12, 28.04, 22.78, 25.36.

Use a bin size of $\Delta x = 1$.
Hint: For each value draw a cross of width Δx and height 1. In this way you do not need to order the measured values, since each cross is drawn within a bin on top of a preceding cross. In this way the bars of the histogram grow while drawing.

6.14 Hints and Solutions

Problem 6.13.1

(a) $E(\mathsf{S}) = E\{\sum a_i \mathsf{x}_i\} = \sum a_i E(\mathsf{x}_i) = \widehat{x} \sum a_i = \widehat{x}$
if and only if
$\sum a_i = 1$.

$$
\begin{aligned}
\text{(b)}\quad \sigma^2 &= \sum \left(\frac{\partial \mathsf{S}}{\partial x_i}\right)^2 \sigma^2 = \sigma^2 \sum a_i^2 \quad , \\
\sigma^2(\mathsf{S}_1) &= \sigma^2 \left(\frac{1}{16} + \frac{1}{16} + \frac{1}{4}\right) = \sigma^2 \cdot \frac{3}{8} = 0.375\sigma^2 \quad , \\
\sigma^2(\mathsf{S}_2) &= \sigma^2 \left(\frac{1}{25} + \frac{4}{25} + \frac{4}{25}\right) = \sigma^2 \cdot \frac{9}{25} = 0.360\sigma^2 \quad , \\
\sigma^2(\mathsf{S}_3) &= \sigma^2 \left(\frac{1}{36} + \frac{1}{9} + \frac{1}{4}\right) = \sigma^2 \cdot \frac{7}{18} = 0.389\sigma^2 \quad .
\end{aligned}
$$

$$
\begin{aligned}
\text{(c)}\quad \sigma^2(\mathsf{S}) &= [a_1^2 + a_2^2 + (1 - (a_1 + a_2))^2]\sigma^2 \quad , \\
\frac{\partial \sigma^2(\mathsf{S})}{\partial a_1} &= [2a_1 - 2(1 - (a_1 + a_2))]\sigma^2 = 0 \quad , \\
\frac{\partial \sigma^2(\mathsf{S})}{\partial a_2} &= [2a_2 - 2(1 - (a_1 + a_2))]\sigma^2 = 0 \quad , \\
a_1 &= 1 - (a_1 + a_2) \quad , \\
a_2 &= 1 - (a_1 + a_2) \quad , \qquad a_1 = a_2 = \tfrac{1}{3} \quad ; \\
\sigma^2(\mathsf{S}) &= \tfrac{1}{3}\sigma^2 = 0.333\sigma^2 \quad .
\end{aligned}
$$

Problem 6.13.2

$$
\begin{aligned}
a &= 20 \quad , \\
\Delta &= \frac{1}{n} \sum \delta_i \\
&= \frac{1}{10}(-2 + 1 + 3 - 1 + 0 + 1 + 0 - 1 + 0 - 3) = -0.2 \quad , \\
\bar{\mathsf{x}} &= 19.8 \quad , \\
\mathsf{s} &= \frac{1}{9}(1.8^2 + 1.2^2 + 3.2^2 + 0.8^2 + 0.2^2 + \\
&\quad + 1.2^2 + 0.2^2 + 0.8^2 + 0.2^2 + 2.8^2) \\
&= \frac{1}{9} \cdot 25.60 = 2.84 \quad , \\
\mathsf{s}_{\bar{\mathsf{x}}}^2 &= 0.284 \quad . \qquad \text{Therefore } \bar{\mathsf{x}} = 19.8 \pm 0.53 \quad .
\end{aligned}
$$

Problem 6.13.3

(a)
$$\begin{aligned}
\bar{x}_1 &= 0.2,\ \bar{x}_2 = 0.5,\ \bar{x}_3 = 0.7 \quad ,\\
\widetilde{x} &= 0.02 + 0.35 + 0.14 = 0.51 \quad ,\\
s_1^2 &= \frac{1}{9}(2 - 10 \cdot 0.2^2) = 0.178 \quad ,\\
s_2^2 &= \frac{1}{9}(5 - 10 \cdot 0.5^2) = 0.278 \quad ,\\
s_3^2 &= \frac{1}{9}(7 - 10 \cdot 0.7^2) = 0.233 \quad ,\\
s_{\bar{x}}^2 &= \frac{0.1^2}{10}0.178 + \frac{0.7^2}{10}0.278 + \frac{0.2^2}{10}0.233\\
&= 0.001 \cdot 0.178 + 0.049 \cdot 0.278 + 0.004 \cdot 0.233\\
&= 0.0147 \quad ,\\
s_{\bar{x}} &= \sqrt{s_{\bar{x}}^2} = 0.12 \quad .
\end{aligned}$$

The result $\widetilde{x} = 0.51 \pm 0.12$ does not favor any one party significantly.

(b)
$$\begin{aligned}
s_1 &= 0.422,\ s_2 = 0.527,\ s_3 = 0.483 \quad ,\\
p_1 s_1 &= 0.0422,\ p_2 s_2 = 0.369,\ p_3 s_3 = 0.0966 \quad ,\\
\sum p_i s_i &= 0.508 \quad ,\\
\frac{n_1}{n} &= 0.083,\ \frac{n_2}{n} = 0.726,\ \frac{n_3}{n} = 0.190 \quad .
\end{aligned}$$

Problem 6.13.4

(a) For simplicity we write $\chi^2 = u$.

$$\begin{aligned}
\mu_3 &= E\{(u - \widehat{u})^3\} = E\{u^3 - 3u^2\widehat{u} + 3u\widehat{u}^2 - \widehat{u}^3\}\\
&= E(u^3) - 3\widehat{u}E(u^2) + 3\widehat{u}^2 E(u) - \widehat{u}^3\\
&= \lambda_3 - 3\widehat{u}E(u^2) + 2\widehat{u}^3 \quad ,\\
E(u^3) &= \lambda_3 = \frac{1}{i^3}\varphi'''(0)\\
&= i\varphi'''(0) = i(-\lambda)(-\lambda - 1)(-\lambda - 2)(-2i)^3\\
&= 8\lambda(\lambda + 1)(\lambda + 2) = 8\lambda^3 + 24\lambda^2 + 16\lambda \quad ,
\end{aligned}$$

$$\begin{aligned}
3\widehat{u}E(\mathrm{u}^2) &= 6\lambda(4\lambda^2+4\lambda) = 24\lambda^3+24\lambda^2 \quad , \\
2\widehat{u}^3 &= 2\cdot(2\lambda)^3 = 16\lambda^3 \quad , \\
\mu_3 &= 16\lambda \quad , \\
\gamma &= \mu_3/\sigma^3 = 16\lambda/8\lambda^{\frac{3}{2}} = 2\lambda^{-\frac{1}{2}} \quad .
\end{aligned}$$

(b) Obviously $\gamma = 2/\sqrt{\lambda} \to 0$ holds for $\lambda = \frac{1}{2}n \to \infty$.

Problem 6.13.5

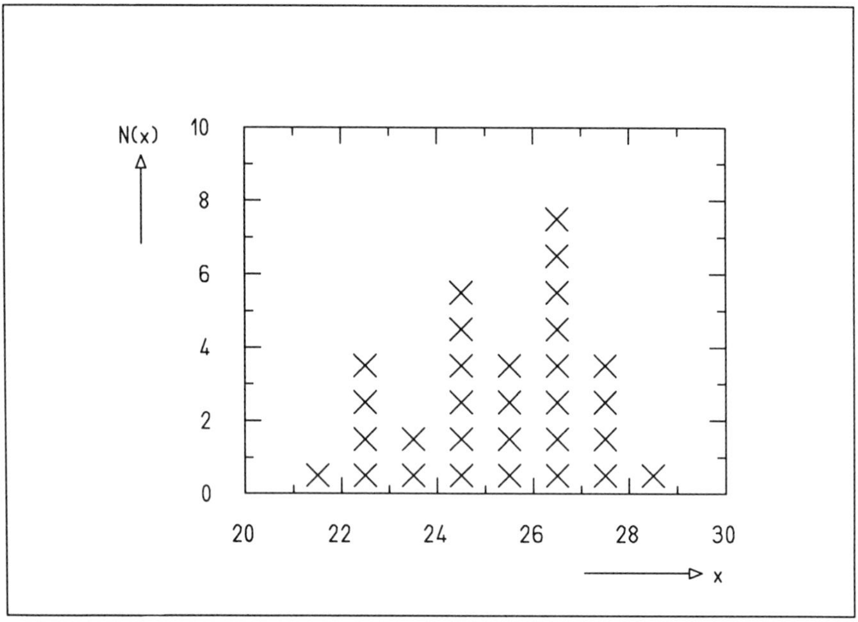

Fig. 6.13. Histogram of the data of Problem 6.13.5.

7. The Method of Maximum Likelihood

7.1 Likelihood Ratio. Likelihood Function

In the last chapter we introduced the concept of parameter estimation. We have also described the desirable properties of estimators, though without specifying how such estimators can be constructed in a particular case. We have derived estimators only for the important quantities expectation value and variance. We now take on the general problem.

In order to specify explicitly the parameters

$$\boldsymbol{\lambda} = (\lambda_1, \lambda_2, \ldots, \lambda_p) \quad ,$$

we now write the probability density of the random variables

$$\mathbf{x} = (\mathrm{x}_1, \mathrm{x}_2, \ldots, \mathrm{x}_n)$$

in the form

$$f = f(\mathbf{x}; \boldsymbol{\lambda}) \quad . \tag{7.1.1}$$

If we now carry out a certain number of experiments, say N, or we draw a sample of size N from a population, then we can give a number to each experiment j:

$$\mathrm{d}P^{(j)} = f(\mathbf{x}^{(j)}; \boldsymbol{\lambda})\,\mathrm{dx} \quad . \tag{7.1.2}$$

The number $\mathrm{d}P^{(j)}$ has the character of an *a posteriori probability*, i.e., given *after* the experiment, how probable it was to find the result $\mathbf{x}^{(j)}$ (within a small interval). The total probability to find exactly all of the events

$$\mathbf{x}^{(1)}, \mathbf{x}^{(2)}, \ldots, \mathbf{x}^{(j)}, \ldots, \mathbf{x}^{(N)}$$

is then the product

$$\mathrm{d}P = \prod_{j=1}^{N} f(\mathbf{x}^{(j)}; \boldsymbol{\lambda})\,\mathrm{d}\mathbf{x} \quad . \tag{7.1.3}$$

This probability still clearly depends on $\boldsymbol{\lambda}$. There are cases where the population is determined by only two possible sets of parameters, $\boldsymbol{\lambda}_1$ and $\boldsymbol{\lambda}_2$. Such cases occur, for example, in nuclear physics, where the parity of a state is necessarily "even" or "odd". One can construct the ratio

$$Q = \frac{\prod_{j=1}^{N} f(\mathbf{x}^{(j)}; \boldsymbol{\lambda}_1)}{\prod_{j=1}^{N} f(\mathbf{x}^{(j)}; \boldsymbol{\lambda}_2)} \tag{7.1.4}$$

and say that the values $\boldsymbol{\lambda}_1$ are "Q times more probable" than the values $\boldsymbol{\lambda}_2$. This factor is called the *likelihood ratio*.*

A product of the form

$$L = \prod_{j=1}^{N} f(\mathbf{x}^{(j)}; \boldsymbol{\lambda}) \tag{7.1.5}$$

is called a *likelihood function*. One must clearly distinguish between a probability density and a likelihood function, which is a function of a sample and is hence a random variable. In particular, the *a posteriori* nature of the probability in (7.1.5) is of significance in many discussions.

Example 7.1: Likelihood ratio

Suppose one wishes to decide whether a coin belongs to type A or B by means of a number of tosses. The coins in question are asymmetric in such a way that A shows heads with a probability of 1/3, and B shows heads with a probability of 2/3.

	A	B
Heads	1/3	2/3
Tails	2/3	1/3

If an experiment yields heads once and tails four times, then one has $L_A = \frac{1}{3} \cdot (\frac{2}{3})^4$ and $L_B = \frac{2}{3} \cdot (\frac{1}{3})^4$,

$$Q = \frac{L_A}{L_B} = 8 \quad .$$

One would therefore tend towards the position that the coin is of type A. ∎

*Although the likelihood ratio Q and the likelihood functions L and ℓ introduced below are random variables, since they are functions of a sample, we do not write them here with a special character type.

7.2 The Method of Maximum Likelihood

The generalization of the likelihood ratio is now clear. One gives the greatest confidence to that choice of the parameters $\boldsymbol{\lambda}$ for which the likelihood function (7.1.5) is a maximum. Figure 7.1 illustrates the situation for various forms of the likelihood function for the case of a single parameter λ.

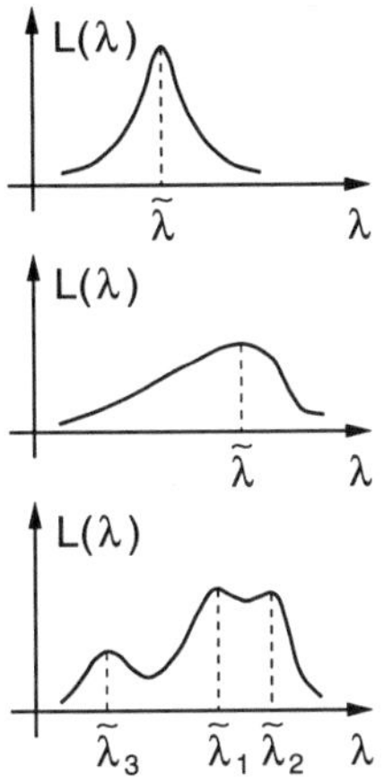

Fig. 7.1. Likelihood functions.

The maximum can be located simply by setting the first derivative of the likelihood function with respect to the parameter λ_i equal to zero. The derivative of a product with many factors is, however, unpleasant to deal with. One first constructs therefore the logarithm of the likelihood function,

$$\ell = \ln L = \sum_{j=1}^{N} \ln f(\mathrm{x}^{(j)}; \boldsymbol{\lambda}) \quad . \tag{7.2.1}$$

The function ℓ is also often called the likelihood function. Sometimes one says explicitly "*log-likelihood function*". Clearly the maxima of (7.2.1) are identical with those of (7.1.5). For the case of a single parameter we now construct

$$\ell' = \mathrm{d}\ell/\mathrm{d}\lambda = 0 \quad . \tag{7.2.2}$$

The problem of estimating a parameter is now reduced to solving this *likelihood equation*. By application of (7.2.1) we can write

$$\ell' = \sum_{j=1}^{N} \frac{\mathrm{d}}{\mathrm{d}\lambda} \ln f(\mathrm{x}^{(j)}; \lambda) = \sum_{j=1}^{N} \frac{f'}{f} = \sum_{j=1}^{N} \varphi(\mathrm{x}^{(j)}; \lambda) \quad , \tag{7.2.3}$$

where

$$\varphi(\mathrm{x}^{(j)}; \lambda) = \left(\frac{\mathrm{d}}{\mathrm{d}\lambda} f(x^{(j)}; \lambda) \right) \Big/ f(\mathrm{x}^{(j)}; \lambda) \tag{7.2.4}$$

is the *logarithmic derivative* of the density f with respect to λ.

In the general case of p parameters the likelihood equation (7.2.2) is replaced by the system of p simultaneous equations,

$$\frac{\partial \ell}{\partial \lambda_i} = 0 \quad , \qquad i = 1, 2, \ldots, p \quad . \tag{7.2.5}$$

Example 7.2: Repeated measurements of differing accuracy

If a quantity is measured with different instruments, then the measurement errors are in general different. The measurements $\mathrm{x}^{(j)}$ are spread about the true value λ. Suppose the errors are normally distributed, so that a measurement corresponds to obtaining a sample from a Gaussian distribution with mean λ and standard deviation σ_j. The *a posteriori* probability for a measured value is then

$$f(\mathrm{x}^{(j)}; \lambda)\,\mathrm{d}x = \frac{1}{\sqrt{2\pi}\sigma_j} \exp\left(-\frac{(\mathrm{x}^{(j)} - \lambda)^2}{2\sigma_j^2}\right) \mathrm{d}x \quad .$$

From all N measurements one obtains the likelihood function

$$L = \prod_{j=1}^{N} \frac{1}{\sqrt{2\pi}\sigma_j} \exp\left(-\frac{(\mathrm{x}^{(j)} - \lambda)^2}{2\sigma_j^2}\right) \tag{7.2.6}$$

with the logarithm

$$\ell = -\frac{1}{2} \sum_{j=1}^{N} \frac{(\mathrm{x}^{(j)} - \lambda)^2}{\sigma_j^2} + \mathrm{const} \quad . \tag{7.2.7}$$

The likelihood equation thus becomes

$$\frac{\mathrm{d}\ell}{\mathrm{d}\lambda} = \sum_{j=1}^{N} \frac{\mathrm{x}^{(j)} - \lambda}{\sigma_j^2} = 0 \quad .$$

It has the solution

$$\tilde{\lambda} = \frac{\sum_{j=1}^{N} \frac{\mathrm{x}^{(j)}}{\sigma_j^2}}{\sum_{j=1}^{N} \frac{1}{\sigma_j^2}} \quad . \tag{7.2.8}$$

Since $\mathrm{d}^2\ell/\mathrm{d}\lambda^2 = -\sum \sigma_j^{-2} < 0$, the solution is, in fact, a maximum. Thus we see that we obtain the maximum likelihood estimator as the mean of the N measurements weighted inversely by the variances of the individual measurements. ■

Example 7.3: Estimation of the parameter N of the hypergeometric distribution

As in the example with coins at the beginning of this chapter, sometimes parameters to be estimated can only take on discrete values. In Example 5.2 we indicated the possibility of estimating zoological population densities by means of tagging and recapture. According to (5.3.1), the probability to catch exactly n fish of which k are tagged out of a pond with an unknown total of N fish, out of which K are tagged, is given by

$$L(k; n, K, N) = \frac{\binom{K}{k}\binom{N-K}{n-k}}{\binom{N}{n}} \quad .$$

We must now find the value of N for which the function L is maximum. For this we use the ratio

$$\frac{L(k; n, k, N)}{L(k; n, k, N-1)} = \frac{(N-n)(N-k)}{(N-n-K+k)N} \begin{cases} > 1\,, & Nk < nK\,, \\ < 1\,, & Nk > nK\,. \end{cases}$$

The function L is thus maximum when N is the integer closest to nK/k. ∎

7.3 Information Inequality. Minimum Variance Estimators. Sufficient Estimators

We now want to discuss once more the quality of an estimator. In Sect. 6.1 we called an estimator unbiased if for every sample the *bias* vanished,

$$B(\lambda) = E(\mathsf{S}) - \lambda = 0 \quad . \tag{7.3.1}$$

Lack of bias is, however, not the only characteristic required of a "good" estimator. More importantly one should require that the variance

$$\sigma^2(\mathsf{S})$$

is small. Here one must often find a compromise, since there is a connection between B and σ^2, described by the *information inequality*†.

One immediately sees that it is easy to achieve $\sigma^2(\mathsf{S}) = 0$ simply by using a constant for S. We consider an estimator $\mathsf{S}(\mathsf{x}^{(1)}, \mathsf{x}^{(2)}, \ldots, \mathsf{x}^{(N)})$ that is a function of the sample $\mathsf{x}^{(1)}, \mathsf{x}^{(2)}, \ldots, \mathsf{x}^{(N)}$. According to (6.1.3) and (6.1.4) the joint probability density of the elements of the sample is

†This inequality was independently found by H. Cramer, M. Fréchet, and C. R. Rao as well as by other authors. It is also called the Cramer–Rao or Fréchet inequality.

$$f(x^{(1)}, x^{(2)}, \ldots, x^{(N)}; \lambda) = f(x^{(1)}; \lambda) f(x^{(2)}; \lambda) \cdots f(x^{(N)}; \lambda) \quad .$$

The expectation value of S is thus

$$E(\mathsf{S}) = \int \mathsf{S}(\mathsf{x}^{(1)}, \ldots, \mathsf{x}^{(N)}) f(x^{(1)}; \lambda) \cdots f(x^{(N)}; \lambda) \times \mathrm{d}x^{(1)}\, \mathrm{d}x^{(2)} \cdots \mathrm{d}x^{(N)} \quad . \tag{7.3.2}$$

According to (7.3.1), however, one also has

$$E(\mathsf{S}) = B(\lambda) + \lambda \quad .$$

We now assume that we can differentiate with respect to λ in the integral. We then obtain

$$1 + B'(\lambda) = \int \mathsf{S} \left(\sum_{j=1}^{N} \frac{f'(x^{(j)}; \lambda)}{f(x^{(j)}; \lambda)} \right) f(x^{(1)}; \lambda) \cdots f(x^{(N)}; \lambda)\, \mathrm{d}x^{(1)} \cdots \mathrm{d}x^{(N)} \quad ,$$

which is equivalent to

$$1 + B'(\lambda) = E\left\{ \mathsf{S} \sum_{j=1}^{N} \frac{f'(x^{(j)}; \lambda)}{f(x^{(j)}; \lambda)} \right\} = E\left\{ \mathsf{S} \sum_{j=1}^{N} \varphi(x^{(j)}; \lambda) \right\}$$

and therefore

$$1 + B'(\lambda) = E\{\mathsf{S}\ell'\} \quad . \tag{7.3.3}$$

One clearly has

$$\int f(x^{(1)}; \lambda) \cdots f(x^{(N)}; \lambda)\, \mathrm{d}x^{(1)} \cdots \mathrm{d}x^{(N)} = 1 \quad .$$

If we also compute the derivative with respect to λ, we obtain

$$\int \sum_{j=1}^{N} \frac{f'(x^{(j)}; \lambda)}{f(x^{(j)}; \lambda)} f(x^{(1)}; \lambda) \cdots f(x^{(N)}; \lambda)\, \mathrm{d}x^{(1)} \cdots \mathrm{d}x^{(N)} = E(\ell') = 0 \quad .$$

By multiplying this equation by $E(\mathsf{S})$ and subtracting the result of (7.3.3) one obtains

$$1 + B'(\lambda) = E\{\mathsf{S}\ell'\} - E(\mathsf{S})E(\ell') = E\{[\mathsf{S} - E(\mathsf{S})]\ell'\} \quad . \tag{7.3.4}$$

In order to see the significance of this expression, we need to use the Cauchy–Schwarz inequality in the following form:

If x and y are random variables and if x^2 and y^2 have finite expectation values, then

$$\{E(\mathsf{xy})\}^2 \le E(\mathsf{x}^2)E(\mathsf{y}^2) \quad . \tag{7.3.5}$$

To prove this inequality we consider the expression

$$E((a\mathsf{x}+\mathsf{y})^2) = a^2E(\mathsf{x}^2) + 2aE(\mathsf{xy}) + E(\mathsf{y}^2) \ge 0 \quad . \tag{7.3.6}$$

This is a non-negative number for all values of a. If we consider for the moment the case of equality, then this is a quadratic equation for a with the solutions

$$a_{1,2} = -\frac{E(\mathsf{xy})}{E(\mathsf{x}^2)} \pm \sqrt{\left(\frac{E(\mathsf{xy})}{E(\mathsf{x}^2)}\right)^2 - \frac{E(\mathsf{y}^2)}{E(\mathsf{x}^2)}} \quad . \tag{7.3.7}$$

The inequality (7.3.6) is then valid for all a if the term under the square root is negative or zero. From this follows the assertion

$$\frac{\{E(\mathsf{xy})\}^2}{\{E(\mathsf{x}^2)\}^2} - \frac{E(\mathsf{y}^2)}{E(\mathsf{x}^2)} \le 0 \quad .$$

If we now apply the inequality (7.3.5) to (7.3.4), one obtains

$$\{1 + B'(\lambda)\}^2 \le E\{[\mathsf{S} - E(\mathsf{S})]^2\}E(\ell'^2) \quad . \tag{7.3.8}$$

We now use (7.2.3) in order to rewrite the expression for $E(\ell'^2)$,

$$\begin{aligned} E(\ell'^2) &= E\left\{\left(\sum_{j=1}^{N} \varphi(\mathsf{x}^{(j)};\lambda)\right)^2\right\} \\ &= E\left\{\sum_{j=1}^{N}(\varphi(\mathsf{x}^{(j)};\lambda))^2\right\} + E\left\{\sum_{i\ne j} \varphi(\mathsf{x}^{(i)};\lambda)\varphi(\mathsf{x}^{(j)};\lambda)\right\} \quad . \end{aligned}$$

All terms on the right-hand side vanish, since for $i \ne j$

$$\begin{aligned} E\{\varphi(\mathsf{x}^{(i)};\lambda)\varphi(\mathsf{x}^{(j)};\lambda)\} &= E\{\varphi(\mathsf{x}^{(i)};\lambda)\}E\{\varphi(\mathsf{x}^{(j)};\lambda)\} \quad , \\ E\{\varphi(\mathsf{x};\lambda)\} &= \int_{-\infty}^{\infty} \frac{f'(x;\lambda)}{f(x;\lambda)} f(x;\lambda)\,\mathrm{d}x = \int f'(x;\lambda)\,\mathrm{d}x \quad , \end{aligned}$$

and

$$\int_{-\infty}^{\infty} f(x;\lambda)\,\mathrm{d}x = 1 \quad .$$

By differentiating the last line with respect to λ one obtains

$$\int_{-\infty}^{\infty} f'(x;\lambda)\,\mathrm{d}x = 0 \quad .$$

Thus one has simply

$$E(\ell'^2) = E\left\{\sum_{j=1}^{N}(\varphi(\mathsf{x}^{(j)};\lambda))^2\right\} = E\left\{\sum_{j=1}^{N}\left(\frac{f'(\mathsf{x}^{(j)};\lambda)}{f(\mathsf{x}^{(j)};\lambda)}\right)^2\right\} \quad .$$

Since the individual terms of the sum are independent, the expectation value of the sum is simply the sum of the expectation values. The individual expectation values do not depend on the elements of the sample. Therefore one has

$$I(\lambda) = E(\ell'^2) = NE\left\{\left(\frac{f'(x;\lambda)}{f(x;\lambda)}\right)^2\right\} \quad .$$

This expression is called the *information of the sample with respect to* λ. It is a non-negative number, which vanishes if the likelihood function does not depend on the parameter λ.

It is sometimes useful to write the information in a somewhat different form. To do this we differentiate the expression

$$E\left(\frac{f'(x;\lambda)}{f(x;\lambda)}\right) = \int_{-\infty}^{\infty}\frac{f'(x;\lambda)}{f(x;\lambda)}f(x;\lambda)\,\mathrm{d}x = 0$$

once more with respect to λ and obtain

$$\begin{aligned} 0 &= \int_{-\infty}^{\infty}\left\{\frac{f'^2}{f} + f\left(\frac{f'}{f}\right)'\right\}\mathrm{d}x = \int_{-\infty}^{\infty}\left\{\left(\frac{f'}{f}\right)^2 + \left(\frac{f'}{f}\right)'\right\}f\,\mathrm{d}x \\ &= E\left\{\left(\frac{f'}{f}\right)^2\right\} + E\left\{\left(\frac{f'}{f}\right)'\right\} \quad . \end{aligned}$$

The information can then be written as

$$I(\lambda) = NE\left\{\left(\frac{f'(x;\lambda)}{f(x;\lambda)}\right)^2\right\} = -NE\left\{\left(\frac{f'(x;\lambda)}{f(x;\lambda)}\right)'\right\}$$

or

$$I(\lambda) = E(\ell'^2) = -E(\ell'') \quad . \qquad (7.3.9)$$

The inequality (7.3.8) can now be written in the following way:

$$\{1 + B'(\lambda)\}^2 \le \sigma^2(\mathsf{S})I(\lambda)$$

or

$$\sigma^2(\mathsf{S}) \ge \frac{\{1 + B'(\lambda)\}^2}{I(\lambda)} \quad . \qquad (7.3.10)$$

This is the *information inequality*. It gives the connection between the bias and the variance of an estimator and the information of a sample. It should be noted that in its derivation no assumption about the estimator was made. The right-hand side of the inequality (7.3.10) is therefore a lower bound for the variance of an estimator. It is called the *minimum variance bound* or *Cramer–Rao bound.* In cases where the bias does not depend on λ, i.e., particularly in cases of vanishing bias, the inequality (7.3.10) simplifies to

$$\sigma^2(\mathsf{S}) \geq 1/I(\lambda) \quad . \tag{7.3.11}$$

This relation justifies using the name information. As the information of a sample increases, the variance of an estimator can be made smaller.

We now ask under which circumstances the minimum variance bound is attained, or explicitly, when the equals sign in the relation (7.3.10) holds. In the inequality (7.3.6), one has equality if $(a\mathsf{x}+\mathsf{y})$ vanishes, since only then does one have $E\{(a\mathsf{x}+\mathsf{y})^2\} = 0$ for all values of a, x, and y. Applied to (7.3.8), this means that

$$\ell' + a(\mathsf{S} - E(\mathsf{S})) = 0$$

or

$$\ell' = A(\lambda)(\mathsf{S} - E(\mathsf{S})) \quad . \tag{7.3.12}$$

Here A means an arbitrary quantity that does not depend on the sample $\mathsf{x}^{(1)}, \mathsf{x}^{(2)}, \ldots, \mathsf{x}^{(N)}$, but may be, however, a function of λ. By integration we obtain

$$\ell = \int \ell' \,\mathrm{d}\lambda = B(\lambda)\mathsf{S} + C(\lambda) + D \tag{7.3.13}$$

and finally

$$L = d\, \exp\{B(\lambda)\mathsf{S} + C(\lambda)\} \quad . \tag{7.3.14}$$

The quantities d and D do not depend on λ.

We thus see that estimators attain the minimum variance bound when the likelihood function is of the special form (7.3.14). They are therefore called *minimum variance estimators.*

For the case of an unbiased minimum variance estimator we obtain from (7.3.11)

$$\sigma^2(\mathsf{S}) = \frac{1}{I(\lambda)} = \frac{1}{E(\ell'^2)} \quad . \tag{7.3.15}$$

By substituting (7.3.12) one obtains

$$\sigma^2(\mathsf{S}) = \frac{1}{(A(\lambda))^2 E\{(\mathsf{S} - E(\mathsf{S}))^2\}} = \frac{1}{(A(\lambda))^2 \sigma^2(\mathsf{S})}$$

or

$$\sigma^2(\mathsf{S}) = \frac{1}{|A(\lambda)|} \quad . \tag{7.3.16}$$

If instead of (7.3.14) only the weaker requirement

$$L = g(\mathsf{S}, \lambda) c(\mathsf{x}^{(1)}, \mathsf{x}^{(2)}, \ldots, \mathsf{x}^{(N)}) \tag{7.3.17}$$

holds, then the estimator S is said to be *sufficient* for λ. One can show [see, e.g., Kendall and Stuart, Vol. 2 (1967)], that no other estimator can contribute information about λ that is not already contained in S, if the requirement (7.3.17) is fulfilled. Hence the name "sufficient estimator" (or statistic).

Example 7.4: Estimator for the parameter of the Poisson distribution
Consider the *Poisson distribution* (5.4.1)

$$f(k) = \frac{\lambda^k}{k!} \mathrm{e}^{-\lambda} \quad .$$

The likelihood function of a sample $\mathsf{k}^{(1)}, \mathsf{k}^{(2)}, \ldots, \mathsf{k}^{(N)}$ is

$$\ell = \sum_{j=1}^{N} \{\mathsf{k}^{(j)} \ln \lambda - \ln(\mathsf{k}^{(j)}!) - \lambda\}$$

and its derivative with respect to λ is

$$\frac{\mathrm{d}\ell}{\mathrm{d}\lambda} = \ell' = \sum_{j=1}^{N} \left\{ \frac{\mathsf{k}^{(j)}}{\lambda} - 1 \right\} = \frac{1}{\lambda} \sum_{j=1}^{N} \{\mathsf{k}^{(j)} - \lambda\} \quad ,$$

$$\ell' = \frac{N}{\lambda} (\bar{\mathsf{k}} - \lambda) \quad . \tag{7.3.18}$$

Comparing with (7.3.12) and (7.3.16) shows that the arithmetic mean $\bar{k}$ is an unbiased minimum variance estimator with variance λ/N. ∎

Example 7.5: Estimator for the parameter of the binomial distribution
The likelihood function of a sample from a *binomial distribution* with the parameters $p = \lambda$, $q = 1 - \lambda$ is given directly by (5.1.3),

$$L(\mathsf{k}, \lambda) = \binom{n}{\mathsf{k}} \lambda^{\mathsf{k}} (1 - \lambda)^{n - \mathsf{k}} \quad .$$

(The result of the sample can be summarized by the statement that in n experiments, the event A occurred k times; see Sect. 5.1.) One then has

$$\begin{aligned} \ell &= \ln L = \mathsf{k} \ln \lambda + (n - \mathsf{k}) \ln(1 - \lambda) + \ln \binom{n}{\mathsf{k}} \quad , \\ \ell' &= \frac{\mathsf{k}}{\lambda} - \frac{n - \mathsf{k}}{1 - \lambda} = \frac{n}{\lambda(1 - \lambda)} \left(\frac{\mathsf{k}}{n} - \lambda \right) \quad . \end{aligned}$$

By comparing with (7.3.12) and (7.3.16) one finds k/n to be a minimum variance estimator with variance $\lambda(1 - \lambda)/n$. ∎

Example 7.6: Law of error combination ("Quadratic averaging of individual errors")

We now return to the problem of Example 7.2 of repeated measurements of the same quantity with varying uncertainties, or expressed in another way, to the problem of drawing a sample from normal distributions with the same mean λ and different but known variances σ_j. From (7.2.7) we obtain

$$\frac{\mathrm{d}\ell}{\mathrm{d}\lambda} = \ell' = \sum_{j=1}^{N} \frac{\mathsf{x}^{(j)} - \lambda}{\sigma_j^2} \quad .$$

We can rewrite this expression as

$$\begin{aligned} \ell' &= \sum \frac{\mathsf{x}^{(j)}}{\sigma_j^2} - \sum \frac{\lambda}{\sigma_j^2} \\ &= \sum_{j=1}^{N} \frac{1}{\sigma_j^2} \left\{ \frac{\sum \frac{\mathsf{x}^{(j)}}{\sigma_j^2}}{\sum \frac{1}{\sigma_j^2}} - \lambda \right\} \quad . \end{aligned}$$

As in Example 7.2 we recognize

$$\mathsf{S} = \widetilde{\lambda} = \frac{\sum \frac{\mathsf{x}^{(j)}}{\sigma_j^2}}{\sum \frac{1}{\sigma_j^2}} \tag{7.3.19}$$

as an unbiased estimator for λ. Comparing with (7.3.12) shows that it is also a minimum variance estimator. From (7.3.16) one determines that its variance is

$$\sigma^2(\widetilde{\lambda}) = \left(\sum_{j=1}^{N} \frac{1}{\sigma_j^2} \right)^{-1} \quad . \tag{7.3.20}$$

The relation (7.3.20) often goes by the name of the *law of error combination* or *quadratic averaging of individual errors*. It could have been obtained by application of the rule of error propagation (3.8.7) to (7.3.19). If we identify $\sigma(\widetilde{\lambda})$ as the error of the estimator $\widetilde{\lambda}$ and σ_j as the error of the jth measurement, then we can write it in its usual form

$$\Delta\widetilde{\lambda} = \left(\frac{1}{(\Delta x_1)^2} + \frac{1}{(\Delta x_2)^2} + \cdots + \frac{1}{(\Delta x_n)^2} \right)^{-\frac{1}{2}} \quad . \tag{7.3.21}$$

If all of the measurements have the same error $\sigma = \sigma_j$, Eqs. (7.3.19), (7.3.20) simplify to

$$\widetilde{\lambda} = \bar{\mathsf{x}} \quad , \qquad \sigma^2(\widetilde{\lambda}) = \sigma^2/n \quad ,$$

which we have already found in Sect. 6.2. ■

7.4 Asymptotic Properties of the Likelihood Function and Maximum-Likelihood Estimators

We can now show heuristically several important properties of the likelihood function and maximum-likelihood estimators for very large data samples, that is, for the limit $N \to \infty$. The estimator $\mathrm{S} = \widetilde{\lambda}$ was defined as the solution to the likelihood equation

$$\ell'(\lambda) = \sum_{j=1}^{N} \left(\frac{f'(\mathrm{x}^{(j)};\lambda)}{f(\mathrm{x}^{(j)};\lambda)} \right)_{\widetilde{\lambda}} = 0 \quad . \tag{7.4.1}$$

Let us assume that $\ell'(\lambda)$ can be differentiated with respect to λ one more time, so that we can expand it in a series around the point $\lambda = \widetilde{\lambda}$,

$$\ell'(\lambda) = \ell'(\widetilde{\lambda}) + (\lambda - \widetilde{\lambda})\ell''(\widetilde{\lambda}) + \cdots \quad . \tag{7.4.2}$$

The first term on the right side vanishes because of Eq. (7.4.1). In the second term one can write explicitly

$$\ell''(\widetilde{\lambda}) = \sum_{j=1}^{N} \left(\frac{f'(\mathrm{x}^{(j)};\lambda)}{f(\mathrm{x}^{(j)};\lambda)} \right)'_{\widetilde{\lambda}} \quad .$$

This expression has the form of the mean value of a sample. For very large N it can be replaced by the expectation value of the population (Sect. 6.2),

$$\ell''(\widetilde{\lambda}) = NE\left\{ \left(\frac{f'(x;\lambda)}{f(x;\lambda)} \right)'_{\widetilde{\lambda}} \right\} \quad . \tag{7.4.3}$$

Using Eq. (7.3.9) we can now write

$$\ell''(\widetilde{\lambda}) = E(\ell''(\widetilde{\lambda})) = -E(\ell'^2(\widetilde{\lambda})) = -I(\widetilde{\lambda}) = -1/b^2 \quad . \tag{7.4.4}$$

In this way we can replace the expression for $\ell''(\widetilde{\lambda})$, which is a function the sample $\mathrm{x}^{(1)}, \mathrm{x}^{(2)}, \ldots, \mathrm{x}^{(N)}$, by the quantity $-1/b^2$, which only depends on the probability density f and the estimator $\widetilde{\lambda}$. If one neglects higher-order terms, Eq. (7.4.2) can be expressed as

$$\ell'(\lambda) = -\frac{1}{b^2}(\lambda - \widetilde{\lambda}) \quad . \tag{7.4.5}$$

By integration one obtains

$$\ell(\lambda) = -\frac{1}{2b^2}(\lambda - \widetilde{\lambda})^2 + c \quad .$$

Inserting $\lambda = \widetilde{\lambda}$ gives $c = \ell(\widetilde{\lambda})$, or

$$\ell(\lambda) - \ell(\widetilde{\lambda}) = -\frac{1}{2}\frac{(\lambda - \widetilde{\lambda})^2}{b^2} \quad . \tag{7.4.6}$$

By exponentiation one obtains

$$L(\lambda) = k \exp\{-(\lambda - \widetilde{\lambda})^2/2b^2\} \quad , \tag{7.4.7}$$

where k is a constant. The likelihood function $L(\lambda)$ has the form of a normal distribution with mean $\widetilde{\lambda}$ and variance b^2. At the values $\lambda = \widetilde{\lambda} \pm b$, where λ is one standard deviation from $\widetilde{\lambda}$, one has

$$-\left(\ell(\lambda) - \ell(\widetilde{\lambda})\right) = \frac{1}{2} \quad . \tag{7.4.8}$$

We can now compare (7.4.7) with Eqs. (7.3.12) and (7.3.16). Since we are estimating the parameter λ, we must write $\mathsf{S} = \widetilde{\lambda}$ and thus $E(\mathsf{S}) = \lambda$. The estimator $\widetilde{\lambda}$ is therefore an unbiased minimum variance estimator with variance

$$\sigma^2(\widetilde{\lambda}) = b^2 = \frac{1}{I(\widetilde{\lambda})} = \frac{1}{E(\ell'^2(\widetilde{\lambda}))} = -\frac{1}{E(\ell''(\widetilde{\lambda}))} \quad . \tag{7.4.9}$$

Since the estimator $\widetilde{\lambda}$ only possesses this property for the limiting case $N \to \infty$, we call it *asymptotically unbiased*. This is equivalent to the statement that the maximum likelihood estimator is consistent (Sect. 6.1). For the same reason the likelihood function is called *asymptotically normal*.

In Sect. 7.2 we interpreted the likelihood function $L(\lambda)$ as a measure of the probability that the true value λ_0 of a parameter is equal to λ. The result of an estimator is often represented in abbreviated form,

$$\lambda = \widetilde{\lambda} \pm \sigma(\widetilde{\lambda}) = \widetilde{\lambda} \pm \Delta\lambda \quad .$$

Since the likelihood function is asymptotically normal, at least in the case of large samples, i.e., many measurements, this can be interpreted by saying that the probability that the true value λ_0 lies in the interval

$$\widetilde{\lambda} - \Delta\widetilde{\lambda} < \lambda_0 < \widetilde{\lambda} + \Delta\widetilde{\lambda}$$

is 68.3 % (Sect. 5.8). In practice the relation above is used for large but finite samples. Unfortunately one cannot construct any general rule for determining when a sample is large enough for this procedure to be reliable. Clearly if N is finite, (7.4.3) is only an approximation, whose accuracy depends not only on N, but also on the particular probability density $f(x; \lambda)$.

Example 7.7: Determination of the mean lifetime from a small number of decays

The probability that a radioactive nucleus, which exists at time $t = 0$, decays in the time interval between t and $t + \mathrm{d}t$ is

$$f(t)\,\mathrm{d}t = \frac{1}{\tau}\exp(-t/\tau)\,\mathrm{d}t \quad .$$

For observed decay times $t_1, t_2, \ldots, t_N$ the likelihood function is

$$L = \frac{1}{\tau^N}\exp\left\{-\frac{1}{\tau}\sum_{i=1}^{N} t_i\right\} = \frac{1}{\tau^N}\exp\left\{-\frac{N}{\tau}\bar{t}\right\} \quad .$$

Its logarithm is

$$\ell = \ln L = -\frac{N}{\tau}\bar{t} - N\ln\tau$$

with the derivative

$$\ell' = \frac{N}{\tau}\left(\frac{\bar{t}}{\tau} - 1\right) = \frac{N}{\tau^2}(\bar{t} - \tau) \quad .$$

Comparing with (7.3.12) we see that $\tilde{\tau} = \bar{t}$ is the maximum likelihood solution, which has a variance of $\sigma^2(\tau) = \tau^2/N$. For $\tau = \tilde{\tau} = \bar{t}$ one obtains $\Delta\tilde{\tau} = \bar{t}/\sqrt{N}$.

For $\tilde{\tau} = \bar{t}$ one has

$$\ell(\tilde{\tau}) = \ell_{\max} = -N(1 + \ln\bar{t}) \quad .$$

We can write

$$-\left(\ell(\tau) - \ell(\tilde{\tau})\right) = N\left(\frac{\bar{t}}{\tau} + \ln\frac{\tau}{\bar{t}} - 1\right) \quad .$$

From this expression for the log-likelihood function one cannot so easily recognize the asymptotic form (7.4.6) for $N \to \infty$. For small values of N it clearly does not have this form. Corresponding to (7.4.8), we want to use the values $\tau_+ = \tilde{\tau} + \Delta_+$ and $\tau_- = \tilde{\tau} - \Delta_-$, where one has

$$-\left(\ell(\tau_\pm) - \ell(\tilde{\tau})\right) = \frac{1}{2}$$

for the *asymmetric errors* Δ_+, Δ_-. Clearly we expect for $N \to \infty$ that Δ_+, $\Delta_- \to \Delta\tilde{\tau} = \sigma(\tilde{\tau})$.

In Fig. 7.2 the N observed decay times t_i are marked as vertical tick marks on the abscissa for various small values of N. In addition the function $-(\ell - \ell_{\max}) = -(\ell(\tau) - \ell(\tilde{\tau}))$ is plotted. The points τ_+ and τ_- are found where a horizontal line intersects $-(\ell - \ell_{\max}) = 1/2$. The point $\tilde{\tau}$ is indicated by an additional mark on the horizontal line. One sees that with increasing N the function $-(\ell - \ell_{\max})$ approaches more and more the symmetric parabolic form and that the errors Δ_+, Δ_-, and $\Delta\tilde{\tau}$ become closer to each other. ■

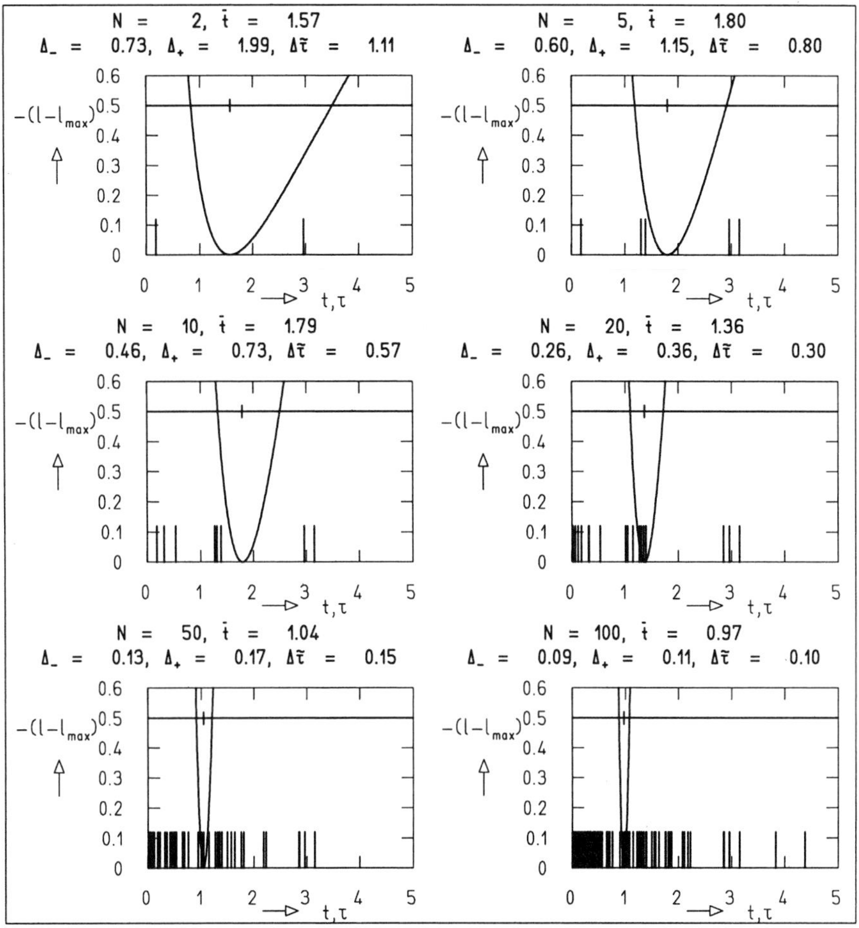

Fig. 7.2. Data and log-likelihood function for Example 7.7.

7.5 Simultaneous Estimation of Several Parameters. Confidence Intervals

We have already given a system of equations (7.2.5) allowing the simultaneous determination of p parameters $\lambda = (\lambda_1, \lambda_2, \ldots, \lambda_p)$. It turns out that it is not the parameter determination but rather the estimation of their errors that becomes significantly more complicated in the case of several parameters. In particular we will need to consider correlations as well as errors of the parameters.

We extend our considerations from Sect. 7.4 on the properties of the likelihood function to the case of several parameters. The log-likelihood function

$$\ell(\mathbf{x}^{(1)}, \mathbf{x}^{(2)}, \ldots, \mathbf{x}^{(N)}; \boldsymbol{\lambda}) = \sum_{j=1}^{N} \ln f(\mathbf{x}^{(j)}; \boldsymbol{\lambda}) \tag{7.5.1}$$

can be expanded in a series about the point

$$\widetilde{\boldsymbol{\lambda}} = (\widetilde{\lambda}_1, \widetilde{\lambda}_2, \ldots, \widetilde{\lambda}_p) \tag{7.5.2}$$

to give

$$\begin{aligned} \ell(\boldsymbol{\lambda}) &= \ell(\widetilde{\boldsymbol{\lambda}}) + \sum_{k=1}^{p} \left(\frac{\partial \ell}{\partial \lambda_k}\right)_{\widetilde{\boldsymbol{\lambda}}} (\lambda_k - \widetilde{\lambda}_k) \\ &\quad + \frac{1}{2} \sum_{\ell=1}^{p} \sum_{m=1}^{p} \left(\frac{\partial^2 \ell}{\partial \lambda_\ell \partial \lambda_m}\right)_{\widetilde{\boldsymbol{\lambda}}} (\lambda_\ell - \widetilde{\lambda}_\ell)(\lambda_m - \widetilde{\lambda}_m) + \cdots \quad . \end{aligned} \tag{7.5.3}$$

Since by the definition of $\widetilde{\boldsymbol{\lambda}}$ one has

$$\left(\frac{\partial \ell}{\partial \lambda_k}\right)_{\widetilde{\boldsymbol{\lambda}}} = 0 \quad , \qquad k = 1, 2, \ldots, p \quad , \tag{7.5.4}$$

which holds for all k, the series simplifies to

$$-\left(\ell(\boldsymbol{\lambda}) - \ell(\widetilde{\boldsymbol{\lambda}})\right) = \frac{1}{2}(\boldsymbol{\lambda} - \widetilde{\boldsymbol{\lambda}})^{\mathrm{T}} A (\boldsymbol{\lambda} - \widetilde{\boldsymbol{\lambda}}) + \cdots \tag{7.5.5}$$

with

$$-A = \begin{pmatrix} \dfrac{\partial^2 \ell}{\partial \lambda_1^2} & \dfrac{\partial^2 \ell}{\partial \lambda_1 \partial \lambda_2} & \cdots & \dfrac{\partial^2 \ell}{\partial \lambda_1 \partial \lambda_p} \\ \dfrac{\partial^2 \ell}{\partial \lambda_1 \partial \lambda_2} & \dfrac{\partial^2 \ell}{\partial \lambda_2^2} & \cdots & \dfrac{\partial^2 \ell}{\partial \lambda_2 \partial \lambda_p} \\ \vdots & & & \\ \dfrac{\partial^2 \ell}{\partial \lambda_1 \partial \lambda_p} & \dfrac{\partial^2 \ell}{\partial \lambda_2 \partial \lambda_p} & \cdots & \dfrac{\partial^2 \ell}{\partial \lambda_p^2} \end{pmatrix}_{\boldsymbol{\lambda} = \widetilde{\boldsymbol{\lambda}}} \quad . \tag{7.5.6}$$

In the limit $N \to \infty$ we can replace the elements of A, which still depend on the specific sample, by the corresponding expectation values,

$$B = E(A) =$$

$$-\begin{pmatrix} E\left(\dfrac{\partial^2 \ell}{\partial \lambda_1^2}\right) & E\left(\dfrac{\partial^2 \ell}{\partial \lambda_1 \partial \lambda_2}\right) & \cdots & E\left(\dfrac{\partial^2 \ell}{\partial \lambda_1 \partial \lambda_p}\right) \\ E\left(\dfrac{\partial^2 \ell}{\partial \lambda_1 \partial \lambda_2}\right) & E\left(\dfrac{\partial^2 \ell}{\partial \lambda_2^2}\right) & \cdots & E\left(\dfrac{\partial^2 \ell}{\partial \lambda_2 \partial \lambda_p}\right) \\ \vdots & & & \\ E\left(\dfrac{\partial^2 \ell}{\partial \lambda_1 \partial \lambda_p}\right) & E\left(\dfrac{\partial^2 \ell}{\partial \lambda_2 \partial \lambda_p}\right) & \cdots & E\left(\dfrac{\partial^2 \ell}{\partial \lambda_p^2}\right) \end{pmatrix}_{\boldsymbol{\lambda} = \widetilde{\boldsymbol{\lambda}}} \quad . \tag{7.5.7}$$

If we neglect higher-order terms, we can give the likelihood function as

$$L = k\,\exp\{-\frac{1}{2}(\boldsymbol{\lambda} - \widetilde{\boldsymbol{\lambda}})^{\mathrm{T}} B (\boldsymbol{\lambda} - \widetilde{\boldsymbol{\lambda}})\} \quad . \tag{7.5.8}$$

Comparing with (5.10.1) shows that this is a p-dimensional normal distribution with mean $\widetilde{\boldsymbol{\lambda}}$ and covariance matrix

$$C = B^{-1} \quad . \tag{7.5.9}$$

The variances of the maximum likelihood estimators $\widetilde{\lambda}_1, \widetilde{\lambda}_2, \ldots, \widetilde{\lambda}_p$ are given by the diagonal elements of the matrix (7.5.9). The off-diagonal elements are the covariances between all possible pairs of estimators,

$$\sigma^2(\widetilde{\lambda}_i) = c_{ii} \quad , \tag{7.5.10}$$

$$\mathrm{cov}(\widetilde{\lambda}_j, \widetilde{\lambda}_k) = c_{jk} \quad . \tag{7.5.11}$$

For the correlation coefficient between the estimators $\widetilde{\lambda}_j, \widetilde{\lambda}_k$ we can define

$$\varrho(\widetilde{\lambda}_j, \widetilde{\lambda}_k) = \frac{\mathrm{cov}(\widetilde{\lambda}_j, \widetilde{\lambda}_k)}{\sigma(\widetilde{\lambda}_j)\sigma(\widetilde{\lambda}_k)} \quad . \tag{7.5.12}$$

As in the case of a single parameter, the square roots of the variances are given as the error or standard deviations of the estimators,

$$\Delta\widetilde{\lambda}_i = \sigma(\widetilde{\lambda}_i) = \sqrt{c_{ii}} \quad . \tag{7.5.13}$$

In Sect. 7.4 we determined that by giving the maximum-likelihood estimator and its error one defines a region that contains the true value of the parameter with a probability of 68.3 %. Since the likelihood function in the several parameter case is asymptotically a Gaussian distribution of several variables, this region is not determined only by the errors, but rather by the entire covariance matrix. In the special case of two parameters this is the covariance ellipse, which we introduced in Sect. 5.10.

The expression (7.5.8) has (with the replacement $\mathbf{x} = \boldsymbol{\lambda}$) exactly the form of (5.10.1). We can therefore use it for all of the results of Sect. 5.10. For the exponent one has

$$-\frac{1}{2}(\boldsymbol{\lambda} - \widetilde{\boldsymbol{\lambda}})^{\mathrm{T}} B (\boldsymbol{\lambda} - \widetilde{\boldsymbol{\lambda}}) = -\frac{1}{2} g(\boldsymbol{\lambda}) = -\left\{\ell(\boldsymbol{\lambda}) - \ell(\widetilde{\boldsymbol{\lambda}})\right\} \quad . \tag{7.5.14}$$

In the parameter space spanned by $\lambda_1, \ldots, \lambda_p$, the covariance ellipsoid of the distribution (7.5.8) is then determined by the condition

$$g(\boldsymbol{\lambda}) = 1 = 2\left\{\ell(\boldsymbol{\lambda}) - \ell(\widetilde{\boldsymbol{\lambda}})\right\} \quad . \tag{7.5.15}$$

For other values of $g(\boldsymbol{\lambda})$ one obtains the confidence ellipsoids introduced in Sect. 5.10. For smaller values of N, the series (7.5.3) cannot be truncated and the approximation (7.5.7) is not valid. Nevertheless, the solution (7.5.4) can clearly still be computed. For a given probability W one obtains instead of a confidence ellipsoid a *confidence region*, contained within the hypersurface

$$g(\boldsymbol{\lambda}) = 2\left\{\ell(\boldsymbol{\lambda}) - \ell(\widetilde{\boldsymbol{\lambda}})\right\} = \text{const} \quad . \qquad (7.5.16)$$

The value of g is determined in the same way as for the confidence ellipsoid as in Sect. 5.10.

In Example 7.7 we computed the region $\widetilde{\lambda} - \Delta_- < \lambda < \widetilde{\lambda} + \Delta_+$ for the case of a single variable. This corresponds to a confidence region with the probability content 68.3 %.

Example 7.8: Estimation of the mean and variance of a normal distribution

We want to determine the mean λ_1 and standard deviation λ_2 of a normal distribution using a sample of size N. This problem occurs, for example, in the measurement of the range of α-particles in matter. Because of the statistical nature of the energy loss through a large number of independent individual collisions, the range of the individual particles is Gaussian distributed about some mean value. By measuring the range $\mathrm{x}^{(j)}$ of N different particles, the mean λ_1 and "straggling constant" $\lambda_2 = \sigma$ can be estimated. We obtain the likelihood function

$$L = \prod_{j=1}^{N} \frac{1}{\lambda_2\sqrt{2\pi}} \exp\left(-\frac{(\mathrm{x}^{(j)} - \lambda_1)^2}{2\lambda_2^2}\right)$$

and

$$\ell = -\frac{1}{2}\sum_{j=1}^{N} \frac{(\mathrm{x}^{(j)} - \lambda_1)^2}{\lambda_2^2} - N\ln\lambda_2 - \text{const} \quad .$$

The system of likelihood equations is

$$\begin{aligned} \frac{\partial\ell}{\partial\lambda_1} &= \sum_{j=1}^{N} \frac{\mathrm{x}^{(j)} - \lambda_1}{\lambda_2^2} = 0 \quad , \\ \frac{\partial\ell}{\partial\lambda_2} &= \frac{1}{\lambda_2^3}\sum_{j=1}^{N}(\mathrm{x}^{(j)} - \lambda_1)^2 - \frac{N}{\lambda_2} = 0 \quad . \end{aligned}$$

Its solution is

$$\begin{aligned} \widetilde{\lambda}_1 &= \frac{1}{N}\sum_{j=1}^{N}\mathrm{x}^{(j)} \quad , \\ \widetilde{\lambda}_2 &= \sqrt{\frac{\sum_{j=1}^{N}(\mathrm{x}^{(j)} - \widetilde{\lambda}_1)^2}{N}} \quad . \end{aligned}$$

For the estimator of the mean, the maximum-likelihood method leads to the arithmetic mean of the individual measurements. For the variance it gives the quantity s'^2 (6.2.4), which has a small bias, and not s^2, the unbiased estimator (6.2.6).

Let us now determine the matrix B. The second derivatives are

$$\begin{aligned} \frac{\partial^2 \ell}{\partial \lambda_1^2} &= -\frac{N}{\lambda_2^2} \quad , \\ \frac{\partial^2 \ell}{\partial \lambda_1 \partial \lambda_2} &= -\frac{2\sum(\mathrm{x}^{(j)} - \lambda_1)}{\lambda_2^3} \quad , \\ \frac{\partial^2 \ell}{\partial \lambda_2^2} &= -\frac{3\sum(\mathrm{x}^{(j)} - \lambda_1)^2}{\lambda_2^4} + \frac{N}{\lambda_2^2} \quad . \end{aligned}$$

We use the procedure of (7.5.7), substitute λ_1, λ_2 by $\widetilde{\lambda}_1, \widetilde{\lambda}_2$ and find

$$B = \begin{pmatrix} N/\widetilde{\lambda}_2^2 & 0 \\ 0 & 2N/\widetilde{\lambda}_2^2 \end{pmatrix}$$

or for the covariance matrix

$$C = B^{-1} = \begin{pmatrix} \widetilde{\lambda}_2^2/N & 0 \\ 0 & \widetilde{\lambda}_2^2/2N \end{pmatrix} \quad .$$

We interpret the diagonal elements as the errors of the corresponding parameters, i.e.,

$$\Delta\widetilde{\lambda}_1 = \widetilde{\lambda}_2/\sqrt{N} \quad , \qquad \Delta\widetilde{\lambda}_2 = \widetilde{\lambda}_2/\sqrt{2N} \quad .$$

The estimators for λ_1 and λ_2 are not correlated. ■

Example 7.9: Estimators for the parameters of a two-dimensional normal distribution

To conclude we consider a population described by a two-dimensional normal distribution (Sect. 5.10)

$$\begin{aligned} f(x_1, x_2) = {} & \frac{1}{2\pi\sigma_1\sigma_2\sqrt{1-\varrho^2}} \exp\left[-\frac{1}{2(1-\varrho^2)}\right. \\ & \times \left.\left\{\frac{(x_1-a_1)^2}{\sigma_1^2} + \frac{(x_2-a_2)^2}{\sigma_2^2} - 2\varrho\frac{(x_1-a_1)(x_2-a_2)}{\sigma_1\sigma_2}\right\}\right] \quad . \end{aligned}$$

By constructing and solving a system of five simultaneous likelihood equations for the five parameters $a_1, a_2, \sigma_1^2, \sigma_2^2, \varrho$ we obtain for the maximum-likelihood estimators

$$\bar{\mathrm{x}}_1 = \frac{1}{N}\sum_{j=1}^{N} \mathrm{x}_1^{(j)} \quad , \qquad \bar{\mathrm{x}}_2 = \frac{1}{N}\sum_{j=1}^{N} \mathrm{x}_2^{(j)} \quad ,$$

$$\mathrm{s}_1'^2 = \frac{1}{N}\sum_{j=1}^{N} (\mathrm{x}_1^{(j)} - \bar{\mathrm{x}}_1)^2 \quad , \qquad \mathrm{s}_2'^2 = \frac{1}{N}\sum_{j=1}^{N} (\mathrm{x}_2^{(j)} - \bar{\mathrm{x}}_2)^2 \quad ,$$

$$\mathrm{r} = \frac{\sum_{j=1}^{N} (\mathrm{x}_1^{(j)} - \bar{\mathrm{x}}_1)(\mathrm{x}_2^{(j)} - \bar{\mathrm{x}}_2)}{N \mathrm{s}_1' \mathrm{s}_2'} \quad . \tag{7.5.17}$$

Exactly as in Example 7.8, the estimators for the variances $\mathrm{s}_1'^2$ and $\mathrm{s}_2'^2$ are biased. This also holds for the expression (7.5.17), the *sample correlation coefficient* r. Like all maximum likelihood estimators, r is consistent, i.e., it provides a good estimation of ϱ for very large samples. For $N \to \infty$ the probability density of the random variable r becomes a normal distribution with mean ϱ and variance

$$\sigma^2(\mathrm{r}) = (1 - \varrho^2)^2 / N \quad . \tag{7.5.18}$$

For finite samples the distribution is asymmetric. It is therefore important to have a sufficiently large sample before applying Eq. (7.5.17). As a rule of thumb, $N \geq 500$ is usually recommended. ■

7.6 Example Programs

The source files of the example programs are available in the subdirectory `\DATAN\EXASRC`.

7.6.1 Program `E1ML` to Compute the Mean Lifetime and Asymmetric Errors from a Small Number of Radioactive Decays

The program performs the computations and the graphical display for the problem described in Example 7.7. First by Monte Carlo method a total of N decay times t_i of radioactive nuclei with a mean lifetime of $\tau = 1$ are simulated. The number N of decays and also the seeds for the random number generator are entered interactively. (The choice of seeds allows for the generation of different sets of decay times.)

7.6.2 Program E2ML to Compute the Maximum-Likelihood Estimates of the Parameters of a Bivariate Normal Distribution from a Simulated Sample

The program first asks the user for the number $n_{\rm exp}$ of experiments to simulate (i.e., of the samples to be treated consecutively), for the size $n_{\rm pt}$ of each sample and for the means a_1, a_2, the standard deviations σ_1, σ_2, and the correlation coefficient ρ of a bivariate Gaussian distribution.

The covariance matrix C of the normal distribution is calculated and the generator of random numbers from a multivariate Gaussian distribution is initialized by calling RNMNPR. Each sample is generated and then analyzed, i.e., the quantities $\bar{x}_1$, $\bar{x}_2$, s'_1, s'_2, and r are computed, which are estimates of a_1, a_2, σ_1, σ_2, and ρ [cf. (7.5.17)]. The quantities are displayed for each sample.

Suggestions: Choose $n_{\rm exp} = 20$, keep all other parameters fixed, and study the statistical fluctuations of r for $n_{\rm pt} = 5, 50, 500$. Use the values $\rho = 0,\ 0.5,\ 0.95$.

7.7 Programming Problems

The source files of the example solutions are available in the subdirectory `\DATAN\SOLSRC`.

7.7.1 Distribution of Lifetimes Determined from a Small Number of Radioactive Decays

In Example Program 7.6.1, an estimate $\bar{t}$ of the mean lifetime τ and its asymmetric errors Δ_- and Δ_+ are found from a single small sample. In all cases the program yields $\Delta_- < \Delta_+$. Write a program that simulates a large number $n_{\rm exp}$ of experiments, in each of which N radioactive decays of mean lifetime $\tau = 1$ are measured. Compute for each experiment the estimate $\bar{t}$ and construct a histogram of the quantity $\bar{t}$ for all experiments. Allow for the interactive choice between the presentation of this histogram $N_i(\bar{t}_i \le \bar{t} < \bar{t}_i + \Delta\bar{t})$ and of the cumulative frequency distribution $h_i = (1/n_{\rm exp})\sum_{\bar{t}<\bar{t}_i} N_i$. Allow for interactive input of $n_{\rm exp}$ and N. Demonstrate that the distributions are asymmetric for small N and that they become symmetric for large N. Show that for small N the value $\bar{t} = 1$ is not the most probable value, but that it is the expectation value of $\bar{t}$. Determine for a fixed value of N, e.g., $N = 4$, limits Δ_- and Δ_+ in such a way that $\bar{t} < 1 - \Delta_-$ holds with the probability $0.683/2$ and that with the same probability one has $\bar{t} > 1 + \Delta_+$. Compare the results found with a series of simulated experiments from the program E1ML, Example Program 7.6.1.

7.7.2 Distribution of the Sample Correlation Coefficient

Modify the program E2ML so that instead of numerical output, a histogram of the correlation coefficient r is presented (Fig. 7.3). Produce histograms for $\rho = 0$ and $\rho = 0.95$, each for $n_{pt} = 5, 50, 500$. Under what circumstances is the distribution asymmetric and why? Is this asymmetry in contradiction to the Central Limit theorem?

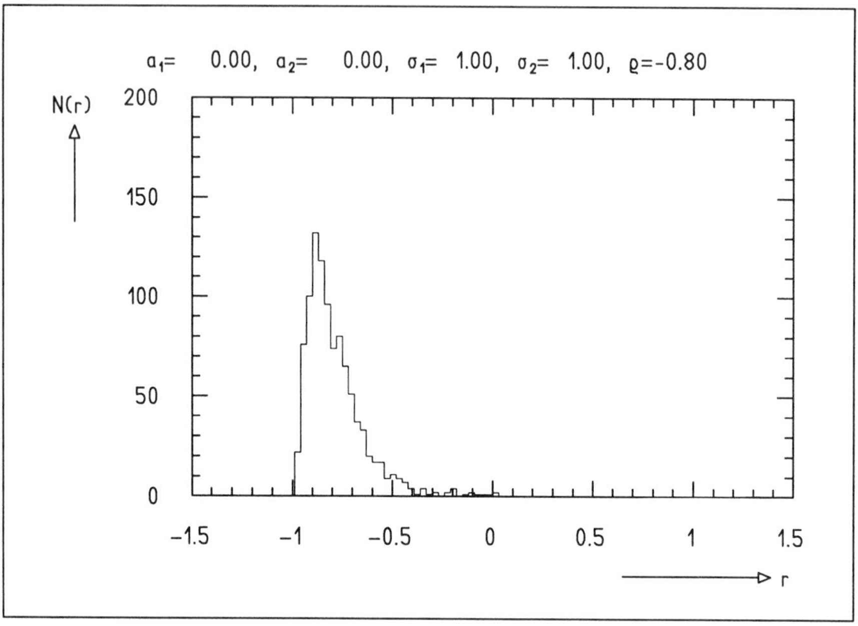

Fig. 7.3. Histogram of the sample correlation coefficient computed for 1000 samples of size 10 from a bivariate Gaussian distribution with the correlation coefficient $\rho = -0.8$.

7.8 Problems

7.8.1 Maximum-Likelihood Estimates

(a) Suppose it is known that a random variable x follows the uniform distribution $f(x) = 1/b$ for $0 < x < b$. The parameter b is to be estimated from a sample. Show that $\mathrm{S} = \widetilde{b} = \mathrm{x}_{max}$ is the maximum-likelihood estimator of b. (Hint: This result cannot be obtained by differentiation, but from a simple consideration about the likelihood function).

(b) Write down the likelihood equations for the two parameters a and Γ of the Lorentz distribution (see Example 3.5). Show that these do not necessarily have unique solutions. You can, however, easily convince yourself that for $|\mathsf{x}^{(j)} - a| \ll \Gamma$ the arithmetic mean $\bar{\mathsf{x}}$ is an estimator of a.

7.8.2 Information

(a) Determine the information $I(\lambda)$ of a sample of size N that was obtained from a normal distribution of known variance σ^2 but unknown mean $\lambda = a$.

(b) Determine the information $I(\lambda)$ of a sample of size N which was drawn from a normal distribution of known mean a but unknown variance $\lambda = \sigma^2$. Show that the maximum-likelihood estimator of σ^2 is given by

$$\mathsf{S} = \frac{1}{N} \sum_{j=1}^{N} (\mathsf{x}^{(j)} - a)^2$$

and that the estimator is unbiased, i.e., $B(\mathsf{S}) = E(\mathsf{S}) - \lambda = 0$.

7.8.3 Variance of an Estimator

(a) Use the information inequality to obtain a lower limit on the variance of the estimator of the mean in Problem 7.8.2 (a). Show that this limit is equal to the minimum variance that was determined in Problem 6.13.1 (c).

(b) Use the information inequality to obtain a lower limit on the variance of S in Problem 7.8.2 (b).

(c) Show using Eq. (7.3.12) that S is a minimum variance estimator with the same variance the lower limit found in (b).

7.9 Hints and Solutions

Problem 7.8.1

(a) $L = \prod_{j=1}^{N} \frac{1}{b} = \frac{1}{b^N}$; obviously one has $L = \max$ for $b = \mathsf{x}_{\max}$.

(b) $\lambda_1 = a$, $\lambda_2 = \Gamma$,

$$\begin{aligned}
L &= \prod_{j=1}^{N} \frac{2}{\pi\lambda_2} \frac{\lambda_2^2}{4(\mathsf{x}^{(j)} - \lambda_1)^2 + \lambda_2^2} \\
&= \left(\frac{2\lambda_2}{\pi}\right)^N \prod_{j=1}^{N} \frac{1}{4(\mathsf{x}^{(j)} - \lambda_1)^2 + \lambda_2^2} \quad , \\
\ell &= N(\ln 2 - \ln\pi + \ln\lambda_2) - \sum_{j=1}^{N} \ln[4(\mathsf{x}^{(j)} - \lambda_1)^2 + \lambda_2^2] \quad , \\
\frac{\partial\ell}{\partial\lambda_1} &= \sum_{j=1}^{N} \frac{8(\mathsf{x}^{(i)} - \lambda_1)}{4(\mathsf{x}^{(j)} - \lambda_1)^2 + \lambda_2^2} = 0 \quad , \\
\frac{\partial\ell}{\partial\lambda_2} &= \frac{N}{\lambda_2} - 2\lambda_2 \sum_{j=1}^{N} \frac{1}{4(\mathsf{x}^{(j)} - \lambda_1)^2 + \lambda_2^2} = 0 \quad .
\end{aligned}$$

There is not necessarily a unique solution since the equations are not linear in λ_1 and λ_2. For $|\mathsf{x}^{(j)} - \lambda_1| \ll \lambda_2$, however, we may write

$$\frac{8}{\lambda_2^2} \sum_{j=1}^{N} (\mathsf{x}^{(j)} - \lambda_1) = 0 \quad , \qquad N\lambda_1 = \sum_{j=1}^{N} \mathsf{x}^{(j)} \quad , \qquad \lambda_1 = a = \bar{\mathsf{x}} \quad .$$

Problem 7.8.2

(a)

$$\begin{aligned}
L &= \prod_{j=1}^{N} \frac{1}{\sqrt{2\pi}\sigma} \exp[-(\mathsf{x}^{(j)} - \lambda)^2/2\sigma^2] \\
&= (\sqrt{2\pi}\sigma)^{-N} \prod_{j=1}^{N} \exp[-(\mathsf{x}^{(j)} - \lambda)^2/2\sigma^2] \quad , \\
\ell &= -N\ln(\sqrt{2\pi}\sigma) - \frac{1}{2\sigma^2} \sum_{j=1}^{N} (\mathsf{x}^{(j)} - \lambda)^2 \quad , \\
\ell' &= \frac{1}{\sigma^2} \sum_{j=1}^{N} (\mathsf{x}^{(j)} - \lambda) \quad , \\
\ell'' &= -N/\sigma^2 \quad , \\
I(\lambda) &= -E(\ell'') = N/\sigma^2 \quad .
\end{aligned}$$

(b)

$$\begin{aligned}
L &= \prod_{j=1}^{N} \frac{1}{\sqrt{2\pi}\sqrt{\lambda}} \exp\left(-\frac{(\mathsf{x}^{(j)} - a)^2}{2\lambda}\right) \\
&= (2\pi)^{-N/2} \lambda^{-N/2} \prod_{j=1}^{N} \exp\left(-\frac{(\mathsf{x}^{(j)} - a)^2}{2\lambda}\right) \quad ,
\end{aligned}$$

$$
\begin{aligned}
\ell &= -\frac{N}{2}\ln(2\pi) - \frac{N}{2}\ln\lambda - \frac{1}{2\lambda}\sum_{j=1}^{N}(\mathsf{x}^{(j)} - a)^2 \quad , \\
\ell' &= -\frac{N}{2\lambda} + \frac{1}{2\lambda^2}\sum_{j=1}^{N}(\mathsf{x}^{(j)} - a)^2 \quad , \\
\ell'' &= \frac{N}{2\lambda^2} - \frac{1}{\lambda^3}\sum_{j=1}^{N}(\mathsf{x}^{(j)} - a)^2 \quad , \\
I(\lambda) &= -E(\ell'') = -\frac{N}{2\lambda^2} + \frac{1}{\lambda^3}E\left\{\sum_{j=1}^{N}(\mathsf{x}^{(j)} - a)^2\right\} \\
&= -\frac{N}{2\lambda^2} + \frac{1}{\lambda^3}N\lambda \quad , \\
I(\lambda) &= \frac{1}{2\lambda^2}(-N + 2N) = \frac{N}{2\lambda^2} \quad , \\
E(\mathsf{S}) &= \frac{1}{N}E\left\{\sum_{j=1}^{N}(\mathsf{x}^{(j)} - a)^2\right\} = \sigma^2 = \lambda \quad .
\end{aligned}
$$

Problem 7.8.3

(a) $\sigma^2(\mathsf{S}) \geq \dfrac{1}{I(\lambda)} = \dfrac{\sigma^2}{N}$,

In Problem 6.13.1 (c) we had $\sigma^2(\bar{\mathsf{x}}) = 1/3\sigma^2$ for $N = 3$.

(b) $\sigma^2(\mathsf{S}) \geq \dfrac{1}{I(\lambda)} = \dfrac{2\lambda^2}{N}$.

(c) From Problem 7.8.2 we take

$$
\ell' = -\frac{N}{2\lambda} + \frac{1}{2\lambda^2}N\mathsf{S} = \frac{N}{2\lambda^2}(\mathsf{S} - \lambda) = \frac{N}{2\lambda^2}(\mathsf{S} - E(\mathsf{S})) \quad .
$$

Thus, $\sigma^2(\mathsf{S}) = \dfrac{2\lambda^2}{N}$.

8. Testing Statistical Hypotheses

8.1 Introduction

Often the problem of a statistical analysis does not involve determining an originally unknown parameter, but rather, one already has a preconceived opinion about the value of the parameter, i.e., a *hypothesis*. In a sample taken for quality control, for example, one might initially assume that certain critical values are normally distributed within tolerance levels around their nominal values. One would now like to test this hypothesis. To elucidate such test procedures, called *statistical tests*, we will consider such an example and for simplicity make the hypothesis that a sample of size 10 originates from a standard normal distribution.

Suppose the analysis of the sample resulted in the arithmetic mean $\bar{\mathrm{x}} = 0.154$. Under the assumption that our hypothesis is correct, the random variable $\bar{\mathrm{x}}$ is normally distributed with mean 0 and standard deviation $\frac{1}{\sqrt{10}}$. We now ask for the probability to observe a value $|\bar{\mathrm{x}}| \geq 0.154$ from such a distribution. From (5.8.5) and Table I.3 this is

$$P(|\bar{\mathrm{x}}| \geq 0.154) = 2\{1 - \psi_0(0.154\sqrt{10})\} = 0.62 \quad .$$

Thus we see that even if our hypothesis is correct, there is a probability of 62 % that a sample of size 10 will lead to a sample mean that differs from the population mean by 0.154 or more.

We now find ourselves in the difficult situation of having to answer the simple question: "Is our hypothesis true or false?" A solution to this problem is provided by the concept of the *significance level*: One specifies before the test a (small) test probability α. Staying with our previous example, if $P(|\bar{\mathrm{x}}| \geq 0.154) < \alpha$, then one would regard the occurrence of $\bar{\mathrm{x}} = 0.154$ as improbable. That is, one would say that $\bar{\mathrm{x}}$ differs significantly from the hypothesized value and the hypothesis would be rejected. The converse is, however, not true. If P does not fall below α, we cannot say that "the hypothesis is true", but rather "it is not contradicted by the result of the sample". The choice of the significance level depends on the problem being considered. For quality control of pencils one might be satisfied with 1 %. If, however, one wishes to determine insurance premiums such that the probability for the company to go bankrupt is less than α, then one would probably still regard 0.01 % as too high. In the analysis of scientific data α values of 5 %, 1 %, or 0.1 % are typically used. From Table I.3 we can obtain limiting values for $|\bar{x}|$ such that a deviation in excess of these values corresponds to the given probabilities. These are

$$\begin{aligned} 0.05 &= 2\{1-\psi_0(1.96)\} = 2\{1-\psi_0(0.62\sqrt{10})\} \quad , \\ 0.01 &= 2\{1-\psi_0(2.58)\} = 2\{1-\psi_0(0.82\sqrt{10})\} \quad , \\ 0.001 &= 2\{1-\psi_0(3.29)\} = 2\{1-\psi_0(1.04\sqrt{10})\} \quad . \end{aligned}$$

At these significance levels the value $|\bar{x}|$ would have to exceed the values 0.62, 0.82, 1.04 before we could reject the hypothesis.

In some cases the sign of $\bar{x}$ is important. In many production processes, deviations in a positive and negative direction are of different importance. (If a baker's rolls are too heavy, this reduces profits; if they are too light they cost the baker his license.) If one sets, e.g.,

$$P(\bar{x} \geq x'_\alpha) < \alpha \quad ,$$

then this is referred to as a *one-sided test* in contrast to the *two-sided test*, which we have already considered (Fig. 8.1).

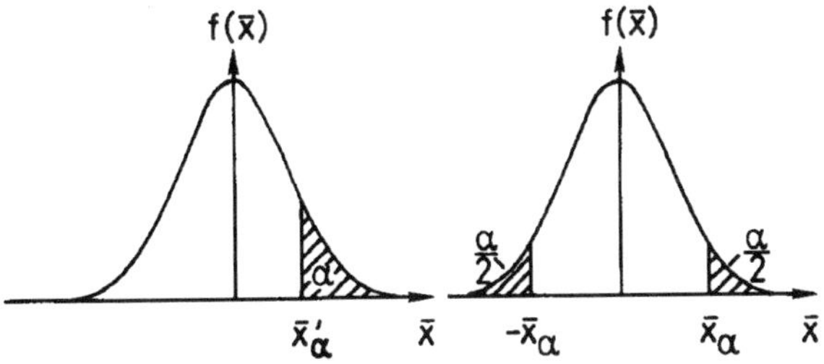

Fig. 8.1. One-sided and two-sided tests.

For many tests one does not construct the sample mean but rather a certain function of the sample called a *test statistic*, which is particularly suited for tests of certain hypotheses. As above one specifies a certain significance level α and determines a region U in the space of possible values of the test statistic T in which

$$P_H(T \in U) = \alpha \quad .$$

The index H means that the probability was computed under the assumption that the hypothesis H is valid. One then obtains a sample, which results in a particular value T' for the test statistic. If T' is in the region U, the *critical region* of the test, then the hypothesis is rejected.

In the next sections we will discuss some important tests in detail and then turn to a more rigorous treatment of test theory.

8.2 *F*-Test on Equality of Variances

The problem of comparing two variances occurs frequently in the development of measurement techniques or production procedures. Consider two populations with the same expectation value; e.g., one measures the same quantity with two different devices without systematic error. One may then ask if they also have the same variance.

To test this hypothesis we take samples of size N_1 and N_2 from both populations, which we assume to be normally distributed. We construct the sample variance (6.2.6) and consider the ratio

$$F = \mathrm{s}_1^2/\mathrm{s}_2^2 \quad . \tag{8.2.1}$$

If the hypothesis is true, then F will be near unity. It is known from Sect. 6.6 that for every sample we can construct a quantity that follows a χ^2-distribution:

$$\begin{aligned} X_1^2 &= \frac{(N_1-1)\mathrm{s}_1^2}{\sigma_1^2} = \frac{f_1 \mathrm{s}_1^2}{\sigma_1^2} \quad , \\ X_2^2 &= \frac{(N_2-1)\mathrm{s}_2^2}{\sigma_2^2} = \frac{f_2 \mathrm{s}_2^2}{\sigma_2^2} \quad . \end{aligned}$$

The two distributions have $f_1 = (N_1 - 1)$ and $f_2 = (N_2 - 1)$ degrees of freedom.

Under the assumption that the hypothesis ($\sigma_1^2 = \sigma_2^2$) is true, one has

$$F = \frac{f_2}{f_1}\frac{X_1^2}{X_2^2} \quad .$$

The probability density of a χ^2-distribution with f degrees of freedom is [see (6.6.10)]

$$f(\chi^2) = \frac{1}{\Gamma(\frac{1}{2}f)2^{\frac{1}{2}f}}(\chi^2)^{\frac{1}{2}(f-2)}\mathrm{e}^{-\frac{1}{2}\chi^2} \quad .$$

We now compute the probability*

$$W(Q) = P\left(\frac{X_1^2}{X_2^2} < Q\right)$$

that the ratio X_1^2/X_2^2 is smaller than Q:

$$W(Q) = \frac{1}{\Gamma(\frac{1}{2}f_1)\Gamma(\frac{1}{2}f_2)2^{\frac{1}{2}(f_1+f_2)}} \int\int_{\substack{x>0\\ y>0\\ x/y>Q}} x^{\frac{1}{2}f_1-1}\mathrm{e}^{-\frac{1}{2}x} y^{\frac{1}{2}f_2-1}\mathrm{e}^{-\frac{1}{2}y}\,\mathrm{d}x\,\mathrm{d}y \quad .$$

*We use here the symbol W for a distribution function in order to avoid confusion with the ratio F.

Calculating the integral gives

$$W(Q) = \frac{\Gamma(\frac{1}{2}f)}{\Gamma(\frac{1}{2}f_1)\Gamma(\frac{1}{2}f_2)} \int_0^Q t^{\frac{1}{2}f_1 - 1}(t+1)^{-\frac{1}{2}f}\, dt \quad , \tag{8.2.2}$$

where

$$f = f_1 + f_2 \quad .$$

Finally if one sets

$$F = Q\, f_2/f_1 \quad ,$$

then the distribution function of the ratio F can be obtained from (8.2.2),

$$W(F) = P\left(\frac{s_1^2}{s_2^2} < F\right) \quad .$$

This is called the Fisher F-distribution.† It depends on the parameters f_1 and f_2. The probability density for the F-distribution is

$$f(F) = \left(\frac{f_1}{f_2}\right)^{\frac{1}{2}f_1} \frac{\Gamma(\frac{1}{2}(f_1+f_2))}{\Gamma(\frac{1}{2}f_1)\Gamma(\frac{1}{2}f_2)} F^{\frac{1}{2}f_1 - 1}\left(1 + \frac{f_1}{f_2}F\right)^{-\frac{1}{2}(f_1+f_2)} \quad . \tag{8.2.3}$$

This is shown in Fig. 8.2 for fixed values of f_1 and f_2. The distribution is reminiscent of the χ^2-distribution; it is only non-vanishing for $F \geq 0$, and has a long tail for $F \to \infty$. Therefore it cannot be symmetric. One can easily show that for $f_2 > 2$ the expectation value is simply

$$E(F) = f_2/(f_2 - 2) \quad .$$

We can now determine a limit F'_α with the requirement

$$P\left(\frac{s_1^2}{s_2^2} > F'_\alpha\right) = \alpha \quad . \tag{8.2.4}$$

This expression means that the limit F'_α is equal to the *quantile* $F_{1-\alpha}$ of the F-distribution (see Sect. 3.3) since

$$P\left(\frac{s_1^2}{s_2^2} < F'_\alpha\right) = P\left(\frac{s_1^2}{s_2^2} < F_{1-\alpha}\right) = 1 - \alpha \quad . \tag{8.2.5}$$

If this limit is exceeded, then we say that $\sigma_1^2 > \sigma_2^2$ with the significance level α. The quantiles $F_{1-\alpha}$ for various pairs of values (f_1, f_2) are given in Table I.8. In general one applies a two-sided test, i.e., one tests whether the ratio F is between two limits F''_α and F'''_α, which are determined by

†This is also often called the v^2-distribution, ω^2-distribution, or Snedecor distribution.

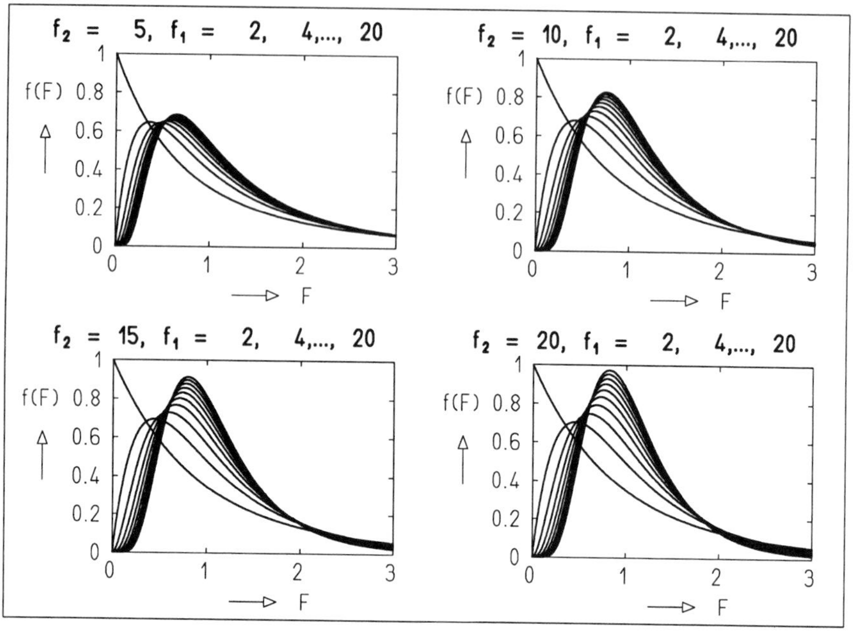

Fig. 8.2. Probability density of the F-distribution for fixed values of $f_1 = 2, 4, \ldots, 20$. For $f_1 = 2$ one has $f(F) = \mathrm{e}^{-F}$. For $f_1 > 2$ the function has a maximum which increases for increasing f_1.

$$P\left(\frac{s_1^2}{s_2^2} > F''_\alpha\right) = \frac{1}{2}\alpha \quad , \qquad P\left(\frac{s_1^2}{s_2^2} < F'''_\alpha\right) = \frac{1}{2}\alpha \quad . \tag{8.2.6}$$

Because of the definition of F as a ratio, the inequality

$$s_1^2/s_2^2 < F'''_\alpha(f_1, f_2)$$

clearly has the same meaning as

$$s_2^2/s_1^2 > F'''_\alpha(f_2, f_1) \quad .$$

Here the first argument gives the number of degrees of freedom in the numerator, and the second in the denominator. The requirement (8.2.6) can also be written:

$$P\left(\frac{s_1^2}{s_2^2} > F''_\alpha(f_1, f_2)\right) = \frac{1}{2}\alpha \quad , \qquad P\left(\frac{s_2^2}{s_1^2} > F''_\alpha(f_2, f_1)\right) = \frac{1}{2}\alpha \quad . \tag{8.2.7}$$

Table I.8 can therefore be used for the one-sided as well as the two-sided F-test.

A glance at Table I.8 also shows that $F_{1-\alpha/2} > 1$ for all reasonable values of α. Therefore one needs only to find the limit for the ratio

$$s_g^2/s_k^2 > F_{1-\frac{1}{2}\alpha}(f_g, f_k) \quad . \tag{8.2.8}$$

Here the indices g and k give the larger and smaller values of the two variances, i.e., $s_g^2 > s_k^2$. If the inequality (8.2.8) is satisfied, then the hypothesis of equal variances must be rejected.

Example 8.1: F-test of the hypothesis of equal variance of two series of measurements

A standard length (100 μm) is measured using two traveling microscopes. The measurements and computations are summarized in Table 8.1. From Table I.8 we find for the two-sided F-test with a significance level of 10 %,

$$F''_{0.1}(6,9) = F_{0.95}(6,9) = 3.37 \quad .$$

The hypothesis of equal variances cannot be rejected. ■

Table 8.1. F-Test on the equality of variances. Data from Example 8.1.

	Measurement with	
Measurement number	instrument 1 [μm]	instrument 2 [μm]
1	100	97
2	101	102
3	103	103
4	98	96
5	97	100
6	98	101
7	102	100
8	101	
9	99	
10	101	
Mean	100	99.8
Degrees of freedom	9	6
s^2	$34/9 = 3.7$	$39/6 = 6.5$
$F = 6.5/3.7 = 1.8$		

8.3 Student's Test. Comparison of Means

We now consider a population that follows a standard Gaussian distribution. Let $\bar{\mathrm{x}}$ be the arithmetic mean of a sample of size N. According to (6.2.3) the variance of $\bar{\mathrm{x}}$ is related to the population variance by

$$\sigma^2(\bar{\mathrm{x}}) = \sigma^2(\mathrm{x})/N \quad . \tag{8.3.1}$$

If N is sufficiently large, then from the Central Limit Theorem, $\bar{\mathrm{x}}$ will be normally distributed with mean $\widehat{x}$ and variance $\sigma^2(\bar{\mathrm{x}})$. That is,

$$\mathrm{y} = (\bar{\mathrm{x}} - \widehat{x})/\sigma(\bar{\mathrm{x}}) \tag{8.3.2}$$

will be described by a standard normal distribution. The quantity $\sigma(\mathrm{x})$ is, however, not known. We only know the estimate for $\sigma^2(\mathrm{x})$,

$$\mathrm{s}_x^2 = \frac{1}{N-1}\sum_{j=1}^{N}(\mathrm{x}_j - \bar{\mathrm{x}})^2 \quad . \tag{8.3.3}$$

Then with (8.3.1) we can also estimate $\sigma^2(\bar{\mathrm{x}})$ to be

$$\mathrm{s}_{\bar{\mathrm{x}}}^2 = \frac{1}{N(N-1)}\sum_{j=1}^{N}(\mathrm{x}_j - \bar{\mathrm{x}})^2 \quad . \tag{8.3.4}$$

We now ask to what extent (8.3.2) differs from the standard Gaussian distribution if $\sigma(\bar{\mathrm{x}})$ is replaced by $\mathrm{s}_{\bar{\mathrm{x}}}$. By means of a simple translation of coordinates we can always have $\widehat{x} = 0$. We therefore only consider the distribution of

$$\mathrm{t} = \bar{\mathrm{x}}/\mathrm{s}_{\bar{\mathrm{x}}} = \bar{\mathrm{x}}\sqrt{N}/\mathrm{s}_{\mathrm{x}} \quad . \tag{8.3.5}$$

Since $(N-1)\,\mathrm{s}_{\mathrm{x}}^2 = f\,\mathrm{s}_{\mathrm{x}}^2$ follows a χ^2-distribution with $f = N-1$ degrees of freedom, we can write

$$\mathrm{t} = \bar{\mathrm{x}}\sqrt{N}\sqrt{f}/\chi \quad .$$

The distribution function of t is given by

$$F(t) = P(\mathrm{t} < t) = P\left(\frac{\bar{\mathrm{x}}\sqrt{N}\sqrt{f}}{\chi} < t\right) \quad . \tag{8.3.6}$$

After a somewhat lengthy calculation one finds

$$F(t) = \frac{\Gamma(\frac{1}{2}(f+1))}{\Gamma(\frac{1}{2}f)\sqrt{\pi}\sqrt{f}}\int_{-\infty}^{t}\left(1+\frac{t^2}{f}\right)^{-\frac{1}{2}(f+1)} \mathrm{d}t \quad .$$

The corresponding probability density is

$$f(t) = \frac{\Gamma(\frac{1}{2}(f+1))}{\Gamma(\frac{1}{2}f)\sqrt{\pi}\sqrt{f}}\left(1+\frac{t^2}{f}\right)^{-\frac{1}{2}(f+1)} \quad . \tag{8.3.7}$$

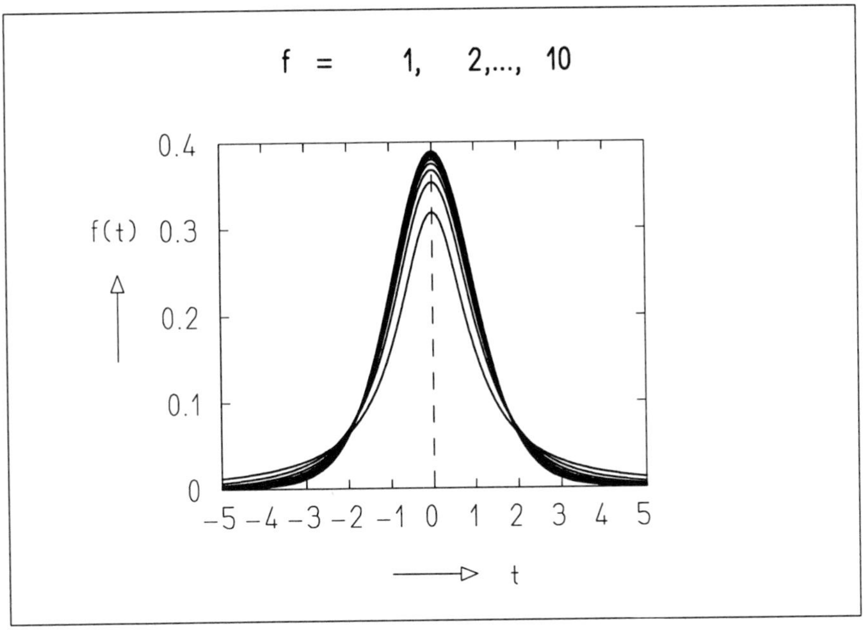

Fig. 8.3. Student's distribution $f(t)$ for $f = 1, 2, \ldots, 10$ degrees of freedom. For $f = 1$ the maximum is lowest and the tails are especially prominent.

Figure 8.3 shows the function $f(t)$ for various degrees of freedom $f = N - 1$. A comparison with Fig. 5.7 shows that for $f \to \infty$, the distribution (8.3.7) becomes the standard normal distribution $\phi_0(t)$, as expected. Like $\phi_0(t)$, $f(t)$ is symmetric about 0 and has a bell shape. Corresponding to (5.8.3) one has

$$P(|t| \le t) = 2F(|t|) - 1 \quad . \tag{8.3.8}$$

By requiring

$$\int_0^{t'_\alpha} f(t)\,\mathrm{d}t = \frac{1}{2}(1 - \alpha) \tag{8.3.9}$$

we can again determine limits $\pm t'_\alpha$ at a given significance level α, where

$$t'_\alpha = t_{1-\frac{1}{2}\alpha} \quad .$$

The quantiles $t_{1-\frac{1}{2}\alpha}$ are given in Table I for various values of α and f.

The application of Student's test[‡] can be described in the following way: One has a hypothesis λ_0 for the population mean of a normal distribution. A sample of size N yields the sample mean $\bar{\mathrm{x}}$ and sample variance s_x^2. If the inequality

[‡]The t-distribution was introduced by W. S. Gosset and published under the pseudonym "Student".

$$|t| = \frac{|\bar{\mathsf{x}} - \lambda_0|\sqrt{N}}{\mathsf{s}_\mathsf{x}} > t'_\alpha = t_{1-\frac{1}{2}\alpha} \qquad (8.3.10)$$

is fulfilled for a given significance level α, then the hypothesis must be rejected.

This is clearly a two-sided test. If deviations only in one direction are important, then the corresponding test at the significance level α is

$$t = \frac{(\bar{\mathsf{x}} \pm \lambda_0)\sqrt{N}}{\mathsf{s}_\mathsf{x}} > t'_{2\alpha} = t_{1-\alpha} \quad . \qquad (8.3.11)$$

We can make the test more general and apply it to the problem of comparing two mean values. Suppose samples of size N_1 and N_2 have been taken from two populations X and Y. We wish to find a measure of correctness for the hypothesis that the expectation values are equal,

$$\widehat{x} = \widehat{y} \quad .$$

Because of the Central Limit Theorem, the mean values are almost normally distributed. Their variances are

$$\sigma^2(\bar{\mathsf{x}}) = \frac{1}{N_1}\sigma^2(\mathsf{x}) \quad , \qquad \sigma^2(\bar{\mathsf{y}}) = \frac{1}{N_2}\sigma^2(\mathsf{y}) \qquad (8.3.12)$$

and the estimates for these quantities are

$$\begin{aligned} \mathsf{s}^2_{\bar{\mathsf{x}}} &= \frac{1}{N_1(N_1-1)}\sum_{j=1}^{N_1}(\mathsf{x}-\bar{\mathsf{x}})^2 \quad , \\ \mathsf{s}^2_{\bar{\mathsf{y}}} &= \frac{1}{N_2(N_2-1)}\sum_{j=1}^{N_2}(\mathsf{y}-\bar{\mathsf{y}})^2 \quad . \end{aligned} \qquad (8.3.13)$$

According to the discussion in Example 5.10, the difference

$$\Delta = \bar{\mathsf{x}} - \bar{\mathsf{y}} \qquad (8.3.14)$$

also has an approximate normal distribution with

$$\sigma^2(\Delta) = \sigma^2(\bar{\mathsf{x}}) + \sigma^2(\bar{\mathsf{y}}) \quad . \qquad (8.3.15)$$

If the hypothesis of equal means is true, i.e., $\widehat{\Delta} = 0$, then the ratio

$$\Delta/\sigma(\Delta) \qquad (8.3.16)$$

follows the standard normal distribution. But only s_Δ is known. The corresponding ratio

$$\Delta/\mathsf{s}_\Delta \qquad (8.3.17)$$

will in general be somewhat larger.

Usually the hypothesis $\widehat{x} = \widehat{y}$ implies that $\bar{\mathsf{x}}$ and $\bar{\mathsf{y}}$ come from the same population. Then $\sigma^2(\mathsf{x})$ and $\sigma^2(\mathsf{y})$ are equal, and we can use the weighted mean of s_x^2 and s_y^2 as the corresponding estimator. The weights are given by $(N_1 - 1)$ and $(N_2 - 1)$:

$$\mathsf{s}^2 = \frac{(N_1 - 1)\mathsf{s}_\mathsf{x}^2 + (N_2 - 1)\mathsf{s}_\mathsf{y}^2}{(N_1 - 1) + (N_2 - 1)} \quad . \tag{8.3.18}$$

From this we construct

$$\mathsf{s}_{\bar{\mathsf{x}}}^2 = \frac{\mathsf{s}^2}{N_1} \quad , \qquad \mathsf{s}_{\bar{\mathsf{y}}}^2 = \frac{\mathsf{s}^2}{N_2} \quad ,$$

and

$$\mathsf{s}_\Delta^2 = \mathsf{s}_{\bar{\mathsf{x}}}^2 + \mathsf{s}_{\bar{\mathsf{y}}}^2 = \frac{N_1 + N_2}{N_1 N_2}\mathsf{s}^2 \quad . \tag{8.3.19}$$

It can be shown (see [8]) that the ratio (8.3.17) follows the Student's t-distribution with $f = N_1 + N_2 - 2$ degrees of freedom. With this one can now perform *Student's difference test*:

The quantity (8.3.17) is computed from the results of two samples. This value is compared to a quantile of Student's distribution with $f = N_1 + N_2 - 2$ degrees of freedom with a significance level α. If

$$|t| = \frac{|\Delta|}{\mathsf{s}_\Delta} = \frac{|\bar{\mathsf{x}} - \bar{\mathsf{y}}|}{\mathsf{s}_\Delta} \geq t'_\alpha = t_{1-\frac{1}{2}\alpha} \quad , \tag{8.3.20}$$

then the hypothesis of equal means must be rejected. Instead one would assume $\widehat{x} > \widehat{y}$ or $\widehat{x} < \widehat{y}$, depending on whether one has $\bar{\mathsf{x}} > \bar{\mathsf{y}}$ or $\bar{\mathsf{x}} < \bar{\mathsf{y}}$.

Example 8.2: Student's test of the hypothesis of equal means of two series of measurements

Column x of Table 8.2 contains measured values (in arbitrary units) of the concentration of neuraminic acid in the red blood cells of patients suffering from a certain blood disease. Column y gives the measured values for a group of healthy persons. From the mean values and variances of the two samples one finds

$$\begin{aligned} |\Delta| &= |\bar{\mathsf{x}} - \bar{\mathsf{y}}| = 1.3 \quad , \\ \mathsf{s}^2 &= \frac{15\mathsf{s}_\mathsf{x}^2 + 6\mathsf{s}_\mathsf{y}^2}{21} = 9.15 \quad , \\ \mathsf{s}_\Delta^2 &= \frac{23}{112}\mathsf{s}^2 = 1.88 \quad . \end{aligned}$$

For $\alpha = 5\,\%$ and $f = 21$ we find $t_{1-\alpha/2} = 2.08$. We must therefore conclude that the experimental data is not sufficient to determine an influence of the disease on the concentration. ■

Table 8.2. Student's difference test on the equality of means. Data from Example 8.2.

x	y
21	16
24	20
18	22
19	19
25	18
17	19
18	19
22	
21	
23	
18	
13	
16	
23	
22	
24	
$N_1 = 16$	$N_2 = 7$
$\bar{\mathrm{x}} = 20.3$	$\bar{\mathrm{y}} = 19.0$
$s_x^2 = 171.8/15$	$s_y^2 = 20/6$

8.4 Concepts of the General Theory of Tests

The test procedures discussed so far have been obtained more or less intuitively and without rigorous justification. In particular we have not given any specific reasons for the choice of the critical region. We now want to deal with the theory of statistical tests in a somewhat more critical way. A complete treatment of this topic would, however, go beyond the scope of this book.

Each sample of size N can be characterized by N points in the sample space of Sect. 2.1. For simplicity we will limit ourselves to a continuous random variable x, so that the sample can be described by N points $(\mathrm{x}^{(1)}, \mathrm{x}^{(2)}, \ldots, \mathrm{x}^{(N)})$ on the x axis. In the case of r random variables we would have N points in an r-dimensional space. The result of such a sample, however, could also be specified by a single point in a space of dimension rN. A sample of size 2 with a single variable could, for example, be depicted as a point in a two-dimensional plane, spanned by the axes $\mathrm{x}^{(1)}, \mathrm{x}^{(2)}$. We will call such a space the E space. Every *hypothesis* H consists of an assumption about the probability density

$$f(x; \lambda_1, \lambda_2, \ldots, \lambda_p) = f(x; \boldsymbol{\lambda}) \quad . \tag{8.4.1}$$

The hypothesis is said to be *simple* if the function f is completely specified, i.e., if the hypothesis gives the values of all of the parameters λ_i. It is said to be *composite* if the general mathematical form of f is known, but the exact value of at least one parameter remains undetermined. A simple hypothesis could, for example, specify a standard Gaussian distribution. A Gaussian distribution with a mean of zero but an undetermined variance, however, is a composite hypothesis. The hypothesis H_0 is called the *null hypothesis*. Sometimes we will write explicitly

$$H_0(\boldsymbol{\lambda} = \boldsymbol{\lambda}_0) = H_0(\lambda_1 = \lambda_{10}, \lambda_2 = \lambda_{20}, \ldots, \lambda_p = \lambda_{p0}) \quad . \tag{8.4.2}$$

Other possible hypotheses are called *alternative hypotheses*, e.g.,

$$H_1(\boldsymbol{\lambda} = \boldsymbol{\lambda}_1) = H_1(\lambda_1 = \lambda_{11}, \lambda_2 = \lambda_{21}, \ldots, \lambda_p = \lambda_{p1}) \quad . \tag{8.4.3}$$

Often one wants to test a null hypothesis of the type (8.4.2) against a composite alternative hypothesis

$$H_1(\lambda \neq \lambda_0) = H_1(\lambda_1 \neq \lambda_{10}, \lambda_2 \neq \lambda_{20}, \ldots, \lambda_p \neq \lambda_{p0}) \quad . \tag{8.4.4}$$

Since the null hypothesis makes a statement about the probability density in the sample space, it also predicts the probability for observing a point $X = (\mathsf{x}^{(1)}, \mathsf{x}^{(2)}, \ldots, \mathsf{x}^{(N)})$ in the E space.[§] We now define a *critical region* S_c with the significance level α by the requirement

$$P(X \in S_c | H_0) = \alpha \quad , \tag{8.4.5}$$

i.e., we determine S_c such that the probability to observe a point X within S_c is α, under the assumption that H_0 is true. If the point X from the sample actually falls into the region S_c, then the hypothesis H_0 is rejected. One must note that the requirement (8.4.5) does not necessarily determine the critical region S_c uniquely.

Although using the E space is conceptually elegant, it is usually not very convenient for carrying out tests. Instead one constructs a *test statistic*

$$T = T(X) = T(\mathsf{x}^{(1)}, \mathsf{x}^{(2)}, \ldots, \mathsf{x}^{(N)}) \tag{8.4.6}$$

and determines a region U of the variable T such that it corresponds to the critical region S_c, i.e., one performs the mapping

$$X \to T(X) \quad , \qquad S_c(X) \to U(X) \quad . \tag{8.4.7}$$

The null hypothesis is rejected if $T \in U$.

[§] Although X and the function $T(x)$ introduced below are random variables, we do not use for them a special character type.

Because of the statistical nature of the sample, it is clearly possible that the null hypothesis could be true, even though it was rejected since $X \in S_c$. The probability for such an error, an *error of the first kind*, is equal to α. There is in addition another possibility to make a wrong decision, if one does not reject the hypothesis H_0 because X was not in the critical region S_c, even though the hypothesis was actually false and an alternative hypothesis was true. This is an *error of the second kind*. The probability for this,

$$P(X \notin S_c|H_1) = \beta \quad , \tag{8.4.8}$$

depends of course on the particular alternative hypotheses H_1. This connection provides us with a method to specify the critical region S_c. A test is clearly most reasonable if for a given significance level α the critical region is chosen such that the probability β for an error of the second kind is a minimum. The critical region and therefore the test itself naturally depend on the alternative hypothesis under consideration.

Once the critical region has been determined, we can consider the probability for rejecting the null hypothesis as a function of the "true" hypothesis, or rather as a function of the parameters that describe it. In analogy to (8.4.5), this is

$$M(S_c, \boldsymbol{\lambda}) = P(X \in S_c|H) = P(X \in S_c|\boldsymbol{\lambda}) \quad . \tag{8.4.9}$$

This probability is a function of S_c and of the parameters $\boldsymbol{\lambda}$. It is called the *power function* of a test. The complementary probability

$$L(S_c, \boldsymbol{\lambda}) = 1 - M(S_c, \boldsymbol{\lambda}) \tag{8.4.10}$$

is called the *acceptance probability* or the *operating characteristic function* of the test. It gives the probability to accept[¶] the null hypothesis. One clearly has

$$\begin{aligned} M(S_c, \boldsymbol{\lambda}_0) &= \alpha \quad , & M(S_c, \boldsymbol{\lambda}_1) &= 1 - \beta \quad , \\ L(S_c, \boldsymbol{\lambda}_0) &= 1 - \alpha \quad , & L(S_c, \boldsymbol{\lambda}_1) &= \beta \quad . \end{aligned} \tag{8.4.11}$$

The *most powerful test* of a simple hypothesis H_0 with respect to the simple alternative hypothesis is defined by the requirement

$$M(S_c, \boldsymbol{\lambda}_1) = 1 - \beta = \max \quad . \tag{8.4.12}$$

Sometimes there exists a *uniformly most powerful test*, for which the requirement (8.4.12) holds for all possible alternative hypotheses.

A test is said to be *unbiased* if its power function is greater than or equal to α for all alternative hypotheses:

$$M(S_c, \boldsymbol{\lambda}_1) \geq \alpha \quad . \tag{8.4.13}$$

[¶]We use here the word "acceptance" of a hypothesis, although more precisely one should say, "There is no evidence to reject the hypothesis."

This definition is reasonable, since the probability to reject the null hypothesis is then a minimum if the null hypothesis is true. An *unbiased most powerful test* is the most powerful of all the unbiased tests. Correspondingly one can define a *unbiased uniformly most powerful test.* In the next sections we will learn the rules which sometimes allow one to construct tests with such desirable properties. Before turning to this task, we will first give an example to illustrate the definitions just introduced.

Example 8.3: Test of the hypothesis that a normal distribution with given variance σ^2 has the mean $\lambda = \lambda_0$

We wish to test the hypothesis $H_0(\lambda = \lambda_0)$. As a test statistic we use the arithmetic mean $\bar{\mathrm{x}} = \frac{1}{n}(\mathrm{x}_1 + \mathrm{x}_2 + \ldots + \mathrm{x}_n)$. (We will see in Example 8.4 that this is the most appropriate test statistic for our purposes.) From Sect. 6.2 we know that $\bar{\mathrm{x}}$ is normally distributed with mean λ and variance σ^2/n, i.e., that the probability density for $\bar{\mathrm{x}}$ for the case $\lambda = \lambda_0$ is given by

$$f(\bar{x}; \lambda_0) = \frac{\sqrt{n}}{\sqrt{2\pi}\sigma} \exp\left(-\frac{n}{2\sigma^2}(\bar{x} - \lambda_0)^2\right) \quad . \tag{8.4.14}$$

This is shown in Fig. 8.4 together with four different critical regions, all of which have the same significance level α.

These are the regions

$$\begin{array}{ll}
U_1 : \bar{x} < \lambda^{\mathrm{I}} \text{ and } \bar{x} > \lambda^{\mathrm{II}} & \text{with } \int_{-\infty}^{\lambda^{\mathrm{I}}} f(\bar{x})\,\mathrm{d}\bar{x} = \int_{\lambda^{\mathrm{II}}}^{\infty} f(\bar{x})\,\mathrm{d}\bar{x} = \frac{1}{2}\alpha \ , \\
U_2 : \bar{x} > \lambda^{\mathrm{III}} & \text{with } \int_{\lambda^{\mathrm{III}}}^{\infty} f(\bar{x})\,\mathrm{d}\bar{x} = \alpha \ , \\
U_3 : \bar{x} < \lambda^{\mathrm{IV}} & \text{with } \int_{-\infty}^{\lambda^{\mathrm{IV}}} f(\bar{x})\,\mathrm{d}\bar{x} = \alpha \ , \\
U_4 : \lambda^{\mathrm{V}} \le \bar{x} < \lambda^{\mathrm{VI}} & \text{with } \int_{\lambda^{\mathrm{V}}}^{\lambda_0} f(\bar{x})\,\mathrm{d}\bar{x} = \int_{\lambda_0}^{\lambda^{\mathrm{VI}}} f(\bar{x})\,\mathrm{d}\bar{x} = \frac{1}{2}\alpha \ .
\end{array}$$

In order to obtain the power functions for each of these regions, we must vary the mean value λ. The probability density of $\bar{\mathrm{x}}$ for an arbitrary value of λ is, in analogy to (8.4.14), given by

$$f(\bar{x}; \lambda) = \frac{\sqrt{n}}{\sqrt{2\pi}\sigma} \exp\left[-\frac{n}{2\sigma^2}(\bar{x} - \lambda)^2\right] \quad . \tag{8.4.15}$$

The dashed curve in Fig. 8.4b represents the probability density for $\lambda = \lambda_1 = \lambda_0 + 1$. The power function (8.4.9) is now simply

$$P(\bar{\mathrm{x}} \in U|\lambda) = \int_U f(\bar{x}; \lambda)\,\mathrm{d}\bar{x} \quad . \tag{8.4.16}$$

The power functions obtained in this way for the critical regions U_1, U_2, U_3, U_4 are shown in Fig. 8.4c for $n = 2$ (solid curve) and $n = 10$ (dashed curve).

We can now compare the effects of the four tests corresponding to the various critical regions. From Fig. 8.4c we immediately see that U_1 corresponds

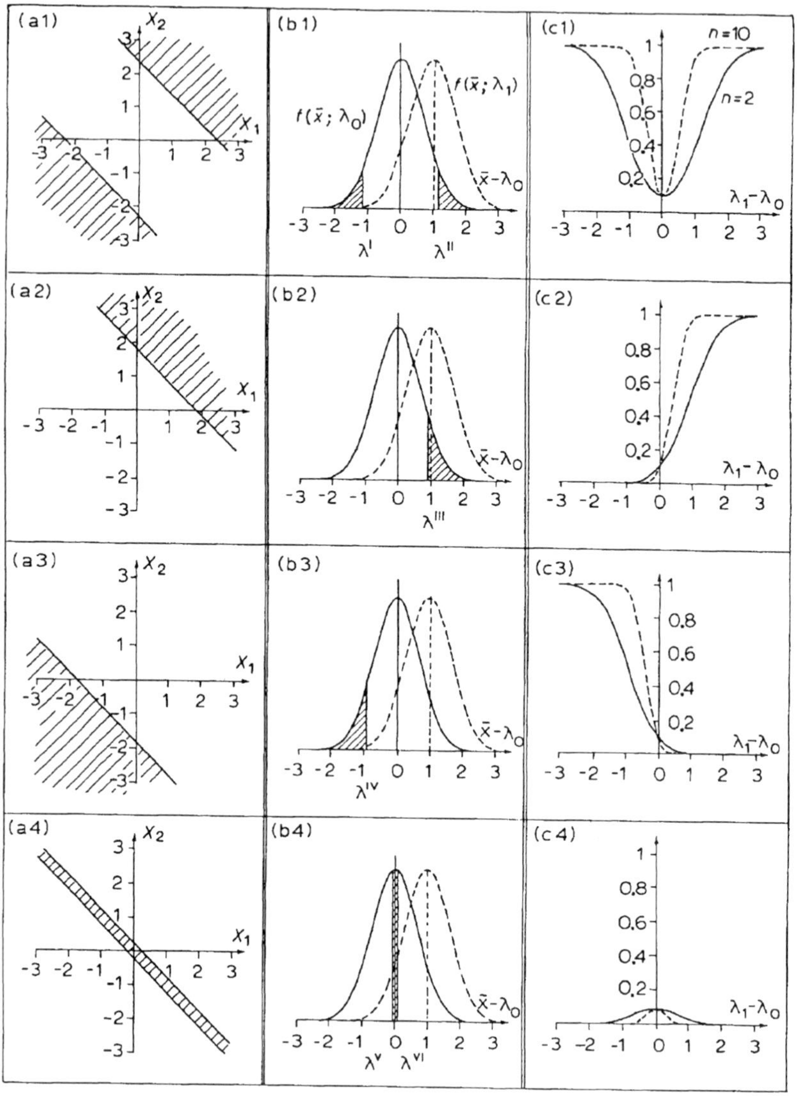

Fig. 8.4. (a) Critical regions in E space, (b) critical region of the test function, and (c) power function of the test from Example 8.3.

to an unbiased test, since the requirement (8.4.13) is clearly fulfilled. On the other hand, the test with the critical region U_2 is more powerful for the alternative hypothesis $H_1(\lambda_1 > \lambda_0)$, but is not good for $H_1(\lambda_1 < \lambda_0)$. For the test with U_3, the situation is exactly the opposite. Finally, the region U_4 provides a test for which the rejection probability is a maximum if the null hypothesis is true. Clearly this is very undesirable. The test was only constructed for demonstration purposes. If we compare the first three tests, we see that none of them are more powerful than the other two for all values of λ_1. Thus we have not succeeded in finding a uniformly most powerful test. In Example 8.4, where we will continue the discussion of the present example, we will determine that for this problem there does not exist a uniformly most powerful test. ■

8.5 The Neyman–Pearson Lemma and Applications

In the last section we introduced the E space, in which a sample is represented by a single point X. The probability to observe a point X within the critical region S_c – providing the null hypothesis H_0 is true – was defined in (8.4.5),

$$P(X \in S_c|H_0) = \alpha \quad . \tag{8.5.1}$$

We now define a conditional probability in E space,

$$f(X|H_0) \quad .$$

One clearly has

$$\int_{S_c} f(X|H_0)\,\mathrm{d}X = P(X \in S_c|H_0) = \alpha \quad . \tag{8.5.2}$$

The NEYMAN–PEARSON lemma states the following:

> A test of the simple hypothesis H_0 with respect to the simple alternative hypothesis H_1 is a most powerful test if the critical region S_c in E space is chosen such that
>
> $$\frac{f(X|H_0)}{f(X|H_1)} \begin{cases} \leq c \text{ for all } X \in S_c & , \\ \geq c \text{ for all } X \notin S_c & . \end{cases} \tag{8.5.3}$$
>
> Here c is a constant which depends on the significance level.

We will prove this by considering another region S along with S_c. It may partially overlap with S_c, as sketched in Fig. 8.5. We choose the size of the region S such that it corresponds to the same significance level, i.e.,

$$\int_S f(X|H_0)\,\mathrm{d}X = \int_{S_c} f(X|H_0)\,\mathrm{d}X = \alpha \quad .$$

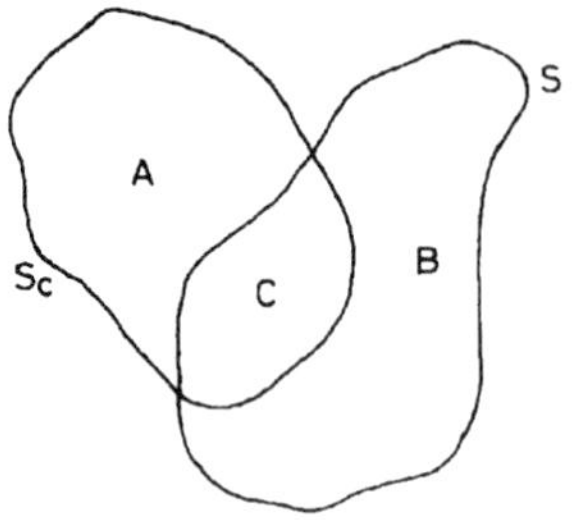

Fig. 8.5. The regions S and S_c.

Using the notation of Fig. 8.5, we can write

$$\begin{aligned}\int_A f(X|H_0)\,\mathrm{d}X &= \int_{S_c} f(X|H_0)\,\mathrm{d}X - \int_C f(X|H_0)\,\mathrm{d}X \\ &= \int_S f(X|H_0)\,\mathrm{d}X - \int_C f(X|H_0)\,\mathrm{d}X \\ &= \int_B f(X|H_0)\,\mathrm{d}X \quad .\end{aligned}$$

Since A is contained in S_c, we can use (8.5.3), i.e.,

$$\int_A f(X|H_0)\,\mathrm{d}X \le c\int_A f(X|H_1)\,\mathrm{d}X \quad .$$

Correspondingly, since B is outside of S_c, one has

$$\int_B f(X|H_0)\,\mathrm{d}X \ge c\int_B f(X|H_1)\,\mathrm{d}X \quad .$$

We can now express the power function (8.4.9) using these integrals:

$$\begin{aligned}M(S_c,\boldsymbol{\lambda}_1) &= \int_{S_c} f(X|H_1)\,\mathrm{d}X = \int_A f(X|H_1)\,\mathrm{d}X + \int_C f(X|H_1)\,\mathrm{d}X \\ &\ge \frac{1}{c}\int_A f(X|H_0)\,\mathrm{d}X + \int_C f(X|H_1)\,\mathrm{d}X \\ &\ge \int_B f(X|H_1)\,\mathrm{d}X + \int_C f(X|H_1)\,\mathrm{d}X \\ &\ge \int_S f(X|H_1)\,\mathrm{d}X = M(S,\boldsymbol{\lambda}_1)\end{aligned}$$

or directly,

$$M(S_c,\boldsymbol{\lambda}_1) \ge M(S,\boldsymbol{\lambda}_1) \quad . \tag{8.5.4}$$

This is exactly the condition (8.4.12) for a uniformly most powerful test. Since we have not made any assumptions about the alternative hypothesis $H_1(\lambda = \lambda_1)$ or the region S, we have proven that the requirement (8.5.3) provides a uniformly most powerful test when it is fulfilled by the alternative hypothesis.

Example 8.4: Most powerful test for the problem of Example 8.3

We now continue with the ideas from Example 8.3, i.e., we consider tests with a sample of size N, obtained from a normal distribution with known variance σ^2 and unknown mean λ. The conditional probability density of a point $X = (\mathsf{x}^{(1)}, \mathsf{x}^{(2)}, \ldots, \mathsf{x}^{(N)})$ in E space is the joint probability density of the $\mathsf{x}^{(j)}$ for given values of λ, i.e.,

$$f(X|H_0) = \left(\frac{1}{\sqrt{2\pi}\sigma}\right)^N \exp\left[-\frac{1}{2\sigma^2}\sum_{j=1}^{N}(\mathsf{x}^{(j)} - \lambda_0)^2\right] \tag{8.5.5}$$

for the null hypothesis and

$$f(X|H_1) = \left(\frac{1}{\sqrt{2\pi}\sigma}\right)^N \exp\left[-\frac{1}{2\sigma^2}\sum_{j=1}^{N}(\mathsf{x}^{(j)} - \lambda_1)^2\right] \tag{8.5.6}$$

for the alternative hypothesis. The ratio (8.5.3) takes on the form

$$\begin{aligned} Q = \frac{f(X|H_0)}{f(X|H_1)} &= \exp\left[-\frac{1}{2\sigma^2}\left\{\sum_{j=1}^{N}(\mathsf{x}^{(j)} - \lambda_0)^2 - \sum_{j=1}^{N}(\mathsf{x}^{(j)} - \lambda_1)^2\right\}\right] \\ &= \exp\left[-\frac{1}{2\sigma^2}\left\{N(\lambda_0^2 - \lambda_1^2) - 2(\lambda_0 - \lambda_1)\sum_{j=1}^{N}\mathsf{x}^{(j)}\right\}\right] \quad . \end{aligned}$$

The expression

$$\exp\left[-\frac{N}{2\sigma^2}(\lambda_0^2 - \lambda_1^2)\right] = k \geq 0$$

is a non-negative constant. The condition (8.5.3) thus has the form

$$k \exp\left[\frac{\lambda_0 - \lambda_1}{\sigma^2}\sum_{j=1}^{N}\mathsf{x}^{(j)}\right] \begin{cases} \leq c, & X \in S_c \quad , \\ \geq c, & X \notin S_c \quad . \end{cases}$$

This is the same as

$$(\lambda_0 - \lambda_1)\bar{\mathsf{x}} \begin{cases} \leq c', & X \in S_c \quad , \\ \geq c', & X \notin S_c \quad . \end{cases} \tag{8.5.7}$$

Here c' is a constant different from c. Equation (8.5.7) is, however, not only a condition for S_c, but also specifies directly that $\bar{\mathsf{x}}$ should be used as the test variable. For each given λ_1, i.e., for every simple alternative hypothesis $H_1(\lambda = \lambda_1)$, (8.5.7) gives a clear prescription for the choice of S_c or U, i.e., for the critical region and the test variable $\bar{\mathsf{x}}$.

For the case $\lambda_1 < \lambda_0$, the relation (8.5.7) becomes

$$\bar{\mathsf{x}} \begin{cases} \leq c'', & X \in S_c \quad , \\ \geq c'', & X \notin S_c \quad . \end{cases}$$

This corresponds to the situation in Fig. 8.4 (b3) with $c'' = \lambda^{IV}$. Similarly, for every alternative hypothesis with $\lambda_1 > \lambda_0$, the critical region of the most powerful test is given by

$$\bar{\mathsf{x}} \geq c'''$$

(see Fig. 8.4 (b2) with $c''' = \lambda'''$). There does not exist a uniformly most powerful test, since the factor $(\lambda_0 - \lambda_1)$ in Eq. (8.5.7) changes sign at $\lambda_1 = \lambda_0$. ■

8.6 The Likelihood-Ratio Method

The Neyman–Pearson lemma gave the condition for a uniformly most powerful test. Such a test did not exist if the alternative hypothesis included parameter values that could be both greater and less than that of the null hypothesis. We determined this in Example 8.4; it can be shown, however, that it is true in general. The question thus arises as to what test should be used when no uniformly most powerful test exists. Clearly this question is not formulated precisely enough to allow a unique answer. We would like in the following to give a prescription that allows us to construct tests that have desirable properties and that have the advantage of being relatively easy to use.

We consider from the beginning the general case with p parameters $\boldsymbol{\lambda} = (\lambda_1, \lambda_2, \ldots, \lambda_p)$. The result of a sample, i.e., the point $X = (\mathsf{x}^{(1)}, \mathsf{x}^{(2)}, \ldots, \mathsf{x}^{(N)})$ in E space is to be used to test a given hypothesis. The (composite) null hypothesis is characterized by a given region for each parameter. We can use a p-dimensional space, with the $\lambda_1, \lambda_2, \ldots, \lambda_p$ as coordinate axes, and consider the region allowed by the null hypothesis as a region in this parameter space, called ω. We denote the region in this space representing all possible values of the parameters by Ω. The most general alternative hypothesis is then the part of Ω that does not contain ω. We denote this by $\Omega - \omega$. Recall now from Chapter 7 the maximum-likelihood estimator $\widetilde{\lambda}$ for a parameter λ. It is that value of λ for which the likelihood function is a maximum. In Chapter 7 we tacitly assumed that one searched for the maximum in the entire allowable parameter space. In the following we will consider maxima in a restricted region (e.g., in ω). We write in this case $\widetilde{\lambda}^{(\omega)}$. The *likelihood-ratio test* defines a test statistic

$$T = \frac{f(\mathsf{x}^{(1)}, \mathsf{x}^{(2)}, \ldots, \mathsf{x}^{(N)}; \widetilde{\lambda}_1^{(\Omega)}, \widetilde{\lambda}_2^{(\Omega)}, \ldots, \widetilde{\lambda}_p^{(\Omega)})}{f(\mathsf{x}^{(1)}, \mathsf{x}^{(2)}, \ldots, \mathsf{x}^{(N)}; \widetilde{\lambda}_1^{(\omega)}, \widetilde{\lambda}_2^{(\omega)}, \ldots, \widetilde{\lambda}_p^{(\omega)})} \quad . \tag{8.6.1}$$

Here $f(\mathrm{x}^{(1)}, \mathrm{x}^{(2)}, \ldots, \mathrm{x}^{(N)}; \lambda_1, \lambda_2, \ldots, \lambda_p)$ is the joint probability density of the $\mathrm{x}^{(j)}$ $(j = 1, 2, \ldots, N)$, i.e., the likelihood function (7.1.5). The procedure of the likelihood ratio test prescribes that we reject the null hypothesis if

$$T > T_{1-\alpha} \quad . \tag{8.6.2}$$

Here $T_{1-\alpha}$ is defined by

$$P(T > T_{1-\alpha}|H_0) = \int_{T_{1-\alpha}}^{\infty} g(T|H_0)\,\mathrm{d}T \quad , \tag{8.6.3}$$

and $g(T|H_0)$ is the conditional probability density for the test statistic T. The following theorem by WILKS [9] concerns the distribution function of T (or actually $-2\ln T$) in the limiting case of very large samples:

> If a population is described by the probability density $f(x; \lambda_1, \lambda_2, \ldots, \lambda_p)$ that satisfies reasonable requirements of continuity, and if r of the p parameters of the null hypothesis
>
> $$H_0(\lambda_1 = \lambda_{10}, \lambda_2 = \lambda_{20}, \ldots, \lambda_r = \lambda_{r0}) \quad , \qquad r \le p \quad ,$$
>
> are fixed then the statistic $-2\ln T$ follows a χ^2-distribution with $p-r$ degrees of freedom for very large samples, i.e., for $N \to \infty$. For the case of a simple null hypothesis, i.e., $r = p$, then the number of degrees of freedom is equal to one.

We now apply this method to the problem of Examples 8.3 and 8.4, i.e., we consider tests with samples from a normally distributed population with known variance and unknown mean.

Example 8.5: Power function for the test from Example 8.3

For the simple hypothesis $H_0(\lambda = \lambda_0)$, the region ω of the parameter space is reduced to the point $\lambda = \lambda_0$. We have thus

$$\widetilde{\lambda}^{(\omega)} = \lambda_0 \quad . \tag{8.6.4}$$

If we consider the most general alternative hypothesis $H_1(\lambda = \lambda_1 \neq \lambda_0)$, then we obtain as the maximum-likelihood estimator of λ the sample mean $\bar{\mathrm{x}}$. The likelihood ratio (8.6.1) thus becomes

$$T = \frac{f(\mathrm{x}^{(1)}, \mathrm{x}^{(2)}, \ldots, \mathrm{x}^{(N)}; \bar{\mathrm{x}})}{f(\mathrm{x}^{(1)}, \mathrm{x}^{(2)}, \ldots, \mathrm{x}^{(N)}; \lambda_0)} \quad . \tag{8.6.5}$$

The joint probability density is given by (7.2.6),

$$f(x^{(1)}, x^{(2)}, \ldots, x^{(N)}) = \left(\frac{1}{\sqrt{2\pi}\sigma}\right)^N \exp\left[-\frac{1}{2\sigma^2}\sum_{j=1}^{N}(x^{(j)} - \lambda)^2\right] \quad . \tag{8.6.6}$$

Therefore,

$$\begin{aligned} T &= \exp\left[\frac{1}{2\sigma^2}\left\{-\sum_{j=1}^{N}(\mathrm{x}^{(j)} - \bar{\mathrm{x}})^2 + \sum_{j=1}^{N}(\mathrm{x}^{(j)} - \lambda_0)^2\right\}\right] \\ &= \exp\left[\frac{1}{2\sigma^2}\sum_{j=1}^{N}(\bar{\mathrm{x}} - \lambda_0)^2\right] \\ &= \exp\left[\frac{N}{2\sigma^2}(\bar{\mathrm{x}} - \lambda_0)^2\right] \quad . \end{aligned}$$

We must now calculate $T_{1-\alpha}$ and reject the hypothesis H_0 if the inequality (8.6.2) is fulfilled. Since the logarithm of T is a monotonic function of T, we can use

$$T' = 2\ln T = \frac{N}{\sigma^2}(\bar{\mathrm{x}} - \lambda_0)^2 \tag{8.6.7}$$

as the test statistic and reject H_0 if

$$T' > T'_{1-\alpha}$$

with

$$\int_{T'_{1-\alpha}}^{\infty} h(T'|H_0)\,\mathrm{d}T' = \alpha \quad .$$

In order to calculate the probability density $h(T'|H_0)$ of T', we start with the density $f(\bar{x})$ for the sample mean with the condition $\lambda = \lambda_0$,

$$f(\bar{x}|H_0) = \sqrt{\frac{N}{2\pi\sigma^2}} \exp\left(-\frac{N}{2\sigma^2}(\bar{x} - \lambda_0)^2\right) \quad .$$

In order to carry out the transformation of variables (3.7.1), we need in addition the derivative,

$$\left|\frac{\mathrm{d}\bar{x}}{\mathrm{d}T'}\right| = \frac{1}{2}\sqrt{\frac{\sigma^2}{N}}T'^{-1/2} \quad ,$$

which can be easily obtained from (8.6.7). One then has

$$h(T'|H_0) = \left|\frac{\mathrm{d}\bar{x}}{\mathrm{d}T'}\right| f(\bar{x}|H_0) = \frac{1}{\sqrt{2\pi}}T'^{-1/2}\mathrm{e}^{-T'/2} \quad . \tag{8.6.8}$$

This is indeed a χ^2-distribution for one degree of freedom. Thus in our example, WILKS' theorem holds even for finite N. We see, therefore, that the likelihood-ratio test yields the unbiased test of Fig. 8.4 (b1). The test

$$T' = \frac{N}{\sigma^2}(\bar{x} - \lambda_0)^2 > T'_{1-\alpha}$$

is equivalent to

$$\left(\frac{N}{\sigma^2}\right)^{1/2} |\bar{x} - \lambda_0| < \lambda' \quad , \qquad \left(\frac{N}{\sigma^2}\right)^{1/2} |\bar{x} - \lambda_0| > \lambda'' \tag{8.6.9}$$

with

$$-\lambda' = \lambda'' = (T'_{1-\alpha})^{1/2} = (\chi^2_{1-\alpha})^{1/2} = \chi_{1-\alpha} \quad .$$

We can use this result to compute explicitly the power function of our test. For a given value of the population mean λ, the probability density for the sample mean is

$$f(\bar{x};\lambda) = \left(\frac{N}{2\pi\sigma^2}\right)^{1/2} \exp\left[-\frac{N(\bar{x}-\lambda)^2}{2\sigma^2}\right] = \phi_0\left(\frac{\bar{x}-\lambda}{\sigma/\sqrt{N}}\right) \quad .$$

Using (8.4.9) and (8.6.9) we obtain

$$\begin{aligned} M(S_c;\lambda) &= \int_{-\infty}^{A} f(\bar{x};\lambda)\,\mathrm{d}\bar{x} + \int_{B}^{\infty} f(\bar{x};\lambda)\,\mathrm{d}\bar{x} \qquad (8.6.10) \\ &= \psi_0\left(\chi_{1-\alpha} - \frac{\lambda-\lambda_0}{\sigma/\sqrt{N}}\right) + \psi_0\left(\chi_{1-\alpha} + \frac{\lambda-\lambda_0}{\sigma/\sqrt{N}}\right) \quad , \\ A &= -\chi_{1-\alpha}\sigma/\sqrt{N} - \lambda_0 \quad , \qquad B = \chi_{1-\alpha}\sigma/\sqrt{N} - \lambda_0 \quad . \end{aligned}$$

Here ϕ_0 and ψ_0 are the probability density and distribution function of the standard normal distribution. The power function (8.6.10) is shown in Fig. 8.6 for $\alpha = 0.05$ and various values of N/σ^2. ■

Example 8.6: Test of the hypothesis that a normal distribution of unknown variance has the mean value $\lambda = \lambda_0$

In this case the null hypothesis $H_0(\lambda = \lambda_0)$ is composite, i.e., it makes no statement about σ^2. From Example 7.8 we know the maximum-likelihood estimator in the full parameter space,

$$\begin{aligned} \tilde{\lambda}^{(\Omega)} &= \bar{\mathsf{x}} \quad , \\ \tilde{\sigma}^{2(\Omega)} &= \frac{1}{N}\sum_{j=1}^{N}(\mathsf{x}^{(j)} - \bar{\mathsf{x}})^2 = \mathsf{s}'^2 \quad . \end{aligned}$$

In the parameter space of the null hypothesis we have

$$\begin{aligned} \tilde{\lambda}^{(\omega)} &= \lambda_0 \quad , \\ \tilde{\sigma}^{2(\omega)} &= \frac{1}{N}\sum_{j=1}^{N}(\mathsf{x}^{(j)} - \lambda_0)^2 \quad . \end{aligned}$$

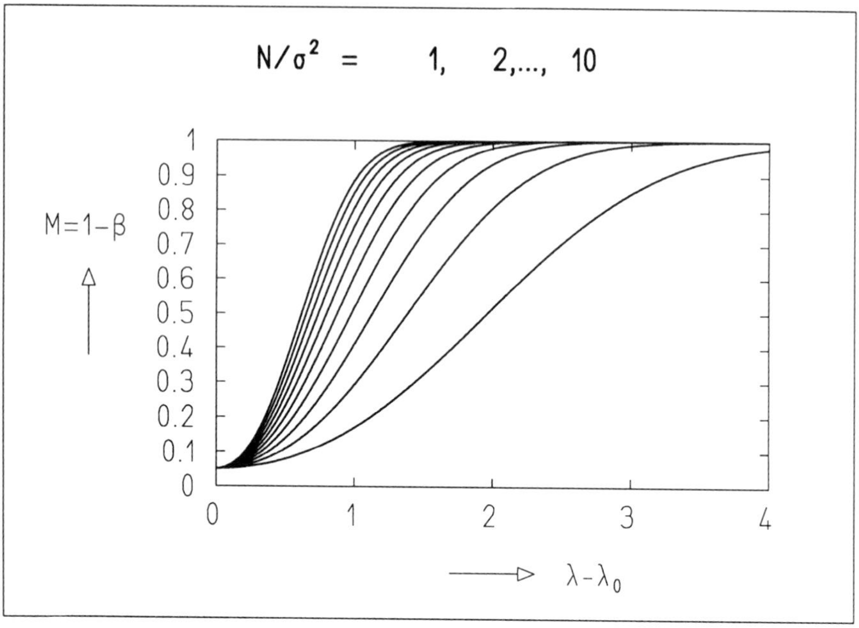

Fig. 8.6. Power function of the test from Example 8.5. The right-most curve corresponds to $N/\sigma^2 = 1$.

The likelihood ratio (8.6.1) is then

$$\begin{aligned} T &= \left(\frac{\sum(\mathrm{x}^{(j)}-\lambda_0)^2}{\sum(\mathrm{x}^{(j)}-\bar{\mathrm{x}})^2}\right)^{N/2} \exp\left(-\frac{N}{2}\frac{\sum(\mathrm{x}^{(j)}-\bar{\mathrm{x}})^2}{\sum(\mathrm{x}^{(j)}-\bar{\mathrm{x}})^2}+\frac{N}{2}\frac{\sum(\mathrm{x}^{(j)}-\lambda_0)^2}{\sum(\mathrm{x}^{(j)}-\lambda_0)^2}\right) \\ &= \left(\frac{\sum(\mathrm{x}^{(j)}-\lambda_0)^2}{\sum(\mathrm{x}^{(j)}-\bar{\mathrm{x}})^2}\right)^{N/2} . \end{aligned}$$

We transform again to a different test statistic T' that is a monotonic function of T,

$$\begin{aligned} T' &= T^{2/N} = \frac{\sum(\mathrm{x}^{(j)}-\lambda_0)^2}{\sum(\mathrm{x}^{(j)}-\bar{\mathrm{x}})^2} = \frac{\sum(\mathrm{x}^{(j)}-\bar{\mathrm{x}})^2+N(\bar{\mathrm{x}}-\lambda_0)^2}{\sum(\mathrm{x}^{(j)}-\bar{\mathrm{x}})^2} , \qquad (8.6.11) \\ T' &= 1+\frac{t^2}{N-1} , \end{aligned}$$

where

$$t = \sqrt{N}\frac{\bar{\mathrm{x}}-\lambda_0}{\left(\frac{\sum(\mathrm{x}^{(j)}-\bar{\mathrm{x}})^2}{N-1}\right)^{1/2}} = \sqrt{N}\frac{\bar{\mathrm{x}}-\lambda_0}{\mathrm{s}_x} = \frac{\bar{\mathrm{x}}-\lambda_0}{\mathrm{s}_{\bar{\mathrm{x}}}} \qquad (8.6.12)$$

is Student's test variable introduced in Sect. 8.3. From (8.6.11) we can compute a value of t for a given sample and reject the null hypothesis if

$$|t| > t_{1-\frac{1}{2}\alpha} \quad .$$

The very generally formulated method of the likelihood ratio has led us to Student's test, which was originally constructed for tests with samples from a normal distribution with known mean and unknown variance. ■

8.7 The χ^2-Test for Goodness-of-Fit

8.7.1 χ^2-test with Maximal Number of Degrees of Freedom

Suppose one has N measured values g_i, $i = 1, 2, \ldots, N$, each with a known measurement error σ_i. The meaning of the measurement error is the following: g_i is a measurement of the (unknown) true quantity h_i. One has

$$g_i = h_i + \varepsilon_i \quad , \qquad i = 1, 2, \ldots, N \quad . \tag{8.7.1}$$

Here the deviation ε_i is a random variable that follows a normal distribution with mean 0 and standard deviation σ_i.

We now want to test the hypothesis specifying the values h_i on which the measurement is based,

$$h_i = f_i \quad , \qquad i = 1, 2, \ldots, N \quad . \tag{8.7.2}$$

If this hypothesis is true, then all of the quantities

$$u_i = \frac{g_i - f_i}{\sigma_i} \quad , \qquad i = 1, 2, \ldots, N \quad , \tag{8.7.3}$$

follow the standard Gaussian distribution. Therefore,

$$T = \sum_{i=1}^{N} u_i^2 = \sum_{i=1}^{N} \left(\frac{g_i - f_i}{\sigma_i} \right)^2 \tag{8.7.4}$$

follows a χ^2-distribution for N degrees of freedom. If the hypothesis (8.7.2) is false, then the individual deviations of the measured values g_i from the values predicted by the hypothesis f_i, normalized by the errors σ_i, (8.7.3), will be greater. For a given significance level α, the hypothesis (8.7.2) is rejected if

$$T > \chi^2_{1-\alpha} \quad , \tag{8.7.5}$$

i.e., if the quantity (8.7.4) is greater than the quantile $\chi^2_{1-\alpha}$ of the χ^2-distribution for N degrees of freedom.

8.7.2 χ^2-test with Reduced Number of Degrees of Freedom

The number of degrees of freedom is reduced when the hypothesis to be tested is less explicit than (8.7.2). For this case we consider the following example. Suppose a quantity g can be measured as a function of an independent *controlled* variable t, which itself can be set without error,

$$g = g(t) \quad .$$

The individual measurements g_i correspond to given fixed values t_i of the independent variable. The corresponding true quantities h_i are given by some function

$$h_i = h(t_i) \quad .$$

A particularly simple hypothesis for this function is the linear equation

$$f(t) = h(t) = at + b \quad . \tag{8.7.6}$$

The hypothesis can in fact include specifying the numerical values for the parameters a and b. In this case, all values f_i in (8.7.2) are exactly known, and the quantity (8.7.4) follows – if the hypothesis is true – a χ^2-distribution for N degrees of freedom.

The hypothesis may, however, only state: There exists a linear relationship (8.7.6) between the controlled variable t and the variable h. The numerical values of the parameters a and b are, however, unknown. In this case one constructs estimators $\widetilde{a}$, $\widetilde{b}$ for the parameters, which are functions of the measurements g_i and the errors σ_i. The hypothesis (8.7.2) is then

$$h_i = h(t_i) = f_i = \widetilde{a}t_i + \widetilde{b} \quad .$$

Since, however, $\widetilde{a}$ and $\widetilde{b}$ are functions of the measurements g_i, the normalized deviations u_i in (8.7.3) are no longer all independent. Therefore the number of degrees of freedom of the χ^2-distribution for the sum of squares (8.7.4) is reduced by 2 to $N - 2$, since the determination of the two quantities $\widetilde{a}$, $\widetilde{b}$ introduces two equations of constraint between the quantities u_i.

8.7.3 χ^2-Test and Empirical Frequency Distribution

Suppose we have a distribution function $F(x)$ and its probability density $f(x)$. The full region of the random variable x can be divided into r intervals

$$\xi_1, \xi_2, \ldots, \xi_i, \ldots, \xi_r \quad ,$$

as shown in Fig. 8.7. By integrating $f(x)$ over the individual intervals we obtain the probability to observe x in ξ_i,

$$p_i = P(\mathrm{x} \in \xi_i) = \int_{\xi_i} f(x)\,\mathrm{d}x \quad ; \qquad \sum_{i=1}^{r} p_i = 1 \quad . \tag{8.7.7}$$

We now take a sample of size n and denote by n_i the number of elements of the sample that fall into the interval ξ_i. An appropriate graphical representation of the sample is a histogram, as described in Sect. 6.3.

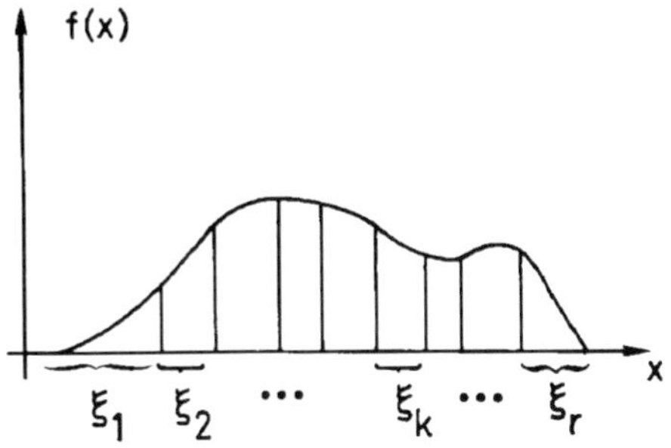

Fig. 8.7. Dividing the range of the variable x into the intervals ξ_k.

One clearly has

$$\sum_{i=1}^{r} n_i = n \quad . \tag{8.7.8}$$

From the (hypothetical) probability density for the population we would have expected the value

$$n\,p_i$$

for n_i. For large values of n_i, the variance of n_i is equal to n_i (Sect. 6.8), and the distribution of the quantity u_i with

$$u_i^2 = \frac{(n_i - n\,p_i)^2}{n_i} \tag{8.7.9}$$

becomes approximately – if the hypothesis is true – a standard Gaussian distribution. This holds also if one uses the expected variances $n\,p_i$ instead of the observed quantities n_i in the denominator of (8.7.9),

$$u_i^2 = \frac{(n_i - n\,p_i)^2}{n\,p_i} \quad . \tag{8.7.10}$$

If we now construct the sum of squares of the u_i for all intervals,

$$X^2 = \sum_{i=1}^{r} u_i^2 \quad , \tag{8.7.11}$$

then we expect (for large n) that this follows a χ^2-distribution if the hypothesis is true. The number of degrees of freedom is $r - 1$, since the u_i are not independent because of (8.7.8). The number of degrees of freedom is reduced to $r - 1 - p$ if, in addition, p parameters are determined from the observations.

Example 8.7: χ^2-test for the fit of a Poisson distribution to an empirical frequency distribution

In an experiment investigating photon-proton interactions, a beam of high energy photons (γ-quanta) impinge on a hydrogen bubble chamber. The processes by which a photon materializes in the chamber, electron-positron pair conversion, are counted in order to obtain a measure of the intensity of the photon beam. The frequency of cases in which 0,1,2,... pairs are observed simultaneously, i.e., in the same bubble-chamber photograph, follows a Poisson distribution (see Example 5.3). Deviations from the Poisson distribution provide information about measurement losses, which are important for uncovering systematic errors. The results of observing $n = 355$ photographs are given in column 2 of Table 8.3 and in Fig. 8.8. From Example 7.4, we know that the maximum-likelihood estimator of the parameter of the Poisson distribution is given by $\tilde{\lambda} = \sum_k k\, n_k / \sum_k n_k$. We find $\tilde{\lambda} = 2.33$. The values p_k of the Poisson distribution with this parameter multiplied by n are given in column 3. By summing the squared terms in column 4 one obtains the value $X^2 = 10.44$. The problem has six degrees of freedom, since $r = 8, p = 1$. We chose $\alpha = 1\,\%$ and find $\chi^2_{0.99} = 16.81$ from Table I.7. We therefore have no reason to reject the hypothesis of a Poisson distribution. ■

Table 8.3. Data for the χ^2-test from Example 8.7.

Number of electron pairs per photograph k	Number of photographs with k electron pairs n_k	Prediction of Poisson distribution $n\,p_k$	$\dfrac{(n_k - n\,p_k)^2}{n\,p_k}$
0	47	34.4	4.61
1	69	80.2	1.56
2	84	93.7	1.00
3	76	72.8	0.14
4	49	42.6	0.96
5	16	19.9	0.76
6	11	7.8	1.31
7	3	2.5	0.10
8	—	(0.7)	
	$n = \sum n_k = 355$		$X^2 = 10.44$

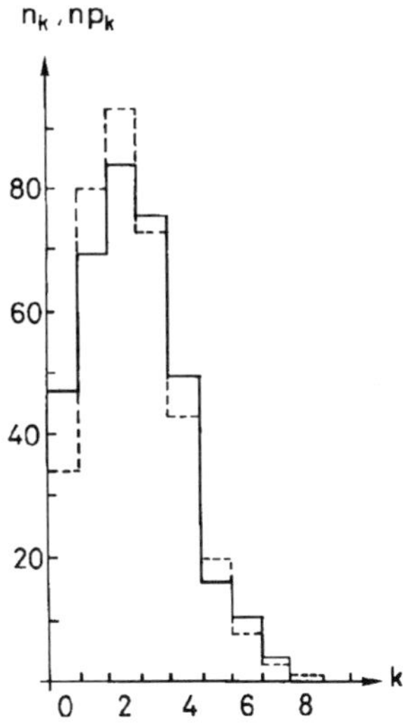

Fig. 8.8. Comparison of the experimental distribution n_k (histogram with solid line) from Example 8.7 with the Poisson distribution np_k (dashed line).

8.8 Contingency Tables

Suppose n experiments have been carried out whose results are characterized by the values of two random variables x and y. We consider the two variables as discrete, being able to take on the values $x_1, x_2, \ldots, x_k$; $y_1, y_2, \ldots, y_\ell$. Continuous variables can be approximated by discrete ones by dividing their range into intervals, as shown in Fig. 8.7. Let the number of times the result $\mathsf{x} = x_i$ and $\mathsf{y} = y_j$ is observed be n_{ij}. One can arrange the numbers n_{ij} in a matrix, called a *contingency table* (Table 8.4).

Table 8.4. Contingency table.

	y_1	y_2	$\ldots$	y_ℓ
x_1	n_{11}	n_{12}	$\ldots$	$n_{1\ell}$
x_2	n_{21}	n_{22}	$\ldots$	$n_{2\ell}$
$\vdots$	$\vdots$	$\vdots$		$\vdots$
x_k	n_{k1}	n_{k2}	$\ldots$	$n_{k\ell}$

We denote by p_i the probability for $\mathsf{x} = x_i$ to occur, and by q_j the probability for $\mathsf{y} = y_j$. If the variables are independent, then the probability to simultaneously observe $\mathsf{x} = x_i$ and $\mathsf{y} = y_j$ is equal to the product $p_i q_j$. The maximum-likelihood estimators for p and q are

$$\widetilde{p}_i = \frac{1}{n}\sum_{j=1}^{\ell} n_{ij} \quad , \qquad \widetilde{q}_j = \frac{1}{n}\sum_{i=1}^{k} n_{ij} \quad .$$

Since

$$\sum_{j=1}^{\ell} \widetilde{q}_j = \sum_{i=1}^{k} \widetilde{p}_i = \frac{1}{n} \sum_{j=1}^{\ell} \sum_{i=1}^{k} n_{ij} = 1 \quad ,$$

one has $k + \ell - 2$ independent estimators $\widetilde{p}_i$, $\widetilde{q}_j$. We can now organize the elements of the contingency table into a single line,

$$n_{11}, n_{12}, \ldots, n_{1\ell}, n_{21}, n_{22}, \ldots, n_{2\ell}, \ldots, n_{k\ell} \quad ,$$

and carry out a χ^2-test. For this we must compute the quantity

$$X^2 = \sum_{i=1}^{k} \sum_{j=1}^{\ell} \frac{(n_{ij} - n\, \widetilde{p}_i\, \widetilde{q}_j)^2}{n\, \widetilde{p}_i\, \widetilde{q}_j} \tag{8.8.1}$$

and compare it to the quantile $\chi^2_{1-\alpha}$ of the χ^2-distribution corresponding to a given significance level α. The number of degrees of freedom is still obtained from the number of intervals minus the number of estimated parameters minus one,

$$f = k\ell - 1 - (k + \ell - 2) = (k-1)(\ell - 1) \quad .$$

If the variables are not independent, then n_{ij} will not, in general, be near $n\, \widetilde{p}_i\, \widetilde{q}_j$, i.e., one will find

$$X^2 > \chi^2_{1-\alpha} \tag{8.8.2}$$

and the hypothesis will be rejected.

8.9 2 × 2 Table Test

The simplest nontrivial contingency table has only two rows and two columns, and is called a 2 × 2 table, as shown in Table 8.5. It is often used in medical studies. (The variables x_1 and x_2 could represent, for example, two different treatment methods, and y_1 and y_2 could represent success and failure of the treatment. One wishes to determine whether success is independent of the treatment.)

One computes the quantity X^2 either according to (8.8.1) or from the formula

Table 8.5. 2 × 2 table.

	y_1	y_2
x_1	$n_{11} = a$	$n_{12} = b$
x_2	$n_{21} = c$	$n_{22} = d$

$$X^2 = \frac{n(ad - bc)^2}{(a+b)(c+d)(a+c)(b+d)} \quad ,$$

which is obtained by rearranging (8.8.1). If the variables x and y are independent, then X^2 follows a χ^2-distribution with one degree of freedom. One rejects the hypothesis of independence at the significance level α if

$$X^2 > \chi^2_{1-\alpha} \quad .$$

In order for the quantity X^2 to actually follow a χ^2-distribution is again necessary that the individual n_{ij} are sufficiently large, (and the hypothesis of independence must be true).

8.10 Example Programs

The source files of the example programs are available in the subdirectory `\DATAN\EXASRC`.

8.10.1 Main Program **E1TEST** to Generate Samples and Test the Equality of their Variances Using the *F*-Test

The program performs a total of $n_{\rm exp}$ simulated experiments. Each experiment consists of the simulation of two samples of sizes N_1 and N_2 from normal distributions with standard deviations σ_1 and σ_2. The variance of each of the samples is computed using the program SMMNVR. The sample variances are called s_g^2 and s_k^2, so that

$$s_g^2 > s_k^2 \quad .$$

From the corresponding sample sizes the numbers of degrees of freedom $f_g = N_g - 1$ and $f_k = N_k - 1$ are computed. Finally, the ratio s_g^2/s_k^2 is compared with the quantile $F_{1-\alpha/2}(f_g, f_k)$ at a given confidence level $\beta = 1 - \alpha$. If the ratio is larger than the quantile, then the hypothesis of equal variances has to be rejected. The program asks for the quantities $n_{\rm exp}$, N_1, N_2, σ_1, σ_2, and β. For each simulated experiment one line of output is displayed.

Suggestions: Choose $n_{\rm exp} = 20$ and $\beta = 0.9$. (a) For $\sigma_1 = \sigma_2$ you would expect the hypothesis to be rejected in 2 out of 20 cases because of an error of the first kind. Note the large statistical fluctuations, which obviously depend on N_1 and N_2, and choose different pairs of values N_1, N_2 for $\sigma_1 = \sigma_2$. (b) Check the power of the test for different variances $\sigma_1 \neq \sigma_2$.

8.10.2 Main Program E2TEST to Generate Samples and Test the Equality of their Means with a Given Value Using Student's Test

This short program performs $n_{\rm exp}$ simulation experiments. In each experiment a sample of size N is drawn from a normal distribution with mean x_0 and width σ. Using SMMNVR the sample mean $\bar{\mathsf{x}}$ and the sample variance s_x^2 are determined. If λ_0 is the population mean specified by the hypothesis, then the quantity

$$|t| = \frac{|\bar{\mathsf{x}} - \lambda_0|\sqrt{N}}{\mathsf{s}_\mathsf{x}}$$

can be used to test the hypothesis. At a given confidence level $\beta = 1 - \alpha$ the hypothesis is rejected if

$$|t| > t_{1-\alpha/2} \quad .$$

Here $t_{1-\alpha/2}$ is the quantile of Student's distribution with $f = N - 1$ degrees of freedom. The program asks for the quantities $n_{\rm exp}$, N, x_0, σ, λ_0, and β. For each simulated experiment one line of output is displayed.

Suggestion: Modify the suggestions at the end of Sect. 8.10.1 to apply to Student's test.

8.10.3 Main Program E3TEST to Generate Samples and Compute the Test Statistic χ^2 for the Hypothesis that the Samples are Taken from a Normal Distribution with Known Parameters

For samples of size N the hypothesis H_0 that they stem from a normal distribution with mean a_0 and standard deviation σ_0 is tested. A total of $n_{\rm exp}$ samples are drawn in simulated experiments from a normally distributed population with mean a and standard deviation σ. For each sample the quantity

$$\mathsf{X}^2 = \sum_{i=1}^{N} \left(\frac{\mathsf{x}_i - a_0}{\sigma_0} \right)^2 \tag{8.10.1}$$

is computed. Here x_i are the elements of the sample. If $a = a_0$ and $\sigma = \sigma_0$, then the quantity X^2 follows a χ^2-distribution for N degrees of freedom. This quantity can therefore be used to perform a χ^2-test on the hypothesis H_0. The program does not, however, perform the χ^2-test, but rather it displays a histogram of the quantity X^2 together with the χ^2-distribution. One observes that for $a = a_0$ and $\sigma = \sigma_0$ the histogram and χ^2-distribution indeed coincide within statistical fluctuations. If, however, $a \neq a_0$ and/or $\sigma \neq \sigma_0$, then deviations appear. These deviations become particularly clear if instead of X^2 the quantity

$$P(\mathsf{X}^2) = 1 - F(\mathsf{X}^2; N) \tag{8.10.2}$$

is displayed. Here $F(\mathsf{X}^2, N)$ is the distribution function (C.5.2) of the χ^2-distribution for N degrees of freedom. $F(\mathsf{X}^2, N)$ is equal to the probability that a random variable drawn from a χ^2-distribution is smaller than X^2. Thus, P is the probability that it is greater than or equal to X^2. If the hypothesis H_0 is true, then F and therefore also P follow uniform distributions between 0 and 1. If, however, H_0 is false, then the distribution of the X^2 is not a χ^2-distribution, and the distribution of the P is not a uniform distribution. The test statistic X^2 often (not completely correctly) is simply called "χ^2" and the quantity P is then called the "χ^2-probability". Large values of X^2 obviously signify that the terms in the sum (8.10.1) are on the average large compared to unity, i.e., that the x_i are significantly different from a_0. For large values of X^2, however, P becomes small, cf. (8.10.2). Large values of "χ^2" therefore correspond to small values of the "χ^2-probability". The hypothesis H_0 is rejected at the confidence level $\beta = 1 - \alpha$ if $\mathsf{X}^2 > \chi^2_{1-\alpha}(N)$. That is equivalent to $F(\mathsf{X}^2, N) > \beta$ or $P < \alpha$.

The program asks for the quantities n_{exp}, N, a, a_0, σ, σ_0, and for a number (1 or 2), which indicates whether X^2 or $P(\mathsf{X}^2)$ is to be displayed as a histogram.

Suggestions: (a) Choose $n_{\text{exp}} = 1000$; $a = a_0 = 0$, $\sigma = \sigma_0 = 1$ and for $N = 1, N = 2$, and $N = 10$ display both X^2 and $P(\mathsf{X}^2)$. (b) Repeat (a), keeping $a = 0$ and choosing $a_0 = 1$ and $a_0 = 5$. Explain the shift of the histogram for $P(\mathsf{X}^2)$. (c) Repeat (a), keeping $\sigma = 1$ fixed and choosing $\sigma_0 = 0.5$ and $\sigma_0 = 2$. Discuss the results. (d) Modify the program so that instead of a_0 and σ_0^2, the sample mean $\bar{\mathsf{x}}$ sample variance s^2 are used for the computation of X^2. The quantity X^2 can be used for a χ^2-test of the hypothesis that the samples were drawn from a normal distribution. Display histograms of X^2 and $P(\mathsf{X}^2)$ and show that X^2 follows a χ^2-distribution with $N - 2$ degrees of freedom.

8.11 Problems

8.11.1 *F*-Test

Two samples are given:

$$\begin{array}{lll} (1) & 21,\ 19,\ 14,\ 27,\ 25,\ 23,\ 22,\ 18,\ 21, & (N_1 = 9)\,, \\ (2) & 16,\ 24,\ 22,\ 21,\ 25,\ 21,\ 18, & (N_2 = 7)\,. \end{array}$$

Does sample (2) have a smaller variance than sample (1) at a significance level of $\alpha = 5\,\%$?

8.11.2 Student's Test

Test the hypothesis that the 30 measurements of Problem 6.13.5 were drawn from a population with mean 25.5. Use a level of significance of $\alpha = 10\,\%$. Assume that the population is normally distributed.

8.11.3 χ^2-Test for Variance

Use the likelihood-ratio method to construct a test of the hypothesis $H_0(\sigma^2 = \sigma_0^2)$ that a sample stems from a normal distribution with unknown mean a and unknown variance σ_0^2. The parameters are $\boldsymbol{\lambda} = (a, \sigma)$. In ω one has $\widetilde{\boldsymbol{\lambda}}^{(\omega)} = (\bar{\mathrm{x}}, \sigma_0)$, in Ω: $\bar{\boldsymbol{\lambda}}^{(\Omega)} = (\bar{\mathrm{x}}, \mathrm{s})$.

(a) Form the likelihood ratio T.

(b) Show that instead of T, the test statistic $T' = N\mathrm{s}'^2/\sigma_0^2$ can be used as well.

(c) Show that T' follows a χ^2-distribution with $N-1$ degrees of freedom so that the test can be performed with the help of Table I.7.

8.11.4 χ^2-Test of Goodness-of-Fit

(a) Determine the mean $\widetilde{a}$ and the variance $\widetilde{\sigma}^2$ for the histogram of Fig. 6.1b, i.e.,

x_k	193	195	197	199	201	203	205	207	209	211
n_k	1	2	9	12	23	25	11	9	6	2

Use the result of Example 7.8 to construct the estimates. Give the estimators explicitly as functions of n_k and x_k.

(b) Perform (at a significance level of $\alpha = 10\,\%$) a χ^2-test on the goodness-of-fit of a normal distribution with mean $\widetilde{a}$ and variance $\widetilde{\sigma}^2$ to the histogram. Use only those bins of the histogram for which $np_k \geq 4$. Determine p_k from the difference of two entries in Table I.2. Give a formula for p_k as function of x_k, Δx, $\widetilde{a}$, $\widetilde{\sigma}^2$, and ψ_0. Construct a table for the computation of χ^2 containing columns for x_k, n_k, np_k, and $(n_k - np_k)^2/np_k$.

8.11.5 Contingency Table

(a) In an immunology experiment [taken from SOKAL and ROHLF, Biometry (Freeman, San Francisco, 1969)] the effect of an antiserum on a particular type of bacteria is studied. 57 mice received a certain dose of bacteria and antiserum, whereas 54 mice received bacteria only. After some time the mice of both groups were counted and the following contingency table constructed.

	dead	alive		sum
bacteria and antiserum	13	44		57
only bacteria	25	29		54
sum	38	73	total	111

Test (at a significance level of $\alpha = 10\,\%$) the hypothesis that the antiserum has no influence on the survival probability.

(b) In the computation of χ^2 in (a) you will have noticed that the numerators in (6.15) all have the same value. Show that generally for 2×2 contingency tables the following relation holds,

$$n_{ij} - n\tilde{p}_i\tilde{q}_j = (-1)^{i+j}\frac{1}{n}(n_{11}n_{22} - n_{12}n_{21}) \quad .$$

8.12 Hints and Solutions

Problem 8.11.1

$$\begin{aligned} \bar{\mathsf{x}}_1 &= 21.11 \quad , \qquad \bar{\mathsf{x}}_2 = 21.00 \quad , \\ \mathsf{s}_1^2 &= 14.86 \quad , \qquad \mathsf{s}_2^2 = 10.00 \quad , \\ T &= \mathsf{s}_1^2/\mathsf{s}_2^2 = 1.486 \quad , \qquad F_{0.95}(8,6) = 4.15 \quad . \end{aligned}$$

The variance of sample (2) is not significantly smaller.

Problem 8.11.2

$$\begin{aligned} \bar{\mathsf{x}} &= 25.142 \quad , \qquad \mathsf{s}^2 = 82.69/29 = 2.85 \quad , \qquad \mathsf{s} = 1.69 \quad , \\ T &= \frac{\bar{\mathsf{x}} - 25.5}{\mathsf{s}/\sqrt{30}} = \frac{25.142 - 25.5}{1.69/5.48} = -\frac{0.358}{0.309} = -1.16 \quad , \\ |T| &< t_{0.95}(f = 29) = 1.70 \quad . \end{aligned}$$

Therefore the hypothesis cannot be rejected.

Problem 8.11.3

(a)
$$f(\mathsf{x}^{(1)},\mathsf{x}^{(2)},\ldots,\mathsf{x}^{(N)},\widetilde{\boldsymbol{\lambda}}^{(\Omega)}) = \prod_{j=1}^{N} \frac{1}{\sqrt{2\pi}\mathsf{s}'} \exp\left(-\frac{(\mathsf{x}^{(j)}-\bar{\mathsf{x}})^2}{2\mathsf{s}'^2}\right)$$
$$= (\sqrt{2\pi}\mathsf{s}')^{-N} \exp\left(-\frac{1}{2\mathsf{s}'^2}\sum_{j=1}^{N}(\mathsf{x}^{(j)}-\bar{\mathsf{x}})^2\right) \quad ,$$
$$f(\mathsf{x}^{(1)},\mathsf{x}^{(2)},\ldots,\mathsf{x}^{(N)},\widetilde{\boldsymbol{\lambda}}^{(\omega)})$$
$$= (\sqrt{2\pi}\sigma_0)^{-N} \exp\left(-\frac{1}{2\sigma_0^2}\sum_{j=1}^{N}(\mathsf{x}^{(j)}-\bar{\mathsf{x}})^2\right) \quad ,$$

$$T = \left(\frac{\sigma_0}{\mathsf{s}'}\right)^N \exp\left[\tfrac{1}{2}N\mathsf{s}'^2\left(\frac{1}{\sigma_0^2}-\frac{1}{\mathsf{s}'^2}\right)\right] \quad ,$$
$$\ln T = N(\ln\sigma_0 - \ln\mathsf{s}') + \tfrac{1}{2}N\mathsf{s}'^2\left(\frac{1}{\sigma_0^2}-\frac{1}{\mathsf{s}'^2}\right)$$
$$= N(\ln\sigma_0 - \ln\mathsf{s}') + \tfrac{1}{2}N\frac{\mathsf{s}'^2}{\sigma_0^2} - \tfrac{1}{2}N \quad .$$

(b) $T' = N\mathsf{s}'^2/\sigma_0^2$ is a monotonically increasing function of s' if $\mathsf{s}' > \sigma_0$, otherwise it is monotonically decreasing. Therefore the test takes on the following form:

$$T' > T'_{1-\frac{1}{2}\alpha} \quad , \qquad T' < T'_{\frac{1}{2}\alpha} \quad \text{for} \quad H_0(\mathsf{s}' = \sigma_0) \quad ,$$
$$T' > T'_{1-\alpha} \quad \text{for} \quad H_0(\mathsf{s}' < \sigma_0) \quad ,$$
$$T' < T'_{\alpha} \quad \text{for} \quad H_0(\mathsf{s}' > \sigma_0) \quad .$$

(c) The answer follows from (6.7.2) and $\mathsf{s}'^2 = (N-1)\mathsf{s}^2/N$.

Problem 8.11.4

(a)
$$\widetilde{a} = \frac{1}{n}\sum_k n_k \mathsf{x}_k = 202.36 \quad ,$$
$$\widetilde{\sigma}^2 = \frac{1}{n-1}\sum_k n_k(\mathsf{x}_k - \widetilde{a})^2 = 13.40 \quad , \qquad \widetilde{\sigma} = 3.66 \quad .$$

(b)
$$p_k(\mathsf{x}_k) = \psi_0\left(\frac{\mathsf{x}_k + \frac{1}{2}\Delta x - \widetilde{a}}{\widetilde{\sigma}}\right) - \psi_0\left(\frac{\mathsf{x}_k - \frac{1}{2}\Delta x - \widetilde{a}}{\widetilde{\sigma}}\right)$$
$$= \psi_{0+} - \psi_{0-} \quad .$$

x_k	n_k	ψ_{0+}	ψ_{0-}	np_k	$\frac{(n_k - np_k)^2}{np_k}$
193	1	0.011	0.002	(0.9)	—
195	2	0.041	0.011	(3.0)	—
197	9	0.117	0.041	7.6	0.271
199	12	0.260	0.117	14.3	0.362
201	23	0.461	0.260	20.1	0.411
203	25	0.673	0.463	21.2	0.679
205	11	0.840	0.673	16.7	1.948
207	9	0.938	0.840	9.8	0.071
209	6	0.982	0.938	4.3	0.647
211	2	0.996	0.982	(1.4)	—
					$\mathrm{X}^2 = 4.389$

The number of degrees of freedom is $7 - 2 = 5$. Since $\chi^2_{0.90}(5) = 9.24$, the test does not reject the hypothesis.

Problem 8.11.5

(a) $$\tilde{p}_1 = \frac{1}{111}(13+44) = \frac{57}{111} \quad , \qquad \tilde{p}_2 = \frac{1}{111}(25+29) = \frac{54}{111} \quad ,$$
$$\tilde{q}_1 = \frac{1}{111}(13+25) = \frac{38}{111} \quad , \qquad \tilde{q}_2 = \frac{1}{111}(44+29) = \frac{73}{111} \quad ,$$

$$\begin{aligned}
\mathrm{X}^2 &= \frac{\left(13 - \frac{57\cdot 38}{111}\right)^2}{\frac{57\cdot 38}{111}} + \frac{\left(44 - \frac{57\cdot 73}{111}\right)^2}{\frac{57\cdot 73}{111}} \\
&\quad + \frac{\left(25 - \frac{54\cdot 38}{111}\right)^2}{\frac{54\cdot 38}{111}} + \frac{\left(29 - \frac{54\cdot 73}{111}\right)^2}{\frac{54\cdot 73}{111}} \\
&= \frac{42.43}{19.51} + \frac{42.43}{37.49} + \frac{42.43}{18.49} + \frac{42.43}{35.51} = 6.78 \quad .
\end{aligned}$$

Since $\chi^2_{0.90} = 2.71$ for $f = 1$, the hypothesis of independence is rejected.

(b) $$\begin{aligned}
& n_{ij} - \frac{1}{n}(n_{i1} + n_{i2})(n_{1j} + n_{2j}) \\
&= \frac{1}{n}[n_{ij}(n_{11} + n_{12} + n_{21} + n_{22}) - (n_{i1} + n_{i2})(n_{1j} + n_{2j})] \quad .
\end{aligned}$$

One can easily show that the expression in square brackets takes the form $(-1)^{i+j}(n_{11}n_{22} - n_{12}n_{21})$ for all i, j.

9. The Method of Least Squares

The method of *least squares* was first developed by LEGENDRE and GAUSS. In the simplest case it consists of the following prescription:

> The repeated measurements y_j can be treated as the sum of the (unknown) quantity x and the measurement error ε_j,
>
> $$\mathsf{y}_j = x + \varepsilon_j \quad .$$
>
> The quantity x should be determined such that the sum of squares of the errors ε_j is a minimum,
>
> $$\sum_j \varepsilon_j^2 = \sum_j (x - \mathsf{y}_j)^2 = \min \quad .$$

We will see that in many cases this prescription can be derived as a result of the principle of maximum likelihood, which historically was developed much later, but that in other cases as well, it provides results with optimal properties. The method of least squares, which is the most widely used of all statistical methods, can also be used in the case where the measured quantities y_j are not directly related to the unknown x, but rather indirectly, i.e., as a linear (or also nonlinear) combination of several unknowns $x_1, x_2, \ldots$. Because of the great practical significance of the method we will illustrate the various cases individually with examples before turning to the most general case.

9.1 Direct Measurements of Equal or Unequal Accuracy

The simplest case of *direct measurements with equal accuracy* has already been mentioned. Suppose one has carried out n measurements of an unknown quantity x. The measured quantities y_j have measurement errors ε_j. We now make the additional assumption that these are normally distributed about zero:

$$\mathsf{y}_j = x + \varepsilon_j \quad , \qquad E(\varepsilon_j) = 0 \quad , \qquad E(\varepsilon_j^2) = \sigma^2 \quad . \tag{9.1.1}$$

This assumption can be justified in many cases by the Central Limit Theorem. The probability to observe the value y_j as the result of a single measurement is thus proportional to

$$f_j\,\mathrm{d}y = \frac{1}{\sigma\sqrt{2\pi}}\exp\left(-\frac{(\mathsf{y}_j - x)^2}{2\sigma^2}\right)\mathrm{d}y \quad .$$

The log-likelihood function for all n measurements is thus (see Example 7.2)

$$\ell = -\frac{1}{2\sigma^2}\sum_{j=1}^{n}(\mathsf{y}_j - x)^2 + \text{const} \quad . \tag{9.1.2}$$

The maximum likelihood condition,

$$\ell = \max \quad ,$$

is thus equivalent to

$$M = \sum_{j=1}^{n}(\mathsf{y}_j - x)^2 = \sum_{j=1}^{n}\varepsilon_j^2 = \min \quad . \tag{9.1.3}$$

This is exactly the least-squares prescription. As we have shown in Examples 7.2 and 7.6, this leads to the result that the best estimator for x is given by the arithmetic mean of the y_j,

$$\widetilde{x} = \bar{\mathsf{y}} = \frac{1}{n}\sum_{j=1}^{n}\mathsf{y}_j \quad . \tag{9.1.4}$$

The variance of this estimator is

$$\sigma^2(\bar{\mathsf{y}}) = \sigma^2/n \quad , \tag{9.1.5}$$

or, if we set the measurement errors and standard deviations equal,

$$\Delta\widetilde{x} = \Delta y/\sqrt{n} \quad . \tag{9.1.6}$$

The more general case of *direct measurements of different accuracy* has also already been treated in Example 7.6. Let us assume again a normal distribution centered about zero for the measurement errors, i.e.,

$$\mathsf{y}_j = x + \varepsilon_j \quad , \qquad E(\varepsilon_j) = 0 \quad , \qquad E(\varepsilon_j^2) = \sigma_j^2 = 1/g_j \quad . \tag{9.1.7}$$

Comparing with (7.2.7) gives the requirement of the maximum-likelihood method

$$M = \sum_{j=1}^{n}\frac{(\mathsf{y}_j - x)^2}{\sigma_j^2} = \sum_{j=1}^{n}g_j(\mathsf{y}_j - x)^2 = \sum_{j=1}^{n}g_j\varepsilon_j^2 = \min \quad . \tag{9.1.8}$$

The individual terms in the sum are now *weighted* with the inverse of the variances. The best estimator for x is then [cf. (7.2.8)]

$$\widetilde{x} = \frac{\sum_{j=1}^{n} g_j \mathrm{y}_j}{\sum_{j=1}^{n} g_j} \quad , \tag{9.1.9}$$

i.e., the weighted mean of the individual measurements. One sees that an individual measurement thus contributes to the final result less when its measurement error is greater. From (7.3.20) we know the variance of $\widetilde{x}$; it is

$$\sigma^2(\widetilde{x}) = \left(\sum_{j=1}^{n} \frac{1}{\sigma_j^2} \right)^{-1} = \left(\sum_{j=1}^{n} g_j \right)^{-1} \quad . \tag{9.1.10}$$

We can use the result (9.1.9) in order to compute the best estimates $\widetilde{\varepsilon}_j$ of the original measurement errors ε_j from (9.1.1) to obtain

$$\widetilde{\varepsilon}_j = \mathrm{y}_j - \widetilde{x} \quad .$$

We expect these quantities to be normally distributed about zero with the variance σ_j^2. That is, the quantities $\widetilde{\varepsilon}_j/\sigma_j$ should follow a standard Gaussian distribution. According to Sect. 6.6, the sum of squares

$$M = \sum_{j=1}^{n} \left(\frac{\widetilde{\varepsilon}_j}{\sigma_j} \right)^2 = \sum_{j=1}^{n} \frac{(\mathrm{y}_j - \widetilde{x})^2}{\sigma_j^2} = \sum_{j=1}^{n} g_j (\mathrm{y}_j - \widetilde{x})^2 \tag{9.1.11}$$

then follows a χ^2-distribution with $n - 1$ degrees of freedom.

This property of the quantity M can now be used to carry out a χ^2-test on the validity of the assumption (9.1.7). If, for a given significance level α, the quantity M exceeds the value $\chi^2_{1-\alpha}$, then we would have to recheck the assumption (9.1.7). Usually one does not doubt that the y_j are in fact measurements of the unknown x. It may be, however, that the errors ε_j are not normally distributed. In particular, the measurements may also be biased, i.e., the expectation value of the errors ε_j may be different from zero. The presence of such systematic errors can often be inferred from the failure of the χ^2-test.

Example 9.1: Weighted mean of measurements of different accuracy

The best values for constants of fundamental significance, such as important constants of nature, are usually obtained as weighted averages of measurements obtained by different experimental groups. For the properties of elementary particles, such mean values are compiled at regular intervals. We will consider as an example somewhat older measurements of the mass of the neutral K-meson (K^0), taken from such a compilation from 1967 [10]. An average was computed from the results of four experiments, all carried out with different techniques. The calculation can be carried out following the scheme of Table 9.1. The resulting value of M is 7.2. If we choose a significance level of 5 %, we find from Table I.7 for three degrees of freedom $\chi^2_{0.95} = 7.82$. At the time of the averaging, one could therefore assume that the result $m_{K^0} = (497.9 \pm 0.2)$ MeV represented the best value for the mass of the K-meson, as long as no further experiments were carried out. More than 20 years later the weighted mean of all measurements was $m_{K^0} = (497.671 \pm 0.030)$ MeV [11]). ■

Table 9.1. Construction of the weighted mean from four measurements of the mass of the neutral K meson. The y_j are the measured values in MeV.

j	y_j	σ_j	$1/\sigma_j^2 = g_j$	$y_j\, g_j$	$y_j - \widetilde{x}$	$(y_j - \widetilde{x})^2 g_j$
1	498.1	0.4	6.3	3038.0	0.2	0.3
2	497.44	0.33	10	4974.4	−0.46	2.1
3	498.9	0.5	4	1995.6	1.0	4.0
4	497.44	0.5	4	1989.8	−0.46	0.8
$\sum$			24.3	11 997.8		7.2
$\widetilde{x} = \sum y_j g_j / \sum g_j = 497.9$, $\Delta\widetilde{x} = (\sum g_j)^{-\frac{1}{2}} = 0.20$						

Let us now consider the case where the χ^2-test fails. As mentioned above, one usually assumes that at least one of the measurements has a systematic error. By investigation of the individual measurements, one can sometimes determine that one or two measurements deviate from the others by a large amount. Such a case is illustrated in Fig. 9.1a, where several different measurements are shown with their errors. (The measured value is plotted along the vertical axis; the horizontal axis merely distinguishes between the different measurements.) Although the χ^2-test would fail if all measurements from Fig. 9.1a are used, this is not the case if measurements 4 and 6 are excluded.

Unfortunately the situation is not always so clear. Figure 9.1b shows a further example where the χ^2-test fails. (According to Chapter 8 one would reject the hypothesis that the measurements are determinations of the same quantity.) There is no single measurement, however, that is responsible for this fact. It would now be mathematically correct to not give any average value at all, and to make no statement about a best value, as long as no further measurements are available. In practice, this is clearly not satisfactory. ROSENFELD et al. [10] have suggested that the individual measurement errors should be increased by a *scale factor* $\sqrt{M/(n-1)}$, i.e., one replaces the σ_j by

$$\sigma'_j = \sigma_j \sqrt{\frac{M}{n-1}} \quad . \tag{9.1.12}$$

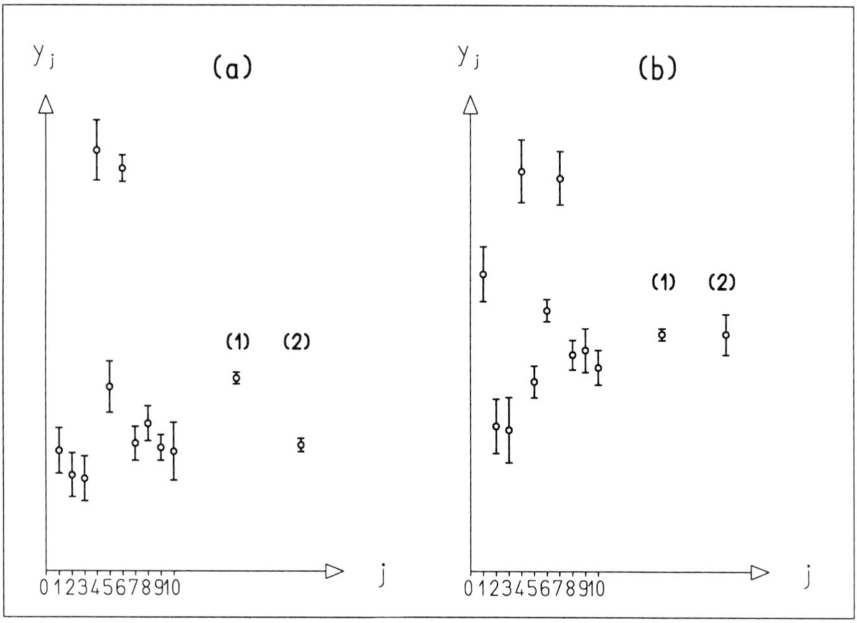

Fig. 9.1. Averaging of 10 measurements where the χ^2-test fails. (a) Anomalous deviation of certain measurements: (1) Averaging of all measurements; (2) averaging without y_4 and y_6. (b) Errors of individual measurements clearly too small: (1) Error of the mean according to (9.1.10); (2) error of the mean according to (9.1.13).

The weighted mean $\tilde{x}$ obtained by using these measurement errors does not differ from that of expression (9.1.9). The variance, however, is different from (9.1.10). It becomes

$$\sigma'^2(\tilde{x}) = \frac{M}{n-1}\left(\sum_{j=1}^{n}\frac{1}{\sigma_j^2}\right)^{-1} \quad . \tag{9.1.13}$$

We now compute the analogous expression to (9.1.11),

$$M' = \frac{n-1}{M}\sum_{j=1}^{n}\frac{(y_j-\tilde{x})^2}{\sigma_j^2} = \frac{n-1}{M}M = n-1 \quad . \tag{9.1.14}$$

This is the expectation value of χ^2 for $n-1$ degrees of freedom. Equation (9.1.14) clearly provided the motivation for relation (9.1.12). We repeat that this relation has no rigorous mathematical basis. It should be used with caution, since it hides the influence of systematic errors. On the other hand it provides reasonable errors for the mean for cases such as those in Fig. 9.1b, while direct application of Eq. (9.1.10) leads to an error that is far to small to reflect the actual dispersion of the individual measurements about the mean. Both solutions for the error of the mean are shown in Fig. 9.1b.

9.2 Indirect Measurements. Linear Case

Let us now consider the more general case of several unknown quantities $x_i\,(i = 1, 2, \ldots, r)$. The unknowns are often not measured directly. Instead, only a set of linear functions of the x_i are measurable,

$$\eta_j = p_{j0} + p_{j1}x_1 + p_{j2}x_2 + \cdots + p_{jr}x_r \quad . \tag{9.2.1}$$

We now write this relation in a somewhat different form,

$$f_j = \eta_j + a_{j0} + a_{j1}x_1 + a_{j2}x_2 + \cdots + a_{jr}x_r = 0 \quad . \tag{9.2.2}$$

We define a column vector,

$$a_j = \begin{pmatrix} a_{j1} \\ a_{j2} \\ \vdots \\ a_{jr} \end{pmatrix} \quad , \tag{9.2.3}$$

and write (9.2.2) in the more compact form,

$$f_j = \eta_j + a_{j0} + \mathbf{a}_j^{\mathrm{T}}\mathbf{x} = 0 \quad , \qquad j = 1, 2, \ldots, n \quad . \tag{9.2.4}$$

If we define in addition

$$\boldsymbol{\eta} = \begin{pmatrix} \eta_1 \\ \eta_2 \\ \vdots \\ \eta_n \end{pmatrix} , \quad \mathbf{a}_0 = \begin{pmatrix} a_{10} \\ a_{20} \\ \vdots \\ a_{n0} \end{pmatrix} , \quad A = \begin{pmatrix} a_{11} & a_{12} & \cdots & a_{1r} \\ a_{21} & a_{22} & \cdots & a_{2r} \\ \vdots & & & \\ a_{n1} & a_{n2} & \cdots & a_{nr} \end{pmatrix} , \tag{9.2.5}$$

then the system of equations (9.2.4) can be written as a matrix equation,

$$\mathbf{f} = \boldsymbol{\eta} + \mathbf{a}_0 + A\mathbf{x} = 0 \quad . \tag{9.2.6}$$

Of course the measured quantities still have measurement errors ε_j, which we assume to be normally distributed. We then have*

$$\begin{aligned} y_j &= \eta_j + \varepsilon_j \quad , \\ E(\varepsilon_j) &= 0 \quad , \\ E(\varepsilon_j^2) &= \sigma_j^2 = 1/g_j \quad . \end{aligned} \tag{9.2.7}$$

Since the y_j are *independent* measurements, we can arrange the variances σ_j^2 in a *diagonal* covariance matrix for y_j or ε_j,

---

*For simplicity of notation we no longer write random variables with a special character type. From context it will always be evident which variables are random.

$$C_y = C_\varepsilon = \begin{pmatrix} \sigma_1^2 & & & 0 \\ & \sigma_2^2 & & \\ & & \ddots & \\ 0 & & & \sigma_n^2 \end{pmatrix} \quad . \tag{9.2.8}$$

In analogy to (9.1.7) we call the inverse of the covariance matrix a *weight matrix*,

$$G_y = G_\varepsilon = C_y^{-1} = C_\varepsilon^{-1} = \begin{pmatrix} g_1 & & & 0 \\ & g_2 & & \\ & & \ddots & \\ 0 & & & g_n \end{pmatrix} \quad . \tag{9.2.9}$$

If we now put the measurements and errors together into vectors, we obtain from (9.2.7)

$$\mathbf{y} = \boldsymbol{\eta} + \boldsymbol{\varepsilon} \quad . \tag{9.2.10}$$

From (9.2.6) one then has

$$\mathbf{y} - \boldsymbol{\varepsilon} + \mathbf{a}_0 + A\mathbf{x} = 0 \quad . \tag{9.2.11}$$

We want to solve this system of equations for the unknowns $\mathbf{x}$ with the maximum-likelihood method. With our assumption (9.2.7), the measurements y_j are normally distributed with the probability density

$$f(y_j) = \frac{1}{\sigma_j\sqrt{2\pi}} \exp\left(-\frac{(y_j-\eta_j)^2}{2\sigma_j^2}\right) = \frac{1}{\sigma_j\sqrt{2\pi}} \exp\left(-\frac{\varepsilon_j^2}{2\sigma_j^2}\right) \quad . \tag{9.2.12}$$

Thus for all measurements one obtains likelihood functions

$$L = \prod_{j=1}^{n} f(y_j) = (2\pi)^{-\frac{1}{2}n} \left(\prod_{j=1}^{n} \sigma_j^{-1}\right) \exp\left(-\frac{1}{2}\sum_{j=1}^{n} \frac{\varepsilon_j^2}{\sigma_j^2}\right) \quad , \tag{9.2.13}$$

$$\ell = \ln L = -\tfrac{1}{2}n \ln 2\pi + \ln\left(\prod_{j=1}^{n} \sigma_j^{-1}\right) - \frac{1}{2}\sum_{j=1}^{n} \frac{\varepsilon_j^2}{\sigma_j^2} \quad . \tag{9.2.14}$$

This expression is clearly a maximum when

$$M = \sum_{j=1}^{n} \frac{\varepsilon_j^2}{\sigma_j^2} = \sum_{j=1}^{n} \frac{(y_j + \mathbf{a}_j^{\mathrm{T}}\mathbf{x} + a_{j0})^2}{\sigma_j^2} = \min \quad . \tag{9.2.15}$$

Using (9.2.9) and (9.2.11), we can rewrite this expression as

$$M = \boldsymbol{\varepsilon}^{\mathrm{T}} G_y \boldsymbol{\varepsilon} = \min \tag{9.2.16}$$

or

$$M = (\mathbf{y} + \mathbf{a}_0 + A\mathbf{x})^{\mathrm{T}} G_y (\mathbf{y} + \mathbf{a}_0 + A\mathbf{x}) = \min \quad , \tag{9.2.17}$$

or in abbreviated form

$$\mathbf{c} = \mathbf{y} + \mathbf{a}_0 \quad , \tag{9.2.18}$$

$$M = (\mathbf{c} + A\mathbf{x})^{\mathrm{T}} G_y (\mathbf{c} + A\mathbf{x}) = \min \quad . \tag{9.2.19}$$

We will simplify this expression further by using the Cholesky decomposition (cf. Sect. A.10) of the positive-definite symmetric weight matrix G_y,

$$G_y = H^{\mathrm{T}} H \quad . \tag{9.2.20}$$

In the frequently occurring case of uncorrelated measurements (9.2.9) one has

$$H = H^{\mathrm{T}} = \begin{pmatrix} 1/\sigma_1 & & & 0 \\ & 1/\sigma_2 & & \\ & & \ddots & \\ 0 & & & 1/\sigma_n \end{pmatrix} \quad . \tag{9.2.21}$$

Using the notation

$$\mathbf{c}' = H\mathbf{c} \quad , \qquad A' = HA \tag{9.2.22}$$

Eq. (9.2.19) takes on the simple form

$$M = (A'\mathbf{x} + \mathbf{c}')^2 = \min \quad . \tag{9.2.23}$$

The method for solving this equation for $\mathbf{x}$ is described in detail in Appendix A, in particular in Sects. A.6 through A.15. The solution can be written in the form

$$\widetilde{\mathbf{x}} = -A'^{+}\mathbf{c}' \quad . \tag{9.2.24}$$

[See (A.11.3).] Here A'^{+} is the pseudo-inverse of the matrix A' (see Sect. A.11). In the Program A.36 in Sect. A.15, a solution of the form of (9.2.24) is indeed found with the help of the singular value decomposition of the matrix A'. This procedure is particularly accurate numerically.

To compute by hand one uses instead of Eq. (9.2.24) the mathematically equivalent expression for the solution of the normal equations [see (A.6.17)],

$$\widetilde{\mathbf{x}} = -(A'^{\mathrm{T}} A')^{-1} A'^{\mathrm{T}} \mathbf{c}' \tag{9.2.25}$$

or, with (9.2.22) in terms of the quantities $\mathbf{c}$ and A,

$$\widetilde{\mathbf{x}} = -(A^{\mathrm{T}} G_y A)^{-1} A^{\mathrm{T}} G_y \mathbf{c} \quad . \tag{9.2.26}$$

The solution includes, of course, the special case of Sect. 9.1. In the case of direct measurements of different accuracy, $\mathbf{x}$ has only one element, $\mathbf{a}_0$ vanishes, and A is simply an n-component column vector whose elements are all equal to -1. One then has

$$\mathbf{c}' = \begin{pmatrix} y_1/\sigma_1 \\ y_2/\sigma_2 \\ \vdots \\ y_n/\sigma_n \end{pmatrix} \quad , \qquad A' = \begin{pmatrix} -1/\sigma_1 \\ -1/\sigma_2 \\ \vdots \\ -1/\sigma_n \end{pmatrix} \quad , \qquad A'^{\mathrm{T}}A = \sum_{j=1}^{n} \frac{1}{\sigma_j^2} \quad ,$$

and (9.2.25) becomes

$$\tilde{\mathbf{x}} = \left(\sum_{j=1}^{n} \frac{1}{\sigma_j^2} \right)^{-1} \sum_{j=1}^{n} \frac{y_j}{\sigma_j^2} \quad ,$$

which is identical to (9.1.9).

The solution of (9.2.26) represents a linear relation between the solution vector $\tilde{\mathbf{x}}$ and the vector of measurements $\mathbf{y}$, since $\mathbf{c} = \mathbf{y} + \mathbf{a}_0$. We can thus apply the error propagation techniques of Sect. 3.8. Using (3.8.2) and (3.8.4) one immediately obtains

$$C_{\tilde{x}} = G_{\tilde{x}}^{-1} = [(A^{\mathrm{T}}G_yA)^{-1}A^{\mathrm{T}}G_y]G_y^{-1}[(A^{\mathrm{T}}G_yA)^{-1}A^{\mathrm{T}}G_y]^{\mathrm{T}} \quad .$$

The matrices G_y, G_y^{-1}, and $(A^{\mathrm{T}}G_yA)$ are symmetric, i.e., they are identical to their transposed matrices. Using the rule (A.1.8), this expression simplifies to

$$\begin{aligned} G_{\tilde{x}}^{-1} &= (A^{\mathrm{T}}G_yA)^{-1}A^{\mathrm{T}}G_yG_y^{-1}G_yA(A^{\mathrm{T}}G_yA)^{-1} \\ &= (A^{\mathrm{T}}G_yA)^{-1}(A^{\mathrm{T}}G_yA)(A^{\mathrm{T}}G_yA)^{-1} \quad , \\ G_{\tilde{x}}^{-1} &= (A^{\mathrm{T}}G_yA)^{-1} = (A'^{\mathrm{T}}A')^{-1} \quad . \end{aligned} \tag{9.2.27}$$

We have thus obtained a simple expression for the covariance matrix of the estimators $\tilde{\mathbf{x}}$ for the unknowns $\mathbf{x}$. The square roots of the diagonal elements of this matrix can be viewed as "measurement errors", although the quantities $\mathbf{x}$ were not directly measured.

We can also use the result (9.2.26) to improve the original measurements $\mathbf{y}$. By substituting (9.2.26) into (9.2.11) one obtains a vector of estimators of the measurement errors $\boldsymbol{\varepsilon}$,

$$\tilde{\boldsymbol{\varepsilon}} = A\tilde{\mathbf{x}} + \mathbf{c} = -A(A^{\mathrm{T}}G_yA)^{-1}A^{\mathrm{T}}G_y\mathbf{c} + \mathbf{c} \quad . \tag{9.2.28}$$

These measurement errors can now be used to compute improved measured values,

$$\begin{aligned} \tilde{\boldsymbol{\eta}} = \mathbf{y} - \tilde{\boldsymbol{\varepsilon}} &= \mathbf{y} + A(A^{\mathrm{T}}G_yA)^{-1}A^{\mathrm{T}}G_y\mathbf{c} - \mathbf{c} \quad , \\ \tilde{\boldsymbol{\eta}} &= A(A^{\mathrm{T}}G_yA)^{-1}A^{\mathrm{T}}G_y\mathbf{c} - \mathbf{a}_0 \quad . \end{aligned} \tag{9.2.29}$$

The $\tilde{\boldsymbol{\eta}}$ are again linear in $\mathbf{y}$. We can again use error propagation to determine the covariance matrix of the improved measurements,

$$
\begin{aligned}
G_{\tilde{\eta}}^{-1} &= [A(A^{\mathrm{T}}G_yA)^{-1}A^{\mathrm{T}}G_y]G_y^{-1}[A(A^{\mathrm{T}}G_yA)^{-1}A^{\mathrm{T}}G_y]^{\mathrm{T}} \quad , \\
G_{\tilde{\eta}}^{-1} &= A(A^{\mathrm{T}}G_yA)^{-1}A^{\mathrm{T}} = A\,G_{\tilde{x}}^{-1}A^{\mathrm{T}} \quad . \qquad (9.2.30)
\end{aligned}
$$

The improved measurements $\tilde{\eta}$ satisfy (9.2.1) if the unknowns are replaced by their estimators $\tilde{x}$.

9.3 Fitting a Straight Line

We will examine in detail the simple but in practice frequently occurring task of fitting a straight line to a set of measurements y_j at various values t_j of a so-called *controlled variable* t. The values of these variables will be assumed to be known exactly, i.e., without error. The variable t_i could be, for example, the time at which an observation y_i is made, or a temperature or voltage that is set in an experiment. (If t also has an error, then fitting a line in the (t, y) plane becomes a nonlinear problem. It will be treated in Sect. 9.10, Example 9.11.)

In the present case the relation (9.2.1) has the simple form

$$\eta_j = y_j - \varepsilon_j = x_1 + x_2 t_j$$

or using vector notation,

$$\boldsymbol{\eta} - x_1 - x_2\mathbf{t} = 0 \quad .$$

We will attempt to determine the unknown parameters

$$\mathbf{x} = \begin{pmatrix} x_1 \\ x_2 \end{pmatrix}$$

from the measurements in Table 9.2.

A comparison of our problem to Eqs. (9.2.2) through (9.2.6) gives $\mathbf{a}_0 = 0$,

Table 9.2. Data for fitting a straight line.

j	1	2	3	4
t_j	0.0	1.0	2.0	3.0
y_j	1.4	1.5	3.7	4.1
σ_j	0.5	0.2	1.0	0.5

$$A = -\begin{pmatrix} 1 & t_1 \\ 1 & t_2 \\ 1 & t_3 \\ 1 & t_4 \end{pmatrix} = -\begin{pmatrix} 1 & 0 \\ 1 & 1 \\ 1 & 2 \\ 1 & 3 \end{pmatrix} \quad , \qquad \mathbf{y} = \mathbf{c} = \begin{pmatrix} 1.4 \\ 1.5 \\ 3.7 \\ 4.1 \end{pmatrix} \quad .$$

The matrices G_y and H are found by substitution of the last line of Table 9.2 into (9.2.9) and (9.2.21),

$$G_y = \begin{pmatrix} 4 & & & \\ & 25 & & \\ & & 1 & \\ & & & 4 \end{pmatrix} \quad , \qquad H = \begin{pmatrix} 2 & & & \\ & 5 & & \\ & & 1 & \\ & & & 2 \end{pmatrix} \quad .$$

One thus has

$$A' = -\begin{pmatrix} 2 & 0 \\ 5 & 5 \\ 1 & 2 \\ 2 & 6 \end{pmatrix} \quad , \qquad \mathbf{c}' = \begin{pmatrix} 2.8 \\ 7.5 \\ 3.7 \\ 8.2 \end{pmatrix} \quad , \qquad A'^{\mathrm{T}}\mathbf{c}' = -\begin{pmatrix} 63.2 \\ 94.1 \end{pmatrix} \quad ,$$

$$\begin{aligned} (A'^{\mathrm{T}}A')^{-1} &= \begin{pmatrix} 34 & 39 \\ 39 & 65 \end{pmatrix}^{-1} \\ &= \frac{1}{689}\begin{pmatrix} 65 & -39 \\ -39 & 34 \end{pmatrix} = \begin{pmatrix} 0.0943 & -0.0556 \\ -0.0566 & 0.0493 \end{pmatrix} \quad . \end{aligned}$$

To invert the 2×2 matrix we use (A.7.8). The solution (9.2.25) is then

$$\widetilde{\mathbf{x}} = \begin{pmatrix} 0.0943 & -0.0566 \\ -0.0566 & 0.0493 \end{pmatrix} \begin{pmatrix} 63.2 \\ 94.1 \end{pmatrix} = \begin{pmatrix} 0.636 \\ 1.066 \end{pmatrix} \quad .$$

The covariance matrix of for $\widetilde{\mathbf{x}}$ is

$$C_{\widetilde{x}} = G_{\widetilde{x}}^{-1} = (A'^{\mathrm{T}}A')^{-1} = \begin{pmatrix} 0.0943 & -0.0566 \\ -0.0566 & 0.0494 \end{pmatrix} \quad .$$

Its diagonal elements are the variances of $\widetilde{x}_1$ and $\widetilde{x}_2$, and their square roots are the errors,

$$\Delta\widetilde{x}_1 = 0.307 \quad , \qquad \Delta\widetilde{x}_2 = 0.222 \quad .$$

For the correlation coefficient between $\widetilde{x}_1$ and $\widetilde{x}_2$ one finds

$$\rho = \frac{-0.0566}{0.307 \cdot 0.222} = -0.830 \quad .$$

The improved measurements are

$$\widetilde{\boldsymbol{\eta}} = -A\widetilde{\mathbf{x}} = \begin{pmatrix} 1 & 0 \\ 1 & 1 \\ 1 & 2 \\ 1 & 3 \end{pmatrix} \begin{pmatrix} 0.636 \\ 1.066 \end{pmatrix} = \begin{pmatrix} 0.636 \\ 1.702 \\ 2.768 \\ 3.834 \end{pmatrix} \quad .$$

They lie on a line given by $\tilde{\boldsymbol{\eta}} = -A\tilde{\mathbf{x}}$, which of course is different in general from the "true" solution. The "residual errors" of $\tilde{\boldsymbol{\eta}}$ can be obtained with (9.2.30),

$$G_{\tilde{\eta}}^{-1} = C_{\tilde{\eta}} = \begin{pmatrix} 1 & 0 \\ 1 & 1 \\ 1 & 2 \\ 1 & 3 \end{pmatrix} \begin{pmatrix} 0.0943 & -0.0566 \\ -0.0566 & 0.0493 \end{pmatrix} \begin{pmatrix} 1 & 1 & 1 & 1 \\ 0 & 1 & 2 & 3 \end{pmatrix}$$

$$= \begin{pmatrix} 0.0943 & 0.0377 & -0.0189 & -0.0755 \\ 0.0377 & 0.0305 & 0.0232 & 0.0160 \\ -0.0189 & 0.0232 & 0.0653 & 0.1074 \\ -0.0755 & 0.0160 & 0.1074 & 0.1988 \end{pmatrix} .$$

The square roots of the diagonal elements are

$$\Delta\tilde{\eta}_1 = 0.31 \quad , \qquad \Delta\tilde{\eta}_2 = 0.17 \quad , \qquad \Delta\tilde{\eta}_3 = 0.26 \quad , \qquad \Delta\tilde{\eta}_4 = 0.45 \quad .$$

The fit procedure, in which more measurements (four) were used than were necessary to determine the two unknowns, has noticeably reduced the individual errors of the measurements in comparison to the original values σ_j.

Finally we will compute the value of the minimum function (9.2.16),

$$M = \tilde{\varepsilon}^{\mathrm{T}} G_y \tilde{\varepsilon} = (\mathbf{y} - \tilde{\boldsymbol{\eta}})^{\mathrm{T}} G_y (\mathbf{y} - \tilde{\boldsymbol{\eta}}) = \left(\sum_{j=1}^{n} \frac{y_j - \tilde{\eta}_j}{\sigma_j} \right)^2 = 4.507 \quad .$$

With this result we can carry out a χ^2-test on the goodness-of-fit of a straight line to our data. Since we started with $n = 4$ measured values and have determined $r = 2$ unknown parameters, we still have $n - r = 2$ degrees of freedom available (cf. Sect. 9.7). If we choose a significance level of 5 %, we find from Table I.7 for two degrees of freedom $\chi^2_{0.95} = 5.99$. There is thus no reason to reject the hypothesis of a straight line.

The results of the fit are shown in Fig. 9.2. The measurements y_j are shown as functions of the variables t. The vertical bars give the measurement errors. They cover the range $y_j \pm \sigma_j$. The plotted line corresponds to the result $\tilde{x}_1$, $\tilde{x}_2$. The improved measurements lie on this line. They are shown in Fig. 9.2b together with the residual errors $\Delta\tilde{\eta}_j$. In order to illustrate the accuracy of the estimates $\tilde{x}_1$, $\tilde{x}_2$, we consider the covariance matrix $C_{\tilde{x}}$. It determines a covariance ellipse (Sect. 5.10) in a plane spanned by the variables x_1, x_2. This ellipse is shown in Fig. 9.2c. Points on the ellipse correspond to fits of equal probability. Each of these points determines a line in the (t, y) plane. Some of the points are indicated in Fig. 9.2c and the corresponding lines are plotted in Fig. 9.2d. The points on the covariance ellipse thus correspond to a bundle of lines. The line determined by the "true" values of the unknowns lies in this bundle with the probability $1 - \mathrm{e}^{-1/2}$; cf. (5.10.18).

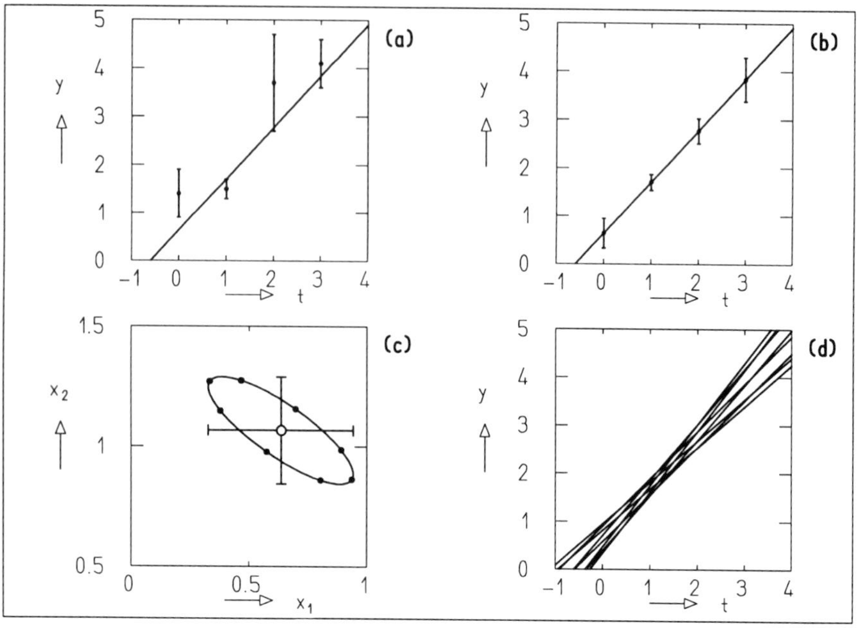

Fig. 9.2. Fit of a straight line to data from Table 9.2. (a) Original measured values and errors; (b) improved measurement values and residual errors; (c) covariance ellipse for the fitted quantities x_1, x_2; (d) various lines corresponding to individual points on the covariance ellipse.

9.4 Programs for Fitting Linear Functions of the Unknowns

The starting point of the ideas in Sect. 9.2 was the assumption (9.2.6) of a linear relationship between the "true" values $\boldsymbol{\eta}$ of the measured quantities $\mathbf{y}$ and the unknowns $\mathbf{x}$. We will write this relation in the form

$$\boldsymbol{\eta} = \mathbf{h}(\mathbf{x}) = -\mathbf{a}_0 - A\mathbf{x} \quad , \tag{9.4.1}$$

or in terms of components,

$$\eta_j = h_j(\mathbf{x}) = -a_{0j} - A_{j1}x_1 - A_{j2}x_2 - \cdots - A_{jr}x_r \quad . \tag{9.4.2}$$

Often it is useful to consider the index j as specifying that the measurement y_j corresponds to a value t_j of a controlled variable, which is taken to be known without error. Then (9.4.2) can be written in the form

$$\eta_j = h(\mathbf{x}, t_j) \quad . \tag{9.4.3}$$

This relation describes a curve in the (t, η) plane. It is characterized by the parameters $\mathbf{x}$. The determination of the parameters is thus equivalent to fitting a curve to the measurements $y_j = y(t_j)$. Usually the individual measurements y_j are uncorrelated. The weight matrix G_y is diagonal; it has the Cholesky decomposition (9.2.20). One then has simply

$$A'_{jk} = A_{jk}/\sigma_j \quad , \qquad c'_j = c/\sigma_j \quad ,$$

see (9.2.22).

9.4.1 Fitting a Polynomial

As a particularly simple but useful example of a function for (9.4.3), let us consider the relation

$$\eta_j = h_j = x_1 + x_2 t_j + x_3 t_j^2 + \cdots + x_r t_j^{r-1} \quad . \tag{9.4.4}$$

This is a polynomial in t_j, but is linear in the unknowns **x**. The special case $r = 2$ has been treated in detail in Sect. 9.3.

A comparison of (9.4.4) with (9.4.2) yields directly

$$a_{0j} = 0 \quad , \qquad A_{j\ell} = -t_j^{\ell-1}$$

or more completely,

$$\mathbf{a}_0 = \begin{pmatrix} 0 \\ 0 \\ \vdots \\ 0 \end{pmatrix} \quad , \qquad A = -\begin{pmatrix} 1 & t_1 & t_1^2 & \ldots & t_1^{r-1} \\ 1 & t_2 & t_2^2 & \ldots & t_2^{r-1} \\ \vdots & & & & \\ 1 & t_n & t_n^2 & \ldots & t_n^{r-1} \end{pmatrix} \quad .$$

Program 9.1: Subroutine `LSQPOL` fits a polynomial to measured values

CALLING SEQUENCE: `CALL LSQPOL(T,Y,DELTAY,N,NR,X,CX,R, A,SCRAT,OK)`

INPUT ARGUMENTS: `T` — n-vector of values of the controlled variable t_j
`Y` — n-vector of the corresponding measured values y_j
`DELTAY` — n-vector of measurement errors σ_j
`N` — number of measured values n
`NR` — number of unknowns r ($r = 1$ for a polynomial of order zero, $r = 2$ for a polynomial of order one, ...)
`A` — working array, minimum size $n \times r$
`SCRAT` — working array, minimum size $r \times r$

OUTPUT ARGUMENTS: `X` — r-vector of the unknowns
`CX` — $r \times r$ covariance matrix for the unknowns
`R` — value of the minimum function M
`OK` — logical variable (`.TRUE.` for convergence, `.FALSE.` if procedure does not converge)

```
      SUBROUTINE LSQPOL(T,Y,DELTAY,N,NR,X,CX,R,A,SCRAT,OK)
      IMPLICIT DOUBLE PRECISION (A-H,O-Z)
      PARAMETER(NMAX=1000)
      DIMENSION A(N,NR),T(N),Y(N),DELTAY(N),X(NR),CX(NR,NR)
      DIMENSION SCRAT(NR,NR),C(NMAX)
      COMMON /DASV04/ C
      LOGICAL OK
      PARAMETER(ONE=1.D0)
C Set up matrix A'
      DO 20 I=1,N
        DO 10 K=1,NR
          IF(K.EQ.1) THEN
            A(I,K)=-ONE/DELTAY(I)
          ELSE
            A(I,K)=-(T(I)**(K-1))/DELTAY(I)
          END IF
   10   CONTINUE
   20 CONTINUE
C Set up vector C'
      DO 30 I=1,N
        C(I)=-Y(I)/DELTAY(I)
   30 CONTINUE
C Set up matrix GX
      CALL MTXMAT(A,A,CX,NR,N,NR)
C Determine vector X of unknowns
      CALL MTXSVD(A,C,X,R,N,NR,1,DBLE(0.),OK)
      IF(OK) THEN
C Determine covariance matrix CX by inverting GX
        CALL MTXCHI(CX,SCRAT,NR)
      END IF
      END
```

Example 9.2: Fitting of various polynomials

It is often interesting to fit polynomials of various orders to measured data, as long as the number of degrees of freedom of the fit is at least one, i.e., $n > \ell + 1$. As a numerical example we will use measurements from an elementary particle physics experiment. Consider an investigation of elastic scattering of negative K mesons on protons with a fixed K meson energy. The distribution of the cosine of the scattering angle Θ in the center-of-mass system of the collision is characteristic of the angular momentum of possible intermediate states in the collision process. If, in particular, the distribution is considered as a polynomial in $\cos\Theta$, the order of the polynomial can be used to determine the spin quantum numbers of such intermediate states.

The measured values $y_j (j = 1, 2, \ldots, 10)$ are simply the numbers of collisions for which $\cos\Theta$ was observed in a small interval around $t_j = \cos\Theta_j$. As measurement errors the statistical errors were used, i.e., the square roots of the number of observations. The data are given in Table 9.3. The results of the fit of polynomials of various orders are summarized in Table 9.4 and Fig. 9.3. With the χ^2-test we can check successively whether a polynomial of order zero, one, ... gives a good fit to the data.

We see that the first two hypotheses (a constant and a straight line) are not in agreement with the experimental data. This can be seen in Fig. 9.3

Table 9.3. Data for Example 9.2. One has $\sigma_j = \sqrt{y_j}$.

j	$t_j = \cos\Theta_j$	y_j
1	−0.9	81
2	−0.7	50
3	−0.5	35
4	−0.3	27
5	−0.1	26
6	0.1	60
7	0.3	106
8	0.5	189
9	0.7	318
10	0.9	520

Table 9.4. Summary of results from Example 9.2 ($n = 10$ measured points, r parameters, $f = n - r$ degrees of freedom).

r	$\widetilde{x}_1$	$\widetilde{x}_2$	$\widetilde{x}_3$	$\widetilde{x}_4$	$\widetilde{x}_5$	$\widetilde{x}_6$	f	M
1	57.85						9	833.55
2	82.66	99.10					8	585.45
3	47.27	185.96	273.61				7	36.41
4	37.94	126.55	312.02	137.59			6	2.85
5	39.62	119.10	276.49	151.91	52.60		5	1.68
6	39.88	121.39	273.19	136.58	56.90	16.72	4	1.66

and is also reflected in the values of the minimum function. The hypothesis $r = 3$, a second-order polynomial, gives qualitative agreement. Most of the measurements, however, do not fall on the fitted parabola within the error bars. The χ^2-test fails with a significance level of 0.0001. For the hypotheses $r = 4$, 5, and 6, however, the agreement is very good. The fitted curves go through the error bars and are almost identical. The χ^2-test does not call for a rejection of the hypothesis even at $\alpha = 0.5$. We can therefore conclude that a third order polynomial is sufficient to describe the data. An even more careful investigation of the question as to what order a polynomial must be used to describe the data is possible with orthogonal polynomials; cf. Sect. 12.1. ■

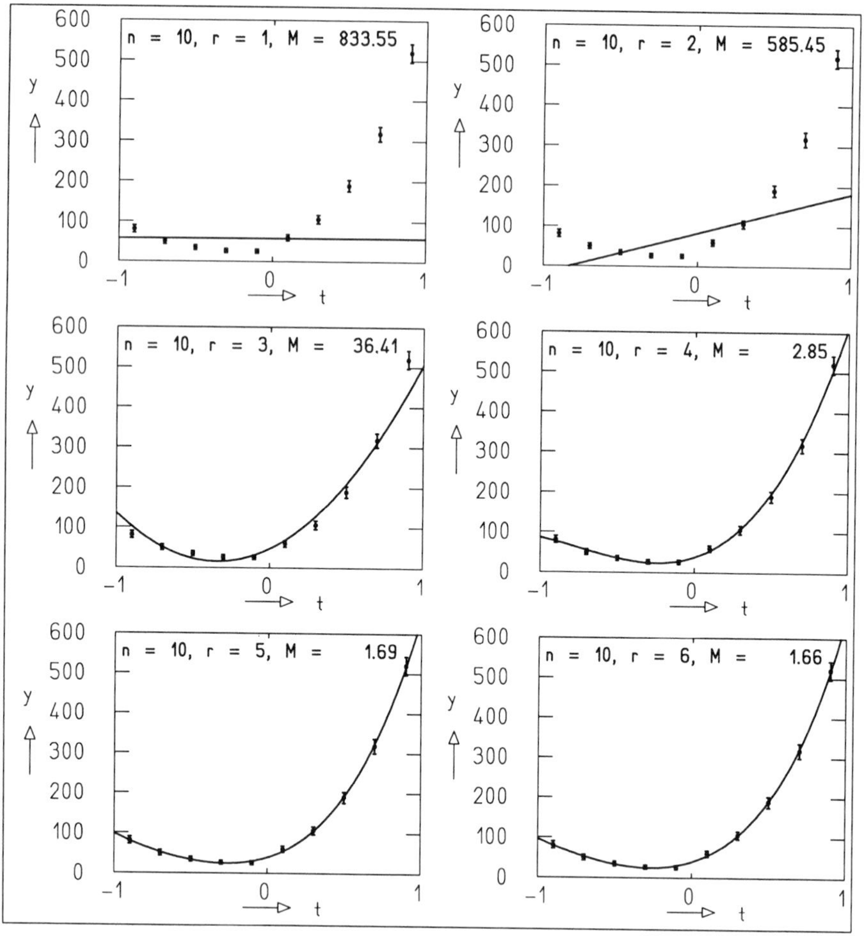

Fig. 9.3. Fit of polynomials of various orders $(0, 1, \ldots, 5)$ to the data from Example 9.2.

9.4.2 Fit of an Arbitrary Linear Function

The matrix A and the vector $\mathbf{c}$ enter into the solution of the problem of Sect. 9.2. They depend on the form of the function to be fitted and must therefore be provided by the user. (In Sect. 9.4.1 the function was known, so the user did not have to worry about computing A and $\mathbf{c}$.)

Program 9.2: Subroutine `LSQLIN` fits a linear function of the unknowns to measured data

CALLING SEQUENCE: `CALL LSQLIN(T,C,DELTAY,N,NR,X,CX,R, A,SCRAT,OK)`

INPUT ARGUMENTS: `T` — n-vector of values of the controlled variable t_j

`C` — n-vector $\mathbf{c} = \mathbf{y} + \mathbf{a}_0$

`DELTAY` — n-vector of measurement errors σ_j

`N` — number of measurements n

`NR` — number of unknowns r

`A` — $n \times r$ matrix A

`SCRAT` — working array, minimum size $r \times r$

OUTPUT ARGUMENTS: `X` — r-vector of unknowns

`CX` — $r \times r$ covariance matrix of the unknowns

`R` — value of the minimum function M

`OK` — logical variable (`.TRUE.` for convergence, `.FALSE.` if procedure does not converge)

```
      SUBROUTINE LSQLIN(T,C,DELTAY,N,NR,X,CX,R,A,SCRAT,OK)
      IMPLICIT DOUBLE PRECISION (A-H,O-Z)
      DIMENSION A(N,NR),T(N),C(N),DELTAY(N),X(NR),CX(NR,NR)
      DIMENSION SCRAT(NR,NR)
      LOGICAL OK
C compute matrix A' from A
      DO 20 I=1,N
        DO 10 K=1,NR
          A(I,K)=A(I,K)/DELTAY(I)
   10   CONTINUE
   20 CONTINUE
C Set up vector C'
      DO 30 I=1,N
        C(I)=C(I)/DELTAY(I)
   30 CONTINUE
C Set up matrix GX
      CALL MTXMAT(A,A,CX,NR,N,NR)
C Determine vector X of unknowns
      CALL MTXSVD(A,C,X,R,N,NR,1,DBLE(0.),OK)
      IF(OK) THEN
C Determine covariance matrix CX by inverting GX
        CALL MTXCHI(CX,SCRAT,NR)
      END IF
      END
```

Example 9.3: Fitting a proportional relation

Suppose from the construction of an experiment it is known that the true value η_j of the measurement y_j is directly proportional to the value of the controlled variable t_j:

$$\eta_j = x_1 t_j \quad .$$

The constant of proportionality x_1 is to be determined from the measurements. This relation is simpler than a first-order polynomial, which contains two constants. A comparison with (9.4.2) gives

$$a_{0j} = 0 \quad , \qquad A_{j1} = -t_j$$

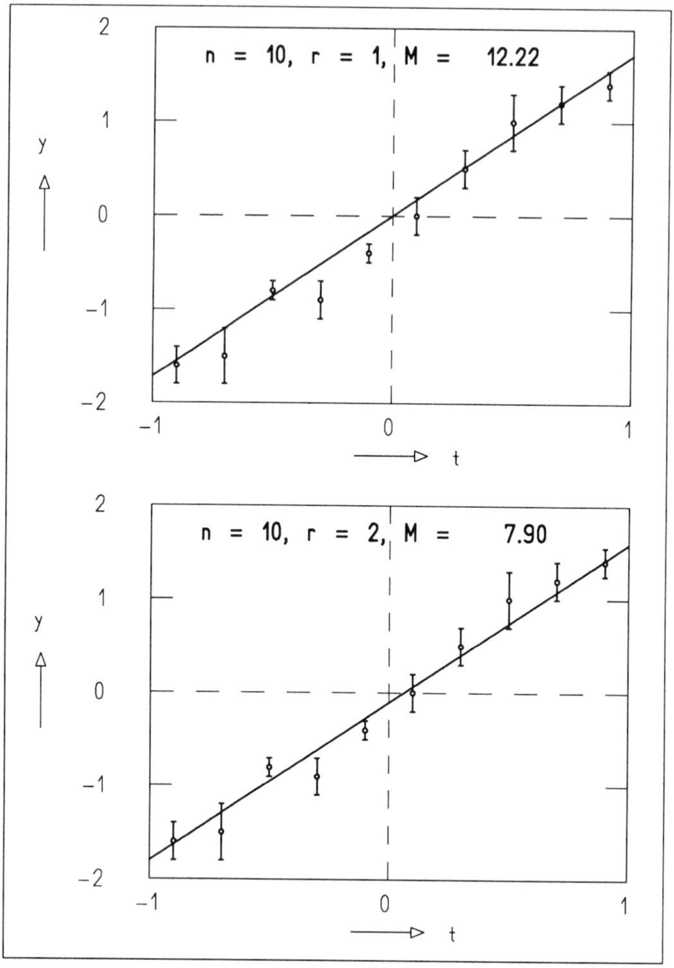

Fig. 9.4. Fit of a proportional relation (above) and a first-order polynomial (below) to data.

and thus

$$\mathbf{c} = \begin{pmatrix} y_1 \\ \vdots \\ y_n \end{pmatrix} , \qquad A = - \begin{pmatrix} t_1 \\ \vdots \\ t_n \end{pmatrix} .$$

Shown in Fig. 9.4 are the results of the fit of a line through the origin, i.e., a proportionality, and the fit of a first-order polynomial. The value of the minimum function is clearly smaller in the case of the general line, and the fit is visibly better. The number of degrees of freedom, however, is less, and the desired constant of proportionality is not determined. ∎

9.5 Indirect Measurements. Nonlinear Case

If the relation between the n-vector $\boldsymbol{\eta}$ of true values of the measured quantities $\mathbf{y}$ and the r-vector of unknowns is given by a function,

$$\boldsymbol{\eta} = \mathbf{h}(\mathbf{x}) \quad ,$$

that is not – as (9.4.1) – linear in $\mathbf{x}$, then our previous procedure fails in the determination of the unknowns.

Instead of (9.2.2) one has

$$f_j(\mathbf{x}, \boldsymbol{\eta}) = \eta_j - h_j(\mathbf{x}) = 0 \quad , \tag{9.5.1}$$

or in vector notation,

$$\mathbf{f}(\mathbf{x}, \boldsymbol{\eta}) = 0 \quad . \tag{9.5.2}$$

We can, however, relate this situation to the linear case if we expand the f_j in a Taylor series and keep only the first term. We carry out the expansion about the point $\mathbf{x}_0 = (x_{10}, x_{20}, \ldots, x_{r0})$, which is a first approximation for the unknowns, which has been obtained in some way,

$$f_j(\mathbf{x}, \boldsymbol{\eta}) = f_j(\mathbf{x}_0, \boldsymbol{\eta}) + \left(\frac{\partial f_j}{\partial x_1}\right)_{\mathbf{x}_0}(x_1 - x_{10}) + \cdots + \left(\frac{\partial f_j}{\partial x_r}\right)_{\mathbf{x}_0}(x_r - x_{r0}) \quad . \tag{9.5.3}$$

If we now define

$$\boldsymbol{\xi} = \mathbf{x} - \mathbf{x}_0 = \begin{pmatrix} x_1 - x_{10} \\ x_2 - x_{20} \\ \vdots \\ x_r - x_{r0} \end{pmatrix} \quad , \tag{9.5.4}$$

$$a_{j\ell} = \left(\frac{\partial f_j}{\partial x_\ell}\right)_{x_0} = -\left(\frac{\partial h_j}{\partial x_\ell}\right)_{\mathbf{x}_0} \quad , \qquad A = \begin{pmatrix} a_{11} & a_{12} & \cdots & a_{1r} \\ a_{21} & a_{22} & \cdots & a_{2r} \\ \vdots & & & \\ a_{n1} & a_{n2} & \cdots & a_{nr} \end{pmatrix} \quad , \tag{9.5.5}$$

$$c_j = f_j(\mathbf{x}_0, \mathbf{y}) = y_j - h_j(\mathbf{x}_0) \quad , \qquad \mathbf{c} = \begin{pmatrix} c_1 \\ c_2 \\ \vdots \\ c_n \end{pmatrix} \quad , \tag{9.5.6}$$

and use the relation (9.2.10), then we obtain

$$f_j(\mathbf{x}_0, \boldsymbol{\eta}) = f_j(\mathbf{x}_0, \mathbf{y} - \boldsymbol{\varepsilon}) = f_j(\mathbf{x}_0, \mathbf{y}) - \boldsymbol{\varepsilon} \quad . \tag{9.5.7}$$

Thus we can now write the system of equations (9.5.2) in the form

$$\mathbf{f} = A\boldsymbol{\xi} + \mathbf{c} - \boldsymbol{\varepsilon} = 0 \quad , \qquad \boldsymbol{\varepsilon} = A\boldsymbol{\xi} + \mathbf{c} \quad . \tag{9.5.8}$$

The least-squares condition (9.2.16) is then

$$M = (\mathbf{c} + A\boldsymbol{\xi})^{\mathrm{T}} G_y (\mathbf{c} + A\boldsymbol{\xi}) = \min \tag{9.5.9}$$

in complete analogy to (9.2.19). Using the quantities defined in (9.2.20) through (9.2.22) we can obtain the solution directly from (9.2.24),

$$\widetilde{\boldsymbol{\xi}} = -A'^{+}\mathbf{c}' \quad . \tag{9.5.10}$$

Corresponding to (9.5.4) one can use $\widetilde{\boldsymbol{\xi}}$ to find a better approximation

$$\mathbf{x}_1 = \mathbf{x}_0 + \widetilde{\boldsymbol{\xi}} \tag{9.5.11}$$

and compute new values of A and $\mathbf{c}$ at the point $\mathbf{x}_1$. A solution $\widetilde{\boldsymbol{\xi}}$ (9.5.10) for (9.5.9) with these values gives $\mathbf{x}_2 = \mathbf{x}_1 + \widetilde{\boldsymbol{\xi}}$, etc. This iterative procedure can be terminated when the minimum function in the last step has not significantly decreased in comparison to the result of the previous step.

There is no guaranty, however, for the convergence of this procedure. Heuristically it is clear, however, that the expectation for convergence is greater when the Taylor series, truncated after the first term, is a good approximation in the region over which $\mathbf{x}$ is varied in the procedure. This is at least the region between $\mathbf{x}_0$ and the solution $\widetilde{\mathbf{x}}$. (Intermediate steps can also lie outside of this region.) It is therefore important, particularly in highly nonlinear problems, to start from a good first approximation.

If the solution $\widetilde{\mathbf{x}} = \mathbf{x}_n = \mathbf{x}_{n-1} + \widetilde{\boldsymbol{\xi}}$ is reached in n steps, then it can be expressed as a linear function of $\widetilde{\boldsymbol{\xi}}$. Using error propagation one finds that the covariance matrices of $\widetilde{\mathbf{x}}$ and $\widetilde{\boldsymbol{\xi}}$ are then identical and one finds that

$$C_{\widetilde{x}} = G_{\widetilde{x}}^{-1} = (A^{\mathrm{T}} G_y A)^{-1} = (A'^{\mathrm{T}} A')^{-1} \quad . \tag{9.5.12}$$

The covariance matrix loses its validity, however, if the linear approximation (9.5.3) is not a good description in the region $\widetilde{x}_i \pm \Delta x_i$, $i = 1, \ldots, r$. (Here $\Delta x_i = \sqrt{C_{ii}}$.)

9.6 Programs for Fitting Nonlinear Functions

It is sometimes useful to set one or several of the r unknown parameters equal to given values, i.e., to treat them as constants and not as adjustable parameters. This can clearly be done by means of a corresponding definition of $\mathbf{f}$ in (9.5.1). For the user, however, it is more convenient to write only one subprogram that computes the function $\mathbf{f}$ for the r original parameters and when needed, to communicate to the program that the number of adjustable parameters is to be reduced from r to r'. Of course a list with r elements ℓ_i must also be given, in which one specifies which parameters x_i are to be held constant ($\ell_i = 0$) and which should remain adjustable ($\ell_i = 1$).

Two more difficulties come up when implementing the considerations of the previous section in a program:

On the one hand, the elements of the matrix A must be found by constructing the function to be fitted and differentiating it with respect to the parameters. Of course it is particularly convenient for the user if he does not have to program these derivatives himself, but rather can turn this task over to a routine for numerical differentiation. In Sect. E.1 we provide such a subroutine, which we will call from the program discussed below. Numerical differentiation implies, however, a loss of precision and an increase in computing time. In addition, the method can fail. Our programs communicate such an occurrence by means of an output parameter. Thus in some cases the user will be forced to program the derivatives by hand.

The second difficulty is related to the fact that the minimum function

$$M = (\mathbf{y} - \mathbf{h}(\mathbf{x}))^{\mathrm{T}} G_y (\mathbf{y} - \mathbf{h}(\mathbf{x})) \tag{9.6.1}$$

is no longer a simple quadratic form of the unknowns like (9.2.17) and (9.2.23). One consequence of this is that the position of the minimum $\widetilde{\mathbf{x}}$ cannot be reached in a single step. In addition, the convergence of the iterative procedure strongly depends on whether the first approximation $\mathbf{x}_0$ is in a region where the minimum function is sufficiently similar to a quadratic form. The determination of a good first approximation must be handled according to the given problem. Some examples are given below. If one constructs the iterative procedure as indicated in the previous section, then it can easily happen that the minimum function does not decrease with every step. In order to ensure convergence despite this, two methods can be applied, which are described in Sects. 9.6.1 and 9.6.2. The first (reduction of step size) is simpler and faster. The second (the Marquardt procedure) has, however, a larger region of convergence. We give programs for both methods, but recommend applying Marquardt procedure in cases of doubt.

9.6.1 Iteration with Step-Size Reduction

As mentioned, the inequality

$$M(\mathbf{x}_i) = M(\mathbf{x}_{i-1} + \widetilde{\boldsymbol{\xi}}) < M(\mathbf{x}_{i-1}) \tag{9.6.2}$$

does not hold in every case for the result $\mathbf{x}_i$ of step i. The following consideration helps us in handling such steps. Let us consider the expression $M(\mathbf{x}_{i-1} + s\widetilde{\boldsymbol{\xi}})$ as a function of the quantity s with $0 \le s \le 1$. If we replace $\widetilde{\boldsymbol{\xi}}$ by $s\widetilde{\boldsymbol{\xi}}$ in (9.5.9), then we obtain

$$M = (\mathbf{c} + sA\widetilde{\boldsymbol{\xi}})^{\mathrm{T}} G_y (\mathbf{c} + sA\widetilde{\boldsymbol{\xi}}) = (\mathbf{c}' + s\mathbf{A}'\widetilde{\boldsymbol{\xi}})^2 \quad .$$

Differentiating with respect to s gives

$$M' = 2(\mathbf{c}' + sA'\widetilde{\boldsymbol{\xi}})^{\mathrm{T}} A'\widetilde{\boldsymbol{\xi}}$$

or with $\mathbf{c}' = -A'\widetilde{\boldsymbol{\xi}}$, cf. (9.5.10),

$$M' = 2(s-1)\widetilde{\boldsymbol{\xi}}^{\mathrm{T}} A'^{\mathrm{T}} A'\widetilde{\boldsymbol{\xi}} \quad ,$$

and thus

$$M'(s=0) < 0 \quad ,$$

if $A'^{\mathrm{T}}A' = A^{\mathrm{T}}G_yA$ is positive definite. (This is always the case near the minimum.) The matrix $A'^{\mathrm{T}}A'$ gives the curvature of the function M in the space spanned by the unknowns $x_1, \ldots, x_r$. For only one such unknown, the region of positive curvature is the region between the points of inflection around the minimum (cf. Fig. 10.1). Since the function M is continuous in s, there exists a value $\lambda > 0$ such that

$$M'(s) < 0 \quad , \qquad 0 \le s \le \lambda \quad .$$

After an iteration $i+1$, for which (9.6.2) does not hold, one multiplies $\widetilde{\boldsymbol{\xi}}$ by a number s, e.g., $s = 1/2$, and checks whether

$$M(x_{i-1} + s\widetilde{\boldsymbol{\xi}}) < M(x_{i-1}) \quad .$$

If this is the case, then one sets $\mathbf{x}_i = \mathbf{x}_{i-1} + s\widetilde{\boldsymbol{\xi}}$. If it is not the case, then one multiplies again with s, and so forth.

The following program operates according to this procedure of iteration with step-size reduction. As in the linear case we consider the measurements as being dependent on a controlled variable t, i.e., $y_j = y_j(t_j)$. For the true values η_j corresponding to the measurements one has

$$\eta_j = h(\mathbf{x}, t_j)$$

or [cf. (9.6.1)]

$$\eta = h(\mathbf{x}, t) \quad .$$

This must be computed by a function subprogram provided by the user. Its name can be chosen freely, but must be declared as `EXTERNAL` in the calling program.

Program 9.3: Subroutine `LSQNON` fits a nonlinear function to measured data using an iterative procedure with step-size reduction

CALLING SEQUENCE: `CALL LSQNON(USERFN,T,Y,DELTAY,N,NR, NRED,LIST,X,CX,R,A,SCRAT,NSTEP)`

INPUT ARGUMENTS: `USERFN` — name of user function (see also comments below)
`T` — n-vector of values of the controlled variable t_j
`Y` — n-vector of measured values y_j
`DELTAY` — n-vector of measurement errors σ_j
`N` — number of measurements n
`NR` — number of unknowns r
`NRED` — reduced number of parameters not fixed r'
`LIST` — r-vector containing the elements of a list specifying which of the r unknowns are fixed (list element $= 0$) or variable (list element $= 1$)
`X` — r-vector of the first approximations for the unknowns
`A` — working array, minimum size $n \times r'$
`SCRAT` — working array, minimum size $r' \times r'$
`NSTEP` — maximum number of iterative steps. If negative, the covariance matrix `CX` is not computed.

OUTPUT ARGUMENTS: `X` — r-vector of the unknowns
`CX` — $r' \times r'$ covariance matrix of the unknowns
`R` — value of the minimum function M
`NSTEP` — number of steps needed. If an error occurs, `NSTEP` is set negative (-2: no convergence, -3: numerical differentiation failed, -4: step-size reduction not successful).

DEFAULT VALUES: If the input value of `NSTEP` is zero, then a maximum of 100 steps are carried out. For the two particularly frequent cases `NRED = NR` and `NRED = 0`, the vector `LIST` does not have to be set explicitly. In all cases, however, an r-vector `LIST` must be provided as a working array.

REMARKS: The user must provide a function subprogram (called `USERFN` above). It has three input arguments, `X` – r-vector $\mathbf{x}$ of the unknowns, `NR` – the number r of unknowns, and `T` – the value of the controlled variables. It must compute the function $\eta = h(\mathbf{x}, t)$.

```
      SUBROUTINE LSQNON(USERFN,T,Y,DELTAY,N,NR,NRED,LIST,X,CX,R,A,
     +SCRAT,NSTEP)
      IMPLICIT DOUBLE PRECISION (A-H,O-Z)
      PARAMETER(NMAX=1000)
      DIMENSION A(N,NRED),T(N),Y(N),DELTAY(N),X(NR)
      DIMENSION SCRAT(NRED,NRED),CX(NRED,NRED),C(NMAX)
      DIMENSION X2(NMAX),X1RED(NMAX)
      DIMENSION LIST(NR)
      COMMON /DASV02/ C,X2,X1RED
      PARAMETER(MAXSTP=100,EPSILN=1.D-8,TT=1.D-15,FACT=0.5D0)
      PARAMETER(ZERO=0.0D0)
      LOGICAL OK,COVMAT
      EXTERNAL USERFN
      COVMAT=.TRUE.
      IF(NSTEP.LT.0) THEN
        COVMAT=.FALSE.
        NSTEP=ABS(NSTEP)
      END IF
      IF(NSTEP.LT.1) NSTEP=MAXSTP
C Initial value of minimum function
      R=ZERO
      DO 10 I=1,N
        R=R+((Y(I)-USERFN(X,NR,T(I)))/DELTAY(I))**2
   10 CONTINUE
C For NRED=0 : set NSTEP=1 and return
      IF(NRED.LE.0) THEN
        NSTEP=1
        GO TO 130
      END IF
C For NR=NRED :  set LIST
      IF(NR.EQ.NRED) THEN
        DO 20 I=1,NR
          LIST(I)=1
   20   CONTINUE
      END IF
C start iteration
      DO 120 ISTEP=1,NSTEP
C Numerical Derivatives
        CALL AUXDRI(USERFN,X,T,N,NR,NRED,LIST,A,OK)
        IF(.NOT.OK) THEN
          NSTEP=-3
          GO TO 130
        END IF
        DO 40 I=1,N
          DO 30 K=1,NRED
            A(I,K)=A(I,K)/DELTAY(I)
   30     CONTINUE
   40   CONTINUE
        DO 50 I=1,N
          C(I)=(Y(I)-USERFN(X,NR,T(I)))/DELTAY(I)
   50   CONTINUE
        CALL MTXSVD(A,C,X1RED,R1,N,NRED,1,DBLE(0.),OK)
        IF(.NOT.OK) THEN
          NSTEP=-1
          GO TO 130
        END IF
        CALL MTXZRV(X2,NR)
        CALL MTXPSV(X2,X1RED,NR,NRED,LIST)
        CALL MTXADV(X,X2,X2,NR)
        R1=ZERO
        DO 60 I=1,N
```

```
            R1=R1+((Y(I)-USERFN(X2,NR,T(I)))/DELTAY(I))**2
   60     CONTINUE
          DO 80 K=1,30
            IF(R1.GT.R*(DBLE(1.)+EPSILN)+TT) THEN
C minimum function has increased: reduce step by factor of two
              CALL MTXMSV(X1RED,X1RED,FACT,NRED)
              CALL MTXZRV(X2,NR)
              CALL MTXPSV(X2,X1RED,NR,NRED,LIST)
              CALL MTXADV(X,X2,X2,NR)
              R1=ZERO
              DO 70 I=1,N
                R1=R1+((Y(I)-USERFN(X2,NR,T(I)))/DELTAY(I))**2
   70         CONTINUE
            ELSE
              GO TO 90
            END IF
   80     CONTINUE
          NSTEP=-4
          GO TO 130
   90     CONTINUE
          CALL MTXZRV(X2,NR)
          CALL MTXPSV(X2,X1RED,NR,NRED,LIST)
          CALL MTXADV(X,X2,X,NR)
C test for break-off criterion
          IF(ABS(R-R1) .LT. EPSILN*ABS(R)+TT) THEN
            R=R1
C Compute covariance matrix CX of unknowns
            IF(COVMAT) THEN
              CALL AUXDRI(USERFN,X,T,N,NR,NRED,LIST,A,OK)
              IF(.NOT.OK) THEN
                NSTEP=-3
                GO TO 130
              END IF
              DO 110 I=1,N
                DO 100 K=1,NRED
                  A(I,K)=A(I,K)/DELTAY(I)
  100           CONTINUE
  110         CONTINUE
              CALL MTXMAT(A,A,CX,NRED,N,NRED)
              CALL MTXCHI(CX,SCRAT,NRED)
              NSTEP=ISTEP
            END IF
            GO TO 130
          ELSE
            R=R1
          END IF
  120 CONTINUE
      NSTEP=-2
  130 RETURN
      END
```

Example 9.4: Fitting a Gaussian curve

In many experiments one has signals $y(t)$ that have the form of a Gaussian curve,

$$y(t) = x_1 \exp(-(t - x_2)^2/2x_3^2) \quad . \tag{9.6.3}$$

One wishes to determine the parameters x_1, x_2, x_3 that give the amplitude, position of the maximum, and the width of the signal. To perform this task with the program LSQNON one must write a short function subprogram, whose name is then the first argument of LSQNON. We will call this user routine LSQGSS.

Program 9.4: Function LSQGSS computes a Gaussian function for three given parameters

CALLING SEQUENCE: F=LSQGSS(X,NR,T)

INPUT ARGUMENTS: X — r-vector of the parameters x_1, x_2, x_3
NR — number of parameters r (must be set to three before function called)
T — value of the controlled variable t

OUTPUT ARGUMENTS: function value LSQGSS given by (9.6.3)

REMARKS: Note that the magnitude of the argument of the exponential function is checked. For large values of the argument (also when negative) the exponential function would fail with some compilers.

```
DOUBLE PRECISION FUNCTION LSQGSS(X,NR,T)
IMPLICIT DOUBLE PRECISION (A-H,O-Z)
DIMENSION X(NR)
PARAMETER (ZERO=0.D0,BIG=700.D0)
ARG=(X(2)-T)**2/(DBLE(2.)*X(3)**2)
IF (ARG.GT.BIG) THEN
  LSQGSS=ZERO
ELSE
  LSQGSS=X(1)*EXP(-ARG)
END IF
END
```

Figure 9.5 shows the result of the fit to data. The values $x_1 = 0.5, x_2 = 1.5$, $x_3 = 0.2$ were used as a first approximation. In fact we could have estimated significantly better initial values directly from the plot of the measurements, i.e., $x_1 \approx 1$ for the amplitude, $x_2 \approx 1.25$ for the position of the maximum, and $x_3 \approx 0.4$ for the width.

Here we have just mentioned a particularly useful aid in determining the initial approximations: the analysis of a plot of the data by eye. One can also proceed more formally, however, and consider the logarithm of (9.6.3),

$$\begin{aligned}\ln y(t) &= \ln x_1 - \frac{(t-x_2)^2}{2x_3^2} \\ &= \left(\ln x_1 - \frac{x_2^2}{2x_3^2}\right) + t\left(\frac{x_2}{x_3^2}\right) - t^2\left(\frac{1}{2x_3^2}\right) \quad .\end{aligned}$$

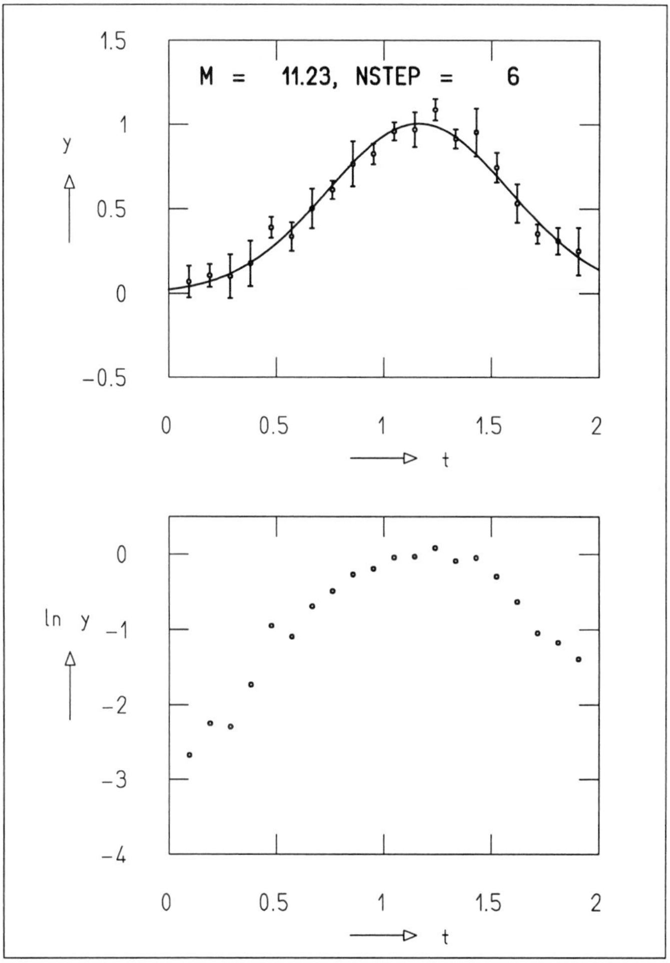

Fig. 9.5. Measured values and fitted Gaussian curve (above); logarithm of the measured values (below).

This is a polynomial in t, which is linear in the three terms in parentheses,

$$a_1 = \ln x_1 - \frac{x_2^2}{2x_3^2} \quad , \qquad a_2 = \frac{x_2}{x_3^2} \quad , \qquad a_3 = -\frac{1}{2x_3^2} \quad .$$

By taking the logarithm, however, the distribution for the measurement errors, originally Gaussian, has been changed, so that strictly speaking one cannot fit a polynomial to determine a_1, a_2, and a_3. For purposes of determining the first approximation we can disregard this difficulty and set the errors of the $\ln y(t_i)$ equal to one. We then determine the quantities a_1, a_2, a_3, and from them the values x_1, x_2, x_3, and use these as initial values.

The logarithms of the measured values y_i are shown in the lower part of Fig. 9.5. One can see that fit range for the parabola must be restricted to points in the bell-shaped region of the curve, since the points in the tails are subject to large fluctuations. ■

Example 9.5: Fit of an exponential function

In studies of radioactivity, for example, a function of the form

$$y(t) = x_1 \exp(-x_2 t) \tag{9.6.4}$$

must be fitted to measured values $y_i(t_i)$. The program to be provided here by the user can have the following form.

Program 9.5: Function `LSQEXP` computes an exponential function for two given parameters

CALLING SEQUENCE: `F=LSQEXP(X,NR,T)`

INPUT ARGUMENTS: `X` — r-vector of the parameters x_1, x_2

`NR` — number of parameters r (must be set to two before function called)

`T` — value of the controlled variable t

OUTPUT ARGUMENTS: function value `LSQEXP` given by (9.6.4)

```
DOUBLE PRECISION FUNCTION LSQEXP(X,NR,T)
IMPLICIT DOUBLE PRECISION (A-H,O-Z)
DIMENSION X(NR)
PARAMETER (BIG=700.D0,ZERO=0.D0)
ARG=X(2)*T
IF(ARG.GT.BIG) THEN
  LSQEXP=ZERO
ELSE
  LSQEXP=X(1)*EXP(-ARG)
END IF
END
```

In Fig. 9.6 the result of a fit to data is shown. The determination of the first approximation for the unknowns can again be obtained by fitting a straight line (graphically or numerically) to the logarithm of the function $y(t)$,

$$\ln y(t) = \ln x_1 - x_2 t \quad .$$

Here one usually uses only the values at small t, since these points have smaller fluctuations. ■

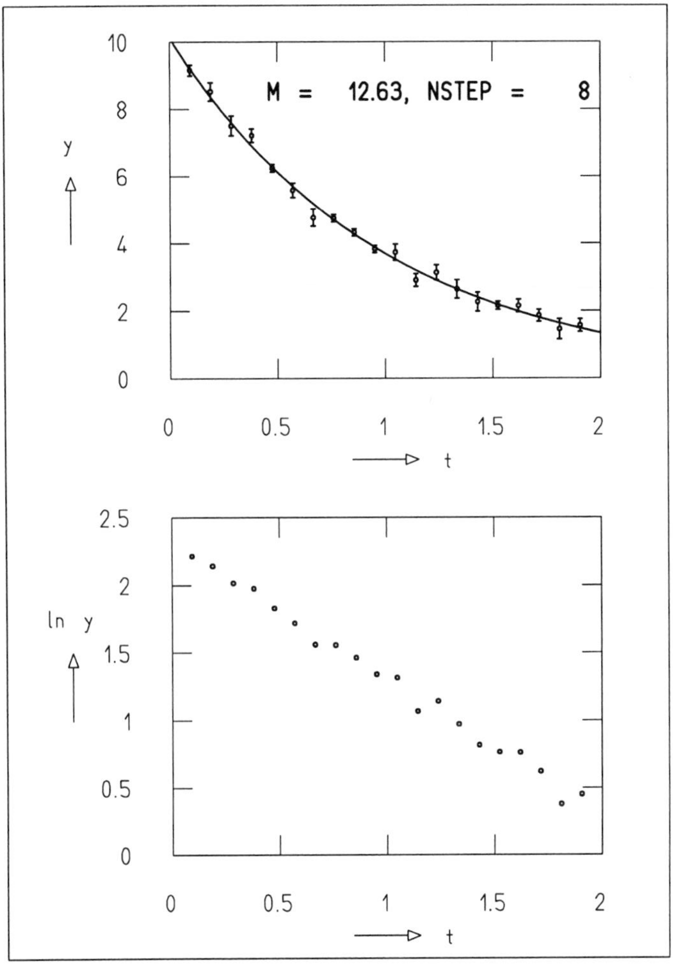

Fig. 9.6. Measured values and fitted exponential function (above); logarithms of the measured values (below).

Example 9.6: Fitting a sum of exponential functions

A radioactive substance often consists of a mixture of components with different decay times. One must therefore fit a sum of several exponential functions. We will consider the case of two functions

$$y(t) = x_1 \exp(-x_2 t) + x_3 \exp(-x_4 t) \quad . \tag{9.6.5}$$

The function that must be written by the user is then:

Program 9.6: Function `LSQ2EX` computes the sum of two exponential functions for four given parameters

CALLING SEQUENCE: F=LSQ2EX(X,NR,T)

INPUT ARGUMENTS: X — r-vector of the parameters x_1, x_2, x_3, x_4
NR — number of parameters r (must be set to 4 before routine is called)
T — value of the controlled variable t

OUTPUT ARGUMENTS: function value LSQ2EX given by (9.6.5)

```
DOUBLE PRECISION FUNCTION LSQ2EX(X,NR,T)
IMPLICIT DOUBLE PRECISION (A-H,O-Z)
DIMENSION X(NR)
PARAMETER (BIG=700.D0,ZERO=0.D0)
ARG=X(2)*T
IF(ARG.GT.BIG) THEN
  E1=ZERO
ELSE
  E1=X(1)*EXP(-ARG)
END IF
ARG=X(4)*T
IF(ARG.GT.BIG) THEN
  E2=ZERO
ELSE
  E2=X(3)*EXP(-ARG)
END IF
LSQ2EX=E1+E2
END
```

Figure 9.7 shows the result of fitting a sum of two exponential functions to the data. A first approximation can be determined by fitting two different lines to $\ln y_i(t_i)$ for the regions of smaller and larger t_i. ■

9.6.2 Marquardt Iteration

The procedure with step-size reduction discussed in Sect. 9.6.1 leads to a minimum when $\mathbf{x}$ is already in a region where $A'^{\mathrm{T}}A'$ is positive definite, i.e., in a region around the minimum where the function $M(\mathbf{x})$ has a positive curvature. (In the one-dimensional case of Fig. 10.1, this is the region between the two points of inflection on either side of the minimum.) It is clear, however, that it must be possible to extend the region of convergence to the region between the two maxima surrounding the minimum. This possibility is offered by the Marquardt procedure, which is presented in Sect. 10.15 in a somewhat more general form. It is therefore recommended that readers wanting to understand the workings of the following program read the present section as well as the introductory sections of Chapter 10, and also Sect. A.18.

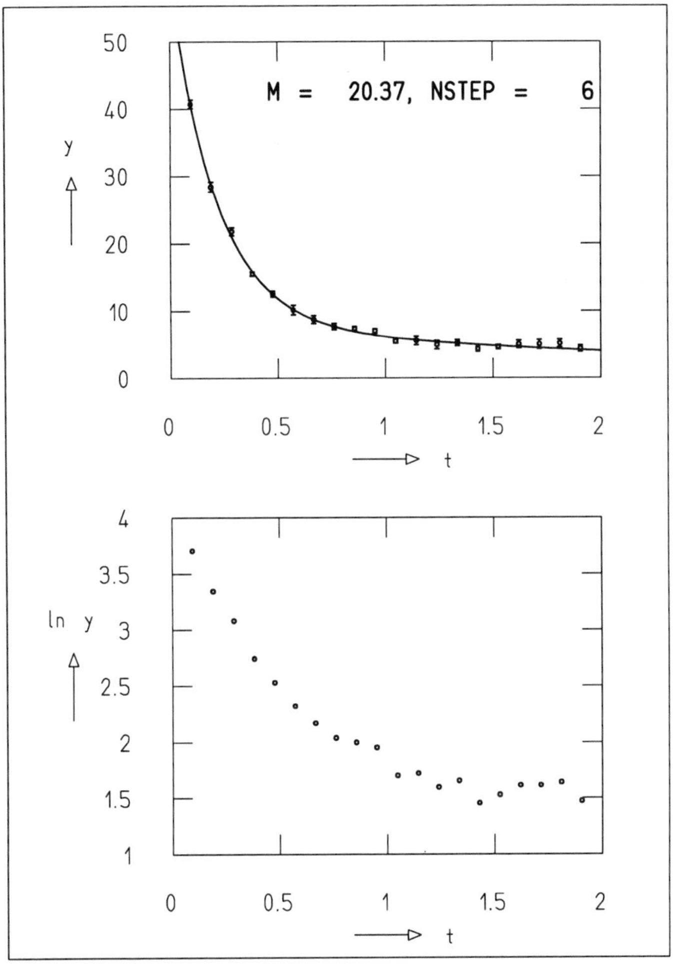

Fig. 9.7. Measured values with the fitted sum of two exponential functions (above); logarithm of the measured values (below).

Program 9.7: Subroutine `LSQMAR` fits a nonlinear function of unknown parameters to data using the Marquardt procedure

CALLING SEQUENCE: `CALL LSQMAR(USERFN,T,Y,DELTAY,N,NR, NRED,LIST,X,CX,R,A,SCRAT,NSTEP)`

INPUT ARGUMENTS: as in `LSQNON`, Program 9.3

OUTPUT ARGUMENTS: as in `LSQNON`

DEFAULT VALUES: as in `LSQNON`

REMARKS: The matrix A is computed with numerical differentiation with Program E.2. If the numerical differentiation fails, the user must provide a routine with the name AUXDRI and the argument list of E.2 such that the matrix A is computed with analytic differentiation. This is also recommended if the problem is to be repeated many times with different data, since the derivatives can be evaluated faster analytically than numerically.

```
      SUBROUTINE LSQMAR(USERFN,T,Y,DELTAY,N,NR,NRED,LIST,X,CX,R,A,
     +SCRAT,NSTEP)
      IMPLICIT DOUBLE PRECISION (A-H,O-Z)
      PARAMETER(NMAX=1000)
      DIMENSION A(N,NRED),T(N),Y(N),DELTAY(N),X(NR)
      DIMENSION SCRAT(NRED,NRED),CX(NRED,NRED)
      DIMENSION C(NMAX),X1(NMAX),X2(NMAX),X1RED(NMAX),X2RED(NMAX)
      DIMENSION LIST(NR)
      COMMON /DASV02/ C,X1,X2,X1RED,X2RED
      PARAMETER(MAXSTP=100,EPSILN=1.D-8,TT=1.D-15,ALAMS=1.D-3)
      LOGICAL NEW,OK,FINAL,COVMAT
      EXTERNAL USERFN
      COVMAT=.TRUE.
      IF(NSTEP.LT.0) THEN
        COVMAT=.FALSE.
        NSTEP=ABS(NSTEP)
      END IF
      IF(NSTEP.LT.1) NSTEP=MAXSTP
      ALAM=ALAMS
C Initial value of minimum function
      R=DBLE(0.)
      DO 10 I=1,N
        R=R+((Y(I)-USERFN(X,NR,T(I)))/DELTAY(I))**2
   10 CONTINUE
      IF(NRED.LE.0) THEN
        NSTEP=1
        GO TO 100
      END IF
C For NR=NRED :  set LIST
      IF(NR.EQ.NRED) THEN
        DO 20 I=1,NR
          LIST(I)=1
   20   CONTINUE
      END IF
C start iteration
      FINAL=.FALSE.
      DO 90 ISTEP=1,NSTEP
C Numerical Derivatives
        CALL AUXDRI(USERFN,X,T,N,NR,NRED,LIST,A,OK)
        IF(.NOT.OK) THEN
          NSTEP=-3
          GO TO 100
        END IF
        DO 40 I=1,N
          DO 30 K=1,NRED
            A(I,K)=A(I,K)/DELTAY(I)
   30     CONTINUE
   40   CONTINUE
        DO 50 I=1,N
          C(I)=(Y(I)-USERFN(X,NR,T(I)))/DELTAY(I)
```

```
   50   CONTINUE
        IF(FINAL) THEN
C Final Step
          CALL MTXSVD(A,C,X1RED,R,N,NRED,1,DBLE(0.),OK)
          IF(.NOT.OK) THEN
            NSTEP=-1
            GO TO 100
          END IF
          CALL MTXZRV(X1,NR)
          CALL MTXPSV(X1,X1RED,NR,NRED,LIST)
          CALL MTXADV(X,X1,X,NR)
C Compute covariance matrix CX of unknowns
          IF(COVMAT) THEN
            CALL AUXDRI(USERFN,X,T,N,NR,NRED,LIST,A,OK)
            IF(.NOT.OK) THEN
              NSTEP=-3
              GO TO 100
            END IF
            DO 70 I=1,N
              DO 60 K=1,NRED
                A(I,K)=A(I,K)/DELTAY(I)
   60         CONTINUE
   70       CONTINUE
            CALL MTXMAT(A,A,CX,NRED,N,NRED)
            CALL MTXCHI(CX,SCRAT,NRED)
          END IF
          NSTEP=ISTEP
          GO TO 100
        END IF
C compute minimum function for two values of lambda (ALAM)
        CALL MTXMAR(A,C,ALAM,X1RED,X2RED,N,NRED,DBLE(0.),OK)
        IF(.NOT.OK) THEN
          NSTEP=-1
          GO TO 100
        END IF
        CALL MTXZRV(X1,NR)
        CALL MTXZRV(X2,NR)
        CALL MTXPSV(X1,X1RED,NR,NRED,LIST)
        CALL MTXPSV(X2,X2RED,NR,NRED,LIST)
        CALL MTXADV(X1,X,X1,NR)
        CALL MTXADV(X2,X,X2,NR)
        R1=DBLE(0.)
        R2=DBLE(0.)
        DO 80 I=1,N
          R1=R1+((Y(I)-USERFN(X1,NR,T(I)))/DELTAY(I))**2
          R2=R2+((Y(I)-USERFN(X2,NR,T(I)))/DELTAY(I))**2
   80   CONTINUE
        IF(.NOT.OK) THEN
          NSTEP=-1
          GO TO 100
        END IF
C evaluate results
        IF(R2.LE.R+TT) THEN
C reduce lambda and accept new point
          ALAM=ALAM*DBLE(0.1)
          CALL MTXCPV(X2,X,NR)
          RMIN=R2
          NEW=.TRUE.
        ELSE IF (R2.GT.R+TT .AND. R1.LE.R+TT) THEN
C keep current value of lambda and accept new point
          CALL MTXCPV(X1,X,NR)
```

```
            RMIN=R1
            NEW=.TRUE.
          ELSE
C increase lambda and reject new point
            ALAM=ALAM*DBLE(10.)
            NEW=.FALSE.
          END IF
          IF(NEW) THEN
C test for break-off criterion
            IF(ABS(R-RMIN) .LT. EPSILN*ABS(RMIN)+TT) THEN
              FINAL=.TRUE.
            ELSE
              R=RMIN
            END IF
          END IF
   90 CONTINUE
      NSTEP=-2
  100 RETURN
      END
```

Example 9.7: Fitting a sum of two Gaussian functions and a polynomial

In practice it is usually not so easy to find the amplitude, position, and width of signals as was shown in Example 9.4. One usually has more than one signal, lying on a background that varies slowly with the controlled variable t. Since in general this background is not well known, it is approximated by a line or a second-order polynomial. We will consider the sum of such a polynomial and two Gaussian distributions, i.e., a function of nine unknown parameters,

$$\begin{aligned} h(\mathbf{x}, t) &= x_1 + x_2 t + x_3 t^2 \\ &\quad + x_4 \exp\{-(x_5 - t)^2/2x_6^2\} + x_7 \exp\{-(x_8 - t)^2/2x_9^2\} \quad . \end{aligned} \tag{9.6.6}$$

The derivatives (9.5.5) are

$$\begin{aligned} -a_{j1} &= \frac{\partial h_j}{\partial x_1} = 1 \quad , \\ -a_{j2} &= \frac{\partial h_j}{\partial x_2} = t_j \quad , \\ -a_{j3} &= \frac{\partial h_j}{\partial x_3} = t_j^2 \quad , \\ -a_{j4} &= \frac{\partial h_j}{\partial x_4} = \exp\{-(x_5 - t_j)^2/2x_6^2\} \quad , \\ -a_{j5} &= \frac{\partial h_j}{\partial x_5} = 2x_4 \exp\{-(x_5 - t_j)^2/2x_6^2\} \frac{t_j - x_5}{2x_6^2} \quad , \\ -a_{j6} &= \frac{\partial h_j}{\partial x_6} = x_4 \exp\{-(x_5 - t_j)^2/2x_6^2\} \frac{(t_j - x_5)^2}{x_6^3} \quad , \\ &\vdots \quad . \end{aligned}$$

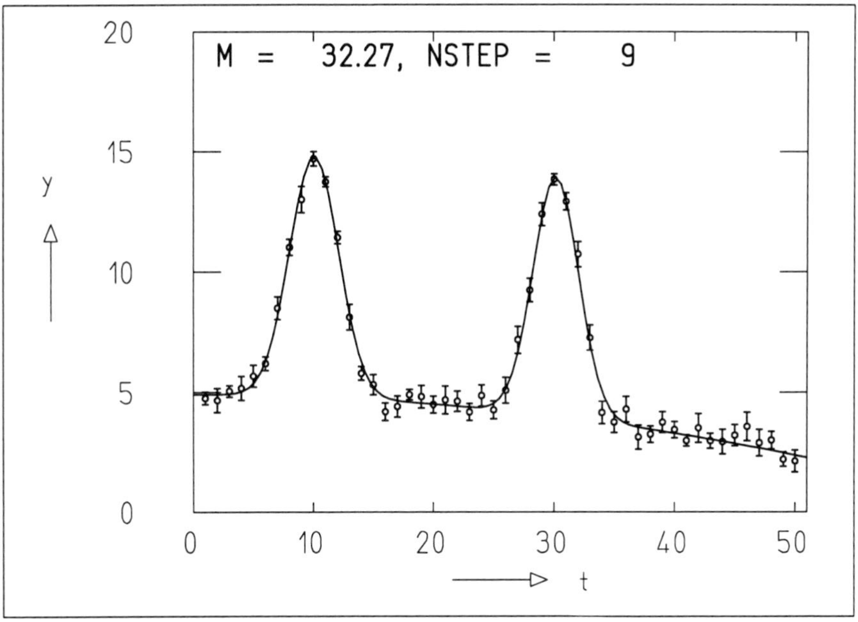

Fig. 9.8. Measured values and fitted sum of a second-order polynomial and two Gaussian functions.

If the numerical differentiation fails, the derivatives must be computed with a specially written version of the routine AUXDRI.

For the numerical example shown in Fig. 9.8, however, this is not necessary. The user must supply, of course, a function to compute (9.6.6). This is:

Program 9.8: Function LSQP2G computes the sum of a polynomial and two Gaussian functions

CALLING SEQUENCE: F=LSQP2G(X,NR,T)

INPUT ARGUMENTS: X — r-vector of parameters $x_1, x_2, \ldots, x_9$
NR — number of parameters r (must be set to 9 before routine called)
T — value of the controlled variable t

OUTPUT ARGUMENT: function value LSQP2G given by (9.6.6)

```
DOUBLE PRECISION FUNCTION LSQP2G(X,NR,T)
IMPLICIT DOUBLE PRECISION (A-H,O-Z)
DIMENSION X(NR)
PARAMETER (BIG=700.D0,TWO=2.D0,ZERO=0.D0)
BACK=X(1)+X(2)*T+X(3)*T**2
ARG1=(X(5)-T)**2/(TWO*X(6)**2)
```

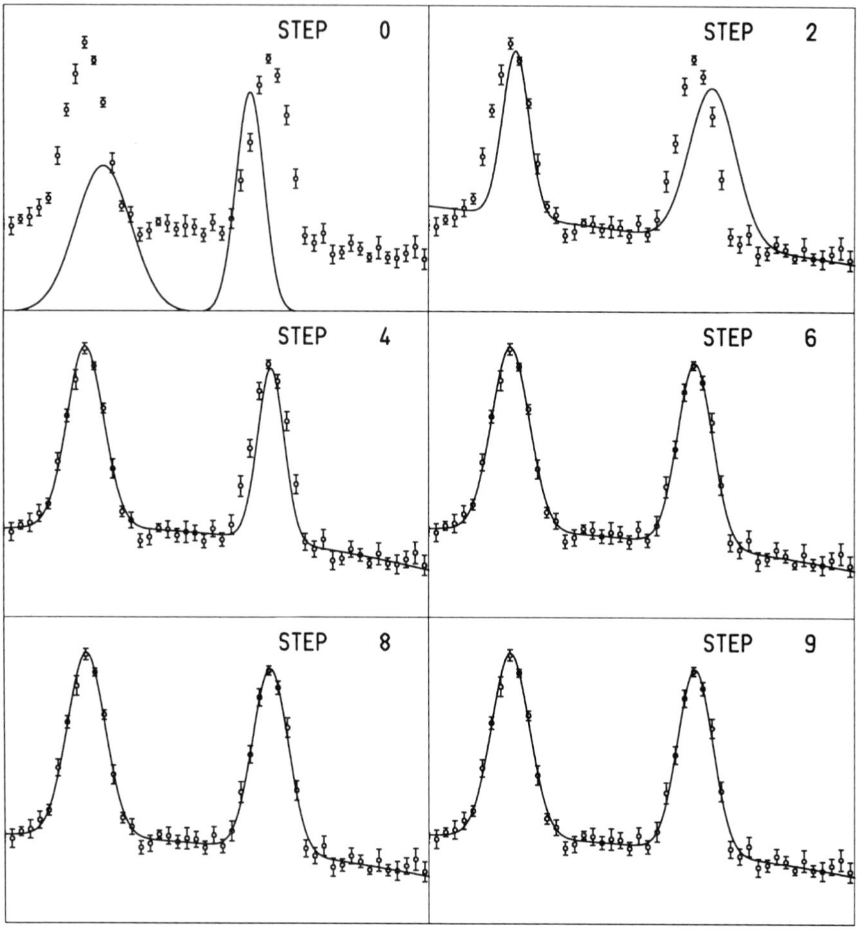

Fig. 9.9. Successive approximations of the fit function to the measurements.

```
ARG2=(X(8)-T)**2/(TWO*X(9)**2)
IF(ARG1.GT.BIG) THEN
  GAUSS1=ZERO
ELSE
  GAUSS1=X(4)*EXP(-ARG1)
END IF
IF(ARG2.GT.BIG) THEN
  GAUSS2=ZERO
ELSE
  GAUSS2=X(7)*EXP(-ARG2)
END IF
LSQP2G=BACK+GAUSS1+GAUSS2
END
```

Figure 9.8 shows the result of fitting to a total of 50 measurements. It is perhaps interesting to look at some of the intermediate steps in Fig. 9.9.

As a first approximation the parameters of the polynomial were set to zero ($x_1 = x_2 = x_3 = 0$). For both of the clearly visible signals, rough estimates of the amplitude, position, and width were used for the initial values. The function (9.6.6) with the parameters of this first approximation is shown in the first frame (`STEP 0`). In the following frames one sees the change of the function after 2, 4, 6, and 8 steps, and finally at convergence of the program `LSQMAR` after a total of 9 steps. ■

9.7 Properties of the Least-Squares Solution. χ^2-Test

Up to now the method of least squares has merely been an application of the maximum-likelihood method to a linear problem. The prescription of least squares (9.2.15) was obtained directly from the maximization of the likelihood function (9.2.14). In order to be able to specify this likelihood function in the first place, it was necessary to know the distribution of the measurement errors. We assumed a normal distribution. But also when there is no exact knowledge of the error distribution, one can still apply the relation (9.2.15) and with it the remaining formulas of the last sections. Such a procedure seems to lack a theoretical justification. The Gauss–Markov theorem, however, states that in this case as well, the method of least squares provides results with desirable properties. Before entering into this, let us list once more the properties of the maximum-likelihood solution.

(a) The solution $\widetilde{\mathbf{x}}$ is asymptotically unbiased, i.e., for very large samples,

$$E(\widetilde{x}_i) = x_i \quad , \qquad i = 1, 2, \ldots, r \quad .$$

(b) It is a minimum variance estimator, i.e.,

$$\sigma^2(\widetilde{x}_i) = E\{(\widetilde{x}_i - x_i)^2\} = \min \quad .$$

(c) The quantity (9.2.16)

$$M = \boldsymbol{\varepsilon}^{\mathrm{T}} G_y \boldsymbol{\varepsilon}$$

follows a χ^2-distribution with $n - r$ degrees of freedom.

The properties (a) and (b) are familiar from Chapter 7. We will demonstrate the validity of (c) for the simple case of direct measurements ($r = 1$), for which the matrix G_y is diagonal:

$$G_y = \begin{pmatrix} 1/\sigma_1^2 & & & 0 \\ & 1/\sigma_2^2 & & \\ & & \ddots & \\ 0 & & & 1/\sigma_n^2 \end{pmatrix} \quad .$$

The quantity M then simply becomes a sum of squares,

$$M = \sum_{j=1}^{n} \varepsilon_j^2/\sigma_j^2 \quad . \tag{9.7.1}$$

Since each ε_j comes from a normally distributed population with mean of zero and variance σ_j^2, the quantities $\varepsilon_j^2/\sigma_j^2$ are described by a standard Gaussian distribution. Thus the sum of squares follows a χ^2-distribution with $n-1$ degrees of freedom.

If the distribution of the errors ε_j is not known, then the least-squares solution has the following properties:

(a) The solution is unbiased.

(b) Of all solutions $\mathbf{x}^*$ that are unbiased estimators for $\mathbf{x}$ and are linear combinations of the measurements $\mathbf{y}$, the least-squares solution has the smallest variance. (This is the GAUSS–MARKOV theorem.)

(c) The expectation value of

$$M = \boldsymbol{\varepsilon}^{\mathrm{T}} G_y \boldsymbol{\varepsilon}$$

is

$$E(M) = n - r \quad .$$

(This is exactly the expectation value of a χ^2 variable for $n-r$ degrees of freedom.)

The quantity M is often simply called χ^2, although it does not necessarily follow a χ^2-distribution. Together with the matrices $C_{\tilde{x}}$ and $C_{\tilde{\eta}}$ it provides a convenient measure of the quality of a fit with the least-squares method. If the value of M obtained from the measurements comes out much larger than $n-r$, then the assumptions of the calculation must be carefully checked. The result should not simply be accepted without critical examination.

The number $f = n - r$ is called the *number of degrees of freedom of a fit* or the *number of equations of constraint of the fit*. It is clear from the beginning (see Appendix A), that the problem of least squares can only be solved for $f \geq 0$. Only for $f > 0$, however, is the quantity M meaningfully defined and usable for testing the quality of a fit.

If it is known that the errors are normally distributed, then a χ^2-test can be done in conjunction with the fit. One rejects the result of the fit if

$$M = \boldsymbol{\varepsilon}^{\mathrm{T}} G_y \boldsymbol{\varepsilon} > \chi^2_{1-\alpha}(n-r) \quad , \tag{9.7.2}$$

i.e., if the quantity M exceeds the quantile of the χ^2-distribution with $n-r$ degrees of freedom at a significance level of α. A larger value of M can be caused be the following reasons (and also by an error of the first kind):

(a) The assumed functional dependence $\mathbf{f}(\mathbf{x}, \boldsymbol{\eta}) = 0$ between the measured values η and the unknown parameters x is false. Either the function $\mathbf{f}(\mathbf{x}, \boldsymbol{\eta})$ is completely wrong, or some of the parameters taken to be known are not correct.

(b) The function $\mathbf{f}(\mathbf{x}, \boldsymbol{\eta})$ is correct, but the series expansion with only one term is not a sufficiently good approximation in the region of parameter space covered in the computation.

(c) The initial approximation $\mathbf{x}_0$ is too far from the true value $\mathbf{x}$. Better values for $\mathbf{x}_0$ could lead to more acceptable values of M. This point is clearly related to (b).

(d) The covariance matrix of the measurements C_y, which is often only based on rough estimates or assumptions, is not correct.

These four points must be carefully taken into consideration if the method of least squares is to be applied successfully. In many cases the least-squares computation is repeated many times for different data sets. One can then look at the empirical frequency distribution of the quantity M and compare it to a χ^2-distribution with the corresponding number of degrees of freedom. Such a comparison is particularly useful for a good estimate of C_y. At the start of a new experiment, the apparatus that provides the measured values $\mathbf{y}$ is usually checked by measuring known quantities. In this way the parameters $\mathbf{x}$ for several data sets are determined. The covariance matrix of the measurements can then be determined such that the distribution (i.e., the histogram) of M agrees as well as possible with a χ^2-distribution. This investigation is particularly illustrative when one considers the distribution of $F(M)$ instead of that of M, where F is the distribution function (6.6.11) of the χ^2-distribution. If M follows a χ^2-distribution, then $F(M)$ follows a uniform distribution.

9.8 Confidence Regions and Asymmetric Errors in the Nonlinear Case

We begin this section with a brief review of the meaning of the covariance matrix $C_{\widetilde{x}}$ of the unknown parameters for the linear least-squares case. The probability that the true value of the unknowns is given by $\mathbf{x}$ is given by a normal distribution of the form (5.10.1),

$$\phi(\mathbf{x}) = k \exp\{-\tfrac{1}{2}(\mathbf{x} - \widetilde{\mathbf{x}})^{\mathrm{T}} B(\mathbf{x} - \widetilde{\mathbf{x}})\} \quad , \tag{9.8.1}$$

where $B = C_{\widetilde{x}}^{-1}$. The exponent (multiplied by -2),

$$g(\mathbf{x}) = (\mathbf{x} - \widetilde{\mathbf{x}})^{\mathrm{T}} B(\mathbf{x} - \widetilde{\mathbf{x}}) \quad , \tag{9.8.2}$$

for $g = 1 =$ const describes the covariance ellipsoid (cf. Sect. 5.10). For other values of $g(\mathbf{x}) =$ const one obtains confidence ellipsoids for a given probability W [see (5.10.19)]. In this way confidence ellipsoids can easily be constructed that have the point $\mathbf{x} = \widetilde{\mathbf{x}}$ at the center and which contain the true value $\mathbf{x}$ of the unknown with probability $W = 0.95$.

We can now find a relation between the confidence ellipsoid and the minimum function. For this we use the expression (9.2.19), which is exactly true in the linear case, and we compute the difference between the minimum function $M(\mathbf{x})$ at a point $\mathbf{x}$ and the minimum function $M(\widetilde{\mathbf{x}})$ at the point $\widetilde{\mathbf{x}}$ where it is a minimum,

$$M(\mathbf{x}) - M(\widetilde{\mathbf{x}}) = (\mathbf{x} - \widetilde{\mathbf{x}})^{\mathrm{T}} A^{\mathrm{T}} G_y A (\mathbf{x} - \widetilde{\mathbf{x}}) \quad . \tag{9.8.3}$$

According to (9.2.27), one has

$$B = C_{\widetilde{x}}^{-1} = G_{\widetilde{x}} = A^{\mathrm{T}} G_y A \quad . \tag{9.8.4}$$

So the difference (9.8.3) is exactly the function introduced above, $g(\mathbf{x})$.

Thus the covariance ellipsoid is in fact the hypersurface in the space of the r variables $x_1, \ldots, x_r$, where the function $M(\mathbf{x})$ has the value $M(\mathbf{x}) = M(\widetilde{\mathbf{x}}) + 1$. Correspondingly, the confidence ellipsoid with probability W is the hypersurface for which

$$M(\mathbf{x}) = M(\widetilde{\mathbf{x}}) + g \quad , \tag{9.8.5}$$

where the constant g is the quantile of the χ^2-distribution with $f = n - r$ degrees of freedom according to (5.10.21),

$$g = \chi^2_W(f) \quad . \tag{9.8.6}$$

In the nonlinear case of least squares, our considerations remain approximately valid. The approximation becomes better when the nonlinear deviations of the expression (9.5.3) are less in the region where the unknown parameters are varied. The deviations are not only small when the derivatives are almost constant, i.e., the function is almost linear, but also when the variation of the unknown parameters is small. A measure of the variation of an unknown is, however, its error. Because of error propagation, the errors of the unknowns are small when the original errors of the measurements are small. In the nonlinear case, therefore, the covariance matrix retains the same interpretation as in the linear case as long as the measurement errors are small.

If the measurement errors are large we retain the interpretation (9.8.5), i.e., we determine the hypersurface with (9.8.5) and state that the true value $\mathbf{x}$ lies with probability W within the *confidence region* around the point. This region is, however, no longer an ellipsoid.

For only one parameter x, the curve $M = M(x)$ can easily be plotted. The confidence region is simply a segment of the x axis. For two parameters x_1, x_2 the boundary of the confidence region is a curve in the (x_1, x_2) plane. It is the contour line (9.8.5) of the function $M = M(\mathbf{x})$. The graphical representation corresponds thus to the contours produced by Program F.15. For more then two parameters, a graphical representation is only possible in the form of slices of the confidence region in the (x_i, x_j) plane in the $(x_1, x_2, \ldots, x_r)$ space through the point $\widetilde{\mathbf{x}} = (\widetilde{x}_1, \widetilde{x}_2, \ldots, \widetilde{x}_r)$, where x_i, x_j can be any possible pair of variables. These contour lines are plotted by the following procedure in a way corresponding to that of Program F.15. To do this the function $M(\mathbf{x})$ must be computed at many points in the plane described above. Since at each of these points all of the parameters $\mathbf{x}$ are fixed, no iteration is carried out, but rather $M(\mathbf{x})$ is computed.

Program 9.9: Subroutine LSQCON plots the confidence region in a plane spanned by two parameters, whose values were determined in the nonlinear case of the least-squares method

CALLING SEQUENCE: CALL LSQCON(IX,JX,XI0,XJ0,DXI,DXJ, NXI,NXJ,FCONT,X0,T,Y,DELTAY,N,NR,USERFN)

INPUT ARGUMENTS: IX,JX — indices i, j, labeling the elements x_i, x_j of the n-vector $\mathbf{x}$, which span the plane of the confidence region
XI0,DXI,NXI — initial value x_{i0}, interval width Δx_i, and number of intervals n_{xi} of the variables x_i
XJ0,DXJ,NXJ — same as above for x_j
FCONT — value of the minimum function at the boundary of the confidence region
X0 — n-vector of the variables at the minimum $\widetilde{\mathbf{x}}$
T,Y,DELTAY,N,NR,USERFN — as in LSQNON

```
      SUBROUTINE LSQCON(IX,JX,XI0,XJ0,DXI,DXJ,NXI,NXJ,FCONT,X0,T,Y,
     +DELTAY,N,NR,USERFN)
      IMPLICIT DOUBLE PRECISION (A-H,O-Z)
      DIMENSION X0(NR),T(N),Y(N),DELTAY(N)
      DIMENSION XI(2),XJ(2),F(2,2),X0SAV(100)
      EXTERNAL USERFN
      CALL MTXCPV(X0,X0SAV,NR)
      DO 50 I=0,NXI-1
        XI(1)=XI0+I*DXI
        XI(2)=XI(1)+DXI
        DO 40 J=0,NXJ-1
          XJ(1)=XJ0+J*DXJ
          XJ(2)=XJ(1)+DXJ
          DO 30 IS=1,2
            DO 20 JS=1,2
              CALL MTXCPV(X0SAV,X0,NR)
              X0(IX)=XI(IS)
```

```
              XO(JX)=XJ(JS)
              R=0.D0
              DO 10 L=1,N
                R=R+((Y(L)-USERFN(XO,NR,T(L)))/DELTAY(L))**2
10            CONTINUE
              F(IS,JS)=R
20          CONTINUE
30        CONTINUE
          CALL GRPXCT(XI(1),XI(2),XJ(1),XJ(2),F(1,1),F(1,2),F(2,1),
     +    F(2,2),FCONT)
40      CONTINUE
50    CONTINUE
      END
```

Example 9.8: The influence of large measurement errors on the confidence region of the parameters for fitting an exponential function

The results of fitting an exponential function as in Example 9.5 to data points with errors of different sizes are shown in Fig. 9.10. The two fitted parameters x_1 and x_2 and their errors and correlation coefficient are also shown in the figure. The quantities Δx_1, Δx_2, and ρ were computed directly from the elements of the covariance matrix $C_{\tilde{x}}$. As expected, these errors increase for increasing measurement errors.

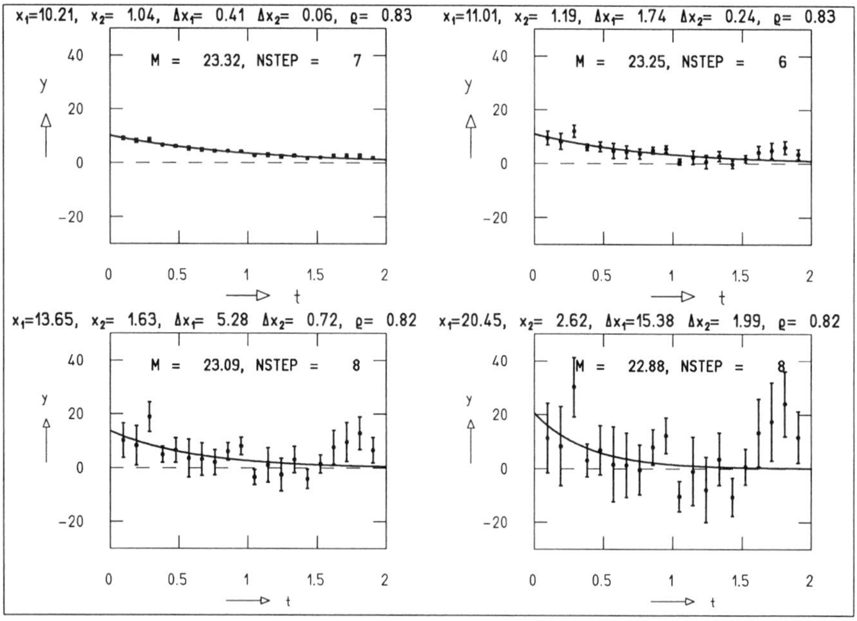

Fig. 9.10. Fit of an exponential function $\eta = x_1 \exp(-x_2 t)$ to measurements with errors of different sizes.

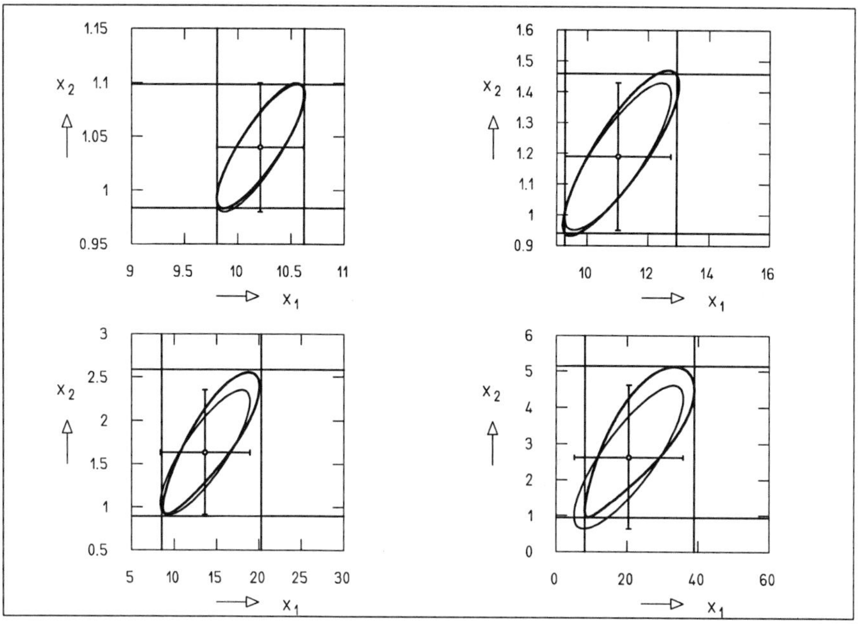

Fig. 9.11. Results of the fit from Fig. 9.10 in the parameter space x_1, x_2. Solution $\widetilde{\mathbf{x}}$ (small circle), symmetric errors (error bars), covariance ellipse, asymmetric error limits (horizontal and vertical lines), confidence region corresponding to the probability of the covariance ellipse (dark contour).

The covariance ellipses for this fit are shown as thin lines in the various frames of Fig. 9.11. The small circle in the center indicates the solution $\widetilde{\mathbf{x}}$. In addition the vertical bars through the point $\widetilde{\mathbf{x}}$ mark the regions $\widetilde{x}_1 \pm \Delta x_1$ and $\widetilde{x}_2 \pm \Delta x_2$. As a thicker line a contour is indicated which surrounds the confidence region $M(\mathbf{x}) = M(\widetilde{\mathbf{x}}) + 1$. We see that for small measurement errors the confidence region and covariance ellipse are practically identical, while for large measurement errors they clearly differ from each other.

The contours and confidence regions were computed and plotted with Program 9.9. Here the solution obtained from LSQNON was used as the argument X0 when calling LSQCON. For the user function LSQEXP was used. ■

Computing and plotting the confidence regions requires, to the extent that they differ from the confidence ellipsoids, considerable effort. It is therefore important to be able to decide whether a clear difference between the two exists. For clarity we will stay with the case of two variables and consider again Example 9.8 and in particular Fig. 9.11. The errors $\Delta x_i = \sigma_i = \sqrt{C_{\widetilde{x}}(i,i)}$ obtained from the covariance matrix $C_{\widetilde{x}}$ have the following property. The lines

$x_i = \widetilde{x}_i \pm \Delta x_i$ are tangent to the covariance ellipse. If the covariance ellipse must be replaced by a confidence region of less regular form, then we can nevertheless find the horizontal and vertical tangents, and at those places in Fig. 9.11 one has

$$x_i = \widetilde{x}_i + \Delta x_{i+} \quad , \qquad x_i = \widetilde{x}_i - \Delta x_{i-} \quad . \tag{9.8.7}$$

Because of the loss of symmetry, the errors Δx_+ and Δx_- are in general different. One speaks of *asymmetric errors*. For r variables one has tangent hypersurfaces of dimension $r - 1$ instead of tangent lines.

We now give a procedure to compute the asymmetric errors $\Delta x_{i\pm}$. One is only interested in asymmetric confidence regions when the asymmetric errors are significantly different from the symmetric ones. The values $x_{i\pm} = \widetilde{x}_i \pm \Delta x_{i\pm}$ have the property that the minimum of the function M for a fixed value of $x_i = x_{i\pm}$ has the value

$$\min\{M(\mathbf{x};\, x_i = x_{i\pm})\} = M(\widetilde{\mathbf{x}}) + g \tag{9.8.8}$$

with $g = 1$. For other values of g one obtains corresponding *asymmetric confidence boundaries*. If we bring all the terms in (9.8.8) to the left-hand side,

$$\min\{M(\mathbf{x};\, x_i = x_{i\pm})\} - M(\widetilde{\mathbf{x}}) - g = 0 \quad , \tag{9.8.9}$$

then we see the problem to be solved is a combination of minimization and zero-finding. We find the zeros using an iterative procedure as in Sect. E.2. The zero, i.e., the point x_i fulfilling (9.8.9), is first bracketed by the values x_{small} and x_{big}, corresponding to negative and positive values of (9.8.9). Next, this interval is reduced by successively dividing it in half until the expression (9.8.9) differs from zero by no more than $g/100$. The minimum in (9.8.9) is found by calling the routines `LSQNON` or `LSQMAR`, where x_i is fixed. The routine `LSQASN` given below can be called directly after `LSQNON`, with which $\widetilde{\mathbf{x}}$, $C_{\widetilde{x}}$, and $M(\widetilde{\mathbf{x}})$ are determined. Most of the many input arguments are input or output arguments of `LSQNON`.

The asymmetric errors for Example 9.8 are shown in Fig. 9.11.

Program 9.10: Subroutine `LSQASN` computes asymmetric errors (or confidence limits) for the nonlinear least-squares problem using the routine `LSQNON`

CALLING SEQUENCE: `CALL LSQASN(USERFN,T,Y,DELTAY,N,NR, NRED,LIST,X,CX,R,W,DXPLUS,DXMINS,A,SCRAT,NSTEP)`

INPUT ARGUMENTS: `USERFN,T,Y,DELTAY,N,NR,NRED,LIST,X, CX,R,A,SCRAT` — as in Program 9.3, `LSQNON`

W — probability W, for which the confidence limits are computed; see (9.8.8). If W is set equal to 0, asymmetric errors rather than confidence limits are computed
NSTEP — maximum number of iteration steps

OUTPUT ARGUMENTS: DXPLUS — r'-vector of errors or confidence limits Δx_{i+}
DXMINS — r'-vector of errors or confidence limits Δx_{i-}
NSTEP — number of steps needed. If an error occurs, NSTEP is set to a negative value (-1: minimization failed, -2: failure to find zero).

DEFAULT VALUES: If the input value of W ≤ 0, then asymmetric errors (or confidence limits) are computed. If the input value of NSTEP ≤ 0, then NSTEP = 100 is taken.

REMARKS: The square roots of the diagonal elements of CX serve as the first approximation for the asymmetric errors.

```
      SUBROUTINE LSQASN(USERFN,T,Y,DELTAY,N,NR,NRED,LIST,X0,CX,R,W,
     +DXPLUS,DXMINS,A,SCRAT,NSTEP)
      IMPLICIT DOUBLE PRECISION (A-H,O-Z)
      PARAMETER(NMAX=1000)
      DIMENSION A(N,NRED),T(N),Y(N),DELTAY(N),X0(NR)
      DIMENSION DXPLUS(NRED),DXMINS(NRED)
      DIMENSION SCRAT(NRED,NRED),CX(NRED,NRED)
      DIMENSION X(NMAX),LIST(NR),DX(NMAX)
      COMMON /DASV03/ X,DX
      PARAMETER(MAXSTP=100,EPS=1.D-2,EPSILN=1.D-4,TT=1.D-6,
     +FACT=1.5D0)
      PARAMETER(ZERO=0.0D0,ONE=1.D0,HALF=0.5D0,TWO=2.D0)
      LOGICAL LENCLS
      EXTERNAL USERFN
      IF(NSTEP.LE.0) NSTEP=MAXSTP
      IF(NR.EQ.NRED) THEN
        DO 10 I=1,NR
          LIST(I)=1
   10   CONTINUE
      END IF
      IF(W.LE.EPSILN) THEN
        G=ONE
      ELSE
        G=SQCHI2(W,NRED)
      END IF
      RTARG=R+G
      DO 20 I=1,NRED
        DX(I)=SQRT(CX(I,I))
   20 CONTINUE
      NRED1=NRED-1
      DO 60 IVAR=1,NR
        IF(LIST(IVAR).EQ.0) GO TO 60
C fix variable IVAR
        LIST(IVAR)=0
        DO 50 ISIGN=-1,1,2
          SIGNUM=DBLE(ISIGN)
```

```
          DEL=DX(IVAR)
          LENCLS=.FALSE.
C set XSMALL to x at minimum position
          XSMALL=X0(IVAR)
          XIVAR=X0(IVAR)+SIGNUM*DEL
          DO 30 I=1,NSTEP
            NSTEPL=-NSTEP
            CALL MTXCPV(X0,X,NR)
            X(IVAR)=XIVAR
            CALL LSQNON(USERFN,T,Y,DELTAY,N,NR,NRED1,LIST,X,CX,
     +      R1,A,SCRAT,NSTEPL)
            IF(NSTEPL.LT.1) THEN
              NSTEP=-1
              GO TO 70
            END IF
C test for convergence
            IF(ABS(R1-RTARG).LT.EPS*G) THEN
              IF(ISIGN.LT.0) THEN
                DXMINS(IVAR)=X0(IVAR)-X(IVAR)
              ELSE
                DXPLUS(IVAR)=X(IVAR)-X0(IVAR)
              END IF
              GO TO 40
            ELSE
              IF(.NOT.LENCLS) THEN
C zero was not yet enclosed in last step
                IF(R1.GT.RTARG) THEN
C zero is now enclosed, perform first interval halving
                  LENCLS=.TRUE.
                  XBIG=X(IVAR)
                  X(IVAR)=HALF*(XBIG+XSMALL)
                ELSE
C zero not enclosed, widen range
                  DEL=DEL*TWO
                  X(IVAR)=XSMALL+SIGNUM*DEL
                END IF
              ELSE
C continue interval halving
                IF(R1.GT.RTARG) THEN
                  XBIG=X(IVAR)
                ELSE
                  XSMALL=X(IVAR)
                END IF
                X(IVAR)=HALF*(XBIG+XSMALL)
              END IF
              XIVAR=X(IVAR)
            END IF
   30     CONTINUE
          NSTEP=-2
          GO TO 70
   40     CONTINUE
   50   CONTINUE
C unfix variable IVAR
        LIST(IVAR)=1
   60 CONTINUE
   70 RETURN
      END
```

If LSQMAR rather than LSQNON is to be used, then the routine LSQASM computes the asymmetric errors. We present here only the calling sequence, since the program itself is almost identical to LSQASN.

Program 9.11: Subroutine LSQASM performs the same task as LSQASN, but using the routine LSQMAR

CALLING SEQUENCE: `CALL LSQASM(USERFN,T,Y,DELTAY,N,NR, NRED,LIST,X,CX,R,W,DXPLUS,DXMINS,A,SCRAT,NSTEP)`

INPUT ARGUMENTS: as in LSQASN

OUTPUT ARGUMENTS: as in LSQASN

DEFAULT VALUES: as in LSQASN

9.9 Constrained Measurements

We now return to the case of Sect. 9.1, where the quantities of interest were directly measured. The N measurements are, however, no longer completely independent, but rather are related to each other by q equations of constraint. One could measure, for example, the three angles of a triangle. The equation of constraint says that their sum is equal to 180°. We again ask for the best estimates $\tilde{\eta}_j$ for the quantities η_j. The measurements give instead the values

$$y_j = \eta_j + \varepsilon_j \quad , \qquad j = 1, 2, \ldots, n \quad . \tag{9.9.1}$$

As above let us assume a normal distribution about zero for the measurement errors ε_j:

$$E(\varepsilon_j) = 0 \quad , \qquad E(\varepsilon_j^2) = \sigma_j^2 \quad .$$

The q equations of constraint have the form

$$f_k(\boldsymbol{\eta}) = 0 \quad , \qquad k = 1, 2, \ldots, q \quad . \tag{9.9.2}$$

Let us first consider the simple case of linear equations of constraint. The equations (9.9.2) are then of the form

$$\begin{aligned} b_{10} + b_{11}\eta_1 + b_{12}\eta_2 + \cdots + b_{1n}\eta_n &= 0 \quad , \\ b_{20} + b_{21}\eta_1 + b_{22}\eta_2 + \cdots + b_{2n}\eta_n &= 0 \quad , \\ &\vdots \\ b_{q0} + b_{q1}\eta_1 + b_{q2}\eta_2 + \cdots + b_{qn}\eta_n &= 0 \quad , \end{aligned} \tag{9.9.3}$$

or in matrix notation,

$$B\boldsymbol{\eta} + \mathbf{b}_0 = 0 \quad . \tag{9.9.4}$$

9.9.1 The Method of Elements

Although not well suited for automatic processing of data, an illustrative procedure is the *method of elements*. We can use the q equations (9.9.3) to eliminate q of the n quantities η. The remaining $n-q$ quantities α_i $(i = 1,2,\ldots,n-q)$ are called *elements*. They can be chosen arbitrarily from the original η or they can be a linear combination of them. We can then express the full vector η as a set of linear combinations of these elements,

$$\eta_j = f_{j0} + f_{j1}\alpha_1 + f_{j2}\alpha_2 + \cdots + f_{j,n-q}\alpha_{n-q} \quad , \qquad j = 1,2,\ldots,n \quad , \tag{9.9.5}$$

or

$$\boldsymbol{\eta} = F\boldsymbol{\alpha} + \mathbf{f}_0 \quad . \tag{9.9.6}$$

Equation (9.9.6) is of the same type as (9.2.2). The solution must thus be of the form of (9.2.26), i.e.,

$$\tilde{\boldsymbol{\alpha}} = (F^{\mathrm{T}} G_y F)^{-1} F^{\mathrm{T}} G_y (\mathbf{y} - \mathbf{f}_0) \tag{9.9.7}$$

describes the estimation of the elements $\boldsymbol{\alpha}$ according to the method of least squares. The corresponding covariance matrix is

$$G_{\tilde{\alpha}}^{-1} = (F^{\mathrm{T}} G_y F)^{-1} \quad , \tag{9.9.8}$$

cf. (9.2.27). The improved measurements are obtained by substituting (9.9.7) into (9.9.6):

$$\tilde{\boldsymbol{\eta}} = F\tilde{\boldsymbol{\alpha}} + \mathbf{f}_0 = F(F^{\mathrm{T}} G_y F)^{-1} F^{\mathrm{T}} G_y (\mathbf{y} - \mathbf{f}_0) + \mathbf{f}_0 \quad . \tag{9.9.9}$$

By error propagation the covariance matrix is found to be

$$G_{\tilde{\eta}}^{-1} = F(F^{\mathrm{T}} G_y F)^{-1} F^{\mathrm{T}} = F G_{\tilde{\alpha}}^{-1} F^{\mathrm{T}} \quad . \tag{9.9.10}$$

Example 9.9: Constraint between the angles of a triangle

Suppose measurements of the angles of a triangle have yielded the values $y_1 = 89°$, $y_2 = 31°$, $y_3 = 61°$, i.e.,

$$\mathbf{y} = \begin{pmatrix} 89 \\ 31 \\ 62 \end{pmatrix} \quad .$$

The linear equation of constraint is

$$\eta_1 + \eta_2 + \eta_3 = 180 \quad .$$

It can be written as

$$B\boldsymbol{\eta} + \mathbf{b}_0 = 0$$

with

$$B = (1, 1, 1) \quad , \qquad \mathbf{b}_0 = b_0 = -180 \quad .$$

As elements we choose η_1 and η_2. The system (9.9.5) then becomes

$$\eta_1 = \alpha_1 \quad , \qquad \eta_2 = \alpha_2 \quad , \qquad \eta_3 = 180 - \alpha_1 - \alpha_2$$

or

$$\boldsymbol{\eta} = \begin{pmatrix} 1 & 0 \\ 0 & 1 \\ -1 & -1 \end{pmatrix} \boldsymbol{\alpha} + \begin{pmatrix} 0 \\ 0 \\ 180 \end{pmatrix} \quad ,$$

i.e.,

$$F = \begin{pmatrix} 1 & 0 \\ 0 & 1 \\ -1 & -1 \end{pmatrix} \quad , \qquad \mathbf{f}_0 = \begin{pmatrix} 0 \\ 0 \\ 180 \end{pmatrix} \quad .$$

We assume a measurement error for the angle of 1°, i.e.,

$$C_y = \begin{pmatrix} 1 & 0 & 0 \\ 0 & 1 & 0 \\ 0 & 0 & 1 \end{pmatrix} = I \quad , \qquad G_y = C_y^{-1} = I \quad .$$

Using these in (9.9.7) gives

$$\begin{aligned} \tilde{\boldsymbol{\alpha}} &= \left[\begin{pmatrix} 1 & 0 & -1 \\ 0 & 1 & -1 \end{pmatrix} I \begin{pmatrix} 1 & 0 \\ 0 & 1 \\ -1 & -1 \end{pmatrix} \right]^{-1} \begin{pmatrix} 1 & 0 & -1 \\ 0 & 1 & -1 \end{pmatrix} I \begin{pmatrix} 89 \\ 31 \\ -118 \end{pmatrix} \\ &= \begin{pmatrix} 2 & 1 \\ 1 & 2 \end{pmatrix}^{-1} \begin{pmatrix} 207 \\ 149 \end{pmatrix} = \frac{1}{3} \begin{pmatrix} 2 & -1 \\ -1 & 2 \end{pmatrix} \begin{pmatrix} 207 \\ 149 \end{pmatrix} \\ &= \begin{pmatrix} 88\frac{1}{3} \\ 30\frac{1}{3} \end{pmatrix} \quad . \end{aligned}$$

Using (9.9.9) one finally obtains

$$\tilde{\boldsymbol{\eta}} = F\tilde{\boldsymbol{\alpha}} + \mathbf{f}_0 = \begin{pmatrix} 1 & 0 \\ 0 & 1 \\ -1 & -1 \end{pmatrix} \begin{pmatrix} 88\frac{1}{3} \\ 30\frac{1}{3} \end{pmatrix} + \begin{pmatrix} 0 \\ 0 \\ 180 \end{pmatrix} = \begin{pmatrix} 88\frac{1}{3} \\ 30\frac{1}{3} \\ 61\frac{1}{3} \end{pmatrix} \quad .$$

This result was clearly to be expected. The "excess" in the measured sum of angles of 2° is subtracted equally from the three measurements. This would not have been the case, however, if the individual measurements had errors of different magnitudes. The reader can easily repeat the computation for such a case. The residual errors of the improved measurements can be determined by application of (9.9.10),

$$
\begin{aligned}
G_{\tilde{\eta}}^{-1} \\
&= \begin{pmatrix} 1 & 0 \\ 0 & 1 \\ -1 & -1 \end{pmatrix} \left[\begin{pmatrix} 1 & 0 & 1 \\ 0 & 1 & -1 \end{pmatrix} I \begin{pmatrix} 1 & 0 \\ 0 & 1 \\ -1 & -1 \end{pmatrix} \right]^{-1} \begin{pmatrix} 1 & 0 & 1 \\ 0 & 1 & -1 \end{pmatrix} \\
&= \frac{1}{3} \begin{pmatrix} 1 & 0 \\ 0 & 1 \\ -1 & -1 \end{pmatrix} \begin{pmatrix} 2 & -1 \\ -1 & 2 \end{pmatrix} \begin{pmatrix} 1 & 0 & 1 \\ 0 & 1 & -1 \end{pmatrix} \\
&= \frac{1}{3} \begin{pmatrix} 2 & -1 & -1 \\ -1 & 2 & -1 \\ -1 & -1 & 2 \end{pmatrix} \quad .
\end{aligned}
$$

The residual error of each angle is thus equal to $\sqrt{2/3} \approx 0.82$. ■

At this point we want to make a general statement on measurements related by equations of constraint. Although in the statistical methods used so far we have found no way of dealing with systematic errors, equations of constraint offer such a possibility in many cases. If, for example, the sum of angles in many measurements is observed to be greater than 180° more frequently than less, then one can conclude that the measurement apparatus has a systematic error.

9.9.2 The Method of Lagrange Multipliers

Instead of elements, one more frequently uses the method of *Lagrange multipliers*. Although both methods clearly give the same results, the latter has the advantage that all unknowns are treated in the same way and thus the user is spared having to choose elements. The method of Lagrange multipliers is a well-known procedure in differential calculus for determination of extrema in problems with additional constraints.

We begin again from the linear system of equations of constraint (9.9.4)

$$B\boldsymbol{\eta} + \mathbf{b}_0 = 0$$

and we recall that the measured quantities are the sum of the true value η and the measurement error ε,

$$\mathbf{y} = \boldsymbol{\eta} + \boldsymbol{\varepsilon} \quad .$$

Thus one has

$$B\mathbf{y} - B\boldsymbol{\varepsilon} + \mathbf{b}_0 = 0 \quad . \tag{9.9.11}$$

Since $\mathbf{y}$ is known from the measurement, and $\mathbf{b}_0$ and B are also constructed from known quantities, we can construct a column vector with q elements,

$$\mathbf{c} = B\mathbf{y} + \mathbf{b}_0 \quad , \tag{9.9.12}$$

which does not contain any unknowns. Thus (9.9.11) can be written in the form

$$\mathbf{c} - B\boldsymbol{\varepsilon} = 0 \quad . \tag{9.9.13}$$

We now introduce an additional column vector with q elements, whose elements, not yet known, are the *Lagrange multipliers*,

$$\boldsymbol{\mu} = \begin{pmatrix} \mu_1 \\ \mu_2 \\ \vdots \\ \mu_q \end{pmatrix} \quad . \tag{9.9.14}$$

Using this we extend the original minimum function (9.2.16)

$$M = \boldsymbol{\varepsilon}^{\mathrm{T}} G_y \boldsymbol{\varepsilon}$$

to

$$L = \boldsymbol{\varepsilon}^{\mathrm{T}} G_y \boldsymbol{\varepsilon} + 2\mu^{\mathrm{T}}(\mathbf{c} - B\boldsymbol{\varepsilon}) \quad . \tag{9.9.15}$$

The function L is called the *Lagrange function*. The requirement

$$M = \min$$

with the constraint

$$\mathbf{c} - B\boldsymbol{\varepsilon} = 0$$

is then fulfilled when the total differential of the Lagrange function vanishes, i.e., when

$$\mathrm{d}L = 2\boldsymbol{\varepsilon}^{\mathrm{T}} G_y \, \mathrm{d}\boldsymbol{\varepsilon} - 2\boldsymbol{\mu}^{\mathrm{T}} B \, \mathrm{d}\boldsymbol{\varepsilon} = 0 \quad .$$

This is equivalent to

$$\boldsymbol{\varepsilon}^{\mathrm{T}} G_y - \boldsymbol{\mu}^{\mathrm{T}} B = 0 \quad . \tag{9.9.16}$$

The system (9.9.16) consists of n equations containing a total of $n + q$ unknowns, $\varepsilon_1, \varepsilon_2, \ldots, \varepsilon_n$ and $\mu_1, \mu_2, \ldots, \mu_q$. In addition we have the q equations of constraint (9.9.13). We transpose (9.9.16) and obtain

$$G_y \boldsymbol{\varepsilon} = B^{\mathrm{T}} \boldsymbol{\mu} \quad , \qquad \boldsymbol{\varepsilon} = G_y^{-1} B^{\mathrm{T}} \boldsymbol{\mu} \quad . \tag{9.9.17}$$

By substitution into (9.9.13) we obtain

$$\mathbf{c} - B G_y^{-1} B^{\mathrm{T}} \boldsymbol{\mu} = 0 \quad ,$$

which can easily be solved for $\boldsymbol{\mu}$:

$$\widetilde{\boldsymbol{\mu}} = (B G_y^{-1} B^{\mathrm{T}})^{-1} \mathbf{c} \quad . \tag{9.9.18}$$

With (9.9.17) we thus have the least-squares estimators of the measurement errors,

$$\widetilde{\boldsymbol{\varepsilon}} = G_y^{-1}B^{\mathrm{T}}(BG_y^{-1}B^{\mathrm{T}})^{-1}\mathbf{c} \quad . \tag{9.9.19}$$

The estimators of the unknowns are then given by (9.9.1),

$$\widetilde{\boldsymbol{\eta}} = \mathbf{y} - \widetilde{\boldsymbol{\varepsilon}} = \mathbf{y} - G_y^{-1}B^{\mathrm{T}}(BG_y^{-1}B^{\mathrm{T}})^{-1}\mathbf{c} \quad . \tag{9.9.20}$$

With the abbreviation

$$G_B = (BG_y^{-1}B^{\mathrm{T}})^{-1}$$

this becomes

$$\widetilde{\boldsymbol{\eta}} = \mathbf{y} - G_y^{-1}B^{\mathrm{T}}G_B\mathbf{c} \quad . \tag{9.9.21}$$

The covariance matrices of $\widetilde{\boldsymbol{\mu}}$ and $\widetilde{\boldsymbol{\eta}}$ are easily obtained by applying error propagation to the linear system of equations (9.9.18) and (9.9.19),

$$G_{\widetilde{\mu}}^{-1} = (BG_y^{-1}B^{\mathrm{T}})^{-1} = G_B \quad , \tag{9.9.22}$$

$$G_{\widetilde{\eta}}^{-1} = G_y^{-1} - G_y^{-1}B^{\mathrm{T}}G_BBG_y^{-1} \quad . \tag{9.9.23}$$

Example 9.10: Application of the method of Lagrange multipliers to Example 9.9

We apply the method of Lagrange multipliers to the problem of Example 9.9. We then have

$$\mathbf{c} = B\mathbf{y} + \mathbf{b}_0 = (1,1,1)\begin{pmatrix} 89 \\ 31 \\ 62 \end{pmatrix} - 180 = 182 - 180 = 2 \quad ,$$

and in addition,

$$G_B = (BG_yB^{\mathrm{T}})^{-1} = \left[(1,1,1)I\begin{pmatrix} 1 \\ 1 \\ 1 \end{pmatrix}\right]^{-1} = 3^{-1} = \frac{1}{3}$$

and

$$G_yB^{\mathrm{T}} = I\begin{pmatrix} 1 \\ 1 \\ 1 \end{pmatrix} = \begin{pmatrix} 1 \\ 1 \\ 1 \end{pmatrix} \quad .$$

We can now compute (9.9.21),

$$\widetilde{\boldsymbol{\eta}} = \begin{pmatrix} 89 \\ 31 \\ 62 \end{pmatrix} - \begin{pmatrix} 1 \\ 1 \\ 1 \end{pmatrix}\frac{2}{3} = \begin{pmatrix} 88\frac{1}{3} \\ 30\frac{1}{3} \\ 61\frac{1}{3} \end{pmatrix} \quad .$$

The covariance matrices are then

$$G_{\tilde{\mu}}^{-1} = \frac{1}{3} \quad ,$$

$$G_{\tilde{\eta}}^{-1} = I - I\begin{pmatrix}1\\1\\1\end{pmatrix}\frac{1}{3}(1,1,1)I$$

$$= I - \frac{1}{3}\begin{pmatrix}1&1&1\\1&1&1\\1&1&1\end{pmatrix} = \frac{1}{3}\begin{pmatrix}2&-1&-1\\-1&2&-1\\-1&-1&2\end{pmatrix} \quad . \blacksquare$$

We now generalize the method of Lagrange multipliers for the case of nonlinear equations of constraint having the general form (9.9.2), i.e.,

$$f_k(\boldsymbol{\eta}) = 0 \quad , \qquad k = 1, 2, \ldots, q \quad .$$

These equations can be expanded in a series about $\boldsymbol{\eta}_0$,

$$f_k(\boldsymbol{\eta}) = f_k(\boldsymbol{\eta}_0) + \left(\frac{\partial f_k}{\partial \eta_1}\right)_{\boldsymbol{\eta}_0} (\eta_1 - \eta_{10}) - \cdots + \left(\frac{\partial f_k}{\partial \eta_n}\right)_{\boldsymbol{\eta}_0} (\eta_n - \eta_{n0}) \quad . \qquad (9.9.24)$$

Here the $\boldsymbol{\eta}_0$ are the first approximations for the true values $\boldsymbol{\eta}$. Using the definitions

$$b_{kl} = \left(\frac{\partial f_k}{\partial \eta_l}\right)_{\boldsymbol{\eta}_0} \quad , \qquad B = \begin{pmatrix} b_{11} & b_{12} & \cdots & b_{1n} \\ b_{21} & b_{22} & \cdots & b_{2n} \\ \vdots & & & \\ b_{q1} & b_{q2} & \cdots & b_{qn} \end{pmatrix} \quad ,$$

$$c_k = f_k(\boldsymbol{\eta}_0) \quad , \qquad \mathbf{c} = \begin{pmatrix} c_1 \\ c_2 \\ \vdots \\ c_q \end{pmatrix} \quad ,$$

$$\delta_k = (\eta_k - \eta_{k0}) \quad , \qquad \boldsymbol{\delta} = \begin{pmatrix} \delta_1 \\ \delta_2 \\ \vdots \\ \delta_n \end{pmatrix} \quad ,$$

we can write (9.9.24) in the form

$$B\boldsymbol{\delta} + \mathbf{c} = 0 \quad . \qquad (9.9.25)$$

In the present case it is not difficult to find a first approximation for $\boldsymbol{\eta}_0$, since the quantities $\mathbf{y}$ are measurements of $\boldsymbol{\eta}$. We can thus use them directly as the first approximation and set

$$\boldsymbol{\eta}_0 = \mathbf{y} \quad . \qquad (9.9.26)$$

The vector $\boldsymbol{\delta}$ is then

$$\boldsymbol{\delta} = \boldsymbol{\eta} - \boldsymbol{\eta}_0 = \boldsymbol{\eta} - \mathbf{y} = -\boldsymbol{\varepsilon} \quad . \tag{9.9.27}$$

Equation (9.9.25) thus takes on the form

$$\mathbf{c} - B\boldsymbol{\varepsilon} = 0 \quad ,$$

which is identical to (9.9.13). The solution $\widetilde{\boldsymbol{\varepsilon}}$ is given by (9.9.19). In the nonlinear case one is usually not satisfied with a single step, but will rather continue iterating. Starting from the second step one no longer has the simple relation $\boldsymbol{\delta} = -\boldsymbol{\varepsilon}$. We solve this problem in the same way as in the linear case, by requiring that the expression

$$M' = \boldsymbol{\delta}^{\mathrm{T}} G_y \boldsymbol{\delta} \tag{9.9.28}$$

be a minimum under the constraint (9.9.25). This requirement is equivalent to saying that the weighted sum of squares of the differences between the two approximations of $\boldsymbol{\eta}$ is a minimum. The solution is then, in analogy to (9.9.19), given by

$$\widetilde{\boldsymbol{\delta}} = -G_y^{-1} B^{\mathrm{T}} (B G_y^{-1} B^{\mathrm{T}})^{-1} \mathbf{c} \quad . \tag{9.9.29}$$

The result of each individual step is used to determine a better approximation for $\boldsymbol{\eta}$. If the iteration is terminated after s steps, one has

$$\widetilde{\boldsymbol{\eta}} = \boldsymbol{\eta}_s = \boldsymbol{\eta}_{s-1} + \widetilde{\boldsymbol{\delta}} \quad . \tag{9.9.30}$$

The corresponding covariance matrix is still given by (9.9.23), i.e.,

$$G_{\widetilde{\eta}}^{-1} = G_y^{-1} - G_y B^{\mathrm{T}} G_B B G_y^{-1} \quad , \tag{9.9.31}$$

where the values computed in the last step are used as elements of the matrix B. The best estimates of the measurement errors can then be computed from

$$\widetilde{\boldsymbol{\varepsilon}} = \mathbf{y} - \widetilde{\boldsymbol{\eta}} \quad . \tag{9.9.32}$$

One finally obtains from them the minimum function

$$\widetilde{M} = \widetilde{\boldsymbol{\varepsilon}}^{\mathrm{T}} G_y \widetilde{\boldsymbol{\varepsilon}} \quad .$$

This quantity can in turn be used for a χ^2-test with q degrees of freedom.

Although the method of Lagrange multipliers is mathematically elegant, in programs provided here we use the method of orthogonal transformations (see Sect. A.19). We do not give any program here, since the problems of this section can be done with the more general routine `LSQGEN`, which will be described in Sect. 9.11 (see also Example 9.11).

9.10 The General Case of Least-Squares Fitting

After the preparation of the previous sections we can now take up the general case of fitting with the method of least squares.

We first recall the notation. The r unknown parameters are placed in a vector $\mathbf{x}$. The quantities to be measured form an n-vector $\boldsymbol{\eta}$. The values $\mathbf{y}$ actually measured differ from $\boldsymbol{\eta}$ by the errors $\boldsymbol{\varepsilon}$. We will assume a normal distribution for the individual errors ε_j $(j = 1, 2, \ldots, n)$, i.e., a normal distribution for the n variables ε_j with the null vector as the vector of expectation values, and a covariance matrix $C_y = G_y^{-1}$. The vectors $\mathbf{x}$ and $\boldsymbol{\eta}$ are related by m functions

$$f_k(\mathbf{x}, \boldsymbol{\eta}) = f_k(\mathbf{x}, \mathbf{y} - \boldsymbol{\varepsilon}) = 0 \quad , \qquad k = 1, 2, \ldots, m \quad . \tag{9.10.1}$$

We will further assume that we have already obtained in some way a first approximation for the unknowns $\mathbf{x}_0$. As a first approximation for $\boldsymbol{\eta}$ we use $\boldsymbol{\eta}_0 = \mathbf{y}$ as in Sect. 9.9. Finally we require that the functions f_k can be approximated by linear functions in the range of variability of our problem, i.e., in the region around $(\mathbf{x}_0, \boldsymbol{\eta}_0)$, which is given by the differences $\mathbf{x} - \mathbf{x}_0$ and $\boldsymbol{\eta} - \boldsymbol{\eta}_0$. We can then write

$$\begin{aligned} f_k(\mathbf{x}, \boldsymbol{\eta}) = {} & f_k(\mathbf{x}_0, \boldsymbol{\eta}_0) \\ & + \left(\frac{\partial f_k}{\partial x_1}\right)_{\mathbf{x}_0, \boldsymbol{\eta}_0} (x_1 - x_{10}) + \cdots + \left(\frac{\partial f_k}{\partial x_r}\right)_{\mathbf{x}_0, \boldsymbol{\eta}_0} (x_r - x_{r0}) \\ & + \left(\frac{\partial f_k}{\partial \eta_1}\right)_{\mathbf{x}_0, \boldsymbol{\eta}_0} (\eta_1 - \eta_{10}) + \cdots + \left(\frac{\partial f_k}{\partial \eta_n}\right)_{\mathbf{x}_0, \boldsymbol{\eta}_0} (\eta_n - \eta_{n0}) \quad . \end{aligned} \tag{9.10.2}$$

With the abbreviations

$$a_{k\ell} = \left(\frac{\partial f_k}{\partial x_\ell}\right)_{\mathbf{x}_0, \boldsymbol{\eta}_0} \quad , \qquad A = \begin{pmatrix} a_{11} & a_{12} & \cdots & a_{1r} \\ a_{21} & a_{22} & \cdots & a_{2r} \\ \vdots & & & \\ a_{m1} & a_{m2} & \cdots & a_{mr} \end{pmatrix} \quad , \tag{9.10.3}$$

$$b_{k\ell} = \left(\frac{\partial f_k}{\partial \eta_\ell}\right)_{\mathbf{x}_0, \boldsymbol{\eta}_0} \quad , \qquad B = \begin{pmatrix} b_{11} & b_{12} & \cdots & b_{1n} \\ b_{21} & b_{22} & \cdots & b_{2n} \\ \vdots & & & \\ b_{m1} & b_{m2} & \cdots & b_{mn} \end{pmatrix} \quad , \tag{9.10.4}$$

$$c_k = f_k(\mathbf{x}_0, \boldsymbol{\eta}_0) \quad , \qquad \mathbf{c} = \begin{pmatrix} c_1 \\ c_2 \\ \vdots \\ c_m \end{pmatrix} \quad , \tag{9.10.5}$$

$$\boldsymbol{\xi} = \mathbf{x} - \mathbf{x}_0 \quad , \qquad \boldsymbol{\delta} = \boldsymbol{\eta} - \boldsymbol{\eta}_0 \tag{9.10.6}$$

the system of equations (9.10.2) can be written as follows:

$$A\boldsymbol{\xi} + B\boldsymbol{\delta} + \mathbf{c} = 0 \quad . \tag{9.10.7}$$

The Lagrange function is

$$L = \boldsymbol{\delta}^{\mathrm{T}} G_y \boldsymbol{\delta} + 2\boldsymbol{\mu}^{\mathrm{T}}(A\boldsymbol{\xi} + B\boldsymbol{\delta} + \mathbf{c}) \quad . \tag{9.10.8}$$

Here $\boldsymbol{\mu}$ is an m-vector of the Lagrange multipliers. We require that the total derivative of (9.10.8) with respect to $\boldsymbol{\delta}$ vanishes. This is equivalent to requiring

$$G_y \boldsymbol{\delta} + B^{\mathrm{T}} \boldsymbol{\mu} = 0$$

or

$$\boldsymbol{\delta} = -G_y^{-1} B^{\mathrm{T}} \boldsymbol{\mu} \quad . \tag{9.10.9}$$

Substitution into (9.10.7) gives

$$A\boldsymbol{\xi} - BG_y^{-1}B^{\mathrm{T}}\boldsymbol{\mu} + \mathbf{c} = 0 \tag{9.10.10}$$

or

$$\boldsymbol{\mu} = G_B(A\boldsymbol{\xi} + \mathbf{c}) \quad , \tag{9.10.11}$$

where

$$G_B = (BG_y^{-1}B^{\mathrm{T}})^{-1} \quad . \tag{9.10.12}$$

With (9.10.9) we can now write

$$\boldsymbol{\delta} = -G_y^{-1} B^{\mathrm{T}} G_B(A\boldsymbol{\xi} + \mathbf{c}) \quad . \tag{9.10.13}$$

Since the Lagrange function L is a minimum also with respect to $\boldsymbol{\xi}$, the total derivative of (9.10.8) with respect to $\boldsymbol{\xi}$ must also vanish, i.e.,

$$2\boldsymbol{\mu}^{\mathrm{T}} A = 0 \quad .$$

By transposing and substituting (9.10.11) one obtains

$$2A^{\mathrm{T}} G_B(A\boldsymbol{\xi} + \mathbf{c}) = 0$$

or

$$\widetilde{\boldsymbol{\xi}} = -(A^{\mathrm{T}} G_B A)^{-1} A^{\mathrm{T}} G_B \mathbf{c} \quad . \tag{9.10.14}$$

Substituting (9.10.14) into (9.10.13) and (9.10.11) immediately gives the estimates of the deviations $\boldsymbol{\delta}$ and the Lagrange multipliers $\boldsymbol{\mu}$,

$$\widetilde{\boldsymbol{\delta}} = -G_y^{-1}B^{\mathrm{T}}G_B(\mathbf{c} - A(A^{\mathrm{T}}G_BA)^{-1}A^{\mathrm{T}}G_B\mathbf{c}) \quad , \tag{9.10.15}$$

$$\widetilde{\boldsymbol{\mu}} = G_BA(\mathbf{c} - A(A^{\mathrm{T}}G_BA)^{-1}A^{\mathrm{T}}G_B\mathbf{c}) \quad . \tag{9.10.16}$$

The estimates for the parameters $\mathbf{x}$ and for the improved measurements $\boldsymbol{\eta}$ are

$$\widetilde{\mathbf{x}} = \mathbf{x}_0 + \widetilde{\boldsymbol{\xi}} \quad , \tag{9.10.17}$$

$$\widetilde{\boldsymbol{\eta}} = \boldsymbol{\eta}_0 + \widetilde{\boldsymbol{\delta}} \quad . \tag{9.10.18}$$

From (9.10.14), (9.10.4), and (9.10.5) we obtain for the matrix of derivatives of the elements of $\widetilde{\boldsymbol{\xi}}$ with respect to the elements of $\mathbf{y}$

$$\frac{\partial \widetilde{\boldsymbol{\xi}}}{\partial \mathbf{y}} = -(A^{\mathrm{T}}G_BA)^{-1}A^{\mathrm{T}}G_B\frac{\partial \mathbf{c}}{\partial \mathbf{y}} = -(A^{\mathrm{T}}G_BA)^{-1}A^{\mathrm{T}}G_BB \quad .$$

Using error propagation one obtains the covariance matrix

$$G_{\widetilde{x}}^{-1} = G_{\widetilde{\xi}}^{-1} = (A^{\mathrm{T}}G_BA)^{-1} \quad . \tag{9.10.19}$$

Correspondingly one finds

$$G_{\widetilde{\eta}}^{-1} = G_y^{-1} - G_y^{-1}B^{\mathrm{T}}G_BBG_y^{-1} + G_y^{-1}B^{\mathrm{T}}G_BA(A^{\mathrm{T}}G_BA)^{-1}A^{\mathrm{T}}G_BBG_y^{-1} \quad . \tag{9.10.20}$$

One can show that under the assumed conditions, i.e., sufficient linearity of (9.10.2) and normally distributed measurement errors, the minimum function M, which can also be written in the form

$$M = (B\widetilde{\varepsilon})^{\mathrm{T}}G_B(B\widetilde{\varepsilon}) \quad , \tag{9.10.21}$$

follows a χ^2-distribution with $m - r$ degrees of freedom.

If the equations (9.10.1) are linear, then the relations (9.10.17) through (9.10.20) already give the final solution. In nonlinear cases they are first considered as a better approximation; i.e., one would replace $\mathbf{x}_0$, $\boldsymbol{\eta}_0$ by $\widetilde{\mathbf{x}}$, $\widetilde{\boldsymbol{\eta}}$, compute again the matrices (9.10.3) through (9.10.5), and obtain better approximations by using them in the relations (9.10.14) through (9.10.18). The iterative procedure can be repeated until a satisfactory solution is found. It is in general not easy to determine when a solution is satisfactory. The simplest procedure consists of examining the value M after every step and terminating the procedure when M changes by less than a given fractional amount from one step to the next. Naturally it is also possible that M does not decrease at all, i.e., that the iterative procedure diverges. In this case one must consider points (a) through (d) mentioned at the end of Sect. 9.7. More detailed criteria for the assumption of a solution will depend on the specific problem at hand.

9.11 Program for the General Case of Least Squares

In the program for treating the general case of least squares we do not use the method of Lagrange multipliers but rather the procedure of Sect. A.19, which is based on orthogonal transformations.

At every step of the iteration we must determine the r-vector $\boldsymbol{\xi}$ and the n-vector $\boldsymbol{\delta}$. We combine both into an $(r+n)$-vector $\mathbf{u}$,

$$\mathbf{u} = \begin{pmatrix} \boldsymbol{\xi} \\ \boldsymbol{\delta} \end{pmatrix} \quad . \tag{9.11.1}$$

The $m \times r$ matrix A and the $m \times n$ matrix B are also combined into an $m \times (r+n)$ matrix E,

$$E = (A, B) \quad . \tag{9.11.2}$$

The vector containing the solutions $\mathbf{u}$ must satisfy the constraint (9.10.7), i.e.,

$$E\mathbf{u} = \mathbf{d} \quad , \qquad \mathbf{d} = -\mathbf{c} \quad . \tag{9.11.3}$$

We now consider the minimum function in the ith iterative step. It depends on

$$\boldsymbol{\eta} = \mathbf{y} + \sum_{\ell=1}^{i} \boldsymbol{\delta}_\ell = \mathbf{y} + \sum_{\ell=1}^{i-1} \boldsymbol{\delta}_\ell + \boldsymbol{\delta}_i = \mathbf{y} + \mathbf{s} + \boldsymbol{\delta} \quad , \qquad \boldsymbol{\delta} = \boldsymbol{\delta}_i \quad , \tag{9.11.4}$$

where

$$\mathbf{s} = \sum_{\ell=1}^{i-1} \boldsymbol{\delta}_\ell \tag{9.11.5}$$

is the result of all of the previous steps that changed $\mathbf{y}$. One then has

$$M = (\boldsymbol{\eta} - \mathbf{y})^{\mathrm{T}} G_y (\boldsymbol{\eta} - \mathbf{y}) = (\boldsymbol{\delta} + \mathbf{s})^{\mathrm{T}} G_y (\boldsymbol{\delta} + \mathbf{s}) = \min \quad . \tag{9.11.6}$$

We now extend the $n \times n$ matrix G_y to an $(r+n) \times (r+n)$ matrix

$$G = \begin{pmatrix} 0 & 0 \\ 0 & G_y \end{pmatrix} \begin{matrix} \}r \\ \}n \end{matrix} \quad , \tag{9.11.7}$$

for which we find a Cholesky decomposition according to Sect. A.10,

$$G = F^{\mathrm{T}} F \quad . \tag{9.11.8}$$

Then (9.11.6) becomes

$$\begin{aligned} M &= (\mathbf{u} + \mathbf{t})^{\mathrm{T}} G (\mathbf{u} + \mathbf{t}) \\ &= (F\mathbf{u} + F\mathbf{t})^2 = \min \end{aligned}$$

or

$$(F\mathbf{u} - \mathbf{b})^2 = \min \qquad (9.11.9)$$

with

$$\mathbf{t} = \begin{pmatrix} \mathbf{0} \\ \mathbf{s} \end{pmatrix} \begin{matrix} \}r \\ \}n \end{matrix} \quad , \qquad \mathbf{b} = -F\mathbf{t} \quad . \qquad (9.11.10)$$

Now one must merely solve the problem (9.11.9) with the constraint (9.11.3), e.g., with the procedure of Sect. A.19. With the solution

$$\tilde{\mathbf{u}} = \begin{pmatrix} \tilde{\boldsymbol{\xi}} \\ \tilde{\boldsymbol{\delta}} \end{pmatrix}$$

or rather with the vectors $\tilde{\boldsymbol{\xi}}$, $\tilde{\boldsymbol{\delta}}$ one finds improved values for $\mathbf{x}$ [cf. (9.10.17)], $\boldsymbol{\eta}$ [cf. (9.10.18)], and for $\mathbf{s}$, as well as for $\mathbf{t}$ [cf. (9.11.5) and (9.11.10)], with which an additional iterative step can be carried out.

The procedure can be regarded as having converged and terminated when the minimum function (9.11.9) in two successive steps only changes by an insignificant amount, or it can be terminated without success if after a given number of steps convergence has not been reached. In the case of convergence, the covariance of the unknowns can be computed according to (9.10.19). The calculation of the covariance matrix of the "improved" measurements $\tilde{\boldsymbol{\eta}}$ according to (9.10.20) is possible as well. It is, however, rarely of interest.

Finally, the value of M obtained in the last step can be used for a χ^2-test of the goodness-of-fit with $m - r$ degrees of freedom.

Program 9.12: Subroutine `LSQGEN` treats the general case of fitting with least squares

CALLING SEQUENCE: `CALL LSQGEN(Y,CY,FY,F,E,M,N,NR,NRED, LIST,X,CX,R,A2,NSTEP)`

INPUT ARGUMENTS: `Y` — n-vector of the measurements $\mathbf{y}$

`CY` — $n \times n$ covariance matrix for the measurements C_y

`FY` — working array, minimum size $n \times n$

`F` — working array, minimum size $(n + r) \times (n + r)$

`E` — working array, minimum size $m \times (n + r)$

`M` — number of equations of constraint m

`N` — number of measurements n

`NR` — number of unknowns r

`NRED` — reduced number of unknowns not fixed r'

`LIST` — r-vector containing the elements of a list which specify which of the r unknowns are fixed (list element $= 0$), or variable (list element $= 1$).

X — r-vector of the first approximations (or of the fixed components) of $\mathbf{x}$
A2 — working array, minimum size $(n + r) \times (n + r - m)$
NSTEP — The value of NSTEP is the maximum number of steps. If negative, the covariance matrix CX or CY is not computed.

OUTPUT ARGUMENTS: X — r-vector of unknowns $\tilde{\mathbf{x}}$
Y — n-vector of the improved measurements $\tilde{\boldsymbol{\eta}}$
CX — $r' \times r'$ covariance matrix for the unknowns $C_{\tilde{x}}$ (for $r' > 0$)
CY — $n \times n$ covariance matrix of the improved measurements $C_{\tilde{\eta}}$ (for $r' = 0$)
R — value of the minimum function M
NSTEP — number of steps needed. If an error occurs, NSTEP is set negative (-1: algorithm MTXLSC did not converge, -2: no convergence within the given number of steps, -3: numerical differentiation failed).

DEFAULT VALUES: If the input value of NSTEP is zero, then a maximum of 100 steps are carried out. For the particularly common cases NRED = NR and NRED = 0, the vector LIST does not need to be set explicitly. In any case, however, one must provide the r-vector LIST as a working array.

REMARKS: The user must provide a function subprogram with the name LSQGFN, corresponding to the description of the function 9.13. If the matrix E should (or must) be computed analytically, then in addition the routine E.3 (AUXDRG) must be replaced by a user routine of the same name and argument list, which provides the derivatives.

```
      SUBROUTINE LSQGEN(Y,CY,FY,F,E,M,N,NR,NRED,LIST,X,CX,R,
     +                  A2,NSTEP)
      IMPLICIT DOUBLE PRECISION (A-H,O-Z)
      DOUBLE PRECISION LSQGFN
      PARAMETER(LMAX=1050)
      DIMENSION Y(N),CY(N,N),X(NR),CX(NRED,NRED),LIST(NR)
      DIMENSION E(M,N+NRED),FY(N,N),F(N+NRED,N+NRED),A2(N+NRED,*)
      DIMENSION D(LMAX),T(LMAX),B(LMAX),U(LMAX)
      COMMON /DASV03/ D,T,B,U
      PARAMETER(MAXSTP=100,EPSILN=1.D-8,TT=1.D-15,ZERO=0.D0)
      LOGICAL OK,COVMAT
C general case of least squares fitting
      OK=.TRUE.
      COVMAT=.TRUE.
      IF(NSTEP.LT.0) THEN
        COVMAT=.FALSE.
        NSTEP=ABS(NSTEP)
      END IF
      IF(NSTEP.LT.1) NSTEP=MAXSTP
      IF(NR.GT.0) THEN
```

```
C For NR=NRED :  set LIST
        IF(NR.EQ.NRED) THEN
          DO 10 I=1,NR
            LIST(I)=1
   10     CONTINUE
        END IF
      END IF
      L=N+NRED
      CALL MTXZRV(T,L)
      CALL MTXTRA(CY,F,N,N)
      CALL MTXCHI(CY,FY,N)
      CALL MTXCHL(CY,FY,N)
      CALL MTXTRA(F,CY,N,N)
C start iteration
      R=ZERO
      DO 50 ISTEP=1,NSTEP
        RLST=R
        CALL MTXZER(F,L,L)
        CALL MTXPSM(F,FY,L,L,N,N,NRED+1,NRED+1)
        DO 20 K=1,M
          D(K)=-LSQGFN(Y,X,N,NR,K)
   20   CONTINUE
C Numerical Derivatives
        CALL AUXDRG(X,Y,M,N,NR,NRED,LIST,E,OK)
        IF(.NOT.OK) THEN
          NSTEP=-3
          GO TO 60
        END IF
        CALL MTXCHM(F,T,B,L,1)
        CALL MTXMSV(B,B,DBLE(-1.),L)
        CALL MTXLSC(F,B,E,D,U,R,A2,L,L,M,DBLE(0.),OK)
        IF(.NOT.OK) THEN
          NSTEP=-1
          GO TO 60
        END IF
        IF(NRED.GT.0) THEN
          IRED=0
          DO 30 I=1,NR
            IF(LIST(I).NE.0) THEN
              IRED=IRED+1
              X(I)=X(I)+U(IRED)
            END IF
   30     CONTINUE
        END IF
        DO 40 I=1,N
          Y(I)=Y(I)+U(I+NRED)
          T(I+NRED)=T(I+NRED)+U(I+NRED)
   40   CONTINUE
C test for convergence
        IF(ISTEP.GT. 1.AND. ABS(R-RLST).LT.EPSILN*R+TT) THEN
          NSTEP=ISTEP
          IF(COVMAT) THEN
C compute matrix GB
            CALL AUXDRG(X,Y,M,N,NR,NRED,LIST,E,OK)
            IF(.NOT.OK) THEN
              NSTEP=-3
              GO TO 60
            END IF
            CALL MTXGSM(E,A2,M,L,M,N,1,NRED+1)
            CALL MTXMBT(CY,A2,F,N,N,M)
            CALL MTXMLT(A2,F,FY,M,N,M)
```

```
          CALL MTXCHI(FY,F,M)
C array FY now contains matrix GB
          IF(NRED.GT.0) THEN
            CALL MTXGSM(E,A2,M,L,M,NRED,1,1)
            CALL MTXMAT(A2,FY,F,NRED,M,M)
            CALL MTXMLT(F,A2,CX,NRED,M,NR)
            CALL MTXCHI(CX,F,NRED)
C array CX now contains covariance matrix of unknowns
          ELSE
            CALL MTXMLT(A2,CY,F,M,N,N)
            CALL MTXMLT(FY,F,A2,M,M,N)
            CALL MTXMAT(F,A2,FY,N,M,N)
            CALL MTXSUB(CY,FY,CY,N,N)
C array CY now contains covariance matrix of 'improved'
C measurements
          END IF
        END IF
        GO TO 60
      END IF
   50 CONTINUE
      NSTEP=-2
   60 RETURN
      END
```

Example 9.11: Fitting a line to points with measurement errors in both the abscissa and ordinate

Suppose a number of measured points (t_i, s_i) in the (t, s) plane are given. Each point has measurement errors Δt_i, Δs_i, which can, in general, be correlated. The covariance between the measurement errors Δt_i and Δs_i is c_i. We identify t_i and s_i with elements of the n-vector $\mathbf{y}$ of measured quantities

$$y_1 = t_1 \quad , \qquad y_2 = s_1 \quad , \qquad \ldots \quad , \qquad y_{n-1} = t_{n/2} \quad , \qquad y_n = s_{n/2} \quad .$$

The covariance matrix is

$$C_y = \begin{pmatrix} (\Delta t_1)^2 & c_1 & 0 & 0 & \\ c_1 & (\Delta s_1)^2 & 0 & 0 & \\ 0 & 0 & (\Delta t_2)^2 & c_2 & \\ 0 & 0 & c_2 & (\Delta s_2)^2 & \\ & & & & \ddots \end{pmatrix} \quad .$$

A straight line in the (t, s) plane is described by the equation $s = x_1 + x_2 t$. For the assumption of such a line through the measured points, the equations of constraint (9.10.1) take on the form

$$f_k(\mathbf{x}, \boldsymbol{\eta}) = \eta_{2k} - x_1 - x_2\eta_{2k-1} = 0 \quad , \qquad k = 1, 2, \ldots, n/2 \quad .$$

Because of the term $x_2\eta_{2k-1}$, the equations are not linear. The derivatives with respect to η_{2k-1} depend on x_2 and those with respect to x_2 depend on η_{2k-1}. The user routine LSQGFN for this example is then of the form given below. The program description holds for all user routines for the general least-squares case.

Program 9.13: Function LSQGFN computes the functions $f_k(\mathbf{x}, \boldsymbol{\eta})$ for the general least-squares problem

CALLING SEQUENCE: F=LSQGFN(ETA,X,N,NR,K)

INPUT ARGUMENTS: ETA — n-vector of the measurements $\boldsymbol{\eta}$
X — r-vector of the unknowns $\mathbf{x}$
N, NR — dimensions n, r
K – index k

OUTPUT ARGUMENT: Function value LSQGFN — $f_k(\mathbf{x}, \boldsymbol{\eta})$

REMARKS: A function with the name LSQGFN and argument list given above must be provided by the user. The routine given below is an example. It treats the problem of fitting a line to points with measurement errors in both the abscissa and ordinate (Example 9.11).

```
      DOUBLE PRECISION FUNCTION LSQGFN(ETA,X,N,NR,K)
      IMPLICIT DOUBLE PRECISION (A-H,O-Z)
      DIMENSION X(NR),ETA(N)
C computes constraint equations for fit of straight
C line to measurements with errors in abscissa and ordinate
      LSQGFN=ETA(2*K)-X(1)-X(2)*ETA(2*K-1)
      END
```

The results of fitting to four measured points are shown in Fig. 9.12. The examples in the two individual plots differ only in the correlation coefficient of the third measurement, $\rho_3 = 0.5$ and $\rho_3 = -0.5$. One can see a noticeable effect of the sign of ρ_3 on the fit result and in particular on the value of the minimum function. ■

Example 9.12: Fixing parameters

In Fig. 9.13 the results of fitting a line to measured points with errors in the abscissa and ordinate are shown, where in each plot one parameter of the line was held fixed. In the upper plot the intercept of the vertical axis x_1 was fixed, and in the lower plot the slope x_2 was fixed. ■

9.12 Applying the Program for the General Case to Constrained Measurements

If all of the variables $x_1, \ldots, x_r$ are fixed, then there are no more unknowns in the least-squares problem.

In the equations of constraint (9.10.2) only the components of $\boldsymbol{\eta}$ are variable. Thus all terms containing the matrix A vanish from the formulas of

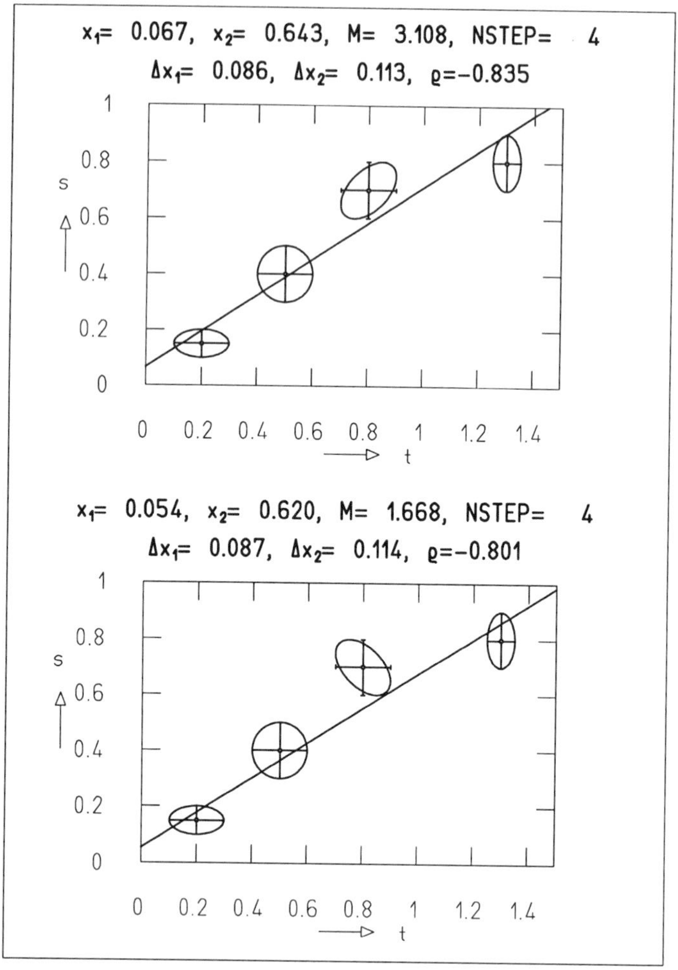

Fig. 9.12. Fitting a line to four measured points in the (t, s) plane. The points are shown with measurement errors (in t and s) and covariance ellipses. The individual plots show the results of the fit, the errors, and correlation coefficients.

the previous sections. As previously, however, the improved measurements $\tilde{\boldsymbol{\eta}}$ can be computed. The quantity M can in addition be used for a χ^2-test with m degrees of freedom for how well the equations of constraint are satisfied by the measurements. Finally, the covariance matrix of the improved measurements $C_{\tilde{\eta}}$ can be determined.

Mathematically – as well as when using the program LSQGEN – it does not matter whether all variables are fixed ($r \neq 0$, $r' = 0$) or whether from the start the equations of constraint and thus the program LSQGFN do not depend on the variables $\mathbf{x}$ ($r = 0$). In both cases LSQGEN gives the same solution.

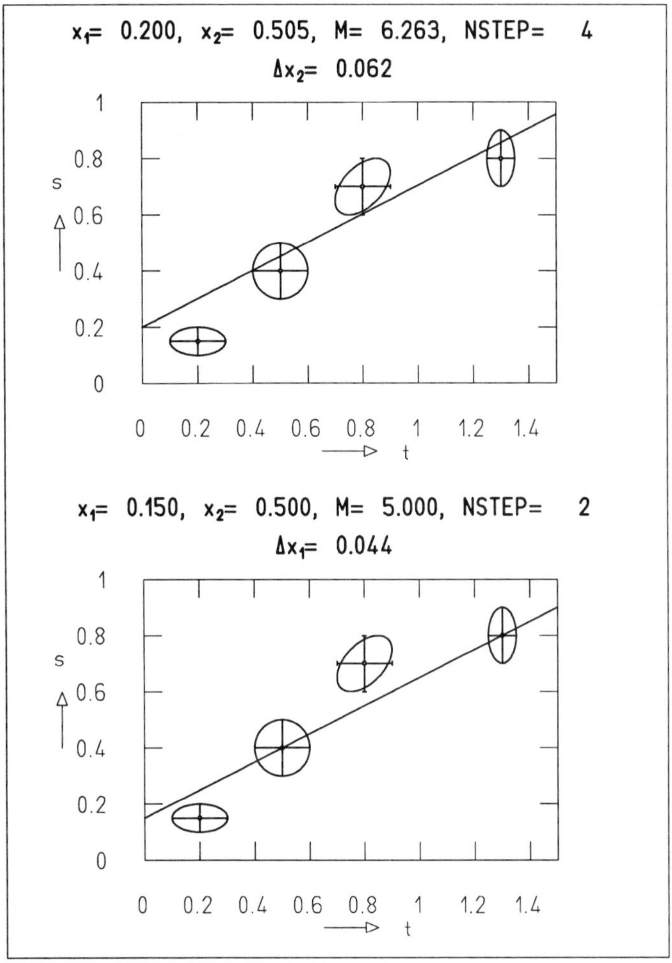

Fig. 9.13. Fitting a line to the same points as in the first plot of Fig. 9.12. The intercept with the vertical axis x_1 was held constant in the upper plot, and below the slope x_2 was held constant.

Example 9.13: χ^2-test of the description of measured points with errors in abscissa and ordinate by a given line

We use the same measurements as used already in Examples 9.11 and 9.12. The results of the analysis of these measurements with LSQGEN with fixed parameters are shown in Fig. 9.14. For the upper plot, x_1 and x_2 were fixed to the values obtained from fitting with two adjustable parameters in Example 9.11, Fig. 9.12. Clearly we also obtain the same value of M as previously. For the lower plot, arbitrarily crude estimates ($x_1 = 0$, $x_2 = 0.5$) were used. They give a significantly higher value of M. This value would lead to rejection of the

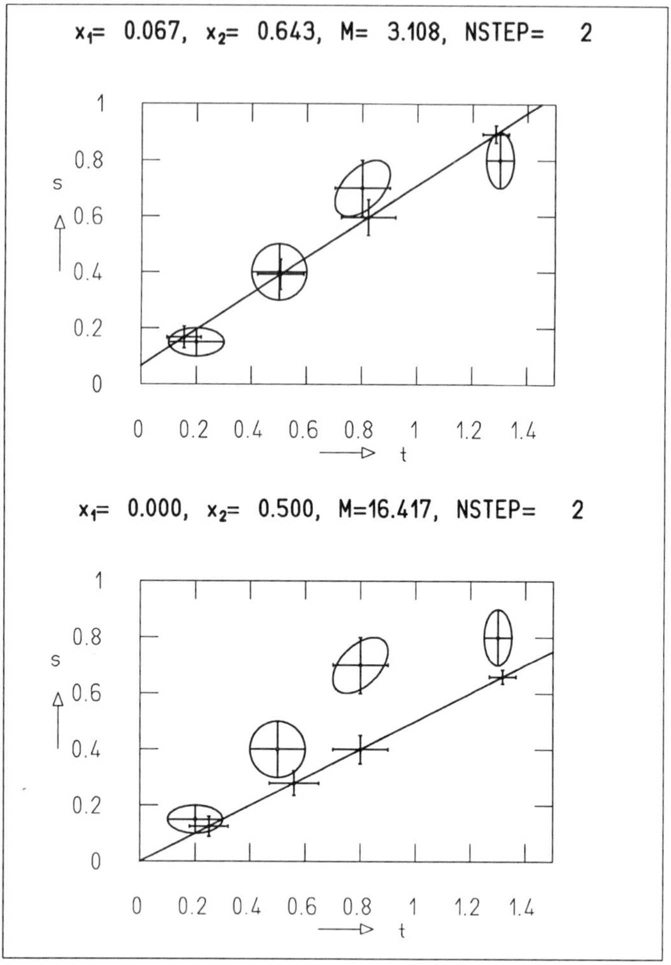

Fig. 9.14. Constrained measurements. The hypothesis was tested whether the true values of the measured points, indicated by their covariance ellipses, lie on the line $s = x_1 + x_2 t$. With the numerical values of M, a χ^2-test with four degrees of freedom can be carried out. Shown as well are the improved measurements, which lie on the line, and their errors.

hypothesis with a confidence level of 99 % that the data points are described by a linear relation with these parameter values ($\chi^2_{0.99} = 13.28$ for four degrees of freedom).

It is also interesting to consider the improved measurements $\tilde{\eta}$ and their errors, which are shown in Fig. 9.14. The improved measurements naturally lie on the line. The measurement errors are the square roots of the diagonal elements of $C_{\tilde{\eta}}$. The correlations between the errors of a measured point in s and t are obtained from the corresponding off-diagonal elements $C_{\tilde{\eta}}$. They

are exactly equal to one. The covariance ellipses of the individual improved measurements collapse to line segments which lie on the line given by x_1, x_2.
■

9.13 Confidence Region and Asymmetric Errors in the General Case

The results obtained in Sect. 9.8 on confidence regions and asymmetric errors are also valid for the general case. We can therefore limit ourselves to presenting a program and to the discussion of an example.

Program 9.14: Subroutine LSQASG computes asymmetric errors (or confidence limits) for the general case of least squares using LSQGEN

CALLING SEQUENCE: CALL LSQASG(Y,CY,FY,F,E,M,N,NR,NRED, LIST,X,CX,R,W,DXPLUS,DXMINS,A2,NSTEP)

INPUT ARGUMENTS: Y,CY — as for call to LSQGEN
M,N,NR,NRED,LIST,X,CX,R,FY,F — as after call to LSQGEN
A2 — working array, minimum size $(n+r) \times (n+r-m)$
NSTEP — maximum number of steps for each call to LSQGEN
W — probability W, for which the confidence limits are computed [cf. (9.8.6)]. If W is set equal to 0, the asymmetric errors rather than confidence limits are computed

OUTPUT ARGUMENTS: DXPLUS — r'-vector of errors (or confidence limits) Δx_{i+}
DXMINS — r'-vector of errors (or confidence limits) Δx_{i-}
NSTEP — number of steps needed for the last call to LSQGEN. If an error occurs, NSTEP is set negative (-1: minimization failed, -2: failure to find zero).

DEFAULT VALUES: If the input value of W ≤ 0, then asymmetric errors are computed. Otherwise, asymmetric confidence limits are given. If the input value of NSTEP ≤ 0, NSTEP = 100 is assumed.

```
      SUBROUTINE LSQASG(Y,CY,FY,F,E,M,N,NR,NRED,LIST,X0,CX,R,W,
     +DXPLUS,DXMINS,A2,NSTEP)
      IMPLICIT DOUBLE PRECISION (A-H,O-Z)
      PARAMETER(NMAX=1000)
      DIMENSION Y(N),CY(N,N),X0(NR),CX(NRED,NRED),LIST(NR)
      DIMENSION E(M,N+NRED),FY(N,N),F(N+NRED,N+NRED),A2(N+NRED,*)
      DIMENSION X(NMAX),DXPLUS(NRED),DXMINS(NRED),DX(NMAX)
      DIMENSION YSAV(NMAX)
      COMMON /DASV04/ X,DX,YSAV
```

```
      PARAMETER(MAXSTP=100,EPS=1.D-2,EPSILN=1.D-4,TT=1.D-6,
     +FACT=1.5D0)
      PARAMETER(ZERO=0.0D0,ONE=1.D0,HALF=0.5D0,TWO=2.D0)
      LOGICAL LENCLS
      IF(NSTEP.LE.0) NSTEP=MAXSTP
      IF(NR.EQ.NRED) THEN
        DO 10 I=1,NR
          LIST(I)=1
   10   CONTINUE
      END IF
      IF(W.LE.EPSILN) THEN
        G=ONE
      ELSE
        G=SQCHI2(W,NRED)
      END IF
      RTARG=R+G
      DO 20 I=1,NRED
        DX(I)=SQRT(CX(I,I))
   20 CONTINUE
      NRED1=NRED-1
      CALL MTXCPV(Y,YSAV,N)
      DO 60 IVAR=1,NR
        IF(LIST(IVAR).EQ.0) GO TO 60
C fix variable IVAR
        LIST(IVAR)=0
        DO 50 ISIGN=-1,1,2
          SIGNUM=DBLE(ISIGN)
          DEL=DX(IVAR)
          LENCLS=.FALSE.
C set XSMALL to x at minimum position
          XSMALL=X0(IVAR)
          XIVAR=X0(IVAR)+SIGNUM*DEL
          DO 30 I=1,NSTEP
            NSTEPL=NSTEP
            CALL MTXCPV(X0,X,NR)
            X(IVAR)=XIVAR
            CALL MTXCPV(YSAV,Y,N)
            CALL LSQGEN(Y,CY,FY,F,E,M,N,NR,NRED1,LIST,X,CX,R1,A2,
     +      NSTEPL)
            IF(NSTEPL.LT.1) THEN
              NSTEP=-1
              GO TO 70
            END IF
C test for convergence
            IF(ABS(R1-RTARG).LT.EPS*G) THEN
              IF(ISIGN.LT.0) THEN
                DXMINS(IVAR)=X0(IVAR)-X(IVAR)
              ELSE
                DXPLUS(IVAR)=X(IVAR)-X0(IVAR)
              END IF
              GO TO 40
            ELSE
              IF(.NOT.LENCLS) THEN
C zero was not yet enclosed in last step
                IF(R1.GT.RTARG) THEN
C zero is now enclosed, perform first interval halving
                  LENCLS=.TRUE.
                  XBIG=X(IVAR)
                  X(IVAR)=HALF*(XBIG+XSMALL)
                ELSE
C zero not enclosed, widen range
```

```
                  DEL=DEL*TWO
                  X(IVAR)=XSMALL+SIGNUM*DEL
                END IF
              ELSE
C continue interval halving
                IF(R1.GT.RTARG) THEN
                  XBIG=X(IVAR)
                ELSE
                  XSMALL=X(IVAR)
                END IF
                X(IVAR)=HALF*(XBIG+XSMALL)
              END IF
              XIVAR=X(IVAR)
            END IF
   30     CONTINUE
          NSTEP=-2
          GO TO 70
   40     CONTINUE
   50   CONTINUE
C unfix variable IVAR
        LIST(IVAR)=1
   60 CONTINUE
   70 RETURN
      END
```

Program 9.15: Subroutine LSQCOG draws the confidence region in a plane spanned by two parameters whose values have been determined for the general least-squares case

CALLING SEQUENCE: CALL LSQCOG(IX,JX,XI0,XJ0,DXI,DXJ, NXI,NXJ,FCONT,Y,CY,GY,F,E,M,N,NR,X0,A2)

INPUT ARGUMENTS:

IX,JX,XI0,XJ0,DXI,DXJ,NXI,NXJ,FCONT,X0 — as in LSQCON, Program 9.9

Y,CY,FY,F,E,M,N,NR,A2 — as in LSQGEN, Program 9.12

REMARKS: As in LSQGEN, Program 9.12

```
      SUBROUTINE LSQCOG(IX,JX,XI0,XJ0,DXI,DXJ,NXI,NXJ,FCONT,Y,CY,
     +GY,F,E,M,N,NR,X0,A2)
      IMPLICIT DOUBLE PRECISION (A-H,O-Z)
      DIMENSION Y(N),CY(N,N),X0(NR),E(M,N),GY(N,N),F(N,N),A2(N,*)
      DIMENSION XI(2),FUN(2,2)
      PARAMETER(NRMAX=100,NMAX=1000)
      DIMENSION YSAV(NMAX),LIST(NRMAX)
      COMMON /DASV04/ YSAV
      DO 5 I=1,NR
        LIST(I)=0
    5 CONTINUE
      CALL MTXCPV(Y,YSAV,N)
      DO 40 I=0,NXI-1
        XI(1)=XI0+I*DXI
        XI(2)=XI(1)+DXI
        DO 30 J=0,NXJ
```

```
              XJ=XJ0+J*DXJ
              DO 20 IS=1,2
                CALL MTXCPV(YSAV,Y,N)
                X0(IX)=XI(IS)
                X0(JX)=XJ
                NSTEP=-100
                NRED=0
                CALL LSQGEN(Y,CY,GY,F,E,M,N,NR,NRED,LIST,X0,CX,R,A2,
     +          NSTEP)
                FUN(IS,2)=R
   20         CONTINUE
              IF (J.GT.0)
     +          CALL GRPXCT(XI(1),XI(2),XJ0,XJ,FUN(1,1),FUN(1,2),
     +                      FUN(2,1),FUN(2,2),FCONT)
              XJ0=XJ
              FUN(1,1)=FUN(1,2)
              FUN(2,1)=FUN(2,2)
   30       CONTINUE
   40 CONTINUE
      CALL MTXCPV(YSAV,Y,N)
      END
```

Example 9.14: Asymmetric errors and confidence region for fitting a straight line to measured points with errors in the abscissa and ordinate

Figure 9.15 shows the result of fitting to four points with large measurement errors. From Sect. 9.8 we already know that large measurement errors can lead to asymmetric errors in the fitted parameters. In fact, we see highly asymmetric errors and large differences between the covariance ellipse and the corresponding confidence region. ■

9.14 Example Programs

The source files of the example programs are available in the subdirectory \DATAN\EXASRC.

9.14.1 Main Program `E1LSQ` to Demonstrate Subprogram `LSQPOL`

The short program contains the data of Table 9.3 in two `DATA` statements. It computes vectors $\tilde{\mathbf{x}}$ of coefficients and their covariance matrix C_x for $r = 1, 2, 3, 4$. Here $r - 1$ is the degree of the polynomial, i.e., r is the number of elements in $\mathbf{x}$. The results are presented numerically.

Suggestions: (a) Modify the program (by modifying a single statement) so that the cases $r = 1, 2, \ldots, 10$ are treated. Which peculiarity do you expect for $r = 10$? (b) Instead of the data of Table 9.3 use different data determined without error by a polynomial, e.g., $y = t^2$, and let the program determine the parameters of the polynomial from the data. Use different (although obviously

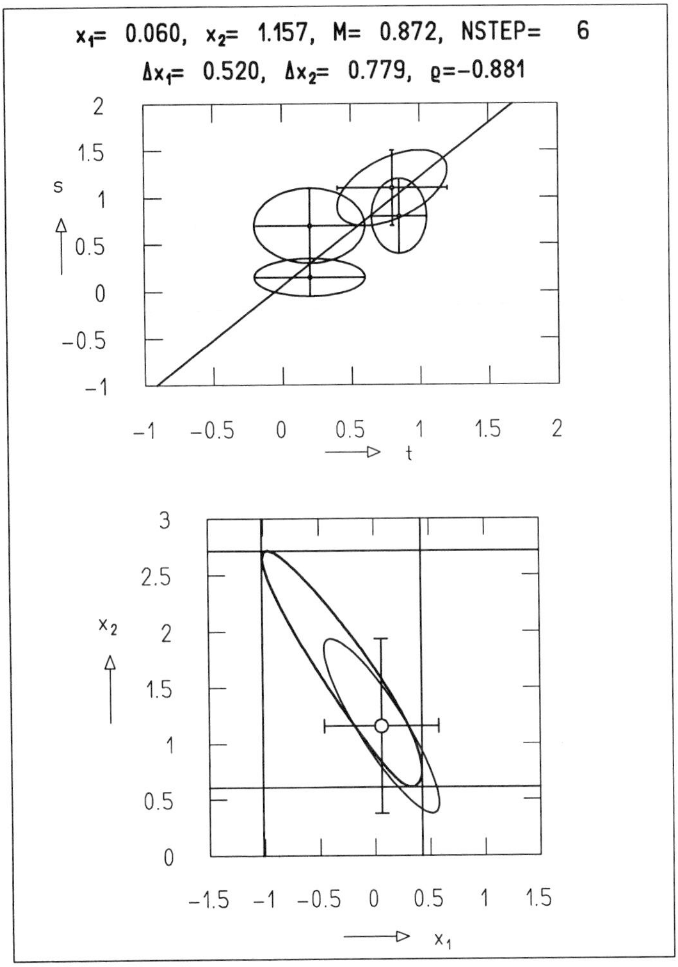

Fig. 9.15. (Above) Measured points with covariance ellipses and fitted line. (Below) Result of the fit, given in the plane spanned by the parameters x_1, x_2. Shown are the fitted parameter values (circle), symmetric errors (crossed bars), covariance ellipse, asymmetric errors (horizontal and vertical lines), and the confidence region (dark contour).

incorrect) sets of values for the errors Δy_i, e.g., $\Delta y_i = \sqrt{y_i}$ for one run through the program and $\Delta y_i = 1$ for another run. What is the influence of the choice of Δy_i on the coefficients $\tilde{\mathbf{x}}$, on the minimum function, and on the covariance matrix?

9.14.2 Main Program E2LSQ to Demonstrate Subprogram LSQLIN

The program contains the data of Fig. 9.4 in three DATA statements. The matrix A and the vector $\mathbf{c}$ are set up; these are needed for the fit of a proportional relation $y = x_1 t$ to the data. Next the fit is performed by calling LSQLIN. The results are displayed numerically.

Suggestions: (a) Modify the program so that in addition a first degree polynomial is fitted. Set up the matrix A yourself, i.e., do not use LSQPOL. Compare your result with Fig. 9.4. (b) Display the results graphically as in Fig. 9.4.

9.14.3 Main Program E3LSQ to Demonstrate Subprogram LSQNON

The program solves the following problem. First 20 pairs of values (t_i, y_i) are generated. The values t_i of the controlled variables are $1/21$, $2/21$, ..., $20/21$. The values y_i are given by

$$y_i = x_1 \exp(-(t - x_2)^2 / 2x_3^2) + \varepsilon_i \quad .$$

Here ε_i is an error taken from a normal distribution with expectation value zero and width σ_i. The width σ_i is different from point to point. It is taken from a uniform distribution with the limits $\sigma/2$ and $3\sigma/2$. Thus, the y_i are points scattered within their errors around a Gaussian curve defined by the parameters $\mathbf{x} = (x_1, x_2, x_3)$. The widths of the error distributions are known, i.e., $\Delta y_i = \sigma_i$. The data points are generated for the parameter values $x_1 = 1$, $x_2 = 1.2$, $x_3 = 0.4$. (They are identical with those shown in Fig. 9.5.) The program now successively performs four different fits of a Gaussian curve to the data. In a first step all three parameters in the fit procedure are considered variable, NRED = 3. Then successively one, two, and finally all parameters are fixed. Before each fit, first approximations for the variables are set which, of course, are not modified during the fitting procedure for the fixed parameters. The results are presented numerically.

Suggestions: (a) Choose different first approximations for the non-fixed parameters and observe the influence of the choice on the results. (b) Obtain first approximations by computation, e.g., by the procedure described in Example 9.4, in which a parabola is fitted to the logarithms of the data. (c) Modify the program by adding graphical output corresponding to Fig. 9.5.

9.14.4 Main Program E4LSQ to Demonstrate Subprogram LSQMAR

The program solves the problem of Example 9.7. First a set of 50 data points (t_i, y_i) is generated. They are scattered according to their errors around a curve corresponding to the sum of a second degree polynomial and two Gaussians, cf. (9.6.6). The 9 parameters of this function are combined to form the vector $\mathbf{x}$. The measurement errors Δy_i are generated by the procedure described in Sect. 9.14.3. The data points are generated for predefined values of $\mathbf{x}$. Next, by calling LSQMAR (with significantly different values $\mathbf{x}$ as first approximations) solutions $\widetilde{\mathbf{x}}$ are obtained by a fit to the data points. The results are presented numerically.

Suggestions: (a) Fix all parameters except for x_5 and x_8, which determine the mean values of the two Gaussians used for generating the data. Allow for interactive input of x_5 and x_8. For different input values of (x_5, x_8), e.g., $(10, 30)$, $(19, 20)$, $(19, 15)$, $(10, 11)$, try to determine whether you can still separate the two Gaussians by the fit, i.e., whether you obtain significantly different values for $\widetilde{x}_5$ and $\widetilde{x}_8$, considering their errors $\Delta\widetilde{x}_5$, $\Delta\widetilde{x}_8$. (b) Repeat (a), but for smaller measurement errors. Choose, e.g., $\sigma = 0.1$ or 0.01 instead of $\sigma = 0.4$. (c) Supplement the program by a graphical display as in Fig. 9.8. (d) In a copy of the file E4LSQ.FOR, delete the subprogram AUXDRI. By compiling and linking this version of E4LSQ an executable program is produced which computes the derivatives numerically. Compare the computing time needed for the execution of this version of the program to that of the original program.

9.14.5 Main Program **E5LSQ** to Demonstrate Subprogram **LSQASN**

The program solves the problem of Example 9.8. First, pairs (t_i, y_i) of data are produced. The y_i are scattered according to their measurement errors around a curve given by the function $y_i(t_i) = x_1 \exp(-x_2 t)$, cf. (9.6.4). The measurement errors Δy_i are generated by the procedure already used in Sect. 9.14.3. Starting from a given first approximation for the parameters $\mathbf{x}_1$, values $\widetilde{\mathbf{x}}$ of these parameters are fitted to the data using LSQNON. Finally the asymmetric errors are found using LSQASN.

Suggestions: (a) The data points generated by the program are identical to those in Figs. 9.10 and 9.11 (lower right section of each figure). Compare the results of the fit (minimum function M, parameters $\widetilde{x}_1$, $\widetilde{x}_2$, symmetric and asymmetric errors) with values read off the figures. (b) Modify the program so that the value σ determining the size of the measurement errors (FORTRAN variable SIGMA), can be entered interactively. For consecutive runs of the program use $\sigma = 0.5$, 2, 5, 10 and compare the results with the four plots in Figs. 9.10 and 9.11.

9.14.6 Main Program E6LSQ to Demonstrate Subprogram LSQCON

The program solves the same problem as E5LSQ, but in addition computes and displays as a contour the confidence region in the (x_1, x_2) plane of the parameters. The asymmetric errors are shown as horizontal and vertical lines. In addition the plot contains the solution $(\tilde{x}_1, \tilde{x}_2)$ and its covariance ellipse. The plot obtained corresponds to the upper right-hand plot of Fig. 9.11.

Suggestions: (a) Supplement the program so that the value of σ can be entered interactively. Choose $\sigma = 0.5$, 2, 5, 10 as in the suggestions given in the preceding section and compare the results with Fig. 9.11. Note that the scales of the plots have to be chosen properly in each case. Devise a method to perform that choice automatically using the errors found by the fit. (b) Draw the confidence regions corresponding to the confidence levels 90 %, 95 %, 99 %. Use Eq. (9.8.5).

9.14.7 Main Program E7LSQ to Demonstrate Subprogram LSQGEN

The program fits a straight line to points with measurement errors in the abscissa and ordinate, i.e., it solves the problem of Example 9.11. The measured values, their errors and covariances are contained in DATA statements. Using this information the vector **y** and the covariance matrix C_y are set up. To determine the two parameters x_1 (ordinate intercept) and x_2 (slope) with LSQGEN the program LSQGFN is used in the form shown in Sect. 9.11 and which is contained in the program library DALIB. In a loop spanning nearly the entire program, three cases are treated consecutively, which differ only by the number NRED of the non-fixed parameters. The results are presented numerically. For NRED = 2 the first approximations of x_1 and x_2 are obtained by constructing a straight line through the outermost two points. The results correspond to the values shown in Fig. 9.12 (above). For NRED = 1 the value of x_1 is fixed at 0.2. The first approximation for x_2 is still the slope of the straight line through the two outermost points. The result corresponds to Fig. 9.13 (top). For NRED = 0 no fit is performed, but a χ^2-test is done (cf. Example 9.13). The result corresponds to Fig. 9.14 (below).

Suggestions: Supplement the program so that it treats the situations of all plots from Figs. 9.12, 9.13, and 9.14.

9.14.8 Main Program E8LSQ to Demonstrate Subprograms LSQASG and LSQCOG

First, the program solves the same problem as Program E7LSQ, described in the preceding section, i.e., it fits a straight line to points with measurement errors in abscissa and ordinate. (The program only treats the case where both x_1 and x_2 are determined.) The input data and the result of the fit are presented in Fig. 9.15 (above). Then the asymmetric errors are computed by calling LSQASG, and the results are displayed numerically. Finally (if requested by the user) the result is displayed graphically. It shows a section of the (x_1, x_2) plane containing the solution $(\widetilde{x}_1, \widetilde{x}_2)$, the symmetric errors as crossed bars, the covariance ellipse, the asymmetric errors, and the confidence region as in Fig. 9.15 (below). Note: The computation of the contour of the confidence region requires considerable computer time since for each pixel in the section of the plane the program LSQGEN has to be called.

9.15 Programming Problems

The source files of the example solutions are available in the subdirectory \DATAN\SOLSRC.

9.15.1 Fit of a First-Degree Polynomial to Data that Correspond to a Second-Degree Polynomial

In experimental or empirical studies one is often confronted with a large number of measurements or objects of the same kind (animals, elementary particle collisions, industrial products from a given production process, ...). The outcomes of the measurements performed on each object are described by some law. Certain assumptions are made about that law, which are to be checked by experiment.

Consider the following example. A series of measurements may contain n_{exp} experiments. Each experiment yields the measurements $y_i = x_1 + x_2 t_i + x_3 t_i^2 + \varepsilon_i$ for 10 values $t_i = 1, 2 \ldots, 10$ of the controlled variable t. The ε_i are taken from a normal distribution with mean zero and width σ. In the analysis of the experiments it is assumed, however, that the true values η_i underlying the measurements y_i can be described by a first-degree polynomial $\eta_i = x_1 + x_2 t$. As result of the fit we obtain a minimum function M from which we can compute the "χ^2-probability" $P = 1 - F(M, n - r)$. Here $F(M, f)$ is the distribution function of a χ^2 distribution with f degrees of freedom, n is the number of data points, and r the number of parameters determined in the fit. If $P < \alpha$, then the fit of a first-degree polynomial to the data is rejected at a confidence level of $\beta = 1 - \alpha$.

Write a program performing the following steps:

(i) Interactive input of n_{exp}, x_1, x_2, x_3, σ, Δy.

(ii) Generation of $n_{\rm exp}$ sets of data $(t_i, y_i, \Delta y)$, fit of a first-degree polynomial to each set of data and computation of P. Entry of P into a histogram.

(iii) Graphical representation of the histogram.

Suggestions:(a) Choose $n_{\rm exp} = 1000$, $x_1 = x_2 = 1$, $x_3 = 0$, $\sigma = \Delta y = 1$. As expected you will obtain a flat distribution for P. (b) Choose (keeping the other input quantities as above) different values $x_3 \neq 0$. You will observe a shift of the distribution towards small P values, cf. Fig. 9.16. Determine approximately the smallest positive value of x_3 such that the hypothesis of a first-degree polynomial is rejected at 90 % confidence level in 95 % of all experiments. (c) Choose $x_3 = 0$, but $\sigma \neq \Delta y$. You will again observe a shift in the distribution, e.g., towards larger P values for $\Delta y > \sigma$. (d) From the experience gained in (a), (b), and (c), one might conclude that if erroneously too large measurement errors are assumed ($\Delta y > \sigma$) then a flat P distribution would result. In this way one would get the impression that a first-degree polynomial could describe the data. Begin with $n_{\rm exp} = 1000$, $x_1 = x_2 = 1$, $x_3 = 0.2$, $\sigma = 1$, $\Delta y = 1$ and increase Δy in steps of 0.1 up to $\Delta y = 2$.

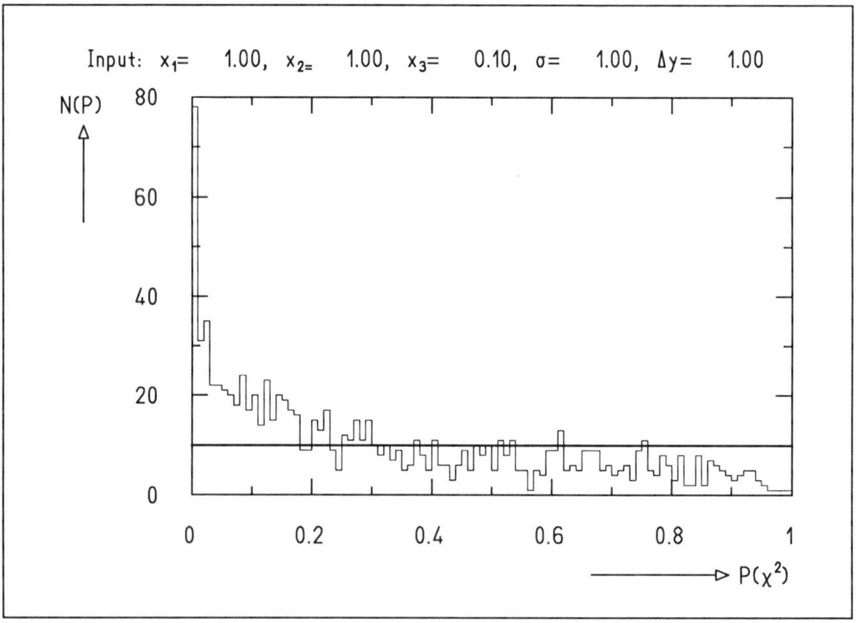

Fig. 9.16. Histogram of the χ^2-probability for fits of a first-degree polynomial to 1000 data sets generated according to a second-degree polynomial.

9.15.2 Fit of a Power Law (Linear Case)

A power law

$$\eta = xt^w$$

is linear in the parameter x if w is a constant. This function is to be fitted to measurements (t_i, y_i) given by

$$\begin{aligned} t_i &= t_0 + (i-1)\Delta t \quad , \qquad i = 1, \ldots, n \quad , \\ y_i &= xt_i^w + \varepsilon_i \quad . \end{aligned}$$

Here the ε_i follow a normal distribution centered about zero with width σ.

Write a program performing the following steps:

(i) Interactive input of n, t_0, Δt, x, w, σ.

(ii) Generation of measured points.

(iii) Fit of the power law.

(iv) Graphical display of the data and the fitted function, cf. Fig. 9.17.

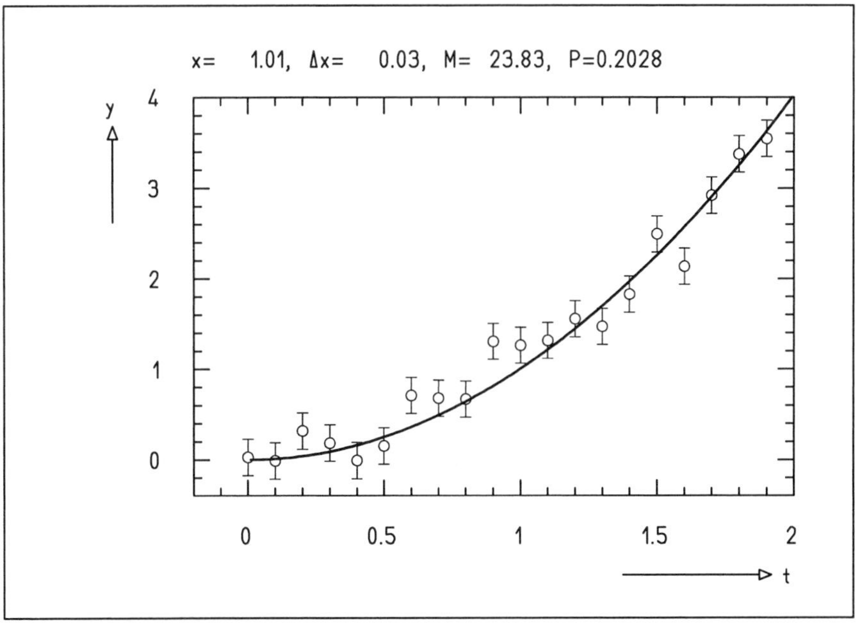

Fig. 9.17. Result of the fit of a parabola $y = xt^2$ to 20 measured points.

9.15.3 Fit of a Power Law (Nonlinear Case)

If the power law has the form

$$\eta = x_1 t^{x_2} \quad ,$$

i.e., if the power itself is an unknown parameter, the problem becomes nonlinear. For the fit of a nonlinear function we have to start from a first approximation of the parameters. We limit ourselves to the case $t_i > 0$ for all i which occurs frequently in practice. Then one has $\ln \eta = \ln x_1 + x_2 \ln t$. If instead of (t_i, y_i) we now use $(\ln t_i, \ln y_i)$ as measured variables, we obtain a linear function in the parameters $\ln x_1$ and x_2. However, in this transformation the errors are distorted so that they are no longer Gaussian. We simply choose all errors to be of equal size and use the result of the linear fit as the first approximation of a nonlinear fit to the (t_i, y_i). We still have to keep in mind (in any case for $x_1 > 0$) that one always has $\eta > 0$ for $t > 0$. Because of measurement errors, however, measured values $y_i < 0$ can occur. Such points of course must not be used for the computation of the first approximation.

Write a program with the following steps:

(i) Interactive input of n, t_0, Δt, x_1, x_2, σ.

(ii) Generation of the n measured points

$$\begin{aligned} t_i &= t_0 + (i-1)\Delta t \quad , \qquad i = 1, \ldots, n \quad , \\ y_i &= x_1 t_i^{x_2} + \varepsilon_i \quad , \end{aligned}$$

where ε_i comes from a normal distribution centered about zero with width σ.

(iii) Computation of first approximations x_1, x_2 by fitting a linear function to $(\ln t_i, \ln y_i)$ with `LSQLIN`.

(iv) Fit of a power law to (t_i, y_i) with `LSQNON`.

(v) Graphical display of the results, cf. Fig. 9.18.

9.15.4 Fit of a Breit–Wigner Function to Data Points with Errors

For the $N = 21$ values $t_i = -3, -2.7, \ldots, 3$ of the controlled variable the measured values

$$y_i = f(t_i) + \varepsilon_i \tag{9.15.1}$$

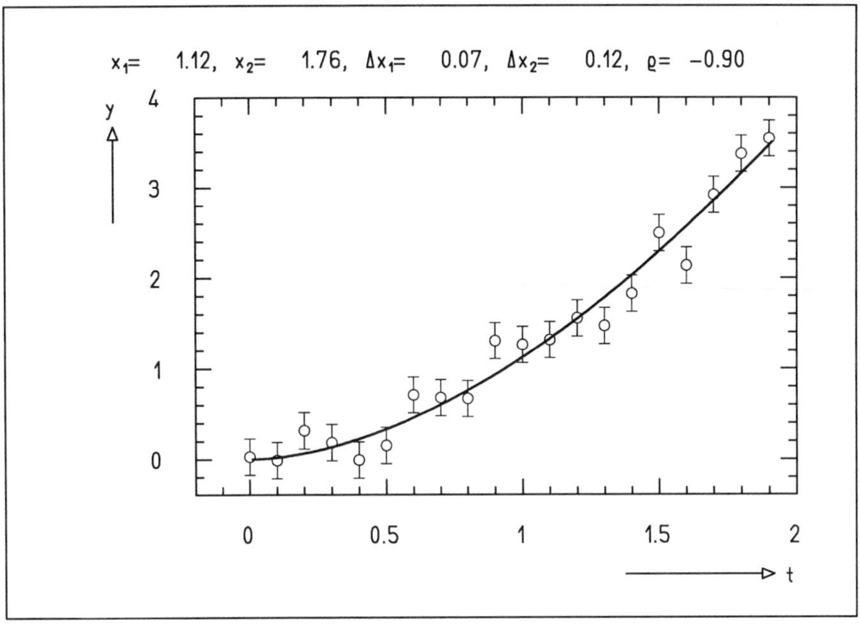

Fig. 9.18. Fit of a function $y = x_1 t^{x_2}$ to 20 data points. The data points are identical to those in Fig. 9.17.

are to be simulated. Here,

$$f(t) = \frac{2}{\pi x_2} \frac{x_2^2}{4(t - x_1)^2 + x_2^2} \tag{9.15.2}$$

is the Breit–Wigner function (3.3.32) with $a = x_1$ and $\Gamma = x_2$. The measurement errors ε_i are to be taken from a normal distribution around zero with width σ. Choose $a = 0$ and $\Gamma = 1$. The points (t_i, y_i) scatter within the measurement errors around a bell-shaped curve with a maximum at $t = a$. A bell-shaped curve with a maximum at the same position, however, could be given by the Gaussian function,

$$f(t) = \frac{1}{x_2\sqrt{2\pi}} \exp\left\{-\frac{(t - x_1)^2}{2x_2^2}\right\} \quad . \tag{9.15.3}$$

Write a program with the following steps:

(i) Interactive input of σ and of a switch `ISWIT`, which determines whether a Breit–Wigner function (`ISWIT = 1`) or a Gaussian (`ISWIT = 2`) is to be fitted.

(ii) Generation of the data, i.e., of the triplets of numbers $(t_i, y_i, \Delta y_i = \varepsilon_i)$.

(iii) Fit of the Breit–Wigner function (9.15.2) or of the Gaussian (9.15.3) to the data and computation of the minimum function M.

(iv) Graphical representation of the measured points with measurement errors and of the fitted function, cf. Fig. 9.19.

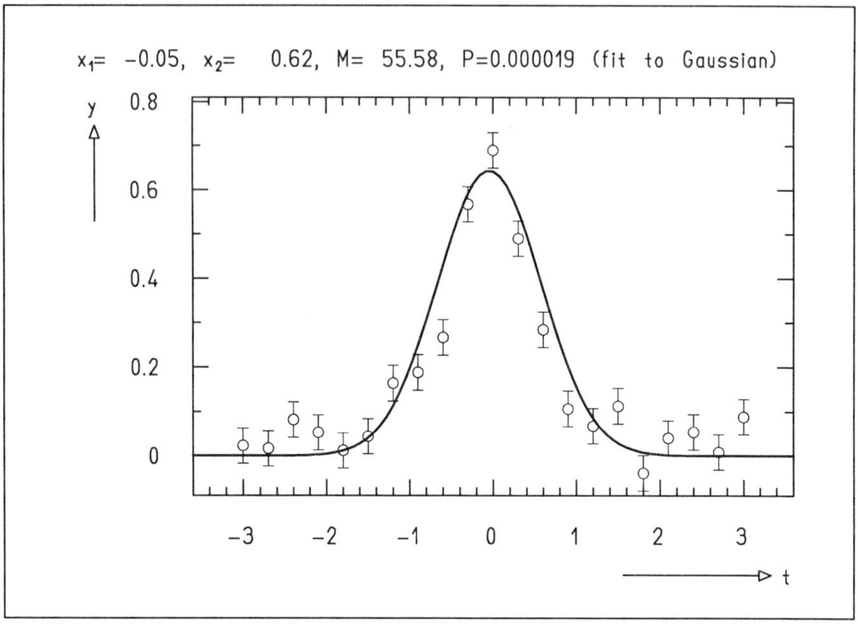

Fig. 9.19. Fit of a Gaussian to data points that correspond to a Breit–Wigner function. The goodness-of-fit is poor.

Run the program using different values of σ and find out for which range of σ the data allow a clear discrimination between the Breit–Wigner and Gaussian functions.

9.15.5 Asymmetric Errors and Confidence Region for the Fit of a Breit–Wigner Function

Modify the program of the preceding section (or rather a copy of that program) such that a graphical presentation of the fitted parameters, their errors, the covariance matrix, the asymmetric errors, and the confidence region is performed similar to the plots in Fig. 9.11.

Discuss the difference in the results you obtain in the fit of a Breit–Wigner function and of a Gaussian, respectively, in each case using $\sigma = 0.1$ and $\sigma = 1$.

9.15.6 Fit of a Breit–Wigner Function to a Histogram

In the Programming Problem 9.15.4 we started from measurements $y_i = f(t_i) + \varepsilon_i$. Here $f(t)$ was a Breit–Wigner function (3.3.32), and the measurement errors ε_i corresponded to a Gaussian distribution centered about zero with width σ.

We now generate a sample of size n_{ev} from a Breit–Wigner distribution with mean $a = 0$ and full width at half maximum $\Gamma = 1$. We represent the sample by a histogram that is again characterized by triplets of numbers $(t_i, y_i, \Delta y_i)$. Now t_i is the center of the ith bin $t_i - \Delta t/2 \le t < t_i + \Delta t/2$, and y_i is the number of sample elements falling into this bin. For not too small y_i the corresponding statistical error is $\Delta y_i = \sqrt{y_i}$. For small values y_i this simple statement is problematic. It is completely wrong for $y_i = 0$. In the fit of a function to a histogram, care is therefore to be taken that empty bins (possibly also bins with few entries) are not to be considered as data points.

The function to be fitted is

$$f(t) = x_3 \frac{2}{\pi x_2} \frac{x_2^2}{4(t - x_1)^2 + x_2^2} \quad . \tag{9.15.4}$$

The function is similar to (9.15.2) but there is an additional parameter x_3.

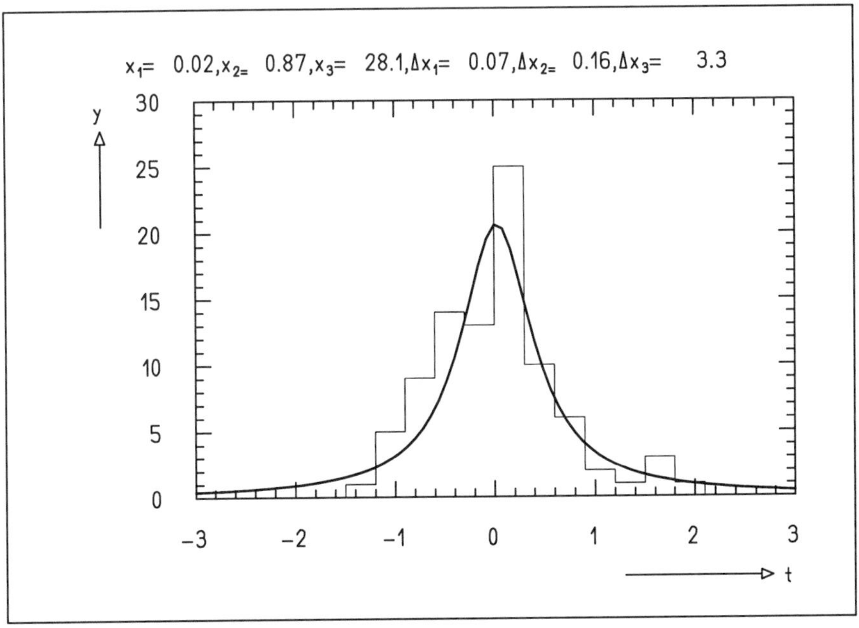

Fig. 9.20. Fit of a Breit–Wigner function to a histogram.

Write a program with the following steps:

(i) Interactive input of $n_{\rm ev}$ (sample size) and n_t (number of histogram bins). The lower limit of the histogram is to be fixed at $t = -3$, the upper limit at $t = 3$.

(ii) Generation of the sample, cf. Programming Problem 4.15.1.

(iii) Construction of the histogram.

(iv) Construction of the triplets $(t_i, y_i, \Delta y_i)$ to be used for the fit.

(v) Fit of the function (9.15.4).

(vi) Output of the results in numerical and graphical form, cf. Fig. 9.20.

Suggestions: Perform consecutive fits for the same sample but different numbers of bins. Try to find an optimum for the number of bins.

9.15.7 Fit of a Circle to Points with Measurement Errors in Abscissa and Ordinate

A total of m data points (s_i, t_i) are given in the (s, t) plane. The measurement errors are defined by 2×2 covariance matrices of the form

$$\begin{pmatrix} \Delta s_i^2 & c_i^2 \\ c_i^2 & \Delta t_i^2 \end{pmatrix} \tag{9.15.5}$$

as in Example 9.11. Here $c_i = \Delta s_i \Delta t_i \rho_i$ and ρ_i is the correlation coefficient between the measurement errors Δs_i, Δt_i. As in Example 9.11, construct the vector $\mathbf{y}$ of measurements from the s_i and t_i and construct the covariance matrix C_y. Set up the equations $f_k(x, \boldsymbol{\eta}) = 0$ assuming that the true positions underlying the measured points lie on a circle with center (x_1, x_2) and radius x_3.

Write a program with the following steps:

(i) Input of m, Δt, Δs, ρ.

(ii) Generation of m measured points (s_i, t_i) using bivariate normal distributions, the means of which are positioned at regular intervals on the unit circle ($x_1 = x_2 = 0$, $x_3 = 1$) and the covariance matrix of which is given by (9.15.5) with $\Delta s_i = \Delta s$, $\Delta t_i = \Delta t$, $c = \Delta s \Delta t \rho$.

(iii) Determination of a first approximation for x_1, x_2, x_3 by computation of the parameters of a circle through the first three measured points.

(iv) Fit to all measured points using `LSQGEN` and using a subprogram `LSQGF` specially written for this problem.

(v) Graphical representation of the measured points and the fitted circle as in Fig. 9.21.

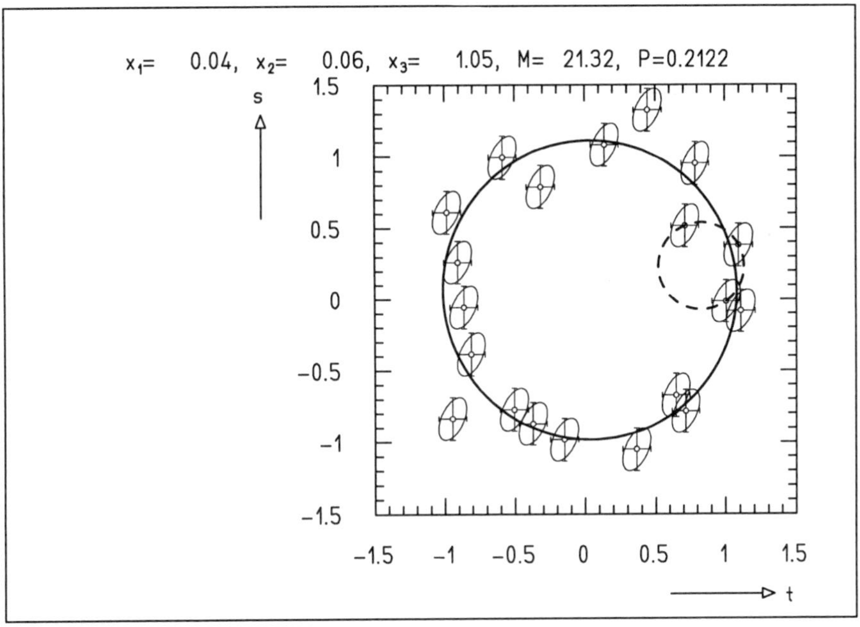

Fig. 9.21. Measured points with covariance ellipses, circle of the first approximation which is given by 3 measured points (broken line), and circle fitted to all measured points.

10. Function Minimization

Locating extreme values (maxima and minima) is particularly important in data analysis. This task occurs in solving the least-squares problem in the form $M(\mathbf{x},\mathbf{y}) = \text{min}$ and in the maximum likelihood problem as $L = \text{max}$. By means of a simple change of sign, the latter problem can also be treated as locating a minimum. We always speak therefore of *minimization*.

10.1 Overview. Numerical Accuracy

We consider first the simple *quadratic form*

$$M(x) = c - bx + \frac{1}{2}Ax^2 \quad . \tag{10.1.1}$$

It has an extremum where the first derivative vanishes,

$$\frac{\mathrm{d}M}{\mathrm{d}x} = 0 = -b + Ax \quad , \tag{10.1.2}$$

that is, at the value

$$x_{\mathrm{m}} = \frac{b}{A} \quad . \tag{10.1.3}$$

With $M(x_{\mathrm{m}}) = M_{\mathrm{m}}$, Eq. (10.1.1) can easily be put into the form

$$M(x) - M_{\mathrm{m}} = \frac{1}{2}A(x - x_{\mathrm{m}})^2 \quad . \tag{10.1.4}$$

Although the function whose minimum we want to find does not usually have the simple form of (10.1.1), it can nevertheless be approximated by a quadratic form in the region of the minimum, where one has the Taylor expansion around the point x_0,

$$M(x) = M(x_0) - b(x - x_0) + \frac{1}{2}A(x - x_0)^2 + \cdots \quad , \tag{10.1.5}$$

where

$$b = -M'(x_0) \quad , \qquad A = M''(x_0) \quad . \tag{10.1.6}$$

In this approximation the minimum is given by the point where the derivative $M'(x)$ is zero, i.e.,

$$x_{\mathrm{mp}} = x_0 + \frac{b}{A} \quad . \tag{10.1.7}$$

This holds only if x_0 is sufficiently close to the minimum so that terms of order higher than quadratic in (10.1.5) can be neglected. The situation is depicted in Fig. 10.1. The function $M(x)$ has a minimum at x_{m}, maxima at x_{M}, and points of inflection at x_{s}. For the second derivative in the region $x > x_{\mathrm{m}}$ one has $M''(x) > 0$ for $x < x_{\mathrm{s}}$ and $M(x'') < 0$ for $x > x_{\mathrm{s}}$. If we now choose x_0 to be in the region $x_{\mathrm{m}} < x_0 < x_{\mathrm{M}}$, then the first derivative $M'(x)$ there is always positive. Therefore x_{mp} lies closer to x_{m} only if $x_0 < x_{\mathrm{s}}$. Clearly the point x_0 is not in general chosen arbitrarily, but rather as close as possible to where the minimum is expected to be. We can call this estimated value the *zeroth approximation* of x_{m}. Various strategies are available to obtain successively better approximations:

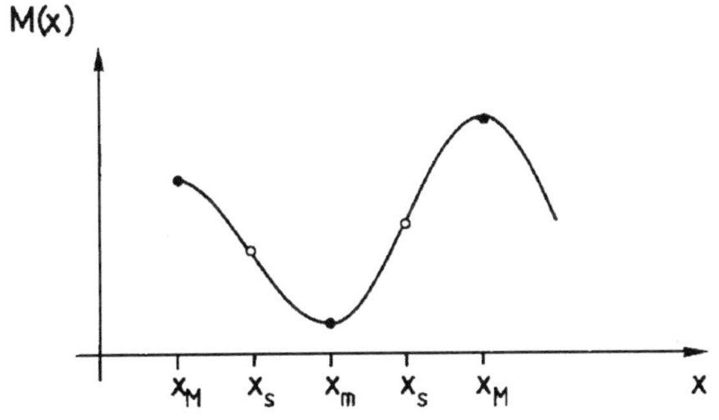

Fig. 10.1. The function $M(x)$ has a minimum at x_{m}, maxima at the points x_{M}, and points of inflection at x_{s}.

(i) *Use of the function and its first and second derivatives at* x_0.

One computes x_{mp} according to (10.1.7), takes x_{mp} as a first approximation, i.e., x_0 is replaced by x_{mp}, and obtains by repeated application of (10.1.7) a second approximation. The procedure is repeated until two successive approximations differ by less than a given value ε. From the discussion above it follows that this procedure does not converge if the zeroth approximation lies outside the points of inflection in Fig. 10.1.

(ii) *Use of the function and its first derivative at* x_0.

The sign of the derivative $M'(x_0) = -b$ at the point x_0 determines the direction in which the function increases. It is assumed that one should search for the minimum in the direction in which the function decreases. The quantity

$$x_1 = x_0 + b \tag{10.1.8}$$

is computed, i.e., one replaces the second derivative by the value unity.

Instead of x_0 one now uses x_1 as the approximation, and so forth. Alternatively one can also use instead of (10.1.8) the rule

$$x_1 = x_0 + cb \quad , \tag{10.1.9}$$

where c is an arbitrary positive constant. Both rules ensure that the step from x_0 to x_1 proceeds in the direction of the minimum. If in addition one chooses c to be small, then the step is small, so that it does not go beyond (or not far beyond) the minimum.

(iii) *Use of the function at various points.*

The procedure (i) can be carried out without knowing the derivative of the function if the function itself is known at three points. One can then uniquely fit a parabola through these three points and take the extreme value as an approximation of the minimum of the function. One speaks of locating the minimum by *quadratic interpolation*. It is, however, by no means certain that the extreme value of the parabola is a minimum and not a maximum. As in procedure (i) it is therefore important that the three chosen points are already in the region of the minimum of the function.

(iv) *Successive reduction of an interval containing the minimum.*

In none of the procedures discussed up to this point were we able to guaranty that the minimum of the function would actually be found. The minimum can be found with certainty provided one knows an interval $x_a < x < x_b$ containing the minimum. If such an interval is known, one can locate the minimum with arbitrary accuracy by successively subdividing and checking in which subinterval the minimum is located.

In Sects. 10.2 to 10.7 we shall examine the minimization of a function of only one variable. In Sect. 10.2 the formula for a parabola determined by three points will be given. In Sect. 10.3 it is shown that the minimization of a function of one variable is equivalent to the minimization of a function of n variables on a line in an n-dimensional space. Section 10.4 describes a procedure for locating an interval containing the minimum. In Sect. 10.5 a minimum search by means of interval division is described. This is combined in Sect. 10.5 with the procedure of quadratic interpolation in a way that the interpolation is only used when it leads quickly to the minimum of the function. If this is not the case, one continues with interval division. In this way we possess a procedure with the certainty of interval division combined with – as much as possible – the speed of quadratic interpolation. The same procedure can also be used for a function of n variables if the search for the minimum is restricted to a line in the n-dimensional space. This problem is addressed in Sect. 10.7.

Next we turn to the task of searching for the minimum of a function of n variables. We begin in Sect. 10.8 with the particularly elegant simplex method. This is followed by the discussion of various procedures of successive minimization along fixed directions in the n-dimensional space. These directions can simply be the directions of the coordinates (Sect. 10.9), or for a function that depends only quadratically on the variables they can be chosen such that the minimum is reached in at most n steps (Sects. 10.10 and 10.11).

Finally we discuss a procedure of n-dimensional minimization which employs aspects of methods (i) and (ii) of the one-dimensional case. If $\mathbf{x}$ is the n-dimensional vector of the variables, then the general quadratic form, i.e., the generalization of (10.1.1) to n variables, is

$$\begin{aligned} M(\mathbf{x}) &= c - \mathbf{b}\cdot\mathbf{x} + \frac{1}{2}\mathbf{x}^{\mathrm{T}}A\mathbf{x} \qquad (10.1.10)\\ &= c - \sum_k b_k x_k + \frac{1}{2}\sum_{k,\ell} x_k A_{k\ell} x_\ell \quad . \end{aligned}$$

Here A is a symmetric matrix, $A_{k\ell} = A_{\ell k}$. The partial derivative with respect to x_i is

$$\begin{aligned} \frac{\partial M}{\partial x_i} &= -b_i + \frac{1}{2}\left(\sum_\ell A_{i\ell}x_\ell + \sum_k x_k A_{ki}\right)\\ &= -b_i + \sum_\ell A_{i\ell}x_\ell \quad . \qquad (10.1.11) \end{aligned}$$

Expressing all of the partial derivatives as a vector $\boldsymbol{\nabla} M$ gives

$$\boldsymbol{\nabla} M = -\mathbf{b} + A\mathbf{x} \quad . \qquad (10.1.12)$$

At the minimum point the vector of derivatives vanishes. The minimum is therefore located at

$$\mathbf{x}_{\mathrm{m}} = A^{-1}\mathbf{b} \quad , \qquad (10.1.13)$$

in analogy to (10.1.3).

Clearly the function $M(\mathbf{x})$ does not in general have the simple form of (10.1.10). We can, however, expand it in a series around the point $\mathbf{x}_0$,

$$M(\mathbf{x}) = M(\mathbf{x}_0) - \mathbf{b}(\mathbf{x}-\mathbf{x}_0) + \frac{1}{2}(\mathbf{x}-\mathbf{x}_0)^{\mathrm{T}}A(\mathbf{x}-\mathbf{x}_0) + \cdots \quad , \qquad (10.1.14)$$

with the negative *gradient*

$$\mathbf{b} = -\boldsymbol{\nabla} M(\mathbf{x}_0) \quad , \qquad \text{i.e.,} \qquad b_i = -\left.\frac{\partial M}{\partial x_i}\right|_{\mathbf{x}=\mathbf{x}_0} \quad , \qquad (10.1.15)$$

and the *Hessian matrix* of second derivatives

$$A_{ik} = \left. \frac{\partial^2 M}{\partial x_i \partial x_k} \right|_{\mathbf{x}=\mathbf{x}_0} \quad . \tag{10.1.16}$$

The series (10.1.14) is the starting point for various minimization procedures:

(i) *Minimization in the direction of the gradient.*

Starting from the point $\mathbf{x}_0$ one searches for the minimum along the direction given by the gradient $\boldsymbol{\nabla} M(\mathbf{x}_0)$ and calls the point where it is found $\mathbf{x}_1$. Starting from $\mathbf{x}_1$ one looks for the minimum along the direction $\boldsymbol{\nabla} M(\mathbf{x}_1)$, and so forth. We will discuss this procedure, called *minimization in the direction of steepest descent*, in Sect. 10.12.

(ii) *Step of given size in the gradient direction.*

One computes in analogy to (10.1.9)

$$\mathbf{x}_1 = \mathbf{x}_0 + c\mathbf{b} \quad , \qquad \mathbf{b} = -\boldsymbol{\nabla} M(\mathbf{x}_0) \quad , \tag{10.1.17}$$

with a given positive c. That is, one takes a *step in the direction of steepest descent* of the function, without, however, searching exactly for the minimum in this direction. Next one computes the gradient at $\mathbf{x}_1$, steps from $\mathbf{x}_1$ in the direction of this gradient, etc. In Sect. 10.13 we will combine this method with the following one.

(iii) *Use of the gradient and the Hessian matrix at* $\mathbf{x}_0$.

If we truncate (10.1.14) after the quadratic term, we obtain a function whose minimum is, according to (10.1.13), given by

$$\mathbf{x}_{mp} = \mathbf{x}_0 + A^{-1}\mathbf{b} \quad . \tag{10.1.18}$$

We take $\mathbf{x}_1 = \mathbf{x}_{mp}$ as the first approximation, compute for this point the gradient and Hessian matrix, obtain by corresponding use of (10.1.18) the second approximation, and so forth. This procedure is discussed in Sect. 10.14. It converges quickly if the zeroth approximation $\mathbf{x}_0$ is sufficiently close to the minimum. If that is not the case, however, then it gives – as for the corresponding one dimensional procedure – no reasonable solution. We will combine it, therefore, in Sect. 10.15 with method (ii), in order to obtain, when possible, the speed of (iii), but when necessary, the certainty of (ii).

In Sects. 10.8 through 10.15 very different methods for solving the same problem, the minimization of a function of n variables, will be discussed. In Sect. 10.16 we give information on how to choose one of the methods appropriate for the problem in question. Section 10.17 is dedicated to considerations of errors. In Sect. 10.18 several examples are discussed in detail.

Before we find the minimum $x_{\rm m}$ of a function, we would like to inquire briefly about the numerical accuracy we expect for $x_{\rm m}$. The minimum is after all almost always determined by a comparison of values of the function at points close in x. If we solve (10.1.4) for $(x - x_{\rm m})$, we obtain

$$(x - x_{\rm m}) = \sqrt{\frac{2[M(x) - M(x_{\rm m})]}{A}} \quad .$$

We assume that A, i.e., the second derivative of the function M, is of order of magnitude unity close to the minimum. (This need only be true approximately. In fact, in numerical calculations one always scales all of the quantities such that they are of order unity, i.e., not something like 10^6 or 10^{-6}.) If we compute the function M with the precision δ then the difference $M(x) - M(x_{\rm m})$ is also known at best with a precision of δ, i.e., two function values can not be considered as being significantly different if they differ by only δ. For the corresponding x values one then has

$$(x - x_{\rm m}) \approx \sqrt{\frac{2\delta}{A}} \approx \sqrt{\delta} \quad . \tag{10.1.19}$$

If the computer has n binary places available for representing the mantissa, then a value x can be represented with the precision (4.2.7)

$$\frac{\Delta x}{x} = 2^{-n} \quad .$$

For computing a value x one chooses therefore a relative precision

$$\varepsilon \geq 2^{-n} \quad ,$$

since it is clearly pointless to try to compute a value with a higher precision than that with which it can be represented. If x is computed iteratively, i.e., one computes a series $x_0, x_1, \ldots$ of approximations for x, then one can truncate this series as soon as

$$\frac{|x_k - x_{k-1}|}{|x_k|} < \varepsilon$$

or

$$|x_k - x_{k-1}| < \varepsilon |x_k| \tag{10.1.20}$$

for a given ε. With this prescription we will have difficulties, however, if $x_k = 0$. We introduce, therefore, in addition to ε a constant $t \neq 0$ and extend (10.1.20) to

$$|x_k - x_{k-1}| < \varepsilon |x_k| + t \quad . \tag{10.1.21}$$

The last task remaining is to choose the numerical values for ε and t. If x is the position of the minimum, then by (10.1.19) a value for ε must be chosen that is greater than or equal to the square root of the relative precision for the representation of a floating point number. With computations using "single precision" most computers have $n = 24$ binary places available for the representation of the mantissa. Then only the values

$$\varepsilon > 2^{-n/2} \approx 2 \cdot 10^{-4}$$

are reasonable. The quantity t corresponds to an absolute precision. Therefore it can be chosen to be considerably smaller. In the programs of the following sections the numerical value of ε is set by a parameter statement.

10.2 Parabola through Three Points

If three points (x_a, y_a), (x_b, y_b), (x_c, y_c) of a function are known, then we can determine the parabola

$$y = a_0 + a_1 x + a_2 x^2 \tag{10.2.1}$$

that passes through these points. Instead of (10.2.1) we can also represent the parabola by

$$y = c_0 + c_1(x - x_b) + c_2(x - x_b)^2 \quad . \tag{10.2.2}$$

This relationship is naturally valid for three given points, i.e.,

$$y_b = c_0$$

and

$$(y_a - y_b) = c_1(x_a - x_b) + c_2(x_a - x_b)^2 \quad ,$$
$$(y_c - y_b) = c_1(x_c - x_b) + c_2(x_c - x_b)^2 \quad .$$

From this we obtain

$$c_1 = C[(x_c - x_b)^2(y_a - y_b) - (x_a - x_b)^2(y_c - y_b)] \quad , \tag{10.2.3}$$

$$c_2 = C[-(x_c - x_b)(y_a - y_b) + (x_a - x_b)(y_c - y_b)] \tag{10.2.4}$$

with

$$C = \frac{1}{(x_a - x_b)(x_c - x_b)^2 - (x_c - x_b)(x_a - x_b)^2}$$

and for the extremum of the parabola

$$x_{\mathrm{mp}} = x_b - \frac{c_1}{2c_2} \quad . \tag{10.2.5}$$

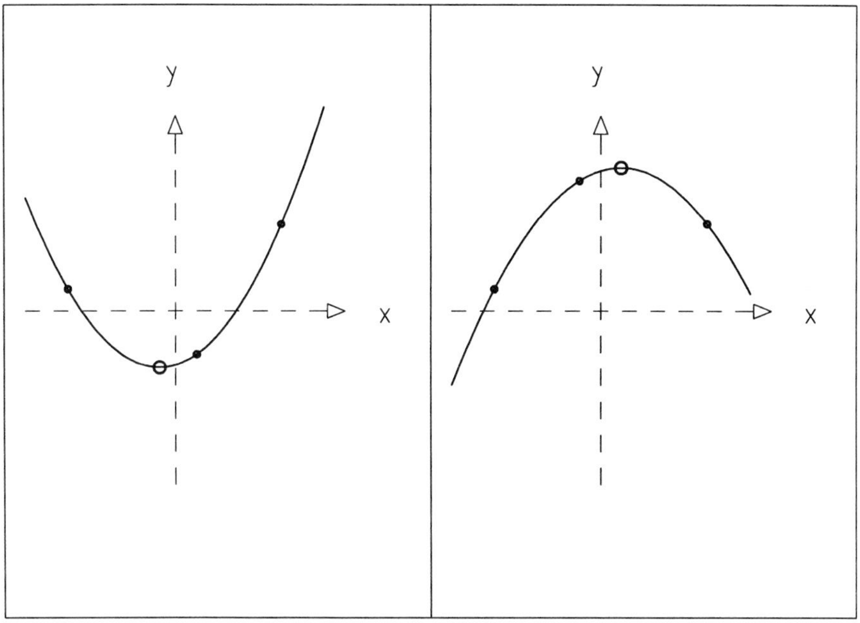

Fig. 10.2. Parabola through three points (small circles) and its extremum (large circle). In the left figure there is a minimum, and on the right a maximum.

Program 10.1: Subroutine MINPRB computes the extremum of a parabola through three points

CALLING SEQUENCE: CALL MINPRB(XMP,XA,YA,XB,YB,XC,YC)

INPUT ARGUMENTS: XA,YA — coordinates of the first point (x_a, y_a)
XB,YB — coordinates of the second point (x_b, y_b)
XC,YC — coordinates of the third point (x_c, y_c)

OUTPUT ARGUMENT: XMP — position of the extreme value $x_{\rm mp}$

```
      SUBROUTINE MINPRB(XMP,XA,YA,XB,YB,XC,YC)
      IMPLICIT DOUBLE PRECISION (A-H,O-Z)
      PARAMETER (EPS=1.D-10)
      XBA=XB-XA
      XBC=XB-XC
      F1=XBA*(YB-YC)
      F2=XBC*(YB-YA)
      ANUM=XBA*F1-XBC*F2
      DEN=F1-F2
      DEN=SIGN(MAX(ABS(DEN),EPS),DEN)
      XMP=XB-ANUM/(2.D0*DEN)
      END
```

We must still determine whether the extremum of the parabola is a minimum or a maximum (cf. Fig. 10.2). One has a minimum if the second derivative of (10.2.2) with respect to $x - x_b$ is positive, i.e., if $c_2 > 0$. We now order the three given points such that

$$x_a < x_b < x_c$$

and find that then

$$1/C = (x_c - x_b)(x_a - x_b)(x_c - x_a) = -(x_c - x_b)(x_b - x_a)(x_c - x_a) < 0 \quad .$$

With this one has for the sign of c_2

$$\operatorname{sign} c_2 = \operatorname{sign}[(x_c - x_b)(y_a - y_b) + (x_b - x_a)(y_c - y_b)] \quad .$$

Both expressions $(x_c - x_b)$ and $(x_b - x_a)$ are positive. Therefore for the extremum to be a minimum it is sufficient that

$$y_a > y_b \quad , \qquad y_c > y_b \quad . \tag{10.2.6}$$

The condition is not necessary, but has the advantage of greater clarity. In the interval $x_a < x < x_c$ one clearly has a minimum if there is a point (x_b, y_b) in the interval where the function has a value smaller than at the two end points. Clearly this statement is also valid if the function is not a parabola. We will make use of this fact in the next section.

10.3 Function of n Variables on a Line in an n-Dimensional Space

Locating the minimum of the function $M(x)$ of a single variable x in an interval of the x axis is equivalent to locating the minimum of a function $M(\mathbf{x})$ of an n-dimensional vector of variables $\mathbf{x} = (x_1, x_2, \dots, x_n)$ with respect to a given line in the n-dimensional space. If $\mathbf{x}_0$ is a fixed point and $\mathbf{d}$ is a fixed vector, then

$$\mathbf{x}_0 + a\mathbf{d} \quad , \qquad -\infty < a < \infty \quad , \tag{10.3.1}$$

describes a fixed line (see Fig. 10.3), and

$$f(a) = M(\mathbf{x}_0 + a\mathbf{d}) \tag{10.3.2}$$

is the value of the function at the point a on this line. For $n = 1$, $\mathbf{x}_0 = 0$, $\mathbf{d} = 1$ and with the change of notation $a = x$, that is, $f(x) = M(x)$, one recovers the original problem.

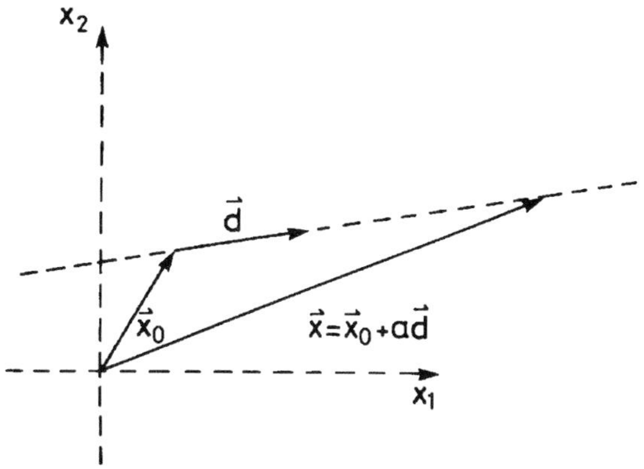

Fig. 10.3. The line given by (10.3.1) in two dimensions.

Program 10.2: Function MINFND computes the value of a function of n variables on a given line in an n-dimensional space

CALLING SEQUENCE: F=MINFND(A,X0,XDIR,N,USERFN)

INPUT ARGUMENTS: A — parameter value a

X0 — n-vector $\mathbf{x}_0$

XDIR — n-vector $\mathbf{d}$

N — number of variables n

USERFN — FORTRAN name of a function subprogram provided by the user which returns the function $M(\mathbf{x})$; must be declared in calling program as EXTERNAL

OUTPUT ARGUMENT: MINFND is the value of the function $M(\mathbf{x})$ at the point $\mathbf{x} = \mathbf{x}_0 + a\mathbf{d}$

REMARKS: The user must provide a function subprogram that returns the function $M(\mathbf{x})$. The name of the function must agree with the name of the last argument of the call of MINFND. The user function itself must have exactly two arguments: the n-vector X and the number of variables N.

```
      DOUBLE PRECISION  FUNCTION MINFND(A,X0,XDIR,N,USERFN)
      IMPLICIT DOUBLE PRECISION (A-H,O-Z)
      PARAMETER(MAXN=50)
      EXTERNAL USERFN
      DIMENSION X0(N),XDIR(N),XLINE(MAXN)
      DO 10 I=1,N
        XLINE(I)=X0(I)+A*XDIR(I)
   10 CONTINUE
      MINFND=USERFN(XLINE,N)
      END
```

In Sects. 10.4 through 10.6 we consider the minimum of a function of a single variable. The programs also treat, however, the case of a minimum of a function of n variables on a line in the n-dimensional space.

10.4 Bracketing the Minimum

For many minimization procedures it is important to know ahead of time that the minimum $x_{\rm m}$ is located in a specific interval,

$$x_a < x_{\rm m} < x_c \quad . \tag{10.4.1}$$

By systematically reducing the interval, the position of the minimum can then be further constrained until finally for a given precision ε one has

$$|x_a - x_c| < \varepsilon \quad . \tag{10.4.2}$$

There is, in fact, a minimum in the interval (10.4.1) if there is an x value x_b such that

$$M(x_b) < M(x_a) \quad , \qquad M(x_b) < M(x_c) \quad , \qquad x_a < x_b < x_c \quad . \tag{10.4.3}$$

The subroutine MINENC below attempts to bracket the minimum of a function by giving values x_a, x_b, x_c with the property (10.4.3). The program is based on a similar subroutine by PRESS et al. [12]. Starting from the input values x_a, x_b, which, if necessary, are relabeled so that $y_b \leq y_a$, a value $x_c = x_b + p(x_b - x_a)$ is computed that presumably lies in the direction of decreasing function value, i.e., closer to the minimum. The factor p in our program is set to $p = 1.618\,034$. In this way the original interval (x_a, x_b) is enlarged by the ratio of the golden section (cf. Sect. 10.5). The goal is reached if $y_c > y_b$. If this is not the case, a parabola is constructed with the subroutine MINPRB through the three points (x_a, y_a), (x_b, y_b), (x_c, y_c), whose minimum is at $x_{\rm m}$.

We now examine the point $(x_{\rm m}, y_{\rm m})$. Here one must distinguish between various cases:

a) $x_b < x_{\rm m} < x_c$:

- a1) $y_{\rm m} < y_c$: $(x_b, x_{\rm m}, x_c)$ is the desired interval.
- a2) $y_b < y_{\rm m}$: $(x_a, x_b, x_{\rm m})$ is the desired interval.
- a3) $y_{\rm m} > y_c$ and $y_{\rm m} < y_b$: There is no minimum. The interval will be extended further to the right.

b) $x_c < x_{\rm m} < x_{\rm end}$ and $x_{\rm end} = x_b + f(x_c - x_b)$ and $f = 10$ in our program.

b1) $y_m > y_c$: (x_b, x_c, x_m) is the desired interval.

b2) $y_m < y_c$: There is no minimum. The interval will be extended further to the right.

c) $x_{end} < x_m$: As a new interval (x_b, x_c, x_{end}) is used.

d) $x_m < x_b$: This result is actually impossible. It can, however, be caused by a rounding error. The interval will be extended further to the right.

If the goal is not reached in the current step, a further step is taken with the new interval.

Program 10.3: Subroutine MINENC brackets the minimum of a function of n variables on a given line in the n-dimensional space

CALLING SEQUENCE: CALL MINENC(XA,XB,XC,YA,YB,YC, USERFN,NSTEP,X0,XDIR,N)

INPUT ARGUMENTS: USERFN — name of a function subprogram for computing the function; cf. Sect. 10.3
N — number of variables n
X0 — n-vector $\mathbf{x}_0$; cf. (10.3.1)
XDIR — n-vector $\mathbf{d}$; cf. (10.3.1)
XA, XB — initial values x_a, x_b of the interval boundaries for the parameter x
NSTEP — maximum number of steps

OUTPUT ARGUMENTS: XA, XB, XC — parameter values x_a, x_b, x_c, which determine the interval containing the minimum
YA, YB, YC — corresponding function values y_a, y_b, y_c
NSTEP — number of steps needed. If the procedure does not succeed within the maximum allowed number of steps, then NSTEP = -1.

DEFAULT VALUES: If the input value of NSTEP is less than or equal to zero, 1000 steps are made.

```
      SUBROUTINE MINENC(XA,XB,XC,YA,YB,YC,USERFN,NSTEP,X0,XDIR,N)
      IMPLICIT DOUBLE PRECISION (A-H,O-Z)
      PARAMETER(ZERO=0.D0,GSPAR=1.618034D0,FACMAG=10.D0,
     +MAXSTP=1000)
      DIMENSION X0(N),XDIR(N)
      DOUBLE PRECISION MINFND
      EXTERNAL USERFN
      YA=MINFND(XA,X0,XDIR,N,USERFN)
      YB=MINFND(XB,X0,XDIR,N,USERFN)
      NST=NSTEP
      IF (NST.LE.0) NST=MAXSTP
```

```
      IF (YB.GT.YA) THEN
C exchange A and B
        BUF=XA
        XA=XB
        XB=BUF
        BUF=YA
        YA=YB
        YB=BUF
      END IF
      XC=XB+GSPAR*(XB-XA)
      YC=MINFND(XC,X0,XDIR,N,USERFN)
      ISTEP=0
   10 IF (YB.GE.YC .AND. ISTEP.LT.NST) THEN
        ISTEP=ISTEP+1
C step was still downwards
        XEND=XB+FACMAG*(XC-XB)
        CALL MINPRB(XM,XA,YA,XB,YB,XC,YC)
        YM=MINFND(XM,X0,XDIR,N,USERFN)
        IF ((XM-XB)*(XC-XM) .GT. ZERO) THEN
C XM is between XB and XC
          IF (YM.LT.YC) THEN
C minimum is between XB and XC
            XA=XB
            YA=YB
            XB=XM
            YB=YM
            GO TO 10
          ELSE IF (YM.GT.YB) THEN
C minimum is between XA and XM
            XC=XM
            YC=YM
            GO TO 10
          END IF
C there was no minimum, go on
          XM=XC+GSPAR*(XC-XB)
        ELSE IF ((XC-XM)*(XM-XEND) .GT. ZERO) THEN
C XM is between XC and XEND
          IF (YM.LT.YC) THEN
            XB=XC
            XC=XM
            XM=XC+GSPAR*(XC-XB)
            YB=YC
            YC=YM
          END IF
        ELSE IF ((XM-XEND)*(XEND-XC) .GE. ZERO) THEN
C XM is beyond XEND
          XM=XEND
        ELSE
          XM=XC+GSPAR*(XC-XB)
        END IF
        XA=XB
        XB=XC
        XC=XM
        YA=YB
        YB=YC
        YC=MINFND(XC,X0,XDIR,N,USERFN)
        GO TO 10
      END IF
      IF(ISTEP.EQ.NST) THEN
        NSTEP=-1
      ELSE
```

```
      NSTEP=ISTEP
   END IF
   END
```

Figure 10.4 shows an example of the individual steps carried out until the bracketing of the minimum is reached.

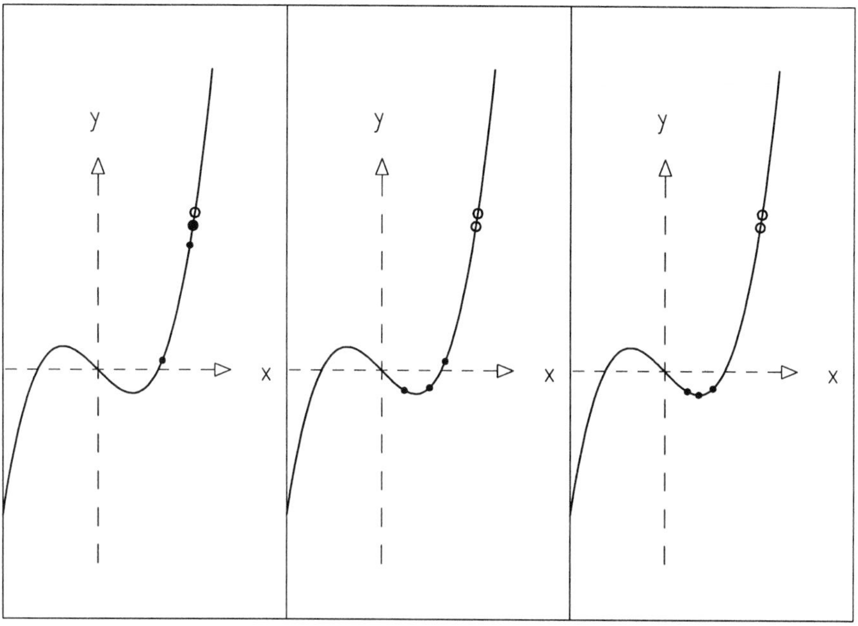

Fig. 10.4. Bracketing of a minimum with three points a, b, c according to (10.4.3). The initial values are shown as larger circles, and the results of the individual steps are shown as small circles.

10.5 Minimum Search with the Golden Section

As soon as the minimum has been enclosed by giving three points x_a, x_b, x_c with the property (10.4.3), the bracketing can easily be tightened further. One chooses a point x inside the larger of the two subintervals (x_a, x_b) and (x_b, x_c). If the value of the function at x is smaller than at x_b, then the subinterval containing x is taken as the new interval. If the value of the function is greater, then x is taken as the endpoint of the new interval.

A particularly clever division of the intervals is possible with the *golden section*. Let us assume (see Fig. 10.5) that

$$g = \frac{\ell}{L} \ , \qquad g > \frac{1}{2} \ , \tag{10.5.1}$$

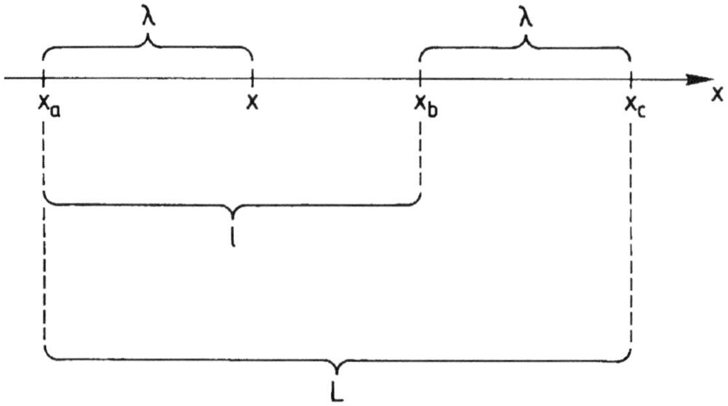

Fig. 10.5. The golden section.

is the length of the subinterval (x_a, x_b) (to be determined later) measured in units of the length of the full interval (x_a, x_c). We now want to be able to divide the subinterval (x_a, x_b) again with a point x corresponding to a fraction g,

$$g = \frac{\lambda}{\ell} \quad . \tag{10.5.2}$$

In addition, the points x and x_b should be situated symmetrically with respect to each other in the interval (x_a, x_c), i.e.,

$$\lambda = L - \ell \quad . \tag{10.5.3}$$

It follows that

$$\frac{\ell}{\lambda + \ell} = \frac{\lambda}{\ell} \quad ,$$

that is,

$$\lambda = \frac{\sqrt{5} - 1}{2} \ell$$

and

$$g = \frac{\sqrt{5} - 1}{2} \approx 0.618\,034 \quad . \tag{10.5.4}$$

As shown at the beginning of this section (for the case shown in Fig. 10.5 $x_b - x_a > x_c - x_b$), the minimum, which was originally only constrained to be in the interval (x_a, x_c), now lies either in the interval (x_a, x_b) or in the interval (x, x_c). By subdividing by the golden section one obtains intervals of equal size.

Program 10.4: Subroutine `MINGLD` determines the minimum of a function of n variables on a given line in the n-dimensional space according to the method of the golden section

CALLING SEQUENCE: CALL MINGLD(A,B,C,EPSILN,USERFN, XMIN,FMIN,NSTEP,X0,XDIR,N)

INPUT ARGUMENTS: A,B,C — parameter values x_a, x_b, x_c, specifying the interval containing the minimum
EPSILN — relative precision ε according to (10.1.21)
USERFN,NSTEP,X0,XDIR,N — as in Program 10.3

OUTPUT ARGUMENTS: XMIN — position of the minimum
FMIN — value of the function at the minimum
NSTEP — as in Program 10.3

DEFAULT VALUES: NSTEP is taken to be 1000 if a value ≤ 0 is given.
EPSILN is taken to be 10^{-8} if a value ≤ 0 is given.

```
      SUBROUTINE MINGLD(A,B,C,EPSILN,USERFN,XMIN,YMIN,NSTEP,X0,
     +XDIR,N)
      IMPLICIT DOUBLE PRECISION(A-H,O-Z)
      PARAMETER(Q=0.61803D0,MAXSTP=1000,EPSDEF=1.D-8,TT=1.D-15)
      EXTERNAL USERFN
      EPS=EPSILN
      IF(EPS.LE.0.)EPS=EPSDEF
      IF(NSTEP.LT.1)NSTEP=MAXSTP
      S=USERFN(B,X0,XDIR,N)
      ISTEP=0
   10 ISTEP=ISTEP+1
      IF(ABS(B-A).GT.ABS(C-B)) THEN
        X=A+Q*(B-A)
        Y=USERFN(X,X0,XDIR,N)
        IF(Y.LT.S) THEN
          C=B
          B=X
          S=Y
        ELSE
          A=X
        END IF
      ELSE
        X=B+Q*(C-B)
        Y=USERFN(X,X0,XDIR,N)
        IF(Y.LT.S) THEN
          A=B
          B=X
          S=Y
        ELSE
          C=X
        END IF
      END IF
      IF (ABS(C-A).GT.EPS*ABS(A)+TT .AND. ISTEP.LE.NSTEP) GO TO 10
      IF(ISTEP.GT.NSTEP) THEN
        NSTEP=-1
      ELSE
        NSTEP=ISTEP
      END IF
      XMIN=B
      YMIN=C
      END
```

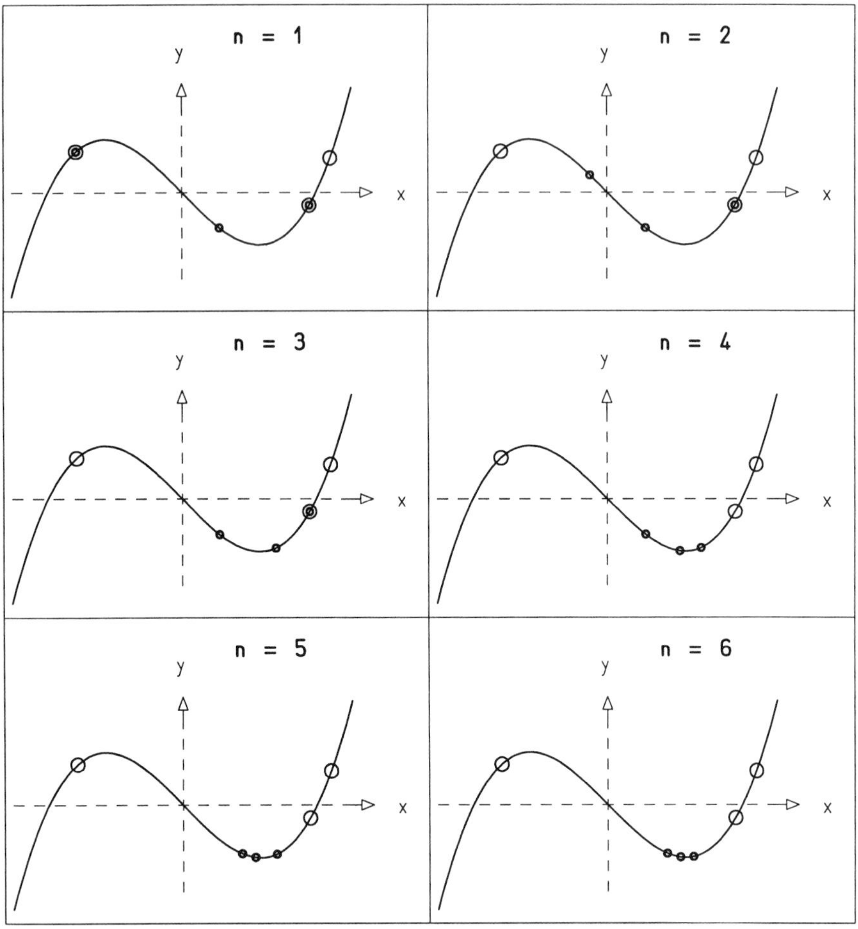

Fig. 10.6. Stepwise bracketing of a minimum by subdividing according to the golden section. The original interval (larger circles) is reduced with every step (small circles).

Figure 10.6 shows the first six steps of an example of minimization with the golden section.

10.6 Minimum Search with Quadratic Interpolation

From the example shown in Fig. 10.6 one sees that the procedure of interval subdivision is quite certain, but works slowly. We now combine it, therefore, with quadratic interpolation.

The program `MINCMB` given here is based on a program developed by Brent [13], who first combined the two methods. The meanings of the most important FORTRAN names in the program are as follows. A and B denote the x values x_a and x_b that contain the minimum. XM is their mean value. X is the point at which the function has its lowest value up to this step, W is the point with the second lowest, and V is the point with the third lowest value of the function. U is the point where the function was last computed. One begins with the two initial values, x_a and x_b, that contain the minimum and adds a point x, that divides the interval (x_a, x_b) according to the golden section. Then in each subsequent iteration parabolic interpolation is attempted as in Sect. 10.2. The result is accepted if it lies in the interval defined by the last step *and* if in this step the change in the minimum is less than half as much as in the previous one. By this condition it is ensured that the procedure converges, i.e., that the steps become smaller on the average, although a temporary increase in step size is tolerated. If both conditions are not fulfilled, a reduction of the interval is carried out according to the golden section.

Numerical questions are handled in a particularly careful way of Brent. Starting from the two parameters ε and t, which define the relative precision, and the current value x for the position of the minimum, an absolute precision $\Delta x = \varepsilon x + t$ = TOL is computed according to (10.1.21). The iterations are continued until the half of the interval width falls below the value TOL (i.e., until the distance from X to XM is not greater than TOL) or until the maximum number of steps is reached. In addition the function is not computed at points that are separated by a distance less than TOL, since such function values would not differ significantly.

Program 10.5: Subroutine MINCMB determines the minimum of a function of n variables on a line in the n-dimensional space according to the combined method by Brent

CALLING SEQUENCE: CALL MINCMB(A,B,EPSILN,USERFN,XMIN, FMIN,NSTEP,X0,XDIR,N)

INPUT ARGUMENTS: as in Program 10.4 (in contrast to MINGLD there is no input argument C)

OUTPUT ARGUMENTS: as in Program 10.4

DEFAULT VALUES: as in Program 10.4

```
      SUBROUTINE MINCMB(A,B,EPSILN,USERFN,XMIN,FMIN,NSTEP,
     +X0,XDIR,N)
      IMPLICIT DOUBLE PRECISION (A-H,O-Z)
      DOUBLE PRECISION MINFND
      LOGICAL PARSTP
      PARAMETER(C=0.381966D0,MAXSTP=1000,EPSDEF=1.D-12,TT=1.D-15)
      PARAMETER(ZERO=0.D0, HALF=0.5D0, ONE=1.D0, TWO=2.D0)
      DIMENSION X0(N),XDIR(N)
```

```
      EXTERNAL USERFN
C if some input values are not given use defaults
      IF(NSTEP.LT.1) NSTEP=MAXSTP
      EPS=EPSILN
      IF(EPS.LE.ZERO) EPS=EPSDEF
C initialize X at golden section position between A and B
      X=A+C*(B-A)
C initialize E,V,W,FX,FV,FW
      E=ZERO
      V=X
      W=X
      FX=MINFND(X,X0,XDIR,N,USERFN)
      FV=FX
      FW=FX
      ISTEP=1
C start iteration
   10 CONTINUE
      IF(ISTEP.GE.NSTEP) THEN
C prepare exit if step number is too large
        NSTEP=-1
        XMIN=X
        FMIN=FX
        GO TO 20
      END IF
      XM=HALF*(A+B)
      TOL=EPS*ABS(X)+TT
      T2=TWO*TOL
      IF(ABS(X-XM).GT.T2-HALF*(B-A)) THEN
        P=ZERO
        Q=ZERO
        R=ZERO
        PARSTP = .FALSE.
        IF(ABS(E).GT.TOL) THEN
C fit parabola
          R=(X-W)*(FX-FV)
          Q=(X-V)*(FX-FW)
          P=(X-V)*Q-(X-W)*R
          Q=TWO*(Q-R)
          IF(Q.GT.ZERO) THEN
            P=-P
          ELSE
            Q=-Q
          END IF
          R=E
          E=D
          IF(ABS(P).LT.ABS(HALF*Q*R) .AND. P.GT.Q*(A-X) .AND. P.LT.
     +    Q*(B-X)) THEN
C use result of parabolic fit
            D=P/Q
            U=X+D
            IF((U-A).LT.T2 .OR. (B-U).LT.T2) THEN
C make sure that U is not too near to A or B
              D=-TOL
              IF(X.LT.XM) D=TOL
            END IF
            PARSTP = .TRUE.
          END IF
        END IF
        IF(.NOT. PARSTP) THEN
C perform golden section step
          IF(X.LT.XM) THEN
```

```
            E=B-X
          ELSE
            E=A-X
          END IF
          D=C*E
        END IF
C determine point U where function is to be computed
C making sure that it is not too close to X
        IF(ABS(D) .GE. TOL) THEN
          U=X+D
        ELSE IF (D.GT.ZERO) THEN
          U=X+TOL
        ELSE
          U=X-TOL
        END IF
        FU=MINFND(U,X0,XDIR,N,USERFN)
C update A,B,V,W,X
        IF(FU.LE.FX) THEN
          IF(U.LT.X) THEN
            B=X
          ELSE
            A=X
          END IF
          V=W
          FV=FW
          W=X
          FW=FX
          X=U
          FX=FU
        ELSE
          IF(U.LT.X) THEN
            A=U
          ELSE
            B=U
          END IF
          IF(FU.LE.FW .OR. W.EQ.X) THEN
            V=W
            FV=FW
            W=U
            FW=FU
          ELSE IF(FU.LE.FV .OR. V.EQ.X .OR. V.EQ.W) THEN
            V=U
            FV=FU
          END IF
        END IF
        ISTEP=ISTEP+1
        GO TO 10
      ELSE
        XMIN=X
        FMIN=FX
        NSTEP=ISTEP
      END IF
   20 CONTINUE
      END
```

The first six steps of an example of minimization according to Brent are shown in Fig. 10.7. Steps according to the golden section are marked by *GS*, and those from quadratic interpolation with *QI*. The comparison with Fig. 10.6 shows the considerably faster convergence achieved by quadratic interpolation.

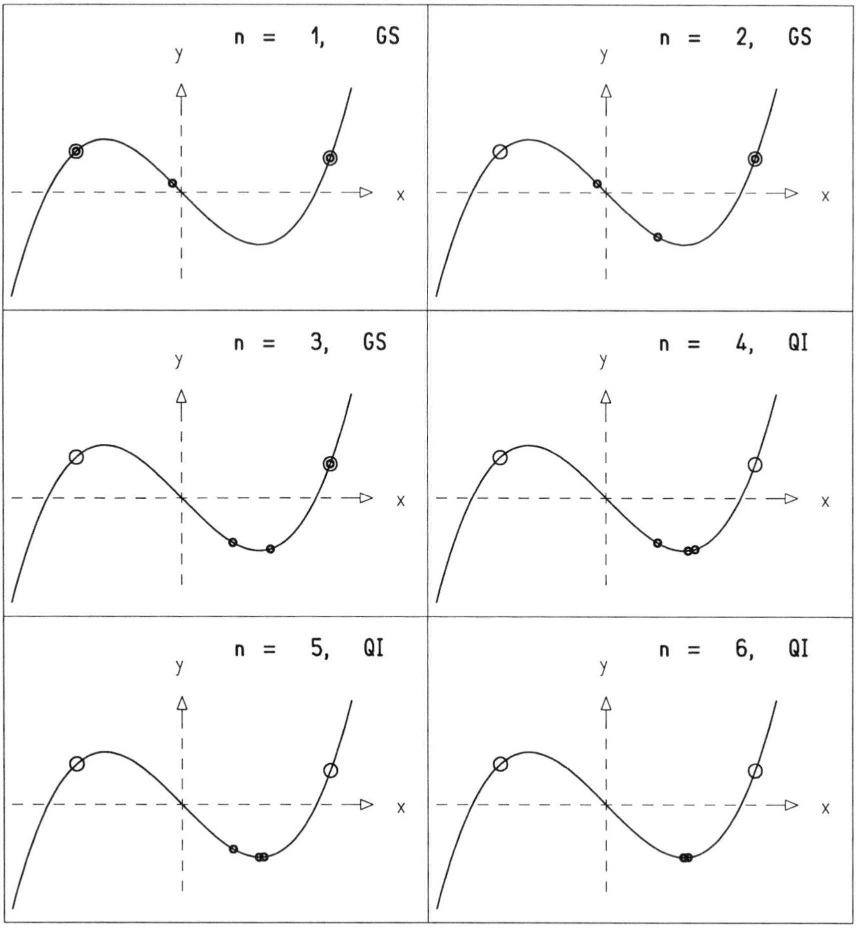

Fig. 10.7. Stepwise bracketing of a minimum with the combined method of Brent. The initial interval (large circles) is reduced with each step (small circles). Steps are carried out according to the golden section (GS) or by quadratic interpolation (QI).

10.7 Minimization along a Direction in n Dimensions

With the following program we compute the minimum of a function of n variables along the line defined in Sect. 10.3 by $\mathbf{x}_0$ and $\mathbf{d}$. We first use `MINENC` to bracket the minimum and then `MINCMB` to locate it exactly.

Program 10.6: Subroutine `MINDIR` determines the minimum of a function of n variables on a line in the n-dimensional space

CALLING SEQUENCE: `CALL MINDIR(X0,XDIR,N,USERFN,FMIN, NSTEP,EPSILN)`

INPUT ARGUMENTS: `X0,XDIR,N,USERFN` — as in Program 10.2
`NSTEP` — maximum number of steps
`EPSILN` — relative precision ε corresponding to (10.1.21)

OUTPUT ARGUMENTS: `X0` — n-vector of the position of the minimum
`FMIN` — value of the function at the minimum
`NSTEP` — number of steps needed (-1, if the procedure does not succeed)

DEFAULT VALUES: `NSTEP` is taken to be 1000 if a value ≤ 0 is given as input.
`EPSILN` is taken to be 10^{-8} if a value ≤ 0 is given as input.

```
      SUBROUTINE MINDIR(X0,DIR,N,USERFN,FMIN,NSTEP,EPSILN)
      IMPLICIT DOUBLE PRECISION (A-H,O-Z)
      PARAMETER(MAXSTP=1000,EPSDEF=1.D-8)
      PARAMETER(DELTAX=1.D-3,CUT=0.1D0,ZERO=0.D0)
      DIMENSION X0(N),DIR(N)
      EXTERNAL USERFN
C initialization
      NST=NSTEP
      IF(NST.LE.0) NST=MAXSTP
      EPS=EPSILN
      IF(EPS.LE.ZERO) EPS=EPSDEF
      XA=0.
      E=ABS(XA)
      IF(E.LT.CUT) E=CUT
      XB=E*DELTAX
      NSTIN=NST
C enclose minimum
      CALL MINENC(XA,XB,XC,YA,YB,YC,USERFN,NST,X0,DIR,N)
      IF(NST.LT.0) THEN
        NSTEP=-1
        GO TO 20
      END IF
      IF(XC.LT.XA)THEN
C reverse order of arguments
        DUM=XC
        XC=XA
        XA=DUM
      END IF
      NST=NSTIN
C locate minimum
      CALL MINCMB(XA,XC,EPS,USERFN,XMIN,FMIN,NST,X0,DIR,N)
      IF(NST.LT.0) THEN
        NSTEP=-1
        GO TO 20
      END IF
      DO 10 I=1,N
        X0(I)=X0(I)+XMIN*DIR(I)
   10 CONTINUE
   20 CONTINUE
      END
```

10.8 Simplex Minimization in n Dimensions

A simple and very elegant (although relatively slow) procedure for determining the minimum of a function of several variables is the simplex method by NELDER and MEAD [14]. The variables $x_1, x_2, \ldots, x_n$ define an n-dimensional space. A *simplex* is defined in this space by $n + 1$ points $\mathbf{x}_i$,

$$\mathbf{x}_i = (x_{1i}, x_{2i}, \ldots, x_{ni}) \quad . \tag{10.8.1}$$

A simplex in two dimensions is a triangle with the corner points $\mathbf{x}_1, \mathbf{x}_2, \mathbf{x}_3$.

We use y_i to designate the value of the function at $\mathbf{x}_i$ and use particular indices for labeling special points $\mathbf{x}_i$. At the point $\mathbf{x}_H$ the value of the function is highest, i.e., $y_H > y_i, i \neq H$. At the point $\mathbf{x}_h$ it is higher than at all other points except $\mathbf{x}_H$ ($y_h > y_i$, $i \neq H$, $i \neq h$) and at the point $\mathbf{x}_\ell$ it is lowest ($y_\ell < y_i, i \neq \ell$). The simplex is now changed step by step. In each substep one of four operations takes place, these being *reflection*, *stretching*, *flattening*, or *contraction* of the simplex (cf. Fig. 10.8).

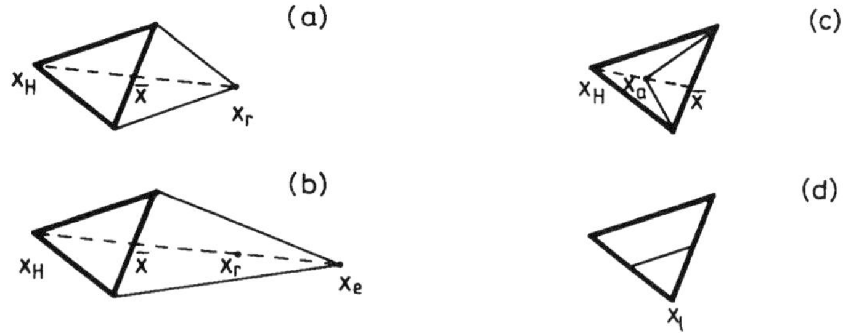

Fig. 10.8. Transformation of a simplex (triangle with thick border) into a new form (triangle with thin border) by means of a reflection (a), stretching (b), flattening (c), and contraction (d).

Using $\bar{\mathbf{x}}$ to designate the center-of-gravity of the (hyper)surface of the simplex opposite to $\mathbf{x}_H$,

$$\bar{\mathbf{x}} = \frac{1}{N-1} \sum_{i \neq H} \mathbf{x}_i \quad , \tag{10.8.2}$$

the *reflection* of $\mathbf{x}_H$ is

$$\mathbf{x}_r = (1 + \alpha)\bar{\mathbf{x}} - \alpha \mathbf{x}_H \tag{10.8.3}$$

with $\alpha > 0$ as the *coefficient of reflection*. The reflected simplex differs from the original one only in that $\mathbf{x}_H$ is replaced by $\mathbf{x}_r$.

A *stretching* of the simplex consists of replacing $\mathbf{x}_H$ by

$$\mathbf{x}_e = \gamma \mathbf{x}_r + (1 - \gamma)\bar{\mathbf{x}} \tag{10.8.4}$$

with the *coefficient of stretching* γ. One therefore chooses (for $\gamma > 1$) a point along the line joining $\mathbf{x}_H$ and $\mathbf{x}_r$, which is still further than the point $\mathbf{x}_r$.

In a *flattening*, $\mathbf{x}_H$ is replaced by a point lying on the line joining $\mathbf{x}_H$ and $\overline{\mathbf{x}}$. This point is located between the two points,

$$\mathbf{x}_a = \beta \mathbf{x}_H + (1 - \beta)\overline{\mathbf{x}} \quad . \tag{10.8.5}$$

The *coefficient of flattening* β is in the range $0 < \beta < 1$.

In the three operations discussed up to now only one point of the simplex is changed, that being the point $\mathbf{x}_H$, which corresponds to the highest value of the function. The point is displaced along the line by $\mathbf{x}_H$ and $\overline{\mathbf{x}}$. After the displacement it either lies on the same side of $\overline{\mathbf{x}}$ (flattening) or on the other side of $\overline{\mathbf{x}}$ (reflection) or even far beyond $\overline{\mathbf{x}}$ (stretching). In contrast to these operations, in a *contraction*, all points but one are replaced. The point $\mathbf{x}_\ell$ where the function has its lowest value is retained. All remaining points are moved to the midpoints of the line segments joining them with $\mathbf{x}_\ell$

$$\mathbf{x}_{ci} = (\mathbf{x}_i + \mathbf{x}_\ell)/2 \quad , \qquad i \neq \ell \quad . \tag{10.8.6}$$

For the original simplex and for each one created by an operation, the points $\overline{\mathbf{x}}$ and $\mathbf{x}_r$ and the corresponding function values $\overline{y}$ and y_r are computed. The next operation is determined as follows:

(a) If $y_r < y_\ell$, a stretching is attempted. If this gives $y_e < y_\ell$, then the stretching is carried out. Otherwise one performs a *reflection*.

(b) For $y_r > y_h$ a reflection is carried out if $y_r < y_H$. Otherwise the simplex is unchanged. In each case this is followed by a *flattening*. If one obtains as a result of the flattening a point $\mathbf{x}_a$ for which the function value is not less than both y_H and $\overline{y}$, the flattening is rejected, and instead a contraction is carried out.

After every step we examine the quantity

$$r = \frac{|\, y_H - y_\ell \,|}{|\, y_H \,| + |\, y_\ell \,|} \quad . \tag{10.8.7}$$

If this falls below a given value, the procedure is terminated and we regard $\mathbf{x}_\ell$ as the point where the function has its minimum.

Program 10.7: Subroutine `MINSIM` determines the minimum of a function of n variables according to the simplex method

CALLING SEQUENCE: `CALL MINSIM(X0,N,NRED,LIST,USERFN, FMIN,EPSILN,NSTEP,SCRAT)`

INPUT ARGUMENTS: `X0` — n-vector of a point $\mathbf{x}_0$ for the initial simplex
`N` — number of variables n
`NRED` — number of non-fixed variables n'
`LIST` — n-vector containing the elements of a list specifying which of the n variables are fixed (list element $= 0$) and which are adjustable (list element $= 1$)
`USERFN` — user function as in Program 10.2
`FMIN` — typical side lengths of the initial simplex
`EPSILN` — relative precision according to (10.1.21)
`NSTEP` — maximum number of steps
`SCRAT` — working array, minimum size $n \times (n+1)$

OUTPUT ARGUMENTS: `X0` — n-vector specifying the position of the minimum
`FMIN` — value of the function at the minimum
`NSTEP` — number of steps required: -1, if the operation was not successful

DEFAULT VALUES: `NSTEP` is taken to be 1000 if a value ≤ 0 is given as input.
`EPSILN` is taken to be 10^{-25} if a value ≤ 0 is given as input.
`FMIN` is taken to be 10^{-5} if a value ≤ 0 is given as input.

REMARKS: The initial simplex is constructed out of n vectors with length `FMIN`, which start from $\mathbf{x}_0$ and are parallel to the coordinate directions.

```
      SUBROUTINE MINSIM(X0,N,NRED,LIST,USERFN,FMIN,EPSILN,NSTEP,X)
      IMPLICIT DOUBLE PRECISION (A-H,O-Z)
      PARAMETER(MAXN=50)
      DIMENSION X0(N),X(NRED,NRED+1)
      DIMENSION Y(MAXN+1),XEXP(MAXN),XR(MAXN),XR2(MAXN),
     +XPRIME(MAXN),X1(MAXN),X2(MAXN),XMX(MAXN),XMN(MAXN)
      DIMENSION LIST(N)
      PARAMETER (EPSDEF=1.D-25,TT=1.D-30,MAXSTP=1000)
      PARAMETER (ALPHA=1.D0,BETA=.5D0,GAMMA=2.D0)
      PARAMETER (ONE=1.D0,HALF=.5D0,ZERO=0.D0)
      EXTERNAL USERFN
C initialization
      EPS=EPSILN
      IF(EPS.LE.ZERO) EPS=EPSDEF
      NST=NSTEP
      IF(NST.LE.0) NST=MAXSTP
      M=NRED+1
C construct initial simplex and compute function at its corners
      Y(1)=USERFN(X0,N)
      CALL MTXGSV(X0,X1,N,NRED,LIST)
      CALL MTXPCL(X,X1,NRED,M,1)
      D=FMIN
      IF(D.LE.ZERO) D=1.D-5
      IRED=0
      DO 10 I=1,N
```

```
        IF(LIST(I).EQ.1) THEN
          IRED=IRED+1
          CALL MTXCPV(X1,X2,NRED)
          X2(IRED)=X2(IRED)+D
          CALL MTXPCL(X,X2,NRED,M,IRED+1)
          CALL MTXCPV(X0,XEXP,N)
          CALL MTXPSV(XEXP,X2,N,NRED,LIST)
          Y(IRED+1)=USERFN(XEXP,N)
        END IF
   10 CONTINUE
      ISTEP=0
C begin iteration
   20 ISTEP=ISTEP+1
C compute indices of corner with largest (MX), next to largest (NE)
C and smallest (MN) function value
      MN=1
      IF(Y(1).GT.Y(2)) THEN
        MX=1
        NE=2
      ELSE
        MX=2
        NE=1
      END IF
      DO 30 I=1,NRED+1
        IF(Y(I).LT.Y(MN)) MN=I
        IF(Y(I).GT.Y(MX)) THEN
          NE=MX
          MX=I
        ELSE IF (Y(I).GT.Y(NE)) THEN
          IF(I.NE.MX) NE=I
        END IF
   30 CONTINUE
C test for break-off condition
      IF(ABS(Y(MX)-Y(MN)).LT.EPS*ABS(Y(MN))+TT .OR. ISTEP.GT.NST)
     +GO TO 60
C compute XPRIME, i.e. point in center of gravity of hyperplane
C opposite to point with largest function value
      CALL MTXGCL(X,XMX,NRED,M,MX)
      CALL MTXGCL(X,XMN,NRED,M,MN)
      CALL MTXZRV(XPRIME,NRED)
      DO 40 I=1,NRED+1
        IF(I.NE.MX) THEN
          CALL MTXGCL(X,X1,NRED,M,I)
          CALL MTXADV(XPRIME,X1,XPRIME,NRED)
        END IF
   40 CONTINUE
      F=ONE/DBLE(NRED)
      CALL MTXMSV(XPRIME,XPRIME,F,NRED)
C construct points by reflection (XR) and extension (XR2)
      CALL MTXMSV(XPRIME,X1,ONE+ALPHA,NRED)
      CALL MTXMSV(XMX,X2,ALPHA,NRED)
      CALL MTXSBV(X1,X2,XR,NRED)
      CALL MTXCPV(X0,XEXP,N)
      CALL MTXPSV(XEXP,XR,N,NRED,LIST)
      YR=USERFN(XEXP,N)
      IF(YR.LE.Y(MN)) THEN
        CALL MTXMSV(XR,X1,GAMMA,NRED)
        CALL MTXMSV(XPRIME,X2,ONE-GAMMA,NRED)
        CALL MTXADV(X1,X2,XR2,NRED)
        CALL MTXCPV(X0,XEXP,N)
        CALL MTXPSV(XEXP,XR2,N,NRED,LIST)
```

```
         YR2=USERFN(XEXP,N)
         IF(YR2.LT.Y(MN)) THEN
C perform extension
           CALL MTXPCL(X,XR2,NRED,M,MX)
           Y(MX)=YR2
         ELSE
C perform reflection
           CALL MTXPCL(X,XR,NRED,M,MX)
           Y(MX)=YR
         END IF
       ELSE IF(YR.GE.Y(NE)) THEN
         IF(YR.LT.Y(MX)) THEN
C perform reflection
           CALL MTXPCL(X,XR,NRED,M,MX)
           Y(MX)=YR
         END IF
         CALL MTXMSV(XMX,X1,BETA,NRED)
         CALL MTXMSV(XPRIME,X2,ONE-BETA,NRED)
         CALL MTXADV(X1,X2,XR2,NRED)
         CALL MTXCPV(X0,XEXP,N)
         CALL MTXPSV(XEXP,XR2,N,NRED,LIST)
         YR2=USERFN(XEXP,N)
         IF(YR2.LT.Y(MX)) THEN
C perform compression
           CALL MTXPCL(X,XR2,NRED,M,MX)
           Y(MX)=YR2
         ELSE
C perform contraction
           DO 50 I=1,NRED+1
             IF(I.NE.MN) THEN
               CALL MTXGCL(X,X1,NRED,M,I)
               CALL MTXADV(X1,XMN,X2,NRED)
               CALL MTXMSV(X2,X2,HALF,NRED)
               CALL MTXPCL(X,X2,NRED,M,I)
               CALL MTXCPV(X0,XEXP,N)
               CALL MTXPSV(XEXP,X2,N,NRED,LIST)
               Y(I)=USERFN(XEXP,N)
             END IF
   50      CONTINUE
         END IF
       ELSE
C perform reflection
         CALL MTXPCL(X,XR,NRED,M,MX)
         Y(MX)=YR
       END IF
       GO TO 20
C Fill output arguments
   60 ISTEP=ISTEP-1
       FMIN=Y(MN)
       CALL MTXGCL(X,X1,NRED,M,MN)
       CALL MTXPSV(X0,X1,N,NRED,LIST)
       IF(ISTEP.GT.NST) THEN
         NSTEP=-1
       ELSE
         NSTEP=ISTEP
       END IF
       END
```

The program is illustrated in the example of Fig. 10.9. The triangle with the dark border is the initial simplex. The sequence of triangles with thin

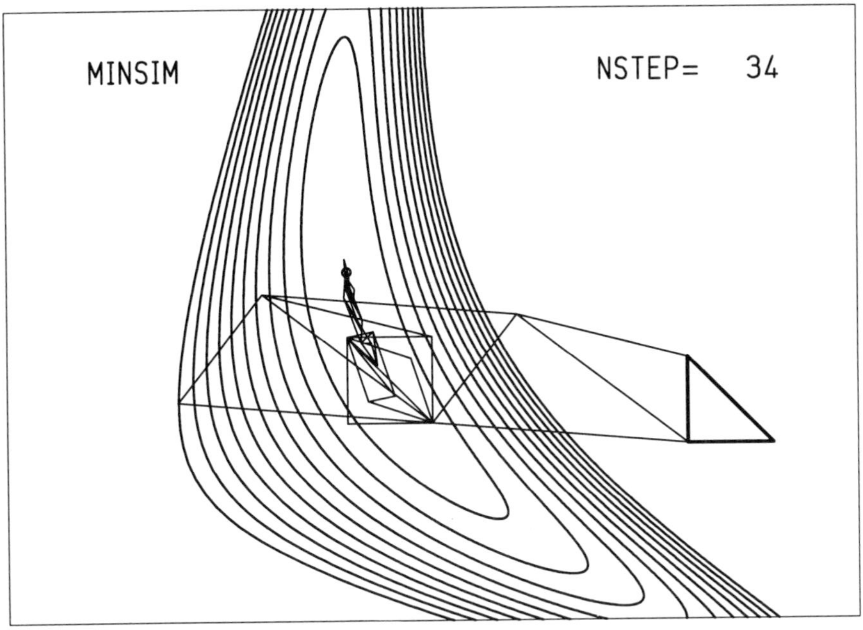

Fig. 10.9. Determining the minimum of a function of two variables with the simplex method. The function is shown by contour lines on which the function is constant. The function is highest on the outermost contour. The minimum is at the position of the small circle within the innermost contour. Each simplex is a triangle. The initial simplex is marked by thicker lines.

borders starting from it correspond to the individual transformations. One clearly recognizes as the first steps: stretching, stretching, reflection, reflection, flattening . . .. The simplex first finds its way into the "valley" of the function and then runs along the bottom of the valley towards the minimum. It is reshaped in the process so that it is longest in the direction in which it is progressing, and in this way it can also pass through narrow valleys. It is worth remarking that this method, which is obtained by considering the problem in two and three dimensions, also works in n dimensions and also in this case possesses a certain descriptiveness.

10.9 Minimization along the Coordinate Directions

Some of the methods for searching for a minimum in an n-dimensional space are based on the following principle. Starting from a point $\mathbf{x}_0$ one searches for the minimum along a given direction in the space. Next one minimizes from there along another direction and finds a new minimum, etc. Within this general framework various *strategies* can be developed to choose the individual directions.

The simplest strategy consists in the choice of the coordinate directions in the space of the n variables x_i. We can label the corresponding basis vectors with $\mathbf{e}_1, \mathbf{e}_2, \ldots, \mathbf{e}_n$, and they are then chosen in order as directions. After $\mathbf{e}_n$ one begins again with $\mathbf{e}_1, \mathbf{e}_2, \ldots$. A partial sequence of minimizations along all coordinate directions starting from $\mathbf{x}_0$ gives the point $\mathbf{x}_1$. After a new partial sequence one obtains $\mathbf{x}_2$, etc.

The procedure is ended successfully when for the values of the function $M(\mathbf{x}_n)$ and $M(\mathbf{x}_{n-1})$ for two successive steps one has

$$M(\mathbf{x}_{n-1}) - M(\mathbf{x}_n) < \varepsilon|M(\mathbf{x}_n)| + t \quad , \tag{10.9.1}$$

i.e., a condition corresponding to (10.1.21) for given values of ε and t. We compare here, however, the value of the function M and not the independent variables $\mathbf{x}$. Otherwise we would have to compute the distance between two points in an n-dimensional space.

Figure 10.10 shows the minimization of the same function as in Fig. 10.9 with the coordinate-direction method. After the first comparatively large step, which leads into the "valley" of the function, the following steps are quite small. The individual directions are naturally perpendicular to each other. The "best" point at each step moves along a staircase-like path along the floor of the valley towards the minimum.

10.10 Conjugate Directions

The slow convergence seen in Fig. 10.10 clearly stems from the fact that when minimizing along one direction, one loses the result of the minimization with respect to another direction. We will now try to choose the directions in such a way that this does not happen. For this we suppose for simplicity that the function is a quadratic form (10.1.10). Its gradient at the point $\mathbf{x}$ is then given by (10.1.12),

$$\nabla M = -\mathbf{b} + A\mathbf{x} \quad . \tag{10.10.1}$$

The change of the gradient in moving by $\Delta\mathbf{x}$ is

$$\Delta(\nabla M) = \nabla M(\mathbf{x}+\Delta\mathbf{x}) - \nabla M(\mathbf{x}) = A\mathbf{x} + A\Delta\mathbf{x} - A\mathbf{x} = A\Delta\mathbf{x} \ . \tag{10.10.2}$$

This expression is a vector that gives the direction of change of the gradient when the argument is moved in the direction $\Delta\mathbf{x}$.

If a minimization has been carried out along a direction $\mathbf{p}$, then the reduction in M with respect to $\mathbf{p}$ is retained when one moves in a direction $\mathbf{q}$ such that the gradient only changes perpendicular to $\mathbf{p}$, i.e., when one has

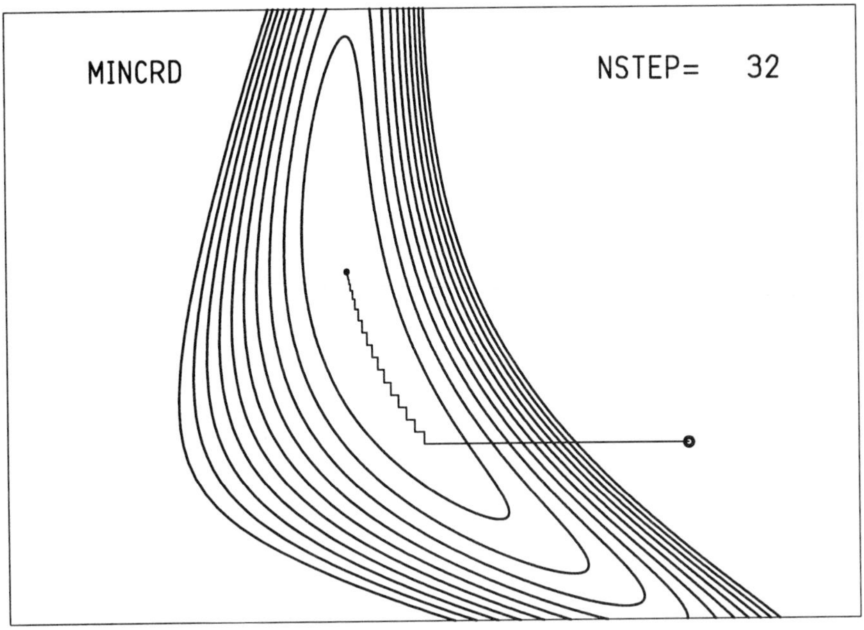

Fig. 10.10. Minimization along the coordinate directions. The starting point is shown by the larger circle, and the end by the smaller circle. The line shows the results of the individual steps.

$$\mathbf{p} \cdot (A\mathbf{q}) = \mathbf{p}^{\mathrm{T}} A\mathbf{q} = 0 \quad . \tag{10.10.3}$$

The vectors $\mathbf{p}$ and $\mathbf{q}$ are said to be *conjugate* to each other with respect to the positive-definite matrix A. If one has n variables, i.e., if A is an $n \times n$ matrix, then one can find in general n conjugate linearly independent vectors.

POWELL [15] has given a method to find a set of conjugate directions for a function described by a quadratic form. For this one chooses as a first set of directions n linearly independent unit vectors $\mathbf{p}_i$, e.g., the coordinate directions $\mathbf{p}_i = \mathbf{e}_i$. Starting from a point $\mathbf{x}_0$ one finds successively the minima in the directions $\mathbf{p}_i$. The results can be labeled by

$$\begin{aligned}
\mathbf{a}_1 &= \alpha_1\mathbf{p}_1 + \alpha_2\mathbf{p}_2 + \cdots + \alpha_n\mathbf{p}_n \quad , \\
\mathbf{a}_2 &= \qquad\quad \alpha_2\mathbf{p}_2 + \cdots + \alpha_n\mathbf{p}_n \quad , \\
&\vdots \\
\mathbf{a}_n &= \qquad\qquad\qquad\quad \alpha_n\mathbf{p}_n \quad .
\end{aligned}$$

Here $\mathbf{a}_1$ is the vector representing all n substeps, $\mathbf{a}_2$ contains all of the steps except the first one, and $\mathbf{a}_n$ is just the last substep. The sum of the n substeps leads then from the point $\mathbf{x}_0$ to

$$\mathbf{x}_1 = \mathbf{x}_0 + \mathbf{a}_1 \ .$$

The direction $\mathbf{a}_1$ describes the average direction of the first n substeps. Therefore we carry out a step in the direction $\mathbf{a}_1$, call the result again $\mathbf{x}_0$, determine a new set of directions

$$\begin{aligned} \mathbf{q}_1 &= \mathbf{p}_2 \quad , \\ \mathbf{q}_2 &= \mathbf{p}_3 \quad , \\ &\vdots \\ \mathbf{q}_{n-1} &= \mathbf{p}_n \quad , \\ \mathbf{q}_n &= \mathbf{a}_1/|\mathbf{a}_1| \quad , \end{aligned}$$

then call these $\mathbf{q}_i$ again $\mathbf{p}_i$ and proceed as above. As was shown by POWELL [15], the directions after n steps, i.e., $n(n+1)$ individual minimizations, are individually conjugate to each other if the function is a quadratic form.

10.11 Minimization along Chosen Directions

The procedure given at the end of the last section contains, however, the danger that the directions $\mathbf{p}_1, \ldots, \mathbf{p}_n$ can become almost linearly dependent, because at every step $\mathbf{p}_1$ is rejected in favor of $\mathbf{a}_1/|\mathbf{a}_1|$, and these directions need not be very different from step to step. Powell has therefore suggested not to replace the direction $\mathbf{p}_1$ by $\mathbf{a}_1/|\mathbf{a}_1|$, but rather to replace the direction $\mathbf{p}_{\max}$, along which the greatest reduction of the function took place. This sounds at first paradoxical, since what is clearly the best direction is replaced by another. But since out of all the directions $\mathbf{p}_{\max}$ gave the largest contribution towards reducing the function, $\mathbf{a}_1/|\mathbf{a}_1|$ has a significant component in the direction $\mathbf{p}_{\max}$. By retaining these two similar directions one would increase the danger of a linear dependence.

In some cases, however, after completing a step we will retain the old directions without change. We denote by $\mathbf{x}_0$ the point before carrying out the step in progress, by $\mathbf{x}_1$ the point obtained by this step, by

$$\mathbf{x}_\mathrm{e} = \mathbf{x}_1 + (\mathbf{x}_1 - \mathbf{x}_0) = 2\mathbf{x}_1 - \mathbf{x}_0 \tag{10.11.1}$$

an extrapolated point that lies in the new direction $\mathbf{x}_1 - \mathbf{x}_0$ from $\mathbf{x}_0$ but is still further than $\mathbf{x}_1$, and by M_0, M_1, and M_e the corresponding function values. If

$$M_\mathrm{e} \geq M_0 \quad , \tag{10.11.2}$$

then the function no longer decreases significantly in the direction $(\mathbf{x}_0 - \mathbf{x}_1)$. We then remain with the previous directions.

Further we denote by ΔM the greatest change in M along a direction in the step in progress and compute the quantity

$$T = 2(M_0 - 2M_1 + M_e)(M_0 - M_1 - \Delta M)^2 - (M_0 - M_e)^2 \Delta M \quad . \quad (10.11.3)$$

If

$$T \geq 0 \quad ,$$

then we retain the old directions. This requirement is satisfied if either the first or second factor in the first term of (10.11.3) becomes large. The first factor $(M_0 - 2M_1 + M_e)$ is proportional to a second derivative of the function M. If it is large (compared to the first derivative in meaningful units), then we are already close to the minimum. The second factor $(M_0 - M_1 - \Delta M)^2$ is large when the reduction of the function $M_0 - M_1$, with the contribution ΔM, does not stem mainly from a single direction.

Program 10.8: Subroutine `MINPOW` determines the minimum of a function of n variables by successive minimization along a chosen direction with the method of Powell

CALLING SEQUENCE: `CALL MINPOW(X0,N,NRED,LIST,USERFN, FMIN,EPSILN,NSTEP,SCRAT)`

INPUT ARGUMENTS: `X0` — n-vector of a starting point
`N` — number of variables n
`NRED,LIST` — as in Program 10.7
`USERFN` — user function as in Program 10.2
`EPSILN` — relative accuracy according to (10.1.21)
`NSTEP` — maximum number of steps
`SCRAT` — working array, minimum size $n \times n$

OUTPUT ARGUMENTS: `X0` — n-vector giving the position of the minimum
`FMIN` — function value at the minimum
`NSTEP` — number of steps needed: -1, if procedure fails

DEFAULT VALUES: `NSTEP` is taken to be 1000 if a value ≤ 0 is given.
`EPSILN` is taken to be 10^{-8} if a value ≤ 0 is given.

REMARKS: The input number of steps `NSTEP` refers both to the maximum length of the sequence $\mathbf{x}_0, \mathbf{x}_1, \ldots$ and to the maximum number of steps for minimization in a given direction. The output number of steps refers to the length of the sequence.

```
      SUBROUTINE MINPOW(X0,N,NRED,LIST,USERFN,FMIN,EPSILN,
     +NSTEP,ALLDIR)
      PARAMETER(MAXN=50)
      IMPLICIT DOUBLE PRECISION (A-H,O-Z)
      DIMENSION X0(N),XOLD(MAXN),XE(MAXN),DIR(MAXN),DIRN(MAXN)
      DIMENSION LIST(N),ALLDIR(N,NRED)
      EXTERNAL USERFN
      PARAMETER(MAXSTP=1000,EPSDEF=1.D-8,T=1.D-15,ONE=1.D0,
     +ZERO=0.D0)
      IF(NSTEP.LT.1) NSTEP=MAXSTP
      EPS=EPSILN
      IF(EPS.LE.0.) EPS=EPSDEF
C Initialize directions along coordinates
      F=USERFN(X0,N)
      IRED=0
      DO 10 I=1,N
        IF(LIST(I).EQ.1) THEN
          IRED=IRED+1
          CALL MTXZRV(DIR,N)
          DIR(I)=ONE
          CALL MTXPCL(ALLDIR,DIR,N,NRED,IRED)
        END IF
   10 CONTINUE
      DO 30 ISTEP=1,NSTEP
        LSTEP=ISTEP
        FOLD=F
        FL=FOLD
        CALL MTXCPV(X0,XOLD,N)
        DELTAF=0.
        IMAX=0
C Loop over complete direction set and find direction which
C yields largest function decrease. Its index is IMAX
        DO 20 I=1,NRED
          CALL MTXGCL(ALLDIR,DIR,N,NRED,I)
          NSTDIR=NSTEP
          CALL MINDIR(X0,DIR,N,USERFN,F,NSTDIR,ZERO)
          IF(NSTDIR.LT.0) GO TO 40
          D=ABS(F-FL)
          IF(D.GT.DELTAF) THEN
            DELTAF=D
            IMAX=I
          END IF
          FL=F
   20   CONTINUE
C Test for break-off criterion
        IF(ABS(FL-FOLD) .LT. EPS*ABS(FL)+T) GO TO 50
C Construct extrapolated point XE and direction DIRN from
C XOLD to X0
        CALL MTXSBV(X0,XOLD,DIRN,N)
        CALL MTXADV(DIRN,X0,XE,N)
        FE=USERFN(XE,N)
C Now there are 3 points (XOLD, X0, XE)
C and their function values (FOLD, F, FE)
        IF(FE.LT.F) THEN
          TEST=2.*(FOLD-2.*F+FE)*(FOLD-F-DELTAF)**2-
     +           (FOLD-FE)**2*DELTAF
          IF(TEST.LT.0.) THEN
            NST=NSTEP
C Find minimum along DIRN and replace X0 by position
C of minimum. Replace direction with index IMAX by DIRN
            CALL MINDIR(X0,DIRN,N,USERFN,F,NST,ZERO)
```

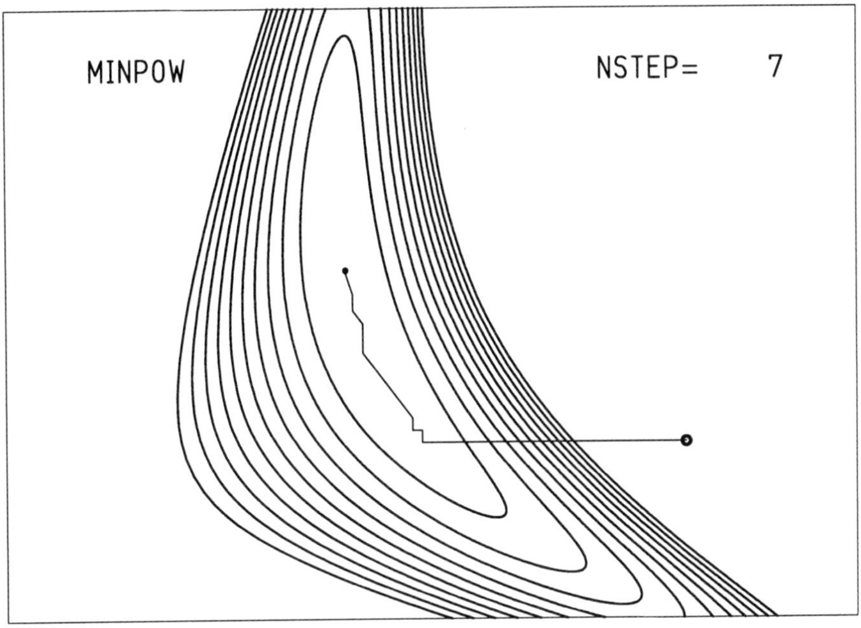

Fig. 10.11. Minimization along a chosen direction with the method of Powell.

```
            IF(NST.LT.0) GO TO 40
            CALL MTXPCL(ALLDIR,DIRN,N,NRED,IMAX)
          END IF
        END IF
   30 CONTINUE
   40 NSTEP=-1
      GO TO 60
   50 NSTEP=LSTEP
      FMIN=FL
   60 END
```

Figure 10.11 shows the advantages of Powell's method. One can see a significantly faster convergence to the minimum compared to that of Fig. 10.10.

10.12 Minimization in the Direction of Steepest Descent

In order to get from a point $\mathbf{x}_0$ to the minimum of a function $M(\mathbf{x})$, it is sufficient to always follow the negative gradient $\mathbf{b}(\mathbf{x}) = -\nabla M(\mathbf{x})$. It is thus obvious that one should look for the minimum along the direction $\nabla M(\mathbf{x}_0)$. One calls this point $\mathbf{x}_1$, searches then for the minimum along $\nabla M(\mathbf{x}_1)$, and so forth, until the termination requirement (10.9.1) is satisfied.

A comparison of the steps taken with this method, shown in Fig. 10.12, with those from minimization along a coordinate direction (Fig. 10.10) shows,

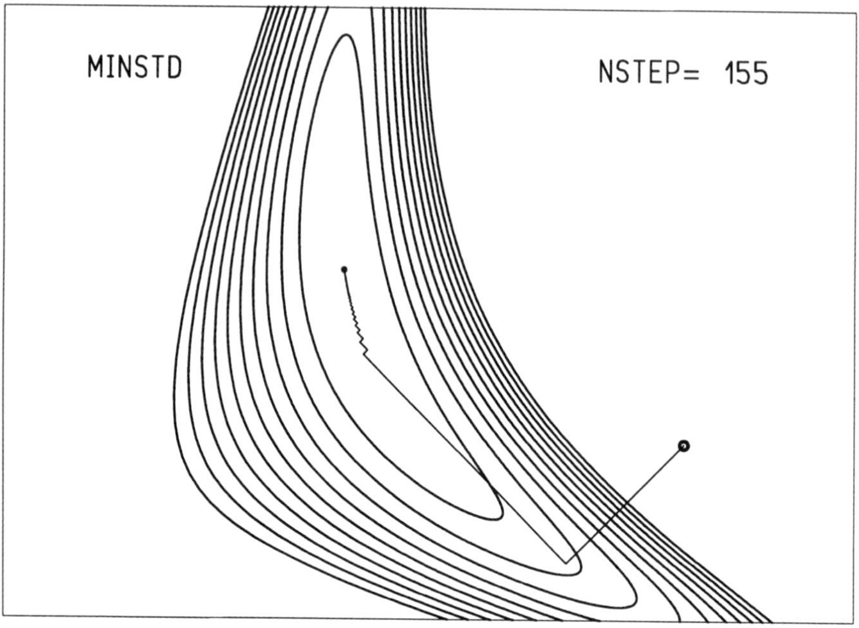

Fig. 10.12. Minimization in the direction of steepest descent.

however, a surprising similarity. In both cases, successive steps are perpendicular to each other. Thus the directions cannot be fitted to the function in the course of the procedure and convergence is slow.

The initially surprising fact that successive gradient directions are perpendicular to each other stems from the construction of the procedure. Searching from $\mathbf{x}_1$ for the minimum in the direction $\mathbf{b}_0$ means that the derivative in the direction $\mathbf{b}_0$ vanishes at the point $\mathbf{x}_1$,

$$\mathbf{b}_0 \cdot \boldsymbol{\nabla} M(\mathbf{x}_1) = 0 \quad ,$$

and thus the gradient is perpendicular to $\mathbf{b}_0$.

10.13 Minimization along Conjugate Gradient Directions

We take up again the idea of conjugate directions from Sect. 10.10. We construct, starting from an arbitrary vector $\mathbf{g}_1 = \mathbf{h}_1$, two sequences of vectors,

$$\mathbf{g}_{i+1} = \mathbf{g}_i - \lambda_i A \mathbf{h}_i \quad , \qquad i = 1, 2, \ldots \quad , \tag{10.13.1}$$

$$\mathbf{h}_{i+1} = \mathbf{g}_{i+1} + \gamma_i \mathbf{h}_i \quad , \qquad i = 1, 2, \ldots \quad , \tag{10.13.2}$$

with

$$\lambda_i = \frac{\mathbf{g}_i^{\mathrm{T}} \mathbf{g}_i}{\mathbf{g}_i^{\mathrm{T}} A \mathbf{h}_i} \quad , \qquad \gamma_i = \frac{\mathbf{g}_{i+1}^{\mathrm{T}} A \mathbf{h}_i}{\mathbf{h}_i^{\mathrm{T}} A \mathbf{h}_i} \quad . \tag{10.13.3}$$

Thus one has

$$\mathbf{g}_{i+1}^{\mathrm{T}} \mathbf{g}_i = 0 \quad , \tag{10.13.4}$$

$$\mathbf{h}_{i+1}^{\mathrm{T}} A \mathbf{h}_i = 0 \quad . \tag{10.13.5}$$

This means that successive vectors **g** are orthogonal, and successive vectors **h** are conjugate to each other.

Rewriting the relation (10.13.3) gives

$$\gamma_i = \frac{\mathbf{g}_{i+1}^{\mathrm{T}} \mathbf{g}_{i+1}}{\mathbf{g}_i^{\mathrm{T}} \mathbf{g}_i} = \frac{(\mathbf{g}_{i+1} - \mathbf{g}_i)^{\mathrm{T}} \mathbf{g}_{i+1}}{\mathbf{g}_i^{\mathrm{T}} \mathbf{g}_i} \quad , \tag{10.13.6}$$

$$\lambda_i = \frac{\mathbf{g}_i^{\mathrm{T}} \mathbf{h}_i}{\mathbf{h}_i^{\mathrm{T}} A \mathbf{h}_i} \quad . \tag{10.13.7}$$

We now try to construct the vectors $\mathbf{g}_i$ and $\mathbf{h}_i$ without explicit knowledge of the Hessian matrix A. To do this we again assume that the function to be minimized is a quadratic form (10.1.10). At a point $\mathbf{x}_i$ we define the vector $\mathbf{g}_i = -\nabla M(\mathbf{x}_i)$. If we now search for the minimum starting from $\mathbf{x}_i$ along the direction $\mathbf{h}_i$, we find it at $\mathbf{x}_{i+1}$ and construct there $\mathbf{g}_{i+1} = -\nabla M(\mathbf{x}_{i+1})$. Then $\mathbf{g}_{i+1}$ and $\mathbf{g}_i$ are orthogonal, since from (10.1.12) one has

$$\mathbf{g}_i = -\nabla M(\mathbf{x}_i) = \mathbf{b} - A\mathbf{x}_i$$

and

$$\mathbf{g}_{i+1} = -\nabla M(\mathbf{x}_{i+1}) = b - A(\mathbf{x}_i + \lambda_i \mathbf{h}_i) = \mathbf{g}_i - \lambda_i A \mathbf{h}_i \quad . \tag{10.13.8}$$

Here λ_i has been chosen such that $\mathbf{x}_{i+1}$ is the minimum along the direction $\mathbf{h}_i$. This means that there the gradient is perpendicular to $\mathbf{h}_i$,

$$\mathbf{h}_i^{\mathrm{T}} \nabla M(\mathbf{x}_{i+1}) = -\mathbf{h}_i^{\mathrm{T}} \mathbf{g}_{i+1} = 0 \quad . \tag{10.13.9}$$

Substituting into (10.13.8) gives indeed

$$0 = \mathbf{h}_i^{\mathrm{T}} \mathbf{g}_{i+1} = \mathbf{h}_i^{\mathrm{T}} \mathbf{g}_i - \lambda_i \mathbf{h}_i^{\mathrm{T}} A \mathbf{h}_i \quad ,$$

in agreement with (10.13.7).

From these results we can now construct the following algorithm. We begin at $\mathbf{x}_0$, construct there the gradient $\nabla M(\mathbf{x}_0)$, and set its negative equal to the two vectors

$$\mathbf{g}_1 = -\nabla M(\mathbf{x}_0) \quad , \qquad \mathbf{h}_1 = -\nabla M(\mathbf{x}_0) \quad .$$

We minimize along $\mathbf{h}_1$. At the point of the minimum $\mathbf{x}_1$ we construct $\mathbf{g}_2 = -\nabla M(\mathbf{x}_1)$ and compute from (10.13.6)

$$\gamma_1 = \frac{(\mathbf{g}_2 - \mathbf{g}_1)^{\mathrm{T}} \mathbf{g}_1}{\mathbf{g}_1^{\mathrm{T}} \mathbf{g}_1}$$

and from (10.13.2)

$$\mathbf{h}_2 = \mathbf{g}_1 + \gamma_1 \mathbf{h}_1 \quad .$$

From $\mathbf{x}_1$ we then minimize along $\mathbf{h}_2$, and so forth.

Program 10.9: Subroutine MINCJG determines the minimum of a function of n variables by successive minimization along conjugate gradient directions

CALLING SEQUENCE: CALL MINCJG(X0,N,NRED,LIST,USERFN, FMIN,EPSILN,NSTEP)

INPUT ARGUMENTS: X0 — n-vector of a starting point
N — number of variables n
NRED,LIST — as in Program 10.7
USERFN — user function as in Program 10.2
EPSILN — relative accuracy according to (10.1.21)
NSTEP — maximum number of steps

OUTPUT ARGUMENTS: X0 — n-vector giving the position of the minimum
FMIN — function value at the minimum
NSTEP — number of steps needed: -1, if procedure fails

DEFAULT VALUES: NSTEP is taken to be 1000 if a value ≤ 0 is given.
EPSILN is taken to be 10^{-8} if a value ≤ 0 is given.

REMARKS: The input number of steps NSTEP refers both to the maximum length of the sequence $\mathbf{x}_0, \mathbf{x}_1, \ldots$ and also to the maximum number of steps for minimization in a given direction. The output number of steps refers to the length of the sequence.

```
      SUBROUTINE MINCJG(X0,N,NRED,LIST,USERFN,FMIN,EPSILN,NSTEP)
      IMPLICIT DOUBLE PRECISION (A-H,O-Z)
      PARAMETER(MAXN=50)
      DIMENSION X0(N),GR(MAXN),G(MAXN),H(MAXN),D(MAXN),LIST(N)
      PARAMETER(MAXSTP=1000,EPSDEF=1.D-8,T=1.D-15,ZERO=0.D0,
     +ONE=1.D0)
      EXTERNAL USERFN
      IF(NSTEP.LT.1) NSTEP=MAXSTP
      EPS=EPSILN
      IF(EPS.LE.ZERO) EPS=EPSDEF
C initialization : directions G and H identical to gradient at X0
      FMINL=USERFN(X0,N)
      IF(NRED.LE.0) THEN
C all variables are fixed
        FMIN=FMINL
        NSTEP=1
        GO TO 40
      END IF
      CALL AUXGRD(X0,GR,N,NRED,LIST,USERFN)
```

```
      CALL MTXMSV(GR,G,-ONE,N)
      CALL MTXCPV(G,H,N)
C start iteration
      DO 10 ISTEP=1,NSTEP
        LSTEP=ISTEP
        IF(ISTEP.GT.1) THEN
          IF(ABS(FMINL-FMIN) .LT. EPS*ABS(FMIN)+T) THEN
            NSTEP=ISTEP
            GO TO 30
          END IF
          FMINL=FMIN
        END IF
        NST=NSTEP
C XO is position of minimum along direction H
        CALL MINDIR(XO,H,N,USERFN,FMIN,NST,ZERO)
        IF(NST.LT.0) GO TO 20
        CALL AUXGRD(XO,GR,N,NRED,LIST,USERFN)
C GR is gradient at XO
C compute next set of directions G and H
        CALL MTXMSV(GR,GR,-ONE,N)
        CALL MTXSBV(GR,G,D,N)
        CALL MTXDOT(D,GR,S,N)
        CALL MTXDOT(G,G,A,N)
        IF(A.EQ.ZERO) GO TO 30
        GAMMA=S/A
        CALL MTXMSV(H,H,GAMMA,N)
        CALL MTXADV(GR,H,H,N)
        CALL MTXCPV(GR,G,N)
   10 CONTINUE
   20 NSTEP=-1
      GO TO 40
   30 NSTEP=LSTEP
   40 CONTINUE
      END
```

A comparison of Fig. 10.13 with Fig. 10.12 shows the superiority of the method of conjugate gradients over determination of direction according to steepest descent, especially near the minimum.

10.14 Minimization with the Quadratic Form

If the function to be minimized $M(\mathbf{x})$ is of the simple form (10.1.10), then the position of the minimum is given directly by (10.1.13). Otherwise one can always expand $M(\mathbf{x})$ about a point $\mathbf{x}_0$,

$$M(\mathbf{x}) = M(\mathbf{x}_0) - b(\mathbf{x} - \mathbf{x}_0) + \frac{1}{2}(\mathbf{x} - \mathbf{x}_0)^{\mathrm{T}} A(\mathbf{x} - \mathbf{x}_0) + \cdots \qquad (10.14.1)$$

with

$$\mathbf{b} = -\boldsymbol{\nabla} M(\mathbf{x}_0) \quad , \qquad A_{ik} = \frac{\partial^2 M}{\partial x_i \partial x_k} \qquad (10.14.2)$$

and obtain as an approximation for the minimum

$$\mathbf{x}_1 = \mathbf{x}_0 + A^{-1}\mathbf{b} \quad . \qquad (10.14.3)$$

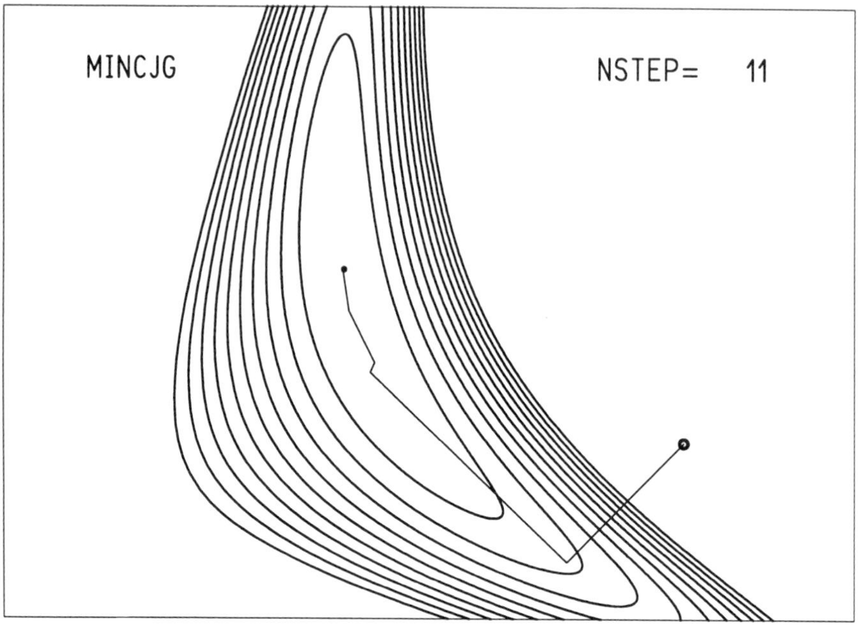

Fig. 10.13. Minimization along conjugate gradient directions.

One can now compute again $\mathbf{b}$ and A as derivatives at the point $\mathbf{x}_1$ and from this obtain a further approximation $\mathbf{x}_2$ according to (10.14.3), and so forth.

For the case where the approximation (10.14.2) gives a good description of the function $M(\mathbf{x})$, the procedure converges quickly, since it tries to jump directly to the minimum. Otherwise it might not converge at all. We have already discussed the difficulties for the corresponding one-dimensional case in Sect. 10.1 with Fig. 10.1.

Program 10.10: Subroutine `MINQDR` determines the minimum of a function of n variables by minimization with a quadratic form

CALLING SEQUENCE: `CALL MINQDR(X0,N,NRED,LIST,USERFN, FMIN,EPSILN,NSTEP,SCRAT)`

INPUT ARGUMENTS: `X0` — n-vector of a starting point
`N` — number of variables n
`NRED,LIST` — as in Program 10.7
`USERFN` — user function as in Program 10.2
`EPSILN` — relative accuracy according to (10.1.21)
`NSTEP` — maximum number of steps
`SCRAT` — working array, minimum size $n' \times n'$

OUTPUT ARGUMENTS: X0 — n-vector giving the position of the minimum
FMIN — function value at the minimum
NSTEP — number of steps needed; -1, if procedure fails

DEFAULT VALUES: NSTEP is set to 100 if a value ≤ 0 is given.
EPSILN is set to 10^{-8} if a value ≤ 0 is given.

```
      SUBROUTINE MINQDR(X0,N,NRED,LIST,USERFN,FMIN,EPSILN,
     +NSTEP,HESSE)
      PARAMETER(MAXN=50)
      IMPLICIT DOUBLE PRECISION (A-H,O-Z)
      DIMENSION X0(N),GRAD(MAXN),HESSE(NRED,NRED),X1(MAXN),
     +X1RED(MAXN)
      DIMENSION LIST(N)
      LOGICAL OK
      PARAMETER(MAXSTP=100,EPSDEF=1.D-8,T=1.D-15,ALAMS=1.D-3)
      PARAMETER(ZERO=0.D0,ONE=1.D0)
      EXTERNAL USERFN
      OK=.TRUE.
      IF(NSTEP.LT.1) NSTEP=MAXSTP
      EPS=EPSILN
      IF(EPS.LE.ZERO) EPS=EPSDEF
      ALAM=ALAMS
C initial value of minimum function
      FMINL=USERFN(X0,N)
      IF(NRED.LE.0) THEN
        FMIN=FMINL
        NSTEP=1
        GO TO 20
      END IF
C start iteration
      DO 10 ISTEP=1,NSTEP
        CALL AUXGRD(X0,GRAD,N,NRED,LIST,USERFN)
        CALL AUXHES(X0,HESSE,N,NRED,LIST,USERFN)
C compute minimum function for two values of lambda (ALAM)
        CALL MTXSVD(HESSE,GRAD,X1RED,R,NRED,NRED,1,ZERO,OK)
        IF(.NOT.OK) THEN
          NSTEP=-4
          GO TO 20
        END IF
        CALL MTXZRV(X1,N)
        CALL MTXPSV(X1,X1RED,N,NRED,LIST)
        CALL MTXSBV(X0,X1,X1,N)
        FMIN=USERFN(X1,N)
C test for break-off criterion
        IF(ABS(FMINL-FMIN) .LT. EPS*ABS(FMIN)+T) THEN
          NSTEP=ISTEP
          GO TO 20
        END IF
        CALL MTXCPV(X1,X0,N)
        FMINL=FMIN
   10 CONTINUE
      NSTEP=-1
   20 CONTINUE
      END
```

Figure 10.14 illustrates the operation of the method. One can observe that the minimum is in fact reached in very few steps.

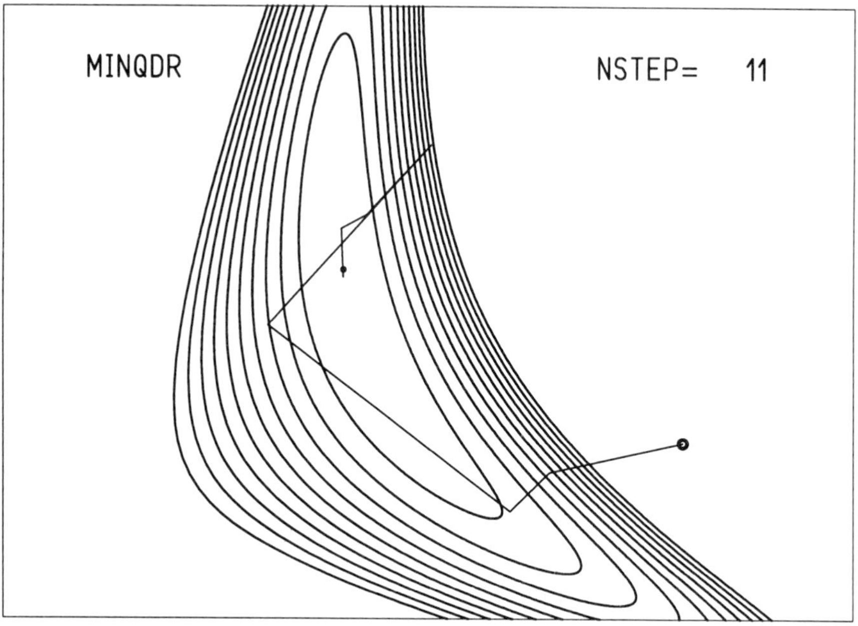

Fig. 10.14. Minimization with quadratic form.

10.15 Marquardt Minimization

MARQUARDT [16] has given a procedure that combines the speed of minimization with the quadratic form near the minimum with the robustness of the method of steepest descent, where one finds the minimum even starting from a point far away. It is based on the following simple consideration.

The prescription (10.14.3), written as a computation of the ith approximation for the position of the minimum,

$$\mathbf{x}_i = \mathbf{x}_{i-1} + A^{-1}\mathbf{b} \quad , \tag{10.15.1}$$

means that one obtains $\mathbf{x}_i$ from the point $\mathbf{x}_{i-1}$ by taking a step along the vector $A^{-1}\mathbf{b}$. Here $\mathbf{b} = -\boldsymbol{\nabla} M(\mathbf{x}_{i-1})$ is the negative gradient, i.e., a vector in the direction of steepest descent of the function M at the point $\mathbf{x}_{i-1}$. If in (10.15.1) one had the unit matrix multiplied by a constant instead of the matrix A, i.e., if instead of (10.15.1) one were to use the prescription

$$\mathbf{x}_i = \mathbf{x}_{i-1} + (\lambda I)^{-1}\mathbf{b} \quad , \tag{10.15.2}$$

then one would step by the vector $\mathbf{b}/\lambda$ from $\mathbf{x}_{i-1}$. This is a step in the direction of steepest descent of the function, which is smaller for larger values of the constant λ. A sufficiently small step in the direction of steepest descent is, however, always a step towards the minimum (at least when one is still in the "approach" to the minimum, i.e., in the one-dimensional case of Fig. 10.1, between the two maxima). The Marquardt procedure consists of interpolating between the prescriptions (10.15.1) and (10.15.2) in such a way that the function M is reduced with every step, and such that the fast convergence of (10.15.1) is exploited as much as possible.

In place of (10.15.1) or (10.15.2) one computes

$$\mathbf{x}_i = \mathbf{x}_{i-1} + (A + \lambda I)^{-1}\mathbf{b} \quad . \tag{10.15.3}$$

Here λ is determined in the following way. One first chooses a fixed number $\nu > 1$ and denotes by $\lambda^{(i-1)}$ the value of λ from the previous step. As an initial value one chooses, e.g., $\lambda^{(0)} = 0.01$. The value obtained from (10.15.3) of $\mathbf{x}_i$ clearly depends on λ. One computes two points $\mathbf{x}_i(\lambda^{(i-1)})$ and $\mathbf{x}_i(\lambda^{(i-1)}/\nu)$, where for λ one chooses the values $\lambda^{(i-1)}$ and $\lambda^{(i-1)}/\nu$, and the corresponding function values $M_i = M(\mathbf{x}_i(\lambda^{(i-1)}))$ and $M_i^{(\nu)} = M(\mathbf{x}_i(\lambda^{(i-1)}/\nu))$. These are compared with the function value $M_{i-1} = M(\mathbf{x}_{i-1})$. The result of the comparison determines what happens next. The following cases are possible:

(i) $M_i^{(\nu)} \leq M_{i-1}$:
One sets $\mathbf{x}_i = \mathbf{x}_i(\lambda^{(i-1)}/\nu)$ and $\lambda^{(i)} = \lambda^{(i-1)}/\nu$.

(ii) $M_i^{(\nu)} > M_{i-1}$ and $M_i \leq M_{i-1}$:
One sets $\mathbf{x}_i = \mathbf{x}_i(\lambda^{(i-1)})$ and $\lambda^{(i)} = \lambda^{(i-1)}$.

(iii) $M_i^{(\nu)} > M_{i-1}$ and $M_i > M_{i-1}$:
One replaces $\lambda^{(i-1)}$ by $\lambda^{(i-1)}\nu$ and repeats the computation of $\mathbf{x}_i(\lambda^{(i-1)}/\nu)$ and $\mathbf{x}_i(\lambda^{(i-1)})$ and the corresponding function values, and repeats the comparisons.

In this way one ensures that the function value in fact decreases with each step and that the value of λ is always as small as possible when adjusted to the local situation. Clearly (10.15.3) becomes (10.15.1) for $\lambda \to 0$, i.e., it describes minimization with the quadratic form. For very large values of λ, Eq. (10.15.3) becomes the relation (10.15.2), which prescribes a small but sure step in the direction of steepest descent.

In Fig. 10.15 one sees the operation of the Marquardt method. It follows a systematic path to the minimum. Comparing with Fig. 10.14 one sees that the method of the quadratic form converged in fewer steps. The Marquardt procedure, however, leads to the minimum in many cases where the method of the quadratic form would fail, e.g., in cases with poorly determined initial values.

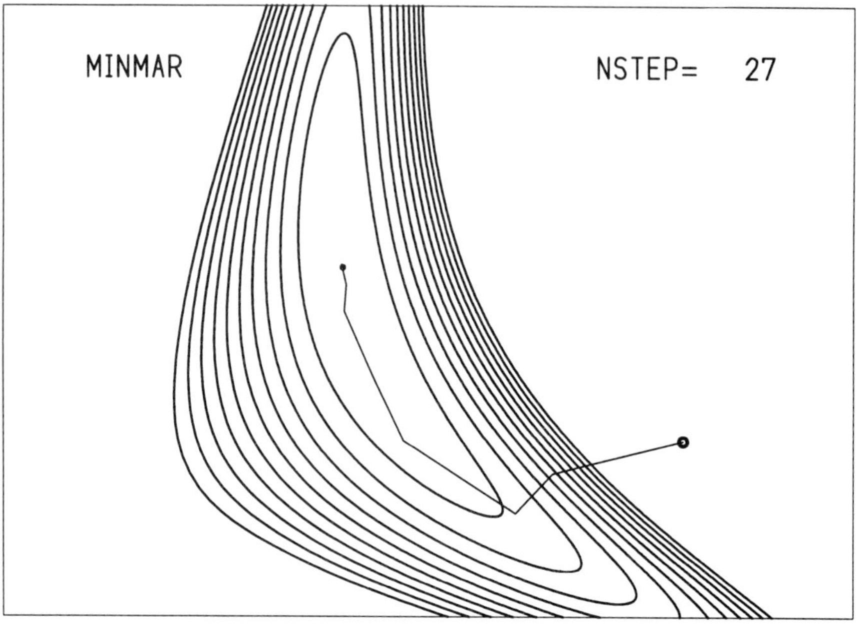

Fig. 10.15. Minimization with the Marquardt procedure.

Program 10.11: Subroutine `MINMAR` determines the minimum of a function of n variables with the Marquardt procedure

CALLING SEQUENCE: `CALL MINMAR(X0,N,NRED,LIST,USERFN, FMIN,EPSILN,NSTEP,SCRAT)`

INPUT ARGUMENTS: as in Program 10.10

OUTPUT ARGUMENTS: as in Program 10.10

DEFAULT VALUES: as in Program 10.10

```
      SUBROUTINE MINMAR(X0,N,NRED,LIST,USERFN,FMIN,EPSILN,
     +NSTEP,HESSE)
      PARAMETER(MAXN=50)
      IMPLICIT DOUBLE PRECISION (A-H,O-Z)
      DIMENSION X0(N),GRAD(MAXN),HESSE(NRED,NRED),X1(MAXN),
     +X2(MAXN),X1RED(MAXN),X2RED(MAXN)
      DIMENSION LIST(N)
      LOGICAL NEW,OK
      PARAMETER(MAXSTP=100,EPSDEF=1.D-8,T=1.D-15,ALAMS=1.D-3)
      PARAMETER(ZERO=0.D0,ONE=1.D0)
      EXTERNAL USERFN
      OK=.TRUE.
      IF(NSTEP.LT.1) NSTEP=MAXSTP
      EPS=EPSILN
```

```
      IF(EPS.LE.ZERO) EPS=EPSDEF
      ALAM=ALAMS
C initial value of minimum function
      FMINL=USERFN(X0,N)
      IF(NRED.LE.0) THEN
        FMIN=FMINL
        NSTEP=1
        GO TO 20
      END IF
C start iteration
      DO 10 ISTEP=1,NSTEP
        CALL AUXGRD(X0,GRAD,N,NRED,LIST,USERFN)
        CALL AUXHES(X0,HESSE,N,NRED,LIST,USERFN)
C compute minimum function for two values of lambda (ALAM)
        CALL MTXMAR(HESSE,GRAD,ALAM,X1RED,X2RED,NRED,NRED,ZERO,OK)
        IF(.NOT.OK) THEN
          NSTEP=-4
          GO TO 20
        END IF
        CALL MTXZRV(X1,N)
        CALL MTXZRV(X2,N)
        CALL MTXPSV(X1,X1RED,N,NRED,LIST)
        CALL MTXPSV(X2,X2RED,N,NRED,LIST)
        CALL MTXSBV(X0,X1,X1,N)
        CALL MTXSBV(X0,X2,X2,N)
        F1=USERFN(X1,N)
        F2=USERFN(X2,N)
C evaluate results
        IF(F2.LE.FMINL) THEN
C reduce lambda and accept new point
          ALAM=ALAM*0.1D0
          CALL MTXCPV(X2,X0,N)
          FMIN=F2
          NEW=.TRUE.
        ELSE IF (F2.GT.FMINL+T .AND. F1.LE.FMINL+T) THEN
C keep current value of lambda and accept new point
          CALL MTXCPV(X1,X0,N)
          FMIN=F1
          NEW=.TRUE.
        ELSE
C increase lambda and reject new point
          ALAM=ALAM*10.D0
          NEW=.FALSE.
        END IF
        IF(NEW) THEN
C test for break-off criterion
          IF(ABS(FMINL-FMIN) .LT. EPS*ABS(FMIN)+T) THEN
            NSTEP=ISTEP
            GO TO 20
          END IF
          FMINL=FMIN
        END IF
   10 CONTINUE
      NSTEP=-1
   20 CONTINUE
      END
```

10.16 On Choosing a Minimization Method

Given the variety of minimization methods, the user is naturally faced with the question of choosing a method appropriate to the task. Before we give recommendations for this, we will first recall the various methods once again.

The *simplex method* (routine `MINSIM`) is particularly robust. Only function values $M(\mathbf{x})$ are computed. The method is slow, however. Faster, but still quite robust is *minimization along a chosen direction* (routine `MINPOW`). This also requires only function values.

The *method of conjugate gradients* (routine `MINCJG`) requires, as the name indicates, the computation of not only the function, but also its gradient. The number of iteration steps is, however, approximately equal to that of `MINPOW`.

For *minimization with the quadratic form* (routine `MINQDR`) and *Marquardt minimization* (routine `MINMAR`) one requires in addition the Hessian matrix of second derivatives. The derivatives are computed numerically with utility routines. The user can replace these utility routines by routines in which the analytic formulas for the derivatives are programmed. If the starting values are sufficiently accurate, `MINQDR` converges after just a few steps. The convergence is slower for `MINMAR`. The method is, however, more robust; it often converges when starting from values from which `MINQDR` would fail.

From these characteristics of the methods we arrive at the following recommendations:

1. For problems that need only to be solved once or not many times, that is, in which the computing time does not play an important role, one should choose `MINSIM` or `MINPOW`.

2. For problems occurring repeatedly (with different numerical values) one should use `MINMAR`. If one always has an accurate starting approximation, `MINQDR` can be used.

3. For repeated problems the derivatives needed for MINMAR or MINQDR should be calculated analytically. Although this entails additional programming work, one gains precision compared to numerical derivatives and saves in many cases computing time.

At this point we should make an additional remark about the comparison of the minimization methods of this chapter with the method of least squares from Chapter 9. The method of least squares is a special case of minimization. The function to be minimized is a sum of squares, e.g., (9.1.8), or the generalization of a sum of squares, e.g., (9.5.9). In this generalized sum of squares there appears the matrix of derivatives A. These are not, however, the derivatives of the function to be minimized, but rather derivatives of a function **f**, which characterizes the problem under consideration; cf. (9.5.2). Second derivatives are never needed. In addition, if one uses the singular value decomposition, as has been done in our programs in Chapter 9 to solve the least-squares problem, one works in numerically critical cases with a considerably higher accuracy than in the computation of sums of squares (cf. Sect. A.14, particularly Example A.4).

Least-squares problems should therefore always be carried out with the routines of Chapter 9. This applies particularly to problems of fitting of functions like those in the examples of Sect. 9.6, when one has many measured points. The matrix A then contains many rows, but few columns. In computing the function to be minimized the product $A^{\mathrm{T}}A$ is encountered, and one is threatened with the above mentioned loss of accuracy in comparison to the singular value decomposition.

10.17 Consideration of Errors

In data analysis minimization procedures are used for determining best estimates $\widetilde{\mathbf{x}}$ for unknown values $\mathbf{x}$. The function to be minimized $M(\mathbf{x})$ is here usually a sum of squares (Chapter 9) or a log-likelihood function (multiplied by -1) as in Chapter 7. In using the equations from Chapter 7 one must note, however, that there the n-vector that was called $\boldsymbol{\lambda}$ is now denoted by $\mathbf{x}$. The variables in Chapter 7 that were called $x^{(i)}$ are the measured values (usually called $\mathbf{y}$ in Chapter 9).

Information on the error of $\widetilde{\mathbf{x}}$ is obtained directly from results of Sects. 9.7, 9.8, and 9.13 by defining

$$H_{ik} = \left(\frac{\partial^2 M}{\partial x_i \partial x_k} \right)_{\mathbf{x}=\widetilde{\mathbf{x}}} \tag{10.17.1}$$

as the elements of the symmetric matrix of second derivatives (the *Hessian matrix*) of the minimization function. Here one must note that the factor f_{QL} takes on the numerical value

$$f_{QL} = 1 \tag{10.17.2}$$

when the function being minimized is a sum of squares. If the function is a log-likelihood function (times -1) then this must be set equal to

$$f_{QL} = 1/2 \quad . \tag{10.17.3}$$

1. Covariance matrix. Symmetric errors. The covariance matrix of $\widetilde{\mathbf{x}}$ is

$$C_{\widetilde{x}} = 2 f_{QL} H^{-1} \quad . \tag{10.17.4}$$

The square roots of the diagonal elements are the (symmetric) *errors*

$$\Delta\tilde{x}_i = \sqrt{c_{ii}} \quad . \tag{10.17.5}$$

It is only meaningful to give the covariance matrix if the measurement errors are small and/or there are many measurements, i.e., the requirements for (7.5.8) are fulfilled.

2. *Confidence ellipsoid. Symmetric confidence limits.* The covariance matrix defines the covariance ellipsoid; cf. Sects. 5.10 and A.12. Its center is given by $\mathbf{x} = \tilde{\mathbf{x}}$. The probability that the true value of $\mathbf{x}$ is contained inside the ellipsoid is given by (5.10.20). The ellipsoid for which this probability has a given value W, the *confidence level*, is given by the *confidence matrix*

$$C_{\tilde{x}}^{(W)} = \chi^2_W(n_f) C_{\tilde{x}} \quad . \tag{10.17.6}$$

Here $\chi^2_W(n_f)$ is the quantile of the χ^2-distribution for n_f degrees of freedom and probability $P = W$; cf. (5.10.19) and (C.5.3). The number of degrees of freedom n_f is equal to the number of measured values minus the number of parameters determined by the minimization. The square roots of the diagonal elements of $C_{\tilde{x}}^{(W)}$ are the distances from the *symmetric confidence limits*,

$$x_{i\pm}^{(W)} = \tilde{x}_i \pm \sqrt{c_{ii}^{(W)}} \quad . \tag{10.17.7}$$

Program 10.12: Subroutine `MINCOV` computes the covariance matrix or confidence matrix for parameters determined by minimization

CALLING SEQUENCE: `CALL MINCOV(X0,N,NRED,LIST,USERFN, FACT,SCRAT1,SCRAT2,CX,OK)`

INPUT ARGUMENTS: `X0` — n-vector $\mathbf{x}_0$ giving the position of the minimum
`N` — number of variables n
`NRED` — reduced number of variables n'
`LIST` — n-vector containing the elements of a list indicating which of the n variables should be considered as fixed (list element $= 0$) or variable (list element $= 1$)
`USERFN` — name of a user function
`FACT` — $2f_{QL}\chi^2_W(n_f)$, cf. (10.17.4), (10.17.6)
`SCRAT1` — working array, minimum size $n' \times n'$
`SCRAT2` — working array, minimum size $n' \times n'$

OUTPUT ARGUMENTS: `CX` — $n' \times n'$ confidence matrix $C_{\tilde{x}}^{(W)}$. For $\chi^2_W(n_f) = 1$, i.e., for `FACT` $= 2f_{QL}$, `CX` is the covariance matrix.
`OK` — logical variable (`.TRUE.` for convergence of the matrix inversion, `.FALSE.` if inversion does not converge)

DEFAULT VALUES: If the input value of `FACT` ≤ 0, it is taken to be 1.

```
      SUBROUTINE MINCOV(X0,N,NRED,LIST,USERFN,FACT,SCRAT1,SCRAT2,
     +CX,OK)
      PARAMETER(MAXN=50)
      IMPLICIT DOUBLE PRECISION(A-H,O-Z)
      DIMENSION X0(*),LIST(*),CX(NRED,*),SCRAT1(NRED,*),
     +SCRAT2(NRED,*)
      DIMENSION R(MAXN)
      PARAMETER (EPS=1.D-10,ONE=1.D0)
      LOGICAL OK
      EXTERNAL USERFN
      FRAC=0.D0
      MS=NRED
      NS=NRED
      LS=NRED
      CALL AUXHES(X0,SCRAT1,N,NRED,LIST,USERFN)
      CALL MTXUNT(SCRAT2,MS)
      CALL MTXSVD(SCRAT1,SCRAT2,CX,R,MS,NS,LS,FRAC,OK)
      IF(FACT.LT.EPS) THEN
        F=ONE
      ELSE
        F=FACT
      END IF
      CALL MTXMSC(CX,CX,F,NRED,NRED)
      END
```

3. Confidence region. If giving a covariance or confidence ellipsoid is not meaningful, it is still possible to give a confidence region with confidence level W. It is determined by the hypersurface

$$M(\mathbf{x}) = M(\widetilde{\mathbf{x}}) + \chi^2_W(n_f) f_{QL} \quad . \tag{10.17.8}$$

With the help of the following routine a contour of a cross section through this hypersurface is drawn in a plane that contains the point $\widetilde{\mathbf{x}}$ and is parallel to (x_i, x_j). Here x_i and x_j are two components of the vector of parameters $\mathbf{x}$.

Program 10.13: Subroutine `MINCNT` draws the confidence region in a plane spanned by two parameters whose values are determined by minimization

CALLING SEQUENCE: `CALL MINCNT(IX,JX,XI0,XJ0,DXI,DXJ, NXI,NXJ,FCONT,X0,N,USERFN)`

INPUT ARGUMENTS: `IX,JX` — indices i, j, label the elements x_i, x_j of the n-vectors $\mathbf{x}$ of the parameters, which span the plane of the covariance region

`XI0,DXI,NXI` — initial value x_{i0}, interval width Δx_i, and number of intervals n_{x_i} of the variables x_i

`XJ0,DXJ,NXJ` — same as for x_j

`FCONT` — value of the minimum function, i.e., the right hand side of (10.17.8) on the contour line to be drawn

`X0` — n-vector $\widetilde{\mathbf{x}}$ of the variables at the minimum

`N` — number of variables n

`USERFN` — name of user function

```
      SUBROUTINE MINCNT(IX,JX,XI0,XJ0,DXI,DXJ,NXI,NXJ,FCONT,X0,N,
     +USERFN)
      PARAMETER(MAXN=50)
      IMPLICIT DOUBLE PRECISION (A-H,O-Z)
      DIMENSION X0(N)
      DIMENSION XI(2),XJ(2),F(2,2),X0SAV(MAXN)
      EXTERNAL USERFN
      CALL MTXCPV(X0,X0SAV,N)
      DO 40 I=0,NXI-1
        XI(1)=XI0+I*DXI
        XI(2)=XI(1)+DXI
        DO 30 J=0,NXJ-1
          XJ(1)=XJ0+J*DXJ
          XJ(2)=XJ(1)+DXJ
          DO 20 IS=1,2
            DO 10 JS=1,2
              CALL MTXCPV(X0SAV,X0,N)
              X0(IX)=XI(IS)
              X0(JX)=XJ(JS)
              F(IS,JS)=USERFN(X0,N)
   10       CONTINUE
   20     CONTINUE
          CALL GRPXCT(XI(1),XI(2),XJ(1),XJ(2),F(1,1),F(1,2),F(2,1),
     +    F(2,2),FCONT)
   30   CONTINUE
   40 CONTINUE
      CALL MTXCPV(X0SAV,X0,N)
      END
```

4. Asymmetric errors and confidence limits. If the confidence region is not an ellipsoid, the asymmetric *confidence limits* for the variable x_i can still be determined by

$$\min\left\{M(\widetilde{\mathbf{x}});\, x_i = x_{i\pm}^{(W)}\right\} = M(\widetilde{\mathbf{x}}) + \chi^2_W(n_f) f_{QL} \quad . \tag{10.17.9}$$

The differences

$$\begin{aligned} \Delta x_{i+}^{(W)} &= x_{i+}^{(W)} - \widetilde{x}_i \quad , \\ \Delta x_{i-}^{(W)} &= \widetilde{x}_i - x_{i-}^{(W)} \end{aligned} \tag{10.17.10}$$

are the (asymmetric) distances from the confidence limits. If one takes $\chi^2_W(n_f) = 1$, the *asymmetric errors* Δx_{i+} and Δx_{i-} are obtained.

Program 10.14: Subroutine `MINASY` computes the asymmetric errors (or distances from the confidence limits) for parameters determined by minimization

CALLING SEQUENCE: `CALL MINASY(USERFN,N,NRED,LIST,X0,CX,FCONT,DXPLUS,DXMINS,XSCRAT,NSTEP)`

INPUT ARGUMENTS: `USERFN` — name of a user function

`N` — number of variables n

`NRED` — number of non-fixed variables n'

LIST — n-vector, containing the elements of a list indicating which of the n variables are fixed (list element = 0) or adjustable (list element = 1).
X0 — n-vector $\tilde{\mathbf{x}}$ of the variables at the minimum
CX — $n' \times n'$ covariance matrix. The diagonal elements (or their estimates) are needed as initial values.
FCONT — value of the right hand side of (10.17.9)
XSCRAT — working array, minimum size $n' \times (n' + 1)$
NSTEP — maximum number of steps for minimization

OUTPUT ARGUMENTS: DXPLUS, DXMINS — n'-vectors $\Delta x_{i+}^{(W)}$, $\Delta x_{x-}^{(W)}$ $(i = 1, 2, \ldots, n')$
NSTEP — positive for convergence of the procedure, negative if it does not converge (−1 failure in individual minimization, −2 failure in establishing confidence limits)

DEFAULT VALUES: If the input value of NSTEP is zero or negative, it is taken to be 1000.

```
      SUBROUTINE MINASY(USERFN,N,NRED,LIST,X0,CX,FCONT,DXPLUS,
     +DXMINS,XR,NSTEP)
      IMPLICIT DOUBLE PRECISION (A-H,O-Z)
      PARAMETER(NMAX=100)
      DIMENSION X(NMAX),DX(NMAX),LIST(N),XR(NRED,NRED+1)
      DIMENSION CX(NRED,NRED),DXPLUS(NRED),DXMINS(NRED)
      DIMENSION X0(N)
      PARAMETER(MAXSTP=1000)
      PARAMETER(EPS=1.D-3,EPSILN=1.D-4,TT=1.D-6,FACT=1.5D0)
      PARAMETER(ZERO=0.0D0,ONE=1.D0,HALF=0.5D0,TWO=2.D0,BIG=1.D20)
      LOGICAL LENCLS
      EXTERNAL USERFN
      IF(NSTEP.LE.0) NSTEP=MAXSTP
      IF(N.EQ.NRED) THEN
        DO 10 I=1,N
          LIST(I)=1
   10   CONTINUE
      END IF
      DXMIN=BIG
      DO 20 I=1,NRED
        DX(I)=SQRT(CX(I,I))
        DXMIN=MIN(DXMIN,DX(I))
   20 CONTINUE
      DXMIN=DBLE(.01)*DXMIN
      NRED1=NRED-1
      DO 60 IVAR=1,N
        IF(LIST(IVAR).EQ.0) GO TO 60
C fix variable IVAR
        LIST(IVAR)=0
        SAV=X0(IVAR)
        DO 50 ISIGN=-1,1,2
          SIGNUM=DBLE(ISIGN)
          DEL=DX(IVAR)
          LENCLS=.FALSE.
```

```
C set XSMALL to x at minimum position
          XSMALL=X0(IVAR)
          XIVAR=X0(IVAR)+SIGNUM*DEL
          DO 30 I=1,NSTEP
            NSTEPL=NSTEP
            CALL MTXCPV(X0,X,N)
            X(IVAR)=XIVAR
            EPSI=ZERO
            R1=DXMIN
            CALL MINSIM(X,N,NRED1,LIST,USERFN,R1,EPSI,NSTEPL,XR)
            IF(NSTEPL.LT.1) THEN
              NSTEP=-1
              GO TO 70
            END IF
C test for convergence
            IF(ABS(R1-FCONT).LT.EPS) THEN
              IF(ISIGN.LT.0) THEN
                DXMINS(IVAR)=X0(IVAR)-X(IVAR)
              ELSE
                DXPLUS(IVAR)=X(IVAR)-X0(IVAR)
              END IF
              GO TO 40
            ELSE
              IF(.NOT.LENCLS) THEN
C zero was not yet enclosed in last step
                IF(R1.GT.FCONT) THEN
C zero is now enclosed, perform first interval halving
                  LENCLS=.TRUE.
                  XBIG=X(IVAR)
                  X(IVAR)=HALF*(XBIG+XSMALL)
                ELSE
C zero not enclosed, widen range
                  DEL=DEL*TWO
                  X(IVAR)=XSMALL+SIGNUM*DEL
                END IF
              ELSE
C continue interval halving
                IF(R1.GT.FCONT) THEN
                  XBIG=X(IVAR)
                ELSE
                  XSMALL=X(IVAR)
                END IF
                X(IVAR)=HALF*(XBIG+XSMALL)
              END IF
              XIVAR=X(IVAR)
            END IF
   30     CONTINUE
          NSTEP=-2
          GO TO 70
   40     CONTINUE
   50   CONTINUE
C unfix variable IVAR
        LIST(IVAR)=1
   60 CONTINUE
   70 RETURN
      END
```

10.18 Examples

Example 10.1: Determining the parameters of a distribution from the elements of a sample with the method of maximum likelihood

Suppose one has N measurements $y_1, y_2, \ldots, y_n$ that can be assumed to come from a normal distribution with expectation value $a = x_1$ and standard deviation $\sigma = x_2$. The likelihood function is

$$L = \prod_{i=1}^{N} \frac{1}{x_2\sqrt{2\pi}} \exp\left\{ -\frac{(y_i - x_1)^2}{2x_2^2} \right\} \tag{10.18.1}$$

and its logarithm is

$$\ell = -\sum_{i=1}^{N} \frac{(y_i - x_1)^2}{2x_2^2} - N \ln\{x_2\sqrt{2\pi}\} \quad . \tag{10.18.2}$$

The task of determining the maximum-likelihood estimators $\widetilde{x}_1, \widetilde{x}_2$ has already been solved in Example 7.8 by setting the analytically calculated derivatives of the function $\ell(\mathbf{x})$ to zero. Here we will accomplish this by means of numerical minimization of $\ell(\mathbf{x})$. For this we must first provide a function that computes the function to be minimized,

$$M(\mathbf{x}) = -\ell(\mathbf{x}) \quad .$$

Program 10.15: Function `MINGLS` computes the negative log-likelihood function for a sample assumed to come from a normally distributed population

CALLING SEQUENCE: `GLS=MINGLS(X,N)`

INPUT ARGUMENTS: `X` — n-vector $\mathbf{x}$ of the parameters of the normal distribution (x_1 — mean, x_2 — standard deviation)
`N` — number of parameters n (must be set to 2 before calling routine)

OUTPUT ARGUMENT: function value `MINGLS` — negative log-likelihood function

REMARKS: The elements of the sample $y_1, y_2, \ldots, y_N$ and the number of elements N are passed as `Y` and `NY` through the common block `/CMINGS/`.

```
      DOUBLE PRECISION FUNCTION MINGLS(X,N)
      IMPLICIT DOUBLE PRECISION (A-H,O-Z)
      DIMENSION Y(1000)
      DIMENSION X(N)
      COMMON /CMINGS/ Y,NY
      PARAMETER(SQR2PI=2.506628275D0,BIG=-1.D10,TWO=2.D0,
     +SMALL=1.D-10)
      A=SQR2PI*ABS(X(2))
      IF(A.LT.SMALL) THEN
        F=BIG
      ELSE
        F=LOG(A)
      END IF
      MINGLS=DBLE(NY)*F
      DO 10 I=1,NY
        F=(Y(I)-X(1))**2/(TWO*X(2)**2)
        MINGLS=MINGLS+F
   10 CONTINUE
      END
```

The results of the minimization of this user function are shown in Fig. 10.16 for two samples. Confidence regions and covariance ellipses agree reasonably well with each other. The agreement is better for the larger sample. ■

Example 10.2: Determination of the parameters of a distribution from the histogram of a sample by maximizing the likelihood

Instead of the original sample $y_1, y_2, \ldots, y_N$ as in Example 10.1, one often uses the corresponding histogram. Denoting by n_i the number of observations that fall into the interval centered about the point t_i with width Δt,

$$t_i - \Delta t/2 \le y < t_i + \Delta t/2 \quad , \tag{10.18.3}$$

the histogram is given by the pairs of numbers

$$(t_i, n_i) \quad , \qquad i = 1, 2, \ldots, n \quad . \tag{10.18.4}$$

If the original sample is taken from a normal distribution with expectation value $x_1 = a$ and standard deviation $x_2 = \sigma$, i.e., from the probability density

$$f(t; x_1, x_2) = \frac{1}{x_2\sqrt{2\pi}} \exp\left\{-\frac{(t - x_1)^2}{2x_2^2}\right\} \quad , \tag{10.18.5}$$

then one might expect the $n_i(t_i)$ (at least in the limit $N \to \infty$) to be given by

$$g_i = N\Delta t f(t_i; x_1, x_2) \quad . \tag{10.18.6}$$

The quantities $n_i(t_i)$ are integers, which are clearly not equal in general to g_i. We can, however, regard each $n_i(t_i)$ as a sample of size one from a Poisson distribution with the expectation value

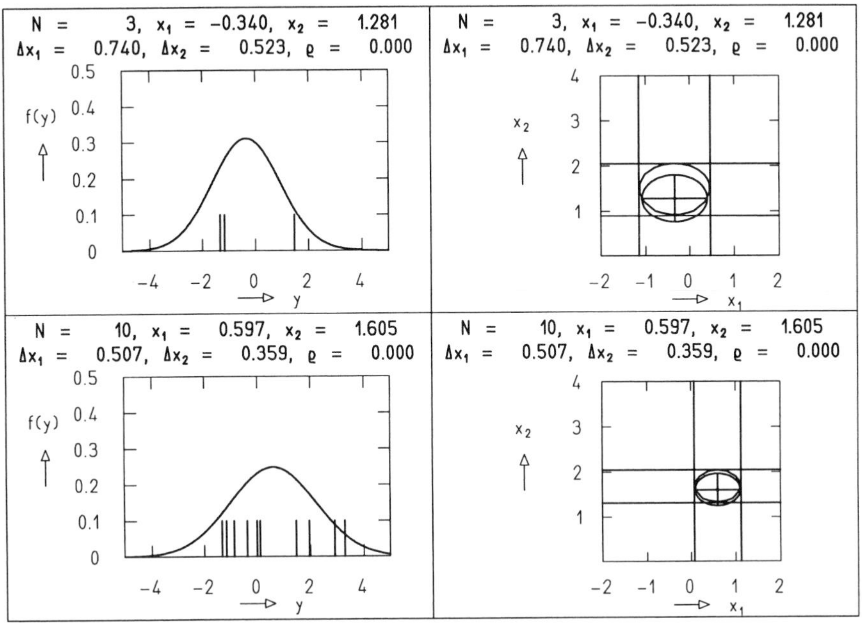

Fig. 10.16. Determination of the parameters x_1 (mean) and x_2 (width) of a Gaussian distribution by maximization of the log-likelihood function of a sample. The plots on the left show two different samples, marked as one-dimensional scatter plots (tick marks) on the y axis. The curves $f(y)$ are Gaussian distributions with the fitted parameters. The right-hand plots show in the (x_1, x_2) plane the values of the parameters obtained with symmetric errors and covariance ellipses, as well as the confidence region for $\chi^2_W = 1$ and the corresponding confidence boundaries (horizontal and vertical lines).

$$\lambda_i = g_i \quad . \tag{10.18.7}$$

The *a posteriori* probability to observe the value $n_i(t_i)$ is clearly

$$\frac{1}{n_i!}\lambda_i^{n_i}\mathrm{e}^{-\lambda_i} \quad . \tag{10.18.8}$$

The likelihood function for the observation of the entire histogram is

$$L = \prod_{i=1}^{n} \frac{1}{n_i!}\lambda_i^{n_i}\mathrm{e}^{-\lambda_i} \quad , \tag{10.18.9}$$

and its logarithm is

$$\ell = -\sum_{i=1}^{n} \ln n_i! + \sum_{i=1}^{n} n_i \ln \lambda_i - \sum_{i=1}^{n} \lambda_i \quad . \tag{10.18.10}$$

If we use for λ_i the notation of (10.18.7) and find the minimum of $-\ell$ with respect to x_1, x_2, then this determines the best estimates of the parameters

x_1, x_2. Of course the same procedure can be applied not only in the case of a simple Gaussian distribution, but for any distribution that depends on parameters. One must simply use the corresponding probability density in (10.18.5) and (10.18.6). In the user function given below one must change only one instruction in order to replace the Gaussian by another distribution. ■

Program 10.16: Function MINGLP computes the negative log-likelihood function for a histogram with the assumption that the sample from which the histogram was obtained stems from a normally distributed population. The Poisson fluctuations of the bin contents of the histogram are taken into consideration.

CALLING SEQUENCE: GLP=MINGLP(X,N)

INPUT ARGUMENTS: X — n-vector **x** of the parameters of the normal distribution (x_1 – mean, x_2 – standard deviation)
N — number of parameters n (must be set to 2)

OUTPUT ARGUMENT: function value MINGLP — negative log-likelihood function

REMARKS: The information about the histogram is passed through the common block /CMINGH/. The meanings of the variables are
T — N-vector $(t_1, t_2, \ldots, t_N)$ of the centers of the bins
HIST — N-vector of bin contents $(n_1, n_2, \ldots, n_N)$
DELTAT — bin width Δt
NT — number of bins N
NEVENT — number of elements n in the sample

```
      DOUBLE PRECISION FUNCTION MINGLP(X,N)
      IMPLICIT DOUBLE PRECISION(A-H,O-Z)
      COMMON /CMINGH/ T(100),HIST(100),DELTAT,NT,NEVENT
      DIMENSION X(N)
      PARAMETER(ZERO=0.D0)
      FNORM=DBLE(NEVENT)*DELTAT
      MINGLP=ZERO
      DO 10 I=1,NT
C GI is the value of the probability density of the population
C at T(I) (by replacing the RHS of the following statement it can
C be changed from normal to any desired distribution)
        GI=SDNORM(T(I),X(1),X(2))
C normalize to number of events in sample
        GI=FNORM*GI
        ALAM=GI
        IF(ALAM.GT.ZERO) THEN
          NI=NINT(HIST(I))
          ALNLAM=LOG(ALAM)
          ALNNI=GLNGAM(DBLE(NI+1))
          MINGLP=MINGLP+ALNNI-HIST(I)*ALNLAM+ALAM
        END IF
   10 CONTINUE
      END
```

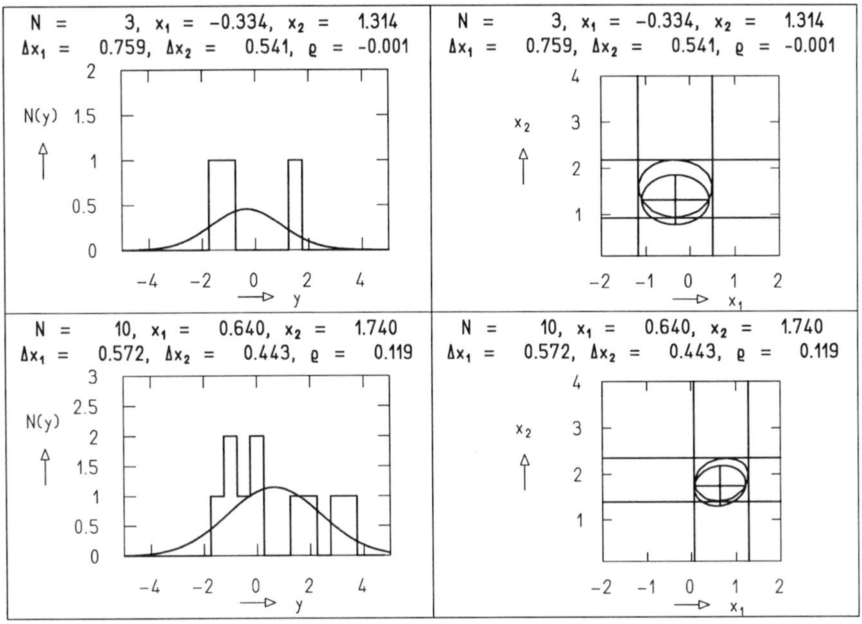

Fig. 10.17. Determination of the parameters x_1 (mean) and x_2 (width) of a Gaussian distribution by maximizing the log-likelihood function of a histogram. The left-hand plots show two histograms corresponding to the samples from Fig. 10.16. The curves are the Gaussian distributions normalized to the histograms. The right-hand plots show the symmetric errors, covariance ellipse, and confidence region represented as in Fig. 10.16.

The results of the minimization with this user function are shown in Fig. 10.17 for two histograms, which are based on the samples from Fig. 10.16. The results are very similar to those from Example 10.1. The errors of the parameters, however, are somewhat larger. This is to be expected, since some information is necessarily lost when constructing a histogram from the sample.

A histogram can be viewed as a sample in a compressed representation. The compression becomes greater as the bin width of the histogram increases. This is made clear in Fig. 10.18. One sees that for the same sample, the errors of the determined parameters increase for greater bin width. The effect is relatively small, however, for the relatively large sample size in this case.

Example 10.3: Determination of the parameters of a distribution from the histogram of a sample by minimization of a sum of squares

If the bin contents n_i of a histogram (10.18.4) are sufficiently large, then the statistical fluctuations of each n_i can be approximated by a Gaussian distribution with the standard deviation

$$\Delta n_i = \sqrt{n_i} \quad . \tag{10.18.11}$$

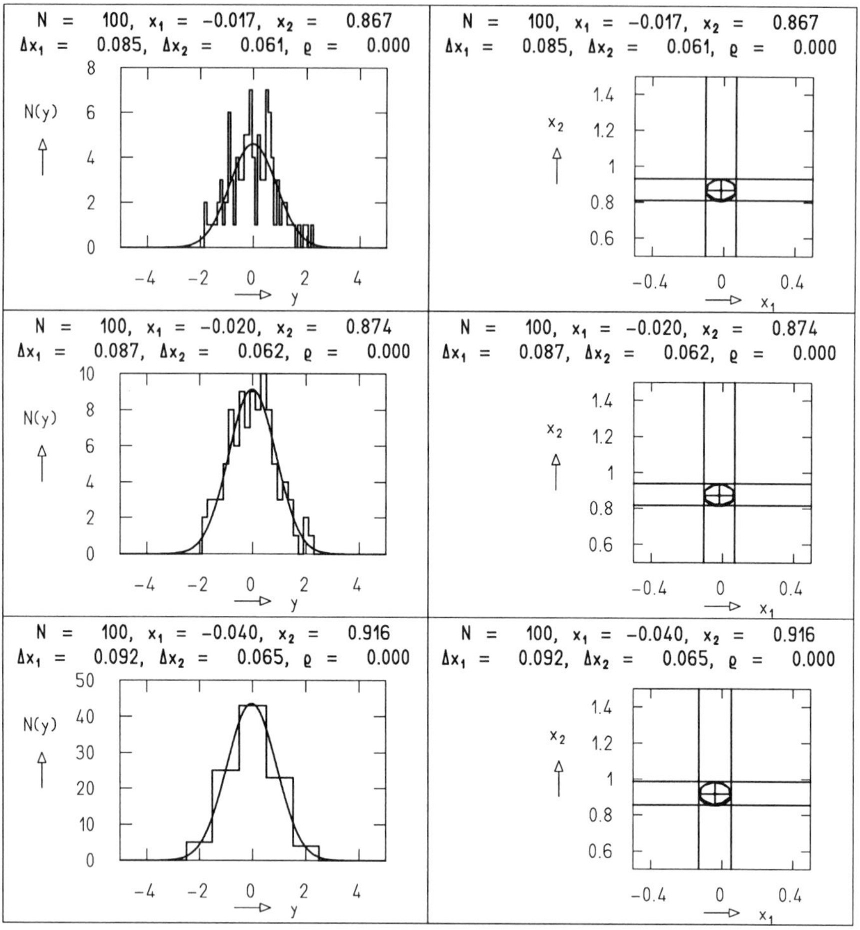

Fig. 10.18. As in Fig. 10.17, but for histograms with different interval widths of the same sample.

(See Sect. 6.8.) The weighted sum of squares describing the deviation of the histogram contents n_i from the expected values g_i from (10.18.6) is then

$$Q = \sum_{i=1}^{N} \frac{(n_i - g_i)^2}{n_i} \quad . \tag{10.18.12}$$

When carrying out the sum one must take care (in contrast to Example 10.2) that empty bins ($n_i = 0$) are not included. Even better is to not include bins in the sum even with low bin contents, e.g., $n_i < 4$.

The following user function computes the sum of squares. Here g_i is given by (10.18.6) as in the previous example. The sum is carried out over all bins with $n_i > 0$.

Program 10.17: Function `MINGSQ` computes the weighted sum of squares for the deviation of a histogram from a distribution determined by the assumption that the sample was taken from a normal distribution. The statistical fluctuations of the bin contents of the histogram are approximated as being distributed according to a Gaussian.

CALLING SEQUENCE: `GSQ=MINGSQ(X,N)`

INPUT ARGUMENTS: as in Program 10.16

OUTPUT ARGUMENT: function value `MINGSQ` — weighted sum of squares

REMARKS: as in Program 10.16

```
      DOUBLE PRECISION FUNCTION MINGSQ(X,N)
      IMPLICIT DOUBLE PRECISION(A-H,O-Z)
      COMMON /CMINGH/ T(100),HIST(100),DELTAT,NT,NEVENT
      DIMENSION X(N)
      PARAMETER(ZERO=0.D0)
      FNORM=DBLE(NEVENT)*DELTAT
      MINGSQ=ZERO
      DO 10 I=1,NT
C GI is the value of the probability density of the population
C at T(I) (by replacing the RHS of the following statement it can
C be changed from normal to any desired distribution)
        GI=SDNORM(T(I),X(1),X(2))
C normalize to number of events in sample
        GI=FNORM*GI
        IF(HIST(I).GT.ZERO) THEN
          MINGSQ=MINGSQ+(HIST(I)-GI)**2/HIST(I)
        END IF
   10 CONTINUE
      END
```

Figure 10.19 shows the results obtained by minimizing the quadratic sum for the histogram in Fig. 10.18. Since the histograms are based on the same sample, the values n_i decrease for decreasing bin width, whereby the requirement for using the sum of squares becomes less well fulfilled. Thus we cannot trust as much the results nor the errors for smaller bin width. One sees, however, that the errors given by the procedure in fact increase for decreasing bin width. ■

We emphasize that the determination of parameters from a histogram by quadratic-sum minimization gives less exact results than those obtained by likelihood maximization. This is because the assumption of a normal distribution of the n_i with the width (10.18.11) is only an approximation, which often requires large bin widths for the histogram and thus implies a loss of information. If, however, enough data, i.e., sufficiently large samples, are available, then the difference between the two procedures is small. One should compare, e.g., Fig. 10.18 (upper right-hand plot) with Fig. 10.19 (lower right-hand plot).

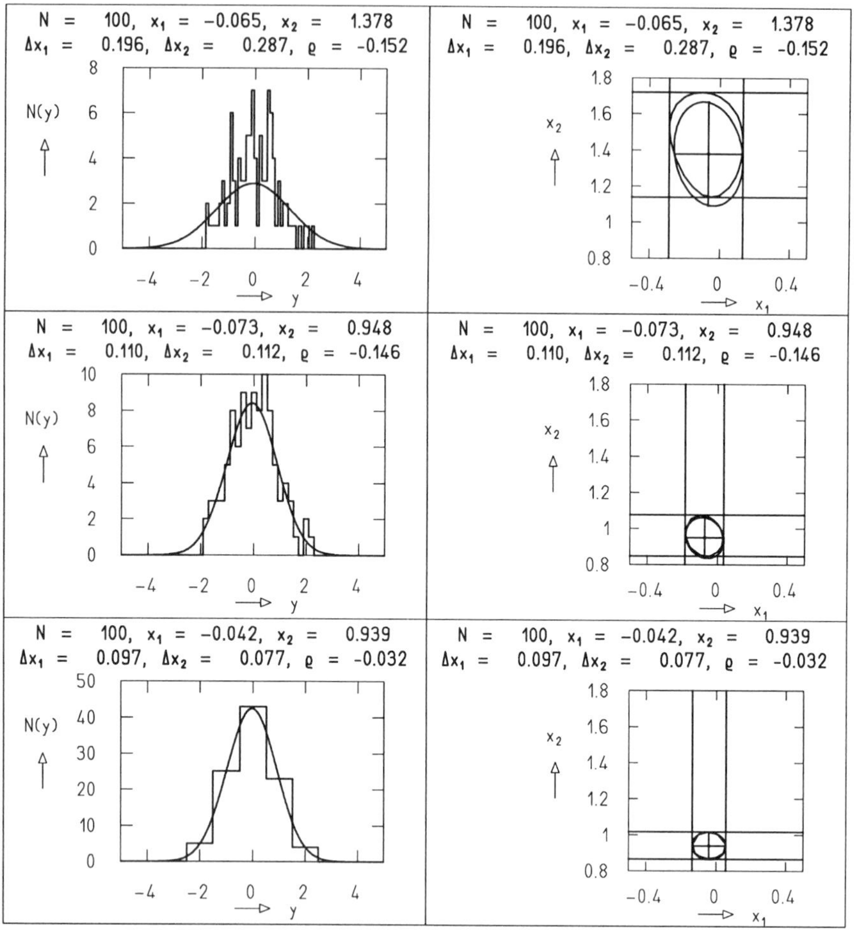

Fig. 10.19. Determination of the parameters x_1 (mean) and x_2 (width) of a Gaussian distribution by minimization of a weighted sum of squares. The histograms on the left are the same as in Fig. 10.18. The fit results and errors, covariance ellipses, and confidence regions are represented as in Fig.10.17 and Fig. 10.18.

10.19 Example Programs

The source files of the example programs are available in the subdirectory \DATAN\EXASRC.

10.19.1 Main Program E1MIN to Demonstrate Subprograms MINSIM, MINPOW, MINCJG, MINQDR, and MINMAR

The program calls one of the five subprograms, as requested by the user, in order to solve the following problem. One wants to find the minimum of the function $f = f(\mathbf{x}) = f(x_1, x_2, x_3)$. The search is started at the point $\mathbf{x}^{(\mathrm{in})} = (x_1^{(\mathrm{in})}, x_2^{(\mathrm{in})}, x_3^{(\mathrm{in})})$, which is also input by the user. The program treats consecutively four different cases:

(i) No variables are fixed (`NRED=3`),

(ii) x_3 is fixed (`NRED=2`),

(iii) x_2 and x_3 are fixed (`NRED=1`),

(iv) all variables are fixed (`NRED=0`).

The user can choose one of the following functions to be minimized:

$$\begin{aligned}
f_1(\mathbf{x}) &= r^2 \ , \quad r = \sqrt{x_1^2 + x_2^2 + x_3^2} \ , \\
f_2(\mathbf{x}) &= r^{10} \ , \\
f_3(\mathbf{x}) &= r \ , \\
f_4(\mathbf{x}) &= -\mathrm{e}^{-r^2} \ , \\
f_5(\mathbf{x}) &= r^6 - 2r^4 + r^2 \ , \\
f_6(\mathbf{x}) &= r^2 \mathrm{e}^{-r^2} \ , \\
f_7(\mathbf{x}) &= -\mathrm{e}^{-r^2} - 10\,\mathrm{e}^{-r_a^2} \ , \quad r_a^2 = (x_1 - 3)^2 + (x_2 - 3)^2 + (x_3 - 3)^2 \ .
\end{aligned}$$

Suggestions: Discuss the functions f_1 through f_7. All of them have a minimum at $x_1 = x_2 = x_3 = 0$. Some possess additional minima. Study the convergence behavior of the different minimization methods for these functions using different starting points and explain this behavior qualitatively.

10.19.2 Main Program E2MIN to Determine the Parameters of a Distribution from the Elements of a Sample and to Demonstrate Subprogram MINCOV

The program solves the problem in Example 10.1. First a sample is drawn from the standard normal distribution. Next the sample is used to estimate the parameters x_1 (mean) and x_2 (standard deviation) of the population by minimizing the negative of the likelihood function (10.18.2). This is done by a call of `MINSIM` using the function subprogram `MINGLS`. The covariance matrix of the parameters is determined by calling `MINCOV`. The results are presented numerically.

The rest of the program performs the graphical display of the sample as a one-dimensional scatter plot and of the fitted function as in Figure 10.16 (left-hand plots).

Suggestion: The seeds are chosen such that you can reproduce directly plots on the left-hand side of Figure 10.16 Run the program also for larger samples.

10.19.3 Main Program `E3MIN` to Demonstrate Subprograms `MINASY` and `MINCNT`

The program solves the problem of the preceding section. In addition it computes the asymmetric errors of the parameters using `MINASY`. Then the solution, the symmetric errors, the covariance ellipse, and the asymmetric errors are displayed graphically in the (x_1, x_2) plane, and with the help of the subprogram `MINCNT` the contour of the confidence region is shown as well. The plot corresponds to Figure 10.16 (right-hand plots).

Suggestion: Reproduce the plots on the right-hand side of Figure 10.16. Note that your plot and that of Figure 10.16 are identical in content but not graphically identical since in `E3MIN` the scales are set automatically. Run the program also for larger samples.

10.19.4 Main Program `E4MIN` to Determine the Parameters of a Distribution from a Histogram of a Sample and to Demonstrate Subprograms `MINGLP` and `MINGSQ`

The program solves the problem of Examples 10.2 and 10.3. First a sample of size n_{ev} is drawn from a standard normal distribution and a histogram with n_t bins between $t_0 = -5.25$ and $t_{\text{max}} = 5.25$ is constructed from the sample. The bin centers are t_i $(i = 1, 2, \ldots, n_t)$, and the bin contents are n_i. As chosen by the user either the likelihood function ℓ given by (10.18.10) is maximized (i.e., $-\ell$ is minimized) by `MINSIM` and `MINGLP`, or the sum of squares Q given by (10.18.12) is minimized, again with `MINSIM`, but using `MINGSQ`.

The results are presented in graphical form. The user has the choice either to show the histogram and the fitted Gaussian (i.e., a plot corresponding to the plots on the left-hand side of Figs. 10.17 or 10.18) or to present the solution in the (x_1, x_2) plane with symmetric and asymmetric errors, covariance ellipse, and confidence region (corresponding to the plots on the right-hand side of the figures).

Suggestions: (a) Choose $n_{\text{ev}} = 100$. Show that for likelihood maximization the errors $\Delta x_1, \Delta x_2$ increase if you decrease the number of bins beginning with $n_t = 100$, but that for the minimization of the sum of squares the number of bins has to be small to get meaningful errors. (b) Show that for $n_{\rm ev} = 1000$ and $n_t = 50$ or $n_t = 20$ there is practically no difference between the results of the methods.

10.20 Programming Problems

The source files of the example solutions are available in the subdirectory `\DATAN\SOLSRC`.

10.20.1 Monte Carlo Minimization to Choose a Good First Approximation

For some functions in Example Program 10.19.1 the choice of the point $\mathbf{x}_0$ defining the first approximation was decisive for success or failure of the minimization. If a function has several minima and if its value is smallest at one of them, that is if an "absolute minimum" exists, the following procedure will work. One uses the Monte Carlo method to determine a first approximation $\mathbf{x}_0$ of the absolute minimum in a larger region of the parameter space by generating points $\mathbf{x} = (x_1, x_2, \ldots, x_n)$ according to a uniform distribution in that region and by choosing that point at which the function has the smallest value.

On the basis of the program `E1MIN` write a program that determines the absolute minimum of the function $f_7(\mathbf{x})$ described in Sect. 10.19.1. A first approximation $\mathbf{x}_0$ within the region

$$-10 < x_{0i} < 10 \quad , \qquad i = 1, 2, 3 \quad ,$$

is to be determined by the Monte Carlo method. Perform the search for the first approximation with N points generated at random and allow for an interactive input of N.

10.20.2 Determination of the Parameters of a Breit–Wigner Distribution from the Elements of a Sample

By modifying a copy of `E2MIN` produce a program that simulates a sample from a Breit–Wigner distribution with mean a and full width at half maximum Γ, and that subsequently determines the numerical values of these parameters by minimization of the negative log-likelihood function of the sample. Allow for interactive input of the sample size and the parameters a and Γ used in the simulation.

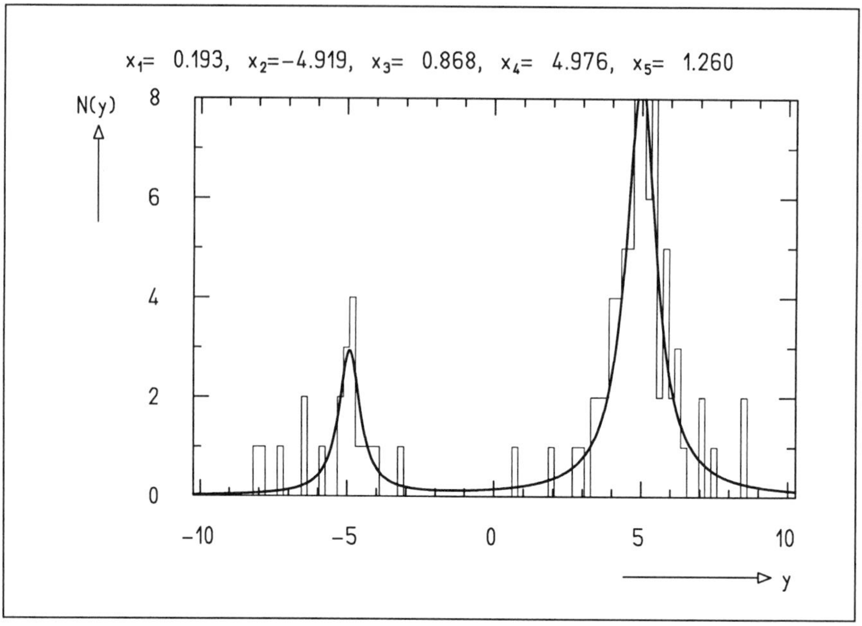

Fig. 10.20. The histogram describes a sample of size 100 which was drawn from a population described by $f(y)$ with the parameters $w = 0.2$, $a_1 = -5$, $\Gamma_1 = 1$, $a_2 = 5$, $\Gamma_2 = 1$. The maximum-likelihood method yields the parameters displayed at the top of the figure. The curve is a distribution with these parameters.

10.20.3 Determination of the Parameters of a Breit–Wigner Distribution from the Histogram of a Sample

Modify a copy of the program E4MIN such that the sample is not taken from a normal distribution but from a Breit–Wigner distribution with mean a and width Γ, and determine these two parameters from the histogram of the sample.

10.20.4 Determination of the Parameters the Sum of Two Breit–Wigner Distributions from the Histogram of a Sample

A sample is to be taken from a population with the probability density

$$f(y) = wg(y; a_1, \Gamma_1) + (1 - w)g(y; a_2, \Gamma_2) \quad .$$

Here

$$g(y; a, \Gamma) = \frac{2}{\pi \Gamma} \frac{\Gamma^2}{4(y - a)^2 + \Gamma^2}$$

is the probability density of a Breit–Wigner distribution with mean a and width Γ. The quantity w, $0 \le w \le 1$, is a weighting factor. From an analysis of a histogram of the sample, the five parameters

$$x_1 = w \quad , \qquad x_2 = a_1 \quad , \qquad x_3 = \Gamma_1 \quad , \qquad x_4 = a_2 \quad , \qquad x_5 = \Gamma_2$$

are to be estimated. In the analysis, as in program E4MIN, allow the user to choose either to minimize the negative of the likelihood function (10.18.10) (using Poisson statistics) or to minimize the sum of squares (10.18.12). Figure 10.20 shows an example. Start from a copy of the program E4MIN. For the computation of errors you can limit yourself to the determination of the covariance matrix using MINCOV. Begin with an interactive input of the sample size and the parameter values for which the sample is generated. Next, ask for the initial values of the parameters to be used in the minimization and for each parameter allow the user to choose whether the parameter is to be kept fixed during the minimization. Running the program for particular values of the parameters you will find that for $a_1 \approx a_2$, convergence of the methods can usually be reached only if Γ_1 and Γ_2 are kept fixed.

11. Analysis of Variance

The analysis of variance (or ANOVA), originally developed by R. A. FISHER, concerns testing the hypothesis of equal means of a number of samples. Such problems occur, for example, in the comparison of a series of measurements carried out under different conditions, or in quality control of samples produced by different machines. One tries to discover what influence the changing of *external variables* (e.g., experimental conditions, the number of a machine) has on a sample. For the simple case of only two samples, this problem can also be solved with Student's difference test (Sect. 8.3).

We speak of *one-way analysis of variance*, or also one-way classification, when only one external variable is changed. The evaluation of a series of measurements of an object micrometer performed with different microscopes can serve as an example. One has a *two- (or more) way analysis of variance* (two-way classification) when several variables are changed simultaneously. In the example above, if different observers carry out the series of measurements with each microscope, then a two-way analysis of variance can investigate influences of both the observer and the instrument on the result.

11.1 One-Way Analysis of Variance

Let us consider a sample of size n, which can be divided into t groups according to a certain criterion A. Clearly the criterion must be related to the sampling or measuring process. We say that the groups are constructed according to the *classification* A. We assume that the populations from which the t subsamples are taken are normally distributed with the same variance σ^2. We now want to test the hypothesis that the mean values of these populations are also equal. If this hypothesis is true, then all of the samples come from the same population. We can then apply the results of Sect. 6.4 (samples from subpopulations). Using the same notation as there, we have t groups of size n_i with

$$n = \sum_{i=1}^{t} n_i$$

and we write the jth element of the ith group as x_{ij}. The sample mean of the ith group is

$$\bar{x}_i = \frac{1}{n_i} \sum_{j=1}^{n_i} x_{ij} \tag{11.1.1}$$

and the mean of the entire sample is

$$\bar{x} = \frac{1}{n} \sum_{i=1}^{t} \sum_{j=1}^{n_i} x_{ij} = \frac{1}{n} \sum_{i=1}^{t} n_i \bar{x}_i \quad . \tag{11.1.2}$$

We now construct the *sum of squares*

$$\begin{aligned} Q &= \sum_{i=1}^{t} \sum_{j=1}^{n_i} (x_{ij} - \bar{x})^2 = \sum_{i=1}^{t} \sum_{j=1}^{n_i} (x_{ij} - \bar{x}_i + \bar{x}_i - \bar{x})^2 \\ &= \sum_{i=1}^{t} \sum_{j=1}^{n_i} (x_{ij} - \bar{x}_i)^2 + \sum_{i=1}^{t} \sum_{j=1}^{n_i} (\bar{x}_i - \bar{x})^2 + 2 \sum_{i=1}^{t} \sum_{j=1}^{n_i} (x_{ij} - \bar{x}_i)(\bar{x}_i - \bar{x}) \quad . \end{aligned}$$

The last term vanishes because of (11.1.1) and (11.1.2). One therefore has

$$\begin{aligned} Q &= \sum_{i=1}^{t} \sum_{j=1}^{n_i} (x_{ij} - \bar{x})^2 = \sum_{i=1}^{t} n_i (\bar{x}_i - \bar{x})^2 + \sum_{i=1}^{t} \sum_{j=1}^{n_i} (x_{ij} - \bar{x}_i)^2 \quad , \\ Q &= Q_A + Q_W \quad . \end{aligned} \tag{11.1.3}$$

The first term is the *sum of squares between the groups* obtained with the classification A. The second term is a sum over the *sums of squares within a group*. The sum of squares Q is decomposed into a sum of two sums of squares corresponding to different "sources" — the variation of means within the classification A and the variation of measurements within the groups. If our hypothesis is correct, then Q is a sum of squares from a normal distribution, i.e., Q/σ^2 follows a χ^2-distribution with $n - 1$ degrees of freedom. Correspondingly, for each group the quantity

$$\frac{Q_i}{\sigma^2} = \frac{1}{\sigma^2} \sum_{j=1}^{n_i} (x_{ij} - \bar{x}_i)^2$$

follows a χ^2-distribution with $n_i - 1$ degrees of freedom. The sum

$$\frac{Q_W}{\sigma^2} = \sum_{i=1}^{t} \frac{Q_i}{\sigma^2}$$

is then described by a χ^2-distribution with $\sum_i (n_i - 1) = n - t$ degrees of freedom (see Sect. 6.6). Finally, Q_A/σ^2 follows a χ^2-distribution with $t - 1$ degrees of freedom.

The expressions

$$\begin{aligned} s^2 &= \frac{Q}{n-1} = \frac{1}{n-1}\sum_i\sum_j(x_{ij}-\bar{x})^2 \quad , \\ s_A^2 &= \frac{Q_A}{t-1} = \frac{1}{t-1}\sum_i n_i(\bar{x}_i-\bar{x})^2 \quad , \\ s_W^2 &= \frac{Q_W}{n-t} = \frac{1}{n-t}\sum_i\sum_j(x_{ij}-\bar{x}_i)^2 \end{aligned} \tag{11.1.4}$$

are unbiased estimators of the population variances. (In Sect. 6.5 we called such expressions mean squares.) The ratio

$$F = s_A^2/s_W^2 \tag{11.1.5}$$

can thus be used to carry out an F-test.

If the hypothesis of equal means is false, then the values $\bar{x}_i$ of the individual groups will be quite different. Thus s_A^2 will be relatively large, while s_W^2, which is the mean of the variances of the individual groups, will not change much. This means that the ratio (11.1.5) will be large. Therefore one uses a one-sided F-test. The hypothesis of equal means is rejected at the significance level α if

$$F = s_A^2/s_W^2 > F_{1-\alpha}(t-1,\, n-t) \quad . \tag{11.1.6}$$

The sums of squares can be computed according to two equivalent formulas,

$$\begin{aligned} Q &= \sum_i\sum_j(x_{ij}-\bar{x})^2 = \sum_i\sum_j x_{ij}^2 - n\bar{x}^2 \quad , \\ Q_A &= \sum_i n_i(\bar{x}_i-\bar{x})^2 = \sum_i n_i\bar{x}_i^2 - n\bar{x}^2 \quad , \\ Q_W &= \sum_i\sum_j(x_{ij}-\bar{x}_i)^2 = \sum_i\sum_j x_{ij}^2 - \sum_i n_i\bar{x}_i^2 \quad . \end{aligned} \tag{11.1.7}$$

The expression on the right of each line is usually easier to compute. Since each sum of squares is obtained by computing the difference of two relatively large numbers, one must pay attention to possible problems with errors in rounding. Although one only needs the sums Q_A and Q_W in order to compute the ratio F, it is recommended to compute Q as well, since one can then perform a check using (11.1.3), i.e., $Q = Q_A + Q_W$. The check is only meaningful when Q is computed with the left-hand form of (11.1.3). Usually the results of an analysis of variance are summarized in a so-called *analysis of variance table*, (or ANOVA table) as shown in Table 11.1.

Before carrying out an analysis of variance one must consider whether the requirements are met under which the procedure has been derived. In particular, one must check the assumption of a normal distribution for the

Table 11.1. ANOVA table for one-way classification.

Source	SS (Sum of squares)	DF (Degrees of freedom)	MS (Mean square)	F
Between the groups	Q_A	$t-1$	$s_A^2 = \frac{Q_A}{t-1}$	
Within the groups	Q_W	$n-t$	$s_W^2 = \frac{Q_W}{n-t}$	$F = \frac{s_A^2}{s_W^2}$
Sum	Q	$n-1$	$s^2 = \frac{Q}{n-1}$	

measurements within each group. This is by no means certain in every case. If, for example, the measured values are always positive (e.g., the length or weight of an object) and if the standard deviation is of a magnitude comparable to the measured values, then the probability density can be asymmetric and thus not Gaussian. If, however, the original measurements (let us denote them for the moment by x') are transformed using a monotonic transformation such as

$$x = a \log(x' + b) \quad , \tag{11.1.8}$$

where a and b are appropriately chosen constants, then a normal distribution can often be sufficiently well approximated. Other transformations sometimes used are $x = \sqrt{x'}$ or $x = 1/x'$.

Example 11.1: One-way analysis of variance of the influence of various drugs

The spleens of mice with cancer are often attacked particularly strongly. The weight of the spleen can thus serve as a measure of the reaction to various drugs. The drugs (I, II, and III) were used to treat ten mice each. Table 11.2 contains the measured spleen weights, which have already been transformed according to $x = \log x'$, where x' is the weight in grams. Most of the calculation is presented in Table 11.2. Table 11.3 contains the resulting ANOVA table. Since even at a significance level of 50 % the F-test gives $F_{0.5}(2, 24) = 3.4$, one cannot reject the hypothesis of equal mean values. The experiment thus showed no significant difference in the effectiveness of the three drugs. ∎

11.2 Two-Way Analysis of Variance

Before we turn to analysis of variance with two external variables, we would like to examine more carefully the results obtained for one-way classification.

Table 11.2. Data for Example 11.1.

Experiment number	Group I	II	III	
1	19	40	32	
2	45	28	26	
3	26	26	30	
4	23	15	17	
5	36	24	23	
6	23	26	24	
7	26	36	29	
8	33	27	20	
9	22	28	—	
10	—	19	—	$\sum_i \sum_j x_{ij}^2 = 20\,607$
$\sum_j x_j$	253	269	201	$\sum_i \sum_j x_{ij} = 723$
n_i	9	10	8	$n = 27$
				$n\bar{x}^2 = 19\,360$
$\bar{x}_i$	28.11	26.90	25.13	$\bar{x} = 26.78$
$\bar{x}_i^2$	790.23	723.61	631.52	$\sum_i n_i \bar{x}_i^2 = 19\,398$

Table 11.3. ANOVA table for Example 11.1.

Source	SS	DF	MS	F
Between the groups	38	2	19.0	0.377
Within the groups	1209	24	50.4	
Sum	1247	26	47.8	

We denoted the jth measurement of the quantity x in group i by x_{ij}. We now assume for simplicity that each group contains the same number of measurements, i.e., $n_i = J$. In addition we denote the total number of groups by I. The classification into individual groups was done according to the criterion A, e.g., the production number of a microscope, by which the groups can be distinguished. The labeling according to measurement and group is illustrated in Table 11.4.

We can write the individual means of the groups in the form

$$\bar{x}_{..} = \bar{x} = \frac{1}{IJ} \sum_i \sum_j x_{ij} \quad ,$$

Table 11.4. One-way classification.

Measurement	Classification A					
number	A_1	A_2	$\dots$	A_i	$\dots$	A_I
1	x_{11}	x_{22}		x_{i1}		x_{I1}
2	x_{12}	x_{22}		x_{i2}		x_{I2}
$\vdots$						
j	x_{1j}	x_{2j}		x_{ij}		x_{Ij}
$\vdots$						
J	x_{1J}	x_{2J}		x_{iJ}		x_{IJ}

$$\begin{aligned} \bar{x}_{i.} &= \frac{1}{J}\sum_j x_{ij} \quad , \\ \bar{x}_{.j} &= \frac{1}{I}\sum_i x_{ij} \quad . \end{aligned} \tag{11.2.1}$$

Here a point denotes summation over the index that it replaces. This notation allows a simple generalization to a larger number of indices. The analysis of variance with one-way classification is based on the assumption that the measurements within a group only differ by the measurement errors, which follow a normal distribution with a mean of zero and variance σ^2. That is, we consider the *model*

$$x_{ij} = \mu_i + \varepsilon_{ij} \quad . \tag{11.2.2}$$

The goal of an analysis of variance was to test the hypothesis

$$H_0(\mu_1 = \mu_2 = \ldots = \mu_I = \mu) \quad . \tag{11.2.3}$$

By choosing measurements out of a certain group i and by applying the maximum-likelihood method to (11.2.2) one obtains the estimator

$$\widetilde{\mu}_i = \bar{x}_{i.} = \frac{1}{J}\sum_j x_{ij} \quad . \tag{11.2.4}$$

If H_0 is true, then one has

$$\widetilde{\mu} = \bar{x} = \frac{1}{IJ}\sum_i\sum_j x_{ij} = \frac{1}{I}\sum_i \widetilde{\mu}_i \quad . \tag{11.2.5}$$

The (composite) alternative hypothesis is that not all of the μ_i are equal. We want, however, to retain the concept of the overall mean and we write

$$\mu_i = \mu + a_i \quad .$$

The model (11.2.2) then has the form

$$x_{ij} = \mu + a_i + \varepsilon_{ij} \quad . \tag{11.2.6}$$

Between the quantities a_i, which represent a measure of the deviation of the mean for the ith group from the overall mean, one has the relation

$$\sum_i a_i = 0 \quad . \tag{11.2.7}$$

The maximum-likelihood estimators for the a_i are

$$\tilde{a}_i = \bar{x}_{i.} - \bar{x} \quad . \tag{11.2.8}$$

The one-way analysis of variance of Sect. 11.1 was derived from the identity

$$x_{ij} - \bar{x} = (\bar{x}_{i.} - \bar{x}) + (x_{ij} - \bar{x}_{i.}) \quad , \tag{11.2.9}$$

which describes the deviation of the individual measurements from the overall mean. The sum of squares Q of these deviations could then be decomposed into the terms Q_A and Q_W; cf. (11.1.3).

After this preparation we now consider a two-way classification, where the measurements are divided into groups according to two criteria, A and B. The measurements x_{ijk} belong to class A_i, which is given by the classification according to A, and also to class B_j. The index k denotes the measurement number within the group that belongs to both class A_i and class B_j.

A two-way classification is said to be *crossed*, when a certain classification B_j has the same meaning for all classes A. If, for example, microscopes are classified by A and observers by B, and if each observer carries out a measurement with each microscope, then the classifications are crossed. If, however, one compares the microscopes in different laboratories, and if therefore in each laboratory a different group of J observers makes measurements with a certain microscope i, then the classification B is said to be *nested* in A. The index j then merely counts the classes B within a certain class A.

The simplest case is a crossed classification with only *one observation*. Since then $k = 1$ for all observations x_{ijk}, we can drop the index k. One uses the model

$$x_{ij} = \mu + a_i + b_j + \varepsilon_{ij} \quad , \qquad \sum_i a_i = 0 \quad , \qquad \sum_j b_j = 0 \quad , \tag{11.2.10}$$

where ε is normally distributed with mean zero and variance σ^2. The null hypothesis says that by classification according to A or B, no deviation from the overall mean occurs. We write this in the form of two individual hypotheses,

$$H_0^{(A)}(a_1 = a_2 = \cdots = a_I = 0) \quad , \qquad H_0^{(B)}(b_1 = b_2 = \cdots = b_J = 0) \quad . \tag{11.2.11}$$

The least-squares estimators for a_i and b_j are

$$\widetilde{a}_i = \bar{x}_{i.} - \bar{x} \quad , \qquad \widetilde{b}_j = \bar{x}_{.j} = \bar{x} \quad .$$

In analogy to Eq. (11.2.9) we can write

$$x_{ij} - \bar{x} = (\bar{x}_{i.} - \bar{x}) + (\bar{x}_{.j} - \bar{x}) + (x_{ij} - \bar{x}_{i.} - \bar{x}_{.j} + \bar{x}) \quad . \tag{11.2.12}$$

In a similar way the sum of squares can be written

$$\sum_i \sum_j (x_{ij} - \bar{x})^2 = Q = Q_A + Q_B + Q_W \quad , \tag{11.2.13}$$

where

$$\begin{aligned} Q_A &= J \sum_i (\bar{x}_{i.} - \bar{x})^2 = J \sum_i \bar{x}_{i.}^2 - IJ\bar{x}^2 \quad , \\ Q_B &= I \sum_j (\bar{x}_{.j} - \bar{x})^2 = I \sum_j \bar{x}_{.j}^2 - IJ\bar{x}^2 \quad , \\ Q_W &= \sum_i \sum_j (x_{ij} - \bar{x}_{i.} - \bar{x}_{.j} + \bar{x})^2 \\ &= \sum_i \sum_j x_{ij}^2 - J \sum_i \bar{x}_{i.}^2 - I \sum_j \bar{x}_{.j}^2 + IJ\bar{x}^2 \quad . \end{aligned} \tag{11.2.14}$$

When divided by the corresponding number of degrees of freedom, these sums are estimators of σ^2, providing that the hypotheses (11.2.11) are correct. The hypotheses $H_0^{(A)}$ and $H_0^{(B)}$ can be tested individually by using the ratio

$$F^{(A)} = (s_A/s_W)^2 \quad , \qquad F^{(B)} = (s_B/s_W)^2 \quad . \tag{11.2.15}$$

Here one uses one-sided F-tests as in Sect. 11.1. The overall situation can be summarized in an ANOVA table (Table 11.5).

If more than one observation is made in each group, then the crossed classification can be generalized in various ways. The most important generalization involves *interaction* between the classes. One then has the model

$$x_{ijk} = \mu + a_i + b_j + (ab)_{ij} + \varepsilon_{ijk} \quad . \tag{11.2.16}$$

The quantity $(ab)_{ij}$ is called the *interaction* between the classes A_i and B_j. It describes the deviation from the group mean that occurs because of the specific interaction of A_i and B_j. The parameters a_i, b_j, $(ab)_{ij}$ are related by

$$\sum_i a_i = \sum_j b_j = \sum_i \sum_j (ab)_{ij} = 0 \quad . \tag{11.2.17}$$

Their maximum-likelihood estimators are

$$\begin{aligned} &\widetilde{a}_i = \bar{x}_{i..} - \bar{x} \quad , \qquad \widetilde{b}_j = \bar{x}_{.j.} - \bar{x} \quad , \\ &\widetilde{(ab)}_{ij} = \bar{x}_{ij.} + \bar{x} - \bar{x}_{i..} - \bar{x}_{.j.} \quad . \end{aligned} \tag{11.2.18}$$

Table 11.5. Analysis of variance table for crossed two-way classification with only one observation.

Source	SS	DF	MS	F
Class. A	Q_A	$I-1$	$s_A^2 = \dfrac{Q_A}{I-1}$	$F^{(A)} = \dfrac{s_A^2}{s_W^2}$
Class. B	Q_B	$J-1$	$s_B^2 = \dfrac{Q_B}{J-1}$	$F^{(B)} = \dfrac{s_B^2}{s_W^2}$
Within groups	Q_W	$(I-1)(J-1)$	$s_W^2 = \dfrac{Q_W}{(I-1)(J-1)}$	
Sum	Q	$IJ-1$	$s^2 = \dfrac{Q}{IJ-1}$	

The null hypothesis can be divided into three individual hypotheses,

$$H_0^{(A)}(a_i = 0;\ i = 1,2,\ldots,I) \quad , \qquad H_0^{(B)}(b_j = 0;\ j = 1,2,\ldots,J) \quad ,$$
$$H_0^{(AB)}((ab)_{ij} = 0;\ i = 1,2,\ldots,I;\ j = 1,2,\ldots,J) \quad , \tag{11.2.19}$$

which can then be tested individually. The analysis of variance is based on the identity

$$\begin{aligned} x_{ijk} - \bar{x} &= (\bar{x}_{i..} - \bar{x}) + (\bar{x}_{.j.} - \bar{x}) \\ &\quad + (\bar{x}_{ij.} + \bar{x} - \bar{x}_{i..} - \bar{x}_{.j.}) + (x_{ijk} - \bar{x}_{ij.}) \quad , \end{aligned} \tag{11.2.20}$$

which allows the decomposition of the sum of squares of deviations into four terms,

$$\begin{aligned} Q &= \sum_i \sum_j \sum_k (x_{ijk} - \bar{x})^2 \\ &= Q_A + Q_B + Q_{AB} + Q_W \quad , \\ Q_A &= JK \sum_i (\bar{x}_{i..} - \bar{x})^2 \quad , \\ Q_B &= IK \sum_j (\bar{x}_{.j.} - \bar{x})^2 \quad , \\ Q_{AB} &= K \sum_i \sum_j (\bar{x}_{ij.} + \bar{x} - \bar{x}_{i..} - \bar{x}_{.j.})^2 \quad , \\ Q_W &= \sum_i \sum_j \sum_k (x_{ijk} - \bar{x}_{ij.})^2 \quad . \end{aligned} \tag{11.2.21}$$

The degrees of freedom and mean squares as well as the F-ratio, which can be used for testing the hypotheses, are given in Table 11.6.

Table 11.6. Analysis of variance table for crossed two-way classification.

Source	SS	DF	MS	F
Class. A	Q_A	$I-1$	$s_A^2 = \dfrac{Q_A}{I-1}$	$F^{(A)} = \dfrac{s_A^2}{s_W^2}$
Class. B	Q_B	$J-1$	$s_B^2 = \dfrac{Q_B}{J-1}$	$F^{(B)} = \dfrac{s_B^2}{s_W^2}$
Interaction	Q_{AB}	$(I-1)(J-1)$	$s_{AB}^2 = \dfrac{Q_{AB}}{(I-1)(J-1)}$	$F^{(AB)} = \dfrac{s_{AB}^2}{s_W^2}$
Within groups	Q_W	$IJ(K-1)$	$s_W^2 = \dfrac{Q_W}{IJ(K-1)}$	
Sum	Q	$IJK-1$	$s^2 = \dfrac{Q}{IJK-1}$	

Finally, we will give the simplest case of a *nested two-way classification.* Because the classification B is only defined within the individual classes of A, the terms b_j and $(ab)_{ij}$ from Eq. (11.2.10) are not defined, since they imply a sum over i for fixed j. Therefore one uses the model

$$x_{ijk} = \mu + a_i + b_{ij} + \varepsilon_{ijk} \tag{11.2.22}$$

with

$$\begin{array}{ll} \sum_i a_i = 0 \quad , & \sum_i \sum_j b_{ij} = 0 \quad , \\ \widetilde{a}_i = \bar{x}_{i..} - \bar{x} \quad , & \widetilde{b}_{ij} = \bar{x}_{ij.} - \bar{x}_{i..} \quad . \end{array}$$

The term b_{ij} is a measure of the deviation of the measurements of class B_j within class A_i from the overall mean of class A_i. The null hypothesis consists of

$$\begin{array}{l} H_0^{(A)}(a_i = 0;\, i = 1, 2, \ldots, I) \quad , \\ H_0^{(B(A))}(b_{ij} = 0;\, i = 1, 2, \ldots, I;\, j = 1, 2, \ldots, J) \quad . \end{array} \tag{11.2.23}$$

An analysis of variance for testing these hypotheses can be carried out with the help of Table 11.7. Here one has

$$\begin{aligned} Q_A &= JK \sum_i (\bar{x}_{i..} - \bar{x})^2 \quad , \\ Q_{B(A)} &= K \sum_i \sum_j (\bar{x}_{ij.} - \bar{x}_{i..})^2 \quad , \\ Q_W &= \sum_i \sum_j \sum_k (x_{ijk} - \bar{x}_{ij.})^2 \quad , \\ Q &= Q_A + Q_{B(A)} + Q_W = \sum_i \sum_j \sum_k (x_{ijk} - \bar{x})^2 \quad . \end{aligned}$$

In a similar way one can construct various models for two-way or multiple classification. For each model the total sum of squares is decomposed into a certain sum of individual sums of squares, which, when divided by the corresponding number of degrees of freedom, can be used to carry out an F-test. With this, the hypotheses implied by the model can be tested.

Some models are, at least formally, contained within others. For example one finds by comparing Tables 11.6 and 11.7 the relation

$$Q_{B(A)} = Q_B + Q_{AB} \quad . \tag{11.2.24}$$

A similar relation holds for the corresponding number of degrees of freedom,

$$f_{B(A)} = f_B + f_{AB} \quad . \tag{11.2.25}$$

We will use these relations in the following program. The routine (AVTBLE) treats both crossed and nested classifications. Only the second routine, AVOUTP, requires the type of classification for printing the results.

Table 11.7. Analysis of variance table for nested two-way classification.

Source	SS	DF	MS	F
Class. A	Q_A	$I-1$	$s_A^2 = \frac{Q_A}{I-1}$	$F^{(A)} = \frac{s_A^2}{s_W^2}$
Class. B				
Within A	$Q_{B(A)}$	$I(J-1)$	$s_{B(A)}^2 = \frac{A_{B(A)}}{I(J-1)}$	$F^{(B(A))} = \frac{s_{B(A)}^2}{s_W^2}$
Within groups	Q_W	$IJ(K-1)$	$s_W^2 = \frac{Q_W}{IJ(K-1)}$	
Sum	Q	$IJK-1$	$s^2 = \frac{Q}{IJK-1}$	

Program 11.1: Subroutine AVTBLE carries out a two-way analysis of variance and returns the individual elements of an ANOVA table

CALLING SEQUENCE: CALL AVTBLE(X,NI,NJ,NK,XB,XBI,XBJ,XBIJ,Q,S,F,NDF,ALPHA)

INPUT ARGUMENTS: X — $I \times J \times K$ matrix of measurements
NI — number I of classes A
NJ — number J of classes B
NK — number K of measurements within each classification

OUTPUT ARGUMENTS: XB — mean value $\bar{x}$

XBI — I-dimensional vector of mean values $\bar{x}_{i..}$

XBJ — J-dimensional vector of mean values $\bar{x}_{.j.}$

XBIJ — $I \times J$ matrix of mean values $\bar{x}_{ij.}$

Q — vector of the six sums of squares Q_A, Q_B, Q_{AB}, $Q_{B(A)}$, Q_W, Q

S — vector of the six variance estimators s^2_A, s^2_B, s^2_{AB}, $s^2_{B(A)}$, s^2_W, s^2

NDF — vector of the six numbers of degrees of freedom corresponding to the six Q values

F — vector of the four F-ratios $F^{(A)}$, $F^{(B)}$, $F^{(AB)}$, $F^{(B(A))}$

ALPHA — vector of four significance levels giving the limits for rejecting the null hypothesis when carrying out F-tests with the F-ratios

```
      SUBROUTINE AVTBLE(X,NI,NJ,NK,XB,XBI,XBJ,XBIJ,Q,S,F,NDF,A)
      IMPLICIT DOUBLE PRECISION (A-H,O-Z)
      DIMENSION X(NI,NJ,NK),XBI(NI),XBJ(NJ),XBIJ(NI,NJ),Q(6),DF(6)
      DIMENSION S(6),F(4),NDF(6),A(4)
      PARAMETER(ZERO=0.D0, ONE=1.D0)
      AI=DBLE(NI)
      AK=DBLE(NK)
      AJ=DBLE(NJ)
C compute means
      XB=ZERO
      DO 10 I=1,NI
        XBI(I)=ZERO
   10 CONTINUE
      DO 20 J=1,NJ
        XBJ(J)=ZERO
        DO 20 I=1,NI
          XBIJ(I,J)=ZERO
   20 CONTINUE
      DO 40 I=1,NI
        DO 40 J=1,NJ
          DO 30 K=1,NK
            XBIJ(I,J) =XBIJ(I,J)+X(I,J,K)
   30     CONTINUE
          XBIJ(I,J) =XBIJ(I,J)/AK
   40 CONTINUE
      DO 60 I=1,NI
        DO 50 J=1,NJ
          XBI(I)=XBI(I)+XBIJ(I,J)
   50   CONTINUE
        XBI(I)=XBI(I)/AJ
   60 CONTINUE
      DO 80 J=1,NJ
        DO 70 I=1,NI
          XBJ(J)=XBJ(J)+XBIJ(I,J)
   70   CONTINUE
        XBJ(J)=XBJ(J)/AI
        XB=XB+XBJ(J)
   80 CONTINUE
      XB=XB/AJ
C compute sums of squares
      DO 90 L=1,6
        Q(L)=ZERO
```

```
   90 CONTINUE
      DO 100 I=1,NI
        Q(1)=Q(1)+(XBI(I)-XB)**2
  100 CONTINUE
      Q(1)=AJ*AK*Q(1)
      DO 110 J=1,NJ
        Q(2)=Q(2)+(XBJ(J)-XB)**2
  110 CONTINUE
      Q(2)=AI*AK*Q(2)
      DO 120 I=1,NI
        DO 120 J=1,NJ
          Q(3)=Q(3)+(XBIJ(I,J)+XB-XBI(I)-XBJ(J))**2
          DO 120 K=1,NK
            Q(5)=Q(5)+(X(I,J,K)-XBIJ(I,J))**2
            Q(6)=Q(6)+(X(I,J,K)-XB)**2
  120 CONTINUE
      Q(3) =AK*Q(3)
      Q(4)=Q(2)+Q(3)
C determine degrees of freedom
      DF(1)=AI-ONE
      DF(2)=AJ-ONE
      DF(3)=(AI-ONE)*(AJ-ONE)
      DF(4)=DF(2)+DF(3)
      DF(5)=AI*AJ*(AK-ONE)
      DF(6)=AI*AJ*AK-ONE
C mean squares
      DO 130 L=1,6
        S(L)=Q(L)/DF(L)
  130 CONTINUE
C F quotients
      DO 140 L=1,4
        F(L)=S(L)/S(5)
  140 CONTINUE
C degrees of freedom as integers
      DO 150 L=1,6
        NDF(L)=NINT(DF(L))
  150 CONTINUE
C levels of significance
      DO 160 L=1,4
        A(L)=ONE-SCFTST(F(L),NDF(L),NDF(5))
  160 CONTINUE
      END
```

Program 11.2: Subroutine AVOUTP prints the results provided by the routine AVTBLE as an ANOVA table

CALLING SEQUENCE: CALL AVOUTP(X,NI,NJ,NK,Q,S,F,NDF,ALPHA,NTYPE)

INPUT ARGUMENTS: X,NI,NJ,NK,Q,S,F,NDF,ALPHA — see program AVTBLE

NTYPE — gives the type of classification, crossed (NTYPE = 0) or nested (NTYPE = 1)

```
      SUBROUTINE AVOUTP(X,NI,NJ,NK,Q,S,F,NDF,ALPHA,NTYPE)
      IMPLICIT DOUBLE PRECISION (A-H,O-Z)
      DIMENSION X(NI,NJ,NK),Q(6),S(6),F(4),NDF(6),SOURCE(6),
     +ALPHA(4)
      CHARACTER*4 SOURCE
      DATA SOURCE /'A   ','B   ','INT.','B(A)','W   ','TTL.'/
      WRITE (*,90)
      DO 10 I=1,NI
        DO 10 J=1,NJ
          WRITE (6,60)I,J,(X(I,J,K),K=1,NK)
   10 CONTINUE
      WRITE (6,70)
      WRITE (6,80) SOURCE(1),Q(1),NDF(1),S(1),F(1),ALPHA(1)
      IF(NTYPE.EQ.0) THEN
        DO 20 L=2,3
          WRITE (6,80) SOURCE(L),Q(L),NDF(L),S(L),F(L),ALPHA(L)
   20   CONTINUE
      ELSE
        WRITE (6,80) SOURCE(4),Q(4),NDF(4),S(4),F(4),ALPHA(4)
      END IF
      DO 30 L=5,6
        WRITE (6,80) SOURCE(L),Q(L),NDF(L),S(L)
   30 CONTINUE
C format statements
   40 FORMAT(4I10)
   50 FORMAT(10F10.5)
   60 FORMAT(' A(',I2,'), B(',I2,') ',(10F10.2))
   70 FORMAT(///10X,'Analysis of Variance Table' //
     +' Source       Sum of     Degrees of    Mean       F Ratio   ',
     +' Alpha'/'                 Squares    Freedom       Square' //)
   80 FORMAT(3X,A4,2X,F10.2,5X,I3,5X,F10.2,F10.5,F15.12)
   90 FORMAT( '                    DATA'/)
      END
```

Example 11.2: Two-way analysis of variance in cancer research

Two groups of rats are injected with the amino acid thymidine containing traces of tritium, a radioactive isotope of hydrogen. In addition, one of the groups receives a certain carcinogen. The incorporation of thymidine into the skin of the rats is investigated as a function of time by measuring the number of tritium decays per unit area of skin. The classifications are crossed since the time dependence is controlled in the same way for both series of test animals. The measurements are compiled in Table 11.8. The numbers are already transformed from the original counting rates x' according to $x = 50 \log x' - 100$. The results of the program are shown in Table 11.9. There is no doubt that the presence or absence of the carcinogen (classification A has an influence on the result, since the ratio $F^{(A)}$ is very large. We now want to test the existence of a time dependence (classification B) and of an interaction between A and B at a significance level of $\alpha = 0.01$. From Table I.8 we find $F_{0.99} = 2.72$. The hypotheses of time independence and vanishing interaction must therefore be rejected. Table 11.9 also contains the values of α for which the hypothesis would not need to be rejected. They are very small. ■

Table 11.8. Data for Example 11.2.

Obs.	Injection	Time after injection (hours)									
no.		4	8	12	16	20	24	28	32	36	48
1		34	54	44	51	62	61	59	66	52	52
2	Thymidine	40	57	52	46	61	70	67	59	63	50
3		38	40	53	51	54	64	58	67	60	44
4		36	43	51	49	60	68	66	58	59	52
1		28	23	42	43	31	32	25	24	26	26
2	Thymidine	32	23	41	48	45	38	27	26	31	27
3	and	34	29	34	36	41	32	27	32	25	27
4	Carcinogen	27	30	39	43	37	34	28	30	26	30

11.3 Example Programs

The source files of the example programs are available in the subdirectory `\DATAN\EXASRC`.

11.3.1 Main Program `E1AV` to Demonstrate Subprograms `AVTBLE` and `AVOUTP`

The short program contains the data of 11.2 in a `DATA` statement. By calling `AVTBLE` and `AVOUTP`, the analysis of variance is performed and the results are output in the form of Table 11.9.

11.3.2 Main Program `E2AV` to Generate Simulated Data and Perform an Analysis of Variance

The program first asks for a parameter σ, for two seeds used by the random number generator, and for the choice between crossed and nested classification. Next it generates data of the simple form

$$x_{ijk} = \varepsilon_{ijk} \quad .$$

Here, the quantities ε_{ijk} are drawn from a normal distribution with mean zero and standard deviation σ. The data correspond to the null hypotheses (11.2.19) and (11.2.23).

An analysis of variance is performed on the data and the results are presented numerically.

Suggestion: Use a fixed value for σ (e.g., $\sigma = 0.1$) and run the program consecutively with different seeds.

Table 11.9. Printout from Example 11.2.

DATA

A(1), B(1)	34.00	40.00	38.00	36.00
A(1), B(2)	54.00	57.00	40.00	43.00
A(1), B(3)	44.00	52.00	53.00	51.00
A(1), B(4)	51.00	46.00	51.00	49.00
A(1), B(5)	62.00	61.00	54.00	60.00
A(1), B(6)	61.00	70.00	64.00	68.00
A(1), B(7)	59.00	67.00	58.00	66.00
A(1), B(8)	66.00	59.00	67.00	58.00
A(1), B(9)	52.00	63.00	60.00	59.00
A(1), B(10)	52.00	50.00	44.00	52.00
A(2), B(1)	28.00	32.00	34.00	27.00
A(2), B(2)	23.00	23.00	29.00	30.00
A(2), B(3)	42.00	41.00	34.00	39.00
A(2), B(4)	43.00	48.00	36.00	43.00
A(2), B(5)	31.00	45.00	41.00	37.00
A(2), B(6)	32.00	38.00	32.00	34.00
A(2), B(7)	25.00	27.00	27.00	28.00
A(2), B(8)	24.00	26.00	32.00	30.00
A(2), B(9)	26.00	31.00	25.00	26.00
A(2), B(10)	26.00	27.00	27.00	30.00

Analysis of Variance Table

Source	Sum of Squares	Degrees of Freedom	Mean Square	F Ratio	Alpha
A	9 945.80	1	9945.80	590.547 25	0.00E-10
B	1 917.50	9	213.06	12.650 50	0.54E-10
INT.	2 234.95	9	248.33	14.744 85	0.03E-10
W	1 010.50	60	16.84		
TTL.	15 108.75	79	191.25		

11.4 Programming Problems

The source files of the example solutions are available in the subdirectory `\DATAN\SOLSRC`.

11.4.1 Two-Way Analysis of Variance with Crossed Classification

The model (11.2.16) for the data in an analysis of variance with crossed classification is

$$x_{ijk} = \mu + a_i + b_j + (ab)_{ij} + \varepsilon_{ijk} \quad .$$

The analysis of variance tests the null hypothesis

$$a_i = b_j = (ab)_{ij} = 0 \quad .$$

Data corresponding to this null hypothesis are generated by the program E2AV, Sect. 11.3.2, and are subsequently analyzed.

Write a program similar to E2AV which generates data x_{ijk} according to the above formula with

$$\begin{aligned} a_i &= \left(i - \frac{I+1}{2}\right) a \quad , \\ b_j &= \left(j - \frac{J+1}{2}\right) b \quad , \\ (ab)_{ij} &= \left[\left(i - \frac{I+1}{2}\right) + \left(j - \frac{J+1}{2}\right)\right] ab \quad . \end{aligned}$$

These relations fulfill the requirements (11.2.7). The ε_{ijk} as in E2AV are to be drawn from a normal distribution with mean zero and standard deviation σ. Allow for interactive input of the quantities a, b, ab, σ, μ. Perform an analysis of variance on the simulated data. Study different cases, e.g., $a = 0$, $b \neq 0$, $ab = 0$; $a \neq 0$, $b = 0$, $ab = 0$; $a = 0$, $b = 0$, $ab \neq 0$; etc.

11.4.2 Two-Way Analysis of Variance with Nested Classification

Modify Programming Problem 11.4.1 for the treatment of a nested classification with data of the form (11.2.22), e.g.,

$$x_{ijk} = \mu + a_i + b_{ij} + \varepsilon_{ijk}$$

and use the relations

$$\begin{aligned} a_i &= \left(i - \frac{I+1}{2}\right) a \quad , \\ b_{ij} &= \left[\left(i - \frac{I+1}{2}\right) + \left(j - \frac{J+1}{2}\right)\right] b \quad . \end{aligned}$$

12. Linear and Polynomial Regression

The fitting of a linear function (or, more generally, of a polynomial) to measured data that depend on a controlled variable is probably the most commonly occurring task in data analysis. This procedure is also referred to as *linear* (or *polynomial*) *regression*. Although we have already treated this problem in Sect. 9.4.1, we take it up again here in greater detail. Here we will use different numerical methods, emphasize the most appropriate choice for the order of the polynomial, treat in detail the question of confidence limits, and also give a procedure for the case where the measurement errors are not known.

12.1 Orthogonal Polynomials

In Sect. 9.4.1 we have already fitted a polynomial of order $r-1$,

$$\eta(t) = x_1 + x_2 t + \cdots + x_r t^{r-1} \quad , \tag{12.1.1}$$

to measured values $y_i(t_i)$, which corresponded to a given value t_i of the controlled variable t. Here the quantities $\eta(t)$ were the true values of the measured quantities $y(t)$. In Example 9.2 we saw that there can be a maximum reasonable order for the polynomial, beyond which a further increase in the order gives no significant improvement in the fit. We want here to pursue the question of how to find the optimal order of the polynomial. Example 9.2 shows that when increasing the order, all of the coefficients $x_1, x_2, \ldots$ change and that all of the coefficients are in general correlated. Because of this the situation becomes difficult to judge. These difficulties are avoided by using *orthogonal polynomials*.

Instead of (12.1.1) one describes the data by the expression

$$\eta(t) = x_1 f_1(t) + x_2 f_2(t) + \cdots + x_r f_r(t) \quad . \tag{12.1.2}$$

Here the quantities f_j are polynomials of order $j-1$,

$$f_j(t) = \sum_{k=1}^{j} b_{jk} t^{k-1} \quad . \tag{12.1.3}$$

These are chosen to satisfy the *condition of orthogonality* (or more precisely condition of orthonormality) with respect to the values t_i and the measurement weights $g_i = 1/\sigma_i^2$,

$$\sum_{i=1}^{N} g_i f_j(t_i) f_k(t_i) = \delta_{jk} \quad . \tag{12.1.4}$$

Defining

$$A_{ij} = f_j(t_i) \quad , \qquad A = \begin{pmatrix} A_{11} & A_{12} & \cdots & A_{1r} \\ \vdots & & & \\ A_{N1} & A_{N2} & \cdots & A_{Nr} \end{pmatrix} \tag{12.1.5}$$

and using the matrix notation (9.2.9)

$$G_y = \begin{pmatrix} g_1 & & & 0 \\ & g_2 & & \\ & & \ddots & \\ 0 & & & g_N \end{pmatrix} \quad , \tag{12.1.6}$$

Eq. (12.1.4) has the simple form

$$A^{\mathrm{T}} G_y A = I \quad . \tag{12.1.7}$$

The least-squares requirement,

$$\begin{aligned} &\sum_{i=1}^{N} g_i \left\{ y_i(t_i) - \sum_{j=1}^{r} x_j f_j(t_i) \right\}^2 \\ &= (\mathbf{y} - A\mathbf{x})^{\mathrm{T}} G_y (\mathbf{y} - A\mathbf{x}) = \min \quad , \end{aligned} \tag{12.1.8}$$

is of the form (9.2.19) and thus has the solution (9.2.26)

$$\widetilde{\mathbf{x}} = -A^{\mathrm{T}} G_y \mathbf{y} \quad , \tag{12.1.9}$$

where we have used (9.2.18) and (12.1.7). Because of (9.2.27) and (12.1.7) one has

$$C_{\widetilde{x}} = I \quad , \tag{12.1.10}$$

i.e., the covariance matrix of the coefficients $\widetilde{x}_1, \ldots, \widetilde{x}_r$ is simply the unit matrix. In particular, the coefficients are uncorrelated.

We now discuss the procedure for determining the matrix elements

$$A_{ij} = f_j(t_i) = \sum_{k=1}^{j} b_{jk} t_i^{k-1} \quad . \tag{12.1.11}$$

For $j = 1$, the orthogonality condition gives

$$\sum_{i=1}^{N} g_i b_{11}^2 = 1 \quad , \qquad b_{11} = 1/\sqrt{\sum_i g_i} \quad . \tag{12.1.12}$$

For $j = 2$ there are two orthogonality conditions. First we obtain

$$\sum_i g_i f_2(t_i) f_1(t_i) = \sum_i g_i (b_{21} + b_{22} t_i) b_{11} = 0$$

and from this

$$b_{21} = -b_{22} \frac{\sum g_i t_i}{\sum g_i} = -b_{22} \bar{t} \quad , \tag{12.1.13}$$

where

$$\bar{t} = \Sigma g_i t_i / \Sigma g_i \tag{12.1.14}$$

is the weighted mean of the values of the controlled variable t_i at which the measurements were made. The second orthogonality condition for $j = 2$ gives

$$\Sigma g_i [f_2(t_i)]^2 = \Sigma g_i (b_{21} + b_{22} t_i)^2 = \Sigma g_i b_{22}^2 (t_i - \bar{t})^2 = 1$$

or

$$b_{22} = 1/\sqrt{\Sigma g_i (t_i - \bar{t})^2} \quad . \tag{12.1.15}$$

Substitution into (12.1.13) gives b_{21}.

For $j > 2$ one can obtain the values of $A_{ij} = f_j(t_i)$ recursively from the quantities for $j - 1$ and $j - 2$. We make the ansatz

$$\gamma f_j(t_i) = (t_i - \alpha) f_{j-1}(t_i) - \beta f_{j-2}(t_i) \quad , \tag{12.1.16}$$

multiply by $g_i f_{j-1}(t_i)$, sum over i, and obtain

$$\begin{aligned} &\gamma \Sigma g_i f_j(t_i) f_{j-1}(t_i) = 0 \\ &= \Sigma g_i t_i [f_{j-1}(t_i)]^2 - \alpha \Sigma g_i [f_{j-1}(t_i)]^2 - \beta \Sigma g_i f_{j-2}(t_i) f_{j-1}(t_i) \quad . \end{aligned}$$

Because of the orthogonality condition, the second term on the right-hand side is equal to unity, and the third term vanishes. Thus one has

$$\alpha = \Sigma g_i t_i [f_{j-1}(t_i)]^2 \quad . \tag{12.1.17}$$

By multiplication of (12.1.16) by $g_i f_{j-2}(t_i)$ and summing one obtains in the same way

$$\beta = \Sigma g_i t_i f_{j-1}(t_i) f_{j-2}(t) \quad . \tag{12.1.18}$$

Finally by computing the expression

$$\Sigma g_i f_j^2(t_i) = 1$$

and substituting $f_j(t_i)$ from (12.1.16) one obtains

$$\gamma^2 = \Sigma g_i [(t_i - \alpha) f_{j-1}(t_i) - \beta f_{j-2}(t_i)]^2 \quad . \tag{12.1.19}$$

Once the quantities α, β, γ have been computed for a given j, then the coefficients b_{jk} are determined to be

$$\begin{aligned}
b_{j1} &= (-\alpha b_{j-1,1} - \beta b_{j-2,1})/\gamma \quad , \\
b_{jk} &= (b_{j-1,k-1} - \alpha b_{j-1,k} - \beta b_{j-2,k})/\gamma \quad , \qquad k = 2, \ldots, j-2 \quad , \\
b_{j,j-1} &= (b_{j-1,j-2} - \alpha b_{j-1,j-1})/\gamma \quad , \\
b_{jj} &= b_{j-1,j-1}/\gamma \quad .
\end{aligned} \tag{12.1.20}$$

With (12.1.3) and (12.1.5) one thus obtains the quantities $A_{ij} = f_j(t_i)$. Since the procedure is recursive, the column j of the matrix A is obtained from the elements of the columns to its left. Be extending the matrix to the right, the original matrix elements are not changed. This has the consequence that by increasing the number of terms in the polynomial (12.1.1) from r to r', the coefficients $\widetilde{x}_1, \ldots, \widetilde{x}_r$ are retained, and one merely obtains additional coefficients. All of the $\widetilde{x}_j$ are uncorrelated and have the standard deviation $\sigma_{\widetilde{x}_j} = 1$. The order of the polynomial sufficient to describe the data can now be determined by the requirement

$$|\widetilde{x}_j| < c \quad , \qquad j > r \quad .$$

With this the contribution of all higher coefficients $\widetilde{x}_j$ is smaller than $c\sigma_{\widetilde{x}_j}$. The simplest choice of c is clearly $c = 1$.

For a given r one can now substitute the values b_{jk} into (12.1.3) and $f_j(t_i)$ into (12.1.2) in order to obtain the best estimates $\widetilde{\eta}_i(t_i)$ of the true values $\eta_i(t_i)$ corresponding to the measurements $y_i(t_i)$. One can also compute the quantity

$$M = \sum_{i=1}^{N} \frac{(\widetilde{\eta}_i(t_i) - y_i(t_i))^2}{\sigma_i^2} \quad .$$

If the data are in fact described by the polynomial of the chosen order, then the quantity M follows a χ^2-distribution with $f = N - r$ degrees of freedom. It can be used for a χ^2-test for the goodness-of-fit of the polynomial.

Program 12.1: Subroutine `REGPOL` carries out a polynomial regression using orthogonal polynomials

CALLING SEQUENCE: `CALL REGPOL(T,Y,DELTA,N,NR,X,B,A, CHI2)`

INPUT ARGUMENTS: `T` — n-vector of the controlled variables t_i
`Y` — n-vector of measured values y_i
`DELTAY` — n-vector of measurement errors $\Delta y_i = \sigma_i$
`N` — number of measurements n
`NR` — maximum number of parameters r

OUTPUT ARGUMENTS: X — r-vector of the parameters $\widetilde{x}_1, \ldots, \widetilde{x}_r$
B — $r \times r$ matrix of coefficients b_{jk}
A — $n \times r$ matrix $A_{ij} = f_j(t_i)$
CHI2 — r-vector of the minimization functions M for fitting a polynomial of order 0, 1, . . ., $r - 1$

```
      SUBROUTINE REGPOL(T,Y,DELTAY,N,NR,X,B,A,CHI2)
      IMPLICIT DOUBLE PRECISION (A-H,O-Z)
      DIMENSION T(N),Y(N),DELTAY(N),X(NR),CHI2(NR)
      DIMENSION B(NR,NR),A(N,NR)
      PARAMETER(MAXN=1000)
      DIMENSION G(MAXN)
      COMMON /DASV04/ G
      PARAMETER (ZERO=0.D0,ONE=1.D0)
C compute weights G and weighted mean TBAR
      SG=ZERO
      TBAR=ZERO
      DO 10 I=1,N
        G(I)=ONE/DELTAY(I)**2
        SG=SG+G(I)
        TBAR=TBAR+G(I)*T(I)
   10 CONTINUE
      TBAR=TBAR/SG
C compute B and A for NR=1
      B(1,1)=ONE/SQRT(SG)
      DO 20 I=1,N
        A(I,1)=B(1,1)
   20 CONTINUE
C compute B and A for NR=2
      IF(NR.GE.2) THEN
        S=ZERO
        DO 30 I=1,N
          S=S+G(I)*(T(I)-TBAR)**2
   30   CONTINUE
        B(2,2)=ONE/SQRT(S)
        B(2,1)=-B(2,2)*TBAR
        DO 40 I=1,N
          A(I,2)=B(2,1)+B(2,2)*T(I)
   40   CONTINUE
      END IF
C compute B and A for NR greater than 2
      IF(NR.GT.2) THEN
        DO 100 J=3,NR
          ALPHA=ZERO
          BETA=ZERO
          GAMMA2=ZERO
          DO 50 I=1,N
            ALPHA=ALPHA+G(I)*T(I)*A(I,J-1)**2
            BETA=BETA+G(I)*T(I)*A(I,J-1)*A(I,J-2)
   50     CONTINUE
          DO 60 I=1,N
            GAMMA2=GAMMA2+G(I)*((T(I)-ALPHA)*A(I,J-1)-
     +                BETA*A(I,J-2))**2
   60     CONTINUE
          GAMMA1=ONE/SQRT(GAMMA2)
          B(J,1)=GAMMA1*(-ALPHA*B(J-1,1)-BETA*B(J-2,1))
          IF(J.GE.4) THEN
            DO 70 K=2,J-2
              B(J,K)=GAMMA1*(B(J-1,K-1)-ALPHA*B(J-1,K)-
```

```
     +                   BETA*B(J-2,K))
   70         CONTINUE
            END IF
            B(J,J-1)=GAMMA1*(B(J-1,J-2)-ALPHA*B(J-1,J-1))
            B(J,J)=GAMMA1*B(J-1,J-1)
            DO 90 I=1,N
              A(I,J)=B(J,1)
              DO 80 K=2,J
                A(I,J)=A(I,J)+B(J,K)*T(I)**(K-1)
   80         CONTINUE
   90       CONTINUE
  100     CONTINUE
        END IF
C compute X and CHI2
        DO 140 J=1,NR
          X(J)=ZERO
          CHI2(J)=ZERO
          DO 110 I=1,N
            X(J)=X(J)+G(I)*A(I,J)*Y(I)
  110     CONTINUE
          DO 130 I=1,N
            S=ZERO
            DO 120 K=1,J
              S=S+A(I,K)*X(K)
  120       CONTINUE
            CHI2(J)=CHI2(J)+G(I)*(Y(I)-S)**2
  130     CONTINUE
  140   CONTINUE
        END
```

Example 12.1: Treatment of Example 9.2 with orthogonal polynomials

Application of the program REGPOL to the data of Example 9.2 yields the results shown in Table 12.1. The numerical values of the minimization function are, of course, exactly the same as in Example 9.2. The fitted polynomials are also the same. The numerical values $\widetilde{x}_1$, $\widetilde{x}_2$, ... are different from those in Example 9.2 since these quantities are now defined differently. We have emphasized that the covariance matrix for $\widetilde{\mathbf{x}}$ is the $r \times r$ unit matrix. We see that the quantities $x_6, x_7, \ldots, x_{10}$ have magnitudes less than unity and thus are no longer significantly different from zero. It is more difficult to judge the significance of $x_5 = 1.08$. In many cases one would not consider this value as being clearly different from zero. This means that a third order polynomial (i.e., $r = 4$) is sufficient to describe the data. ■

12.2 Regression Curve. Confidence Interval

Once the order of the polynomial (12.1.2) in one way or another has been fixed, then every point of the *regression polynomial* or – in reference to the graphical representation – the *regression curve* can be computed. This is done by first substituting the recursively computed values b_{jk} into (12.1.3) for a given t, and using the thus computed $f_j(t)$ and the parameters $\widetilde{\mathbf{x}}$ in (12.1.2):

Table 12.1. Results from the program REGPOL using the data from Example 9.2.

r	$\tilde{x}_r$	M	Degrees of freedom
1	24.05	833.55	9
2	15.75	585.45	8
3	23.43	36.41	7
4	5.79	2.85	6
5	1.08	1.69	5
6	0.15	1.66	4
7	0.85	0.94	3
8	−0.41	0.77	2
9	−0.45	0.57	1
10	0.75	0.00	0

$$\tilde{\eta}(t) = \sum_{j=1}^{r} \tilde{x}_j \left(\sum_{k=1}^{j} b_{jk} t^{k-1} \right) = \mathbf{d}^{\mathrm{T}}(t)\tilde{\mathbf{x}} \quad . \tag{12.2.1}$$

Here **d** is an r-vector with the elements

$$d_j(t) = \sum_{k=1}^{j} b_{jk} t^{k-1} \quad . \tag{12.2.2}$$

After error propagation, (3.8.4), the variance of $\tilde{\eta}(t)$ is

$$\sigma^2_{\tilde{\eta}(t)} = \mathbf{d}^{\mathrm{T}}(t)\mathbf{d}(t) \quad , \tag{12.2.3}$$

since $C_{\tilde{x}} = I$. The reduced variable

$$u = \frac{\tilde{\eta}(t) - \eta(t)}{\sigma_{\tilde{\eta}}(t)} \tag{12.2.4}$$

thus follows a standard Gaussian distribution. This also means that the probability for the position of the true value $\eta(t)$ is distributed according to a Gaussian of width $\sigma_{\tilde{\eta}}(t)$ centered about the value $\tilde{\eta}(t)$.

We can now easily give a *confidence interval* for $\eta(t)$. According to Sect. 5.8 one has with probability P that

$$|u| \geq \Omega'(P) = \Omega\left(\frac{1}{2}(P+1)\right) \quad , \tag{12.2.5}$$

e.g., $\Omega'(0.95) = 1.96$. Thus the confidence limits at the confidence level P (e.g., $P = 0.95$) are

$$\eta(t) = \tilde{\eta}(t) \pm \Omega'(P)\sigma_{\tilde{\eta}(t)} = \tilde{\eta}(t) \pm \delta\tilde{\eta}(t) \quad . \tag{12.2.6}$$

12.3 Regression with Unknown Errors

If the measurement errors $\sigma_i = 1/\sqrt{g_i}$ are not known, but it can be assumed that they are equal,

$$\sigma_i = \sigma \quad , \qquad i = 1, \ldots, N \quad , \tag{12.3.1}$$

then one can easily obtain an estimate s for σ directly from the regression itself. The quantity

$$\sum_{i=1}^{N} \frac{(\tilde{\eta}(t_i) - y_i(t_i))^2}{\sigma^2} = M \tag{12.3.2}$$

follows a χ^2-distribution with $f = N - r$ degrees of freedom. More precisely, if many similar experiments with N measurements each were to be carried out, then the values of M computed for each experiment would be distributed like a χ^2-variable with $f = N - r$ degrees of freedom. Its expectation value is simply $f = N - r$. If the value of σ in (12.3.2) is not known, then it can be replaced by an estimate s such that on the right-hand side one has the expectation value of M,

$$s^2 = \sum (\tilde{\eta}(t_i) - y_i(t_i))^2 / (N - r) \quad . \tag{12.3.3}$$

Here all of the steps necessary to compute $\tilde{\eta}(t_i)$ must be carried out using the formulas of the last section and with the value $\sigma_i = 1$. (In fact, for the case (12.3.1) the quantities σ_i and g_i could be removed completely from the formulas for computing $\tilde{\mathbf{x}}$ and $\tilde{\eta}(t)$.)

If we define

$$\bar{s}^2_{\tilde{\eta}(t)} = \mathbf{d}^{\mathrm{T}}(t)\mathbf{d}(t)$$

as the value of the variance of $\tilde{\eta}(t)$ obtained for $\sigma_i = 1$, then

$$s^2_{\tilde{\eta}(t)} = s^2 \bar{s}^2_{\tilde{\eta}(t)} \tag{12.3.4}$$

is clearly the estimate of the variance of $\tilde{\eta}(t)$, which is based on the estimate (12.3.3) for the variance of the measured quantities. Replacing $\sigma_{\tilde{\eta}(t)}$ by $s_{\tilde{\eta}(t)}$ in (12.2.4), we obtain the variable

$$v = \frac{\tilde{\eta}(t) - \eta(t)}{s_{\tilde{\eta}(t)}} \quad , \tag{12.3.5}$$

which no longer follows a standard normal distribution but rather Student's t-distribution for $f = N - r$ degrees of freedom (cf. Sect. 8.3). For the confidence limits at the confidence level $P = 1 - \alpha$ one now has

$$\eta(t) = \tilde{\eta}(t) \pm t_{1-\alpha/2} s_{\tilde{\eta}(t)} = \tilde{\eta}(t) \pm \delta\tilde{\eta}(t) \quad , \tag{12.3.6}$$

where $t_{1-\alpha/2}$ is the quantile of the t-distribution for $f = N - r$ degrees of freedom.

It must be emphasized at this point that in the case where the errors are not known, one loses the possibility of applying the χ^2-test. The goodness-of-fit for such a polynomial cannot be tested. Thus the procedure described at the end of Sect. 12.1 for determining the order of the polynomial is no longer valid. One must rely on *a priori* knowledge about the order of the polynomial. Therefore one almost always refrains from treating anything beyond linear dependences between η and t, i.e., the case $r = 2$.

Program 12.2: Subroutine REGCON computes a point on the regression curve and its distance from the confidence limits

CALLING SEQUENCE: CALL REGCON(X,B,T,CHI2,P,NR,NF,ETA, CONETA)

INPUT ARGUMENTS: X — r-dimensional vector $\widetilde{\mathbf{x}}$
B — $r \times r$ matrix of coefficients b_{jk}
T — value of the controlled variable t
CHI2 — value of the minimization function M for fitting a polynomial of order $r - 1$ under the assumption that all of the errors are given by $\sigma_i = 1$. If the input value CHI2 ≤ 0, the σ_i were known.
P — probability P for the confidence limits
NR — number of parameters r in the polynomial
NF — number of degrees of freedom $f = N - r$

OUTPUT ARGUMENTS: ETA — regression value $\widetilde{\eta}(t)$
CONETA — distance $\delta\widetilde{\eta}(t)$ of the confidence limits from the value $\widetilde{\eta}(t)$

```
      SUBROUTINE REGCON(X,B,T,CHI2,P,NR,NF,ETA,CONETA)
      IMPLICIT DOUBLE PRECISION (A-H,O-Z)
      DIMENSION X(NR),B(NR,NR)
      PARAMETER (ZERO=0.D0,ONE=1.D0,HALF=0.5D0)
      PPRIME=HALF*(P+1)
      IF(CHI2.LE.ZERO) THEN
        S=ONE
        FACT=SQSTNR(PPRIME)
      ELSE
        S=SQRT(CHI2/DBLE(NF))
        FACT=SQSTUD(PPRIME,NF)
      END IF
      ETA=ZERO
      CONETA=ZERO
      DO 20 J=1,NR
        D=B(J,1)
        IF(J.GT.1) THEN
          DO 10 K= 2,J
            D=D+B(J,K)*T**(K-1)
   10     CONTINUE
        END IF
        ETA=ETA+X(J)*D
        CONETA=CONETA+D*D
   20 CONTINUE
      CONETA=FACT*S*SQRT(CONETA)
      END
```

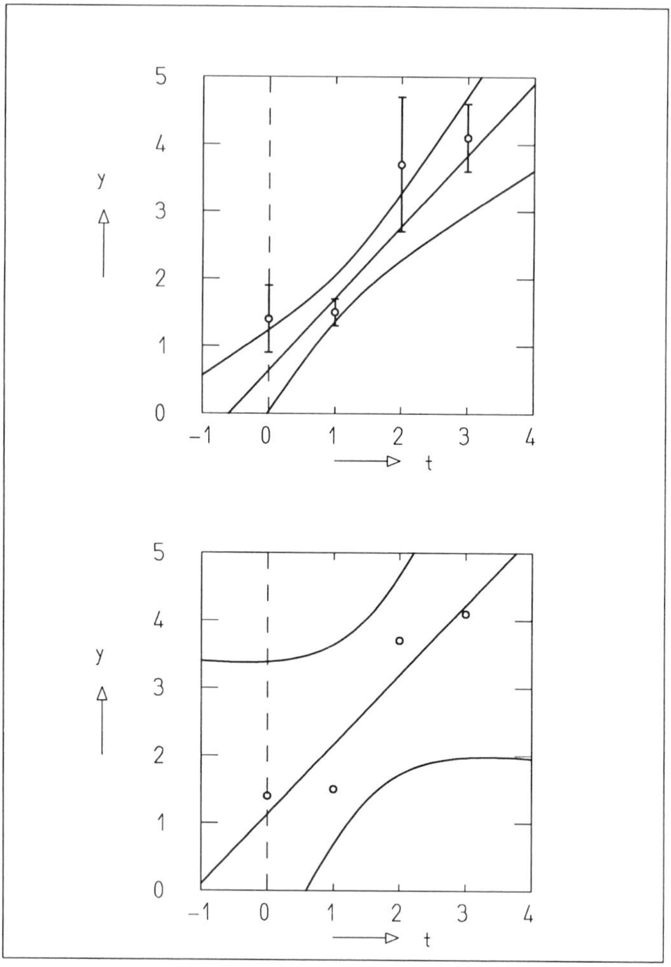

Fig. 12.1. (Above) Measurements with errors and regression line with 95 % confidence limits. (Below) Measured points with errors assumed to be of unknown but equal magnitude, regression line, and 95 % confidence limits.

Example 12.2: Confidence limits for linear regression

In the upper plot of Fig. 12.1, four measured points with their errors, the corresponding regression line, and the limits at a confidence level of 95 % are shown. The measured points are taken from the example of Sect. 9.3. The confidence limits clearly become wider the more one leaves the region of the measured points. The narrowest region is close to the measured point with the smallest error. The plot is similar to Fig. 9.2d. It is easy to convince oneself that the envelope of the lines of Fig. 9.2d corresponds to the 68.3 % confidence limit. In the lower plot of Fig. 12.1 the same measured points were used, but the errors were treated as equal but of unknown magnitude. The regression line and 95 % confidence limits are shown for this assumption. ■

12.4 Example Programs

The source files of the example programs are available in the subdirectory `\DATAN\EXASRC`.

12.4.1 Main Program **E1REG** to Demonstrate Subprogram **REGPOL**

The short program contains the data of Examples 9.2 and 12.1 in `DATA` statements. As measurement errors the statistical errors $\Delta y_i = \sqrt{y_i}$ are taken. By calling `REGPOL` the polynomial regression with $r = 10$ parameters is performed. The results are presented numerically (see also Table 12.1).

12.4.2 Main Program **E2REG** to Graphically Display the Data and All Polynomials for a Polynomial Regression

The first part of the program is identical to `E1REG`. Next, the number of a graphics workstation is asked for as well as the maximum number n_{pol} of terms in the polynomials which are to be displayed. The coordinates of points of the polylines `XPL(I,J)`, `YPL(I,J)` are computed, which are graphs of the functions

$$y_j(t_i) = \sum_{k=1}^{j} x_k f_k(t_i) \quad , \qquad j = 1, \ldots, n_{\mathrm{pol}} \quad .$$

These polylines are displayed together with the data points by calling `GRDTMC`.

Suggestion: Choose consecutively $n_{\mathrm{pol}} = 2, 3, 4, 6, 10$. Try to explain why the curves for $n_{\mathrm{pol}} = 3$ or 4 seem to be particularly convincing although for $n_{\mathrm{pol}} = 10$ all data points lie exactly on the graph of the polynomial.

12.4.3 Main Program **E3REG** to Demonstrate Subprogram **REGCON** and Display the Fitted Polynomial and its Confidence Limits

This program also begins with a repetition of `E1REG`. After the input of the number of a graphics workstation, of the parameter n_{pol}, and of the confidence level needed later in the program, `REGPOL` is called again with `NPOL` as one of the arguments. This call is necessary to ensure that the matrix `B` in memory has `NPOL` rows and `NPOL` columns, which is assumed in a subsequent call of

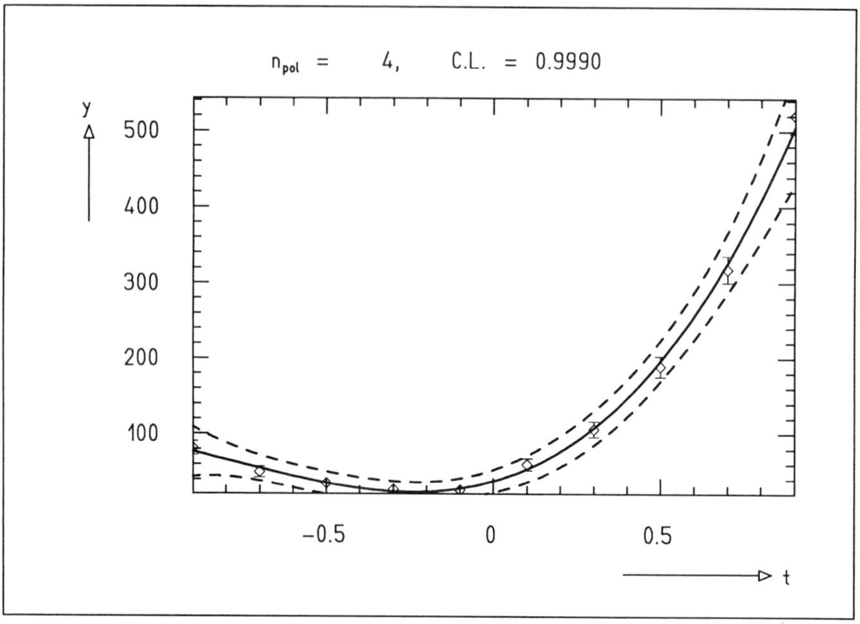

Fig. 12.2. Data points with errors, fitted polynomial, and confidence limits at a confidence level of 0.999.

REGCON. This call is situated in a loop over all points of three polylines. The first polyline contains the regression curve, the other two contain the upper and lower confidence limits at the given confidence level. By a call of GRDTMC, each of the three polylines are displayed. The regression line is shown as a solid line and the confidence limits as dashed lines (Fig. 12.2).

12.4.4 Main Program E4REG to Simulate Data with Measurement Errors and Perform a Polynomial Regression

The program simulates data points which – apart from their measurement errors – are given by a polynomial of degree $m-1$, i.e., a polynomial with m terms. A total of n data points are generated corresponding to the t values $t_1 = -1, t_2 = -1 + \Delta t, \ldots, t_n = +1$ with $\Delta t = 2/(n-1)$,

$$y_i = x_1 + x_2 t_i + x_3 t_i^2 + \cdots + x_m t_i^{m-1} + \varepsilon_i \quad .$$

The measurement errors ε_i follow a normal distribution centered about zero with the standard deviation σ.

The quantities n, σ, m, and the coefficients x_j are entered interactively. Following the simulation of the data, a polynomial regression as in Example Program 12.4.1 is performed including the alphanumeric output.

Suggestion: Convince yourself that as long as the measurement errors are sufficiently small and the input values of x_m are sufficiently different from zero, the degree of the polynomial in the analysis is always recognizable by the fact that $|x_m| \gg 1$ while $|x_{m+1}| \approx 1$, where the coefficients x_j are now the values yielded by the regression. These are not to be confused with the input values.

12.5 Programming Problems

The source files of the example solutions are available in the subdirectory `\DATAN\SOLSRC`.

12.5.1 Simulation of Data and Display of Regression Polynomials of Different Degree

Write a program that simulates data corresponding to Sect. 12.4.4, performs a regression analysis, and finally displays the data and the regression polynomials as described in Sect. 12.4.2. You can use large sections of the programs `E4REG` and `E2REG`. An example of a graphical display is shown in Fig. 12.3.

12.5.2 Simulation of Data and Display of a Regression Line with Confidence Limits

Write a program that generates data corresponding to Sect. 12.4.4, performs a regression and finally displays a regression polynomial of the desired degree together with confidence limits as in Sect. 12.4.3. Allow the user to choose whether in the computation of the confidence limits a knowledge of σ is to be assumed or if only the scattering of the data points around the regression line is to be used. An example of the graphical output is shown in Fig. 12.4.

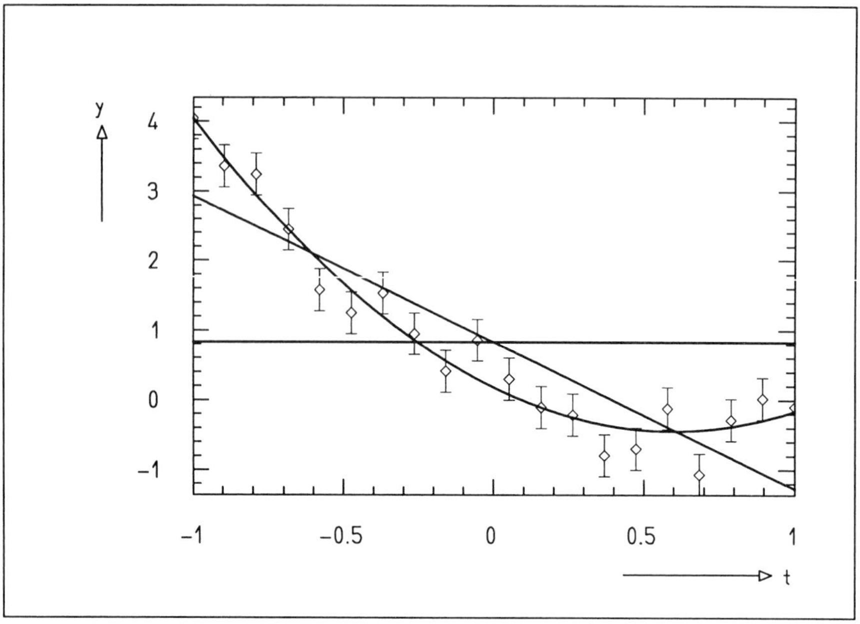

Fig. 12.3. Data points with errors and regression polynomials of different degrees.

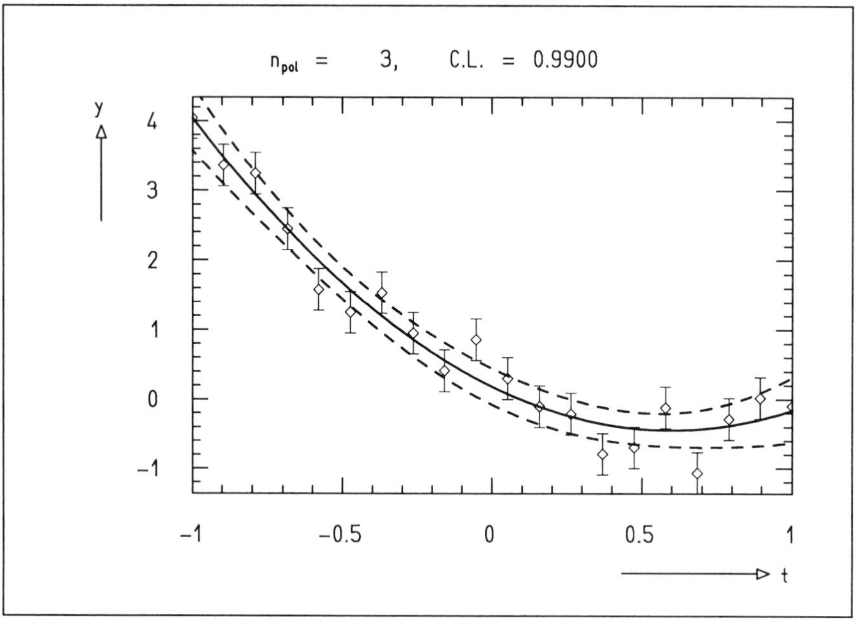

Fig. 12.4. Data points, regression polynomial (solid line), and confidence limits (dashed lines).

13. Time Series Analysis

13.1 Time Series. Trend

In the previous chapter we considered the dependence of a random variable y on a controlled variable t. As in that case we will assume here that y consists of two parts, the true value of the measured quantity η and a measurement error ε,

$$y_i = \eta_i + \varepsilon_i \quad , \qquad i = 1, 2, \ldots, n \quad . \tag{13.1.1}$$

In Chapter 12 we assumed that η_i was a polynomial in t. The measurement error ε_i was considered to be normally distributed about zero.

We now want to make less restrictive assumptions about η. In this chapter we call the controlled variable t "time", although in many applications it could be something different. The method we want to discuss is called *time series analysis* and is often applied in economic problems. It can always be used where one has little or no knowledge about the functional relationship between η and t. In considering time series problems it is common to observe the y_i at equally spaced points in time,

$$t_i - t_{i-1} = \Delta t = \text{const} \quad , \tag{13.1.2}$$

since this leads to a significant simplification of the formulas.

An example of a time series is shown in Fig. 13.1. If we look first only at the measured points, we notice strong fluctuations from point to point. Nevertheless they clearly follow a certain trend. In the left half of the plot they are mostly positive, and in the right half, mostly negative. One could qualitatively obtain the average time dependence by drawing a smooth curve by hand through the points. Since, however, such curves are not free from personal influences, and are thus not reproducible, we must try to develop an objective method.

We use the notation from (13.1.1) and call η_i the *trend* and ε_i the *random component* of the *measurement* y_i. In order to obtain a smoother function of t, one can, for example, construct for every value of y_i the expression

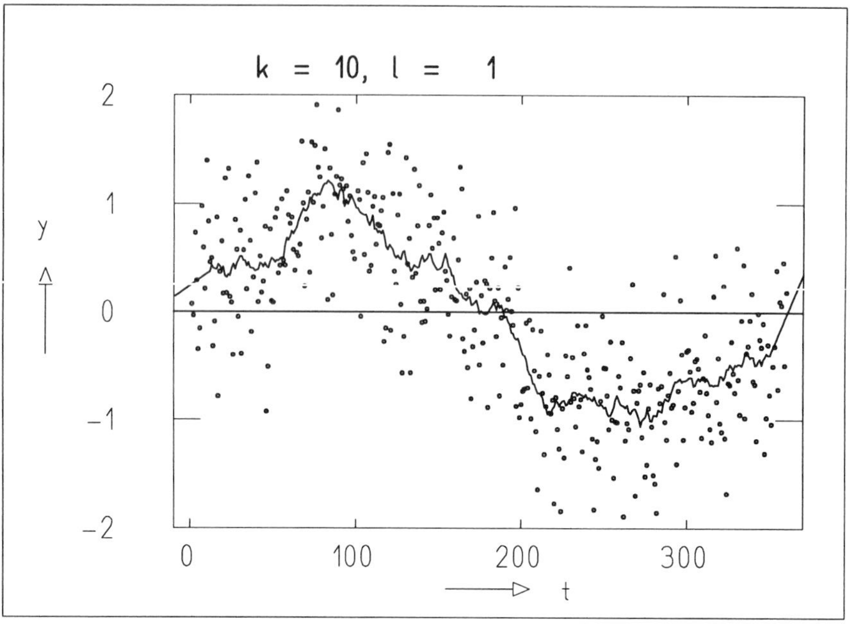

Fig. 13.1. Data points (circles) and moving average (joined by line segments).

$$u_i = \frac{1}{2k+1} \sum_{j=i-k}^{i+k} y_j \quad , \tag{13.1.3}$$

i.e., the unweighted mean of the measurements for the times

$$t_{i-k},\ t_{i-k+1},\ \ldots,\ t_{i-1},\ t_i,\ t_{i+1},\ \ldots,\ t_{i+k} \quad .$$

The expression (13.1.3) is called a *moving average* of y.

13.2 Moving Averages

Of course the moving average (13.1.3) is a very simple construction. We will show later (in Example 13.1) that the use of a moving average of this form is equivalent to the assumption that η is a linear function of time in the interval considered,

$$\eta_j = \alpha + \beta t_j \quad , \qquad j = -k, -k+1, \ldots, k \quad . \tag{13.2.1}$$

Here α and β are constants. They can be estimated from the data by linear regression.

Instead of restricting ourselves to the linear case, we will assume more generally that η can be a polynomial of order ℓ.

In the averaging interval, t takes on the values

$$t_j = t_i + j\Delta t \quad , \qquad j = -k, -k+1, \ldots, k \quad . \tag{13.2.2}$$

Because η is a polynomial in t,

$$\eta_j = a_1 + a_2 t_j + a_3 t_j^2 + \cdots + a_{\ell+1} t_j^\ell \quad , \tag{13.2.3}$$

it is also a polynomial in j,

$$\eta_j = x_1 + x_2 j + x_3 j^2 + \cdots + x_{\ell+1} j^\ell \quad , \tag{13.2.4}$$

since (13.2.2) describes a linear transformation between t_j and j, i.e., it is merely a change of scale. We now want to obtain the coefficients $x_1, x_2, \ldots, x_{\ell+1}$ from the data by fitting with least squares. This task has already been treated in Sect. 9.4.1. We assume (in the absence of any better knowledge) that all of the measurements are of the same accuracy. Thus the matrix $G_y = aI$ is simply a multiple of the unit matrix I. According to (9.2.26), the vector of coefficients is thus given by

$$\widetilde{\mathbf{x}} = -(A^{\mathrm{T}}A)^{-1}A^{\mathrm{T}}\mathbf{y} \quad , \tag{13.2.5}$$

where A is a $(2k+1) \times (\ell+1)$ matrix,

$$A = -\begin{pmatrix} 1 & -k & (-k)^2 & \ldots & (-k)^\ell \\ 1 & -k+1 & (-k+1)^2 & \ldots & (-k+1)^\ell \\ \vdots & & & & \\ 1 & k & k^2 & \ldots & k^\ell \end{pmatrix} \quad . \tag{13.2.6}$$

For the trend $\widetilde{\eta}_0$ at the center of the averaging interval ($j=0$) we obtain from (13.2.4) the estimate

$$\widetilde{\eta}_0 = \widetilde{x}_1 \quad . \tag{13.2.7}$$

It is equal to the first coefficient of the polynomial. According to (13.2.5), $\widetilde{x}_1$ is obtained by multiplication of the column vector of measurements $\mathbf{y}$ on the left with the row vector

$$\mathbf{a} = (-(A^{\mathrm{T}}A)^{-1}A^{\mathrm{T}})_1 \quad , \tag{13.2.8}$$

i.e., with the first row of the matrix $-(A^{\mathrm{T}}A)^{-1}A^{\mathrm{T}}$. We obtain

$$\widetilde{\eta}_0 = \mathbf{a}\mathbf{y} = a_{-k}y_{-k} + a_{-k+1}y_{-k+1} + \cdots + a_0 y_0 + \cdots + a_k y_k \quad . \tag{13.2.9}$$

Table 13.1. Components of the vector **a** for computing moving averages.

$$\mathbf{a} = (a_{-k}, a_{-k+1}, \ldots, a_k) = \frac{1}{A}(\alpha_{-k}, \alpha_{-k+1}, \ldots, \alpha_k)$$

$$\alpha_{-j} = \alpha_j$$

$\ell = 2$ and $\ell = 3$

k	A	α_{-7}	α_{-6}	α_{-5}	α_{-4}	α_{-3}	α_{-2}	α_{-1}	α_0
2	35						−3	12	17
3	21					−2	3	6	7
4	231				−21	14	39	54	59
5	429			−36	9	44	69	84	89
6	143		−11	0	9	16	21	24	25
7	1105	−78	−13	42	87	122	147	162	167

$\ell = 4$ and $\ell = 5$

k	A	α_{-7}	α_{-6}	α_{-5}	α_{-4}	α_{-3}	α_{-2}	α_{-1}	α_0
3	231					5	−30	75	131
4	429				15	−55	30	135	179
5	429			18	−45	−10	60	120	143
6	2431		110	−198	−135	110	390	600	677
7	46 189	2145	−2860	−2937	−165	3755	7500	10 125	11 063

This is a linear function of the measurements within the averaging interval. Here the vector **a** does not depend on the measurements, but rather only on ℓ and k, i.e., on the order of the polynomial and on the length of the interval. Clearly one must choose

$$\ell < 2k + 1 \quad ,$$

since otherwise there would not remain any degrees of freedom for the least-squares fit. The components of **a** for small values of ℓ and k can be obtained from Table 13.1.

Equation (13.2.9) describes the moving average corresponding to the assumed polynomial (13.2.4). Once the vector **a** is determined, the moving averages

$$u_i = \tilde{\eta}_0(i) = \mathbf{a}\mathbf{y}(i) = a_1 y_{i-k} + a_2 y_{i-k+1} + \cdots + a_{2k+1} y_{i+k} \qquad (13.2.10)$$

can easily be computed for each value of i.

Example 13.1: Moving average with linear trend

In the case of a linear trend function,

$$\eta_j = x_1 + x_2 j \quad ,$$

the matrix A becomes simply

$$A = -\begin{pmatrix} 1 & -k \\ 1 & -k+1 \\ \vdots & \\ 1 & k \end{pmatrix} \quad .$$

One then has

$$\begin{aligned} A^{\mathrm{T}}A &= \begin{pmatrix} 1 & 1 & \dots & 1 \\ -k & -k+1 & \dots & k \end{pmatrix} \begin{pmatrix} 1 & -k \\ 1 & -k+1 \\ \vdots & \vdots \\ 1 & k \end{pmatrix} \\ &= \begin{pmatrix} 2k+1 & 0 \\ 0 & k(k+1)(2k+1)/3 \end{pmatrix} \quad , \\ (A^{\mathrm{T}}A)^{-1} &= \begin{pmatrix} \dfrac{1}{2k+1} & 0 \\ 0 & \dfrac{3}{k(k+1)(2k+1)} \end{pmatrix} \quad , \\ \mathbf{a} &= (-(A^{\mathrm{T}}A)^{-1}A^{\mathrm{T}})_1 = \frac{1}{2k+1}(1, 1, \dots, 1) \quad . \end{aligned}$$

In this case the moving average is simply the unweighted mean (13.1.3). ∎

For more complicated models one can obtain the vectors **a** either by solving (13.2.8) or simply from Table 13.1. Because of the symmetry of A, one can show that polynomials of odd order (i.e., $\ell = 2n$ with n an integer) have the same values of **a** as those of polynomials of the next lower order $\ell = 2n - 1$. One can also easily show that **a** has the symmetry

$$a_j = a_{-j} \quad , \qquad j = 1, 2, \dots, k \quad . \qquad (13.2.11)$$

13.3 Edge Effects

Of course the moving average (13.2.10) can be used to estimate the trend only for points i that have at least k neighboring points both to the right and left, since the averaging interval covers $2k+1$ points. This means that for the first and last k points of a time series one must use a different estimator. One obtains the most obvious generalization of the estimator by extrapolating the polynomial (13.2.4) rather than using it only at the center of an interval. One then obtains the estimators

$$\begin{aligned}
\widetilde{\eta}_i = u_i &= \widetilde{x}_1^{(k+1)} + \widetilde{x}_2^{(k+1)}(i-k-1) + \widetilde{x}_3^{(k+1)}(i-k-1)^2 + \cdots \\
&\quad + \widetilde{x}_{\ell+1}^{(k+1)}(i-k-1)^\ell \quad , \qquad i \le k \quad , \\
\widetilde{\eta}_i = u_i &= \widetilde{x}_1^{(n-k)} + \widetilde{x}_2^{(n-k)}(i+k-n) + \widetilde{x}_3^{(n-k)}(i+k-n)^2 + \cdots \\
&\quad + \widetilde{x}_{\ell+1}^{(n-k)}(i+k-n) \quad , \qquad i > n-k \quad . \qquad (13.3.1)
\end{aligned}$$

Here the notation $\widetilde{x}^{(k+1)}$ and $\widetilde{x}^{(n-k)}$ indicates that the coefficients $\widetilde{x}$ were determined for the first and last intervals of the time series for which the centers are at $(k+1)$ and $(n-k)$.

The estimators are now defined even for $i < 1$ and $i > n$. They thus offer the possibility to continue the time series (e.g., into the future). Such extrapolations must be treated with great care for two reasons:

(i) Usually there is no theoretical justification for the assumption that the trend is described by a polynomial. It merely simplifies the computation of the moving average. Without a theoretical understanding for a trend model, the meaning of extrapolations is quite unclear.

(ii) Even in cases where the trend can rightly be described by a polynomial, the confidence limits quickly diverge from the estimated polynomial in the extrapolated region. The extrapolation becomes very inaccurate.

Whether point (i) is correct must be carefully checked in each individual case. The more general point (ii) is already familiar from the linear regression (cf. Fig. 12.1). We will investigate this in detail in the next section.

13.4 Confidence Intervals

We first consider the confidence interval for the moving average u_i from Eq. (13.2.10). The errors of the measurements y_i are unknown and must therefore first be estimated. From (12.3.2) one obtains for the sample variance of the y_j in the interval of length $2k+1$,

$$s_y^2 = \frac{1}{2k-\ell} \sum_{j=-k}^{k} (y_j - \widetilde{\eta}_j)^2 \quad , \tag{13.4.1}$$

where $\widetilde{\eta}_j$ is given by

$$\widetilde{\eta}_j = \widetilde{x}_1 + \widetilde{x}_2 j + \widetilde{x}_3 j^2 + \cdots + \widetilde{x}_{\ell+1} j^\ell \quad . \tag{13.4.2}$$

The covariance matrix for the measurements can then be estimated by

$$G_y^{-1} \approx s_y^2 I \quad . \tag{13.4.3}$$

The covariance matrix of the coefficients **x** is then given by (9.2.27),

$$G_{\widetilde{x}}^{-1} \approx (A^{\mathrm{T}} G_y A)^{-1} = s_y^2 (A^{\mathrm{T}} A)^{-1} \quad . \tag{13.4.4}$$

Since $u_i = \widetilde{\eta}_0 = \widetilde{x}_1$, we thus have for an estimator of the variance of u_i

$$s_{\widetilde{x}_1}^2 = (G_{\widetilde{x}}^{-1})_{11} = s_y^2 ((A^{\mathrm{T}} A)^{-1})_{11} = s_y^2 a_0 \quad . \tag{13.4.5}$$

From (13.2.6), (13.2.7), and (13.2.8) one easily obtains that $(A^{\mathrm{T}} A)_{11}^{-1} = a_0$, since the middle row of A is $-(1, 0, 0, \ldots, 0)$.

Using the same reasoning as in Sect. 12.3 we obtain at a confidence level of $1 - \alpha$

$$\frac{|\widetilde{\eta}_0(i) - \eta_0(i)|}{s_y a_0} \le t_{1-\frac{1}{2}\alpha} \quad . \tag{13.4.6}$$

For a given α we can give the confidence limits as

$$\eta_0^{\pm}(i) = \widetilde{\eta}_0(i) \pm a_0 s_y t_{1-\frac{1}{2}\alpha} \quad . \tag{13.4.7}$$

Here $t_{1-\frac{1}{2}\alpha}$ is a quantile of Student's distribution for $2k - \ell$ degrees of freedom. The true value of the trend lies within these limits with a confidence level of $1 - \alpha$.

Completely analogous, although more difficult computationally, is the determination of confidence limits at the ends of the time series. The moving average is now given by (13.3.1). Labeling the arguments in the expressions (13.3.1) $j = i - k - 1$ and $j = i + k - n$, we obtain

$$\widetilde{\eta} = \mathbf{T}\mathbf{x} \quad . \tag{13.4.8}$$

Here **T** is a row vector of length $\ell + 1$,

$$\mathbf{T} = (1, j, j^2, \ldots, j^\ell) \quad . \tag{13.4.9}$$

According to the law of error propagation we obtain

$$G_{\widetilde{\eta}}^{-1} = T G_{\widetilde{x}}^{-1} T^{\mathrm{T}} \quad . \tag{13.4.10}$$

With (13.4.4) we finally have

$$G_{\tilde{\eta}}^{-1} \approx s_{\tilde{\eta}}^2 = s_y^2 T(A^{\mathrm{T}}A)^{-1}T^{\mathrm{T}} \quad , \tag{13.4.11}$$

where s_y^2 is again given by (13.4.1).

The quantity $s_{\tilde{\eta}}^2$ can now be computed for every value of j, even for values lying outside of the time series itself. Thus we obtain the confidence limits

$$\eta^{\pm}(i) = \tilde{\eta}(i) \pm s_{\tilde{\eta}} t_{1-\frac{1}{2}\alpha} \quad . \tag{13.4.12}$$

13.5 Program for Time Series Analysis. Examples

The following program carries out all necessary computations for a time series analysis.

Program 13.1: Subroutine `TIMSER` carries out a time series analysis

CALLING SEQUENCE: `CALL TIMSER(Y,N,K,L,P,ETA,CONETA,A, ATA1,ATA1AT,SCRAT)`

INPUT ARGUMENTS: `Y` — n-vector of the elements y_i of the time series
`N` — number of elements n
`K` — parameter k giving the length $2k+1$ of the averaging interval
`L` — order of the averaging polynomial ℓ
`P` — probability P for computing the confidence limits
`SCRAT` — working array, minimum size $(\ell+1) \times (\ell+1)$

OUTPUT ARGUMENTS: `ETA` — $(n+2k)$-vector of moving averages; elements in the order $\tilde{\eta}_{1-k}, \tilde{\eta}_{1-k+1}, \ldots, \tilde{\eta}_{n+k}$
`CONETA` — $(n+2k)$-vector of distances of the confidence limits for the moving averages, given in the same order
`A` — $(2k+1) \times (\ell+1)$ matrix A
`ATA1` — $(\ell+1) \times (\ell+1)$ matrix $(A^{\mathrm{T}}A)^{-1}$
`ATA1AT` — $(\ell+1) \times (2k+1)$ matrix $(A^{\mathrm{T}}A)^{-1}A^{\mathrm{T}}$

```
      SUBROUTINE TIMSER(Y,N,K,L,P,ETA,CONETA,A,ATA1,ATA1AT,SCRAT)
      IMPLICIT DOUBLE PRECISION (A-H,O-Z)
      DIMENSION Y(N),ETA(N+2*K),CONETA(N+2*K),A(2*K+1,L+1)
      DIMENSION ATA1(L+1,L+1),ATA1AT(L+1,2*K+1),SCRAT(L+1,L+1)
      PARAMETER (MAX=1000, ZERO=0.D0, ONE=1.D0, HALF=0.5D0)
      DIMENSION X(MAX),YTMP(MAX),ETATMP(MAX),T(MAX)
      COMMON /DADV03/ X,YTMP,ETATMP,T
C quantile of Student's distribution
      PPRIME=HALF*(P+1)
      NF=2*K-L
      TALPHA=SQSTUD(PPRIME,NF)
```

```
C compute matrices depending only on K and L
      K21=2*K+1
      L1=L+1
      DO 20 I=1,K21
        DO 10 J=1,L1
          IF(J.EQ.1) THEN
            A(I,J)=-ONE
          ELSE
            A(I,J)=A(I,J-1)*DBLE(I-K-1)
          END IF
   10   CONTINUE
   20 CONTINUE
      CALL MTXMAT(A,A,ATA1,L1,K21,L1)
      CALL MTXCHI(ATA1,SCRAT,L1)
      CALL MTXMBT(ATA1,A,ATA1AT,L1,L1,K21)
      CALL MTXMSC(ATA1AT,ATA1AT,-ONE,L1,K21)
C loop over inner part of time series
      IA=2*K+1
      IB=N
      DO 60 I=IA,IB
C moving averages and confidence limits for inner part
        CALL MTXGSM(Y,YTMP,N,1,K21,1,I-IA+1,1)
        CALL MTXMLT(ATA1AT,YTMP,X,L1,K21,1)
        ETA(I)=X(1)
        CALL MTXMLT(A,X,ETATMP,K21,L1,1)
        CALL MTXADD(YTMP,ETATMP,ETATMP,K21,1)
        CALL MTXMAT(ETATMP,ETATMP,SY2,1,K21,1)
        SY2=SY2/DBLE(NF)
        A0=SQRT(ABS(ATA1(1,1)))
        CONETA(I)=A0*SQRT(SY2)*TALPHA
C moving averages and confidence limits for end sections
        IF(I.EQ.IA .OR. I.EQ.IB) THEN
          IF(I.EQ.IA) THEN
            IADD=IA
            IS=-1
          ELSE
            IADD=IB
            IS=1
          END IF
          DO 50 I1=1,2*K
            J=IS*I1
            DO 40 I2=1,L1
              DO 30 I3=1,I2
                IF(I3.EQ.1) THEN
                  T(I2)=ONE
                ELSE
                  T(I2)=T(I2)*J
                END IF
   30         CONTINUE
   40       CONTINUE
            CALL MTXMBT(ATA1,T,SCRAT,L1,L1,1)
            CALL MTXMLT(T,SCRAT,SETA2,1,L1,1)
            SETA2=SY2*SETA2
            CALL MTXMLT(T,X,ETAI,1,L1,1)
            CONETA(IADD+J)=SQRT(ABS(SETA2))*TALPHA
            ETA(IADD+J)=ETAI
   50     CONTINUE
        END IF
   60 CONTINUE
      END
```

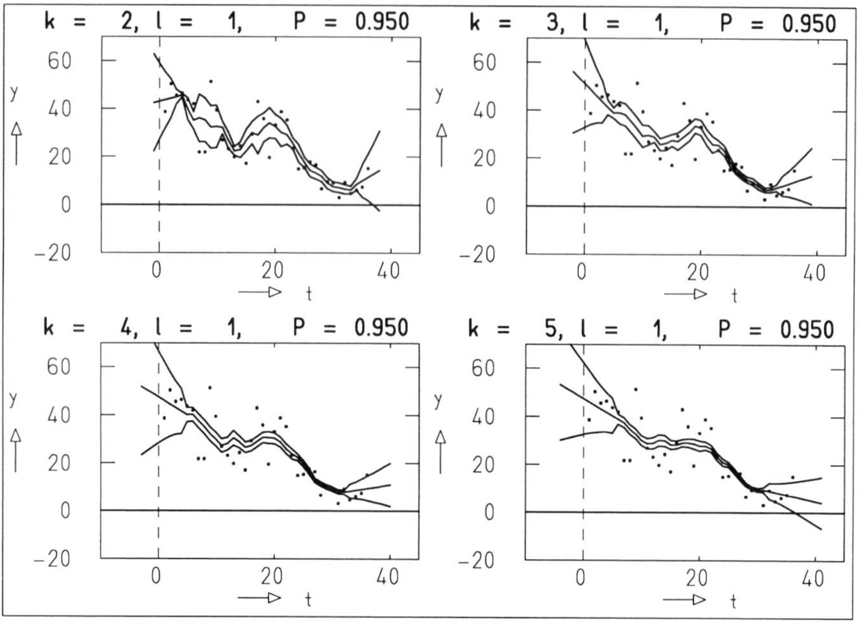

Fig. 13.2. Time series analyses of the same data with fixed ℓ and various values of k.

Caution is always recommended in interpreting the results of a time series analysis. This is particularly true for two reasons:

1. There is usually no *a priori* justification for the mathematical model on which the time series analysis is based. One has simply chosen a convenient procedure in order to "separate out" statistical fluctuations.

2. The user has considerable freedom in choosing the parameters k and ℓ, which, however, can have a significant influence on the results. The following example gives an impression of the magnitude of such influences.

Example 13.2: Time series analysis of the same set of measurements using different averaging intervals and polynomials of different orders

Figures 13.2 and 13.3 contain time series analyses of the average number of sun spots observed in the 36 months from January 1962 through December 1964. Various values of k and ℓ were used. The individual plots in Fig. 13.2 show $\ell = 1$ (linear averaging) but different interval lengths ($2k + 1 = 5, 7, 9,$ and 11). One can see that the curve of moving averages becomes smoother and the confidence interval becomes narrower when k increases, but that then the mean deviation of the individual observations from the curve also increases.

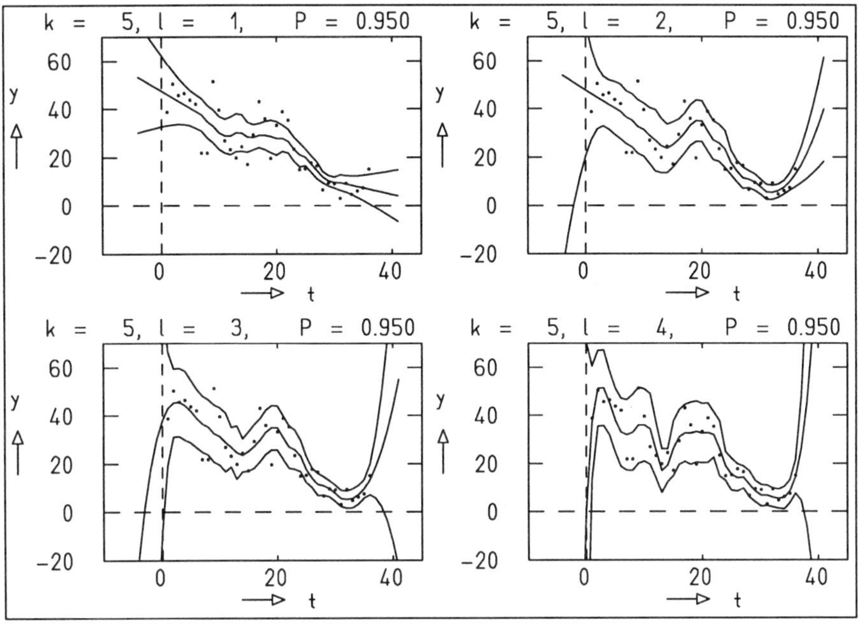

Fig. 13.3. Time series analyses of the same data with fixed k and various values of ℓ.

The extrapolation outside the range of measured points is, of course, a straight line. (For $\ell = 0$ we would have obtained the same moving averages for the inner points. The outer and extrapolated points would lie, however, on a horizontal line, since a polynomial of order zero is a constant.) The plots in Fig. 13.3 correspond to the interval lengths $2k + 1 = 7$ and $\ell = 1, 2, 3, 4$. The moving averages lie closer to the data points and the confidence interval becomes larger when the value of ℓ increases. ■

From these observations we can derive the following qualitative rules:

1. The averaging interval should not be chosen larger than the region where one expects that the data can be well described by a polynomial of the given order. That is, for $\ell = 1$ the interval $2k + 1$ should be chosen such that the expected nonlinear effects within the interval remain small.

2. On the other hand, the smoothing effect becomes stronger as the length of the averaging interval increases. As a rule of thumb, the smoothing becomes more effective for increasing $2k + 1 - \ell$.

3. Caution is required in extrapolation of a time series, especially in nonlinear cases.

The art of time series analysis is, of course, much more highly developed than what we have been able to describe in this short chapter. The interested reader is referred to the specialized literature, where, for example, smoothing functions other than polynomials or multidimensional analyses are treated.

13.6 Example Programs

The source files of the example programs are available in the subdirectory `\DATAN\EXASRC`.

13.6.1 Main Program **E1TIM** to Demonstrate Subprogram **TIMSER**

The short program contains the data of Example 13.2 in a `DATA` statement. After setting some parameters it performs a time series analysis by a call of `TIMSER` with $k = 2$, $\ell = 2$. The data, moving averages, and distances to the confidence limits (at a confidence level of 90 %) are output numerically.

13.6.2 Main Program **E2TIM** to Perform a Time Series Analysis and Display the Output

The program starts with the same data as `E1TIM`. It asks for the parameters k and ℓ and for the confidence level P and then performs a time series analysis with the subroutine `TIMSER`. Subsequently a plot is produced in which the data are displayed as small circles. The floating averages are shown as a polyline. Polylines (in a different color) indicate the confidence limits.

Suggestion: Produce the individual plots of Figures 13.2 and 13.3.

13.7 Programming Problems

The source files of the example solutions are available in the subdirectory `\DATAN\SOLSRC`.

13.7.1 Extrapolation in a Time Series Analysis

In Sect. 13.3 we have stressed that one must be very cautious with the interpretation of the results of a time series analysis at the edge of a time series, and of the extrapolation in regions outside the time series. In particular, we found that the extrapolation yields meaningless results if the data are not at least approximately described by a polynomial. The degree of this polynomial must be smaller than or equal to the degree of the polynomial used for the time series analysis.

Study these statements by simulating a number of time series and analyzing them. Write a program – starting from E2TIM – that for $n = 200$, $i = 1, 2, \ldots, n$, generates data of the form

$$y_i = t_i^m + \varepsilon_i \quad , \qquad t_i = \frac{i - 100}{50} \quad .$$

Here the ε_i are to be generated according to a normal distribution with mean zero and standard deviation σ. Allow for the interactive input of m, σ, k, ℓ, and P. After generating the data perform a time series analysis and produce a plot as in E2TIM. Study different combinations of m, k, and ℓ, and for each combination use small values of σ (e.g., $\sigma = 0.001$) and large values of σ (e.g., $\sigma = 0.1$).

13.7.2 Discontinuities in Time Series

In the development of time series analysis it was assumed that the measurements, apart from their statistical fluctuations, are continuous functions of time. We therefore expect unreliable results in regions where the measurements are discontinuous. Write a program that generates the following three types of time series, analyzes them, and displays the results graphically. One of them is continuous; the other two contain discontinuities.

Sine function:

$$y_i = \sin(\pi t_i/180) + \varepsilon_i \quad , \qquad t_i = i \quad , \qquad i = 1, 2, \ldots, n \quad .$$

Step function:

$$y_i = \begin{cases} -1 + \varepsilon_i \, , & t_i \le 100 \\ 1 + \varepsilon_i \, , & t_i > 100 \end{cases} \quad , \qquad t_i = i \bmod 200 \quad , \qquad i = 1, 2, \ldots, n \quad .$$

Sawtooth function:

$$y_i = (t_i - 50)/100 + \varepsilon_i \quad , \qquad t_i = i \bmod 100 \quad , \qquad i = 1, 2, \ldots, n \quad .$$

The ε_i are again to be generated according to normal distribution with mean zero and standard deviation σ.

Allow for the choice of one of the functions and for the interactive input of n, σ, k, ℓ, and P. Study the time series using different values for the parameters and discuss the results. Figure 13.4 shows an example.

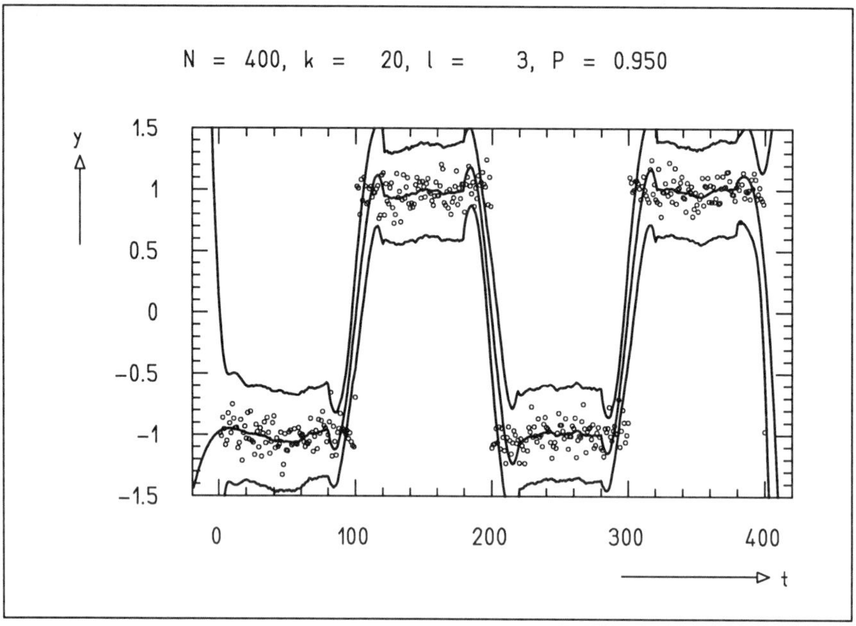

Fig. 13.4. Time series corresponding to a step function with moving averages and confidence limits.

A. Matrix Calculations

The solution of the over-determined linear system of equations $A\mathbf{x} \approx \mathbf{b}$ is of central significance in data analysis. This can be solved in an optimal way with the singular value decomposition, first developed in the late 1960's. In this appendix the elementary definitions and calculation rules for matrices and vectors are summarized in Sects. A.1 and A.2. In Sect. A.3 programs for simple operations are given. In Sect. A.4 orthogonal transformations are introduced, in particular the Givens and Householder transformations, which provide the key to the singular value decomposition.

After a few remarks on determinants (Sect. A.5) there follows in Sect. A.6 a discussion of various cases of matrix equations and a theorem on the orthogonal decomposition of an arbitrary matrix, which is of central importance in this regard. The classical procedure of normal equations, which, however, is inferior to the singular value decomposition, is described here.

Sections A.7 to A.9 concern the particularly simple case of exactly determined, non-singular matrix equations. In this case the inverse matrix A^{-1} exists, and the solution to the problem $A\mathbf{x} = \mathbf{b}$ is $\mathbf{x} = A^{-1}\mathbf{b}$. Methods and programs for finding the solution are given. The important special case of a positive-definite symmetric matrix is treated in Sect. A.10.

In Sect. A.11 we define the pseudo-inverse matrix A^+ of an arbitrary matrix A. After introducing eigenvectors and eigenvalues in Sect. A.12, the singular value decomposition is presented in Sects. A.13 and A.14. Computer routines are given in Sect. A.15. Modifications of the procedure and the consideration of constraints are the subject of Sects. A.16 through A.19.

It has been attempted to make the presentation illustrative rather than mathematically rigorous. Proofs in the text are only indicated in a general way, or are omitted entirely. As mentioned, the singular value decomposition is not yet widespread. An important goal of this appendix is to make its use possible. For readers with a basic knowledge of matrix calculations, the material covered in Sects. A.4, A.13, A.14, A.15.1, and A.19 is sufficient for this task. Sections A.15.2 through A.15.5 contain technical details on carrying out the singular value decomposition and can be omitted by hurried users.

A.1 Definitions. Simple Operations

By a *vector* in m dimensions (an m-vector) $\mathbf{a}$ we mean an m-tuple of real numbers, which are the *components* of $\mathbf{a}$. The arrangement of the components in the form

$$\mathbf{a} = \begin{pmatrix} a_1 \\ a_2 \\ \vdots \\ a_m \end{pmatrix} \qquad \text{(A.1.1)}$$

is called a *column vector.*

A $m \times n$ *matrix* is a rectangular arrangement of $m \times n$ numbers in m rows and n columns,

$$A = \begin{pmatrix} A_{11} & A_{12} & \cdots & A_{1n} \\ A_{21} & A_{22} & \cdots & A_{2n} \\ \vdots & \vdots & & \vdots \\ A_{m1} & A_{m2} & \cdots & A_{mn} \end{pmatrix} \quad . \qquad \text{(A.1.2)}$$

It can be viewed to be composed of n column vectors. By *transposition* of the $m \times n$ matrix A one obtains an $n \times m$ matrix A^{T} with elements

$$A^{\mathrm{T}}_{ik} = A_{ki} \quad . \qquad \text{(A.1.3)}$$

Under transposition a column vector becomes a *row vector,*

$$\mathbf{a}^{\mathrm{T}} = (a_1, a_2, \ldots, a_m) \quad . \qquad \text{(A.1.4)}$$

A column vector is an $m \times 1$ matrix; a row vector is a $1 \times m$ matrix.

For matrices one has the following elementary rules for *addition, subtraction* and *multiplication by a constant*

$$A \pm B = C \quad , \qquad C_{ik} = A_{ik} \pm B_{ik} \quad , \qquad \text{(A.1.5)}$$

$$\alpha A = B \quad , \qquad B_{ik} = \alpha A_{ik} \quad . \qquad \text{(A.1.6)}$$

The *product AB of two matrices* is only defined if the number of columns of the first matrix is equal to the number of rows of the second, e.g., $A = A_{m\times\ell}$ and $B = B_{\ell\times m}$. One has then

$$AB = C \quad , \qquad C_{ik} = \sum_{j=1}^{\ell} A_{ij} B_{jk} \quad . \qquad \text{(A.1.7)}$$

Since

$$C^{\mathrm{T}}_{ik} = C_{ki} = \sum_{j=1}^{\ell} A_{kj} B_{ji} = \sum_{j=1}^{\ell} A^{\mathrm{T}}_{jk} B^{\mathrm{T}}_{ij} = \sum_{j=1}^{\ell} B^{\mathrm{T}}_{ij} A^{\mathrm{T}}_{jk} \quad ,$$

one has

$$C^{\mathrm{T}} = (AB)^{\mathrm{T}} = B^{\mathrm{T}} A^{\mathrm{T}} \quad . \tag{A.1.8}$$

With (A.1.7) one can also define the product of a row vector $\mathbf{a}^{\mathrm{T}}$ with a column vector $\mathbf{b}$, if both have the same number of elements m,

$$\mathbf{a} \cdot \mathbf{b} = \mathbf{a}^{\mathrm{T}} \mathbf{b} = c \quad , \qquad c = \sum_{j=1}^{m} a_j b_j \quad . \tag{A.1.9}$$

The result is then a number, i.e., a *scalar*. The product (A.1.9) is called the *scalar product*. It is usually written without indicating the transposition simply as $\mathbf{a} \cdot \mathbf{b}$. The vectors $\mathbf{a}$ and $\mathbf{b}$ are *orthogonal* to each other if their scalar product vanishes. Starting from (A.1.9) one obtains the following useful property of the matrix product (A.1.7). The element C_{ik}, which is located at the intersection of the ith row and the kth column of the product matrix C, is equal to the scalar product of the ith row of the first matrix A with the kth column of the second matrix B.

The *diagonal elements* of a matrix (A.1.2) are the elements A_{ii}. They form the *main diagonal* of the matrix A. If all of the non-diagonal elements vanish, $A_{ij} = 0, i \neq j$, then A is a *diagonal matrix*. An $n \times n$ diagonal matrix all of whose diagonal elements are unity is the n-dimensional *unit* matrix $I_n = I$,

$$\begin{pmatrix} 1 & 0 & \dots & 0 \\ 0 & 1 & \dots & 0 \\ \vdots & & & \\ 0 & 0 & \dots & 1 \end{pmatrix} = \begin{pmatrix} 1 & & & 0 \\ & 1 & & \\ & & \ddots & \\ 0 & & & 1 \end{pmatrix} = I \quad . \tag{A.1.10}$$

The *null matrix* has only zeros as elements:

$$0 = \begin{pmatrix} 0 & 0 & \dots & 0 \\ 0 & 0 & \dots & 0 \\ \vdots & & & \\ 0 & 0 & \dots & 0 \end{pmatrix} \quad . \tag{A.1.11}$$

A null matrix with only one column is the *null vector* $\mathbf{0}$.

We will now mention several more special types of square matrices.

A square matrix is *symmetric* if

$$A_{ik} = A_{ki} \quad . \tag{A.1.12}$$

If

$$A_{ik} = -A_{ki} \quad , \tag{A.1.13}$$

then the matrix is *antisymmetric*.

A *bidiagonal matrix* B possesses non-vanishing elements only on the main diagonal (b_{ii}) and on the parallel diagonal directly above it ($b_{i,i+1}$).

A *tridiagonal matrix* possesses in a addition non-vanishing elements directly below the main diagonal. A *lower triangular matrix* has non-vanishing elements only on and below the main diagonal, an *upper triangular matrix* only on and above the main diagonal.

The *Euclidian norm* or the *absolute value* of a vector is

$$|\mathbf{a}| = \|\mathbf{a}\|_2 = a = \sqrt{\mathbf{a}^{\mathrm{T}}\mathbf{a}} = \sqrt{\sum_j a_j^2} \quad . \tag{A.1.14}$$

A vector with unit norm is called a *unit vector*. We write this in the form

$$\hat{\mathbf{a}} = \mathbf{a}/a \quad .$$

More general *vector norms* are

$$\begin{aligned} \|\mathbf{a}\|_p &= (\sum_j |a_j|^p)^{1/p} \quad , \qquad 1 \le p < \infty \quad , \\ \|\mathbf{a}\|_\infty &= \max_j |a_j| \quad . \end{aligned} \tag{A.1.15}$$

For every vector norm $\|\mathbf{x}\|$ one defines a *matrix norm* $\|A\|$ as

$$\|A\| = \max_{\mathbf{x} \neq 0} \|A\mathbf{x}\|/\|\mathbf{x}\| \quad . \tag{A.1.16}$$

Matrix norms have the following properties:

$$\|A\| > 0 \quad , \quad A \neq 0 \quad ; \quad \|A\| = 0 \quad , \quad A = 0 \quad , \tag{A.1.17}$$

$$\|\alpha A\| = |\alpha| \|A\| \quad , \qquad \alpha \quad \text{real} \, , \tag{A.1.18}$$

$$\|A + B\| \le \|A\| + \|B\| \quad , \tag{A.1.19}$$

$$\|A\, B\| \le \|A\| \, \|B\| \quad . \tag{A.1.20}$$

A.2 Vector Space, Subspace, Rank of a Matrix

An n*-dimensional vector space* is the set of all n-dimensional vectors. If $\mathbf{u}$ and $\mathbf{v}$ are vectors in this space, then $\alpha\mathbf{u}$ and $\mathbf{u} + \mathbf{v}$ are also in the space, i.e., the vector space is *closed* under vector addition and under multiplication with a scalar α. The vectors $\mathbf{a}_1, \mathbf{a}_2, \ldots, \mathbf{a}_k$ are *linearly independent* if

$$\sum_{j=1}^{k} \alpha_j \mathbf{a}_j \neq 0 \tag{A.2.1}$$

for all α_j except for $\alpha_1 = \alpha_2 = \cdots = \alpha_k = 0$. Otherwise they are *linearly dependent*. The maximum number $k_{\max}$ of vectors that can be linearly independent is equal to the *dimension* of the vector space n. An arbitrary set of n linearly independent vectors $\mathbf{a}_1, \mathbf{a}_2, \ldots, \mathbf{a}_n$ forms a *basis* of the vector space. Any vector $\mathbf{a}$ can be expressed as a *linear combination* of the basis vectors,

$$\mathbf{a} = \sum_{j=1}^{n} \alpha_j \mathbf{a}_j \quad . \tag{A.2.2}$$

A special basis is

$$\mathbf{e}_1 = \begin{pmatrix} 1 \\ 0 \\ 0 \\ \vdots \\ 0 \end{pmatrix} , \quad \mathbf{e}_2 = \begin{pmatrix} 0 \\ 1 \\ 0 \\ \vdots \\ 0 \end{pmatrix} , \quad \ldots , \quad \mathbf{e}_n = \begin{pmatrix} 0 \\ 0 \\ 0 \\ \vdots \\ 1 \end{pmatrix} . \tag{A.2.3}$$

These basis vectors are *orthonormal*, i.e.,

$$\mathbf{e}_i \cdot \mathbf{e}_j = \delta_{ij} = \begin{cases} 1, & i = j \quad , \\ 0, & i \neq j \quad . \end{cases} \tag{A.2.4}$$

The component a_j of the vector $\mathbf{a}$ is the scalar product of $\mathbf{a}$ with the basis vector $\mathbf{e}_j$,

$$\mathbf{a} \cdot \mathbf{e}_j = a_j \quad , \tag{A.2.5}$$

cf. (A.1.1), (A.1.9).

For $n \leq 3$ vectors can be visualized geometrically. A vector $\mathbf{a}$ can be represented as an arrow of length a. The basis vectors (A.2.3) are perpendicular to each other and are of unit length. The perpendicular projections of $\mathbf{a}$ onto the directions of the basis vectors are the components (A.2.5), as shown in Fig. A.1.

A subset T of a vector space S is called a *subspace* if it is itself closed under vector addition and multiplication with a scalar. The greatest possible number of linearly independent vectors in T is the dimension of T. The product of an $m \times n$ matrix A with an n-vector $\mathbf{a}$ is an m-vector $\mathbf{b}$,

$$\mathbf{b} = A\mathbf{a} \quad . \tag{A.2.6}$$

The relation (A.2.6) can be regarded as a *mapping* or *transformation* of the vector $\mathbf{a}$ onto the vector $\mathbf{b}$.

The *span* of a set of vectors $\mathbf{a}_1, \ldots, \mathbf{a}_k$ is the vector space defined by the set of all linear combinations $\mathbf{u}$ of these vectors,

$$\mathbf{u} = \sum_{j=1}^{k} \alpha_j \mathbf{a}_j \quad . \tag{A.2.7}$$

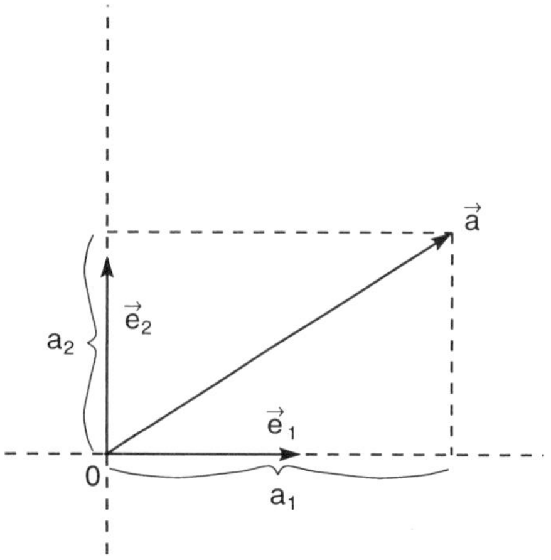

Fig. A.1. The vector **a** in the system of orthonormal basis vectors $\mathbf{e}_1, \mathbf{e}_2$.

It has a dimension $m \leq k$. The *column space* of an $m \times n$ matrix A is the span of the n column vectors of A; in this case the m-vectors $\mathbf{u}$ have the form $\mathbf{u} = A\mathbf{x}$ with arbitrary n-vectors $\mathbf{x}$. Clearly the dimension of the column space is $\leq \min(m, n)$. Similarly, the *row space* of A is the span of the m row vectors.

The *null space* or *kernel* of A consists of the set of vectors $\mathbf{x}$ for which

$$A\mathbf{x} = 0 \quad . \tag{A.2.8}$$

Column and row spaces of an $m \times n$ matrix have the same dimension. This is called the *rank* of the matrix. An $m \times n$ has *full rank* if

$$\mathrm{Rank}(A) = \min(m, n) \quad . \tag{A.2.9}$$

Otherwise it has *reduced rank*. An $n \times n$ matrix with $\mathrm{Rank}(A) < n$ is said to be *singular*.

A vector **a** is orthogonal to a subspace T if it is orthogonal to every vector $t \in T$. (A trivial example is $\mathbf{t} = t_1\mathbf{e}_1 + t_2\mathbf{e}_2$, $\mathbf{a} = a\mathbf{e}_3$, $\mathbf{a} \cdot \mathbf{t} = 0$.) A subspace U is orthogonal to the subspace T if for every pair of vectors $\mathbf{u} \in U$, $\mathbf{t} \in T$ one has $\mathbf{u} \cdot \mathbf{t} = 0$. The set of all vectors $\mathbf{u} + \mathbf{t}$ forms a vector space V, called the *direct sum* of T and U,

$$V = T \oplus U \quad . \tag{A.2.10}$$

Its dimension is

$$\dim(V) = \dim(T) + \dim(U) \quad . \tag{A.2.11}$$

If (A.2.10) holds, then T and U are subspaces of V. They are called *orthogonal complements*, $T = U^\perp$, $U = T^\perp$. If T is a subspace of S, then there always exists an orthogonal complement $T^\perp$, such that $S = T \oplus T^\perp$. Every vector

$\mathbf{a} \in S$ can then be uniquely decomposed into the form $\mathbf{a} = \mathbf{t} + \mathbf{u}$ with $\mathbf{t} \in T$ and $\mathbf{u} \in T^{\perp}$. For the norms of the vectors the relation $a^2 = t^2 + u^2$ holds.

If T is an $(n-1)$-dimensional subspace of an n-dimensional vector space S and if $\mathbf{s}$ is a fixed vector in S, then the set of all vectors $\mathbf{h} = \mathbf{s} + \mathbf{t}$ with $\mathbf{t} \in T$ forms an $(n-1)$-dimensional *hyperplane* H in S. If H is given and $\mathbf{h}_0$ is an arbitrary fixed vector in H, then T is the set of all vectors $\mathbf{t} = \mathbf{h} - \mathbf{h}_0$, $\mathbf{h} \in H$, as shown in Fig. A.2. If $\hat{\mathbf{u}}$ is a unit vector in the one-dimensional subspace $T = H^{\perp}$ orthogonal to H, then the scalar product

$$\hat{\mathbf{u}} \cdot \mathbf{h} = d \tag{A.2.12}$$

has the same value for all $\mathbf{h} \in H$, where d is the distance of the hyperplane from the origin (see Fig. A.3).

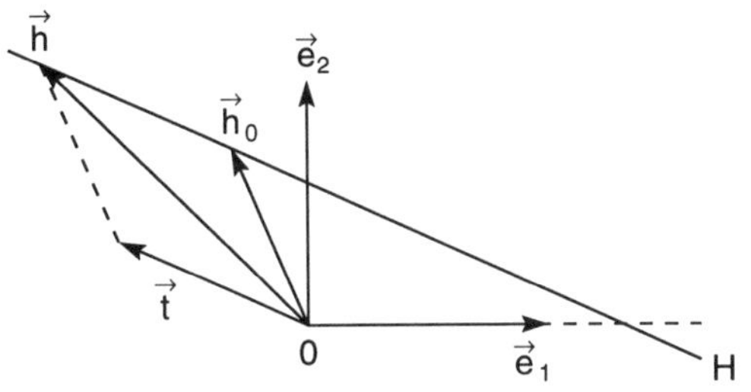

Fig. A.2. Hyperplane H in a two-dimensional vector space.

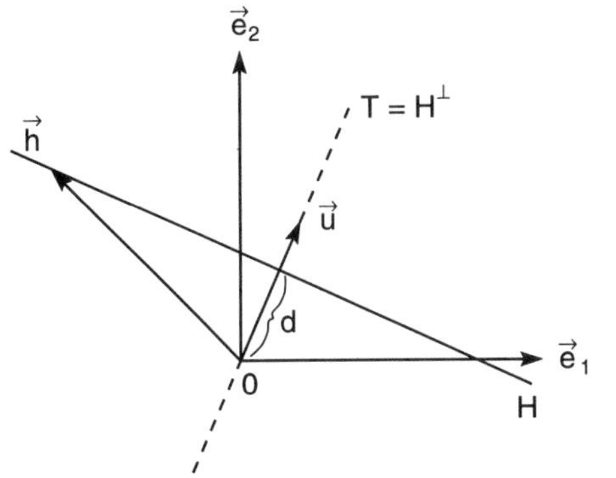

Fig. A.3. Hyperplane H and complementary one-dimensional vector space T.

For a given $\hat{\mathbf{u}}$ and d, Eq. (A.2.12) defines a hyperplane H. It divides the n-dimensional vector space into two *half spaces*, which consist of the set of vectors $\mathbf{x}$ for which $\hat{\mathbf{u}} \cdot \mathbf{x} < 0$ and $\hat{\mathbf{u}} \cdot \mathbf{x} > 0$.

A.3 Programs for Simple Matrix and Vector Operations

In this section we present a number of simple subprograms for matrix and vector operations. They do not require individual explanation. The user must note, however, the following technical points.

The vectors and matrices in the argument lists of the subprograms have *variable dimension*. The declaration of the dimension is contained in the argument list itself. For these vectors and matrices, sufficiently large arrays must be defined in the calling program.

Program A.1: Subroutine MTXTRA copies an $m \times n$ matrix

CALLING SEQUENCE: CALL MTXTRA(A,R,M,N)

INPUT ARGUMENTS: A — $m \times n$ matrix A
M,N — dimensions m, n

OUTPUT ARGUMENTS: R — copy $R = A$ of the matrix A

```
      SUBROUTINE MTXTRA(A,R,M,N)
      IMPLICIT DOUBLE PRECISION (A-H,O-Z)
      DIMENSION A(M,N),R(M,N)
      DO 20 J=1,N
        DO 10 I=1,M
          R(I,J)=A(I,J)
   10   CONTINUE
   20 CONTINUE
      END
```

Program A.2: Subroutine MTXWRT writes the $m \times n$ matrix A line by line to the standard output device

CALLING SEQUENCE: CALL MTXWRT(A,M,N)

INPUT ARGUMENTS: A — $m \times n$ matrix A
M,N — dimensions m, n

```
      SUBROUTINE MTXWRT(A,M,N)
      IMPLICIT DOUBLE PRECISION (A-H,O-Z)
      DIMENSION A(M,N)
      DO 20 I=1,M
        WRITE(*,10)(A(I,J),J=1,N)
   10 FORMAT(1X,4G20.13)
   20 CONTINUE
      END
```

Program A.3: Subroutine MTXADD adds two $m \times n$ matrices

CALLING SEQUENCE: CALL MTXADD(A,B,R,M,N)

INPUT ARGUMENTS: A — $m \times n$ matrix A
B — $m \times n$ matrix B
M,N — dimensions m, n

OUTPUT ARGUMENTS: R — $m \times n$ matrix $R = A + B$

```
      SUBROUTINE MTXADD(A,B,R,M,N)
      IMPLICIT DOUBLE PRECISION (A-H,O-Z)
      DIMENSION A(M,N),B(M,N),R(M,N)
      DO 20 J=1,N
        DO 10 I=1,M
          R(I,J)=A(I,J)+B(I,J)
   10   CONTINUE
   20 CONTINUE
      END
```

Program A.4: Subroutine MTXSUB subtracts an $m \times n$ matrix B from an $m \times n$ matrix A

CALLING SEQUENCE: CALL MTXSUB(A,B,R,M,N)

INPUT ARGUMENTS: A — $m \times n$ matrix A
B — $m \times n$ matrix B
M,N — dimensions m, n

OUTPUT ARGUMENTS: R — $m \times n$ matrix $R = A - B$

```
      SUBROUTINE MTXSUB(A,B,R,M,N)
      IMPLICIT DOUBLE PRECISION (A-H,O-Z)
      DIMENSION A(M,N),B(M,N),R(M,N)
      DO 20 J=1,N
        DO 10 I=1,M
          R(I,J)=A(I,J)-B(I,J)
   10   CONTINUE
   20 CONTINUE
      END
```

Program A.5: Subroutine MTXMLT multiplies an $m \times \ell$ matrix A with an $\ell \times n$ matrix B

CALLING SEQUENCE: CALL MTXMLT(A,B,R,M,L,N)

INPUT ARGUMENTS: A — $m \times \ell$ matrix A
B — $\ell \times n$ matrix B
M,L,N — dimensions m, ℓ, n

OUTPUT ARGUMENTS: R — $m \times n$ matrix $R = AB$

```
      SUBROUTINE MTXMLT(A,B,R,M,L,N)
      IMPLICIT DOUBLE PRECISION (A-H,O-Z)
      DIMENSION A(M,L),B(L,N),R(M,N)
      DO 30 J=1,N
        DO 20 I=1,M
          R(I,J)=0.
          DO 10 LL=1,L
            R(I,J)=R(I,J)+A(I,LL)*B(LL,J)
   10     CONTINUE
   20   CONTINUE
   30 CONTINUE
      END
```

Program A.6: Subroutine MTXMBT multiplies an $m \times \ell$ matrix A with the transpose of the $n \times \ell$ matrix B

CALLING SEQUENCE: CALL MTXMBT(A,B,R,M,L,N)

INPUT ARGUMENTS: A — $m \times \ell$ matrix A
B — $n \times \ell$ matrix B
M,L,N — dimensions m, ℓ, n

OUTPUT ARGUMENTS: R — $m \times n$ matrix $R = AB^{\mathrm{T}}$

```
      SUBROUTINE MTXMBT(A,B,R,M,L,N)
      IMPLICIT DOUBLE PRECISION (A-H,O-Z)
      DIMENSION A(M,L),B(N,L),R(M,N)
      DO 30 J=1,N
        DO 20 I=1,M
          R(I,J)=0.
          DO 10 LL=1,L
            R(I,J)=R(I,J)+A(I,LL)*B(J,LL)
   10     CONTINUE
   20   CONTINUE
   30 CONTINUE
      END
```

Program A.7: Subroutine MTXMAT multiplies the transpose of an $\ell \times m$ matrix A with an $\ell \times n$ matrix B

CALLING SEQUENCE: CALL MTXMAT(A,B,R,M,L,N)

INPUT ARGUMENTS: A — $\ell \times m$ matrix A
B — $\ell \times n$ matrix B
M,L,N — dimensions m, ℓ, n

OUTPUT ARGUMENTS: R — $m \times n$ matrix $R = A^{\mathrm{T}}B$

```
      SUBROUTINE MTXMAT(A,B,R,M,L,N)
      IMPLICIT DOUBLE PRECISION (A-H,O-Z)
      DIMENSION A(L,M),B(L,N),R(M,N)
      DO 30 J=1,N
        DO 20 I=1,M
          R(I,J)=0.
          DO 10 LL=1,L
            R(I,J)=R(I,J)+A(LL,I)*B(LL,J)
   10     CONTINUE
   20   CONTINUE
   30 CONTINUE
      END
```

Program A.8: Subroutine MTXUNT prepares an $n \times n$ unit matrix

CALLING SEQUENCE: CALL MTXUNT(R,N)

INPUT ARGUMENTS: N — dimension n

OUTPUT ARGUMENTS: R — $n \times n$ unit matrix $R = I_n$

```
      SUBROUTINE MTXUNT(R,N)
      IMPLICIT DOUBLE PRECISION (A-H,O-Z)
      DIMENSION R(N,N)
      DO 20 J=1,N
        DO 10 I=1,N
          IF(I.EQ.J) THEN
            R(I,J)=1.D0
          ELSE
            R(I,J)=0.D0
          END IF
   10   CONTINUE
   20 CONTINUE
      END
```

Program A.9: Subroutine MTXZER prepares an $m \times n$ null matrix

CALLING SEQUENCE: CALL MTXZER(R,M,N)

INPUT ARGUMENTS: M, N — dimensions m, n

OUTPUT ARGUMENTS: R — $m \times n$ null matrix R

```
      SUBROUTINE MTXZER(R,N,M)
      IMPLICIT DOUBLE PRECISION (A-H,O-Z)
      DIMENSION R(N,M)
      DO 20 J=1,M
        DO 10 I=1,N
          R(I,J)=0.
   10   CONTINUE
   20 CONTINUE
      END
```

Program A.10: Subroutine MTXMSC multiplies an $m \times n$ matrix A by a scalar factor

CALLING SEQUENCE: CALL MTXMSC(A,R,S,M,N)

INPUT ARGUMENTS: A — $m \times n$ matrix A
S — scalar factor s
M, N — dimensions m, n

OUTPUT ARGUMENTS: R — $m \times n$ matrix $R = sA$

```
      SUBROUTINE MTXMSC(A,R,S,M,N)
      IMPLICIT DOUBLE PRECISION (A-H,O-Z)
      DIMENSION A(M,N),R(M,N)
      DO 20 J=1,N
        DO 10 I=1,M
          R(I,J)=S*A(I,J)
   10   CONTINUE
   20 CONTINUE
      END
```

Program A.11: Subroutine MTXTRP transposes an $m \times n$ matrix A

CALLING SEQUENCE: CALL MTXTRP(A,R,M,N)

INPUT ARGUMENTS: A — $m \times n$ matrix A
M,N — dimensions m, n

OUTPUT ARGUMENTS: R — $n \times m$ matrix $R = A^{\mathrm{T}}$

```
      SUBROUTINE MTXTRP(A,R,M,N)
      IMPLICIT DOUBLE PRECISION (A-H,O-Z)
      DIMENSION A(M,N),R(N,M)
      DO 20 J=1,N
        DO 10 I=1,M
          R(J,I)=A(I,J)
   10   CONTINUE
   20 CONTINUE
      END
```

Program A.12: Subroutine MTXGSM extracts a $k \times \ell$ submatrix S from an $m \times n$ matrix A

CALLING SEQUENCE: CALL MTXGSM(A,S,M,N,K,L,M1,N1)

INPUT ARGUMENTS: A — $m \times n$ matrix A
M,N,K,L — dimensions m, n, k, ℓ
M1,N1 — indices m_1, n_1 in A such that $A_{m_1,n_1} = S_{1,1}$

OUTPUT ARGUMENTS: S — $k \times \ell$ matrix S

```
      SUBROUTINE MTXGSM(A,S,M,N,K,L,M1,N1)
      IMPLICIT DOUBLE PRECISION (A-H,O-Z)
      DIMENSION A(M,N), S(K,L)
      DO 20 I=1,K
        DO 10 J=1,L
          S(I,J)=A(M1-1+I,N1-1+J)
   10   CONTINUE
   20 CONTINUE
      END
```

Program A.13: Subroutine MTXPSM replaces a $k \times \ell$ submatrix S within an $m \times n$ matrix A

CALLING SEQUENCE: CALL MTXPSM(A,S,M,N,K,L,M1,N1)

INPUT ARGUMENTS: A — $m \times n$ matrix A
S — $k \times \ell$ matrix S
M,N,K,L — dimensions m, n, k, ℓ
M1,N1 — indices m_1, n_1 in A such that $A_{m_1,n_1} = S_{1,1}$

OUTPUT ARGUMENTS: A — $m \times n$ matrix A, in which S is inserted as a submatrix such that $A_{m_1,n_1} = B_{1,1}$

```
      SUBROUTINE MTXPSM(A,S,M,N,K,L,M1,N1)
      IMPLICIT DOUBLE PRECISION (A-H,O-Z)
      DIMENSION A(M,N), S(K,L)
      DO 20 I=1,K
        DO 10 J=1,L
          A(M1-1+I,N1-1+J)=S(I,J)
10      CONTINUE
20    CONTINUE
      END
```

Program A.14: Subroutine MTXGCL extracts the ith column vector **c** from an $m \times n$ matrix A

CALLING SEQUENCE: CALL MTXGCL(A,C,M,N,I)

INPUT ARGUMENTS: A — $m \times n$ matrix A
M,N — dimensions m, n
I — column index i of **c** within A

OUTPUT ARGUMENTS: C — m-vector **c**

```
      SUBROUTINE MTXGCL(A,C,M,N,I)
      IMPLICIT DOUBLE PRECISION (A-H,O-Z)
      DIMENSION A(M,N), C(M)
      DO 10 J=1,M
        C(J)=A(J,I)
10    CONTINUE
      END
```

Program A.15: Subroutine MTXPCL replaces the ith column vector by the vector **c** in an $m \times n$ matrix A

CALLING SEQUENCE: CALL MTXPCL(A,C,M,N,I)

INPUT ARGUMENTS: A — $m \times n$ matrix A
C — m-vector **c**
M,N — dimensions m, n
I — column index i of **c** within A

OUTPUT ARGUMENTS: A — $m \times n$ matrix A, whose ith column is replaced by **c**.

```
      SUBROUTINE MTXPCL(A,C,M,N,I)
      IMPLICIT DOUBLE PRECISION (A-H,O-Z)
      DIMENSION A(M,N), C(M)
      DO 10 J=1,M
        A(J,I)=C(J)
   10 CONTINUE
      END
```

Program A.16: Subroutine MTXGRW extracts the ith row vector **r** from an $m \times n$ matrix A

CALLING SEQUENCE: CALL MTXGRW(A,R,M,N,I)

INPUT ARGUMENTS: A — $m \times n$ matrix A
M,N — dimensions m, n
I — row index i of **r** within A

OUTPUT ARGUMENTS: R — n-vector **r**

```
      SUBROUTINE MTXGRW(A,R,M,N,I)
      IMPLICIT DOUBLE PRECISION (A-H,O-Z)
      DIMENSION A(M,N), R(N)
      DO 10 J=1,N
        R(J)=A(I,J)
   10 CONTINUE
      END
```

Program A.17: Subroutine MTXPRW replaces the ith row vector by the vector **r** in an $m \times n$ matrix A

CALLING SEQUENCE: CALL MTXPRW(A,R,M,N,I)

INPUT ARGUMENTS: A — $m \times n$ matrix A
R — n-vector **r**
M,N — dimensions m, n
I — row index i of **r** within A

OUTPUT ARGUMENTS: A — $m \times n$ matrix A, whose ith row is replaced by **r**.

```
      SUBROUTINE MTXPRW(A,R,M,N,I)
      IMPLICIT DOUBLE PRECISION (A-H,O-Z)
      DIMENSION A(M,N), R(N)
      DO 10 J=1,N
        A(I,J)=R(J)
   10 CONTINUE
      END
```

Program A.18: Subroutine MTXCPV copies an n-vector

CALLING SEQUENCE: CALL MTXCPV(U,V,N)

INPUT ARGUMENTS: U — n-vector $\mathbf{u}$
N — dimension n

OUTPUT ARGUMENTS: V — copy of the n-vector $\mathbf{u}$

```
      SUBROUTINE MTXCPV(U,V,N)
      IMPLICIT DOUBLE PRECISION (A-H,O-Z)
      DIMENSION U(N),V(N)
      DO 10 I=1,N
        V(I)=U(I)
   10 CONTINUE
      END
```

Program A.19: Subroutine MTXADV adds two n-vectors

CALLING SEQUENCE: CALL MTXADV(U,V,W,N)

INPUT ARGUMENTS: U, V — n-vectors $\mathbf{u}$, $\mathbf{v}$
N — dimension n

OUTPUT ARGUMENTS: W — sum vector $\mathbf{w} = \mathbf{u} + \mathbf{v}$

```
      SUBROUTINE MTXADV(U,V,W,N)
      IMPLICIT DOUBLE PRECISION (A-H,O-Z)
      DIMENSION U(N),V(N),W(N)
      DO 10 I=1,N
        W(I)=U(I)+V(I)
   10 CONTINUE
      END
```

Program A.20: Subroutine MTXSBV subtracts an n-vector $\mathbf{v}$ from an n-vector $\mathbf{u}$

CALLING SEQUENCE: CALL MTXSBV(U,V,W,N)

INPUT ARGUMENTS: U, V — n-vectors $\mathbf{u}$, $\mathbf{v}$
N — dimension n

OUTPUT ARGUMENTS: W — difference vector $\mathbf{w} = \mathbf{u} - \mathbf{v}$

```
      SUBROUTINE MTXSBV(U,V,W,N)
      IMPLICIT DOUBLE PRECISION (A-H,O-Z)
      DIMENSION U(N),V(N),W(N)
      DO 10 I=1,N
        W(I)=U(I)-V(I)
   10 CONTINUE
      END
```

Program A.21: Subroutine MTXDOT computes the scalar product of two n-vectors

CALLING SEQUENCE: CALL MTXDOT(U,V,S,N)

INPUT ARGUMENTS: U, V — n-vectors $\mathbf{u}$, $\mathbf{v}$
N — dimension n

OUTPUT ARGUMENTS: S — scalar product $s = \mathbf{u} \cdot \mathbf{v}$

```
      SUBROUTINE MTXDOT(U,V,S,N)
      IMPLICIT DOUBLE PRECISION (A-H,O-Z)
      DIMENSION U(N),V(N)
      S=0.
      DO 10 I=1,N
        S=S+U(I)*V(I)
   10 CONTINUE
      END
```

Program A.22: Subroutine MTXNRV computes the Euclidian norm of an n-vector

CALLING SEQUENCE: CALL MTXNRV(U,S,N)

INPUT ARGUMENTS: U — n-vector $\mathbf{u}$
N — dimension n

OUTPUT ARGUMENTS: S — norm $s = \sqrt{\mathbf{u} \cdot \mathbf{u}}$

```
      SUBROUTINE MTXNRV(U,S,N)
      IMPLICIT DOUBLE PRECISION (A-H,O-Z)
      CALL MTXDOT(U,U,S,N)
      S= SQRT(ABS(S))
      END
```

Program A.23: Subroutine MTXMSV multiplies an n-vector by a scalar factor

CALLING SEQUENCE: CALL MTXMSV(U,V,S,N)

INPUT ARGUMENTS: U — n-vector $\mathbf{u}$
N — dimension n
S — scalar factor s

OUTPUT ARGUMENTS: V — n-vector $\mathbf{v} = s\mathbf{u}$

```
      SUBROUTINE MTXMSV(U,V,S,N)
      IMPLICIT DOUBLE PRECISION (A-H,O-Z)
      DIMENSION U(N),V(N)
      DO 10 I=1,N
        V(I)=S*U(I)
   10 CONTINUE
      END
```

Program A.24: Subroutine `MTXZRV` prepares an n-dimensional null vector

CALLING SEQUENCE: `CALL MTXZRV(U,N)`

INPUT ARGUMENTS: `N` — dimension n

OUTPUT ARGUMENTS: `U` — null vector $\mathbf{u} = \mathbf{0}$

```
      SUBROUTINE MTXZRV(U,N)
      IMPLICIT DOUBLE PRECISION (A-H,O-Z)
      DIMENSION U(N)
      DO 10 I=1,N
        U(I)=0.
   10 CONTINUE
      END
```

Program A.25: Subroutine `MTXGSV` extracts an m-dimensional subvector $\mathbf{v}$ from an n-dimensional vector $\mathbf{u}$

CALLING SEQUENCE: `CALL MTXGSV(U,V,N,M,LIST)`

INPUT ARGUMENTS: `U` — n-vector $\mathbf{u}$

`N,M` — dimensions n, m with $m \leq n$

`LIST` — list of n elements having the values 0 or 1. If `LIST(I)` has the value 1, then `U(I)` becomes an element of `V`.

OUTPUT ARGUMENTS: `V` — m-vector $\mathbf{v}$

```
      SUBROUTINE MTXGSV(U,V,N,NRED,LIST)
      IMPLICIT DOUBLE PRECISION (A-H,O-Z)
      DIMENSION U(N),V(NRED),LIST(N)
      IRED=0
      DO 10 I=1,N
        IF(LIST(I).EQ.1) THEN
          IRED=IRED+1
          V(IRED)=U(I)
        END IF
   10 CONTINUE
      END
```

Program A.26: Subroutine `MTXPSV` replaces elements of the n-dimensional vector $\mathbf{u}$ by the elements of the m-dimensional subvector $\mathbf{v}$

CALLING SEQUENCE: `CALL MTXPSV(U,V,N,M,LIST)`

INPUT ARGUMENTS: `U` — n-vector $\mathbf{u}$

`V` — m-vector $\mathbf{v}$

`N,M` — dimensions n, m with $m \leq n$

`LIST` — list of n elements, having the values 0 or 1. The elements `U(I)` for which `LIST(I)` has the value 1 are replaced in order by the elements of `V`.

OUTPUT ARGUMENTS: `U` — n-vector $\mathbf{u}$

```
      SUBROUTINE MTXPSV(U,V,N,NRED,LIST)
      IMPLICIT DOUBLE PRECISION (A-H,O-Z)
      DIMENSION U(N),V(NRED),LIST(N)
      IRED=0
      DO 10 I=1,N
        IF(LIST(I).EQ.1) THEN
          IRED=IRED+1
          U(I)=V(IRED)
        END IF
   10 CONTINUE
      END
```

A.4 Orthogonal Transformations

According to (A.2.6), the mapping of an n-vector $\mathbf{a}$ onto an n-vector $\mathbf{b}$ is performed by multiplication by a square $n \times n$ matrix, $\mathbf{b} = Q\mathbf{a}$. If the length of the vector (A.1.14) remains unchanged, one speaks of an *orthogonal transformation*. For such a case one has $b = a$ or $b^2 = a^2$, i.e.,

$$\mathbf{b}^{\mathrm{T}}\mathbf{b} = \mathbf{a}^{\mathrm{T}}Q^{\mathrm{T}}Q\mathbf{a} = \mathbf{a}^{\mathrm{T}}\mathbf{a} \quad ,$$

and thus

$$Q^{\mathrm{T}}Q = I \quad . \tag{A.4.1}$$

A square matrix Q that fulfills (A.4.1) is said to be *orthogonal*.

It is clear that transformations are orthogonal when the transformed vector $\mathbf{b}$ is obtained from $\mathbf{a}$ by means of a spatial rotation and/or reflection. We will examine some orthogonal transformations that are important for the applications of this appendix.

A.4.1 Givens Transformation

The *Givens rotation* is a transformation that affects only components in a plane spanned by two orthogonal basis vectors. For simplicity we will first consider only two-dimensional vectors.

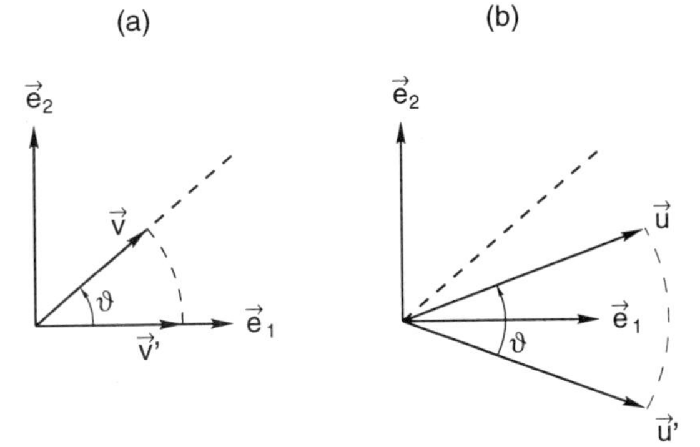

Fig. A.4. Application of the Givens transformation (a) to the vector **v** that defines the transformation and (b) to an arbitrary vector **u**.

A vector **v** can be represented as

$$\mathbf{v} = \begin{pmatrix} v_1 \\ v_2 \end{pmatrix} = v \begin{pmatrix} c \\ s \end{pmatrix} \quad , \qquad c = \cos\vartheta = v_1/v \quad , \qquad s = \sin\vartheta = v_2/v \quad . \tag{A.4.2}$$

A rotation by an angle $-\vartheta$ transforms the vector **v** into the vector $\mathbf{v}' = G\mathbf{v}$, whose second component vanishes, as in Fig. A.4a. Clearly one has

$$G = \begin{pmatrix} c & s \\ -s & c \end{pmatrix} \quad , \qquad \mathbf{v}' = G\mathbf{v} = \begin{pmatrix} v \\ 0 \end{pmatrix} \quad . \tag{A.4.3}$$

Of course the transformation thus defined with the vector **v** can also be applied to any other vector **u**, as shown in Fig. A.4b. In n dimensions the Givens rotation leads to the transformation

$$\mathbf{v} \to \mathbf{v}' = G\mathbf{v} \quad ,$$

such that $v'_k = 0$. The components $v'_\ell = v_\ell$, $\ell \neq k$, $\ell \neq i$, remain unchanged and v'_i is determined such that the norm of the vector remains unchanged, $v' = v$. This is clearly given by

$$G = \begin{pmatrix} 1 & & & & & & & & \\ & \ddots & & & & & & & \\ & & 1 & & & & & & \\ & & & c & & s & & & \\ & & & & 1 & & & & \\ & & & & & \ddots & & & \\ & & & & & & 1 & & \\ & & & -s & & c & & & \\ & & & & & & 1 & & \\ & & & & & & & \ddots & \\ & & & & & & & & 1 \end{pmatrix} \begin{matrix} \leftarrow 1 \\ \\ \\ \leftarrow i \\ \\ \\ \\ \leftarrow k \\ \\ \\ \leftarrow n \end{matrix} \quad . \tag{A.4.4}$$

In practical applications, however, the full matrix is not needed.

Program A.27: Subroutine MTXGVD defines a Givens rotation

CALLING SEQUENCE: CALL MTXGVD(V1,V2,C,S)

INPUT ARGUMENTS: V1,V2 — vector components defining the Givens rotation

OUTPUT ARGUMENTS: C,S — constants corresponding to (A.4.2)

```
      SUBROUTINE MTXGVD(V1,V2,C,S)
      IMPLICIT DOUBLE PRECISION (A-H,O-Z)
      A1=ABS(V1)
      A2=ABS(V2)
      IF(A1.GT.A2) THEN
        W=V2/V1
        Q=SQRT(1.+W*W)
        C=1./Q
        IF(V1.LT.0.) C=-C
        S=C*W
      ELSE
        IF(V2.NE.0.) THEN
          W=V1/V2
          Q=SQRT(1.+W*W)
          S=1./Q
          IF(V2.LT.0.) S=-S
          C=S*W
        ELSE
          C=1.
          S=0.
        END IF
      END IF
      END
```

Program A.28: Subroutine MTXGVA defines a Givens transformation and applies it to the vector by which it is defined

CALLING SEQUENCE: CALL MTXGVA(V1,V2,C,S)

INPUT ARGUMENTS: V1,V2 — vector components that define the Givens transformation

OUTPUT ARGUMENTS: C,S — constants corresponding to (A.4.2).
V1,V2 — components of the transformed vector; in particular, V2 = 0

```
SUBROUTINE MTXGVA(V1,V2,C,S)
IMPLICIT DOUBLE PRECISION (A-H,O-Z)
A1=ABS(V1)
A2=ABS(V2)
IF(A1.GT.A2) THEN
  W=V2/V1
  Q=SQRT(1.+W*W)
  C=1./Q
  IF(V1.LT.0.) C=-C
  S=C*W
  V1=A1*Q
  V2=0.
ELSE
  IF(V2.NE.0.) THEN
    W=V1/V2
    Q=SQRT(1.+W*W)
    S=1./Q
    IF(V2.LT.0.) S=-S
    C=S*W
    V1=A2*Q
    V2=0.
  ELSE
    C=1.
    S=0.
  END IF
END IF
END
```

Program A.29: Subroutine MTXGVT applies a Givens rotation to two components of a vector

CALLING SEQUENCE: CALL MTXGVT(Z1,Z2,C,S)

INPUT ARGUMENTS: C,S — constants determined by calling MTXGVD or MTXGVA
Z1,Z2 — components of a vector to be transformed

OUTPUT ARGUMENTS: Z1,Z2 — components of the transformed vector

```
      SUBROUTINE MTXGVT(Z1,Z2,C,S)
      IMPLICIT DOUBLE PRECISION(A-H,O-Z)
      W=Z1*C+Z2*S
      Z2=-Z1*S+Z2*C
      Z1=W
      END
```

A.4.2 Householder Transformation

The Givens rotation is used to transform a vector in such a way that a given vector component vanishes. A more general transformation is the *Householder transformation*. If the original vector is

$$\mathbf{v} = \begin{pmatrix} v_1 \\ v_2 \\ \vdots \\ v_n \end{pmatrix} \quad , \tag{A.4.5}$$

then for the transformed vector we want to have

$$\mathbf{v}' = H\mathbf{v} = \begin{pmatrix} v_1 \\ \vdots \\ v_{p-1} \\ v'_p \\ v_{p+1} \\ \vdots \\ v_{\ell-1} \\ 0 \\ \vdots \\ 0 \end{pmatrix} \quad . \tag{A.4.6}$$

That is, the components v'_ℓ, $v'_{\ell+1}$, ..., v'_n should vanish. The remaining components should (with the exception of v'_p) remain unchanged. The component v'_p must be changed in such a way that one has $v = v'$. From the Pythagorean theorem in $n - \ell + 1$ dimensions one has

$$v'^2_p = v^2_H = v^2_p + \sum_{i=\ell}^{n} v_i^2$$

or

$$v'_p = -\sigma v_H = -\sigma \sqrt{v^2_p + \sum_{i=\ell}^{n} v_i^2} \tag{A.4.7}$$

with $\sigma = \pm 1$. We choose

$$\sigma = \mathrm{sign}(v_p) \quad . \tag{A.4.8}$$

We now construct the matrix H of (A.4.6). To do this we decompose the vector $\mathbf{v}$ into a sum,

$$\mathbf{v} = \mathbf{v}_H + \mathbf{v}_{H^\perp} \quad , \qquad \mathbf{v}_H = \begin{pmatrix} 0 \\ \vdots \\ 0 \\ v_p \\ 0 \\ \vdots \\ 0 \\ v_\ell \\ \vdots \\ v_n \end{pmatrix} \quad , \qquad \mathbf{v}_{H^\perp} = \begin{pmatrix} v_1 \\ \vdots \\ v_{p-1} \\ 0 \\ v_{p+1} \\ \vdots \\ v_{\ell-1} \\ 0 \\ \vdots \\ 0 \end{pmatrix} \quad . \tag{A.4.9}$$

Here the vector $\mathbf{v}_H$ is in the subspace spanned by the basis vectors $\mathbf{e}_p$, $\mathbf{e}_\ell$, $\mathbf{e}_{\ell+1}$, ..., $\mathbf{e}_n$, and $\mathbf{v}_{H^\perp}$ is in the subspace orthogonal to it. We now construct

$$\mathbf{u} = \mathbf{v}_H + \sigma v_H \mathbf{e}_p \tag{A.4.10}$$

and

$$H = I_n - \frac{2\mathbf{u}\mathbf{u}^{\mathrm{T}}}{u^2} \quad . \tag{A.4.11}$$

If we now decompose an arbitrary vector $\mathbf{a}$ into a sum of vectors parallel and perpendicular to $\mathbf{u}$,

$$\mathbf{a} = \mathbf{a}_{\parallel} + \mathbf{a}_{\perp} \quad ,$$

with

$$\mathbf{a}_{\parallel} = \frac{\mathbf{u}\mathbf{u}^{\mathrm{T}}}{u^2}\mathbf{a} = \frac{\mathbf{u}}{u^2}(\mathbf{u}\cdot\mathbf{a}) = \hat{\mathbf{u}}(\hat{\mathbf{u}}\cdot\mathbf{a}) \quad , \qquad \mathbf{a}_{\perp} = \mathbf{a} - \mathbf{a}_{\parallel} \quad ,$$

then one has

$$\mathbf{a}'_{\parallel} = H\mathbf{a}_{\parallel} = -\mathbf{a}_{\parallel} \quad , \qquad \mathbf{a}'_{\perp} = H\mathbf{a}_{\perp} = \mathbf{a}_{\perp} \quad .$$

Thus we see that the transformation is a reflection in the subspace that is orthogonal to the vector $\mathbf{u}$, as in Fig. A.5. One can easily verify that H in fact yields the transformation (A.4.6).

The transformation is uniquely determined by the vector $\mathbf{u}$. According to (A.4.10) one has for the components of this vector $u_p = v_p + \sigma v_H$, $u_\ell = v_\ell$, $u_{\ell+1} = v_{\ell+1}$, ..., $u_n = v_n$, and $u_i = 0$ for all other i. If the vector $\mathbf{v}$ and the indices p and ℓ are given, then only u_p must be computed. The quantity u^2 appearing in (A.4.11) is then

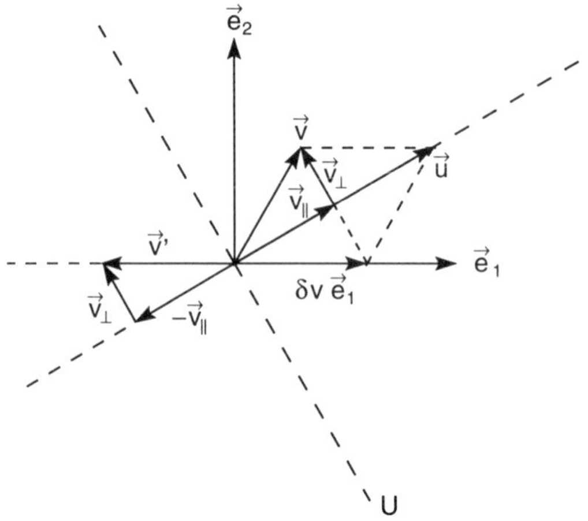

Fig. A.5. The vector **v** is mapped onto **v**′ according to a Householder transformation such that $v_2' = 0$. The mapping corresponds to a reflection in the subspace U that is orthogonal to the auxiliary vector **u**.

$$
\begin{aligned}
u^2 &= u_p^2 + \sum_{i=\ell}^{n} u_i^2 = (v_p + \sigma v_H)^2 + \sum_{i=\ell}^{n} v_i^2 \\
&= v_p^2 + \sum_{i=\ell}^{n} v_i^2 + v_H^2 + 2\sigma v_H v_p \\
&= 2v_H^2 + 2\sigma v_H v_p = 2v_H(v_H + \sigma v_p) = 2v_H u_p \quad .
\end{aligned}
$$

We can thus write (A.4.11) in the form

$$
H = I_n - b\mathbf{u}\mathbf{u}^{\mathrm{T}} \quad , \qquad b = (v_H u_p)^{-1} \quad . \tag{A.4.12}
$$

Program A.30: Subroutine `MTXHSD` defines a Householder transformation

CALLING SEQUENCE: `CALL MTXHSD(V,UP,B,N,LP,L)`

INPUT ARGUMENTS: `V` — n-vector **v**
`N,LP,L` — vector indices n, p, ℓ

OUTPUT ARGUMENTS: `UP` — vector element u_p
`B` — coefficient b

```
      SUBROUTINE MTXHSD(V,UP,B,N,LP,L)
      IMPLICIT DOUBLE PRECISION (A-H,O-Z)
      DIMENSION V(N)
      C=ABS(V(LP))
      DO 10 I=L,N
        C=MAX(ABS(V(I)),C)
   10 CONTINUE
      IF(C.LE.0.D0) GO TO 30
      C1=1./C
      SD=(V(LP)*C1)**2
      DO 20 I=L,N
```

```
      SD=SD+(V(I)*C1)**2
   20 CONTINUE
      VPPRIM=SD
      VPPRIM=C*SQRT(ABS(VPPRIM))
      IF(V(LP).GT.0.D0) VPPRIM=-VPPRIM
      UP=V(LP)-VPPRIM
      B=1./(VPPRIM*UP)
   30 CONTINUE
      END
```

The matrix H, however, is not needed explicitly to compute a transformed vector,

$$\mathbf{c}' = H\mathbf{c} \quad .$$

It is sufficient to know the vector **u** and the constant b. Since, however, **u** only differs from **v** in the element u_p (and in the vanishing elements), it is sufficient, starting from **v**, to first compute the quantities u_p and b and when applying the transformation to use in addition the elements v_ℓ, $v_{\ell+1}$, ..., v_n. These are at the same time the corresponding elements of **u**.

Program A.31: Subroutine `MTXHST` applies a Householder transformation to a vector

CALLING SEQUENCE: `CALL MTXHST(V,UP,B,C,N,LP,L)`

INPUT ARGUMENTS: `V,N,LP,L` — as in `MTXHSD`
`UP,B` — must be determined by a call to `MTXHSD`
`C` — the n-vector to be transformed

OUTPUT ARGUMENT: `C` — transformed n-vector

```
      SUBROUTINE MTXHST(V,UP,B,C,N,LP,L)
      IMPLICIT DOUBLE PRECISION (A-H,O-Z)
      DIMENSION V(N),C(N)
      DUP=UP
      S=C(LP)*DUP
      DO 10 I=L,N
        S=S+C(I)*V(I)
   10 CONTINUE
      S=S*B
      C(LP)=C(LP)+S*DUP
      DO 20 I=L,N
        C(I)=C(I)+S*V(I)
   20 CONTINUE
      END
```

A.4.3 Sign Inversion

If the diagonal element I_{ii} of the unit matrix is replaced by -1, then one obtains a symmetric orthogonal matrix,

$$R^{(i)} = \begin{pmatrix} 1 & & & & \\ & \ddots & & & \\ & & -1 & & \\ & & & \ddots & \\ & & & & 1 \end{pmatrix} \quad .$$

Applying this to the vector **a**,

$$\mathbf{a}' = R^{(i)}\mathbf{a} \quad ,$$

changes the sign of the element a_i and leaves all the other elements unchanged. Clearly $R^{(i)}$ is a Householder matrix which produces a reflection in the subspace orthogonal to the basis vector $\mathbf{e}_i$. This can be seen immediately by substituting $\mathbf{u} = \mathbf{e}_i$ in (A.4.11).

A.4.4 Permutation Transformation

The $n \times n$ unit matrix (A.1.10) can easily be written as an arrangement of the basis vectors (A.2.3) in a square matrix,

$$I_n = (\mathbf{e}_1, \mathbf{e}_2, \ldots, \mathbf{e}_n) = \begin{pmatrix} 1 & 0 & \cdots & 0 \\ 0 & 1 & \cdots & 0 \\ \vdots & & & \\ 0 & 0 & \cdots & 1 \end{pmatrix} \quad .$$

It is clearly an orthogonal matrix. The orthogonal transformation $I\mathbf{a} = \mathbf{a}$ leaves the vector **a** unchanged. If we now exchange two of the basis vectors $\mathbf{e}_i$ and $\mathbf{e}_k$, then we obtain the symmetric orthogonal matrix P^{ik}. As an example we show this for $n = 4$, $i = 2$, $k = 4$,

$$P^{(ik)} = \begin{pmatrix} 1 & 0 & 0 & 0 \\ 0 & 0 & 0 & 1 \\ 0 & 0 & 1 & 0 \\ 0 & 1 & 0 & 0 \end{pmatrix} \quad .$$

The transformation $\mathbf{a}' = P^{(ik)}\mathbf{a}$ leads to an exchange of the elements a_i and a_k. All other elements remain unchanged:

$$\mathbf{a} = \begin{pmatrix} a_1 \\ \vdots \\ a_i \\ \vdots \\ a_k \\ \vdots \\ a_n \end{pmatrix} \quad , \qquad \mathbf{a}' = P^{(ik)}\mathbf{a} = \begin{pmatrix} a_1 \\ \vdots \\ a_k \\ \vdots \\ a_i \\ \vdots \\ a_n \end{pmatrix} \quad .$$

Multiplication of an $n \times m$ matrix A on the left with $P^{(ik)}$ permutes the lines i and k of A. Multiplication of an $m \times n$ matrix A from the right exchanges the columns i and k. If D is an $n \times n$ diagonal matrix, then the elements D_{ii} and D_{kk} are exchanged by the operation

$$D' = P^{(ik)} D \, P^{(ik)} \quad .$$

A.5 Determinants

To every $n \times n$ matrix A one can associate a number, its *determinant*, det A. A determinant, like the corresponding matrix, is written as a square arrangement of the matrix elements, but is enclosed, however, by vertical lines. The determinants of orders two and three are defined by

$$\det A = \begin{vmatrix} A_{11} & A_{12} \\ A_{21} & A_{22} \end{vmatrix} = A_{11}A_{22} - A_{12}A_{21} \tag{A.5.1}$$

and

$$\begin{aligned} \det A &= \begin{vmatrix} A_{11} & A_{12} & A_{13} \\ A_{21} & A_{22} & A_{23} \\ A_{31} & A_{32} & A_{33} \end{vmatrix} \\ &= \begin{array}{ll} \ A_{11}A_{22}A_{33} & - \ A_{11}A_{23}A_{32} \\ + \ A_{12}A_{23}A_{31} & - \ A_{12}A_{21}A_{33} \\ + \ A_{13}A_{21}A_{32} & - \ A_{13}A_{22}A_{31} \end{array} \end{aligned} \tag{A.5.2}$$

or, written in another way,

$$\det A = \begin{array}{ll} \ A_{11}(A_{22}A_{33} & - \ A_{23}A_{32}) \\ - \ A_{12}(A_{21}A_{33} & - \ A_{23}A_{31}) \\ + \ A_{13}(A_{21}A_{32} & - \ A_{22}A_{31}) \end{array} \quad . \tag{A.5.3}$$

A general determinant of order n is written in the form

$$\det A = \begin{vmatrix} A_{11} & A_{12} & \dots & A_{1n} \\ A_{21} & A_{22} & \dots & A_{2n} \\ \vdots & & & \\ A_{n1} & A_{n2} & \dots & A_{nn} \end{vmatrix} \quad . \tag{A.5.4}$$

The *cofactor* $A^{\dagger}_{ij}$ of the element A_{ij} of a matrix is a determinant of order $(n-1)$, calculated from the matrix obtained by deleting the ith row and jth column of the original matrix, and multiplying by $(-1)^{i+j}$,

$$A^\dagger_{ij} = (-1)^{i+j} \begin{vmatrix} A_{11} & A_{12} & \dots & A_{1,j-1} & A_{1,j+1} & \dots & A_{1n} \\ A_{21} & A_{22} & \dots & A_{2,j-1} & A_{2,j+1} & \dots & A_{2n} \\ \vdots & & & & & & \\ A_{i-1,1} & A_{i-1,2} & \dots & A_{i-1,j-1} & A_{i-1,j+1} & \dots & A_{i-1,n} \\ A_{i+1,1} & A_{i+1,2} & \dots & A_{i+1,j-1} & A_{i+1,j+1} & \dots & A_{i+1,n} \\ \vdots & & & & & & \\ A_{n1} & A_{n2} & \dots & A_{n,j-1} & A_{n,j+1} & \dots & A_{nn} \end{vmatrix} \quad . \tag{A.5.5}$$

Determinants of higher order can be written as the sum of all elements of any row or column multiplied with the corresponding cofactors,

$$\det A = \sum_{k=1}^{n} A_{ik} A^\dagger_{ik} = \sum_{k=1}^{n} A_{kj} A^\dagger_{kj} \quad . \tag{A.5.6}$$

One can easily show that the result is independent of the choice of row i or column j. Equation (A.5.3) already shows that (A.5.6) is correct for $n = 3$. Determinants of arbitrary order can be computed by decomposing them according to their cofactors until one reaches, for example, the order two. A singular matrix, i.e., a square matrix whose rows or columns are not linearly independent, has *determinant zero*.

From A we can construct a further matrix by replacing each element ij by the cofactor of the element ji. In this way we obtain the *adjoint matrix* of A,

$$A^\dagger = \begin{pmatrix} A^\dagger_{11} & A^\dagger_{21} & \dots & A^\dagger_{n1} \\ A^\dagger_{12} & A^\dagger_{22} & \dots & A^\dagger_{n2} \\ \vdots & & & \\ A^\dagger_{1n} & A^\dagger_{2n} & \dots & A^\dagger_{nn} \end{pmatrix} \quad . \tag{A.5.7}$$

For determinants the following rules hold:

$$\det A = \det A^{\mathrm{T}} \quad , \tag{A.5.8}$$

$$\det AB = \det A \det B \quad . \tag{A.5.9}$$

For an orthogonal matrix Q one has $Q\,Q^{\mathrm{T}} = I$, i.e.,

$$\det I = 1 = \det Q \det Q$$

and thus

$$\det Q = \pm 1 \quad . \tag{A.5.10}$$

A.6 Matrix Equations. Least Squares

A system of m linear equations with n unknowns $x_1, x_2, \dots, x_n$ has the general form

$$
\begin{aligned}
a_{11}x_1 + a_{12}x_2 + \cdots + a_{1n}x_n - b_1 &= 0 \quad , \\
a_{21}x_1 + a_{22}x_2 + \cdots + a_{2n}x_n - b_2 &= 0 \quad , \\
\vdots \\
a_{m1}x_1 + a_{m2}x_2 + \cdots + a_{mn}x_n - b_m &= 0
\end{aligned}
\tag{A.6.1}
$$

or in matrix notation

$$A\mathbf{x} - \mathbf{b} = 0 \quad . \tag{A.6.2}$$

In finding the solution to this equation $\mathbf{x}$ we must distinguish between various cases, which can be characterized by the values of m, n, and

$$k = \text{Rank}(A) \quad .$$

The vector $A\mathbf{x}$ is in the column space of A, which is of dimension k. Since $\mathbf{b}$ is an m-vector, the equation can in general only be fulfilled if $k = m$, i.e., for $k = n = m$ and for $k = m < n$, since $k \leq \min(m, n)$. For $k = n = m$ one has n independent equations (A.6.1) with n unknowns, which have a *unique* solution. If $k = m < n$, then there are arbitrarily many n-vectors $\mathbf{x}$ that can be mapped on to the m-vector $A\mathbf{x} = \mathbf{b}$ such that (A.6.2) is fulfilled. The system of equations is *underdetermined.* The solution is not unique.

For $k = \text{Rank}(A) \neq m$ there is, in general, no solution of (A.6.2). In this case we look for a vector $\tilde{\mathbf{x}}$, for which the left-hand side of (A.6.2) is a vector of minimum Euclidian norm. That is, we replace Eq. (A.6.2) by

$$r^2 = (A\mathbf{x} - \mathbf{b})^2 = \min \quad , \tag{A.6.3}$$

i.e., we look for a vector $\tilde{\mathbf{x}}$ for which the mapping $A\tilde{\mathbf{x}}$ differs as little as possible from $\mathbf{b}$.

Given a m-vector $\mathbf{c}$, only for $k = \text{Rank}(A) = n$ does there exist only one n-vector $\mathbf{x}$, such that $A\mathbf{x} = \mathbf{c}$. Therefore, only for the case $k = n$ is there a unique solution $\tilde{\mathbf{x}}$. Thus there exists for $\text{Rank}(A) = n$ and $m \geq n$ a *unique solution* $\tilde{\mathbf{x}}$ of (A.6.3). For $n = m$ one has $r = 0$.

The relation (A.6.3) is also often written simply in the form

$$A\mathbf{x} - \mathbf{b} \approx 0 \tag{A.6.4}$$

or even, not entirely correctly, in the form (A.6.2). One calls the solution vector $\tilde{\mathbf{x}}$ the *least-squares solution of* (A.6.4). In Table A.1 we list again the various cases that result from different relationships between m, n, and k.

We now want to state more formally what we have determined in this section with respect to the solution of (A.6.4).

Theorem on the orthogonal decomposition of a matrix:
Every $m \times n$ matrix A of rank k can be written in the form

$$A = H R K^{\mathrm{T}} \quad , \tag{A.6.5}$$

Table A.1. Behavior of the solutions of (A.6.3) for various cases. m is the row number, n is the column number and k is the rank of the matrix A.

Case		Rank(A)	Residual	Solution unique
1a	$m = n$	$k = n$	$r = 0$	yes
1b		$k < n$	$r \geq 0$	no
2a	$m > n$	$k = n$	$r \geq 0$	yes
2b		$k < n$	$r \geq 0$	no
3a	$m < n$	$k = m$	$r = 0$	no
3b		$k < m$	$r \geq 0$	no

where H is an $m \times m$ orthogonal matrix, K is an $n \times n$ orthogonal matrix, and R is an $m \times n$ matrix of the form

$$R = \begin{pmatrix} R_{11} & 0 \\ 0 & 0 \end{pmatrix} \tag{A.6.6}$$

and where R_{11} is a $k \times k$ matrix of rank k.

Substituting (A.6.5) into (A.6.4) and multiplying from the left by H^{T} leads to

$$R K^{\mathrm{T}} \mathbf{x} \approx H^{\mathrm{T}} \mathbf{b} \quad . \tag{A.6.7}$$

We define

$$H^{\mathrm{T}}\mathbf{b} = \mathbf{g} = \begin{pmatrix} \mathbf{g}_1 \\ \mathbf{g}_2 \end{pmatrix} \begin{matrix} \}k \\ \}m-k \end{matrix} \tag{A.6.8}$$

and

$$K^{\mathrm{T}}\mathbf{x} = \mathbf{p} = \begin{pmatrix} \mathbf{p}_1 \\ \mathbf{p}_2 \end{pmatrix} \begin{matrix} \}k \\ \}n-k \end{matrix} \quad , \tag{A.6.9}$$

so that (A.6.7) takes on the form

$$R\mathbf{p} \approx \mathbf{g} \quad . \tag{A.6.10}$$

Because of (A.6.6), this breaks into two independent relations

$$R_{11}\mathbf{p}_1 = \mathbf{g}_1 \tag{A.6.11}$$

and

$$0 \cdot \mathbf{p}_2 \approx \mathbf{g}_2 \quad . \tag{A.6.12}$$

If $m = k$ and/or $n = k$, then the corresponding lower partial vectors are absent in (A.6.8) and/or (A.6.9) and the corresponding lower matrices are absent in (A.6.6). Since in (A.6.11) R_{11} is a $k \times k$ matrix of rank k, and $\mathbf{p}_1$ and $\mathbf{g}_1$ are k-vectors, there exists a solution vector $\widetilde{\mathbf{p}}_1$ for which equality holds. Because of the null matrix on the left-hand side of (A.6.12), we cannot derive any information about $\mathbf{p}_2$ from this relation.

Theorem on the solutions of $A\mathbf{x} \approx \mathbf{b}$: If $\tilde{\mathbf{p}}_1$ is the unique solution vector of (A.6.11), then the following statements hold:

(i) All solutions of (A.6.3) have the form

$$\widehat{\mathbf{x}} = K \begin{pmatrix} \tilde{\mathbf{p}}_1 \\ \tilde{\mathbf{p}}_2 \end{pmatrix} \quad . \tag{A.6.13}$$

Here $\tilde{\mathbf{p}}_2$ is an arbitrary $(n-k)$-vector, i.e., the solution is unique for $k = n$. There is always, however, a *unique solution of minimum absolute value*:

$$\tilde{\mathbf{x}} = K \begin{pmatrix} \tilde{\mathbf{p}}_1 \\ 0 \end{pmatrix} \quad . \tag{A.6.14}$$

(ii) All solutions $\widehat{\mathbf{x}}$ have the same residual vector

$$\mathbf{r} = \mathbf{b} - A\widehat{\mathbf{x}} = H \begin{pmatrix} 0 \\ \mathbf{g}_2 \end{pmatrix} \tag{A.6.15}$$

with the absolute value $r = g_2 = |\mathbf{g}_2|$. The residual vanishes for $k = m$.

The problem $A\mathbf{x} \approx \mathbf{b}$ should always be handled with orthogonal decompositions, preferably with the singular value decomposition and singular value analysis described in Sects. A.13 and A.14. Numerically the results are at least as accurate as with other methods, and are often more accurate (cf. Sect. A.14, Example A.4).

Nevertheless we will briefly present as well the method of *normal equations*. The method is very transparent compared to the orthogonal decomposition and is therefore always described in textbooks.

We consider the square (A.6.3) of the residual vector

$$\begin{aligned} r^2 &= (A\mathbf{x} - \mathbf{b})^{\mathrm{T}}(A\mathbf{x} - \mathbf{b}) = \mathbf{x}^{\mathrm{T}} A^{\mathrm{T}} A\mathbf{x} - 2\mathbf{b}^{\mathrm{T}} A\mathbf{x} + \mathbf{b}^{\mathrm{T}}\mathbf{b} \\ &= \sum_{i=1}^{m}\sum_{j=1}^{n}\sum_{\ell=1}^{n} A_{ij}A_{i\ell}x_j x_\ell - 2\sum_{i=1}^{m}\sum_{j=1}^{n} b_i A_{ij} x_j + \sum_{i=1}^{m} b_i^2 \quad . \end{aligned}$$

The requirement $r^2 = \min$ leads to

$$\frac{\partial r^2}{\partial x_k} = 2\sum_{i=1}^{m}\sum_{\ell=1}^{n} A_{ik}A_{i\ell}x_\ell - 2\sum_{i=1}^{m} b_i A_{ik} = 0 \quad , \qquad k = 1, \ldots, n \quad .$$

These n linear equations are called normal equations. They can be arranged in a matrix,

$$A^{\mathrm{T}} A\mathbf{x} = A^{\mathrm{T}}\mathbf{b} \quad . \tag{A.6.16}$$

This is a system of n equations with n unknowns. If the equations are linearly independent, then the $n \times n$ matrix $(A^{\mathrm{T}}A)$ is of full rank n; it is not singular. According to Sect. A.7 there then exists an inverse $(A^{\mathrm{T}}A)^{-1}$, such that $(A^{\mathrm{T}}A)^{-1}(A^{\mathrm{T}}A) = I$. Thus the desired solution to (A.6.3) is

$$\widetilde{\mathbf{x}} = (A^{\mathrm{T}}A)^{-1}A^{\mathrm{T}}\mathbf{b} \quad . \tag{A.6.17}$$

This simple prescription is, however, useless if $A^{\mathrm{T}}A$ is singular or nearly singular (cf. Example A.4 in Sect. A.14).

A.7 Inverse Matrix

For every non-singular $n \times n$ matrix A, the *inverse matrix* A^{-1} is defined by

$$A\,A^{-1} = I_n = A^{-1}\,A \quad . \tag{A.7.1}$$

It is also an $n \times n$ matrix.

If A^{-1} is known, then the solution of the matrix equation

$$A\mathbf{x} = \mathbf{b} \tag{A.7.2}$$

is given simply by

$$\mathbf{x} = A^{-1}\mathbf{b} \quad . \tag{A.7.3}$$

As we will show, A^{-1} only exists for non-singular square matrices, so that (A.7.3) only gives the solution for the case 1a of Table A.1.

In order to determine A^{-1}, we set $A^{-1} = X$ and express the column vectors of X as $\mathbf{x}_1$, $\mathbf{x}_2$, ..., $\mathbf{x}_n$. Equation (A.7.1) is then decomposed into n equations:

$$A\mathbf{x}_i = \mathbf{e}_i \quad , \qquad i = 1, 2, \ldots, n \quad . \tag{A.7.4}$$

The right-hand sides are the basis vectors (A.2.3). For the case $n = 2$ we can write the system (A.7.4) in various equivalent ways, e.g.,

$$\begin{pmatrix} A_{11} & A_{12} \\ A_{21} & A_{22} \end{pmatrix} \begin{pmatrix} X_{11} \\ X_{21} \end{pmatrix} = \begin{pmatrix} 1 \\ 0 \end{pmatrix} \quad , \quad \begin{pmatrix} A_{11} & A_{12} \\ A_{21} & A_{22} \end{pmatrix} \begin{pmatrix} X_{12} \\ X_{22} \end{pmatrix} = \begin{pmatrix} 0 \\ 1 \end{pmatrix} \quad , \tag{A.7.5}$$

or alternatively

$$\begin{pmatrix} A_{11} & A_{12} \\ A_{21} & A_{22} \end{pmatrix} \begin{pmatrix} X_{11} & X_{12} \\ X_{21} & X_{22} \end{pmatrix} = \begin{pmatrix} 1 & 0 \\ 0 & 1 \end{pmatrix} \quad ,$$

or as a system of four equations with four unknowns,

$$
\begin{aligned}
A_{11}X_{11} + A_{12}X_{21} &= 1 \quad , \\
A_{21}X_{11} + A_{22}X_{21} &= 0 \quad , \\
A_{11}X_{12} + A_{12}X_{22} &= 0 \quad , \\
A_{21}X_{12} + A_{22}X_{22} &= 1 \quad .
\end{aligned}
\tag{A.7.6}
$$

By elimination and substitution one easily finds

$$
\begin{aligned}
X_{11} &= \frac{A_{22}}{A_{11}A_{22} - A_{12}A_{21}} \quad , \\
X_{12} &= \frac{-A_{12}}{A_{11}A_{22} - A_{12}A_{21}} \quad , \\
X_{21} &= \frac{-A_{21}}{A_{11}A_{22} - A_{12}A_{21}} \quad , \\
X_{22} &= \frac{A_{11}}{A_{11}A_{22} - A_{12}A_{21}}
\end{aligned}
\tag{A.7.7}
$$

or in matrix notation

$$
X = A^{-1} = \frac{1}{\det A} \begin{pmatrix} A_{22} & -A_{12} \\ -A_{21} & A_{11} \end{pmatrix} \quad . \tag{A.7.8}
$$

The matrix on the right-hand side is the adjoint of the original matrix, i.e.,

$$
A^{-1} = \frac{A^{\dagger}}{\det A} \quad . \tag{A.7.9}
$$

One can show that this relation holds for square matrices of arbitrary order. From (A.7.9) it is clear that the inverse of a singular matrix, i.e., of a matrix with vanishing determinant, is not determined.

In practice the inverse matrix is not computed using (A.7.9) but rather as in our example of the 2×2 matrix by elimination and substitution from the system (A.7.4). This system consists of n sets of equations of the form

$$
A\mathbf{x} = \mathbf{b} \quad , \tag{A.7.10}
$$

each consisting of n equations with n unknowns. Here A is a non-singular $n \times n$ matrix. We will first give the solution algorithm for square non-singular matrices A, but we will later remove these restrictions.

A.8 Gaussian Elimination

We will write Eq. (A.7.10) for an $n \times n$ matrix A in components,

$$
\begin{aligned}
&A_{11}x_1 + A_{12}x_2 + \cdots + A_{1n}x_n = b_1 \quad , \\
&A_{21}x_1 + A_{22}x_2 + \cdots + A_{2n}x_n = b_2 \quad , \\
&\vdots \\
&A_{n1}x_1 + A_{n2}x_2 + \cdots + A_{nn}x_n = b_n \quad ,
\end{aligned}
\tag{A.8.1}
$$

and we will solve the system by *Gaussian elimination*. For this we define $n-1$ multipliers

$$m_{i1} = \frac{A_{i1}}{A_{11}} \quad , \qquad i = 2, 3, \ldots, n \quad , \tag{A.8.2}$$

multiply the first equation by m_{21}, and subtract it from the second. We then multiply the first equation by m_{31} and subtract it from the third, and so forth. We obtain the system

$$\begin{aligned} A_{11}^{(1)} x_1 \; + \; A_{12}^{(1)} x_2 + \cdots + A_{1n}^{(1)} x_n &= b_1^{(1)} \quad , \\ A_{22}^{(2)} x_2 + \cdots + A_{2n}^{(2)} x_n &= b_2^{(2)} \quad , \\ &\vdots \\ A_{n2}^{(2)} x_2 + \cdots + A_{nn}^{(2)} x_n &= b_n^{(2)} \quad , \end{aligned} \tag{A.8.3}$$

where the unknown x_1 has disappeared from all the equations except the first. The coefficients $A_{ij}^{(2)}$, $b_i^{(2)}$ are given by the equations

$$\begin{aligned} A_{ij}^{(2)} &= A_{ij}^{(1)} - m_{i1} A_{1j}^{(1)} \quad , \\ b_i^{(2)} &= b_i^{(1)} - m_{i1} b_1^{(1)} \quad . \end{aligned}$$

The procedure is then repeated with the last $n-1$ equations by defining

$$m_{i2} = \frac{A_{i2}^{(2)}}{A_{22}^{(2)}} \quad , \qquad i = 3, 4, \ldots, n \quad ,$$

multiplying the second equation of the system (A.8.3) with the corresponding m_{i2}, and then subtracting it from the third, fourth, ..., nth equation. In the kth step of the procedure, the multipliers

$$m_{ik} = \frac{A_{ik}^{(k)}}{A_{kk}^{(k)}} \quad , \qquad i = k+1,\ k+2,\ \ldots, n \quad , \tag{A.8.4}$$

are used and the new coefficients

$$\begin{aligned} A_{ij}^{(k+1)} &= A_{ij}^{(k)} - m_{ik} A_{kj}^{(k)} \quad , \\ b_i^{(k+1)} &= b_i^{(k)} - m_{ik} b_k^{(k)} \end{aligned} \tag{A.8.5}$$

are computed. After $n-1$ steps we have produced the following *triangular system of equations*:

$$\begin{aligned} A_{11}^{(1)} x_1 + \; A_{12}^{(1)} x_2 + \cdots + \; A_{1n}^{(1)} x_n &= b_1^{(1)} \quad , \\ A_{22}^{(2)} x_2 + \cdots + \; A_{2n}^{(2)} x_n &= b_2^{(2)} \quad , \\ \vdots \qquad &\quad\ \vdots \\ A_{nn}^{(n)} x_n &= b_n^{(n)} \quad . \end{aligned} \tag{A.8.6}$$

The last equation contains only x_n. By substituting into the next higher equation one obtains x_{n-1}, i.e., in general

$$x_i = \frac{1}{A_{ii}^{(i)}} \left\{ b_i^{(i)} - \sum_{\ell=i+1}^{n} A_{i\ell}^{(i)} x_\ell \right\} \quad , \qquad i = n, n-1, \ldots, 1 \quad . \tag{A.8.7}$$

One should note that the change of the elements of the matrix A does not depend on the right-hand side **b**. Therefore one can reduce several systems of equations with different right-hand sides simultaneously. That is, instead of $A\mathbf{x} = \mathbf{b}$, one can solve the more general system

$$AX = B \tag{A.8.8}$$

at the same time, where B and X are $n \times m$ matrices, the matrix X being unknown.

Example A.1: Inversion of a 3×3 matrix
As a numerical example let us consider the inversion of a 3×3 matrix, i.e., $B = I$. The individual computations for the example

$$\begin{pmatrix} 1 & 2 & 3 \\ 2 & 1 & -2 \\ 1 & 1 & 2 \end{pmatrix} X = I$$

are shown in Table A.2. The result is

$$X = \frac{1}{5} \begin{pmatrix} -4 & 1 & 7 \\ 6 & 1 & -8 \\ -1 & -1 & 3 \end{pmatrix} \quad . \blacksquare$$

Following the individual steps of the calculation one sees that division is carried out in two places, namely in Eqs. (A.8.4) and (A.8.7). The denominator is in both cases a coefficient

$$A_{ii}^{(i)} \quad , \qquad i = 1, 2, \ldots, n-1 \quad ,$$

i.e., the upper left-hand coefficient of the system, the so-called *pivot* for the step $i-1$ of the reduction process. Our procedure fails if this coefficient is equal to zero. In such a case one can simply exchange the ith line of the system with some other lower line whose first coefficient is not zero. The system of equations itself is clearly not changed by exchanging two equations. The procedure still fails if all of the coefficients of a column vanish. In this case the matrix A is singular, and there is no solution. In practice (at least when using computers where the extra work is negligible) it is advantageous to always carry out a reshuffling so that the pivot is the coefficient with the largest absolute value among those in the first column of the reduced system. One then always has the largest possible denominator. In this way rounding errors are kept as small as possible. The following program is based on *Gaussian elimination with pivoting* as just described.

Table A.2. Application of Gaussian elimination to Example A.1.

Reduction

	Matrix A			Matrix B			Multiplier
	1	2	3	1	0	0	–
Step 0	2	1	−2	0	1	0	2
	1	1	2	0	0	1	1
		−3	−8	−2	1	0	–
Step 1		−1	−1	−1	0	1	$\frac{1}{3}$
Step 2			$\frac{5}{3}$	$-\frac{1}{3}$	$-\frac{1}{3}$	1	–

Substitution

	$j = 1$
x_{3j}	$-\frac{1}{5}$
x_{2j}	$-\frac{1}{3}(-2 - 8 \times \frac{1}{5}) = \frac{6}{5}$
x_{1j}	$1 - 2 \times \frac{6}{5} + 3 \times \frac{1}{5} = -\frac{4}{5}$
	$j = 2$
x_{3j}	$-\frac{1}{5}$
x_{2j}	$-\frac{1}{3}(1 - \frac{8}{5}) = \frac{1}{5}$
x_{1j}	$-2 \times \frac{1}{5} + 3 \times \frac{1}{5} = \frac{1}{5}$
	$j = 3$
x_{3j}	$\frac{3}{5}$
x_{2j}	$-\frac{1}{3}(0 + 8 \times \frac{3}{5}) = -\frac{8}{5}$
x_{1j}	$2 \times \frac{8}{5} - 3 \times \frac{3}{5} = \frac{7}{5}$

Program A.32: Subroutine MTXEQU for solving an exactly determined matrix equation $AX = B$ using Gaussian elimination with pivoting

CALLING SEQUENCE: CALL MTXEQU(A,B,N,M)

INPUT ARGUMENTS: A — non-singular $n \times n$ matrix A
B — $n \times m$ matrix for the right-hand side of $AX = B$
N,M — n, m

OUTPUT ARGUMENT: B — contains the $n \times m$ solution matrix X

REMARKS: The input matrix A is overwritten. It is *not* checked whether A is singular or nearly singular.

```
      SUBROUTINE MTXEQU(A,B,N,M)
      IMPLICIT DOUBLE PRECISION (A-H,O-Z)
      DIMENSION A(N,N),B(N,M)
      DO 70 K=1,N-1
        AMAX=0.
        KK=K
        DO 10 L=K,N
          IF(ABS(AMAX).LT.ABS(A(L,K))) THEN
            AMAX=A(L,K)
            KK=L
          END IF
 10     CONTINUE
        IF(KK.NE.K) THEN
          DO 20 J=K,N
            SAVE=A(K,J)
            A(K,J)=A(KK,J)
            A(KK,J)=SAVE
 20       CONTINUE
          DO 30 I=1,M
            SAVE=B(K,I)
            B(K,I)=B(KK,I)
            B(KK,I)=SAVE
 30       CONTINUE
        END IF
        DO 60 I=K+1,N
          DO 40 J=K+1,N
            A(I,J)=A(I,J)-A(K,J)*A(I,K)/A(K,K)
 40       CONTINUE
          DO 50 J=1,M
            B(I,J)=B(I,J)-B(K,J)*A(I,K)/A(K,K)
 50       CONTINUE
 60     CONTINUE
 70   CONTINUE
      DO 100 J=1,M
        B(N,J)=B(N,J)/A(N,N)
        IF(N.GT.1) THEN
          DO 90 I1=1,N-1
            I=N-I1
            DO 80 L=I+1,N
              B(I,J)=B(I,J)-A(I,L)*B(L,J)
 80         CONTINUE
            B(I,J)=B(I,J)/A(I,I)
 90       CONTINUE
```

```
      END IF
  100 CONTINUE
      END
```

A.9 LR-Decomposition

For the elements $A_{ij}^{(k)}$ transformed by Gaussian elimination [cf. (A.8.6)], one has on and above the main diagonal

$$A_{ij}^{(n)} = A_{ij}^{(n-1)} = \cdots = A_{ij}^{(i)} \quad , \qquad i \leq j \quad , \tag{A.9.1}$$

and below the main diagonal

$$A_{ij}^{(n)} = A_{ij}^{(n-1)} = \cdots = A_{ij}^{(j+1)} = 0 \quad , \qquad i > j \quad . \tag{A.9.2}$$

The transformation (A.8.5) is thus only carried out for $k = 1, 2, \ldots, r$ with $r = \min(i-1,\ j)$,

$$A_{ij}^{(k+1)} = A_{ij}^{(k)} - m_{ik} A_{kj}^{(k)} \quad . \tag{A.9.3}$$

Summation over k gives

$$\sum_{k=1}^{r} A_{ij}^{(k+1)} - \sum_{k=1}^{r} A_{ij}^{(k)} = A_{ij}^{(r+1)} - A_{ij} = -\sum_{k=1}^{r} m_{ik} A_{kj}^{(k)} \quad .$$

Thus the elements $A_{ij}^{(1)} = A_{ij}$ of the original matrix A can be written in the form

$$\begin{aligned} A_{ij} &= A_{ij}^{(i)} + \textstyle\sum_{k=1}^{i-1} m_{ik} A_{kj}^{(k)} \quad , \qquad i \leq j \quad , \\ A_{ij} &= 0 + \textstyle\sum_{k=1}^{j} m_{ik} A_{kj}^{(k)} \quad , \qquad i > j \quad . \end{aligned} \tag{A.9.4}$$

Noting that the multipliers m_{ik} have only been defined up to now for $i > k$, we can in addition set

$$m_{ii} = 1 \quad , \qquad i = 1, 2, \ldots, n \quad ,$$

and reduce the two relations (A.9.4) to one,

$$A_{ij} = \sum_{k=1}^{p} m_{ik} A_{kj}^{(k)} \quad , \qquad 1 \leq i,\ j \leq n \quad , \qquad p = \min(i, j) \quad .$$

The equation shows directly that the matrix A can be represented as the product of two matrices L and R. Here L is a lower and R an upper triangular matrix,

$$A = LR \quad , \tag{A.9.5}$$

$$L = \begin{pmatrix} m_{11} & & & \\ m_{21} & m_{22} & & \\ \vdots & & & \\ m_{n1} & m_{n2} & \dots & m_{nn} \end{pmatrix} \quad , \qquad m_{ii} = 1 \quad ,$$

$$R = \begin{pmatrix} r_{11} & r_{12} & \dots & r_{1n} \\ & r_{22} & \dots & r_{2n} \\ & & & \vdots \\ & & & r_{nn} \end{pmatrix} = \begin{pmatrix} A_{11}^{(1)} & A_{12}^{(1)} & \dots & A_{1n}^{(1)} \\ & A_{22}^{(2)} & \dots & A_{2n}^{(2)} \\ & & & \vdots \\ & & & A_{nn}^{(n)} \end{pmatrix} \quad .$$

The original system of equations (A.7.10)

$$A\mathbf{x} = LR\mathbf{x} = \mathbf{b}$$

is thus equivalent to two triangular systems of equations,

$$L\mathbf{y} = \mathbf{b} \quad , \qquad R\mathbf{x} = \mathbf{y} \quad .$$

Instead of resorting to the formulas from Sect. A.8, we can compute the elements of L and R directly from (A.9.5). We obtain

$$\left. \begin{array}{ll} r_{kj} = A_{kj} - \sum_{\ell=1}^{k-1} m_{k\ell} r_{\ell j} \, , & j = k,\, k+1,\, \dots,\, n \\ m_{ik} = \left(A_{ik} - \sum_{\ell=1}^{k-1} m_{i\ell} r_{\ell k} \right) / r_{kk} \, , & i = k+1,\, \dots,\, n \end{array} \right\} k = 1, \dots, n \quad , \tag{A.9.6}$$

i.e., for $k = 1$ one computes the first row of R (which is equal to the first row of A) and the first column of L; for $k = 2$ one computes the second row of R and the second column of L. In a computer program the elements of R and L can overwrite the original matrix A with the same indices, since when computing r_{kj} one only needs the element A_{kj} and other elements of R and L that have already been computed according to (A.9.6). The corresponding consideration holds for the computation of m_{ik}.

The algorithm (A.9.6) is called the Doolittle *LR-decomposition*. By including pivoting this is clearly equivalent to Gaussian elimination.

A.10 Cholesky Decomposition

If A is a real symmetric positive-definite matrix (cf. Sect. A.12), then it can be uniquely expressed as

$$A = U^{\mathrm{T}} U \quad . \tag{A.10.1}$$

Here U is a real upper triangular matrix with positive diagonal elements. The $n \times n$ matrix A has the required property for the validity of (A.10.1), in particular in those cases where it is equal to the product of an arbitrary real $n \times m$ matrix B (with $m \geq n$ and full rank) with its transpose B^{T}, $A = B\,B^{\mathrm{T}}$.

To determine U we first carry out the Doolittle LR-decomposition, define a diagonal matrix D whose diagonal elements are equal to those of R,

$$D = \mathrm{diag}(r_{11}, r_{22}, \ldots, r_{nn}) \quad , \tag{A.10.2}$$

and introduce the matrices

$$D^{-1} = \mathrm{diag}(r_{11}^{-1}, r_{22}^{-1}, \ldots, r_{nn}^{-1}) \tag{A.10.3}$$

and

$$D^{-1/2} = \mathrm{diag}(r_{11}^{-1/2}, r_{22}^{-1/2}, \ldots, r_{nn}^{-1/2}) \quad . \tag{A.10.4}$$

We can then write

$$A = L\,R = L\,D\,D^{-1}\,R = L\,D\,R' \quad , \qquad R' = D^{-1}\,R \quad . \tag{A.10.5}$$

Because of the assumed symmetry one has

$$A = A^{\mathrm{T}} = (R')^{\mathrm{T}} D\,L^{\mathrm{T}} \quad . \tag{A.10.6}$$

Comparing with (A.10.5) gives $L^{\mathrm{T}} = R'$, i.e.,

$$L = R^{\mathrm{T}}\,D^{-1} \quad .$$

With

$$U = D^{-1/2} R = D^{1/2} L^{\mathrm{T}} \tag{A.10.7}$$

Eq. (A.10.1) is indeed fulfilled. The relation (A.10.7) means for the elements u_{ij} of U

$$u_{ij} = r_{ii}^{-1/2} r_{ij} = r_{ii}^{1/2} m_{ji} \quad .$$

Thus the values r and m can be eliminated in favor of u from (A.9.6), and we obtain with

$$\left.\begin{array}{l} u_{kk} = \left(A_{kk} - \sum_{\ell=1}^{k-1} u_{\ell k}^2\right)^{1/2} \\ u_{kj} = \left(A_{kj} - \sum_{\ell=1}^{k-1} u_{\ell k} u_{\ell j}\right) / u_{kk} \,, \quad j = k+1, \ldots, n \end{array}\right\} k = 1, \ldots, n$$

the algorithm for the *Cholesky decomposition* of a real positive-definite symmetric matrix A. Since for such matrices all of the diagonal elements are different from zero, one does not need in principle any pivoting. One can also show that it would bring no advantage in numerical accuracy.

Program A.33: Subroutine MTXCHL performs the Cholesky decomposition of a symmetric positive-definite square matrix

CALLING SEQUENCE: CALL MTXCHL(A,U,N)

INPUT ARGUMENTS: A — $n \times n$ matrix A
N — dimension of A

OUTPUT ARGUMENT: U — upper triangular matrix $U_{n\times n}$, such that $U^{\mathrm{T}}U = A$

```
      SUBROUTINE MTXCHL(A,U,N)
      IMPLICIT DOUBLE PRECISION(A-H,O-Z)
      DIMENSION A(N,N),U(N,N)
      CALL MTXZER(U,N,N)
      DO 30 K=1,N
        S=0.
        DO 20 J=K,N
          IF(K.GT.1) THEN
            S=0.
            DO 10 L=1,K-1
              S=S+U(L,K)*U(L,J)
   10       CONTINUE
          END IF
          U(K,J)=A(K,J)-S
          IF(K.EQ.J) THEN
            U(K,J)=SQRT(ABS(U(K,J)))
          ELSE
            U(K,J)=U(K,J)/U(K,K)
          END IF
   20   CONTINUE
   30 CONTINUE
      END
```

Positive-definite symmetric matrices play an important role as weight and covariance matrices. In such cases one often requires the Cholesky decomposition $A = U^{\mathrm{T}}U$ as well as multiplication of U by a matrix. It is easier computationally if this multiplication is done by a special subroutine which takes the triangular form of U into account.

Program A.34: Subroutine MTXCHM multiplies an upper triangular matrix U of dimension $m \times m$ with an $m \times n$ matrix A

CALLING SEQUENCE: CALL MTXCHM(U,A,R,M,N)

INPUT ARGUMENTS: U — $m \times m$ matrix U
A — $m \times n$ matrix A
M,N — dimensions m, n

OUTPUT ARGUMENT: R — $m \times n$ product matrix $R = UA$

```
      SUBROUTINE MTXCHM(U,A,R,M,N)
      IMPLICIT DOUBLE PRECISION (A-H,O-Z)
      DIMENSION U(M,M),A(M,N),R(M,N)
      PARAMETER(Z=0.D0)
      DO 30 I=1,M
        DO 20 K=1,N
          R(I,K)=Z
          DO 10 L=I,M
            R(I,K)=R(I,K)+U(I,L)*A(L,K)
10        CONTINUE
20      CONTINUE
30    CONTINUE
      END
```

Of particular interest is the inversion of a symmetric positive-definite $n \times n$ matrix A. For this we will solve the n matrix equations (A.7.4) with the previously described Cholesky decomposition of A,

$$A\mathbf{x}_i = U^{\mathrm{T}} U \mathbf{x}_i = U^{\mathrm{T}} \mathbf{y}_i = \mathbf{e}_i \quad , \qquad i = 1, \ldots, n \quad . \tag{A.10.8}$$

We denote the ℓth component of the three vectors $\mathbf{x}_i$, $\mathbf{y}_i$, and $\mathbf{e}_i$ by $x_{i\ell}$, $y_{i\ell}$, and $e_{i\ell}$. Clearly one has $x_{i\ell} = (A^{-1})_{i\ell}$ and $e_{i\ell} = \delta_{i\ell}$, since $e_{i\ell}$ is the element (i, ℓ) of the unit matrix.

We now determine $y_{i\ell}$ by means of *forward substitution*. From (A.10.8) it follows directly that

$$\sum_{k=1}^{\ell} U_{k\ell} y_{ik} = e_{i\ell} = \delta_{i\ell} \quad ,$$

i.e.,

$$\begin{aligned} U_{11} y_{i1} &= \delta_{i1} \quad , \\ U_{12} y_{i1} + U_{22} y_{i2} &= \delta_{i2} \quad , \\ &\vdots \end{aligned}$$

and thus

$$\begin{aligned} y_{i1} &= \delta_{i1}/U_{11} \quad , \\ &\vdots \\ y_{i\ell} &= \frac{1}{U_{\ell\ell}} \left(\delta_{i\ell} - \sum_{k=1}^{\ell-1} U_{k\ell} y_{ik} \right) \quad . \end{aligned}$$

Since, however, $\delta_{i\ell} = 0$ for $i \neq \ell$, this expression simplifies to

$$\begin{aligned} y_{i\ell} &= 0 \quad , \qquad \ell < i \quad , \\ y_{i\ell} &= \frac{1}{U_{\ell\ell}} \left(\delta_{i\ell} - \sum_{k=i}^{\ell-1} U_{k\ell} y_{ik} \right) \quad , \qquad \ell \geq i \quad . \end{aligned}$$

We can now obtain $x_{i\ell}$ by means of *backward substitution* of $y_{i\ell}$ into

$$U\mathbf{x}_i = \mathbf{y}_i$$

or, written in components,

$$\sum_{k=\ell}^{n} U_{\ell k} x_{ik} = y_{i\ell}$$

or

$$\begin{aligned} U_{11}x_{i1} + U_{12}x_{i2} + \cdots + U_{1n}x_{in} &= y_{i1} \quad , \\ U_{22}x_{i2} + \cdots + U_{2n}x_{in} &= y_{i2} \quad , \\ &\vdots \\ U_{nn}x_{in} &= y_{in} \quad . \end{aligned}$$

We obtain

$$\begin{aligned} x_{in} &= y_{in}/U_{nn} \quad , \\ &\vdots \\ x_{i\ell} &= \frac{1}{U_{\ell\ell}} \left(y_{i\ell} - \sum_{k=\ell+1}^{n} U_{\ell k} x_{ik} \right) \quad . \end{aligned}$$

If we compute $x_{i\ell}$ only for $\ell \geq i$, then by backward substitution one only encounters the elements $y_{i\ell}$ for $\ell \geq i$, i.e., only non-vanishing $y_{i\ell}$. The vanishing elements $y_{i\ell}$ thus do not need to be stored. The elements $x_{i\ell}$ for $\ell < i$ follow simply from the symmetry of the original matrix, $x_{i\ell} = x_{\ell i}$.

In the following program the Cholesky decomposition of A is first performed. Then a loop is carried out over the various right-hand sides e_i, $i = 1, \ldots, n$, of (A.10.8). As the result of running through the loop once one obtains the elements x_{in}, $x_{i,n-1}$, …, x_{ii} of row i. They are stored as the corresponding elements of the output matrix A. The elements of the vector $\mathbf{y}$, which is only used for intermediate results, can be stored in the last row of A. Finally the elements below the main diagonal are filled by copies of their mirror images.

Program A.35: Subroutine `MTXCHI` inverts a positive-definite matrix A by means of the Cholesky decomposition

CALLING SEQUENCE: `CALL MTXCHI(A,SCRAT,N)`

INPUT ARGUMENTS: `A` — positive-definite symmetric $n \times n$ matrix A
`SCRAT` — working array, minimum size $n \times n$
`N` — dimension of A

OUTPUT ARGUMENT: `A` — inverse A^{-1} of the input matrix A

```
      SUBROUTINE MTXCHI(A,U,N)
      IMPLICIT DOUBLE PRECISION (A-H,O-Z)
      DIMENSION A(N,N),U(N,N)
C Step 1: Cholesky decomposition
      CALL MTXCHL(A,U,N)
      DO 50 I=1,N
C Step 2: Forward Substitution
        DO 20 L=I,N
          IF(L.EQ.I) THEN
            A(N,L)=DBLE(1.)/U(L,L)
          ELSE
            A(N,L)=DBLE(0.)
            DO 10 K=I,L-1
              A(N,L)=A(N,L)-U(K,L)*A(N,K)
   10       CONTINUE
            A(N,L)=A(N,L)/U(L,L)
          END IF
   20   CONTINUE
C Step 3: Back Substitution
        DO 40 L=N,I,-1
          IF(L.EQ.N) THEN
            A(I,L)=A(N,L)/U(L,L)
          ELSE
            A(I,L)=A(N,L)
            DO 30 K=N,L+1,-1
              A(I,L)=A(I,L)-U(L,K)*A(I,K)
   30       CONTINUE
            A(I,L)=A(I,L)/U(L,L)
          END IF
   40   CONTINUE
   50 CONTINUE
C Fill lower triangle symmetrically
      IF(N.GT.1) THEN
        DO 70 I=1,N
          DO 60 L=1,I-1
            A(I,L)=A(L,I)
   60     CONTINUE
   70   CONTINUE
      END IF
      END
```

A.11 Pseudo-inverse Matrix

We now return to the problem of Sect. A.6, that of solving the equation (A.6.4)

$$A\mathbf{x} \approx \mathbf{b} \tag{A.11.1}$$

for an arbitrary $m \times n$ matrix A of rank k. According to (A.6.14), the unique solution of minimum norm is

$$\widetilde{\mathbf{x}} = K \begin{pmatrix} \mathbf{p}_1 \\ 0 \end{pmatrix} \quad .$$

The vector $\mathbf{p}_1$ is the solution of Eq. (A.6.11) and therefore

$$\widetilde{\mathbf{p}}_1 = R_{11}^{-1} \mathbf{g}_1 \quad ,$$

since R_{11} is non-singular. Because of (A.6.8) one has finally

$$\tilde{\mathbf{x}} = K \begin{pmatrix} R_{11}^{-1} & 0 \\ 0 & 0 \end{pmatrix} H^{\mathrm{T}}\mathbf{b} \quad . \tag{A.11.2}$$

In analogy to (A.7.3) we write

$$\tilde{\mathbf{x}} = A^{+}\mathbf{b} \tag{A.11.3}$$

and call the $n \times m$ matrix

$$A^{+} = K \begin{pmatrix} R_{11}^{-1} & 0 \\ 0 & 0 \end{pmatrix} H^{\mathrm{T}} \tag{A.11.4}$$

the *pseudo-inverse* of the $m \times n$ matrix A.

The matrix A^{+} is uniquely determined by A and does not depend on the particular orthogonal decomposition (A.6.5). This can easily be seen if one denotes the jth column vector of A^{+} by $\mathbf{a}_j^{+} = A^{+}\mathbf{e}_j$, with $\mathbf{e}_j$ the jth column vector of the m-dimensional unit matrix. According to (A.11.3) the vector $\mathbf{a}_j^{+}$ is the minimum-length solution of the equation $A\mathbf{a}_j^{+} = \mathbf{e}_j$, and is therefore unique.

A.12 Eigenvalues and Eigenvectors

We now consider the *eigenvalue equation*

$$G\mathbf{x} = \lambda\mathbf{x} \tag{A.12.1}$$

for the $n \times n$ matrix G. If this is fulfilled, then the scalar λ is called the *eigenvalue* and the n-vector $\mathbf{x}$ the *eigenvector* of G. Clearly the eigenvector $\mathbf{x}$ is only determined up to an arbitrary factor. One can choose this factor such that $|\mathbf{x}| = 1$.

We consider first the particularly simple case where G is a diagonal matrix with non-negative diagonal elements,

$$G = S^{\mathrm{T}}S = S^2 = \begin{pmatrix} s_1^2 & & & \\ & s_2^2 & & \\ & & \ddots & \\ & & & s_n^2 \end{pmatrix} \quad , \tag{A.12.2}$$

which can be expressed as the product of an arbitrary diagonal matrix S with itself,

$$S = \begin{pmatrix} s_1 & & & \\ & s_2 & & \\ & & \ddots & \\ & & & s_n \end{pmatrix} \quad . \tag{A.12.3}$$

The eigenvalue equation $S^2\mathbf{x} = \lambda\mathbf{x}$ then has the eigenvalues $s_i^2 = \lambda_i$ and the normalized eigenvectors are the basis vectors $\mathbf{x}_i = \mathbf{e}_i$.

In place of S we now set

$$A = U\,S\,V^{\mathrm{T}} \tag{A.12.4}$$

with orthogonal matrices U and V and

$$G = A^{\mathrm{T}}\,A = V\,S^{\mathrm{T}}\,S\,V^{\mathrm{T}} \quad . \tag{A.12.5}$$

We can write the eigenvalue equation of G in the form

$$G\mathbf{x} = \lambda\mathbf{x} \tag{A.12.6}$$

or

$$V\,S^{\mathrm{T}}\,S\,V^{\mathrm{T}}\,\mathbf{x} = \lambda\mathbf{x} \quad .$$

Multiplying on the left with V^{T},

$$S^{\mathrm{T}}S\,V^{\mathrm{T}}\mathbf{x} = \lambda V^{\mathrm{T}}\mathbf{x} \quad ,$$

and comparing with (A.12.1), (A.12.2) shows that G has the same eigenvalues $\lambda_i = s_i^2$ as S^2, but has the orthogonally transformed eigenvectors

$$\mathbf{x}_i = \mathbf{e}'_i = V\mathbf{e}_i \quad . \tag{A.12.7}$$

One can clearly find the eigenvalues and eigenvectors of G if one knows the orthogonal matrix V that transforms G into a diagonal matrix,

$$V^{\mathrm{T}}\,G\,V = S^{\mathrm{T}}\,S = S^2 \quad . \tag{A.12.8}$$

The transformation (A.12.8) is called a *principal axis transformation*. The name becomes clear by considering the equation

$$\mathbf{r}^{\mathrm{T}}G\mathbf{r} = 1 \quad . \tag{A.12.9}$$

We are interested in the geometrical position of all points $\mathbf{r}$ that fulfill (A.12.9).

For the vector $\mathbf{r}$ the following two representations are completely equivalent:

$$\mathbf{r} = \sum_i r_i\mathbf{e}_i \tag{A.12.10}$$

and

$$\mathbf{r} = \sum_i r'_i \mathbf{e}'_i \quad , \tag{A.12.11}$$

with the components r_i and r'_i taken with respect to the basis vectors $\mathbf{e}_i$ and $\mathbf{e}'_i$, respectively. With the representation (A.12.11) and by using (A.12.5) and (A.12.7) we obtain

$$\begin{aligned} 1 &= \mathbf{r}^{\mathrm{T}} G \mathbf{r} = \mathbf{r}^{\mathrm{T}} V S^2 V^{\mathrm{T}} \mathbf{r} \\ &= \sum_i (r'_i \mathbf{e}'_i)^{\mathrm{T}} V S^2 V^{\mathrm{T}} \sum_j (r'_j \mathbf{e}'_j) \\ &= \sum_i (r'_i \mathbf{e}_i^{\mathrm{T}} V^{\mathrm{T}}) V S^2 V^{\mathrm{T}} \sum_j (r'_j V \mathbf{e}_j) \\ &= \sum_i (r'_i \mathbf{e}_i^{\mathrm{T}}) S^2 \sum_j (\mathbf{r}'_j \mathbf{e}_j) \end{aligned}$$

and finally

$$\sum_{i=1}^{n} r'^2_i s^2_i = \sum_{i=1}^{n} r'^2_i / a^2_i = 1 \quad . \tag{A.12.12}$$

This is clearly the equation of an ellipsoid in n dimensions with half-diameters in the directions $\mathbf{e}'_i$ and having the lengths

$$a_i = \frac{1}{s_i} \quad . \tag{A.12.13}$$

The vectors

$$\mathbf{a}_i = \mathbf{e}'_i / s_i = V \mathbf{e}_i / s_i \tag{A.12.14}$$

are the *principal axes* of the ellipsoid. They have the directions of the eigenvectors of G. Their lengths $a_i = 1/\sqrt{s_i^2}$ are determined by the eigenvalues s_i^2.

The matrix

$$C = G^{-1} = (A^{\mathrm{T}} A)^{-1} = V (S^2)^{-1} V^{\mathrm{T}}$$

clearly has the same eigenvectors as G, but has the eigenvalues $1/s_i^2$. The lengths of the half-diameters of the ellipsoid described above are then directly equal to the square roots of the eigenvalues of C.

The matrix C is called the *unweighted covariance matrix* of A. The ellipsoid is called the *unweighted covariance ellipsoid.*

Please note that all considerations were done for matrices of the type (A.12.5) which, by construction, are symmetric and *non-negative definite*, i.e., they have real, non-negative eigenvalues. Ellipses with finite semiaxes are obtained only with *positive-definite* matrices. If the same eigenvalue occurs several times then the ellipsoid has several principal axes of the same length. In this case there is a certain ambiguity in the determination of the principal axes which, however, can always be chosen orthogonal to each other.

Up to now we have not given any prescription for finding the eigenvalues. The eigenvalue equation is $G\mathbf{x} = \lambda\mathbf{x}$ or

$$(G - \lambda I)\mathbf{x} = 0 \quad . \tag{A.12.15}$$

Written in this way it can be considered as a linear system of equations for determining $\mathbf{x}$, for which the right-hand side is the null vector. Corresponding to (A.7.5) and (A.7.9) there is a non-trivial solution only for

$$\det(G - \lambda I) = 0 \quad . \tag{A.12.16}$$

This is the *characteristic equation* for determining the eigenvalues of A. For $n = 2$ this is

$$\begin{vmatrix} g_{11} - \lambda & g_{12} \\ g_{21} & g_{22} - \lambda \end{vmatrix} = (g_{11} - \lambda)(g_{22} - \lambda) - g_{12}g_{21} = 0$$

and has the solutions

$$\lambda_{1,2} = \frac{g_{11} + g_{22}}{2} \pm \sqrt{g_{12}g_{21} + \frac{(g_{11} - g_{22})^2}{4}} \quad .$$

If G is symmetric, as previously assumed, i.e., $g_{12} = g_{21}$, then the eigenvalues are real. If G is positive definite, they are positive.

The characteristic equation of an $n \times n$ matrix has n solutions. In practice, however, for $n > 2$ one does not find the eigenvalues by using the characteristic equation, but rather by means of an iterative procedure such as the singular value decomposition (see Sect. A.13).

A.13 Singular Value Decomposition

We now consider a particular orthogonal decomposition of the $m \times n$ matrix A,

$$A = U\,S\,V^{\mathrm{T}} \quad . \tag{A.13.1}$$

Here and in Sects. A.14 and A.15 we assume that $m \geq n$. If this is not the case, then one can simply extend the matrix A with further rows whose elements are all zero until it has n rows, so that $m = n$. The decomposition (A.13.1) is a special case of the decomposition (A.6.5). The $m \times n$ matrix S, which takes the place of R, has the special form

$$S = \begin{pmatrix} D & 0 \\ 0 & 0 \end{pmatrix} \quad , \tag{A.13.2}$$

and D is a $k \times k$ diagonal matrix with $k = \mathrm{Rank}(A)$. The diagonal elements of S are called the *singular values* of A. If A is of full rank, then $k = n$ and all

$s_i \neq 0$. For reduced rank $k < n$ one has $s_i = 0$ for $i > k$. We will see below that U and V can be determined such that all s_i are positive and ordered to be non-increasing,

$$s_1 \geq s_2 \geq \ldots \geq s_k \quad . \tag{A.13.3}$$

The singular values of A have a very simple meaning. From Sect. A.12 one has directly that the s_i are the square roots of the eigenvalues of $G = A^{\mathrm{T}}A$. Thus the half-diameters of the covariance ellipsoid of G are $a_i = 1/s_i$. If G is singular, then A has the reduced rank $k < n$, and the $n - k$ singular values $s_{k+1}, \ldots, s_n$ vanish. In this case the determinant

$$\det G = \det U \; \det S^2 \; \det V = \det S^2 = s_1^2 s_2^2 \cdots s_n^2 \tag{A.13.4}$$

also vanishes.

With the substitutions $H \to U$, $K \to V$, $R \to S$, $R_{11} \to D$ we obtain from Sect. A.6

$$A\mathbf{x} = U\,S\,V^{\mathrm{T}}\mathbf{x} \approx \mathbf{b} \quad , \tag{A.13.5}$$

$$S\,V^{\mathrm{T}}\mathbf{x} \approx U^{\mathrm{T}}\mathbf{b} \quad . \tag{A.13.6}$$

With

$$V^{\mathrm{T}}\mathbf{x} = \mathbf{p} = \begin{pmatrix} \mathbf{p}_1 \\ \mathbf{p}_2 \end{pmatrix} \begin{matrix} \} & k \\ \} & n-k \end{matrix} \quad , \qquad U^{\mathrm{T}}\mathbf{b} = \mathbf{g} = \begin{pmatrix} \mathbf{g}_1 \\ \mathbf{g}_2 \end{pmatrix} \begin{matrix} \} & k \\ \} & m-k \end{matrix} \tag{A.13.7}$$

one has

$$S\mathbf{p} = \mathbf{g} \quad , \tag{A.13.8}$$

i.e.,

$$D\mathbf{p}_1 = \mathbf{g}_1 \tag{A.13.9}$$

and

$$0 \cdot \mathbf{p}_2 = \mathbf{g}_2 \quad . \tag{A.13.10}$$

These have the solutions

$$\widetilde{\mathbf{p}}_1 = D^{-1}\mathbf{g}_1 \quad , \tag{A.13.11}$$

i.e.,

$$p_\ell = g_\ell / s_\ell \quad , \qquad \ell = 1, 2, \ldots, k \quad ,$$

for arbitrary $\mathbf{p}_2$. The solution with minimum absolute value is

$$\widetilde{\mathbf{x}} = V \begin{pmatrix} \widetilde{\mathbf{p}}_1 \\ 0 \end{pmatrix} \quad . \tag{A.13.12}$$

The residual vector has the form

$$\mathbf{r} = \mathbf{b} - A\widetilde{\mathbf{x}} = U \begin{pmatrix} 0 \\ \mathbf{g}_2 \end{pmatrix} \tag{A.13.13}$$

and the absolute value

$$r = |\mathbf{g}_2| = \left(\sum_{i=k+1}^{m} g_i^2 \right)^{1/2} . \tag{A.13.14}$$

A.14 Singular Value Analysis

The rank k of the matrix A plays a decisive role in finding the solution of $A\mathbf{x} \approx \mathbf{b}$. Here k characterizes the transition from non-zero to vanishing singular values, $s_k > 0$, $s_{k+1} = 0$. How should one judge the case of very small values of s_k, i.e., $s_k < \varepsilon$ for a given small ε?

Example A.2: Almost vanishing singular values
As a simple example we consider the case $m = n = 2$, $U = V = I$. One then has

$$A\mathbf{x} = U\,S\,V^{\mathrm{T}}\mathbf{x} = S\mathbf{x} = \begin{pmatrix} s_1 & 0 \\ 0 & s_2 \end{pmatrix} \begin{pmatrix} x_1 \\ x_2 \end{pmatrix} = \begin{pmatrix} b_1 \\ b_2 \end{pmatrix} = \mathbf{b} \quad , \tag{A.14.1}$$

that is,

$$x_1 = b_1/s_1 \quad , \qquad x_2 = b_2/s_2 \quad . \tag{A.14.2}$$

If we now take $s_2 \to 0$ in (A.14.2), then $|x_2| \to \infty$. At first glance one obtains a completely different picture if one sets $s_2 = 0$ in (A.14.1) directly. This gives

$$s_1 x_1 = b_1 \quad , \qquad 0 \cdot x_2 = b_2 \quad . \tag{A.14.3}$$

Thus $x_1 = b_1/s_1$ as in (A.14.2), but x_2 is completely undetermined. The solution $\tilde{\mathbf{x}}$ of minimum absolute value is obtained by setting $x_2 = 0$. The question is now, "What is right? $x_2 = \infty$ or $x_2 = 0$?" The answer is that one should set

$$x_2 = \begin{cases} b_2/s_2 \, , & s_2 \geq \varepsilon \\ 0 \, , & s_2 < \varepsilon \end{cases}$$

and choose the parameter ε such that the expression b_2/ε is still numerically well-defined. This means that one must have $\varepsilon/|b_2| \gg 2^{-m}$ if m binary digits are available for the representation of a floating point number (cf. Sect. 4.2). Thus a finite value of b_2 is computed as long as the numerical determination of this value is reasonable. If this is not the case then one approaches the situation $s_2 = 0$, where b_2 is completely undetermined, and one sets $b_2 = 0$. ■

Example A.3: Point of intersection of two almost parallel lines

We consider the two lines shown in Fig. A.6, which are described by

$$\begin{aligned} -\alpha x_1 + x_2 &= 1 - \alpha \quad , \\ \alpha x_1 + x_2 &= 1 + \alpha \quad . \end{aligned}$$

For the vector **x** of the intersection point one has $A\mathbf{x} = \mathbf{b}$ with

$$A = \begin{pmatrix} -\alpha & 1 \\ \alpha & 1 \end{pmatrix} \quad , \qquad \mathbf{b} = \begin{pmatrix} 1-\alpha \\ 1+\alpha \end{pmatrix} \quad .$$

One can easily check that $A = U\,S\,V^{\mathrm{T}}$ holds with

$$U = \frac{1}{\sqrt{2}} \begin{pmatrix} -1 & -1 \\ -1 & 1 \end{pmatrix} \quad , \qquad S = \sqrt{2} \begin{pmatrix} 1 & 0 \\ 0 & \alpha \end{pmatrix} \quad , \qquad V = \begin{pmatrix} 0 & 1 \\ -1 & 0 \end{pmatrix} \quad ,$$

i.e., $s_1 = \sqrt{2}$, $s_2 = \alpha\sqrt{2}$. Using

$$\mathbf{g} = U^{\mathrm{T}}\mathbf{b} = \sqrt{2} \begin{pmatrix} -1 \\ \alpha \end{pmatrix} = S\mathbf{p} = \sqrt{2} \begin{pmatrix} p_1 \\ \alpha p_2 \end{pmatrix} \quad , \qquad \mathbf{p} = \begin{pmatrix} -1 \\ 1 \end{pmatrix}$$

one obtains

$$\mathbf{x} = V\mathbf{p} = \begin{pmatrix} 1 \\ 1 \end{pmatrix}$$

independent of α.

If, however, one has $s_2 = \alpha\sqrt{2} < \varepsilon$ and then sets $s_2 = 0$, then one obtains

$$\mathbf{p} = \begin{pmatrix} -1 \\ 0 \end{pmatrix} \quad , \qquad \mathbf{x} = \begin{pmatrix} 0 \\ 1 \end{pmatrix} \quad .$$

From Fig. A.6 one can see that for $\alpha \to 0$ the two lines come together to a single line described by $x_2 = 1$. The x_1-coordinate of the "intersection point" is completely undetermined. It is set equal to zero, since the solution vector **x** has minimum length [cf. (A.6.14)].

As in the case of "indirect measurements" in Chapter 9 we now assume that the vector **b** is equal to the vector of measurements **y**, which characterize the two lines, and that their measurement errors are given by the covariance matrix $C_y = G_y^{-1}$. In the simplest case of equal uncorrelated errors one has

$$C_y = G_y^{-1} = \sigma^2 I \quad .$$

The covariance matrix for the unknowns $\widetilde{\mathbf{x}}$ is according to (9.2.27)

$$C_x = (A^{\mathrm{T}} G_y A)^{-1} \quad ,$$

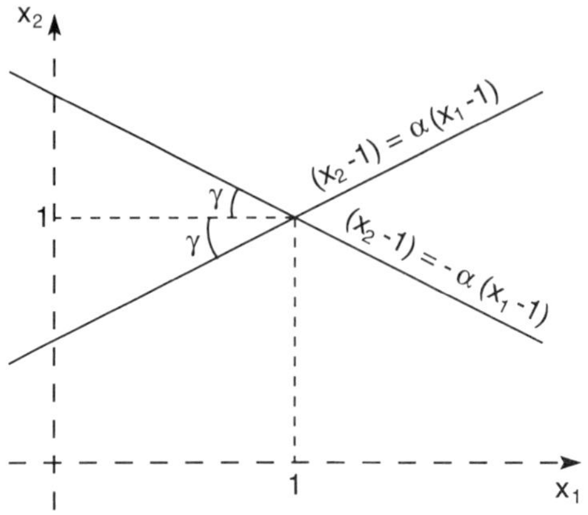

Fig. A.6. Two lines intersect at the point (1,1) with an angle $2\gamma = 2\arctan\alpha$.

and thus in the case of uncorrelated measurement errors,

$$C_x = \sigma^2(A^{\mathrm{T}}A)^{-1} = \sigma^2 C \quad .$$

Up to the factor σ^2 it is thus equal to the unweighted covariance matrix $C = (A^{\mathrm{T}}A)^{-1}$ for the matrix A. For our matrix A one has

$$C = (A^{\mathrm{T}}A)^{-1} = \begin{pmatrix} 1/2\alpha^2 & 0 \\ 0 & 1/2 \end{pmatrix} \quad .$$

The corresponding ellipse has the half-diameters $\mathbf{e}_1/\alpha\sqrt{2}$ and $\mathbf{e}_2/\sqrt{2}$. The covariance ellipse of $\widetilde{\mathbf{x}}$ then has for the case of equal uncorrelated measurements equal half-diameters, multiplied, however, by the factor σ. They have the lengths $\sigma_{x_1} = \sigma/\alpha\sqrt{2}$, $\sigma_{x_2} = \sigma/\sqrt{2}$. Clearly one then sets $x_1 = 0$ if the inequality $x_1 \ll \sigma_{x_1}$ would hold for a finite fixed x_1, i.e., $\alpha\sqrt{2} \ll \sigma$. ■

The decision as to whether a small singular value should be set equal to zero thus depends on numerical considerations and on a consideration of the measurement errors. The following fairly general procedure of *singular value analysis* has proven to be useful in practice.

1. With a computer program one carries out the singular value decomposition, which yields, among other things, the ordered singular values

$$s_1 \geq s_2 \geq \cdots \geq s_k \quad .$$

2. Depending on the problem at hand one chooses a positive factor $f \ll 1$.

3. All singular values for which $s_i < f s_1$ are set equal to zero. In place of k one has thus $\ell \leq k$ such that $s_i = 0$ for $i > \ell$.

4. With the replacements described above ($k \to \ell$, $s_{\ell+1} = \cdots = s_k = 0$) the formulas of Sect. A.13 retain their validity.

In place of the residual (A.13.14) one obtains a somewhat larger value, since in the sum in the expression

$$r = \left(\sum_{i=\ell+1}^{m} g_i^2 \right)^{1/2} \tag{A.14.4}$$

one has more terms than in (A.13.14).

The procedure implies that in some cases one has an effective reduction of the rank k of the matrix A to a value $\ell < k$. If A has its full rank, then this can be reduced. This has the great advantage that numerical difficulties with small singular values are avoided. In contrast to, for example, the Gaussian or Cholesky procedures, the user does not have to worry about whether $G = A^{\mathrm{T}}A$ is singular or nearly singular.

Although the singular value analysis always gives a solution of minimum absolute value $\tilde{\mathbf{x}}$ for the problem $A\mathbf{x} \approx \mathbf{b}$, caution is recommended for the case $\ell < n$ (regardless of whether $k < n$ or $\ell < k = n$). In such a case one has an (almost) linear dependence of the unknowns $x_1, \ldots, x_n$. The solution $\tilde{\mathbf{x}}$ is not the only solution of the problem, but rather is simply the solution with the smallest absolute value out of the many possible solutions.

We have already remarked in Sect. A.6 that the singular value decomposition is advantageous also with respect to numerical precision compared to other methods, especially to the method of normal equations. A detailed discussion (see, e.g., [18]) is beyond the scope of this book. Here we limit ourselves to giving an example.

Example A.4: Numerical superiority of the singular value decomposition compared to the solution of normal equations

Consider the problem $A\mathbf{x} \approx \mathbf{b}$ for

$$A = \begin{pmatrix} 1 & 1 \\ \beta & 0 \\ 0 & \beta \end{pmatrix} \quad .$$

The singular values of A are the square roots of the eigenvalues of

$$G = A^{\mathrm{T}}A = \begin{pmatrix} 1+\beta^2 & 1 \\ 1 & 1+\beta^2 \end{pmatrix}$$

and are determined by (A.12.16) to be

$$s_1 = \sqrt{2+\beta^2} \quad , \qquad s_2 = |\beta| \quad .$$

This was done with singular value decomposition without using the matrix G. If the computing precision is ε and if $\beta^2 < \varepsilon$ but $\beta > \varepsilon$, then one obtains

$$s_1 = \sqrt{2} \quad , \qquad s_2 = |\beta| \quad ,$$

i.e., both singular values remain non-zero. If instead of the singular value decomposition one uses the normal equations

$$A^{\mathrm{T}} A\mathbf{x} = G\mathbf{x} = A^{\mathrm{T}}\mathbf{b} \quad ,$$

then the matrix G appears explicitly. With $\beta^2 < \varepsilon$ it is numerically represented as

$$G = \begin{pmatrix} 1 & 1 \\ 1 & 1 \end{pmatrix} \quad .$$

This matrix is singular, $\det G = 0$, and cannot be inverted, as foreseen in (A.6.17). This is also reflected in its singular values

$$s_1 = \sqrt{2} \quad , \qquad s_2 = 0 \quad . \blacksquare$$

A.15 Algorithm for Singular Value Decomposition

A.15.1 Strategy

We will now give an algorithm for the singular value decomposition and its implementation as a FORTRAN program. We follow for the most part the presentation of LAWSON and HANSON [18], which is based on the work of GOLUB and KAHN [19], BUSINGER and GOLUB [20], and GOLUB and REINSCH [21]. The strategy of the algorithm is based on the successive application of orthogonal transformations.

In the *first step*, the matrix A is transformed into a bidiagonal matrix C,

$$A = Q\, C\, H^{\mathrm{T}} \quad . \tag{A.15.1}$$

The orthogonal matrices Q and H are themselves products of Householder-transformation matrices.

In *step 2* the matrix C is brought into diagonal form by means of an iterative procedure:

$$C = U'\, S'\, V'^{\mathrm{T}} \quad . \tag{A.15.2}$$

Here the matrices U' and V' are given by products of Givens-transformation matrices and if necessary reflection matrices. The latter ensure that all diagonal elements are non-negative.

In *step 3*, further orthogonal matrices U'' and V'' are determined. They are products of permutation matrices and ensure that the diagonal elements of

$$S = U''^{\mathrm{T}} S' V'' \tag{A.15.3}$$

are ordered so as to be non-increasing as in (A.13.3). As the result of the first three steps one obtains the matrices

$$U = Q U' U'' \quad , \qquad V = H V' V'' \quad , \tag{A.15.4}$$

which produce the singular value decomposition (A.13.1) as well as the diagonal elements $s_1, s_2, \ldots, s_k$.

In *step 4* the singular value analysis is finally carried out. Stated simply, all singular values are set to zero for which

$$s_i < s_{\min} \quad . \tag{A.15.5}$$

Thus one can finally give the solution vector $\mathbf{x}$ to the equation

$$A\mathbf{x} \approx \mathbf{b} \quad . \tag{A.15.6}$$

In practice, Eq. (A.15.6) is often solved simultaneously for various right-hand sides, which can then be arranged in an $m \times \ell$ matrix,

$$B = (\mathbf{b}_1, \mathbf{b}_2, \ldots, \mathbf{b}_\ell) \quad .$$

Instead of the solution vector $\widetilde{\mathbf{x}}$ we thus obtain the $n \times \ell$ solution matrix

$$\widetilde{X} = (\widetilde{\mathbf{x}}_1, \widetilde{\mathbf{x}}_2, \ldots, \widetilde{\mathbf{x}}_\ell)$$

of the equation

$$A X \approx B \quad . \tag{A.15.7}$$

The following subroutine computes this solution. It merely consists of four calls to other subroutines that carry out the four steps described above. These are explained in more detail in the following sections.

Program A.36: Subroutine MTXSVD carries out the singular value decomposition and analysis of an $m \times n$ matrix A

CALLING SEQUENCE: `CALL MTXSVD(A,B,X,R,M,N,L,FRAC,OK)`

INPUT ARGUMENTS: `A` — $m \times n$ matrix A

`B` — $m \times \ell$ matrix B

`M,N,L` — dimensions m, n, ℓ

`FRAC` — fraction f. If $s_{\max}$ is the largest singular value, then $s_{\min} = f \cdot s_{\max}$, and in the singular value analysis all singular values $< s_{\min}$ are set equal to zero.

OUTPUT ARGUMENTS: X — $n \times \ell$ solution matrix
N — number of singular values not set to zero
OK — logical variable (.TRUE. for convergence, .FALSE. for non-convergence of step 2)
R — ℓ-vector of the squares of residuals (A.14.4) for the ℓ columns of the matrix B

```
      SUBROUTINE MTXSVD(A,B,X,R,M,N,NB,FRAC,OK)
      IMPLICIT DOUBLE PRECISION (A-H,O-Z)
      PARAMETER(NMAX=1000)
      DIMENSION A(M,N),B(M,NB),X(N,NB),R(NB),D(NMAX),E(NMAX)
      COMMON /DASV01/ D,E
      LOGICAL OK
C STEP 1: Bidiagonalisation of A
      CALL MTXSV1(A,B,D,E,M,N,NB)
C STEP 2: Diagonalisation of bidiagonal matrix
      CALL MTXSV2(A,B,D,E,M,N,NB,OK)
C STEP 3: Order singular values and perform  permutations
      CALL MTXSV3(A,B,D,M,N,NB)
C STEP 4: Singular value analysis
      CALL MTXSV4(A,B,D,X,R,M,N,NB,FRAC)
      END
```

Usually one is only interested in the solution matrix $\widetilde{X}$ of the problem $AX \approx B$ and possibly in the number of singular values not set to zero. Sometimes, however, one would like explicit access to the matrices U and V and to the singular values. This is made possible by the following program.

Program A.37: Subroutine MTXDEC carries out the singular value decomposition $A = U S V^{\mathrm{T}}$ of an $m \times n$ matrix A and gives explicitly the matrices U, S, V and the pseudo-inverse X

CALLING SEQUENCE: CALL MTXDEC(A,B,X,R,M,N,FRAC,OK,D,U,V)

INPUT ARGUMENTS: A — $m \times n$ matrix A
B — working array, minimum size $m \times m$
M,N — dimensions m, n
FRAC — fraction f. If $s_{\max}$ is the largest singular value, then $s_{\min} = f \cdot s_{\max}$, and in the singular value analysis all singular values $< s_{\min}$ are set equal to zero.

OUTPUT ARGUMENTS: X — $n \times m$ matrix A^+, the pseudo-inverse of A
U — $m \times m$ matrix U
V — $n \times n$ matrix V
D — n-vector of the diagonal elements of S
OK — logical variable (.TRUE. for convergence, .FALSE. for non-convergence)

N — number of singular values not set to zero.

R — ℓ-vector of the squares of the residuals (A.14.4) for the ℓ columns of the matrix B

REMARKS: If the input and output values of N are different, then in D and in the computation of X one or more singular values were set to zero.

```
      SUBROUTINE MTXDEC(A,B,X,R,M,N,FRAC,OK,D,U,V)
      IMPLICIT DOUBLE PRECISION (A-H,O-Z)
      PARAMETER(NMAX=1000)
      DIMENSION A(M,N),B(M,M),X(N,M),R(M)
      DIMENSION D(N),E(NMAX)
      DIMENSION U(M,M),V(N,N)
      COMMON /DASV04/ E
      LOGICAL OK
C STEP 0: Set B equal to (N x N) unit matrix
      CALL MTXUNT(B,M)
C STEP 1: Bidiagonalisation of A
      CALL MTXSV1(A,B,D,E,M,N,M)
C STEP 2: Diagonalisation of bidiagonal matrix
      CALL MTXSV2(A,B,D,E,M,N,M,OK)
C STEP 3: Order singular values and perform permutations
      CALL MTXSV3(A,B,D,M,N,M)
      CALL MTXGSM(A,V,M,N,N,N,1,1)
      CALL MTXTRP(B,U,M,M)
C STEP 4: Singular value analysis
      CALL MTXSV4(A,B,D,X,R,M,N,M,FRAC)
      END
```

A.15.2 Bidiagonalization

The goal of this section is to find the $m \times n$ matrix C in (A.15.1). Here C is of the form

$$C = \begin{pmatrix} C' \\ 0 \end{pmatrix} \quad , \tag{A.15.8}$$

and C' is an $n \times n$ bidiagonal matrix

$$C' = \begin{pmatrix} d_1 & e_2 & & & \\ & d_2 & e_3 & & \\ & & \ddots & \ddots & \\ & & & \ddots & e_n \\ & & & & d_n \end{pmatrix} \quad . \tag{A.15.9}$$

The goal is achieved by multiplication of the matrix A with appropriate Householder matrices (alternating on the right and left):

$$C = Q_n(\cdots((Q_1 A)H_2)\cdots H_n) = Q^{\mathrm{T}} A\, H \quad . \tag{A.15.10}$$

The matrix Q_1 is computed from the elements of the first column of A and is applied to all columns of A (and B). It results in only the element $(1,1)$ of A remaining non-zero. The matrix H_2 is computed from the first row of the matrix $Q_1 A$. It results in the element $(1,1)$ remaining unchanged, the element $(1,2)$ being recomputed and the elements $(1,3), \ldots, (1,n)$ being set equal to zero. It is applied to all rows of the matrix $(Q_1 A)$. One thus obtains

$$Q_1 A = \begin{pmatrix} \bullet & . & . & . \\ 0 & . & . & . \\ 0 & . & . & . \\ 0 & . & . & . \\ 0 & . & . & . \end{pmatrix} , \qquad Q_1 A\, H_2 = \begin{pmatrix} \bullet & \bullet & 0 & 0 \\ 0 & . & . & . \\ 0 & . & . & . \\ 0 & . & . & . \\ 0 & . & . & . \end{pmatrix} .$$

Here "$\bullet$" denotes an element of the final bidiagonal matrix C', and "." an element that will be changed further. Now Q_2 is determined from the second column of $Q_1 A\, H_2$ such that upon application to this column the element 1 remains unchanged and the element 2 and all others are changed such that only element 2 remains non-zero, and the elements 3 through m become zero.

The procedure can be summarized in the following way. The matrix Q_i is applied to the column vectors, it leaves the elements 1 through $i-1$ unchanged and changes the elements i through m. It produces an orthogonal transformation in the subspace of the components i through m. At the time of applying Q_i, however, these elements in the columns 1 through $i-1$ are already zero, so that Q_i must only be applied explicitly to columns i through n. The corresponding considerations hold for the matrix H_i. It acts on the row vectors, leaves elements 1 through $i-1$ unchanged, and produces an orthogonal transformation in the subspace of the components i through n. When it is applied, these components in the rows 1 through $i-2$ are all zero. The matrix H_i must only be applied explicitly to the rows $i-1, \ldots, m-1$. Since only $n-1$ rows of A must be processed, the matrix H_n in (A.15.10) is the unit matrix; Q_n is the unit matrix only if $m = n$.

In addition to the transformed matrix $Q^{\mathrm{T}} A\, H$, the matrix $H = H_2 H_3 \cdots H_{n-1}$ is also stored. For this the information about each matrix H_i is saved. As we determined at the end of Sect. A.4, it is sufficient to store the quantities defined there, u_p and b, and the elements $i+1$ through n of the $(i-1)$th row vectors defining the matrix H_i. This can be done, however, in the array elements of these row vectors themselves, since it is these elements that will be transformed to zero and which therefore do not enter further into the calculation. One needs to declare additional variables only for the quantities u_p and b of each of the matrices H_i. If all of the transformations have been carried out, then the diagonal elements d_i and the next-to-diagonal elements e_i are transferred to the arrays E and D. Finally, the product matrix $H = H_2 H_3 \cdots H_{n-1} I$ is constructed in the first n rows of the array of the original matrix A, in the order $H_{n-1}I$, $H_{n-2}(H_{n-1}I)$, ... Here as well, the procedure is as economical as possible, i.e., the Householder matrix is only applied to those columns of the matrix to the right for which it would actually produce a change. The unit matrix I is constructed to the extent that it is needed, row-by-row in the array of the original matrix A starting from below. For this there is exactly the right amount of space available, which up to this point was necessary for storing information about the Householder transformations that were just applied.

Program A.38: Subroutine MTXSV1 performs bidiagonalization of an $m \times n$ matrix A, such that $A = Q\,C\,H^{\mathrm{T}}$ with C bidiagonal

CALLING SEQUENCE: CALL MTXSV1(A,B,D,E,M,N,L)

INPUT ARGUMENTS: A — $m \times n$ matrix A
B — $m \times \ell$ matrix B
M,N,L — dimensions m, n, ℓ

OUTPUT ARGUMENTS: A — $n \times n$ matrix H (in the first n lines of A)
B — $m \times \ell$ product matrix $Q^{\mathrm{T}}B$
D — vector of the diagonal elements of C
E — vector of the next-to-diagonal elements of C

```
      SUBROUTINE MTXSV1(A,B,D,E,M,N,NB)
      IMPLICIT DOUBLE PRECISION (A-H,O-Z)
      PARAMETER(NMAX=1000)
      DIMENSION A(M,N),B(M,NB),V(NMAX),S(NMAX),UPS(NMAX),D(N),E(N)
      DIMENSION BBS(NMAX)
      COMMON /DASV00/ V,S,UPS,BBS
      DO 50 I=1,N
C set up Householder Transformation Q(I)
        IF(I.LT.N .OR. M.GT.N) THEN
          CALL MTXGCL(A,V,M,N,I)
          CALL MTXHSD(V,UP,BB,M,I,I+1)
C apply Q(I) to A
          DO 10 J=I,N
            CALL MTXGCL(A,S,M,N,J)
            CALL MTXHST(V,UP,BB,S,M,I,I+1)
            CALL MTXPCL(A,S,M,N,J)
   10     CONTINUE
C apply Q(I) to B
          DO 20 K=1,NB
            CALL MTXGCL(B,S,M,NB,K)
            CALL MTXHST(V,UP,BB,S,M,I,I+1)
            CALL MTXPCL(B,S,M,NB,K)
   20     CONTINUE
        END IF
        IF(I.LT.N-1) THEN
C set up Householder Transformation H(I)
          CALL MTXGRW(A,V,M,N,I)
          CALL MTXHSD(V,UP,BB,N,I+1,I+2)
C save H(I)
```

```
          UPS(I)=UP
          BBS(I)=BB
C apply H(I) to A
          DO 40 J=I,M
            CALL MTXGRW(A,S,M,N,J)
            CALL MTXHST(V,UP,BB,S,N,I+1,I+2)
C save elements I+2,... in row J of matrix A
            IF (J.EQ.I) THEN
              DO 30 K=I+2,N
                S(K)=V(K)
   30         CONTINUE
            END IF
            CALL MTXPRW(A,S,M,N,J)
   40     CONTINUE
        END IF
   50 CONTINUE
C copy diagonal of transformed matrix A to D
C and upper parallel A to E
      IF(N.GT.1) THEN
        DO 60 I=2,N
          D(I)=A(I,I)
          E(I)=A(I-1,I)
   60   CONTINUE
      END IF
      D(1)=A(1,1)
      E(1)=0.
C construct product matrix H=H(2)*H(3)*...*H(N), H(N)=I
      DO 90 I=N,1,-1
        IF(I.LE.N-1) CALL MTXGRW(A,V,M,N,I)
        DO 70 K=1,N
          A(I,K)=0.
   70   CONTINUE
        A(I,I)=1.
        IF(I.LT.N-1) THEN
          DO 80 K=I,N
            CALL MTXGCL(A,S,M,N,K)
            CALL MTXHST(V,UPS(I),BBS(I),S,N,I+1,I+2)
            CALL MTXPCL(A,S,M,N,K)
   80     CONTINUE
        END IF
   90 CONTINUE
      END
```

A.15.3 Diagonalization

The bidiagonal matrix C, whose non-vanishing elements are stored in the arrays D and E, is now brought into diagonal form by appropriately chosen Givens transformations. The strategy is chosen such that the lowest non-diagonal element vanishes first and the non-diagonal elements always move up and to the left, until C is finally diagonal. All of the transformations applied to C from the left are also applied to the matrix stored in the array B, and all transformations that act from the right are also applied to the matrix in A. (We denote the matrix to be diagonalized during each step by C.)

Only the upper-left submatrix C_k with k rows and columns is not yet diagonal and must still be considered. The index k is determined such that

$e_k \neq 0$ and $e_j = 0$, $j > k$. This means that the program runs through the loop $k = n, n-1, \ldots, 2$. Before the lower non-diagonal element e_k is systematically made zero by means of an iterative procedure, one checks for two special cases, which allow a shortening of the computation by means of special treatment.

Special case 1, $d_k = 0$ (handled by subroutine `MTXS21`): A Givens matrix W is applied from the right, which also causes e_k to vanish. The matrix w is the product $W_{k-1}W_{k-2}\cdots W_1$. Here W_i acts on the columns i and k of C_k, but of course only on those rows where at least one of these columns has a non-vanishing element. W_{k-1} acts on the row $k-1$, annihilates the element $e_k = C_{k-1,k}$, and changes $C_{k-1,k-1}$. In addition, W_{k-1} acts on the row $k-2$, changes the element $C_{k-2,k-1}$, and produces a non-vanishing element $H = C_{k-2,k}$ in column k. Now the matrix W_{k-2} is applied, which annihilates exactly this element, but produces a new element in row $k-3$ and column k. When the additional element finally makes it to row 1, it can then be annihilated by the transformation W_1. As a result of this treatment of special case 1, C_k decomposes into a $(k-1)\times(k-1)$ submatrix C_{k-1} and a 1×1 null matrix.

Special case 2, C_k decomposes into submatrices (handled by subroutine `MTXS22`): If $e_\ell = 0$ for any value ℓ, $2 \leq \ell \leq k$, then the matrix

$$C_k = \begin{pmatrix} d_1 & e_2 & & & & & & \\ & \ddots & & & & & & \\ & & d_{\ell-1} & 0 & & & & \\ & & & d_\ell & e_{\ell+1} & & & \\ & & & & \ddots & & & \\ & & & & & d_{k-1} & e_k \\ & & & & & & d_k \end{pmatrix} = \begin{pmatrix} C_{\ell-1} & 0 \\ 0 & \bar{C} \end{pmatrix}$$

can be decomposed into two matrices $C_{\ell-1}$ and $\bar{C}$. One can first diagonalize $\bar{C}$ and then $C_{\ell-1}$. In particular, if $\ell = k$, then one obtains

$$C_k = \begin{pmatrix} C_{k-1} & 0 \\ 0 & d_k \end{pmatrix} .$$

Here d_k is the singular value, and the loop index can only be decreased by one, $k \to k-1$. First, however, the case $d_k < 0$ must be treated; see *"change of sign"* below. If $d_{\ell-1} = 0$, but $e_\ell \neq 0$, then C can still be decomposed. For this one uses the transformation matrix $T = T_k T_{k-1}, \cdots, T_{\ell+1}$ acting on the left, where T_i acts on the rows ℓ and i. In particular, $T_{\ell+1}$ annihilates the element $e_\ell = C_{\ell,\ell+1}$ and creates an element $H = C_{\ell,\ell+2}$. $T_{\ell+2}$ annihilates this element and creates in its place $H = C_{\ell,\ell+3}$. Finally T_k annihilates the last created element $H = C_{\ell k}$ by a transformation to $C_{kk} = d_k$.

After it has been checked whether one or both of the special cases were present and such an occurrence has been treated accordingly, it remains only to diagonalize the matrix $\bar{C}$. This consists of the rows and columns ℓ through k of C. If special case 2 was not present, then one has $\ell = 1$. The problem is solved iteratively with the QR-algorithm.

QR-algorithm (carried out in subroutine MTXS23): First we denote the square output matrix $\bar{C}$ by C_1. We then determine orthogonal matrices U_i, V_i and carry out transformations

$$C_{i+1} = U_i^{\mathrm{T}} C_i V_i \quad , \qquad i = 1, 2, \ldots \quad ,$$

which lead to a diagonal matrix S,

$$\lim_{i\to\infty} C_i = S \quad .$$

The following prescription is used for determining U_i and V_i:

(A) One determines the eigenvalues λ_1, λ_2 of the lower-right 2×2 submatrix of $C_i^{\mathrm{T}} C_i$. Here σ_i is the eigenvalue closest to the lower-right element of$C_i^{\mathrm{T}} C_i$.

(B) The matrix V_i is determined such that $V_i^{\mathrm{T}}(C_i^{\mathrm{T}} C_i - \sigma_i I)$ has upper triangular form.

(C) The matrix U_i is determined such that $C_{i+1} = U_i^{\mathrm{T}} C_i V_i$ is again bidiagonal.

The matrix V_i from step (B) exists, according to a theorem by FRANCIS [22], if

(a) $C_i^{\mathrm{T}} C_i$ is tridiagonal with non-vanishing subdiagonal elements,

(b) V_i is orthogonal,

(c) σ_i is an arbitrary scalar,

(d) $V_i^{\mathrm{T}}(C_i^{\mathrm{T}} C_i) V_i$ is tridiagonal, and

(e) the first column of $V_i^{\mathrm{T}}(C_i^{\mathrm{T}} C_i - \sigma_i I)$ except the first element vanishes.

The requirement (a) is fulfilled, since C_i is bidiagonal and the special case 2 has been treated if necessary; (b) is also fulfilled by constructing V_i as the product of Givens matrices. This is done in such a way that simultaneously (d) and (e) are fulfilled. In particular one has

$$V_i = R_1 R_2 \cdots R_{n-1} \quad , \qquad U_i^{\mathrm{T}} = L_{n-1} L_n \cdots L_1 \quad ,$$

where

R_j acts on the columns j and $j+1$ of C,
L_j acts on the rows j and $j+1$ of C,
R_1 is determined such that requirement (e) is fulfilled,
L_1, R_2, L_2, ... are determined such that (e) is fulfilled without violating (d).

For σ_i one obtains

$$\sigma_i = d_n^2 + e_n \left(e_n - \frac{d_{n-1}}{t} \right) \quad .$$

With the definitions

$$\begin{aligned} t &= f + \sqrt{1+f^2} \quad , \qquad f \geq 0 \quad , \\ t &= f - \sqrt{1+f^2} \quad , \qquad f < 0 \quad , \end{aligned}$$

and

$$f = \frac{d_n^2 - d_{n-1}^2 + e_n^2 - e_{n-1}^2}{2 e_n d_{n-1}} \quad , \qquad \gamma = \frac{\delta}{e_n d_{n-1}}$$

the first column of the matrix $(C_i^{\mathrm{T}} C_i - \sigma_i I)$ is

$$\begin{pmatrix} d_1^2 - \sigma \\ d_1 e_2 \\ 0 \\ \vdots \\ 0 \end{pmatrix} \quad .$$

One determines the matrix R_1, which defines a Givens transformation, such that all elements of the first column of $R_1^{\mathrm{T}}(C_i^{\mathrm{T}} C_i - \sigma_i I)$ except the first vanish. Application of R_1 on C_i produces, however, an additional element $H = C_{21}$, so that C_i is no longer bidiagonal,

$$C_i = \begin{pmatrix} \cdot & \cdot & & & \\ & \cdot & \cdot & & \\ & & \cdot & \cdot & \\ & & & \cdot & \cdot \\ & & & & \cdot \end{pmatrix} \quad , \qquad C_i R_1 = \begin{pmatrix} \cdot & \cdot & & & \\ H & \cdot & \cdot & & \\ & & \cdot & \cdot & \\ & & & \cdot & \cdot \\ & & & & \cdot \end{pmatrix} \quad .$$

By application of L_1, this element is projected onto the diagonal. In its place a new element $H = C_{13}$ is created,

$$L_1 C_i R_1 = \begin{pmatrix} \cdot & \cdot & H & & \\ & \cdot & \cdot & & \\ & & \cdot & \cdot & \\ & & & \cdot & \cdot \\ & & & & \cdot \end{pmatrix} \quad .$$

By continuing the procedure, the additional element is moved further down and to the right, and can be completely eliminated in the last step:

$$T_{n-1}\begin{pmatrix} \cdot & \cdot & & & & \\ & \cdot & \cdot & & & \\ & & \cdot & \cdot & & \\ & & & \cdot & \cdot & \\ & & H & & \cdot & \end{pmatrix} = \begin{pmatrix} \cdot & \cdot & & & \\ & \cdot & \cdot & & \\ & & \cdot & \cdot & \\ & & & \cdot & \cdot \\ & & & & \cdot \end{pmatrix} = C_{i+1} \quad .$$

If the lower non-diagonal element of C_{i+1} is already zero, then the lower diagonal element is already a singular value. Otherwise the procedure is repeated, whereby it is first checked whether now one of the two special cases is present. The procedure typically converges in about $2\,k$ steps; (k is the rank of the original matrix A). If convergence has still not been reached after $10\,k$ steps, then the algorithm is terminated without success.

Change of sign. If a singular value d_k has been found, i.e., if $e_k = 0$, then it is checked whether it is negative. If this is the case, then a simple orthogonal transformation is applied that multiplies the element $d_k = C_{kk}$ of C and all elements in the kth column of the matrix contained in `A` by -1. The index k can then be reduced by one. Corresponding to (A.13.6) the matrix B is multiplied on the left by U^{T}. This is done successively with the individual factors making up U^{T}.

Program A.39: Subroutine `MTXSV2` iteratively diagonalizes the bidiagonal matrix C, such that $S' = U'^{\mathrm{T}} C\, V'$ is diagonal

CALLING SEQUENCE: `CALL MTXSV2(A,B,D,E,M,N,L,OK)`

INPUT ARGUMENTS: `A` — $n \times n$ matrix H (in the first n rows of `A`)
`B` — $m \times \ell$ product matrix $Q^{\mathrm{T}} B$
`D` — vector of diagonal elements of C
`E` — vector of the next-to-diagonal elements of C
`M,N,L` — dimensions m, n, ℓ

OUTPUT ARGUMENTS: `A` — $n \times n$ matrix $H\,V'$ (in the first n rows of `A`)
`B` — $m \times \ell$ matrix $U'^{\mathrm{T}} Q^{\mathrm{T}} B$
`D` — vector of the diagonal elements of S'
`OK` — `.TRUE.` for convergence, `.FALSE.` for non-convergence of the procedure

```
      SUBROUTINE MTXSV2(A,B,D,E,M,N,NB,OK)
      IMPLICIT DOUBLE PRECISION (A-H,O-Z)
      DIMENSION A(M,N),B(M,NB),D(N),E(N)
      LOGICAL OK,ELZERO
      PARAMETER(ZERO=0.D0)
      OK=.TRUE.
      NITERM=10*N
      NITER=0
      BMX=D(1)
      IF(N.GT.1) THEN
        DO 10 I =  2, N
          BMX=MAX(ABS(D(I))+ABS(E(I)),BMX)
   10   CONTINUE
      END IF
      DO 60 K=N,1,-1
   20   CONTINUE
        IF(K.NE.1) THEN
          IF((BMX+D(K))-BMX.EQ.ZERO) THEN
C Since D(K).EQ.0. perform Givens transform with result E(K)=0.
            CALL MTXS21(A,D,E,M,N,NB,K)
          END IF
C Find L (2. LE. L .LE. K) so that either E(L)=0. or D(L-1)=0.
C In the latter case transform E(L) to zero. In both cases the
C matrix splits and the bottom right minor begins with row L.
C If no such L is found set L=1
          DO 30 LL=K,1,-1
            L=LL
            IF(L.EQ.1) THEN
              ELZERO=.FALSE.
              GO TO 40
            ELSE IF((BMX-E(L))-BMX.EQ.ZERO) THEN
              ELZERO=.TRUE.
              GO TO 40
            ELSE IF((BMX+D(L-1))-BMX.EQ.ZERO) THEN
              ELZERO=.FALSE.
            END IF
   30     CONTINUE
   40     IF (L.GT. 1.AND. .NOT.ELZERO) THEN
            CALL MTXS22(B,D,E,M,N,NB,K,L)
          END IF
          IF(L.NE.K) THEN
C one more QR pass with order K
            CALL MTXS23(A,B,D,E,M,N,NB,K,L)
            NITER=NITER+1
            IF(NITER.LE.NITERM) GO TO 20
C set flag indicating non-convergence
            OK=.FALSE.
          END IF
        END IF
        IF(D(K).LT.0.) THEN
C for negative singular values perform change of sign
          D(K)=-D(K)
          DO 50 J=1,N
            A(J,K)=-A(J,K)
   50     CONTINUE
        END IF
C order is decreased by one in next pass
   60 CONTINUE
      END
C-----------------------------------------------------------------
      SUBROUTINE MTXS21(A,D,E,M,N,NB,K)
```

```
      IMPLICIT DOUBLE PRECISION (A-H,O-Z)
      DIMENSION A(M,N),D(N),E(N)
      DO 20 I = K- 1,  1, -1
        IF(I.EQ.K-1) THEN
          CALL MTXGVA(D(I),E(I+1),CS,SN)
        ELSE
          CALL MTXGVA(D(I),H,CS,SN)
        END IF
        IF(I.GT.1) THEN
          H=0.
          CALL MTXGVT(E(I),H,CS,SN)
        END IF
        DO 10 J =  1,N
          CALL MTXGVT(A(J,I),A(J,K),CS,SN)
   10   CONTINUE
   20 CONTINUE
      END
C------------------------------------------------------------------
      SUBROUTINE MTXS22(B,D,E,M,N,NB,K,L)
      IMPLICIT DOUBLE PRECISION (A-H,O-Z)
      DIMENSION B(M,NB),D(N),E(N)
      DO 20 I=L,K
        IF(I.EQ.L) THEN
          CALL MTXGVA(D(I),E(I),CS,SN)
        ELSE
          CALL MTXGVA(D(I),H,CS,SN)
        END IF
        IF(I.LT.K) THEN
          H=0.
          CALL MTXGVT(E(I+1),H,CS,SN)
        END IF
        DO 10 J=1,NB
          CALL MTXGVT(CS,SN,B(I,J),B(L-1,J))
   10   CONTINUE
   20 CONTINUE
      END
C------------------------------------------------------------------
      SUBROUTINE MTXS23(A,B,D,E,M,N,NB,K,L)
      IMPLICIT DOUBLE PRECISION (A-H,O-Z)
      DIMENSION A(M,N),B(M,NB),D(N),E(N)
C Determine shift parameter
      F=((D(K-1)-D(K))*(D(K-1)+D(K))+(E(K-1)-E(K))*(E(K-1)+E(K)))/
     +  (2.*E(K)*D(K-1))
      IF(ABS(F).GT.1.E10) THEN
        G=ABS(F)
      ELSE
        G=SQRT(1.+F*F)
      END IF
      IF(F.GE.0.) THEN
        T=F+G
      ELSE
        T=F-G
      END IF
      F=((D(L)-D(K))*(D(L)+D(K))+E(K)*(D(K-1)/T-E(K)))/D(L)
      DO 30 I = L , K-1
        IF(I.EQ.L) THEN
C Define R(L)
          CALL MTXGVD(F,E(I+1),CS,SN)
        ELSE
C Define R(I) , I.NE.L
          CALL MTXGVA(E(I),H,CS,SN)
```

```
          END IF
          CALL MTXGVT(D(I),E(I+1),CS,SN)
          H=0.
          CALL MTXGVT(H,D(I+1),CS,SN)
          DO 10 J =  1, N
            CALL MTXGVT(A(J,I),A(J,I+1),CS,SN)
   10     CONTINUE
C Define T(I)
          CALL MTXGVA(D(I),H,CS,SN)
          CALL MTXGVT(E(I+1),D(I+1),CS,SN)
          IF(I.LT.K-1) THEN
            H=0.
            CALL MTXGVT(H,E(I+2),CS,SN)
          END IF
          DO 20 J =  1, NB
            CALL MTXGVT(B(I,J),B(I+1,J),CS,SN)
   20     CONTINUE
   30 CONTINUE
      END
```

A.15.4 Ordering of the Singular Values and Permutation

In the following subroutine the singular values are put into non-increasing order. This is done by a sequence of permutations of neighboring singular values, carried out if the singular value that follows is larger then the preceding. The matrices stored in A and B are multiplied by a corresponding sequence of permutation matrices; cf. (A.15.4).

Program A.40: Subroutine MTXSV3 carries out a permutation transformation of the diagonal matrix S', such that the diagonal elements of $S'' = U''^{\mathrm{T}} S' V''$ are non-increasing

CALLING SEQUENCE: CALL MTXSV3(A,B,D,M,N,L)

INPUT ARGUMENTS: A — $n \times n$ matrix $H\,V'$ (in the first n rows of A)
B — $m \times \ell$ matrix $U'^{\mathrm{T}} Q^{\mathrm{T}} B$
D — vector of the diagonal elements of S'
M,N,L — dimensions m, n, ℓ

OUTPUT ARGUMENTS: A — $m \times n$ matrix $H\,V'V''$ (in the first n rows of A)
B — $m \times \ell$ matrix $U''^{\mathrm{T}} U'^{\mathrm{T}} Q^{\mathrm{T}} B$
D — vector of the diagonal elements of S''

```
      SUBROUTINE MTXSV3(A,B,D,M,N,NB)
      IMPLICIT DOUBLE PRECISION (A-H,O-Z)
      DIMENSION A(M,N),B(M,NB),D(N)
C Order singular values
      IF(N.GT.1) THEN
   10   DO 20 I=2,N
          IF(D(I).GT.D(I-1)) GO TO 30
   20   CONTINUE
```

```
      RETURN
 30   CONTINUE
      DO 70 I =  2, N
        T=D(I-1)
        K=I-1
        DO 40 J = I , N
          IF(T.LT.D(J)) THEN
            T=D(J)
            K=J
          END IF
 40     CONTINUE
        IF(K.NE.I-1) THEN
C perform permutation on singular values
          D(K)=D(I-1)
          D(I-1)=T
C perform permutation on matrix A
          DO 50 J =  1, N
            T=A(J,K)
            A(J,K)=A(J,I-1)
            A(J,I-1)=T
 50       CONTINUE
C perform permutation on matrix B
          DO 60 J =  1, NB
            T=B(K,J)
            B(K,J)=B(I-1,J)
            B(I-1,J)=T
 60       CONTINUE
        END IF
 70   CONTINUE
      GO TO 10
    END IF
    END
```

A.15.5 Singular Value Analysis

In the last step the singular value analysis is carried out as described in Sect. A.14. For a given factor $f \ll 1$ a value $\ell \leq k$ is determined such that $s_i < f s_1$ for $i > \ell$. The columns of the array B, which now contains the vectors **g**, are transformed in their first ℓ elements into $\tilde{\mathbf{p}}_1$ according to (A.13.11), and the elements $\ell + 1, \ldots, n$ are set equal to zero. Then the solution vectors $\tilde{\mathbf{x}}$, which make up the columns of the solution matrix $\widetilde{X}$, are computed according to (A.13.12).

Program A.41: Subroutine MTXSV4 carries out a singular value analysis

CALLING SEQUENCE: CALL MTXSV4(A,B,D,X,R,M,N,L,FRAC)

INPUT ARGUMENTS: A — $n \times n$ matrix $H V'V''$ (in the first n rows of A)
B — $m \times \ell$ matrix $U''^{\mathrm{T}}U'^{\mathrm{T}}Q^{\mathrm{T}}B$
D — vector of diagonal elements of S'', i.e., vector of the singular values of A in non-increasing order
M,N,L — dimensions m, n, ℓ

FRAC — The absolute value of FRAC is the fraction f. If $s_{\max}$ is the largest singular value, then $s_{\min} = f \cdot s_{\max}$, and in the singular value analysis all singular values $< s_{\min}$ are set to zero. (If $f < 10^{-15}$, e.g., $f = 0$, then the program sets $f = 10^{-15}$.)

OUTPUT ARGUMENTS: X — $n \times \ell$ solution matrix

N — number of remaining singular values

R — ℓ-vector of the squares of the residuals (A.14.4) for the ℓ columns of the matrix B

```
      SUBROUTINE MTXSV4(A,B,D,X,R,M,N,NB,FRAC)
      IMPLICIT DOUBLE PRECISION (A-H,O-Z)
      DIMENSION A(M,N),B(M,NB),D(N),X(N,NB),R(NB)
      PARAMETER (EPSILN=1.D-15)
      FRACT=ABS(FRAC)
      IF(FRACT.LT.EPSILN) FRACT=EPSILN
      SINMAX=0.
      DO 10 I =  1, N
        SINMAX=MAX(SINMAX,D(I))
   10 CONTINUE
      SINMIN=SINMAX*FRACT
      KK=N
      DO 20 I=1,N
        IF(D(I).LE.SINMIN) THEN
          KK=I-1
          GO TO 30
        END IF
   20 CONTINUE
   30 CONTINUE
      DO 60 I =  1, M
        IF(I.LE.KK) THEN
          S1=1./D(I)
          DO 40 J=1,NB
            B(I,J)=B(I,J)*S1
   40     CONTINUE
        ELSE
          DO 50 J=1,NB
            IF(I.EQ.KK+1) THEN
              R(J)=B(I,J)**2
            ELSE
              R(J)=R(J)+B(I,J)**2
            END IF
            IF(I.LE.N) B(I,J)=0.
   50     CONTINUE
        END IF
   60 CONTINUE
      DO 90 I=1,N
        DO 80 J=1,NB
          X(I,J)=0.
          DO 70 K=1,N
            X(I,J)=X(I,J)+A(I,K)*B(K,J)
   70     CONTINUE
   80   CONTINUE
   90 CONTINUE
      N=KK
      END
```

A.16 Least Squares with Weights

Instead of the problem (A.6.3),

$$r^2 = (A\mathbf{x} - \mathbf{b})^{\mathrm{T}}(A\mathbf{x} - \mathbf{b}) = \min \quad , \tag{A.16.1}$$

one often encounters a similar problem that in addition contains a positive-definite symmetric weight-matrix $G_{m\times m}$,

$$r^2 = (A\mathbf{x} - \mathbf{b})^{\mathrm{T}}G(A\mathbf{x} - \mathbf{b}) = \min \quad . \tag{A.16.2}$$

In (A.16.1) one simply has $G = I$. Using the Cholesky decomposition (A.10.1) of G, i.e., $G = U^{\mathrm{T}}U$, one has

$$r^2 = (A\mathbf{x} - \mathbf{b})^{\mathrm{T}}U^{\mathrm{T}}U(A\mathbf{x} - \mathbf{b}) = \min \quad . \tag{A.16.3}$$

With the definitions

$$A' = U\,A \quad , \qquad \mathbf{b}' = U\,\mathbf{b} \tag{A.16.4}$$

Eq. (A.16.3) takes on the form

$$r^2 = (A'\mathbf{x} - \mathbf{b}')^{\mathrm{T}}(A'\mathbf{x} - \mathbf{b}') = \min \quad . \tag{A.16.5}$$

After the replacement (A.16.4), the problem (A.16.2) is thus equivalent to the original one (A.16.1).

In Sect. A.12 we called the $n \times n$ matrix

$$C = (A^{\mathrm{T}}A)^{-1} \tag{A.16.6}$$

the unweighted covariance matrix of the unknowns $\mathbf{x}$ in the Problem (A.16.1). In Problem (A.16.2), the *weighted covariance matrix*

$$C_x = (A'^{\mathrm{T}}A')^{-1} = (A^{\mathrm{T}}G\,A)^{-1} \tag{A.16.7}$$

appears in its place.

A.17 Least Squares with Change of Scale

The goal in solving a problem of the type

$$(A\mathbf{x} - \mathbf{b})^{\mathrm{T}}(A\mathbf{x} - \mathbf{b}) = \min \tag{A.17.1}$$

is the most accurate numerical determination of the solution vector $\widetilde{\mathbf{x}}$ and the covariance matrix C. A change of scale in the elements of A, $\mathbf{b}$, and $\mathbf{x}$ can lead to an improvement in the numerical precision.

Let us assume that the person performing the problem already has an approximate idea of $\widetilde{\mathbf{x}}$ and C, which we call $\mathbf{z}$ and K. The matrix K has the Cholesky decomposition $K = L^{\mathrm{T}}L$. By defining

$$A' = A\,L \quad , \qquad \mathbf{b}' = \mathbf{b} - A\,\mathbf{z} \quad , \qquad \mathbf{x}' = L^{-1}(\mathbf{x} - \mathbf{z}) \quad , \tag{A.17.2}$$

Eq. (A.17.1) becomes

$$(A'\mathbf{x}' - \mathbf{b}')^{\mathrm{T}}(A'\mathbf{x}' - \mathbf{b}) = \min \quad .$$

The meaning of the new vector of unknowns $\mathbf{x}'$ is easily recognizable for the case where K is a diagonal matrix. We write

$$K = \begin{pmatrix} \sigma_1^2 & & & \\ & \sigma_2^2 & & \\ & & \ddots & \\ & & & \sigma_n^2 \end{pmatrix} \quad , \qquad L = \begin{pmatrix} \sigma_1 & & & \\ & \sigma_2 & & \\ & & \ddots & \\ & & & \sigma_n \end{pmatrix} \quad ,$$

where the quantities σ_i^2 are the estimated values of the variances of the unknowns x_i. Thus the ith component of the vector $\mathbf{x}'$ becomes

$$x_i' = \frac{x_i - z_i}{\sigma_i} \quad .$$

If in fact one has $\widetilde{x}_i = z_i$ and the corresponding variance σ_i^2, then $x_i' = 0$ and has a variance of one. If the estimates are at least of the correct order of magnitude, the x_i' are close to zero and their variances are of order unity. In addition, in case the full matrix is, in fact, estimated with sufficient accuracy, the components of $\mathbf{x}'$ are not strongly correlated with each other.

In practice, one carries out the transformation (A.17.2) only in exceptional cases. One must take care, however, that "reasonable" variables are chosen for (A.17.2). This technique is applied in the graphical representation of data. If it is known, for example, that a voltage U varies in the region $\sigma = 10$ millivolts about the value $U_0 = 1$ Volts, then instead of U one would plot the quantity $U' = (U - U_0)/\sigma$, or some similar quantity.

A.18 Modification of Least Squares According to Marquardt

Instead of the problem

$$(A\mathbf{x} - \mathbf{b})^{\mathrm{T}}(A\mathbf{x} - \mathbf{b}) = \min \quad , \tag{A.18.1}$$

which we have also written in the shorter form

$$A\mathbf{x} - \mathbf{b} \approx 0 \quad , \tag{A.18.2}$$

let us now consider the modified problem

$$\begin{matrix} m \\ n \end{matrix} \begin{matrix} \{ \\ \{ \end{matrix} \underbrace{\begin{pmatrix} A \\ \lambda I \end{pmatrix}}_{n} \mathbf{x} \approx \begin{pmatrix} \mathbf{b} \\ 0 \end{pmatrix} \begin{matrix} \} \\ \} \end{matrix} \begin{matrix} m \\ n \end{matrix} \quad . \tag{A.18.3}$$

Here I is the $n \times n$ unit matrix and λ is a non-negative number. The modified problem is of considerable importance for fitting nonlinear functions with the method of least squares (Sect. 9.5) or for minimization (Sect. 10.15). For $\lambda = 0$ Eq. (A.18.3) clearly becomes (A.18.2). If, on the other hand, λ is very large, or more precisely, if it is large compared to the absolute values of the elements of A and $\mathbf{b}$, then the last "row" of (A.18.3) determines the solution $\tilde{\mathbf{x}}$, which is the null vector for $\lambda \to \infty$.

We first ask which direction the vector $\tilde{\mathbf{x}}$ has for large values of λ. The normal equations corresponding to (A.18.3), cf. (A.6.16), are

$$(A^{\mathrm{T}}, \lambda I) \begin{pmatrix} A \\ \lambda I \end{pmatrix} \mathbf{x} = (A^{\mathrm{T}}A + \lambda^2 I)\mathbf{x} = (A^{\mathrm{T}}, \lambda I) \begin{pmatrix} \mathbf{b} \\ 0 \end{pmatrix} = A^{\mathrm{T}}\mathbf{b}$$

with the solution

$$\tilde{\mathbf{x}} = (A^{\mathrm{T}}A + \lambda^2 I)^{-1} A^{\mathrm{T}}\mathbf{b} \quad .$$

For large λ the second term in parentheses dominates, and one obtains simply

$$\tilde{\mathbf{x}} = \lambda^{-2} A^{\mathrm{T}}\mathbf{b} \quad .$$

That is, for large values of λ, the solution vector tends toward the direction of the vector $A^{\mathrm{T}}\mathbf{b}$.

We will now show that for a given λ the solution $\mathbf{x}^{(\lambda)}$ to (A.18.3) can easily be found with the singular value decomposition simultaneously with the determination of the solution $\tilde{\mathbf{x}}$ of (A.18.2). The singular value decomposition (A.13.1) of A is

$$A = U \begin{pmatrix} S \\ 0 \end{pmatrix} V^{\mathrm{T}} \quad .$$

By substituting into (A.18.3) and multiplying on the left we obtain

$$\begin{pmatrix} U^{\mathrm{T}} & 0 \\ 0 & V^{\mathrm{T}} \end{pmatrix} \begin{pmatrix} U \begin{pmatrix} S \\ 0 \end{pmatrix} V^{\mathrm{T}} \\ \lambda I \end{pmatrix} \mathbf{x} = \begin{pmatrix} U^{\mathrm{T}} & 0 \\ 0 & V^{\mathrm{T}} \end{pmatrix} \begin{pmatrix} \mathbf{b} \\ 0 \end{pmatrix}$$

or

$$\begin{pmatrix} S \\ 0 \\ \lambda I \end{pmatrix} \mathbf{p} = \begin{pmatrix} \mathbf{g} \\ 0 \end{pmatrix} \quad , \tag{A.18.4}$$

where, using the notation as in Sect. A.13,

$$\mathbf{p} = V^{\mathrm{T}}\mathbf{x} \quad , \qquad \mathbf{g} = U^{\mathrm{T}}\mathbf{b} \quad .$$

By means of Givens transformations, the matrix on the left-hand side of (A.18.4) can be brought into diagonal form. One obtains

$$\begin{pmatrix} S^{(\lambda)} \\ 0 \\ 0 \end{pmatrix} \mathbf{p} = \begin{pmatrix} \mathbf{g}^{(\lambda)} \\ \mathbf{h}^{(\lambda)} \end{pmatrix}$$

with

$$\begin{aligned} g_i^{(\lambda)} &= \frac{g_i s_i}{s_i^{(\lambda)}} \quad , \qquad i = 1, \ldots, n \quad , \\ g_i^{(\lambda)} &= g_i \quad , \qquad i = n+1, \ldots, m \quad , \\ h_i^{(\lambda)} &= -\frac{g_i^{\lambda}}{s_i^{(\lambda)}} \quad , \qquad i = 1, \ldots, n \quad , \\ s_i^{(\lambda)} &= \sqrt{s_i^2 + \lambda^2} \end{aligned}$$

and thus

$$\begin{aligned} p_i^{(\lambda)} &= \frac{g_i s_i}{s_i^2 + \lambda^2} = p_i^{(0)} \frac{s_i^2}{s_i^2 + \lambda^2} \quad , \qquad i = 1, \ldots, k \quad , \\ p_i^{(\lambda)} &= 0 \quad , \qquad i = k+1, \ldots, n \quad . \end{aligned}$$

The solution $\widetilde{\mathbf{x}}^{(\lambda)}$ of (A.18.3) is then

$$\widetilde{\mathbf{x}}^{(\lambda)} = V\mathbf{p}^{(\lambda)} \quad .$$

The following subroutine computes these solution vectors for two values of λ.

Program A.42: Subroutine MTXMAR computes two solution vectors $\widetilde{\mathbf{x}}_1$ and $\widetilde{\mathbf{x}}_2$ of the problem $A\mathbf{x} \approx \mathbf{b}$

CALLING SEQUENCE: `CALL MTXMAR(A,B,ALAM,X1,X2,M,N,FRAC, OK)`

INPUT ARGUMENTS: `A` — $m \times n$ matrix A
`B` — n-vector **b**
`ALAM` — positive number λ
`M,N` — dimensions m, n
`FRAC` — as in Program A.36

OUTPUT ARGUMENTS: `X1` — solution vector $\widetilde{\mathbf{x}}_1$ for Marquardt factor $\lambda_1 = \lambda$
`X2` — solution vector $\widetilde{\mathbf{x}}_2$ for Marquardt factor $\lambda_2 = \lambda/10$
`N` — number of singular values not set equal to zero
`OK` — logical variables (`.TRUE.` for convergence, `.FALSE.` for non-convergence of the singular value decomposition)

```
      SUBROUTINE MTXMAR(A,B,ALAM,X1,X2,M,N,FRAC,OK)
      IMPLICIT DOUBLE PRECISION (A-H,O-Z)
      PARAMETER(NMAX=1000)
      DIMENSION A(M,N),B(M,1),X1(N),X2(N),D(NMAX),E(NMAX)
      LOGICAL OK
      COMMON /DASV01/ D,E
      PARAMETER(NB=1)
C STEP 1: Bidiagonalisation of A
      CALL MTXSV1(A,B,D,E,M,N,NB)
C STEP 2: Diagonalisation of bidiagonal matrix
      CALL MTXSV2(A,B,D,E,M,N,NB,OK)
C STEP 3: Order singular values and perform  permutations
      CALL MTXSV3(A,B,D,M,N,NB)
C STEP 4: Singular value analysis and appl. of Marquardt method
      CALL MTXSVM(A,B,D,ALAM,X1,X2,M,N,FRAC)
      END
C-------------------------------------------------------------------
      SUBROUTINE MTXSVM(A,B,D,ALAM,X1,X2,M,N,FRAC)
      IMPLICIT DOUBLE PRECISION (A-H,O-Z)
      PARAMETER(NMAX=1000)
      DIMENSION A(M,N),B(M),D(N),X1(N),X2(N),P1(NMAX),P2(NMAX)
      COMMON /DASV00/ P1,P2
      PARAMETER (EPSILN=1.D-15)
      FRACT=ABS(FRAC)
      IF(FRACT.LT.EPSILN) FRACT=EPSILN
      SINMAX=DBLE(0.)
      ALAM2=ALAM**2
      ALAMP=ALAM*DBLE(0.1)
      ALAMP2=ALAMP**2
      DO 10 I =  1, N
        SINMAX=MAX(SINMAX,D(I))
   10 CONTINUE
      SINMIN=SINMAX*FRACT
      KK=N
      DO 20 I=1,N
        IF(D(I).LE.SINMIN) THEN
          KK=I-1
          GO TO 30
        END IF
   20 CONTINUE
   30 CONTINUE
      DO 40 I =  1, M
        G=B(I)
        IF(I.LE.KK) THEN
          DEN1=1./(D(I)**2+ALAM2)
          DEN2=1./(D(I)**2+ALAMP2)
          P1(I)=G*D(I)*DEN1
          P2(I)=G*D(I)*DEN2
        ELSE
          IF(I.LE.N) THEN
            P1(I)=0.
            P2(I)=0.
          END IF
        END IF
   40 CONTINUE
      DO 60 I=1,N
        X1(I)=0.
        X2(I)=0.
        DO 50 K=1,N
          X1(I)=X1(I)+A(I,K)*P1(K)
          X2(I)=X2(I)+A(I,K)*P2(K)
```

```
50    CONTINUE
60 CONTINUE
   N=KK
   END
```

A.19 Least Squares with Constraints

One often encounters the problem (A.6.3)

$$r^2 = (A\mathbf{x} - \mathbf{b})^2 = \min \tag{A.19.1}$$

with the *constraint*

$$E\mathbf{x} = \mathbf{d} \quad . \tag{A.19.2}$$

Here A is as before an $m \times n$ matrix and E is an $\ell \times n$ matrix. We will restrict ourselves to the only case that occurs in practice,

$$\text{Rank}\, E = \ell < n \quad . \tag{A.19.3}$$

The determination of an extreme value with constraints is usually treated in analysis textbooks with the method of Lagrange multipliers. Here as well we rely on orthogonal transformations. The following method is due to LAWSON and HANSON [18]. It uses a basis of the null space of E. First we carry out an orthogonal decomposition of E as in (A.6.5),

$$E = H\,R\,K^{\mathrm{T}} \quad . \tag{A.19.4}$$

Here we regard the orthogonal $n \times n$ matrix K as being constructed out of an $n \times \ell$ matrix K_1 and an $n \times (n - \ell)$ matrix K_2,

$$K = (K_1,\, K_2) \quad . \tag{A.19.5}$$

According to (A.6.6) and (A.6.7), all solutions of (A.19.2) have the form

$$\widehat{\mathbf{x}} = K_1\widetilde{\mathbf{p}}_1 + K_2\mathbf{p}_2 = \widetilde{\mathbf{x}} + K_2\mathbf{p}_2 \quad . \tag{A.19.6}$$

Here $\widetilde{\mathbf{x}}$ is the unique solution of minimum absolute value of (A.19.2). For brevity we will write this in the form $\widetilde{\mathbf{x}} = E^{+}\mathbf{d}$; cf. (A.11.3). $\mathbf{p}_2$ is an arbitrary $(n - \ell)$-vector, since the vectors $K_2\mathbf{p}_2$ form the null space of E,

$$E\,K_2\mathbf{p}_2 = 0 \quad . \tag{A.19.7}$$

The constraint (A.19.2) thus says that the vector $\mathbf{x}$ for which (A.19.1) is a minimum must come from the set of all vectors $\widehat{\mathbf{x}}$, i.e.,

$$\begin{aligned}(A\widehat{\mathbf{x}} - \mathbf{b})^2 &= (A(\widetilde{\mathbf{x}} + K_2\mathbf{p}_2) - \mathbf{b})^2 \\ &= (A\,K_2\mathbf{p}_2 - (\mathbf{b} - A\,\widetilde{\mathbf{x}}))^2 = \min \quad .\end{aligned} \tag{A.19.8}$$

This relation is a least-squares problem without constraints, from which the $(n-\ell)$-vector $\mathbf{p}_2$ can be determined. We write its solution using (A.11.3) in the form

$$\widetilde{\mathbf{p}}_2 = (A\,K_2)^+(\mathbf{b} - A\,\widetilde{\mathbf{x}}) \quad . \tag{A.19.9}$$

By substitution into (A.19.6) we finally obtain

$$\mathbf{x} = \widetilde{\mathbf{x}} + A^+(\mathbf{b} - A\,\widetilde{\mathbf{x}}) = E^+\mathbf{d} + K_2(A\,K_2)^+(\mathbf{b} - A\,E^+\mathbf{d}) \tag{A.19.10}$$

as the solution of (A.19.1) with the constraint (A.19.2).

The following prescription leads to the solution (A.19.10). Its starting point is the fact that one can set $H = I$ because of (A.19.3).

Step 1: One determines an orthogonal matrix $K = (K_1,\, K_2)$ as in (A.19.5) such that

$$E\,K = (E\,K_1,\, E\,K_2) = (\widetilde{E}_1,\, 0)$$

and such that $\widetilde{E}_1$ is a lower triangular matrix. In addition one computes

$$A\,K = (A\,K_1,\, A\,K_2) = (\widetilde{A}_1,\, \widetilde{A}_2) \quad .$$

Step 2: One determines the solution $\widetilde{\mathbf{p}}_1$ of

$$\widetilde{E}_1\mathbf{p}_1 = \mathbf{d} \quad .$$

This is easy, since $\widetilde{E}_1$ is a lower triangular matrix of rank ℓ. Clearly one has $\widetilde{\mathbf{x}} = K_1\widetilde{\mathbf{p}}_1$.

Step 3: One determines the vector

$$\bar{\mathbf{b}} = \mathbf{b} - \widetilde{A}_1\widetilde{\mathbf{p}}_1 = \mathbf{b} - A\,K_1K_1^{\mathrm{T}}\widetilde{\mathbf{x}} = \mathbf{b} - A\,\widetilde{\mathbf{x}} \quad .$$

Step 4: One determines the solution $\widetilde{\mathbf{p}}_2$ to the least-squares problem (A.19.8) (without constraints)

$$(\widetilde{A}_2\mathbf{p}_2 - \bar{\mathbf{b}})^2 = \min \quad .$$

Step 5: From the results of steps 2 and 4 one finds the solution to (A.19.1) with the constraint (A.19.2)

$$\mathbf{x} = K\begin{pmatrix}\widetilde{\mathbf{p}}_1\\ \widetilde{\mathbf{p}}_2\end{pmatrix} = \begin{pmatrix}K_1\widetilde{\mathbf{p}}_1\\ K_2\widetilde{\mathbf{p}}_2\end{pmatrix} \quad .$$

We will now consider a simple example that illustrates both the least-squares problem with constraints as well as the method of solution given above. Following this we give a computer program for solving the problem.

Example A.5: Least squares with constraints

Suppose the relation (A.19.1) has the simple form

$$r^2 = \mathbf{x}^2 = \min$$

for $n = 2$. One then has $m = n = 2$, $A = I$, and $\mathbf{b} = 0$. Suppose the constraint is

$$x_1 + x_2 = 1 \quad ,$$

i.e., $\ell = 1$, $E = (1, 1)$, $\mathbf{d} = 1$.

The problem has been chosen such that it can be solved by inspection without mathematical complications. The function $z = \mathbf{x}^2 = x_1^2 + x_2^2$ corresponds to a paraboloid in the (x_1, x_2, z) space, whose minimum is at $x_1 = x_2 = 0$. We want, however, to find not the minimum in the entire (x_1, x_2) plane, but rather only on the line $x_1 + x_2 = 1$, as shown in Fig. A.7. It clearly lies at the point where the line has its smallest distance from the origin, i.e., at $x_1 = x_2 = 1/2$.

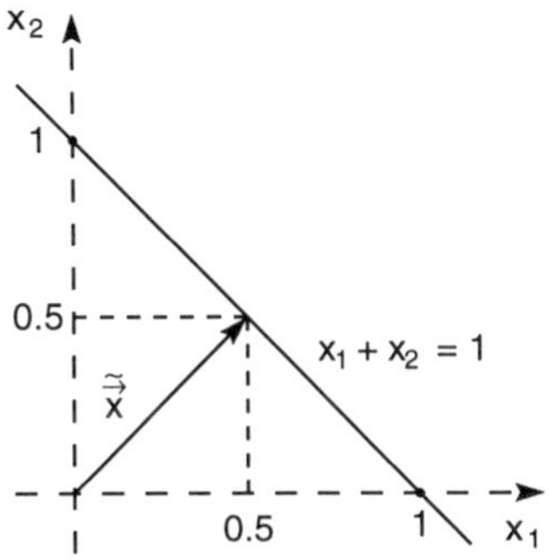

Fig. A.7. The solution $\widetilde{\mathbf{x}}$ to Example A.5 lies on the line given by the constraint $x_1 + x_2 = 1$.

Of course one obtains the same result with the algorithm. With

$$K = \frac{1}{\sqrt{2}} \begin{pmatrix} 1 & -1 \\ 1 & 1 \end{pmatrix}$$

we obtain $\widetilde{E}_1 = \sqrt{2}$, $\widetilde{\mathbf{p}}_1 = 1/\sqrt{2}$,

$$\widetilde{A}_1 = \frac{1}{\sqrt{2}} \begin{pmatrix} 1 \\ 1 \end{pmatrix} \quad , \qquad (\widetilde{A}_2) = \frac{1}{\sqrt{2}} \begin{pmatrix} -1 \\ 1 \end{pmatrix} \quad , \qquad \bar{\mathbf{b}} = -\frac{1}{2} \begin{pmatrix} 1 \\ 1 \end{pmatrix} \quad .$$

We solve the problem $(\widetilde{A}_2 \mathbf{p}_2 - \bar{\mathbf{b}})^2 = \min$ with the normal equations

$$\widetilde{\mathbf{p}}_2 = (\widetilde{A}_2^{\mathrm{T}} \widetilde{A}_2)^{-1} \widetilde{A}_2^{\mathrm{T}} \bar{\mathbf{b}} = (\sqrt{2})^{-1} \cdot 0 = 0 \quad .$$

The full solution is then

$$\mathbf{x} = K \begin{pmatrix} \widetilde{\mathbf{p}}_1 \\ \widetilde{\mathbf{p}}_2 \end{pmatrix} = \frac{1}{\sqrt{2}} \begin{pmatrix} 1 & -1 \\ 1 & 1 \end{pmatrix} \begin{pmatrix} 1/\sqrt{2} \\ 0 \end{pmatrix} = \begin{pmatrix} 1/2 \\ 1/2 \end{pmatrix} \quad . \blacksquare$$

Program A.43: Subroutine `MTXLSC` solves the least-squares problem $(A\mathbf{x} - \mathbf{b})^2 = \min$ with the linear constraint $E\mathbf{x} = \mathbf{d}$

CALLING SEQUENCE: `CALL MTXLSC(A,B,E,D,X,R,A2,M,N,L, FRAC,OK)`

INPUT ARGUMENTS: `A` — $m \times n$ matrix A
`B` — m-vector $\mathbf{b}$
`E` — $\ell \times n$ matrix E
`D` — ℓ-vector $\mathbf{d}$
`A2` — working array, minimum size $m \times (n - \ell)$
`M,N,L` — dimensions m, n, ℓ
`FRAC` — as in Program A.36

OUTPUT ARGUMENTS: `X` — n-vector $\widetilde{\mathbf{x}}$
`R` — value of the minimization function r^2
`OK` — as in Program A.36

```
      SUBROUTINE MTXLSC(A,B,E,D,X,R,A2,M,N,L,FRAC,OK)
      IMPLICIT DOUBLE PRECISION (A-H,O-Z)
      DIMENSION A(M,N),B(M),E(L,N),D(L),X(N)
      LOGICAL OK
      PARAMETER(LMAX=1050,NMAX=1000)
      DIMENSION UP(LMAX),BB(LMAX),A2(M,*),P2(NMAX),S(NMAX),V(NMAX)
      COMMON /DASV02/ UP,BB,P2,S,V
C step 1
      NMINL=N-L
      DO 40 I=1,L
        CALL MTXGRW(E,V,L,N,I)
        CALL MTXHSD(V,UP(I),BB(I),N,I,I+1)
        DO 20 J=I,L
          CALL MTXGRW(E,S,L,N,J)
          CALL MTXHST(V,UP(I),BB(I),S,N,I,I+1)
          IF(J.EQ.I .AND. N.GT.I) THEN
            DO 10 K=I+1,N
              S(K)=V(K)
   10       CONTINUE
          END IF
          CALL MTXPRW(E,S,L,N,J)
   20   CONTINUE
        DO 30 J=1,M
          CALL MTXGRW(A,S,M,N,J)
          CALL MTXHST(V,UP(I),BB(I),S,N,I,I+1)
          CALL MTXPRW(A,S,M,N,J)
   30   CONTINUE
   40 CONTINUE
C step 2
      X(1)=D(1)/E(1,1)
      IF(L.GT.1) THEN
        DO 60 J=2,L
          X(J)=D(J)
          DO 50 K=1,J-1
            X(J)=X(J)-E(J,K)*X(K)
   50     CONTINUE
```

```
          X(J)=X(J)/E(J,J)
   60   CONTINUE
      END IF
C step 3
      DO 80 J=1,M
        DO 70 K=1,L
          B(J)=B(J)-A(J,K)*X(K)
   70   CONTINUE
   80 CONTINUE
C step 4
      L2=1
      CALL MTXGSM(A,A2,M,N,M,NMINL,1,L+1)
      CALL MTXSVD(A2,B,P2,R,M,NMINL,L2,FRAC,OK)
      IF(OK) THEN
        CALL MTXPSM(X,P2,N,1,NMINL,1,L+1,1)
        DO 90 I=L,1,-1
          CALL MTXGRW(E,V,L,N,I)
          CALL MTXHST(V,UP(I),BB(I),X,N,I,I+1)
   90   CONTINUE
      END IF
      END
```

A.20 Example Programs for Simple Matrix and Vector Operations

The source files of the example programs are available in the subdirectory `\DATAN\EXASRC`.

A.20.1 Main Program E1MTX to Demonstrate Simple Operations of Matrix and Vector Algebra with Subprograms MTXTRA, MTXWRT, MTXADD, MTXSUB, MTXMLT, MTXMBT, MTXMAT, MTXUNT, MTXZER, MTXMSC, MTXTRP, MTXCPV, MTXADV, MTXSBV, MTXDOT, MTXMSV, MTXZRV

The matrices

$$A = \begin{pmatrix} 1 & 2 & 3 \\ 2 & 1 & 3 \end{pmatrix} , \qquad B = \begin{pmatrix} 2 & 3 & 1 \\ 1 & 5 & 4 \end{pmatrix} ,$$

$$C = \begin{pmatrix} 1 & 5 \\ 3 & 4 \\ 2 & 3 \end{pmatrix}$$

and the vectors

$$\mathbf{u} = \begin{pmatrix} 0 \\ 3 \\ 4 \end{pmatrix} , \qquad \mathbf{v} = \begin{pmatrix} 3 \\ 1 \\ 2 \end{pmatrix} , \qquad \mathbf{w} = \begin{pmatrix} 5 \\ 2 \end{pmatrix}$$

are made available in DATA statements. With suitable subprograms simple operations are performed with these quantities. The resulting matrices and vectors are output by MTXWRT. The operations are

$$\begin{aligned}
R &= A \quad ,\\
R &= A+B \quad ,\\
R &= A-B \quad ,\\
R &= AC \quad ,\\
S &= AB^{\mathrm{T}} \quad ,\\
T &= A^{\mathrm{T}}B \quad ,\\
R &= I \quad ,\\
R &= 0.5\,A \quad ,\\
R &= A^{\mathrm{T}} \quad ,\\
\mathbf{z} &= \mathbf{w} \quad ,\\
\mathbf{x} &= \mathbf{u}+\mathbf{v} \quad ,\\
\mathbf{x} &= \mathbf{u}-\mathbf{v} \quad ,\\
d &= \mathbf{u}\cdot\mathbf{v} \quad ,\\
s &= |\mathbf{u}| \quad ,\\
\mathbf{x} &= 0.5\,\mathbf{u} \quad ,\\
\mathbf{x} &= 0 \quad .
\end{aligned}$$

A.20.2 Main Program E2MTX to Demonstrate the Manipulation of Submatrices and Subvectors with Subprograms MTXGSM, MTXPSM, MTXGCL, MTXPCL, MTXGRW, MTXPRW, MTXGSV, MTXPSV

The matrices

$$A = \begin{pmatrix} 1 & 2 & 3 \\ 2 & 1 & 3 \end{pmatrix} \quad , \qquad D = \begin{pmatrix} 0 & 2 \\ 1 & 3 \end{pmatrix}$$

and the vectors

$$\mathbf{u} = \begin{pmatrix} 0 \\ 3 \\ 4 \end{pmatrix} \quad , \qquad \mathbf{w} = \begin{pmatrix} 5 \\ 2 \end{pmatrix}$$

are defined in DATA statements.

Next a submatrix S is taken from A, and a submatrix of A is overwritten by D. A column vector and a row vector are taken from A and inserted into A. Finally some elements are taken according to a list from the vector **u** and assembled in a vector **z**, and elements of **w** are put into positions (defined by list) of the vector **u**.

A.21 Example Programs Related to Orthogonal Transformations

A.21.1 Main Program E3MTX to Demonstrate the Givens Transformation with Subprograms MTXGVD, MTXGVT, and MTXGVA

First the two vectors

$$\mathbf{u} = \begin{pmatrix} 3 \\ 4 \end{pmatrix} \quad , \qquad \mathbf{w} = \begin{pmatrix} 1 \\ 1 \end{pmatrix}$$

are defined in a DATA statement. Next, by the use of MTXGVD transformation parameters c and s for the vector $\mathbf{u}$ are computed and displayed. The Givens transformation of $\mathbf{u}$ with these parameters is performed by MTXGVT yielding

$$\mathbf{u}' = \begin{pmatrix} 5 \\ 0 \end{pmatrix} \quad .$$

Finally, by calling MTXGVA the parameters c and s are computed for the vector $\mathbf{w}$ and the transformation is applied to this vector.

A.21.2 Main Program E4MTX to Demonstrate the Householder Transformation with Subprograms MTXHSD and MTXHST

The two vectors

$$\mathbf{v} = \begin{pmatrix} 1 \\ 2 \\ 0 \\ 4 \\ 3 \\ 4 \end{pmatrix} \quad , \qquad \mathbf{c} = \begin{pmatrix} 1 \\ 2 \\ 0 \\ 4 \\ 3 \\ 4 \end{pmatrix}$$

are defined in DATA statements. Moreover, the indices n, p, and ℓ are set to $n = 6$, $p = 3$, $\ell = 5$. By calling MTXHSD the Householder transformation defined by these indices and by the vector $\mathbf{v}$ is initialized. The application of this transformation to the vectors $\mathbf{v}$ and $\mathbf{c}$ is performed by two calls of MTXHST. The results are displayed alphanumerically.

A.22 Example Programs Solving Matrix Equations

A.22.1 Main Program E5MTX to Demonstrate the Gaussian Algorithm with Subprogram MTXEQU

The matrix

$$A = \begin{pmatrix} 1 & 2 & 3 \\ 2 & 1 & -2 \\ 1 & 1 & 2 \end{pmatrix}$$

and the 3×3 unit matrix B are defined in DATA statements. A call of MTXEQU solves the matrix equation $AX = B$, i.e., X is the inverse of A. The matrix A is identical to the one chosen in Example A.1. In that example the algorithm is shown step by step.

A.22.2 Main Program E6MTX to Demonstrate Cholesky Decomposition and Cholesky Inversion with Subprograms MTXCHL, MTXCHI, and MTXCHM

First, the matrices

$$B = \begin{pmatrix} 1 & 7 & 13 \\ 3 & 9 & 17 \\ 5 & 11 & 19 \end{pmatrix} \quad , \qquad C = \begin{pmatrix} 1 & 1 \\ 1 & 1 \\ 1 & 1 \end{pmatrix}$$

are defined in DATA statements and the symmetric, positive-definite matrix $A = B^{\mathrm{T}}B$ is constructed. By calling MTXCHL the Cholesky decomposition $A = U^{\mathrm{T}}U$ is performed and the triangular matrix U is displayed. Multiplication of U^{T} by U yields in fact $R = U^{\mathrm{T}}U = A$. The subprogram MTXCHM is then used to compute $R = UC$. Finally, by Cholesky inversion with MTXCHI, the inverse A^{-1} of A is computed. Multiplication with the original matrix A yields $R = AA^{-1} = I$.

A.22.3 Main Program E7MTX to Demonstrate the Singular Value Decomposition with Subprogram MTXSVD

This short program is nearly identical to E5MTX (cf. Sect. A.22.1). The matrix inversion, however, is now done with the help of MTXSVD. In addition to the inverse X of the 3×3 matrix A, the vector of residues $\mathbf{r}$ is computed and displayed. Since A has full rank one has $\mathbf{r} = 0$.

In the framework of our programs, in particular in the solution of problems of least squares or of minimization, matrix equations are almost always solved by MTXSVD. For a demonstration of the properties of the singular value decomposition with orthogonal transformations, however, the subprogram MTXDEC is more suitable. This program is demonstrated in the next section for nine different cases, all of which can also be solved by MTXSVD.

A.22.4 Main Program E8MTX to Demonstrate the Solution of Matrix Equations in 9 Different Cases by Singular Value Decomposition with Subprogram MTXDEC

In Sect. A.6 we have listed the different cases of the matrix equation $A\mathbf{x} \approx \mathbf{b}$. If A is an $m \times n$ matrix then first we must distinguish between the cases $m = n$ (case 1), $m > n$ (case 2), and $m < n$ (case 3). A further subdivision is brought about by the rank of A. If the rank k of A is $k = \min(m, n)$ then we have the case 1a (or 2a or 3a). If, however, $k < \min(m, n)$ then we are dealing with case 1b (or 2b or 3b). The rank of a matrix is equal to its number of non-vanishing singular values. In numerical calculations, which are always performed with finite accuracy, one obviously has to define more precisely the meaning of "non-vanishing". For this definition we use the method of singular value analysis (Sect. A.14) and set a singular value equal to zero if it is smaller than a fraction f of the largest singular value. The number of finite singular values remaining in this analysis is called the *pseudorank*. In addition to the cases mentioned above we consider as well the cases 1c, 2c, and 3c, in which the matrix A has full rank but not full pseudorank.

The program consists of two nested loops. The outer loop runs through the cases 1, 2, 3, the inner loop through the subcases a, b, c. For each case the matrix A is composed of individual vectors. In the subcase b two of these vectors are chosen to be identical. In the subcase c they are identical except for one element, which in one vector differs by ε compared to the value in the other vector. In case 3 the system of linear equations symbolized by the matrix equation $Ax = b$ has less equations (m) than unknowns (n). This case does not appear in practice and, therefore, is not included in the programs MTXSVD and MTXDEC. It is simulated here in the following way. In case 3 with $m = 2$ and $n = 3$, the matrix A is extended to become a 3×3 matrix by addition of another row the elements, of which are all set to zero, and correspondingly m is set to 3.

For each case, by calling MTXDEC the singular value decomposition of A is performed, $A = USV^{\mathrm{T}}$. The original matrix A, its pseudo-inverse X, the vector $\mathbf{r}$ of residues, the vector $\mathbf{d}$ of the diagonal elements of S, i.e., the vector of the ordered singular values, and the transformation matrices U and V are displayed. As a check of the decomposition, the matrix product USV^{T} is computed as well; this has to be equal to the original matrix A. If the singular value analysis shows that one or several singular values are smaller than the fraction f of the largest singular value then they are set to zero. The product matrix USV^{T} is computed with the correspondingly modified diagonal matrix S.

Studying the displayed quantities for the different cases in detail one easily verifies the statements of Table A.1 about the residues for the different cases. (Ignore the third component of $\mathbf{r}$ in case 3 which would not appear without the added row of A.)

In MTXSVD and MTXDEC, the factor f for the singular value analysis (FORTRAN name FRACT) is assumed to be $f = 10^{-15}$ if the input is set to zero by the user. In this case in our example, the cases 1c, 2c, 3c would be solved as if the matrix A had full rank despite the small value of $\varepsilon = 10^{-12}$. In our example program, however, the singular value analysis reduces the number of singular values since we have set $f = 10^{-10}$.

The numbers mentioned above indicate the high numerical stability of the singular value analysis.

A.22.5 Main Program E9MTX to Demonstrate Subprogram MTXMAR

The program MTXMAR will rarely be used directly. It was written as a subprogram for LSQMAR and MINMAR. For completeness we give here a short main program using it. The main program solves the problem $A\mathbf{x} \approx \mathbf{b}$ modified according to (A.18.3) for $\lambda_1 = \lambda$ and $\lambda_2 = \lambda/10$ for given A, $\mathbf{b}$, and λ and displays the results $\mathbf{x}_1$ and $\mathbf{x}_2$.

A.22.6 Main Program E10MTX to Demonstrate the Solution of a Least-Squares Problem with Constraints by Subprogram MTXLSC

The problem of Examples 9.9 and 9.10 is solved, i.e., the measurement $x_1 = 89$, $x_2 = 31$, $x_3 = 62$ of the three angles of a triangle and the evaluation of these measurements using the constraint $x_1 + x_2 + x_3 = 180$. The evaluation requires the solution of $(A\mathbf{x} - \mathbf{b})^2 = \min$ with the constraint $E\mathbf{x} = \mathbf{d}$ with

$$A = \begin{pmatrix} 1 & 0 & 0 \\ 0 & 1 & 0 \\ 0 & 0 & 1 \end{pmatrix} , \qquad \mathbf{b} = \begin{pmatrix} 89 \\ 31 \\ 62 \end{pmatrix} , \qquad E = (1,1,1) , \qquad \mathbf{d} = 180 .$$

In the program the matrices and vectors A, $\mathbf{b}$, E, $\mathbf{d}$ are provided. The solution is computed by calling MTXLSC. It is, of course, identical to the results found in the previously mentioned examples.

B. Combinatorics

Consider n distinguishable objects a_1, a_2, ..., a_n. We ask for the number of possible ways P_n^k, in which one can place k of them in a given order. Such orderings are called *permutations*. For the example $n = 4$, $k = 2$ these permutations are

$$\begin{array}{cccccc} a_1a_2 & , & a_1a_3 & , & a_1a_4 & , \\ a_2a_1 & , & a_2a_3 & , & a_2a_4 & , \\ a_3a_1 & , & a_3a_2 & , & a_3a_4 & , \\ a_4a_1 & , & a_4a_2 & , & a_4a_3 & , \end{array}$$

i.e., $P_k^n = 12$. The answer for the general problem can be derived from the following scheme. There are n different possible ways to occupy the first place in a sequence. When one of these ways has been chosen, however, there are only $n - 1$ objects left, i.e., there remain $n - 1$ ways to occupy the second place, and so forth. One therefore has

$$P_k^n = n(n-1)(n-2)\cdots(n-k+1) \quad . \tag{B.1}$$

The result can also be written in the form

$$P_k^n = \frac{n!}{(n-k)!} \quad , \tag{B.2}$$

where

$$n! = 1 \cdot 2 \cdots n \quad ; \qquad 0! = 1 \quad , \qquad 1! = 1 \quad . \tag{B.3}$$

Often one is not interested in the order of the k objects within a permutation (the same k objects can be arranged in $k!$ different ways within the sequence), but rather one only considers the number of different ways of choosing of k objects out of a total of n. Such a choice is called a *combination*. The number of possible combinations of k elements out of n is then

$$C_k^n = \frac{P_k^n}{k!} = \frac{n!}{k!(n-k)!} = \binom{n}{k} \quad . \tag{B.4}$$

For the *binomial coefficients* $\binom{n}{k}$ one has the simple recursion relation

n \ k	0	1	2	3	4	5	6	7	8
0	1								
1	1	1							
2	1	2	1						
3	1	3	3	1					
4	1	4	6	4	1				
5	1	5	10	10	5	1			
6	1	6	15	20	15	6	1		
7	1	7	21	35	35	21	7	1	
8	1	8	28	56	70	56	28	8	1
⋮					⋮				

Fig. B.1. Pascal's triangle.

$$\binom{n-1}{k} + \binom{n-1}{k-1} = \binom{n}{k} \quad , \tag{B.5}$$

which can easily be proven by computation:

$$\begin{aligned} & \frac{(n-1)!}{k!(n-k-1)!} + \frac{(n-1)!}{(k-1)!(n-k)!} \\ = & \frac{(n-k)(n-1)! + k(n-1)!}{k!(n-k)!} \\ = & \frac{n!}{k!(n-k)!} \quad . \end{aligned}$$

The recursion formula is the basis for the famous *Pascal's triangle*, which, because of its beauty, is shown in Fig. B.1. The name "binomial coefficient" comes from the well-known *binomial theorem*,

$$(a+b)^n = \sum_{k=0}^{n} \binom{n}{k} a^k b^{n-k} \quad , \tag{B.6}$$

the proof of which (by induction) is left to the reader. We use the theorem in order to derive a very important property of the coefficient $\binom{n}{k}$. For this we write it in the simple form for $b = 1$, i.e.,

$$(a+1)^n = \sum_{k=0}^{n} \binom{n}{k} a^k$$

and we then apply it a second time,

$$(a+1)^{n+m} = (a+1)^n(a+1)^m \quad ,$$

$$\sum_{\ell=0}^{n+m} \binom{n+m}{\ell} a^\ell = \sum_{j=0}^{n} \binom{n}{j} a^j \sum_{k=0}^{m} \binom{m}{k} a^k \quad .$$

If we consider only the term with a^ℓ, then by comparing coefficients we find

$$\binom{n+m}{\ell} = \sum_{j=0}^{\ell} \binom{n}{j} \binom{m}{\ell-j} \quad . \tag{B.7}$$

C. Formulas and Programs for Statistical Functions

C.1 Binomial Distribution

We present here two function subprograms for computing the binomial distribution (5.1.3)

$$W_k^n = \binom{n}{k} p^k (1-p)^{n-k} \tag{C.1.1}$$

and the distribution function

$$P(k < K) = \sum_{k=0}^{K-1} W_k^n \quad . \tag{C.1.2}$$

For reasons of numerical stability the logarithm of Euler's gamma function (Program D.2) is used in the computation.

Program C.1: Function SDBINM computes probability of the Binomial distribution

CALLING SEQUENCE: F=SDBINM(K,N,P)

INPUT ARGUMENTS: K,N,P — k, n, p from (C.1.1)

OUTPUT ARGUMENT: function value SDBINM — probability (C.1.1)

```
      DOUBLE PRECISION FUNCTION SDBINM(K,N,P)
      IMPLICIT DOUBLE PRECISION (A-H,O-Z)
      PARAMETER(ZERO=0.D0,ONE=1.D0,SMALL=.0001D0,BIG=100D0)
      AK=DBLE(K)
      AN=DBLE(N)
      IF(N.LT.1 .OR. K.LT.0 .OR. K.GT.N) THEN
        SDBINM=ZERO
      ELSE
        ALBINC=GLNGAM(AN+ONE)-GLNGAM(AK+ONE)-GLNGAM(AN-AK+ONE)
        IF(P.LE.ZERO) THEN
          IF(K.EQ.0) THEN
            SDBINM=ONE
```

```
          ELSE
            SDBINM=ZERO
          END IF
        ELSE IF(P.GE.ONE) THEN
          IF(K.EQ.N) THEN
            SDBINM=ONE
          ELSE
            SDBINM=ZERO
          END IF
        ELSE
          ALFACT=LOG(P)*AK+LOG(ONE-P)*(AN-AK)
          ARG=ALFACT+ALBINC
          ABSARG=ABS(ARG)
          IF(ABSARG.LT.SMALL) THEN
            SDBINM=ONE
          ELSE IF(ABSARG.GT.BIG) THEN
            SDBINM=ZERO
          ELSE
            SDBINM=EXP(ARG)
          END IF
        END IF
      END IF
      END
```

Program C.2: Function SCBINM computes the distribution function of the binomial distribution

CALLING SEQUENCE: F=SCBINM(K,N,P)

INPUT ARGUMENTS: K,N,P — K, n, p from (C.1.2)

OUTPUT ARGUMENT: function value SCBINM — distribution function (C.1.2)

```
      DOUBLE PRECISION FUNCTION SCBINM(K,N,P)
      IMPLICIT DOUBLE PRECISION(A-H,O-Z)
      S=DBLE(0.)
      IF(K.GT.0) THEN
        DO 10 J=0,K-1
          S=S+SDBINM(J,N,P)
   10   CONTINUE
      END IF
      SCBINM=S
      END
```

C.2 Hypergeometric Distribution

The hypergeometric distribution (5.3.1)

$$W_k = \binom{K}{k}\binom{N-K}{n-k} \Big/ \binom{N}{n} \quad , \qquad n \le N, k \le K \quad , \qquad \text{(C.2.1)}$$

and the corresponding distribution function

$$P(k < k') = \sum_{k=0}^{k'-1} W_k \qquad \text{(C.2.2)}$$

are computed by the two following programs.

Program C.3: Function SDHYPG computes probability of the hypergeometric distribution

CALLING SEQUENCE: F=SDHYPG(K,N,KK,NN)

INPUT ARGUMENTS: K,N,KK,NN — k, n, K, N from (C.2.1)

OUTPUT ARGUMENT: function value SDHYPG — probability (C.2.1)

```
      DOUBLE PRECISION FUNCTION SDHYPG(K,N,KK,NN)
      IMPLICIT DOUBLE PRECISION(A-H,O-Z)
      PARAMETER(ZERO=0.D0,ONE=1.D0)
      AN=DBLE(N)
      ANN=DBLE(NN)
      AK=DBLE(K)
      AKK=DBLE(KK)
      AL=AN-AK
      ALL=ANN-AKK
      IF(AK.LE.AKK .AND. AN.LE.ANN .AND. AL.LE.ALL) THEN
          A=GLNGAM(AKK+ONE)-GLNGAM(AK+ONE)-GLNGAM(AKK-AK+ONE)
          B=GLNGAM(ANN+ONE)-GLNGAM(AN+ONE)-GLNGAM(ANN-AN+ONE)
          C=GLNGAM(ALL+ONE)-GLNGAM(AL+ONE)-GLNGAM(ALL-AL+ONE)
          SDHYPG=EXP(A+C-B)
      ELSE
          SDHYPG=ZERO
      END IF
      END
```

Program C.4: Function SCHYPG computes the distribution function of the hypergeometric distribution

CALLING SEQUENCE: F=SCHYPG(K,N,KK,NN)

INPUT ARGUMENTS: K,N,KK,NN — k', n, K, N from (C.2.1)

OUTPUT ARGUMENT: function value SCHYPG — distribution function (C.2.2)

```
      DOUBLE PRECISION FUNCTION SCHYPG(K,N,KK,NN)
      IMPLICIT DOUBLE PRECISION(A-H,O-Z)
      PARAMETER(ZERO=0.D0)
      SCHYPG = ZERO
      IF(K.GT.0) THEN
        DO 10 J=0,K-1
          SCHYPG = SCHYPG + SDHYPG(J,N,KK,NN)
   10   CONTINUE
      END IF
      END
```

C.3 Poisson Distribution

The Poisson distribution (5.4.1)

$$f(k;\lambda) = \frac{\lambda^k}{k!}\mathrm{e}^{-\lambda} \tag{C.3.1}$$

and the corresponding distribution function

$$P(k < K) = \sum_{k=0}^{K-1} f(k;\lambda) \tag{C.3.2}$$

are computed by the following two programs.

Program C.5: Function `SDPOIS` computes the probability of the Poisson distribution

CALLING SEQUENCE: `F=SDPOIS(K,ALAMBD)`

INPUT ARGUMENTS: `K,ALAMBD` — k, λ from (C.3.1)

OUTPUT ARGUMENT: function value `SDPOIS` — probability (C.3.1)

```
      DOUBLE PRECISION FUNCTION SDPOIS(K,ALAMBD)
      IMPLICIT DOUBLE PRECISION(A-H,O-Z)
      SDPOIS=EXP(-ALAMBD)
      IF (K.GT.0) THEN
        DO 10 J=1,K
          SDPOIS=SDPOIS*ALAMBD/DBLE(J)
   10   CONTINUE
      END IF
      END
```

Program C.6: Function `SCPOIS` computes distribution function of the Poisson distribution

CALLING SEQUENCE: `F=SCPOIS(K,ALAMBD)`

INPUT ARGUMENTS: `K,ALAMBD` — K, λ from (C.3.2) and (C.3.1)

OUTPUT ARGUMENT: function value `SCPOIS` — distribution function (C.3.2)

```
      DOUBLE PRECISION FUNCTION SCPOIS(K,ALAMBD)
      IMPLICIT DOUBLE PRECISION(A-H,O-Z)
      PARAMETER(ZERO=0.D0)
      SCPOIS = ZERO
      IF(K.GT.0) THEN
        DO 10 J=0,K-1
          SCPOIS=SCPOIS+SDPOIS(J,ALAMBD)
   10   CONTINUE
      END IF
      END
```

The quantities $f(k;\lambda)$ and $F(K;\lambda)$ depend not only on the values of the discrete variables k and K, but also on the continuous parameter λ. For a given P there is a certain parameter value λ_P that fulfills Eq. (C.3.2). We can denote this as the quantile

$$\lambda = \lambda_P(K) \tag{C.3.3}$$

of the Poisson distribution.

Program C.7: Function SQPOIS computes a quantile of the Poisson distribution

CALLING SEQUENCE: F=SQPOIS(K,P)

INPUT ARGUMENTS: K — K
P — probability P

OUTPUT ARGUMENT: function value SQPOIS — quantile λ_P; cf. (C.3.3)

```
      DOUBLE PRECISION FUNCTION SQPOIS(K,P)
      IMPLICIT DOUBLE PRECISION(A-H,O-Z)
      PARAMETER(ZERO=0.D0,ONE=1.D0,BIG=1.D10,EPSILN=1.D-8)
      EXTERNAL SZPOIS
C boundary of range
      IF(P.GE.ONE) SQPOIS=BIG
      IF(P.LE.ZERO) SQPOIS=ZERO
C normal range
      IF(P.LT.ONE .AND. P.GT.ZERO) THEN
        X0=ZERO
        X1=DBLE(K)
        CALL AUXZBR(X0,X1,SZPOIS,P,K,0)
        CALL AUXZFN(X0,X1,XZERO,SZPOIS,P,K,0,EPSILN)
        SQPOIS=XZERO
      END IF
      END
C-----------------------------------------------------------------
      DOUBLE PRECISION FUNCTION SZPOIS(ALAMBD,P,K,NDUM2)
C returns P minus cumulative Poisson
      IMPLICIT DOUBLE PRECISION(A-H,O-Z)
      SZPOIS=P-SCPOIS(K,ALAMBD)
      END
```

C.4 Normal Distribution

The probability density of the *standard normal distribution* is

$$\phi_0(x) = \frac{1}{\sqrt{2\pi}} \exp(-x^2/2) \quad . \tag{C.4.1}$$

Program C.8: Function SDSTNR computes the probability density of the standard normal distribution

CALLING SEQUENCE: F=SDSTNR(X)

INPUT ARGUMENT: X — x from (C.4.1)

OUTPUT ARGUMENT: function value SDSTNR — probability density (C.4.1)

```
DOUBLE PRECISION FUNCTION SDSTNR(X)
IMPLICIT DOUBLE PRECISION(A-H,O-Z)
PARAMETER (F=0.39894228D0,HALF=0.5D0, BIG=500.D0)
ARG=-HALF*X**2
IF(ABS(ARG).GE.BIG) THEN
  SDSTNR=DBLE(0.)
ELSE
  SDSTNR=F*EXP(ARG)
END IF
END
```

The *normal distribution* with mean x_0 and variance σ^2,

$$\phi(x) = \frac{1}{\sqrt{2\pi}\sigma} \exp\left(-\frac{(x-x_0)^2}{2\sigma^2}\right) \quad , \tag{C.4.2}$$

can easily be expressed in terms of the standardized variable

$$u = \frac{x - x_0}{\sigma} \tag{C.4.3}$$

using (C.4.1) to be

$$\phi(x) = \frac{1}{\sigma}\phi_0(u) \quad . \tag{C.4.4}$$

Program C.9: Function SDNORM computes the probability density of the normal distribution

CALLING SEQUENCE: F=SDNORM(X,X0,SIGMA)

INPUT ARGUMENTS: X — independent variable x from (C.4.2)
X0,SIGMA — mean x_0 and width σ from (C.4.2)

OUTPUT ARGUMENT: function value SDNORM — probability density (C.4.2)

```
DOUBLE PRECISION FUNCTION SDNORM(X,X0,SIGMA)
IMPLICIT DOUBLE PRECISION(A-H,O-Z)
PARAMETER(ONE=1.D0)
S1=ONE/SIGMA
U=(X-X0)*S1
SDNORM=S1*SDSTNR(U)
END
```

The distribution function of the standard normal distribution

$$\psi_0(x) = \int_{-\infty}^{x} \phi_0(x)\,\mathrm{d}x = \frac{1}{\sqrt{2\pi}} \int_{-\infty}^{x} \exp\left(-\frac{x^2}{2}\right)\,\mathrm{d}x \tag{C.4.5}$$

is an integral that cannot be computed in closed form. We can relate it, however, to the incomplete gamma function described in Sect. D.5.

The distribution function of a normal distribution with mean x_0 and variance σ^2 is found from (C.4.5) to be

$$\psi(x) = \psi_0(u) \quad , \qquad u = \frac{x - x_0}{\sigma} \quad . \tag{C.4.6}$$

We now introduce the *error function*

$$\mathrm{erf}(x) = \frac{2}{\sqrt{\pi}} \int_0^x \mathrm{e}^{-t^2}\,\mathrm{d}t \quad , \qquad x \geq 0 \quad . \tag{C.4.7}$$

Comparing with the definition of the incomplete gamma function (D.5.1) gives

$$\mathrm{erf}(x) = \frac{2}{\Gamma(\frac{1}{2})} \int_{t=0}^{t=x} \mathrm{e}^{-t^2}\,\mathrm{d}t = \frac{1}{\Gamma(\frac{1}{2})} \int_{u=0}^{u=x^2} \mathrm{e}^{-u} u^{-1/2}\,\mathrm{d}u \quad ,$$

$$\mathrm{erf}(x) = P\left(\frac{1}{2}, x^2\right) \quad . \tag{C.4.8}$$

On the other hand there is a more direct connection between (C.4.6) and (C.4.7),

$$\psi_{x_0=0,\sigma=1/\sqrt{2}}(x) = \psi_0(u = \sqrt{2}x) = \frac{1}{2}[1 + \mathrm{sign}(x)\mathrm{erf}(|x|)]$$

or

$$\psi_0(u) = \frac{1}{2}\left[1 + \mathrm{sign}(u)\mathrm{erf}\left(\frac{|u|}{\sqrt{2}}\right)\right] \quad ,$$

i.e.,

$$\psi_0(u) = \frac{1}{2}\left[1 + \mathrm{sign}(u) P\left(\frac{1}{2}, \frac{u^2}{2}\right)\right] \quad . \tag{C.4.9}$$

Program C.10: Function SCSTNR computes the distribution function of the standard normal distribution

CALLING SEQUENCE: F=SCSTNR(X)

INPUT ARGUMENT: X — x from (C.4.5)

OUTPUT ARGUMENT: function value SCSTNR — distribution function (C.4.5)

```
DOUBLE PRECISION FUNCTION SCSTNR(X)
IMPLICIT DOUBLE PRECISION(A-H,O-Z)
PARAMETER (ONE=1.D0, ZERO=0.D0, HALF=0.5D0)
ARG=HALF*(X**2)
S=ONE
IF(X.LT.ZERO) S=-ONE
F=GINCGM(HALF,ARG)
SCSTNR=HALF*(ONE+S*F)
END
```

Program C.11: Function SCNORM computes the distribution function of the normal distribution

CALLING SEQUENCE: F=SCNORM(X,X0,SIGMA)

INPUT ARGUMENTS: X — independent variable x from (C.4.6)
X0, SIGMA — mean x_0 and width σ from (C.4.6)

OUTPUT ARGUMENT: function value SCNORM — distribution function (C.4.6)

```
DOUBLE PRECISION FUNCTION SCNORM(X,X0,SIGMA)
IMPLICIT DOUBLE PRECISION(A-H,O-Z)
U=(X-X0)/SIGMA
SCNORM=SCSTNR(U)
END
```

Finally we compute the quantiles of the standard normal distribution. For a given probability P, the quantile x_p is defined by the relation

$$P = \psi_0(x_p) = \frac{1}{\sqrt{2\pi}} \int_{-\infty}^{x_p} \mathrm{e}^{-x^2/2}\,\mathrm{d}x \quad . \tag{C.4.10}$$

We determine this by finding the zero of the function

$$k(x, P) = P - \psi_0(x) \tag{C.4.11}$$

using the procedure of Sect. E.2.

Program C.12: Function SQSTNR computes a quantile of the standard normal distribution

CALLING SEQUENCE: F=SQSTNR(P)

INPUT ARGUMENT: P — probability P

OUTPUT ARGUMENT: function value SCSTNR — quantile x_P

```
      DOUBLE PRECISION FUNCTION SQSTNR(P)
      IMPLICIT DOUBLE PRECISION(A-H,O-Z)
      PARAMETER(BIG=1.D10,EPSILN=1D-8,ONE=1.D0,ZERO=0.D0,HALF=.5D0)
      EXTERNAL SZSTNR
C boundary of range
      IF(P.GE.ONE) SQSTNR=BIG
      IF(P.LE.ZERO) SQSTNR=-BIG
C normal range
      IF(P.LT.ONE .AND. P.GT.ZERO) THEN
        X0=ZERO
        X1=DBLE(0.1)
        CALL AUXZBR(X0,X1,SZSTNR,P,0,0)
        CALL AUXZFN(X0,X1,XZERO,SZSTNR,P,0,0,EPSILN)
        SQSTNR=XZERO
      END IF
      END
C-----------------------------------------------------------------
      DOUBLE PRECISION FUNCTION SZSTNR(X,P,NDUM1,NDUM2)
C returns P minus cumulative standardized normal of X
      IMPLICIT DOUBLE PRECISION(A-H,O-Z)
      SZSTNR=P-SCSTNR(X)
      END
```

For the quantile x_P for a probability P of the normal distribution with mean x_0 and standard deviation σ one has

$$P = \psi_0(u_P) \quad , \qquad x_P = x_0 + \sigma u_P \quad . \tag{C.4.12}$$

Program C.13: Function SQNORM computes a quantile of the normal distribution

CALLING SEQUENCE: F=SQNORM(P,X0,SIGMA)

INPUT ARGUMENTS: P — probability P
X0,SIGMA — mean x_0 and width σ

OUTPUT ARGUMENT: function value SQNORM — quantile x_P

```
DOUBLE PRECISION FUNCTION SQNORM(P,X0,SIGMA)
IMPLICIT DOUBLE PRECISION(A-H,O-Z)
U=SQSTNR(P)
SQNORM=U*SIGMA+X0
END
```

C.5 χ^2-Distribution

The probability density (6.6.10) of the χ^2-distribution for n degrees of freedom,

$$f(\chi^2) = \frac{1}{2^\lambda \Gamma(\lambda)} (\chi^2)^{\lambda-1} \mathrm{e}^{-\frac{1}{2}\chi^2} \quad , \qquad \lambda = \frac{n}{2} \quad , \tag{C.5.1}$$

is computed by the following program.

Program C.14: Function `SDCHI2` computes the probability density of the χ^2-distribution

CALLING SEQUENCE: `F=SDCHI2(X,N)`

INPUT ARGUMENTS: `X` — independent variable χ^2 from (C.5.1)
`N` — number of degrees of freedom n

OUTPUT ARGUMENT: function value `SDCHI2` — probability density (C.5.1)

```
DOUBLE PRECISION FUNCTION SDCHI2(X,N)
IMPLICIT DOUBLE PRECISION(A-H,O-Z)
PARAMETER (ONE=1.D0, ZERO=0.D0, HALF=0.5D0, TWO=2.D0)
IF(X.NE.ZERO .OR. N.GT.2) THEN
  B=GGAMMA(HALF*DBLE(N))*SQRT(TWO**N)
  C=SQRT(X**(N-2))
  E=EXP(-X*HALF)
  SDCHI2=E*C/B
ELSE
  IF(N.EQ.2) SDCHI2=HALF
  IF(N.EQ.1) SDCHI2=1.D10
END IF
END
```

The distribution function

$$\begin{aligned} F(\chi^2) &= \frac{1}{\Gamma(\lambda)} \int_{u=0}^{u=\chi^2} \frac{1}{2} \left(\frac{u}{2}\right)^{\lambda-1} \mathrm{e}^{-\frac{1}{2}u}\, \mathrm{d}u \\ &= \frac{1}{\Gamma(\lambda)} \int_{t=0}^{t=\chi^2/2} \mathrm{e}^{-t} t^{\lambda-1}\, \mathrm{d}t \end{aligned} \tag{C.5.2}$$

is seen from (D.5.1) to be an incomplete gamma function

$$F(\chi^2) = P\left(\lambda, \frac{\chi^2}{2}\right) = P\left(\frac{n}{2}, \frac{\chi^2}{2}\right) \quad . \tag{C.5.3}$$

Program C.15: Function SCCHI2 computes the distribution function of the χ^2-distribution

CALLING SEQUENCE: F=SCCHI2(X,N)

INPUT ARGUMENTS: X — independent variable χ^2
N — number of degrees of freedom n

OUTPUT ARGUMENT: function value SCCHI2 — distribution function (C.5.2)

```
      DOUBLE PRECISION FUNCTION SCCHI2(X,N)
      IMPLICIT DOUBLE PRECISION(A-H,O-Z)
      PARAMETER (HALF=0.5D0)
      A=HALF*DBLE(N)
      SCCHI2=GINCGM(A,HALF*X)
      END
```

The quantile χ^2_P of the χ^2-distribution for a given probability P, which is given by

$$h(\chi^2_P) = P - F(\chi^2_P) = 0 \quad , \tag{C.5.4}$$

is computed as the zero of the function $h(\chi^2)$.

Program C.16: Function SQCHI2 computes a quantile of the χ^2-distribution

CALLING SEQUENCE: F=SQCHI2(P,N)

INPUT ARGUMENTS: P — probability P
N — number of degrees of freedom n

OUTPUT ARGUMENT: function value SQCHI2 — quantile χ^2_P

```
      DOUBLE PRECISION FUNCTION SQCHI2(P,N)
      IMPLICIT DOUBLE PRECISION(A-H,O-Z)
      PARAMETER (BIG=1D10,ONE=1.D0, ZERO=0.D0)
      EXTERNAL SZCHI2
C boundary of range
      IF(P.GE.ONE) SQCHI2=BIG
      IF(P.LE.ZERO) SQCHI2=ZERO
C normal range
      IF(P.LT.ONE .AND. P.GT.ZERO) THEN
        X1=DBLE(N)
        X0=DBLE(0.5)*X1
        CALL AUXZBR(X0,X1,SZCHI2,P,N,0)
        CALL AUXZFN(X0,X1,XZERO,SZCHI2,P,N,0,EPSILN)
        SQCHI2=XZERO
      END IF
      END
C-----------------------------------------------------------------
      DOUBLE PRECISION FUNCTION SZCHI2(X,P,N,NDUM)
C returns P minus cumulative chisquared distribution of (X,N)
      IMPLICIT DOUBLE PRECISION(A-H,O-Z)
      SZCHI2=P-SCCHI2(X,N)
      END
```

C.6 F-Distribution

The probability density (8.2.3) of the F-distribution with f_1 and f_2 degrees of freedom,

$$f(F) = \left(\frac{f_1}{f_2}\right)^{\frac{1}{2}f_1} \frac{\Gamma(\frac{1}{2}(f_1+f_2))}{\Gamma(\frac{1}{2}f_1)\Gamma(\frac{1}{2}f_2)} F^{\frac{1}{2}f_1-1} \left(1+\frac{f_1}{f_2}F\right)^{-\frac{1}{2}(f_1+f_2)} \quad , \qquad \text{(C.6.1)}$$

is computed as follows.

Program C.17: Function SDFTST computes the probability density of the F distribution

CALLING SEQUENCE: F=SDFTST(X,NF1,NF2)

INPUT ARGUMENTS: X — independent variable F from (C.6.1)
NF1,NF2 — numbers of degrees of freedom f_1, f_2

OUTPUT ARGUMENT: function value SDFTST — probability density (C.6.1)

```
      DOUBLE PRECISION FUNCTION SDFTST(X,NF1,NF2)
      IMPLICIT DOUBLE PRECISION(A-H,O-Z)
      PARAMETER (ONE=1.D0, HALF=0.5D0, ZERO=0.D0)
      AF1=DBLE(NF1)
      AF2=DBLE(NF2)
      AF12=HALF*AF1
      AF22=HALF*AF2
      AFS2=AF12+AF22
      IF(X.LE.ZERO) THEN
        SDFTST=ZERO
      ELSE
        AL=AF12*LOG(AF1/AF2)-AFS2*LOG(ONE+AF1*X/AF2)+(AF12-ONE)*
     +       LOG(X)+GLNGAM(AFS2)-GLNGAM(AF12)-GLNGAM(AF22)
        SDFTST=EXP(AL)
      ENDIF
      END
```

The distribution function

$$F(F) = \frac{\Gamma(\frac{1}{2}(f_1+f_2))}{\Gamma(\frac{1}{2}f_1)\Gamma(\frac{1}{2}f_2)} \left(\frac{f_1}{f_2}\right)^{\frac{1}{2}f_1} \int_0^F F^{\frac{1}{2}f_1-1} \left(1+\frac{f_1}{f_2}F\right)^{-\frac{1}{2}(f_1+f_2)} \mathrm{d}F \qquad \text{(C.6.2)}$$

can be rearranged using

$$t = \frac{f_2}{f_2+f_1F} \quad , \qquad |\mathrm{d}t| = \frac{f_1f_2}{(f_2+f_1F)^2}|\mathrm{d}F|$$

to be

$$\begin{aligned} F(F) &= \frac{1}{B(\frac{1}{2}f_1,\frac{1}{2}f_2)} \int_{t=\frac{f_2}{f_2+f_1F}}^{t=1} (1-t)^{\frac{1}{2}f_1-1} t^{\frac{1}{2}f_2-1}\,\mathrm{d}t \qquad \text{(C.6.3)} \\ &= 1 - I_{f_2/(f_2+f_1F)}\left(\tfrac{1}{2}f_2,\tfrac{1}{2}f_1\right) \quad , \end{aligned}$$

i.e., it is related to the incomplete beta function; cf. (D.6.1).

Program C.18: Function SCFTST computes the distribution function of the F distribution

CALLING SEQUENCE: F=SCFTST(X,NF1,NF2)

INPUT ARGUMENTS: X — independent variable F from (C.6.1)
NF1,NF2 — numbers of degrees of freedom f_1, f_2

OUTPUT ARGUMENT: function value SCFTST — distribution function (C.6.2)

```
      DOUBLE PRECISION FUNCTION SCFTST(X,NF1,NF2)
      IMPLICIT DOUBLE PRECISION(A-H,O-Z)
      PARAMETER (ONE=1.D0, HALF=0.5D0)
      ARG=DBLE(NF2)/(DBLE(NF2)+DBLE(NF1)*X)
      A=HALF*DBLE(NF2)
      B=HALF*DBLE(NF1)
      SCFTST=ONE-GINCBT(A,B,ARG)
      END
```

The quantile F_P of the F-distribution for a given probability P is given by the zero of the function

$$h(F) = P - F(F) \quad . \tag{C.6.4}$$

Program C.19: Function SQFTST computes a quantile of the F distribution

CALLING SEQUENCE: F=SQFTST(P,NF1,NF2)

INPUT ARGUMENTS: P — probability P
NF1,NF2 — numbers of degrees of freedom f_1, f_2

OUTPUT ARGUMENT: function value SQFTST — quantile F_P

```
      DOUBLE PRECISION FUNCTION SQFTST(P,N1,N2)
      IMPLICIT DOUBLE PRECISION(A-H,O-Z)
      PARAMETER (BIG=1D10, EPSILN=1D-6, ONE=1.D0, ZERO=0.D0)
      EXTERNAL SZFTST
C boundary of range
      IF(P.GE.ONE) SQFTST=BIG
      IF(P.LE.ZERO) SQFTST=ZERO
C normal range
      X0=ZERO
      X1=P
      IF(P.LT.ONE .AND. P.GT.ZERO) THEN
        CALL AUXZBR(X0,X1,SZFTST,P,N1,N2)
        CALL AUXZFN(X0,X1,XZERO,SZFTST,P,N1,N2,EPSILN)
        SQFTST=XZERO
      END IF
      END
C-----------------------------------------------------------------
      DOUBLE PRECISION FUNCTION SZFTST(X,P,N1,N2)
C returns P minus cumulative F-distribution of (X,N1,N2)
      IMPLICIT DOUBLE PRECISION(A-H,O-Z)
      SZFTST=P-SCFTST(X,N1,N2)
      END
```

C.7 t-Distribution

The probability density (8.3.7) of Student's t-distribution with f degrees of freedom,

$$f(t) = \frac{\Gamma(\frac{1}{2}(f+1))}{\Gamma(\frac{1}{2}f)\Gamma(\frac{1}{2})\sqrt{f}}\left(1+\frac{t^2}{f}\right)^{-\frac{1}{2}(f+1)} = \frac{1}{B(\frac{1}{2},\frac{f}{2})\sqrt{f}}\left(1+\frac{t^2}{f}\right)^{-\frac{1}{2}(f+1)} , \qquad (C.7.1)$$

is computed with the following program.

Program C.20: Function SDSTUD computes the probability density of Student's distribution

CALLING SEQUENCE: F=SDSTUD(X,N)

INPUT ARGUMENTS: X — independent variable t from (C.7.1)
N — number of degrees of freedom f

OUTPUT ARGUMENT: function value SDSTUD — probability density (C.7.1)

```
DOUBLE PRECISION FUNCTION SDSTUD(X,N)
IMPLICIT DOUBLE PRECISION(A-H,O-Z)
PARAMETER (ONE=1.D0, HALF=0.5D0)
AN=DBLE(N)
AN2=HALF*AN
SQAN=SQRT(AN)
ARG=-AN2-HALF
A=(ONE+X*X/AN)**ARG
BETA=GBETAF(HALF,AN2)
SDSTUD=A/(BETA*SQAN)
END
```

By using the substitution

$$u = \frac{f}{f+t^2} ,$$

the distribution function of the t-distribution can be expressed in terms of the incomplete beta function (D.6.1),

$$F(t) = \frac{1}{B(\frac{1}{2},\frac{f}{2})\sqrt{f}}\int_{-\infty}^{t}\left(1+\frac{t^2}{f}\right)^{-\frac{1}{2}(f+1)} dt = \frac{1}{2}+\frac{\operatorname{sign}(t)}{B(\frac{1}{2},\frac{f}{2})\sqrt{f}}\int_{0}^{|t|}\left(1+\frac{t^2}{f}\right)^{-\frac{1}{2}(f+1)} dt \qquad (C.7.2)$$

$$= \frac{1}{2} + \frac{\operatorname{sign}(t)}{B(\frac{1}{2},\frac{f}{2})\sqrt{f}}\frac{1}{2}\int_{u=f/(f+t^2)}^{u=1} u^{\frac{f}{2}-1}(1-u)^{\frac{1}{2}}\,\mathrm{d}u \quad ,$$

$$F(t) = \frac{1}{2}\left\{1 + \operatorname{sign}(t)\left[1 - I_{f/(f+t^2)}\left(\frac{f}{2},\frac{1}{2}\right)\right]\right\} \quad . \qquad \text{(C.7.3)}$$

Program C.21: Function SCSTUD computes the distribution function of Student's t-distribution

CALLING SEQUENCE: F=SCSTUD(X,N)

INPUT ARGUMENTS: X — independent variable t from (C.7.2)
N — number of degrees of freedom f

OUTPUT ARGUMENT: function value SCSTUD — distribution function (C.7.2)

```
DOUBLE PRECISION FUNCTION SCSTUD(X,N)
IMPLICIT DOUBLE PRECISION(A-H,O-Z)
PARAMETER (ZERO=0.D0, ONE=1.D0, HALF=0.5D0)
AN=DBLE(N)
AN2=HALF*AN
ARG=AN/(AN+X**2)
A=GINCBT(AN2,HALF,ARG)
IF(X.GE.ZERO) THEN
  SCSTUD=ONE-HALF*A
ELSE
  SCSTUD=HALF*A
END IF
END
```

The quantile t_P of the t-distribution for a given probability P is computed by finding the zero of the function

$$h(t) = P - F(t) \quad . \qquad \text{(C.7.4)}$$

Program C.22: Function SQSTUD computes a quantile of Student's t-distribution

CALLING SEQUENCE: F=SQSTUD(P,N)

INPUT ARGUMENTS: P — probability P
N — number of degrees of freedom f

OUTPUT ARGUMENT: function value SQSTUD — quantile t_P

```
      DOUBLE PRECISION FUNCTION SQSTUD(P,N)
      IMPLICIT DOUBLE PRECISION(A-H,O-Z)
      PARAMETER(BIG=1D10,EPSILN=1D-6,ONE=1.D0,ZERO=0.D0,HALF=.5D0)
      EXTERNAL SZSTUD
C boundary of range
      IF(P.GE.ONE) SQSTUD=BIG
      IF(P.LE.ZERO) SQSTUD=-BIG
C normal range
      IF(P.LT.ONE .AND. P.GT.ZERO) THEN
        X0=ZERO
        X1=P
        CALL AUXZBR(X0,X1,SZSTUD,P,N,0)
        CALL AUXZFN(X0,X1,XZERO,SZSTUD,P,N,0,EPSILN)
        SQSTUD=XZERO
      END IF
      END
C-----------------------------------------------------------------
      DOUBLE PRECISION FUNCTION SZSTUD(X,P,N,NDUM)
C returns P minus cumulative Student's distribution of (X,N)
      IMPLICIT DOUBLE PRECISION(A-H,O-Z)
      SZSTUD=P-SCSTUD(X,N)
      END
```

C.8 Example Programs

The source files of the example programs are available in the subdirectory \DATAN\EXASRC.

C.8.1 Main Program E1SD to Demonstrate Subprograms SDBINM, SCBINM, SDHYPG, SCHYPG, SDPOIS, SCPOIS, SQPOIS

The program allows the computation of numerical values from the binomial distribution, the hypergeometric distribution, and the Poisson distribution. The user is asked for the type of function and for the input parameters. The result is presented alphanumerically.

C.8.2 Main Program E2SD to Demonstrate Subprograms SDSTNR, SCSTNR, SQSTNR, SDNORM, SCNORM, SQNORM, SDCHI2, SCCHI2, SQCHI2, SDFTST, SCFTST, SQFTST, SCSTUD, SDSTUD, SQSTUD

The program allows the computation of numerical values (probability density, distribution function, or quantile) of the standard normal distribution, the normal distribution as well as of the χ^2-, F-, and t-distributions. The user is asked for the desired quantity and for input parameters. The result is presented alphanumerically.

C.9 Programming Problems

The source files of the example solutions are available in the subdirectory `\DATAN\SOLSRC`.

C.9.1 Graphical Representation of Statistical Functions of a Discrete Variable

Write a program that displays the binomial, hypergeometric, and Poisson distributions as histograms. As chosen by the user the probabilities or the distribution functions are to be plotted. Allow for an interactive choice of the function and for an interactive input of the parameters. Choose the region of the discrete variables such that for the binomial and hypergeometric distributions, the region covers all reasonable values of the variable, and that for the Poisson distribution, the region is shown in which the probability is significantly different from zero.

C.9.2 Graphical Representation of Statistical Functions of a Continuous Variable

Write a program that produces plots of the standard normal and normal distributions, as well as of the χ^2-, F-, and t-distributions. Allow for the interactive choice of the distribution, display of probability density or distribution function, and input of parameters. Choose scales that are appropriate for the problem at hand.

D. The Gamma Function and Related Functions. Methods and Programs for their Computation

D.1 The Euler Gamma Function

Consider a real number x with $x+1 > 0$. We define the Euler *gamma function* by

$$\Gamma(x+1) = \int_0^\infty t^x \mathrm{e}^{-t}\,\mathrm{d}t \quad . \tag{D.1.1}$$

Integrating by parts gives

$$\int_0^\infty t^x \mathrm{e}^{-t}\,\mathrm{d}t = [-t^x \mathrm{e}^{-t}]_0^\infty + x\int_0^\infty t^{x-1}\mathrm{e}^{-t}\,\mathrm{d}t = x\int_0^\infty t^{x-1}\mathrm{e}^{-t}\,\mathrm{d}t \quad .$$

Thus one has the relation

$$\Gamma(x+1) = x\Gamma(x) \quad . \tag{D.1.2}$$

This is the so-called *recurrence relation* of the gamma function. From (D.1.1) it follows immediately that

$$\Gamma(1) = 1 \quad .$$

With (D.1.2) one then has generally that

$$\Gamma(n+1) = n! \quad , \qquad n = 1, 2, \ldots \quad . \tag{D.1.3}$$

If we now choose in particular $x = -\frac{1}{2}$, we obtain

$$\Gamma(\tfrac{1}{2}) = \sqrt{2}\int_0^\infty \mathrm{e}^{-\frac{1}{2}u^2}\,\mathrm{d}u = \frac{1}{\sqrt{2}}\int_{-\infty}^\infty \mathrm{e}^{-\frac{1}{2}u^2}\,\mathrm{d}u \quad . \tag{D.1.4}$$

The integral can be evaluated in the following way. We consider

$$A = \int_{-\infty}^\infty\int_{-\infty}^\infty \mathrm{e}^{-\frac{1}{2}(x^2+y^2)}\,\mathrm{d}x\,\mathrm{d}y = \int_{-\infty}^\infty \mathrm{e}^{-\frac{1}{2}x^2}\,\mathrm{d}x\int_{-\infty}^\infty \mathrm{e}^{-\frac{1}{2}y^2}\,\mathrm{d}y = 2\{\Gamma(\tfrac{1}{2})\}^2 \quad .$$

The integral A can transformed into polar coordinates:

$$A = \int_0^{2\pi} \int_0^{\infty} \mathrm{e}^{-\frac{1}{2}r^2} r \,\mathrm{d}r \,\mathrm{d}\phi = \int_0^{2\pi} \mathrm{d}\phi \int_0^{\infty} \mathrm{e}^{-\frac{1}{2}r^2} r \,\mathrm{d}r = 2\pi\Gamma(1) = 2\pi \quad .$$

Setting the two expressions for A equal gives

$$\Gamma(\tfrac{1}{2}) = \sqrt{\pi} \quad . \tag{D.1.5}$$

Using (D.1.2) we can thus determine the value of the gamma function for half-integral arguments.

For arguments that are not positive integers or half-integers, the integral (D.1.1) cannot be evaluated in closed form. In such cases one must rely on approximations. We discuss here the approximation of LANCZOS (1964), which is based on the analytic properties of the gamma function. We first extend the definition of the gamma function to negative arguments by means of the reflection formula

$$\Gamma(1-x) = \frac{\pi}{\Gamma(x)\sin(\pi x)} = \frac{\pi x}{\Gamma(1+x)\sin(\pi x)} \quad . \tag{D.1.6}$$

(By relations (D.1.1) and (D.1.6) the gamma function is also defined for an arbitrary complex argument if x is complex.) One sees immediately that the gamma function has poles at zero and at all negative integer values. The approximation of LANCZOS [17],

$$\Gamma(x+1) = \sqrt{2\pi}\left(x+\gamma+\frac{1}{2}\right)^{x+\frac{1}{2}} \exp\left(-x-\gamma-\frac{1}{2}\right)(A_\gamma(x)+\varepsilon) \quad , \tag{D.1.7}$$

takes into account the first few of these poles by the form of the function A_γ,

$$A_\gamma(x) = c_0 + \frac{c_1}{x+1} + \frac{c_2}{x+2} + \cdots + \frac{c_{\gamma+1}}{x+\gamma+1} \quad . \tag{D.1.8}$$

For $\gamma = 5$ one has for the error ε the approximation $|\varepsilon| < 2 \cdot 10^{-10}$ for all points x in the right half of the complex plane.

Program D.1: Function GGAMMA computes the Euler gamma function

CALLING SEQUENCE: F=GGAMMA(X)

INPUT ARGUMENTS: X — argument x of the gamma function

OUTPUT ARGUMENTS: function value GGAMMA — $\Gamma(x)$

```
      DOUBLE PRECISION FUNCTION GGAMMA(X)
      IMPLICIT DOUBLE PRECISION (A-H,O-Z)
      DIMENSION C(6)
      LOGICAL REFLEC
      PARAMETER(RTWOPI=2.506628275D0, PI=3.141592654D0,ONE=1.D0)
      PARAMETER(HALF=0.5D0)
      DATA C/76.18009173D0,-86.50532033D0,24.01409822D0,
     +-1.231739516D0,0.120858003D-2,-0.536382D-5/
      IF(X.GE.ONE) THEN
        REFLEC=.FALSE.
        XX=X-ONE
      ELSE
        REFLEC=.TRUE.
        XX=ONE-X
      END IF
      XH=XX+HALF
      XGH=XX+DBLE(5.5)
      S=ONE
      ANUM=XX
      DO 10 I=1,6
        ANUM=ANUM+ONE
        S=S+C(I)/ANUM
   10 CONTINUE
      S=S*RTWOPI
      G=(XGH**XH)*S/EXP(XGH)
      IF (REFLEC) THEN
        GGAMMA=PI*XX/(G*SIN(PI*XX))
      ELSE
        GGAMMA=G
      END IF
      END
```

The gamma function is plotted in Fig. D.1. For large positive arguments the gamma function grows so quickly that it is difficult to represent its value in a computer. In many expressions, however, there appear ratios of gamma functions which have values in a an unproblematic region. In such cases it is better to use the logarithm of the gamma function.

Program D.2: Function GLNGAM computes the natural logarithm of the Euler gamma function

CALLING SEQUENCE: F=GLNGAM(X)

INPUT ARGUMENTS: X — argument x of the gamma function

OUTPUT ARGUMENTS: function value GLNGAM — $\ln \Gamma(x)$

```
      DOUBLE PRECISION FUNCTION GLNGAM(X)
      IMPLICIT DOUBLE PRECISION (A-H,O-Z)
      DIMENSION C(6)
      LOGICAL REFLEC
      PARAMETER(RTWOPI=2.506628275D0, PI=3.141592654D0, ONE=1.D0)
      PARAMETER(HALF=0.5D0)
      DATA C/76.18009173D0,-86.50532033D0,24.01409822D0,
     +-1.231739516D0,0.120858003D-2,-0.536382D-5/
      IF(X.GE.ONE) THEN
```

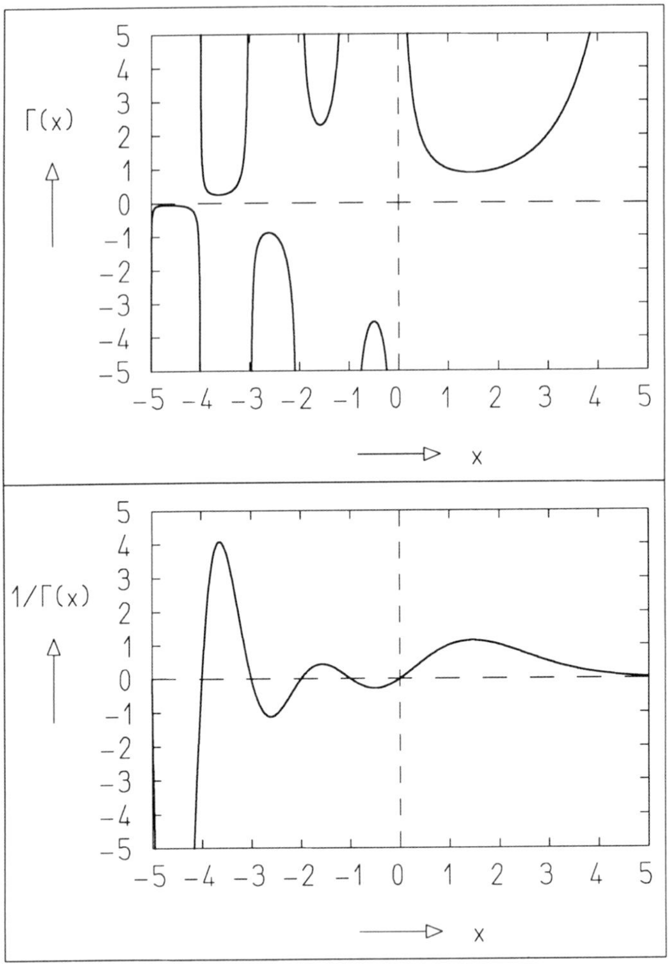

Fig. D.1. The functions $\Gamma(x)$ and $1/\Gamma(x)$.

```
        REFLEC=.FALSE.
        XX=X-ONE
      ELSE
        REFLEC=.TRUE.
        XX=ONE-X
      END IF
      XH=XX+HALF
      XGH=XX+DBLE(5.5)
      S=ONE
      ANUM=XX
      DO 10 I=1,6
        ANUM=ANUM+ONE
        S=S+C(I)/ANUM
   10 CONTINUE
      S=S*RTWOPI
```

```
G=XH*LOG(XGH)+LOG(S)-XGH
IF (REFLEC) THEN
  GLNGAM=LOG(PI*XX)-G-LOG(SIN(PI*XX))
ELSE
  GLNGAM=G
END IF
END
```

D.2 Factorial and Binomial Coefficients

The expression

$$n! = 1 \cdot 2 \cdots n \tag{D.2.1}$$

can either be directly computed as a product or as a gamma function using (D.1.3).

When computing binomial coefficients,

$$\binom{n}{k} = \frac{n!}{k!(n-k)!} = \frac{n}{k} \cdot \frac{n-1}{k-1} \cdots \frac{n-k+1}{1} \quad , \tag{D.2.2}$$

the expression on the right-hand side is preferable for numerical reasons to the expression in the middle.

Program D.3: Function GBINCO computes binomial coefficients

CALLING SEQUENCE: F=GBINCO(N,K)

INPUT ARGUMENTS: N, K — arguments n, k of the binomial coefficient $\binom{n}{k}$

OUTPUT ARGUMENTS: function value GBINCO — $\binom{n}{k}$

```
DOUBLE PRECISION FUNCTION GBINCO(N,K)
IMPLICIT DOUBLE PRECISION (A-H,O-Z)
PARAMETER(ONE=1.DO)
IF(K.EQ.O .OR. N.EQ.O .OR. K.EQ.N) THEN
  GBINCO=ONE
ELSE
  AN=DBLE(N)
  AK=DBLE(K)
  ALBINC=GLNGAM(AN+ONE)-GLNGAM(AK+ONE)-GLNGAM(AN-AK+ONE)
  GBINCO=EXP(ALBINC)
  GBINCO=NINT(GBINCO)
END IF
END
```

D.3 Beta Function

The *beta function* has two arguments and is defined by

$$B(z,w) = \int_0^1 t^{z-1}(1-t)^{w-1}\,\mathrm{d}t \quad . \tag{D.3.1}$$

The integral can be written in a simple way in terms of gamma functions,

$$B(z,w) = B(w,z) = \frac{\Gamma(z)\Gamma(w)}{\Gamma(z+w)} \quad . \tag{D.3.2}$$

Program D.4: Function GBETAF computes the beta function

CALLING SEQUENCE: F=GBETAF(Z,W)

INPUT ARGUMENTS: Z,W — arguments z, w in (D.3.1)

OUTPUT ARGUMENTS: function value GBETAF — $B(z,w)$

```
DOUBLE PRECISION FUNCTION GBETAF(Z,W)
IMPLICIT DOUBLE PRECISION (A-H,O-Z)
PARAMETER(BIG=1.D30, EPSILN=1.D-8)
IF(W.LT.EPSILN) THEN
  GBETAF=BIG
ELSE
  GBETAF=EXP(GLNGAM(Z)+GLNGAM(W)-GLNGAM(Z+W))
END IF
END
```

Figure D.2 shows the beta function as a function of w for several fixed values of z.

D.4 Computing Continued Fractions

In the next two sections there appear *continued fractions*, i.e., expressions of the type

$$f = b_0 + \cfrac{a_1}{b_1 + \cfrac{a_2}{b_2 + \frac{a_3}{b_3 + \cdots}}} \quad , \tag{D.4.1}$$

that can also be written in the typographically simpler form

$$f = b_0 + \frac{a_1}{b_1+}\,\frac{a_2}{b_2+}\,\frac{a_3}{b_3+}\cdots \quad . \tag{D.4.2}$$

If we denote by f_n the value of the fraction (D.4.1) truncated after a finite number of terms up to the coefficients a_n and b_n, then one has

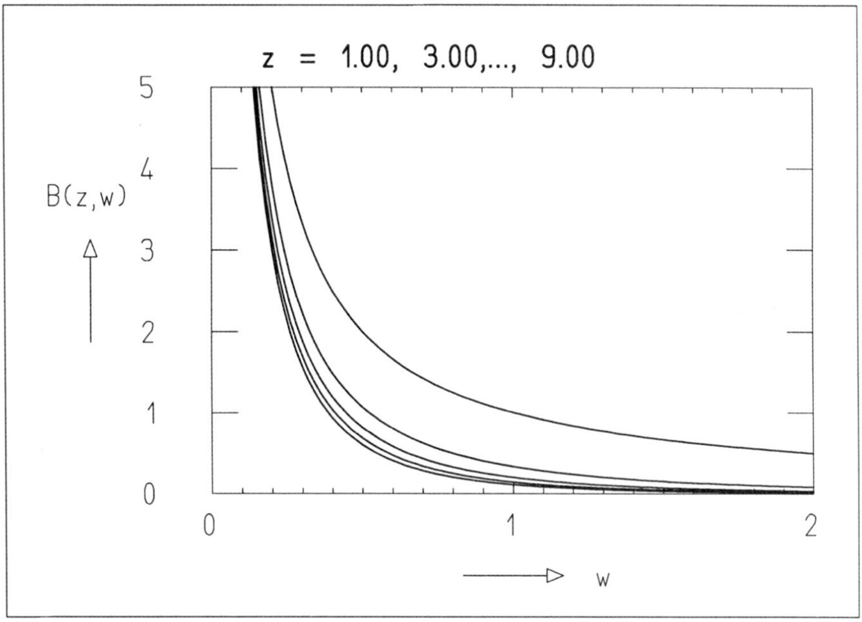

Fig. D.2. The beta function. For increasing z the curves $B(z, w)$ are shifted to the left.

$$f_n = \frac{A_n}{B_n} \quad . \tag{D.4.3}$$

The quantities A_n and B_n can be obtained from the following recursion relation,

$$A_{-1} = 1 \quad , \qquad B_{-1} = 0 \quad , \qquad A_0 = b_0 \quad , \qquad B_0 = 1 \quad , \tag{D.4.4}$$

$$A_j = b_j \, A_{j-1} + a_j \, A_{j-2} \quad , \tag{D.4.5}$$

$$B_j = b_j \, B_{j-1} + a_j \, B_{j-2} \quad . \tag{D.4.6}$$

Since the relations (D.4.5) and (D.4.6) are linear in A_{j-1}, A_{j-2} and B_{j-1}, B_{j-2}, respectively, and since in (D.4.3) only the ratio A_n/B_n appears, one can always multiply the coefficients A_j, A_{j-1}, A_{j-2} and B_j, B_{j-1}, B_{j-2} by an arbitrary normalization factor. One usually chooses for this factor $1/B_j$ and avoids in this way numerical difficulties from very large or very small numbers, which would otherwise occur in the course of the recursion. For steps in which $B_j = 0$, the normalization is not done.

Continued fractions, in a way similar to series expansions, appear in approximations of certain functions. In a region where the approximation for the continued fraction converges, the values of f_{n-1} and f_n for sufficiently large n do not differ much. One can therefore use the following *truncation criterion*. If for a given $\varepsilon \ll 1$ the inequality

$$\left| \frac{f_n - f_{n-1}}{f_n} \right| < \varepsilon$$

holds, then f_n is a sufficiently good approximation of f.

D.5 Incomplete Gamma Function

The *incomplete gamma function* is defined for $a > 0$ by the expression

$$P(a,x) = \frac{1}{\Gamma(a)} \int_0^x \mathrm{e}^{-t} t^{a-1}\, \mathrm{d}t \quad . \tag{D.5.1}$$

It can be expressed as a series expansion

$$P(a,x) = x^a \mathrm{e}^{-x} \sum_{n=0}^{\infty} \frac{x^n}{\Gamma(a+n+1)} = \frac{1}{\Gamma(a)} x^a \mathrm{e}^{-x} \sum_{n=0}^{\infty} \frac{\Gamma(a)}{\Gamma(a+n+1)} x^n \quad . \tag{D.5.2}$$

The sum converges quickly for $x < a + 1$. One uses the right-hand, not the middle form of (D.5.2), since the ratio of the two gamma functions reduces to

$$\frac{\Gamma(a)}{\Gamma(a+n+1)} = \frac{1}{a} \frac{1}{a+1} \cdots \frac{1}{a+n+1} \quad .$$

In the region $x > a + 1$ we use the continued fraction

$$1 - P(a,x) = \frac{1}{\Gamma(a)} \mathrm{e}^{-x} x^a \left(\frac{1}{x+} \frac{1-a}{1+} \frac{1}{x+} \frac{2-a}{1+} \frac{2}{x+} \cdots \right) \quad . \tag{D.5.3}$$

Program D.5: Function `GINCGM` computes the incomplete gamma function

CALLING SEQUENCE: `F=GINCGM(A,X)`

INPUT ARGUMENTS: `A,X` — arguments a, x in (D.5.1)

OUTPUT ARGUMENTS: function value `GINCGM` — $P(a,x)$

```
      DOUBLE PRECISION FUNCTION GINCGM(A,X)
      IMPLICIT DOUBLE PRECISION (A-H,O-Z)
      PARAMETER(BIG=500.D0, EPSILN=1.D-6, ONE=1.D0, ZERO=0.D0)
      ALOGGM=GLNGAM(A)
      IF(X.LE.A+ONE) THEN
C series development
        F=ONE/A
        S=F
        ANUM=A
        DO 10 I=1,100
          ANUM=ANUM+ONE
```

```
          F=X*F/ANUM
          S=S+F
          IF(F.LT.EPSILN) GO TO 20
   10   CONTINUE
   20   IF(X.LT.EPSILN) THEN
          GINCGM=ZERO
        ELSE
          HELP=A*LOG(X)-X-ALOGGM
          IF(ABS(HELP).GE.BIG) THEN
            GINCGM=ZERO
          ELSE
            GINCGM=S*EXP(HELP)
          END IF
        END IF
      ELSE
C continued fraction
        A0=ZERO
        B0=ONE
        A1=ONE
        B1=X
        CF=ONE
        FNORM=ONE
        DO 30 J=1,100
          A2J=DBLE(J)-A
          A2J1=DBLE(J)
          B2J=ONE
          B2J1=X
          A0=(B2J*A1+A2J*A0)*FNORM
          B0=(B2J*B1+A2J*B0)*FNORM
          A1=B2J1*A0+A2J1*A1*FNORM
          B1=B2J1*B0+A2J1*B1*FNORM
          IF(B1.NE.ZERO) THEN
C renormalize and test for convergence
            FNORM=ONE/B1
            CFNEW=A1*FNORM
            IF(ABS(CF-CFNEW)/CF .LT. EPSILN) GO TO 40
            CF=CFNEW
          END IF
   30   CONTINUE
   40   CONTINUE
        HELP=A*LOG(X)-X-ALOGGM
        IF(ABS(HELP).GE.BIG) THEN
          GINCGM=ONE
        ELSE
          GINCGM=ONE-EXP(HELP)*CFNEW
        END IF
      END IF
      END
```

The incomplete gamma function is shown in Fig. D.3 for several values of a. From the figure one sees immediately that

$$P(a,0) = 0 \quad , \qquad \text{(D.5.4)}$$

$$\lim_{x\to\infty} P(a,x) = 1 \quad . \qquad \text{(D.5.5)}$$

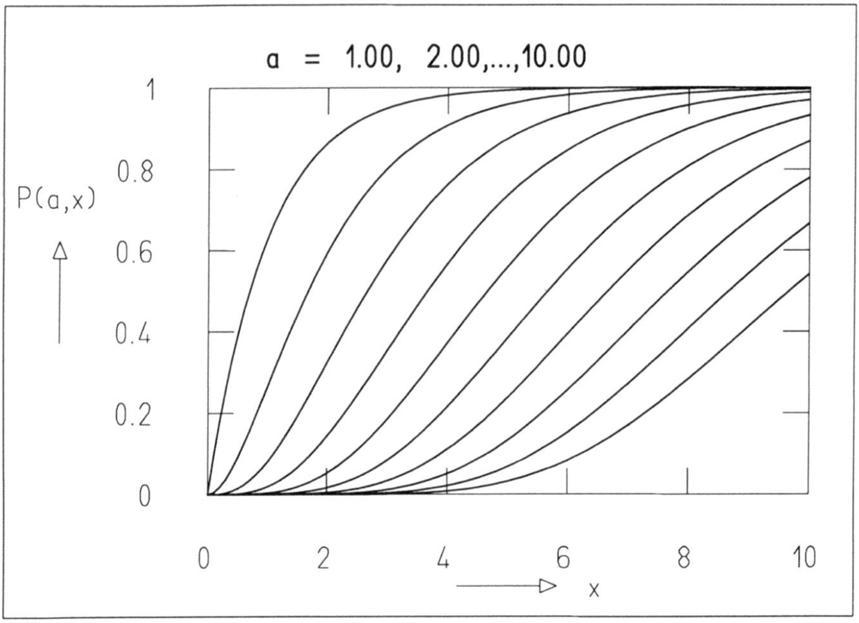

Fig. D.3. The incomplete gamma function. With increasing a the graphs $P(a, x)$ move to the right.

D.6 Incomplete Beta Function

The *incomplete beta function* is defined for $a > 0$, $b > 0$ by the relation

$$I_x(a,b) = \frac{1}{B(a,b)} \int_0^x t^{a-1}(1-t)^{b-1}\,\mathrm{d}t \quad , \qquad x \le 0 \le 1 \quad . \tag{D.6.1}$$

The function obeys the symmetry relation

$$I_x(a,b) = 1 - I_{1-x}(b,a) \quad . \tag{D.6.2}$$

The expression (D.6.1) can be approximated by the following continued fraction:

$$I_x(a,b) = \frac{x^a(1-x)^b}{aB(a,b)} \left\{ \frac{1}{1+} \frac{d_1}{1+} \frac{d_2}{1+} \cdots \right\} \tag{D.6.3}$$

with

$$d_{2m+1} = -\frac{(a+m)(a+b+m)}{(a+2m)(a+2m-1)}\, x \quad ,$$

$$d_{2m} = \frac{m(b-m)}{(a+2m-1)(a+2m)}\, x \quad .$$

The approximation converges quickly for

$$x > \frac{a+1}{a+b+1} \quad . \tag{D.6.4}$$

If this requirement is not fulfilled, then $1 - x$ is greater than the right-hand side of (D.6.4). In this case one computes I_{1-x} as a continued fraction and then uses (D.6.2).

Program D.6: Function GINCBT computes the incomplete beta function

CALLING SEQUENCE: F=GINCBT(A,B,X)

INPUT ARGUMENTS: A,B,X — arguments a, b and index x in (D.6.1)

OUTPUT ARGUMENTS: function value GINCBT — $I_x(a, x)$

```
      DOUBLE PRECISION FUNCTION GINCBT(AA,BB,XX)
      IMPLICIT DOUBLE PRECISION (A-H,O-Z)
      LOGICAL REFLEC
      PARAMETER(EPSILN=1.D-8, ONE=1.D0, ZERO=0.D0, TWO=2.D0)
      XLIM=(AA+ONE)/(AA+BB+ONE)
      IF (XX.LT.XLIM) THEN
        REFLEC=.FALSE.
        A=AA
        B=BB
        X=XX
      ELSE
        REFLEC=.TRUE.
        A=BB
        B=AA
        X=ONE-XX
      END IF
      IF(X.LT.EPSILN) THEN
C function known at end of range
        CF=0.
      ELSE
C continued fraction
        A1=ONE
        B1=ONE
        A2=ONE
        B2=ONE-(A+B)*X/(A+ONE)
        FNORM=ONE/B2
        CF=A2*FNORM
        DO 10 M=1,100
          RM=DBLE(M)
          APL2M=A+TWO*RM
          D2M=RM*(B-RM)*X/((APL2M-ONE)*APL2M)
          D2M1=-(A+RM)*(A+B+RM)*X/(APL2M*(APL2M+1))
          A1=(A2+D2M*A1)*FNORM
          B1=(B2+D2M*B1)*FNORM
          A2=A1+D2M1*A2*FNORM
          B2=B1+D2M1*B2*FNORM
          IF(B2.NE.0.) THEN
C renormalize and test for convergence
            FNORM=ONE/B2
```

```
          CFNEW=A2*FNORM
          IF(ABS(CF-CFNEW)/CF .LT. EPSILN) GO TO 20
          CF=CFNEW
        END IF
 10   CONTINUE
 20   CF=CF*(X**A)*((ONE-X)**B)/(A*GBETAF(A,B))
    END IF
    IF(REFLEC) THEN
      GINCBT=ONE-CF
    ELSE
      GINCBT=CF
    END IF
    END
```

In Fig. D.4 the incomplete beta function is displayed for various values of the parameters a and b. Regardless of these parameters one has

$$I_0(a,b) = 0 \quad , \qquad I_1(a,b) = 1 \quad . \tag{D.6.5}$$

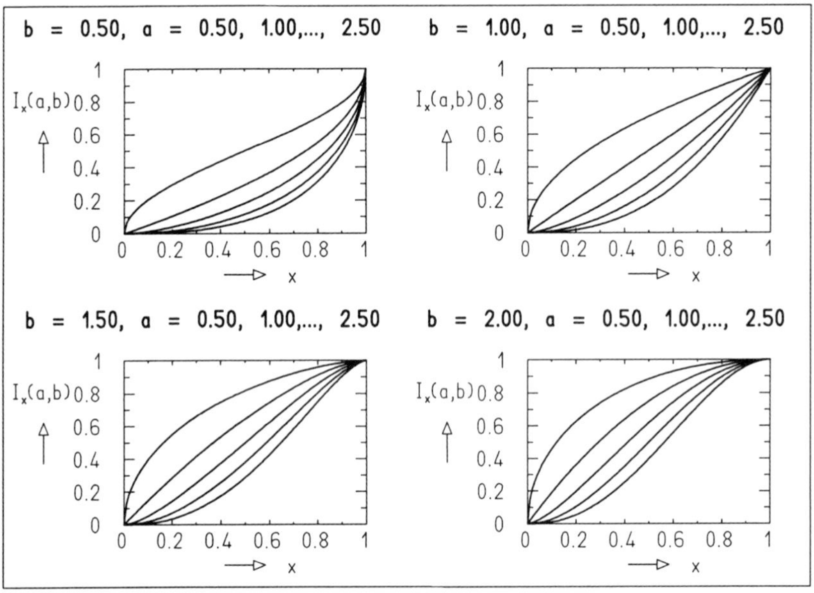

Fig. D.4. The incomplete beta function. With increasing a the curves $I_x(a,b)$ move further to the right.

D.7 Example Programs

The source files of the example programs are available in the subdirectory `\DATAN\EXASRC`.

D.7.1 Main Program `E1GAM` to Demonstrate Subprograms `GGAMMA` and `GLNGAM`

The program asks interactively for the type of function and for the argument x and yields the numerical value of $\Gamma(x)$ or $\ln \Gamma(x)$.

D.7.2 Main Program `E2GAM` to Demonstrate Subprogram `GBINCO`

The program computes and displays the first 12 lines of Pascal's triangle (cf. Fig. B.1).

D.7.3 Main Program `E3GAM` to Demonstrate Subprograms `GBETAF`, `GINCGM`, and `GINCBT`

The program computes values of the beta function $B(z, w)$, the incomplete Gamma function $P(a, x)$, and the incomplete beta function $I_x(a, b)$. The type of function and the arguments are chosen interactively. The result is presented alphanumerically.

D.8 Programming Problems

The source files of the example solutions are available in the subdirectory `\DATAN\SOLSRC`.

D.8.1 Graphical Representations of $\Gamma(x)$ and $1/\Gamma(x)$

Write a program that produces the two plots of Fig. D.1. Note that the graph $\Gamma(x)$ in the left half-plane has to be constructed from separate polylines, since otherwise vertical lines would appear at the position of the poles ($x = 0, -1, -2, \ldots$).

D.8.2 Graphical Representations of the Beta Function, the Incomplete Gamma Function, and the Incomplete Beta Function

Write a program producing graphs of the beta function $B(z, w)$, the incomplete gamma function $P(a, x)$, and the incomplete beta function $I_x(a, b)$. Choose the scales as in Figs. D.2, D.3, and D.4, and display one graph per plot. Allow for the interactive choice of the function and input of parameters.

E. Utility Programs

E.1 Numerical Differentiation

The derivative $\mathrm{d}f(x)/\mathrm{d}x$ of a function $f(x)$ at the point x is given by the limit

$$f'(x) = \lim_{h \to 0} \frac{f(x+h) - f(x)}{h} \quad .$$

Obviously one can approximate $f'(x)$ by

$$\frac{f(x+h) - f(x)}{h}$$

for a small finite value of h. In fact, for this the symmetrical difference ratio

$$\delta(h) = \frac{f(x+h) - f(x-h)}{2h} \tag{E.1.1}$$

is more appropriate. This can be seen from the Taylor expansions at the point x for $f(x+h)$ and $f(x-h)$, which give

$$\delta(h) = f'(x) + \frac{h^2}{3!} f'''(x) + \frac{h^4}{5!} f^{(5)}(x) + \cdots \quad ,$$

in which the leading additional term is already quadratic in h. Nevertheless, the choice of h is still critical, since for very small values of h there occur large rounding errors, and for larger values the approximation may not be valid.

One can compute $\delta(h)$ for a monotonic sequence of values $h = h_0, h_1, h_2, \ldots$. If the sequence $\delta(h_0), \delta(h_1), \ldots$ is monotonic (rising or falling) then this is a sign of convergence of the series to $f'(x)$. According to RUTISHAUSER [27], from the series $\delta(h_0), \delta(h_1)$ one can also obtain others that converge more quickly. The method was modeled after the ROMBERG procedure [28] for numerical integration. Starting from $h_0 = a$ one first chooses the sequence

$$h_0, h_1, \ldots = a, 3a/4, a/2, 3a/8, a/4, \ldots \quad ,$$

sets

$$T_0^{(k)} = \delta(h_k) \quad ,$$

and computes the additional quantities $T_m^{(k)}$

$$T_m^{(k)} = \frac{2^m T_{m-1}^{(k+1)} - 1.125\, T_{m-1}^{(k)}}{2^m - 1.125} \quad ; \qquad m \text{ odd, } k \text{ even} \quad ,$$

$$T_m^{(k)} = \frac{2^m \cdot 1.125\, T_{m-1}^{(k+1)} - T_{m-1}^{(k)}}{2^m \cdot 1.125 - 1} \quad ; \qquad m \text{ odd, } k \text{ odd} \quad ,$$

$$T_m^{(k)} = \frac{2^m T_{m-1}^{(k+1)} - T_{m-1}(k)}{2^m - 1} \quad ; \qquad m \text{ even} \quad .$$

Arranging the quantities $T_m^{(k)}$ in the form of a triangle,

$$\begin{array}{cccc}
T_0^{(0)} & & & \\
 & T_1^{(0)} & & \\
T_0^{(1)} & & T_2^{(0)} & \\
 & T_1^{(1)} & & T_3^{(0)} \\
T_0^{(2)} & & T_2^{(1)} & \\
 & T_1^{(2)} & & \\
T_0^{(3)} & & & \\
\vdots & & &
\end{array}$$

the first column contains the sequence of our original difference ratios. Not only does $T_0^{(k)}$ converge to $f'(x)$, but one has in general

$$\lim_{k\to\infty} T_m^{(k)} = f'(x) \quad , \qquad \lim_{m\to\infty} T_m^{(k)} = f'(x) \quad .$$

The practical significance of the procedure is based on the fact that the columns on the right converge particularly quickly.

In the following program, the sequence $T_0^{(0)}, \ldots, T_0^{(9)}$ is computed starting from $a = 1$. If it is not monotonic, then a is replaced by $a/10$ and a new sequence is computed. After 10 tries without success the procedure is terminated. If, however, a monotonic sequence is found, the triangle scheme is computed and $T_9^{(0)}$ is given as the best approximation for $f'(x)$. The program AUXDER is similar to the program of KOELBIG [29] with the exception of minor changes in the termination criteria.

Program E.1: Subroutine AUXDER computes by numerical differentiation the derivative of a function $f(x)$ at the point x

CALLING SEQUENCE: `CALL AUXDER(F,X,DFDX,OK)`

INPUT ARGUMENTS: `F` — name of a function subprogram that computes the function $f(x)$ supplied by the user
`X` — argument x

OUTPUT ARGUMENTS: `DFDX` — derivative $f'(x)$
`OK` — logical variable (`.TRUE.` for convergence, `.FALSE.` for non-convergence)

```
      SUBROUTINE AUXDER(F,X,DFDX,OK)
C Computes the derivative f'(x) of f(x) at x = X. Based on
C H. Rutishauser, Ausdehnung des Rombergschen Prinzips
C (Extension of Romberg's Principle), Numer. Math. 5 (1963) 48-54
      IMPLICIT DOUBLE PRECISION (A-H,O-Z)
      PARAMETER (EPS = 5.D-8, EPSI=1.D-10, DELTA=10.D0, S=1.D-1)
      DIMENSION DX(0:9),W(0:9,3),T(0:9,0:9),A(0:9)
      LOGICAL LEV(0:9),LMT
      LOGICAL OK
      EXTERNAL F
      DATA DX /0.0256D0, 0.0192D0, 0.0128D0, 0.0096D0, 0.0064D0,
     +0.0048D0, 0.0032D0, 0.0024D0, 0.0016D0, 0.0012D0/
      DATA (LEV(K),K=0,8,2) /5*.TRUE./
      DATA (LEV(K),K=1,9,2) /5*.FALSE./
      DATA W(1,1) /1.33333 33333 333333D+00/
      DATA W(3,1) /1.06666 66666 666667D+00/
      DATA W(5,1) /1.01587 30158 730159D+00/
      DATA W(7,1) /1.00392 15686 274510D+00/
      DATA W(2,1) /3.33333 33333 333333D-01/
      DATA W(4,1) /6.66666 66666 666667D-02/
      DATA W(6,1) /1.58730 15873 015873D-02/
      DATA W(8,1) /3.92156 86274 509804D-03/
      DATA W(0,2) /2.28571 42857 142857D+00/
      DATA W(2,2) /1.16363 63636 363636D+00/
      DATA W(4,2) /1.03643 72469 635628D+00/
      DATA W(6,2) /1.00886 69950 738916D+00/
      DATA W(8,2) /1.00220 21042 329337D+00/
      DATA W(1,2) /1.28571 42857 142857D+00/
      DATA W(3,2) /1.63636 36363 636364D-01/
      DATA W(5,2) /3.64372 46963 562753D-02/
      DATA W(7,2) /8.86699 50738 916256D-03/
      DATA W(9,2) /2.20210 42329 336922D-03/
      DATA W(0,3) /1.80000 00000 000000D+00/
      DATA W(2,3) /1.12500 00000 000000D+00/
      DATA W(4,3) /1.02857 14285 714286D+00/
      DATA W(6,3) /1.00699 30069 930070D+00/
      DATA W(8,3) /1.00173 91304 347826D+00/
      DATA W(1,3) /8.00000 00000 000000D-01/
      DATA W(3,3) /1.25000 00000 000000D-01/
      DATA W(5,3) /2.85714 28571 428571D-02/
      DATA W(7,3) /6.99300 69930 069930D-03/
      DATA W(9,3) /1.73913 04347 826087D-03/
      DEL=DELTA
      DO 40 I=1,10
        DEL=S*DEL
        IF(I.EQ.10 .OR. ABS(X+DEL*DX(9)-X) .LT. EPS) THEN
          OK=.FALSE.
          RETURN
        END IF
        DO 10 K = 0,9
```

```
          H=DEL*DX(K)
          T(K,0)=(F(X+H)-F(X-H))/(H+H)
          A(K)=T(K,0)
10      CONTINUE
        IF(A(0) .GE. A(9)) THEN
          DO 20 K = 0,9
            A(K)=-A(K)
20        CONTINUE
        END IF
        LMT=.TRUE.
        DO 30 K = 1,9
          H=A(K-1)-A(K)
          LMT=LMT .AND. (H.LE.EPSI .OR.
     +                   ABS(H).LE.EPS*ABS(A(K))+EPSI)
30      CONTINUE
        IF(LMT) GO TO 50
40    CONTINUE
50    CONTINUE
      DO 70 M = 1,9
        DO 60 K = 0,9-M
          IF(LEV(M)) THEN
            T(K,M)=W(M-1,1)*T(K+1,M-1)-W(M,1)*T(K,M-1)
          ELSE IF(LEV(K)) THEN
            T(K,M)=W(M-1,2)*T(K+1,M-1)-W(M,2)*T(K,M-1)
          ELSE
            T(K,M)=W(M-1,3)*T(K+1,M-1)-W(M,3)*T(K,M-1)
          END IF
60      CONTINUE
70    CONTINUE
      DFDX=T(0,9)
      OK=.TRUE.
      END
```

We will briefly describe the calling sequences of routines that do not compute single derivatives but rather vectors or matrices of derivatives, as needed in Chapters 9 and 10. To economize on space the programs themselves are not printed.

Program E.2: Subroutine `AUXDRI` computes the matrix of derivatives A, needed for least-squares fitting (indirect measurements, nonlinear case)

CALLING SEQUENCE: `CALL AUXDRI(USERFN,X,T,N,NR,NRED, LIST,A,OK)`

INPUT ARGUMENTS: `USERFN` — name of a user function (examples in Sect. 9.6)
`X,T,N,NR,NRED,LIST` — as in Program 9.3

OUTPUT ARGUMENTS: `A` — as in Program 9.3
`OK` — logical variable (`.TRUE.` for convergence, `.FALSE.` for non-convergence)

Program E.3: Subroutine AUXDRG computes the matrix of derivatives E, needed for the general case of least-squares fitting

CALLING SEQUENCE: CALL AUXDRG(X,Y,M,N,NR,NRED,LIST,E,OK)

INPUT ARGUMENTS: X,Y,M,N,NR,NRED,LIST — as in Program 9.12

OUTPUT ARGUMENTS: E — $m \times (r+n)$ matrix E of derivatives; cf. (9.11.2)
OK — logical variable (.TRUE. for convergence, .FALSE. for non-convergence)

The programs described up to now require considerable computing time. For well behaved functions it is often sufficient to replace the differential ratio by the difference ratio (E.1.1). To compute second derivatives the procedure of difference ratios is extended correspondingly. The following two programs compute the gradient vector and Hessian matrix of second derivatives of a function of n variables, some of which can be treated as fixed if requested.

Program E.4: Subroutine AUXGRD computes the gradient vector of a function of n variables

CALLING SEQUENCE: CALL AUXGRD(X,GRD,N,NRED,LIST,FUNC)

INPUT ARGUMENTS: X — n-vector of variables $\mathbf{x}$
N — number of variables n
NRED — number of variables not fixed n'
LIST — list of n elements giving which variables are fixed (list element 0) and which are adjustable (list element 1)
FUNC — name of a user function; must be declared as EXTERNAL in the calling program

OUTPUT ARGUMENT: GRD — n'-vector of partial derivatives $\partial f/\partial x_i$

REMARKS: The user function must be of the type DOUBLE PRECISION FUNCTION and must have exactly two arguments (X,N).

```
      SUBROUTINE AUXGRD(X,GRD,N,NRED,LIST,FUNC)
      IMPLICIT DOUBLE PRECISION (A-H,O-Z)
      PARAMETER(DELTA=1.D-11,CUT=1.D-9)
      DIMENSION X(*),GRD(*),LIST(*)
      EXTERNAL FUNC
      IL=0
      IF(NRED.GT.0) THEN
        DO 10 I=1,N
          IF(LIST(I).EQ. 1.OR. NRED.EQ.N) THEN
            IL=IL+1
            ARG=ABS(X(I))
```

```
          IF(ARG.LT.CUT) ARG=CUT
          DEL=DELTA*ARG
          SAV=X(I)
          X(I)=SAV+DEL
          FP=FUNC(X,N)
          X(I)=SAV-DEL
          FM=FUNC(X,N)
          X(I)=SAV
          GRD(IL)=(FP-FM)/(DEL+DEL)
        END IF
10    CONTINUE
      END IF
      END
```

Program E.5: Subroutine AUXHES computes the Hessian matrix of second derivatives of a function of n variables

CALLING SEQUENCE: CALL AUXHES(X,HESSE,N,NRED,LIST,FUNC)

INPUT ARGUMENTS: X,N,NRED,LIST,FUNC — as in AUXGRD

OUTPUT ARGUMENT: HESSE — $n' \times n'$ matrix of second derivatives $\partial^2 f/\partial x_i \partial x_k$

REMARKS: as for AUXGRD.

```
      SUBROUTINE AUXHES(X,HESSE,N,NRED,LIST,FUNC)
      IMPLICIT DOUBLE PRECISION (A-H,O-Z)
      PARAMETER(DELTA=1.D-5,CUT=1.D-9)
      DIMENSION X(*),HESSE(NRED,*),LIST(*)
      EXTERNAL FUNC
      IF(NRED.GT.0) THEN
        IL=0
        DO 20 I=1,N
          KL=0
          IF(LIST(I).EQ. 1.OR. N.EQ.NRED) THEN
            IL=IL+1
            DO 10 K=1,N
              IF(LIST(K).EQ. 1.OR. NRED.EQ.N) THEN
                KL=KL+1
                IF(KL.LT.IL) THEN
                  HESSE(IL,KL)=HESSE(KL,IL)
                ELSE
                  E=ABS(X(I))
                  IF(E.LT.CUT) E=CUT
                  DELI=DELTA*E
                  SAVI=X(I)
                  IF (K.EQ.I) THEN
                    F=FUNC(X,N)
                    X(I)=X(I)+DELI
                    FP=FUNC(X,N)
                    X(I)=SAVI-DELI
                    FM=FUNC(X,N)
                    X(I)=SAVI
                    HESSE(IL,IL)=((FP-F)/DELI-(F-FM)/DELI)/DELI
```

```
                  ELSE
                    E=ABS(X(K))
                    IF(E.LT.CUT) E=CUT
                    DELK=DELTA*E
                    SAVK=X(K)
                    X(K)=SAVK+DELK
                    X(I)=SAVI+DELI
                    FP=FUNC(X,N)
                    X(I)=SAVI-DELI
                    FM=FUNC(X,N)
                    DFDXIP=(FP-FM)/(DELI+DELI)
                    X(K)=SAVK-DELK
                    X(I)=SAVI+DELI
                    FP=FUNC(X,N)
                    X(I)=SAVI-DELI
                    FM=FUNC(X,N)
                    X(I)=SAVI
                    X(K)=SAVK
                    DFDXIM=(FP-FM)/(DELI+DELI)
                    HESSE(IL,KL)=(DFDXIP-DFDXIM)/(DELK+DELK)
                  END IF
                END IF
              END IF
   10       CONTINUE
          END IF
   20   CONTINUE
      END IF
      END
```

E.2 Numerical Determination of Zeros

Computing the *quantile* x_P of a distribution function $F(x)$ for a given probability P is equivalent to determining the zero of the function

$$k(x) = P - F(x) \quad . \tag{E.2.1}$$

We treat the problem in two steps. In the first step we determine an interval (x_0, x_1) that contains the zero. In the second step we systematically reduce the interval such that its size becomes smaller than a given value ε.

In the first step we make use of the fact that $k(x)$ is monotonic, since $f(x)$ is monotonic. We begin with initial values for x_0 and x_1. If $f(x_0) \cdot f(x_1) < 0$, i.e., if the function values have different signs, then the zero is contained within the interval. If this is not the case, then we enlarge the interval in the direction where the function has the smallest absolute value, and repeat the procedure with the new values of (x_0, x_1).

Program E.6: Subroutine `AUXZBR` brackets the zero of a function $f(x)$.

CALLING SEQUENCE: `CALL AUXZBR(X0,X1,FUNCT,PAR,NPAR1, NPAR2)`

INPUT ARGUMENTS: X0,X1 — argument values x_0, x_1, believed to be near the zero of $f(x)$
FUNCT — FORTRAN name of a function subprogram that computes the function $f(x)$, supplied by the user
PAR,NPAR1,NPAR2 — parameters p, n_1, n_2, on which the function $f(x)$ can depend in addition to the argument x

OUTPUT ARGUMENTS: X0,X1 — arguments x_0, x_1 which bracket the zero of $f(x)$

REMARKS: The program is reliable for functions that have a zero and are monotonic in x. It should only be used in such cases.

```
      SUBROUTINE AUXZBR(X0,X1,FUNCT,PAR,NPAR1,NPAR2)
      IMPLICIT DOUBLE PRECISION (A-H,O-Z)
      PARAMETER (ZERO=0.D0, ONE=1.D0, TWO=2.D0)
      EXTERNAL FUNCT
      IF(X0.EQ.X1) X1=X0+ONE
      F0=FUNCT(X0,PAR,NPAR1,NPAR2)
      F1=FUNCT(X1,PAR,NPAR1,NPAR2)
      DO 10 I=1,1000
        IF(F0*F1 .GT. ZERO) THEN
          IF(ABS(F0).LE.ABS(F1)) THEN
            XS=X0
            X0=X0+TWO*(X0-X1)
            X1=XS
            F1=F0
            F0=FUNCT(X0,PAR,NPAR1,NPAR2)
          ELSE
            XS=X1
            X1=X1+TWO*(X1-X0)
            X0=XS
            F0=F1
            F1=FUNCT(X1,PAR,NPAR1,NPAR2)
          END IF
        ELSE
          GO TO 20
        END IF
   10 CONTINUE
   20 CONTINUE
      END
```

For localizing the zero within the initial interval (x_0, x_1) we use a comparatively slow but absolutely reliable procedure. The original interval is divided in half and replaced by the half for which the end points are of the opposite sign. The procedure is repeated until the interval width decreases below a given value.

Program E.7: Subroutine AUXZFN determines by successive interval reduction the zeros of a function $f(x)$

CALLING SEQUENCE: CALL AUXZFN(X0,X1,XZERO,FUNCT,PAR, NPAR1,NPAR2,EPSILN)

INPUT ARGUMENTS: X0,X1 — argument values that bracket the zero
FUNCT,PAR,NPAR1,NPAR2 — as in Program E.6
EPSILN — precision ε given such that the output value of the zero differs by less than ε from the true zero

OUTPUT ARGUMENTS: XZERO — zero x_z

```
      SUBROUTINE AUXZFN(X0,X1,XZERO,FUNCT,PAR,NPAR1,NPAR2,EPSILN)
      IMPLICIT DOUBLE PRECISION (A-H,O-Z)
      PARAMETER (ZERO=0.D0, HALF=0.5D0)
      EXTERNAL FUNCT
      XZERO=X0
      DO 10 I=1,2000
        F0=FUNCT(X0,PAR,NPAR1,NPAR2)
        F1=FUNCT(X1,PAR,NPAR1,NPAR2)
        IF(F0.EQ.ZERO) THEN
          XZERO=X0
          GO TO 20
        ELSE IF(F1.EQ.ZERO) THEN
          XZERO=X1
          GO TO 20
        END IF
        XM=HALF*(X0+X1)
        IF(ABS(X0-X1).GE.EPSILN) THEN
          FM=FUNCT(XM,PAR,NPAR1,NPAR2)
          TEST=F0*FM
          IF(TEST .LT. ZERO) THEN
            X1=XM
          ELSE
            X0=XM
          END IF
        ELSE
          XZERO=XM
          GO TO 20
        END IF
   10 CONTINUE
   20 CONTINUE
      END
```

Sections C.4 through C.7 contain examples of the application of AUXZBR and AUXZFN.

F. The Graphics Programming Package GRPACK

F.1 Introductory Remarks

The graphical display of data and of curves of fitted functions has always been an important aid in data analysis. We provide here a simple but powerful system of graphics routines called GRPACK. These routines allow graphics display and output on IBM XT/AT and PS/2 personal computers, provided that these are connected to one of the commonly available display or output devices. Such devices are called *graphics workstations*. The various workstations supported by GRPACK are listed in Sect. F.9.

Since, however, computer graphics is not the subject of this book, the routines are not shown here completely. Their function is only briefly explained. In particular, however, the calling sequences of the individual routines are given explicitly.

The programming package GRPACK consists of a number of user-callable subprograms, and many other routines which remain invisible. All of the subprograms are of the type SUBROUTINE. Their calling sequences and the meanings of their arguments are specified in the following sections. We distinguish between control routines, transformation routines, drawing routines, and utility routines.

The usage of the package GRPACK is described in this appendix which also contains a number of example programs. For many applications, however, it is sufficient to become familiar with the use of only five routines which in turn make extensive use of the subprograms of the package GRPACK. These 5 subprograms generate complete graphical structures by a single call. They are

— the subprogram SMHSGR for displaying a histogram (an example program is E2SM, Sect. 6.12.2)

— The subprogram SMSDGR for displaying a two-dimensional scatter plot (an example program is E3SM, Sect. 6.12.3)

— the subprogram GRHSCV for plotting a histogram and an arbitrary polyline (an example program is E6GR, Sect. F.10.6)

— the subprogram GRDTCV for plotting data points and an arbitrary polyline (an example program is E7GR, Sect. F.10.7)

— the subprogram GRDTMC for plotting data points and several arbitrary polylines (an example program is E8GR, Sect. F.10.8)

F.2 Control Routines

Program F.1: Subroutine GROPEN opens access to the graphics package GRPACK

This routine carries out the necessary initialization and must be called before any of the other routines of GRPACK can be called.

CALLING SEQUENCE: CALL GROPEN

Program F.2: Subroutine GRCLSE terminates the session with the graphics package GRPACK

This routine clears all open workstations and closes them.

CALLING SEQUENCE: CALL GRCLSE

Program F.3: Subroutine GROPWS opens a graphics workstation

The individual procedures for opening depend on the type of workstation; see Sect. F.9.

CALLING SEQUENCE: CALL GROPWS(NWS)

INPUT ARGUMENT: NWS — workstation number

Program F.4: Subroutine GRCLWS closes a graphics workstation

The individual procedures for closing depend on the type of workstation; see Sect. F.9.

CALLING SEQUENCE: CALL GRCLWS(NWS)

INPUT ARGUMENT: NWS — workstation number

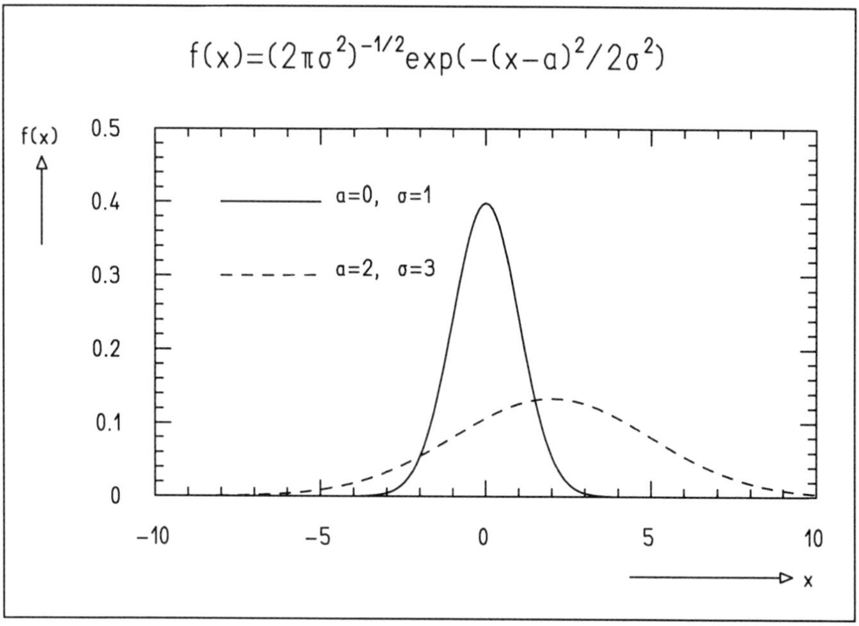

Fig. F.1. Simple example of a plot produced with GRPACK.

F.3 Coordinate Systems and Transformations

F.3.1 Coordinate Systems

World Coordinates (WC) Figure F.1 shows a plot made by GRPACK. Let us imagine for a moment that all of the graphical structures, including text, physically exist, e.g., that they are made out of bent wire. The coordinate system in which this wire structure is described is called the world coordinate system (WC). The coordinates of a point in the WC are denoted by (X, Y).

Computing Coordinates (CC) If we consider the axes in Fig. F.1, we note that the axes designated with x and y have about the same length in world coordinates, but have very different lengths in terms of the numbers shown. The figure shows a plot of a function

$$y = f(x) \quad .$$

Each point (x, y) appears at the point (X, Y). We call the coordinate system of the (x, y) points the computing coordinate system (CC). The transformation between WC and CC is given by Eq. (F.3.1).

Device Coordinates (DC) From the (fictitious) world coordinate system, the plot must be brought onto the working surface of a graphics device (terminal screen or paper). We call the coordinates (u, v) on this surface the device coordinates (DC).

F.3.2 Linear Transformations. Window – Viewport

The concepts defined in this section and the individual transformations are illustrated in Fig. F.2.

Let us assume that the computing coordinates in x cover the range

$$x_a \leq x \leq x_b \quad .$$

The corresponding range in world coordinates is

$$X_a \leq X \leq X_b \quad .$$

A linear transformation $x \to X$ is therefore defined by

$$X = X_a + (x - x_a)\frac{X_b - X_a}{x_b - x_a} \quad . \tag{F.3.1}$$

The transformation for $y \to Y$ is defined in a corresponding way. One speaks of the mapping of the *window* in computing coordinates CC,

$$x_a \leq x \leq x_b \quad , \qquad y_a \leq y \leq y_b \quad , \tag{F.3.2}$$

onto the *viewport* in world coordinates WC,

$$X_a \leq X \leq X_b \quad , \qquad Y_a \leq Y \leq Y_b \quad . \tag{F.3.3}$$

The mapping is in general *distorted*, i.e., a square in CC becomes a rectangle in WC. It is undistorted only if the aspect ratios of the window and viewport are equal,

$$\frac{x_b - x_a}{y_b - y_a} = \frac{X_b - X_a}{Y_b - Y_a} \quad . \tag{F.3.4}$$

The user of the programming package GRPACK computes values first in CC that are to be plotted. By specifying the window and viewport he or she defines the mapping into WC. A possible linear distortion is perfectly acceptable in this step, since it provides a simple means of changing the scale.

Next a mapping onto the physically implemented device coordinates DC must be done. It can of course be defined by again providing a window (in WC) and a viewport (in DC). One wants, however, to avoid an additional distortion. We define, therefore, a viewport

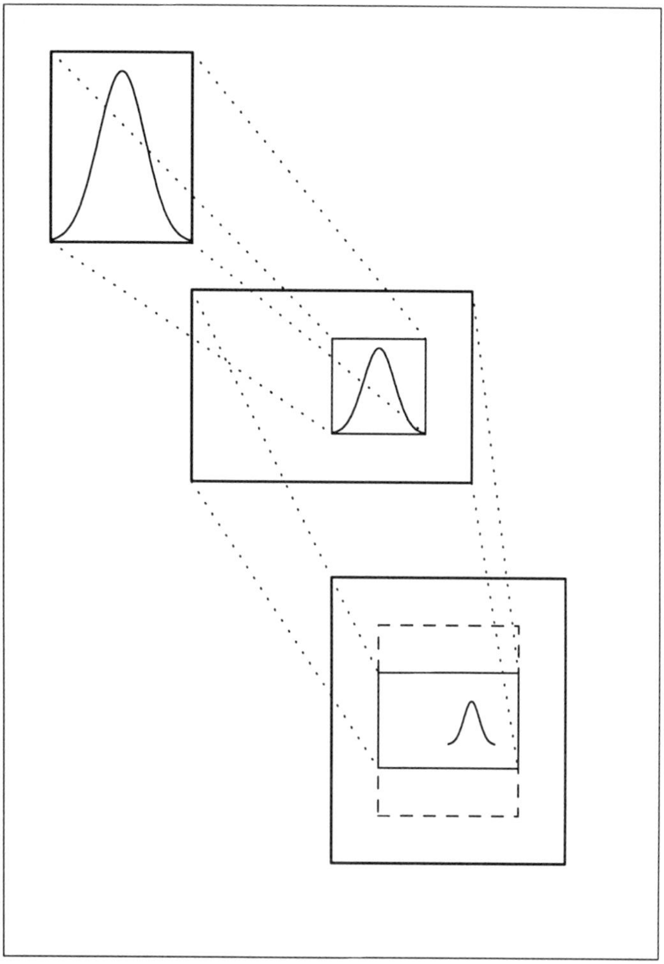

Fig. F.2. The various coordinate systems. Above: window in computing coordinates. Middle: viewport (small rectangle) and window (large rectangle) in world coordinates. Below: preliminary viewport (dashed rectangle), final adjusted viewport (small rectangle), and border of the display surface (large rectangle) in device coordinates. The mappings from computing coordinates to world coordinates and from world coordinates to device coordinates are indicated by dotted lines.

$$u_a \le u \le u_b \quad , \qquad v_a \le v \le v_b \tag{F.3.5}$$

and a window

$$X'_a \le X \le X'_b \quad , \qquad Y'_a \le Y \le Y'_b \quad . \tag{F.3.6}$$

The mapping from the window (F.3.6) onto the viewport (F.3.5) is only carried out if both have the same aspect ratio. Otherwise the viewport (F.3.6) is reduced symmetrically in width to the right and left or in height symmetrically above and below such that the viewport

$$u'_a \le u \le u'_b \quad , \qquad v'_a \le v \le v'_b \tag{F.3.7}$$

has the same aspect ratio as the window (F.3.6). In this way a distortion free mapping is defined between the two.

F.4 Transformation Routines

The user of GRPACK must define the transformations between the various coordinate systems by calling the appropriate routines. The transformations are then applied when drawing a graphical structure without any further intervention.

Transformation CC → WC

This transformation is defined by calling the following two routines.

Program F.5: Subroutine GRWNCC establishes a window in computing coordinates

CALLING SEQUENCE: `CALL GRWNCC(XACC,XBCC,YACC,YBCC)`

INPUT ARGUMENTS: XACC, XBCC, YACC, YBCC have the meaning of x_a, x_b, y_a, y_b in (F.3.2)

DEFAULT VALUES: Until GRWNCC is called, one has $x_a = 0$, $x_b = 1$, $y_a = 0$, $y_b = 1$.

REMARKS: Unreasonable input values are replaced by values that are as reasonable as possible. If XACC = XBCC then the routine sets $x_a = \mathtt{XACC}$, $x_b = x_a + 1$, and correspondingly for y_a, y_b.

Program F.6: Subroutine GRVWWC sets the viewport in world coordinates

CALLING SEQUENCE: `CALL GRVWWC(XAWC,XBWC,YAWC,YBWC)`

INPUT ARGUMENTS: XAWC, XBWC, YAWC, YBWC have the meaning of X_a, X_b, Y_a, Y_b in (F.3.3)

DEFAULT VALUES: Until GRVWWC is called, one has $X_a = 0$, $X_b = 1$, $Y_a = 0$, $Y_b = 1$.

REMARKS: Unreasonable input values are replaced by values that are as reasonable as possible. If XAWC = XBWC, then the routine sets $X_a = \texttt{XAWC}$, $X_b = X_a + 1$.

Transformation WC → DC

This transformation is defined by calling the following two subroutines.

The routine GRSTFR defines the temporary viewport in device coordinates. The final viewport is fitted into the temporary one such that it has the same aspect ratio as does the window defined by the routine GRWNWC in world coordinates.

Program F.7: Subroutine GRSTFR establishes the picture format on the workstation

CALLING SEQUENCE: `CALL GRSTFR(NWS,WIDTH,HEIGHT)`

INPUT ARGUMENTS: NWS — workstation number

(a) If WIDTH > 0., HEIGHT > 0.:
WIDTH — picture width in cm
HEIGHT — picture height in cm

(b) If HEIGHT ≤ 0.:
WIDTH — gives DIN landscape format, e.g., WIDTH = 5 means A5 landscape, corresponding to case (a) WIDTH = 21.0, HEIGHT = 14.8

(c) If WIDTH ≤ 0.:
HEIGHT — gives portrait format

REMARKS: After the workstation has been opened until GRSTFR has been called, the format is taken to be A5 landscape.

If the workstation is a device with a screen, then only the ratio of width and height is considered. The plot always appears as large as possible and centered on the screen. If the workstation is a plotting device or printer, then the measurements in cm only hold inasmuch as a plot with these dimensions can fit onto the paper. If it is larger than the given size, then the plot is reduced until it just fits. In both cases the plot is centered on the paper.

Program F.8: Subroutine GRWNWC establishes a window in world coordinates

CALLING SEQUENCE: CALL GRWNWC(XAPWC,XBPWC,YAPWC,YBPWC)

INPUT ARGUMENTS: XAPWC, XBPWC, YAPWC, YBBWC have the meanings of X'_a, X'_b, Y'_a, Y'_b in (F.3.6)

DEFAULT VALUES: Until GRWNWC has been called, one has $X'_a = X_a$, $X'_b = X_b$, $Y'_a = Y_a$, $Y'_b = Y_b$.

REMARKS: Unreasonable input values are replaced by values that are as reasonable as possible. If XAPWC = XBPWC, then the routine sets $X'_a = \mathtt{XAPWC}$, $X'_b = X'_a + 1$. If XAPWC > XBPWC, then the routine sets $X'_a = \mathtt{XBPWC}$, $X'_b = \mathtt{XAPWC}$, and correspondingly for Y'_a, Y'_b.

Usually the user is interested only in the computing coordinate system, after having defined the transformations onto world and then device coordinates by calls to the routines described in this section.

The graphical structures are not completely drawn under certain circumstances. They are truncated if they extend past the boundary of the so-called *clipping region*. The structures are said to be *clipped*. For polylines, markers, data points, and contour lines (Sect. F.5) the clipping region is in computing coordinates; for text and graphics utility structures the clipping region is the window in world coordinates.

F.5 Drawing Programs

The subroutines given up to this point carry out organizational tasks, but do not, however, produce any graphical structures on the workstation. All of the graphical structures created by GRPACK consists of lines. The lines possess a given line width and/or color. For the workstations supported by GRPACK, line width and/or color can be changed. This takes place by setting the *color index*. The meaning of the color index for various workstations is given in Sect. F.9. Once it has been set, the color index retains its value until it is changed to another.

Program F.9: Subroutine GRSTCL sets the color index

CALLING SEQUENCE: CALL GRSTCL(ICOL)

INPUT ARGUMENT: ICOL — color index

DEFAULT VALUES: The color index is 1 until changed by calling GRSTCL.

REMARKS: The meaning of the color index depends on the type of workstation; cf. Sect. F.9.

The concepts of a polyline and polymarker have been introduced for particularly simple graphics structures. A *polyline* defines a sequence of line segments from the point (x_1, y_1) through the points (x_2, y_2), (x_3, y_3), ... to the point (x_n, y_n). A *polymarker* marks a plotting point with a graphical symbol. The polymarkers available in GRPACK are shown in Fig. F.3.

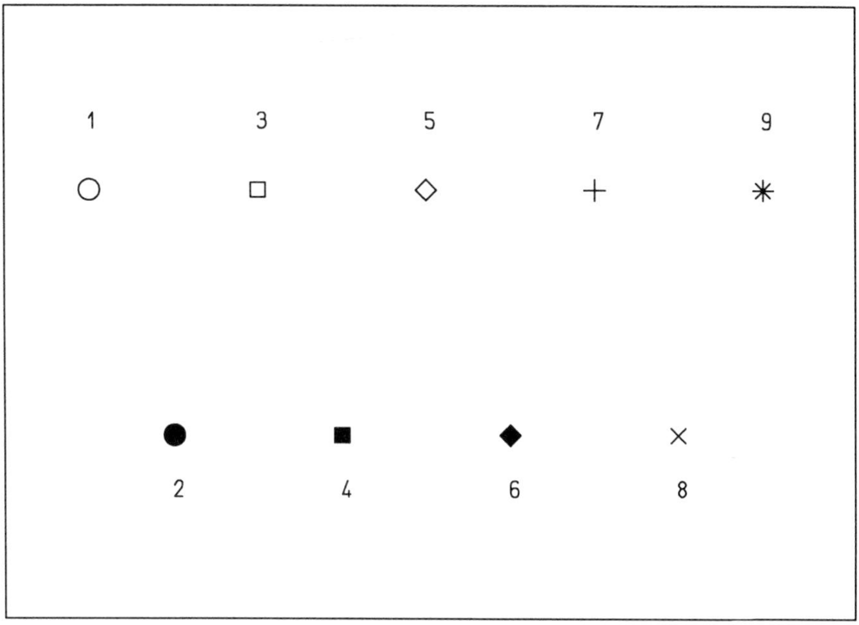

Fig. F.3. Polymarkers.

Program F.10: Subroutine GRPLIN draws a polyline.

CALLING SEQUENCE: CALL GRPLIN(N,XPL,YPL)

INPUT ARGUMENTS: N — number of points defining a polyline
XPL — name of an array containing the x coordinates of the points (in CC)
YPL — name of an array for the y coordinates

REMARKS: The polyline is only drawn inside the CC-window (F.3.2).

One can clearly achieve an arbitrarily good approximation of any graphical structure by means of polylines. For example, *graphs of functions* can be displayed with polylines as long as the individual points are sufficiently close to each other.

Sometimes one wants to draw a polyline not as a solid line but rather *dashed*, *dotted*, or *dot-dashed*.

Program F.11: Subroutine `GRBRPL` draws a polyline with selected line type

CALLING SEQUENCE: `CALL GRBRPL(NTYPE,SCALEF,N,XPL,YPL)`

INPUT ARGUMENTS: `NTYPE` – line type (1: dashed, 2: dotted, 3: dot-dashed)
`SCALEF` — scale factor. The typical length of the dash is one percent of the diagonal of the picture format multiplied by this factor.
`N,XPL,YPL` — as in Program F.10

Program F.12: Subroutine `GRMARK` draws a polymarker

CALLING SEQUENCE: `CALL GRMARK(NMARK,SCALEF,XPM,YPM)`

INPUT ARGUMENTS: `NMARK` — polymarker number; cf. Fig. F.3
`SCALEF` — scale factor. The diameter of the polymarker is one percent of the diagonal of the picture format multiplied by this factor.
`XPM,YPM` — coordinates x, y (in CC) of the point to be plotted with the polymarker.

REMARKS: The polymarkers are only drawn inside the CC window (F.3.2).

Polymarkers are especially suitable for marking data points. If a data point has error bars, then one would like to indicate these errors in one or both coordinates by means of *error bars*. In certain circumstances one would even like to show the complete covariance ellipse. The following routine performs this task. Examples are shown in Fig. F.4.

Program F.13: Subroutine `GRDATP` draws a data point with error bars and covariance ellipse

CALLING SEQUENCE: `CALL GRDATP(NMARK,SCALEF,X,Y,SIGMAX, SIGMAY,COVAR)`

INPUT ARGUMENTS: `NMARK` — polymarker number for plotting the data point; cf. Fig. F.3
`SCALEF` — scale factor as in Program F.12
`X,Y` — computing coordinates (x, y) of the data point
`SIGMAX` — σ_x, error in x
`SIGMAY` — σ_y, error in y
`COVAR` — $\mathrm{cov}(x, y)$, covariance of x and y

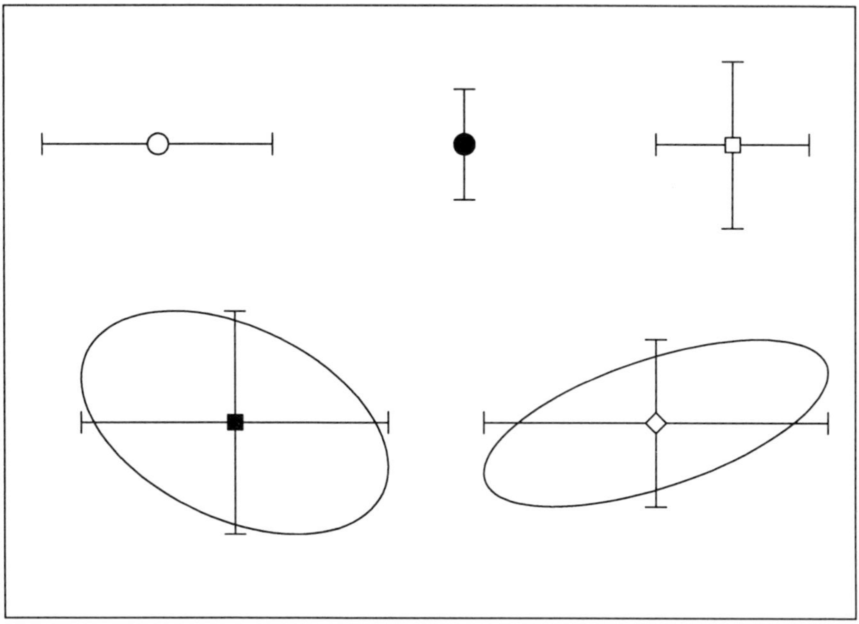

Fig. F.4. Example for plotting data points.

REMARKS: An error bar in the x direction is only drawn if $\sigma_x > 0$, and in the y direction only if $\sigma_y > 0$. The covariance ellipse is only draw if $\sigma_x > 0$, $\sigma_y > 0$, and $\text{cov}(x, y) \neq 0$. Error bars are not drawn if they would lie completely within the polymarker itself. If part of the structure falls outside of the CC window (F.3.2), then this part is not drawn.

The following routine displays data in the form of a histogram. It draws a histogram characterized by the simple data structures defined in Sect. 6.3.

Program F.14: Subroutine `GRHSDR` draws a histogram

CALLING SEQUENCE: `CALL GRHSDR(HIST,X0,DELX,NX)`

INPUT ARGUMENTS: `HIST` — array containing the contents of the first, second, ... bins of the histogram
`X0` — lower boundary of the first bin
`DELX` – bin width
`NX` — number of bins

REMARKS: The array `HIST` will typically have been filled by the routines 6.2 and 6.3.

We conclude our short list of drawing programs with routines for drawing *contour lines*. A function $f = f(x, y)$ defines a surface in a three dimensional (x, y, f) space. One can also get an idea of the function in the (x, y) plane by marking points for which $f(x, y)$ is equal to a given constant c. The set of all such points forms the contour line $f(x, y) = c$. By drawing a set of such contour lines $f(x, y) = c_1, c_2, \ldots$ one obtains (as with a good topographical map) a rather good impression of the function.

Naturally it is impossible to compute the function for all points in the (x, y) plane. We restrict ourselves to a rectangular region in the (x, y) plane, usually the window in computing coordinates, and we break it into a total of $N = n_x n_y$ smaller rectangles. The corner points of these smaller rectangles have x coordinates that are neighboring values in the sequence

$$x_0, x_0 + \Delta x, x_0 + 2\Delta x, \ldots, x_0 + n_x \Delta x \quad .$$

The y coordinates of the corner points are adjacent points of the sequence

$$y_0, y_0 + \Delta y, y_0 + 2\Delta y, \ldots, y_0 + n_y \Delta y \quad .$$

In each rectangle the contour line is approximated linearly. To do this one considers the function $f(x, y) - c$ at the four corner points. If the function is of a different sign at two corner points that are the end points of an edge of the rectangle, then it is assumed that the contour lines intersect the edge. The intersection point is computed with linear interpolation. If the intersection points are on two edges of the rectangle, then they are joined by a line segment. If there are intersection points on more than two edges, then all pairs of such points are joined by line segments.

Clearly the approximation of the contour lines by line segments becomes better for a finer division into small rectangles. With a finer division, of course, the required computing time also becomes longer.

The subroutine GRPLCT defines the large rectangle described above, divides it into smaller rectangles and draws the contour line $f(x, y) = c$.

Program F.15: Subroutine GRPLCT draws the contour line $f(x, y) = c$ of a function f as a line segment in the (x, y) plane

CALLING SEQUENCE: CALL GRPLCT(X0,Y0,DX,DY,NX,NY,FCONT, USERFN)

INPUT ARGUMENTS: X0, DX, NX — initial value x_0, interval width Δx, and number of intervals n_x for dividing the x axis
Y0, DY, NY — same for the y axis
FCONT — function value c of the contour line

REMARKS: The user must supply a short function subprogram with two arguments (x, y) that computes the function $f = f(x, y)$.

The routine GRPLCT successively computes the coordinates of the corner points of all small rectangles and the function values at these points. It calls the routine GRPXCT, which carries out the steps described above for each small rectangle.

Program F.16: Subroutine GRPXCT draws contour line $f(x, y) = c$ approximated as line segment in a small rectangle of the (x, y) plane

CALLING SEQUENCE: CALL GRPXCT(X1,X2,Y1,Y2,F11,F12,F21, F22,FCONT)

INPUT ARGUMENTS: X1,X2,Y1,Y2 — coordinates of the corner points $P_{11} = (x_1, y_1)$, $P_{12} = (x_1, y_2)$, $P_{21} = (x_2, y_1)$, $P_{22} = (x_2, y_2)$
F11,F12,F21,F22 — function value at these points
FCONT — function value of the contour line

An example of a function represented by contour lines is shown in Fig. F.5.

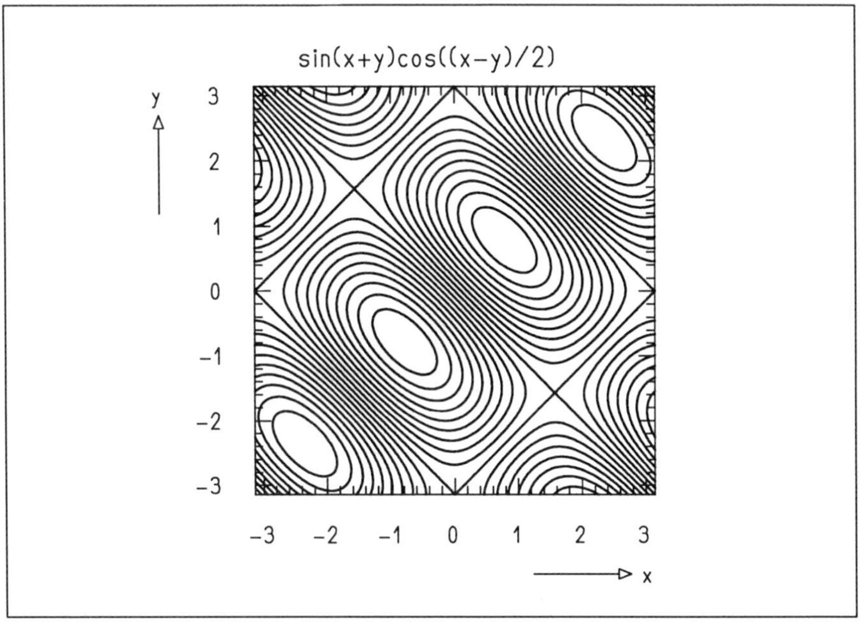

Fig. F.5. Contour lines $f(x, y) = -0.9, -0.8, \ldots, 0.8, 0.9$ of the function $f(x) = \sin(x + y)$ $\cos([x - y]/2)$ in the (x, y) plane.

F.6 Graphics Utility Routines

With the few programs described up to this point, a great variety of complicated plots can be produced. By using the routines of this section and the next, the tasks are made easier for the user, since they help to create graphical structures typically used in conjunction with the plots of data analysis, such as axes, crosses, and explanatory text.

Program F.17: Subroutine GRFRAM draws a frame around the window of the world coordinate system

CALLING SEQUENCE: CALL GRFRAM

Program F.18: Subroutine GRBOUN draws a frame around the computing coordinates window

CALLING SEQUENCE: CALL GRBOUN

Program F.19: Subroutine GRSCLX draws an axis in the x direction.

CALLING SEQUENCE: CALL GRSCLX(STRING,LEN)

INPUT ARGUMENTS: STRING — FORTRAN name of a character string (encoded according to Sect. F.7), whose contents appears below the axis with an arrow in the direction of increasing x values. If the string contains only spaces, neither text nor arrow are displayed.
LEN — length of the character string

REMARKS: Tick marks from the upper and lower edges of the CC window are drawn inside the window. Below the lower edge, numbers are displayed which label some of these marks. It is recommended to first call the subroutine GRBOUN to display lines on the window itself.

Program F.20: Subroutine GRSCLY draws an axis in the y direction

CALLING SEQUENCE: CALL GRSCLY(STRING,LEN)

INPUT ARGUMENTS: STRING — FORTRAN name of a character string (encoded according to Sect. F.7), whose contents appears to the left of the axis with an arrow in the direction of increasing y values. If the string contains only spaces, neither text nor arrow are displayed.
LEN — length of the character string

REMARKS: See Program F.19. The tick marks appear on the left and right edges of the CC window, the numbers are to the left of the window.

The creation of axis divisions and labels with the routines GRSCLX and GRSCLY is usually done automatically without intervention of the user. Sometimes, however, the user will want to influence these operations. This can be done with the following routine.

Program F.21: Subroutine GRSCDF defines the parameters for displaying an axis.

CALLING SEQUENCE: CALL GRSCDF(FIX,DEL,NINTER,SCFTIC, SCFNUM,LNUM)

INPUT ARGUMENTS: FIX — value (in CC) which one wants to mark and label with a number on the axis
DEL — distance between labeled tick marks
NINTER — number of intervals with smaller tick marks between two labeled tick marks
SCFTIC — scale factor for the length of the tick marks
SCFNUM — scale factor for the size of the numbers
LNUM — logical variable defining whether the numbers on the axes are shown in decimal form (LNUM = .TRUE.) or exponential form (LNUM = .FALSE.)

DEFAULT VALUES: For input values ≤ 0, reasonable values are assigned.

REMARKS: A call to GRSCDF only affects the axis created with the next call to GRSCLX or GRSCLY.

Program F.22: Subroutine GRCCRS draws a cross in computing coordinates

CALLING SEQUENCE: CALL GRCCRS

REMARKS: The axes of the computing coordinate system appear as dashed lines within the CC window.

F.7 Text within the Plot

Explanatory text makes plots easier to understand. The subroutines in this section create text superimposed on a plot which can be placed at any location.

The text must be supplied by the user as a FORTRAN character string. Before this text is translated into graphics characters, however, it must first be encoded. The simple encoding system used here allows the user to display simple mathematical formulas. For this there are *three character sets*: Roman, Greek, and mathematics, as shown in Table F.1. The character set is selected by *control characters*. These are the special characters

@ for Roman,
& for Greek,
% for mathematics.

A control character in the text string causes all of the following characters to be produced with the corresponding character set, until another control symbol appears. The default character set is Roman.

In addition there exist the following *positioning symbols:*

^ for superscript (exponent),
_ for subscript (index),
for normal height,
" for backspace.

All characters appear at normal height as long as no positioning symbol has appeared. One can move a maximum of two steps from normal height, e.g., A_{α_β}, $A_{\alpha\beta}$. The positioning symbols ^ and _ remain in effect until the appearance of a #. The symbol " acts only on the character following it. This then appears over the previous character instead of after it. In this way one obtains, e.g., A_α^β instead of $A_\alpha{}^\beta$.

Clearly the character strings supplied by the user will be of differing lengths. In order to allow this despite the awkward handling of strings in FORTRAN, a character variable must be declared in the user program (e.g., with the name TEXT and length *len*) in the following way:

```
CHARACTER*len TEXT
```

Before calling a subroutine that takes a character string as input, the contents of the string must be assigned, e.g., with

```
TEXT = 'f(x) = e^x'
```

For longer strings, only the first 100 characters are used. Blank spaces at the end of a string are suppressed.

Table F.1. The various character sets for producing text.

	Control Characters				Control Characters		
	Roman	Greek	Math		Roman	Greek	Math
Input	@	&	%	Input	@	&	%
A	A	A(ALPHA)	Ä	a	a	α(alpha)	ä
B	B	B(BETA)	B	b	b	β(beta)	b
C	C	X(CHI)	⌉	c	c	χ(chi)	c
D	D	Δ(DELTA)	Δ	d	d	δ(delta)	d
E	E	E(EPSILON)	E	e	e	ϵ(epsilon)	e
F	F	Φ(PHI)	F	f	f	φ(phi)	f
G	G	Γ(GAMMA)	$\neq$	g	g	γ(gamma)	g
H	H	H(ETA)	H	h	h	η(eta)	h
I	I	I(IOTA)	$\int$	i	i	ι(iota)	i
J	J	I(IOTA)	J	j	j	ι(iota)	j
K	K	K(KAPPA)	K	k	k	κ(kappa)	k
L	L	Λ(LAMBDA)	|	l	l	λ(lambda)	l
M	M	M(MU)	$\pm$	m	m	μ(mu)	m
N	N	N(NU)	N	n	n	ν(nu)	n
O	O	Ω(OMEGA)	Ö	o	o	$\acute{\omega}$(omega)	ö
P	P	Π(PI)	Ö	p	p	π(pi)	p
Q	Q	Θ(THETA)	Q	q	q	ϑ(theta)	q
R	R	R(RHO)	∘	r	r	ρ(rho)	r
S	S	Σ(SIGMA)	ß	s	s	σ(sigma)	s
T	T	T(TAU)	¦	t	t	τ(tau)	t
U	U	O(OMICRON)	Ü	u	u	o(omicron)	ü
V	V		Ü	v	v		v
W	W	Ψ(PSI)	$\surd$	w	w	ψ(psi)	w
X	X	Ξ(XI)	X	x	x	ξ(xi)	x
Y	Y	Υ(UPSILON)	Å	y	y	υ(upsilon)	y
Z	Z	Z(ZETA)	Z	z	z	ζ(zeta)	z
~	∼	∼	∼	-	—	—	—
!	!	!	!	=	=	=	≡
$	$	$	$	{	{	{	{
*	*	#	×	}	}	}	}
(	(	↑	←	|	|	|	|
)	)	↓	→	[	[	&	[
+	+	+	+	]	]	@	]
`	‘	‘	‘	\			
1	1	1	1	:	:	:	:
2	2	2	2	;	;	;	;
3	3	3	3	'	’	‘	’
4	4	4	4	<	<	⊂	≤
5	5	5	5	>	>	⊃	≥
6	6	6	6	?	?	§	∼
7	7	7	7	,	,	,	,
8	8	8	8	.	.	.	.
9	9	9	9	/	/	\	%
0	0	0	0				

Program F.23: Subroutine GRTXTC draws text

CALLING SEQUENCE: CALL GRTXTC(SCALEF,STRING,LEN)

INPUT ARGUMENTS: SCALEF — scale factor. The text height is 1.5 % of the diagonal of the window in world coordinates multiplied by the scale factor
STRING — name of the FORTRAN character string containing the encoded text
LEN — length of the character string

REMARKS: The text appears centered just under the upper edge of the window in the world coordinate system.

Sometimes the user wants to write text at a certain place in the plot, e.g., next to an individual curve or data point, and also to choose the text size. This is made possible by the following program.

Program F.24: Subroutine GRTXTF displays text at an arbitrary position

CALLING SEQUENCE: CALL GRTXTF(X,Y,SCALEF,STRING,LEN)

INPUT ARGUMENTS: X,Y — computing coordinates of the point (x, y), which designates the lower left-hand corner of the rectangle surrounding the text
SCALEF — scale factor. The text height is 1.5 % of the diagonal of the window in world coordinates multiplied by the scale factor
STRING — name of the character string containing the encoded text
LEN — length of the character string

F.8 Programs Yielding a Complete Plot

The following three subprograms solve in a very simple way problems which appear frequently in practice.

Program F.25: Subprogram GRHSCV for plotting a histogram and an arbitrary polyline

This problem appears if data exist in the form of a histogram and if the histogram is to be plotted together with a curve (cf. Fig. F.7).

CALLING SEQUENCE: CALL GRHSCV(XPL,YPL,NPL,HIST,X0,
DELX,NX,TX,LTX,TY,LTY,CAPT,LCAPT,NWS)

INPUT ARGUMENTS: XPL, YPL — arrays for x and y coordinates of the points of the polyline
NPL — number of points in the polyline
HIST, X0, DELX, NX — define the histogram (as in Program 6.2)
TX, TY, CAPT — character strings containing text for horizontal and vertical axes, respectively, and text of the caption
LTX, LTY, LCAPT — lengths of the character strings
NWS — number of the graphics workstation

REMARKS: The horizontal axis covers the region of the histogram. The scale on the vertical axis is chosen to match the contents of the histogram. The format of the plot is A5.

Program F.26: Subprogram GRDTCV for plotting data points and an arbitrary polyline

The program allows in a simple way the graphical representation of data points (optionally with error bars and covariance ellipses) and of a curve (cf., e.g., Fig. F.8).

CALLING SEQUENCE: CALL GRDTCV(XPL,YPL,NPL,NMARK, SCALEF,DATX,DATY,DATSX,DATSY,DATCOV,NDAT, TX,LTX,TY,LTY,CAPT,LCAPT,NWS)

INPUT ARGUMENTS: XPL, YPL, NPL — define a polyline as in GRHSCV, Program F.25
NMARK, SCALEF — polymarker number for the data points and scale factor (cf. Program F.13)
DATX, DATY, DATSX, DATSY, DATCOV — arrays (minimum length NDAT) containing the x coordinates, y coordinates, errors in x, errors in y, and the covariance for the data points (cf. Program F.13)
NDAT — number of data points
TX, LTX, TY, LTY, CAPT, LCAPT, NWS — as in GRHSCV, Program F.25

REMARKS: The scales of the abscissa and ordinate are chosen according to the data points. The plot format is A5.

Program F.27: Subprogram GRDTMC for generating a plot with data points and several arbitrary polylines

The program allows in a simple way the representation of data points (optionally with error bars and covariance ellipse) and of several polylines (see, e.g., Fig. F.9).

CALLING SEQUENCE: CALL GRDTMC(XPL,YPL,NPL,NCURVE,NCOL, NMARK,SCALEF,DATX,DATY,DATSX,DATSY,DATCOV,NDAT, TX,LTX,TY,LTY,CAPT,LCAPT,NWS)

INPUT ARGUMENTS: XPL,YPL — arrays containing the points defining the polylines; the point i on the polyline j has the coordinates XPL(I,J), YPL(I,J)
NPL — number of points in each polyline
NCURVE — number of polylines
NCOL — array containing the color indices of the polylines. The color index of polyline j is the modulus of NCOL(J). If NCOL(J) is positive, then the polyline is displayed as a solid line, otherwise as a dashed line.
All other arguments — as in GRDTCV

REMARKS: The program can also be used in cases where only data points (NCURVE = 0) or only polylines (NPL = 0) are to be displayed. If data points exist, the scales of the horizontal and vertical axes are chosen according to the data points. Otherwise the choice of scales is adapted to the polylines. The plot format is A5.

F.9 Graphics Workstations

F.9.1 Types of Workstations. Initialization File. Color Index

Devices on which a graphical representation can be made visible are called in computer terminology *graphics workstations*. The programming package GRPACK supports two types of workstations: computer screens (or windows on computer screens) and files from which at a later time graphics can be produced on paper by a plotter or printer. Information about these devices are transmitted to GRPACK via the initialization file GRINIT.DAT. If the workstation in question does not correspond to the specifications defined in this file, then the file must be changed appropriately.

With the installation (cf. Sect. G.5) a version of the file GRINIT.DAT is provided, which is adapted to your choice of compiler and operating system. It contains information on several devices but only two devices are *activated*. These are

- a computer screen using vga graphics (or equivalent) if you work under DOS (or a graphical window on the computer screen if you work under WINDOWS or Linux and XWindow),
- a PostScript file for later off-line printing.

`GRINIT.DAT` contains device descriptions introduced by `[`*device*`]`, where only certain entries are recognized depending on the operating system and/or compiler. More than one description for one *device* is allowed. Up to 16 valid (recognized) entries for all devices in total can be handled. In most cases you can directly work with this version of the file `GRINIT.DAT`. But you can also edit the file and in this way activate or deactivate devices or change the properties of devices. (If your hardware provides only ega graphics, you may deactivate vga and activate ega. If you want to produce a HP-GL file rather than a PostScript file, you may deactivate PostScript and activate HP-GL.) The file `GRINIT.DAT` is an ASCII file which you can change with any editor. The meaning of all entries is explained by comments within the file. Therefore, here we only list the information given for the workstations without explaining it in detail:

- string describing the workstation,
- detailed technical information (e.g., video mode),
- width and height of workstation surface in internal units (pixels) and/or in centimeters,
- background color,
- colors and/or line widths associated with the 8 *color indices*.

F.9.2 Computer Screen or Window

If you work with the file `GRINIT.DAT` provided at installation time then this type of workstation is addressed by setting the *workstation number* `NWS` equal to `1`. If you have changed the file the workstation number may have changed. You can then get a list of actual workstation numbers using the subroutine `GRNBWS`, cf. Sect. F.9.4.

Screen in Graphical Mode under DOS Under DOS you can choose between different graphic modes:

- vga, ega, cga,
- 8514/A,
- Hercules.

For some compilers the choice is more restricted or extended. But vga or an equivalent is provided as default. For details consult the file `GRINIT.DAT`.

When this workstation is opened the computer screen is cleared and switched from alphanumeric to graphical mode. When it is closed the program pauses until the ENTER key is pressed (to allow inspection of the graphics). Then the screen is cleared again and switched into alphanumeric mode.

By opening the workstation again you again clear the screen, switch to graphical mode, and can produce a second graphics, etc.

Graphical Window under WINDOWS If you use the QuickWin facility provided by Microsoft, then opening such a workstation creates a graphical window into which the program draws graphical output. After closing the workstation the graphical window is not destroyed but remains available for inspection. If you open the workstation again the graphical window is cleared and the program can draw new graphics into it.

If you work under WINDOWS 95 or WINDOWS NT (and do not use the QuickWin facility) then by opening the workstation using GROPWS a resizable child window is opened into which the graphics is drawn. The window remains visible even after a call of GRCLWS. A new call of GROPWS creates an additional child window. Up to ten child windows can be opened. If you call GROPWS while already ten child windows are open, this call is ignored. Calls to GROPWS take effect again if you close one or more of the displayed child windows.

Graphical Window under Linux Under Linux a call of GROPWS opens a graphical window (X11). This window remains visible if the workstation is closed by a call of GRCLWS and even after the application is terminated. To close the graphical window, press the ENTER key while the mouse cursor is within the graphical window. In one application only one graphical window can be opened at a given time. Zoom is possible after calling GRCLWS by defining two corners of a rectangle using the middle or right mouse button. To close the zoom window press the ENTER key with the mouse cursor in that window.

F.9.3 Files for Off-Line Printing or Plotting

If you work with the file GRINIT.DAT provided at installation time then this type of workstation is addressed by setting the *workstation number* NWS equal to 2. If you have changed the file the workstation number may have changed. You can then get a list of actual workstation numbers using the subroutine GRNBWS, cf. Sect. F.9.4.

PostScript Graphical information is written to a file in the form of PostScript commands. It can be displayed later with a PostScript capable printer or display program. When opening the workstation a file with the name `GR???.PS` is created. The three characters symbolized here by `???` serve to number the files. The file is closed when the workstation is closed.

HP-GL The *Hewlett-Packard Graphic Language* HP-GL was originally developed for controlling mechanical plotters. It can also be used with other output devices such as printers which are able to interpret HP-GL commands.

When opening the workstation a file with the name `GR???.HPG` is created. The three characters `???` are used for file numbering. The file is closed when the workstation is closed.

Note that in the initialization file `GRINIT.DAT` you have to specify whether you want to use the `GR???.HPG` files to control a plotter or an HP printer. In the first case the file contains only HP-GL commands. In the second case it contains additional commands at the beginning and the end which switch the printer to HP-GL mode and back to its intrinsic mode.

F.9.4 Determination of the Workstation Numbers

You can display the numbers of the activated workstations (which depend on the initialization file `GRINIT.DAT`) at run time by the following program.

Program F.28: Subroutine `GRNBWS` displays workstation numbers

CALLING SEQUENCE: `CALL GRNBWS`

REMARKS: The program has no input arguments. Output is displayed on the screen.

F.9.5 Error File. File Handling

The program package `GRPACK` accesses simultaneously up to five files. These are the initialization file `GRINIT.DAT`, the files for HP-GL and PostScript output, the file `GRFILE.CNT` mentioned later in this section, and a file `GRERR.LIS`, in which possible error messages are written. This file should be checked by the user after using the routines in case an error is suspected.

If you program in FORTRAN please note: The five above named files are accessed as FORTRAN input/output units 95, 96, 97, 98, and 99. The user of `GRPACK` cannot, therefore, use these unit numbers in his or her own programs.

The following rules hold for the files `GR???.PS`, in which graphics information is stored: If, at the time when the file is to be created, no file of this type exists, then the file GR000.PS is created. The next file is given the name GR001.PS, etc. The character strings used for numbering are stored in a special file GRFILE.CNT, so that after closing GRPACK and opening it again the numbering can be continued. If the file GRFILE.CNT is deleted, then for the next PostScript file the name GR000.PS is used again. If there already exists a file named GR000.PS, it is overwritten.

The scheme described in the last paragraph also holds for the files GR???.HPG.

F.10 Example Programs

F.10.1 Main Program E1GR to Demonstrate Subprograms GRNBWS, GROPEN, GRCLSE, GROPWS, GRCLWS, GRWNCC, GRVWWC, GRWNWC, GRSTFR, GRFRAM, GRBOUN, GRSTCL, GRPLIN, GRBRPL, GRSCLX, GRSCLY, GRTXTF

The program generates the simple plot of Fig. F.1. The program first calls GRNBWS to display the workstation numbers and asks for the number NWS of the graphics workstation to be used. The package GRPACK is initialized by calling GROPEN. The workstation is opened with GROPWS and different coordinate systems are defined by calls of GRWNCC, GRVWWC, GRWNWC, and GRSTFR. The calls of GRFRAM and GRBOUN result in the display of the outer frame (indicating the borders of the selected section of the world coordinate system) and of the inner frame (bounding the window in computing coordinates), respectively. The calls of GRSCLX and GRSCLY ensure that scales will appear on the horizontal and vertical axes. The arrows corresponding to the scales will carry character strings stored in TEXTX and in TEXTY. With GRTXTC a caption of the plot is created, which is stored in CAPT.

With GRSTCL the color index is changed. In a short loop a total of 201 coordinate pairs TX(I), TY(I) are generated corresponding to the points (x_i, y_i). The abscissa values are $x_i = -10, -9.9, -9.8, \ldots, 10$ and the ordinate values are $y_i = f(x_i)$. The function $f(x)$ is the probability density of the standard normal distribution. With GRPLIN this function is displayed as the polyline through the points described above. In a second loop the points of a polyline are computed which corresponds to a normal distribution with mean $a = 2$ and standard deviation $\sigma = 3$. This polyline is displayed as a dashed curve with GRBRPL. Two short straight lines are displayed in the upper left corner of the plot (one as a solid line and one as a dashed line). To the right of these polylines a short text is displayed with GRTXTF, which contains the parameters of the Gaussians displayed as solid and dashed curves. Before termination of the program the workstation is closed by calling GRCLWS, and the package GRPACK is closed by a call of GRCLSE.

F.10.2 Main Program E2GR to Demonstrate Subprogram GRMARK

The short program generates the plot of Fig. F.3 showing the different types of symbols that can be drawn by GRMARK.

F.10.3 Main Program E3GR to Demonstrate Subprogram GRDATP

The program generates the plot Fig. F.4. This plot contains examples for the different possibilities to display data points with errors.

F.10.4 Main Program E4GR to Demonstrate Subprogram GRPLCT

A window of computing coordinates $-\pi \le x \le \pi$, $-\pi \le y \le \pi$ and a square viewport in world coordinates are selected. After creating scales and the caption, input parameters for GRPLCT are prepared. Next by successive calls of GRPLCT in a loop, contours of the function $f(x, y) = \sin(x+y)\cos((x-y)/2)$ are drawn. The function is computed in the function subprogram CONTFN. The result is a plot corresponding to Fig. F.5.

Suggestions: Extend the program such that the parameters NCONT and NSTEP defining the number of contours and the number of intervals in x and y can be set interactively by the user. Study the changes in the plot resulting from very small values of NSTEP.

F.10.5 Main Program E5GR to Demonstrate Subprograms GRSCDF and GRCCRS

The program generates the plots shown in Fig. F.6. It contains four plots which differ only by the design of their scales. The plots are generated in a loop where different viewports in world coordinates are chosen in each step, so that the plots correspond to the upper-left, upper-right, lower-left, and lower-right quadrant of the window in world coordinates. For the upper-left plot the default values for the scale design are used. In the upper-right plot the number of ticks and the lettering of the scale is predefined. In the lower-left plot the size of the symbols used in the lettering of the scales is changed. In the lower-right plot the numbers are written in exponential notation. All plots contain a coordinate cross, which is generated calling GRCCRS, and a curve corresponding to a Gaussian.

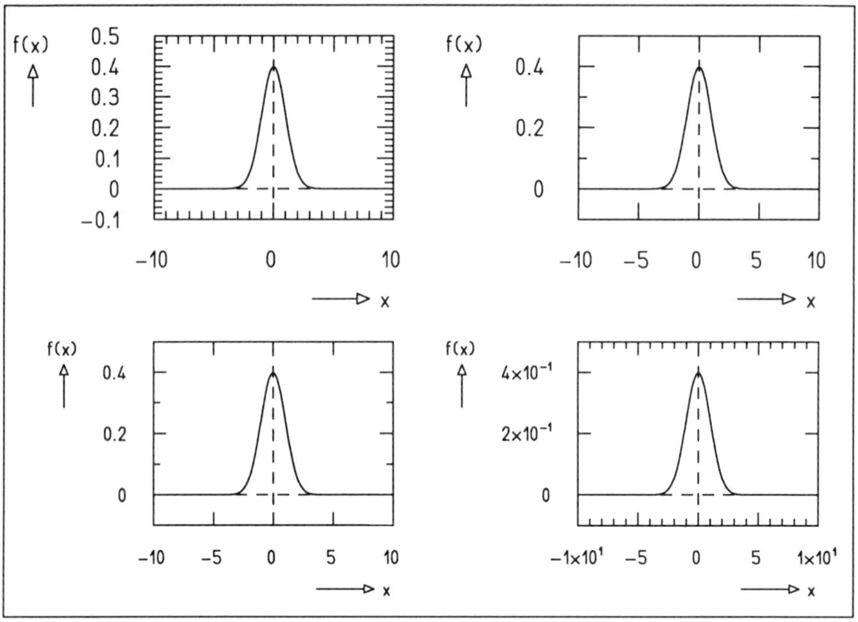

Fig. F.6. Four versions of the same plot with different types of scales.

F.10.6 Main Program E6GR to Demonstrate Subprogram GRHSCV

The program first sets up a histogram which for each bin k contains the Poisson probability $f(k;\lambda)$ for the parameter $\lambda = 10$. Next, points on a polyline are computed corresponding to the probability density of a normal distribution with mean λ and variance λ. Finally the labels of the plot are defined and the complete plot is displayed by a call of GRHSCV (Fig. F.7).

F.10.7 Main Program E7GR to Demonstrate Subprogram GRDTCV

First by calling RNLINE, data points are generated which lie on a straight line $y = at + b$ within the simulated errors. Next, the errors to be presented in the directions of the horizontal and vertical axes and their covariance are defined. The latter two quantities in our example are equal to zero. The polyline defining the straight line consists of only two points. Their computation is trivial. After the definition of the axis labels and the caption, the plot is displayed by calling GRDTCV (Fig. F.8).

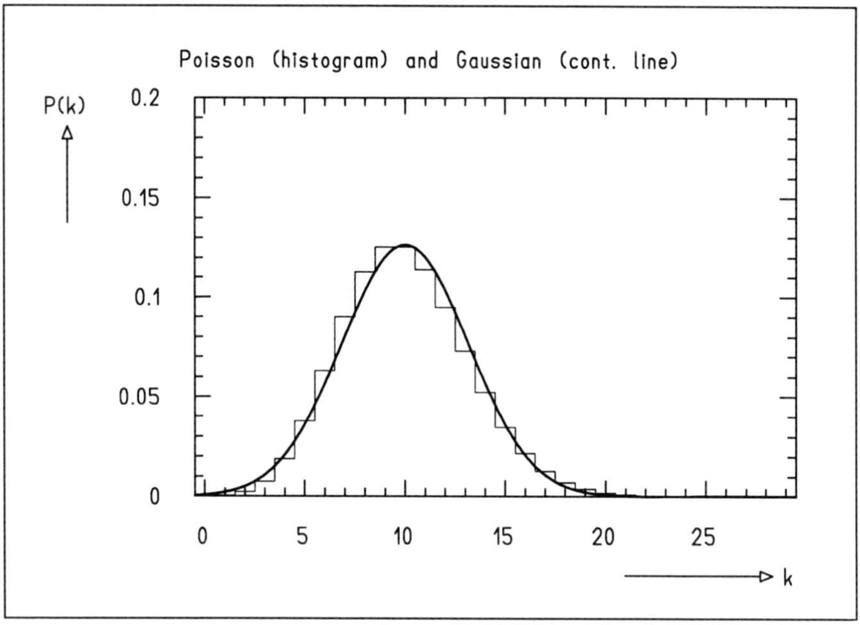

Fig. F.7. A plot generated with GRHSCV containing a histogram and a polyline.

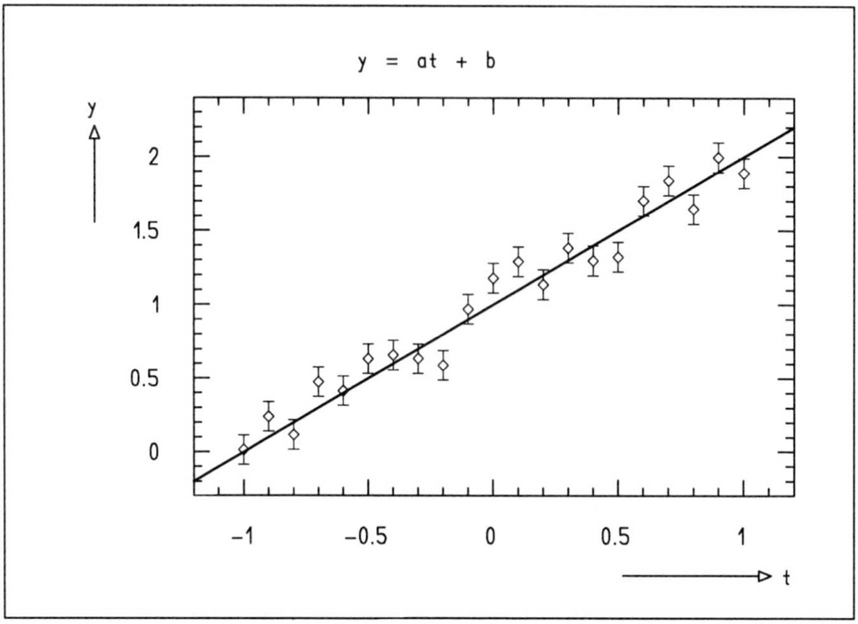

Fig. F.8. A plot with data points and a polyline generated by calling GRDTCV.

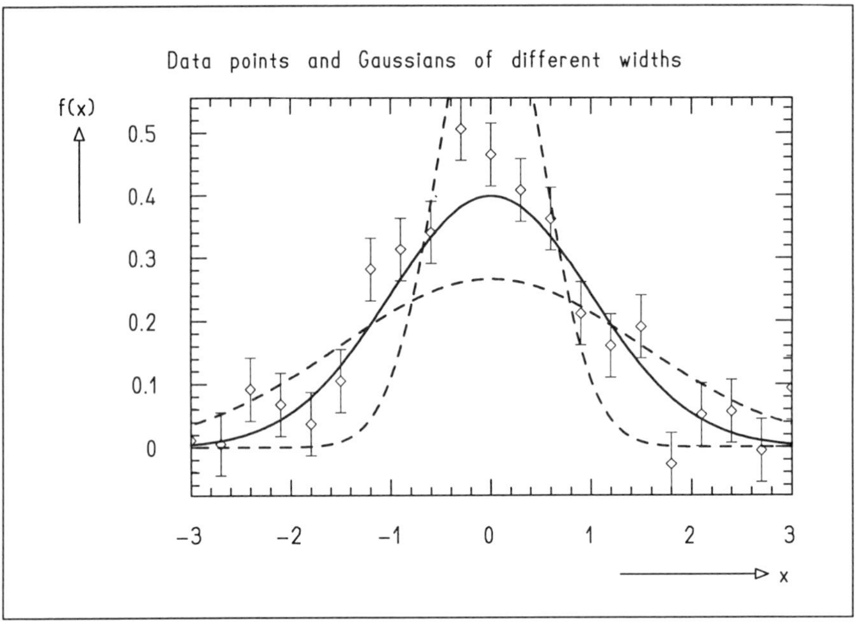

Fig. F.9. A plot with data points and several polylines created by GRDTMC.

F.10.8 Main Program E8GR to Demonstrate Subprogram GRDTMC

The program generates 21 data points which lie within the simulated errors on a Gaussian curve with zero mean and standard deviation $\sigma = 1$, and which span the abscissa region $-3 \leq x \leq 3$. Next, points on three polylines are computed corresponding to Gaussian curves with means of zero and standard deviations $\sigma = 0.5$, $\sigma = 1$, and $\sigma = 1.5$. The polylines span the abscissa region $-10 \leq x \leq 10$. They are displayed in different colors. One polyline is shown as a continuous line, the other two as dashed lines. The user can choose to display only the data points, only the polylines or data points together with polylines (Fig. F.9). In this way the automatic choice of the scales in the different cases is tested.

G. Software Installation and Technical Hints

G.1 The CD-ROM

The program CD-ROM contains

- the FORTRAN and C source files for the programs,
- the FORTRAN and C source files for the example programs,
- the FORTRAN and C source files for solutions of the programming problems,
- a library of graphics routines,
- additional files helpful for compilation and linking,
- executable demo programs for all examples and solutions.

The source files for the graphics routines are not included; the already compiled routines are provided in a library which must be specified during linking. This library exists in several versions on the CD-ROM which are to be used for linking with programs that have been compiled with different FORTRAN or C compilers.

The FORTRAN programs have been written in standard FORTRAN77 independent of any particular computer or compiler and can be used on any computer provided a FORTRAN77 or FORTRAN90 compiler is installed. That is also true for the example programs and for the solutions as long as they do not call graphics routines.

The C programs were essentially obtained by an automatic translation from FORTRAN to C. In this way it was assured that the C routines perform exactly as the FORTRAN ones at the expense, however, that the code is not always very elegant. (For the calling sequences to be used for the C programs and the indexing scheme of vectors and matrices see Sect. G.3.) Again, the C programs, as long as they do not call graphics routines, can be used by any C or C++ compiler.

The files mentioned at the beginning of this section are transferred from the CD-ROM to the hard disk of an IBM compatible Personal computer (cf. installation procedure, Sect. G.5). The source files can then be transferred to any computer. The graphics programs can only be used on an IBM compatible Personal Computer running under DOS, WINDOWS, WINDOWS 95, WINDOWS NT, or Linux, and with certain compilers. These are listed in the next section.

G.2 Different Compilers and Operating Systems

Graphics libraries are provided for a variety of FORTRAN and C compilers and operating systems. Table G.1 gives an overview. During installation (cf. Sect. G.5) you specify the compiler and the operating system you want to work with. The list of compilers comprises the more frequently used 16-bit and 32-bit FORTRAN and C compilers. On the operating systems we make a few short remarks:

DOS For programs running under DOS the whole computer screen is used either for graphics or for alphanumeric output. A mixture of the two on the same screen is not possible.

QuickWin Programming for WINDOWS (cf. Sect. G.4) is somewhat different from programming for DOS. For some Microsoft compilers a facility called QuickWin exists, which provides simple WINDOWS facilities for programs originally written for DOS. We provide a graphics library and files facilitating compilation and linking to produce programs with QuickWin features. Such programs run under WINDOWS (or WINDOWS 95/WINDOWS NT, depending on the compiler) in the following way: A text window is opened for the alphanumeric input and output. If graphical output is required an additional window is opened for it. The advantage of QuickWin is that programs written for DOS need not be changed. On the other hand it offers less flexibility. In our case the graphics cannot be rescaled. To see the full graphics you have to click on the pull-down menu "View" and within this menu on the option "Full screen".

WINDOWS 95/WINDOWS NT Genuine WINDOWS programs have a structure that may be quite different from the structure of DOS programs because a WINDOWS program has to be able to react at run-time to external user actions which may for instance ask for a change in the appearance of a window. The structure of our programs for WINDOWS is briefly described in Sect. G.4. Here we merely sketch the appearance of the computer screen to

Table G.1. Compilers and operating systems for which graphics libraries are available

Compiler	DOS	QuickWin	WINDOWS 95
32-bit:			
Lahey F77L-EM32 FORTRAN version 5.1	X		
Lahey FORTRAN 90 version 3.50a	X		X
Microsoft Power FORTRAN Station version 4.0		X	X
Digital Visual FORTRAN version 5.0		X	X
WATCOM FORTRAN 77^{32} version 9.5	X		
Microsoft Visual C/C++ version 4			X
Borland C/C++ version 5.0			X
16-bit:			
IBM Personal FORTRAN version 1.05	X		
RM FORTRAN version 3	X		
Prospero FORTRAN version iid 2.1	X		
Microsoft FORTRAN version 5.1	X	X	
Microsoft Visual C/C++ version 1.0	X	X	
Microsoft C version 7	X		
Borland C/C++ version 3.1	X		

Linux (32-bit)	XWindow (X11)
GNU FORTRAN version 0.5.19.1	X
GNU C version 2.7.2.1	X

the user. Each program opens a *main window*. In its top part the alphanumeric output of the program is presented. Alphanumeric input to the program is entered in a *command line* at the bottom of the main window. Each line of input has to be terminated by a stroke of the `ENTER` key. It is then copied into the output region of the main window and the command line is cleared. If graphical output is asked for it is placed in a special window called a *child window* within the main window. This window can be manipulated by the user, i.e., changed in size, shrunk to an icon, etc. The child window is treated by the program as a *workstation*. Closing of the workstation does not destroy the window. Several (up to ten) plots can be produced in different child windows.

Linux Programming for Linux does not differ from programming for DOS. (There is one slight exception from this rule: If you write a C program for Linux then programs of the data analysis library and the graphics library have to be referenced with an underscore at the end of the program name, e.g., `mtxadd_` instead of `mtxadd`.) Standard input (normally) comes from the keyboard and is echoed onto the alphanumeric window (e.g., `xterm`) in which the application runs. Standard output also goes to the alphanumeric window. For graphics an additional graphical window (`X11`) is opened by a call of `GROPWS`, cf. Sect. F.9.2.

G.3 Using C Programs

As mentioned in Sect. G.1 the C source programs were obtained by automatic translation of FORTRAN source code. Consequently the C programs have inherited certain features from FORTRAN, in particular conventions for transferring arguments via calling sequences between calling program and subprogram and the indexing conventions for matrices and subroutines.

Arguments in Calling Sequences As in FORTRAN all arguments are referred to *by reference* (in C normally only arrays (i.e., vectors and matrices) are referred to by reference, other variables or constants are referred to *by value*). Therefore, the calling sequence for program `MTXADD` which adds the two $m \times n$ matrices A and B and stores the result in R reads

```
mtxadd (a, b, r, &m, &n);
```

Indexing Conventions for Vectors In FORTRAN the elements of a one-dimensional array V with n elements are addressed as

```
V(1), V(2), ..., V(N).
```

Thus the element v_i of a vector

$$\mathbf{v} = (v_1, v_2, \ldots, v_n)$$

is stored in `V(I)`. Since in C counting starts with zero, the elements of the corresponding array are

```
v[0], v[1], ..., v[n-1]
```

and the vector element v_i is addressed as `v[i-1]`.

Indexing Conventions for Matrices In FORTRAN the elements `A(I,J)` of a matrix are stored columnwise, i.e., the elements of the $m \times n$ matrix A appear in memory in the order

$$a_{11}, a_{21}, \ldots, a_{m1}, a_{12}, a_{22}, \ldots, a_{m2}, \ldots, a_{1n}, a_{2n}, \ldots, a_{mn} \quad .$$

In C a two-dimensional array is stored rowwise. In our programs the FORTRAN conventions are emulated by storing a matrix columnwise in a one-dimensional array,

```
a[0], a[1], ..., a[m * n -1] .
```

The element A_{ij} is then addressed in FORTRAN as `A(I,J)` and in C as `a[(j-1) * n + i - 1]`.

Writing your own C Programs When writing your own C programs you have to comply with the conventions for transferring arguments and referring to vectors and matrices as explained above. Moreover, you have to include the header files `datsrc.h` and `grsrc.h` with the following preprocessor directives:

```
# include "datsrc.h"
# include "grsrc.h"
```

(The second line is needed only if you use graphics routines.)

All programs in the directories `\DATAN\EXASRC` and `\DATAN\SOLSRC` can serve as examples for user-written application programs. Particularly simple examples are `e1mtx.c` (without use of graphics routines), `e6gr.c` (construction of a complete plot by only one call to a graphics routine), and `e1gr.c` (use of several graphics routines).

G.4 Programming for WINDOWS 95/WINDOWS NT

G.4.1 Introductory Remarks

We do not intend here to describe any details of WINDOWS programming. We just provide a framework which a user who is experienced in conventional programming can fill to obtain a WINDOWS program. This is done with the help of five subprograms the use of which is explained further down in this section. WINDOWS programs created in this way handle alphanumeric input and output very much like DOS programs do. Graphical output, however, shows all the flexibility of WINDOWS. (The experienced WINDOWS programmer is free to introduce more WINDOWS features to handle input or output which is specific to his or her problem.) The five subprograms as well as the WINDOWS version of the graphics routines are contained in the *dynamic link library* or DLL named `DATANDLL.DLL`. The complexity of WINDOWS programming is hidden in these library routines. All the user needs to remember can be summarized in two rules:

1. Each application program runs under the control of a main program with the name `WINMAIN` (in FORTRAN) or `WinMain` (in C). The main program is identical for all applications except for the fact that the name of the application program occurs at two places of the main program.

2. The application program (which is a main program under DOS) is a subroutine. It can be programmed as for DOS. Only standard input (input from the keyboard) and standard output (alphanumeric output on the screen) are handled differently.

G.4.2 Main Program

In our examples the main program to run an application and the application program are contained in the same file. We suggest that you study the source file `E1GR` in the directory `\DATAN\EXASRC`. You will find that – apart from the reference to include files (which you need as given) – the main program `WINMAIN` consists essentially of calls to the three subprograms `WNINIT`, `WNRGPR`, and `WNRUNP` which perform initialization, registering of the application program, and running of the application program, respectively. For completeness we still describe these three subprograms:

Program G.1: Subprogram `WNINIT` initializes main program under WINDOWS

CALLING SEQUENCE: `CALL WNINIT(ITEST)`

OUTPUT ARGUMENT: `ITEST` — number characterizing success of initialization (zero if successful, nonzero if not successful)

REMARKS: Unsuccessful initialization usually means that the dynamic link library was not found.

Program G.2: Subprogram `WNRGPR` registers an application program

CALLING SEQUENCE: `CALL WNRGPR(APPROG)`

INPUT ARGUMENT: `APPROG` — name of application program, has to be declared `EXTERNAL` in the main program

Program G.3: Subprogram `WNRUNP` runs registered application program

CALLING SEQUENCE: `CALL WNRUNP`

G.4.3 Application Program

Under WINDOWS the application program is a subroutine without arguments. It is written like a DOS program except for the handling of standard input and standard output. Under DOS input from the keyboard (standard input) and output on the screen (standard output) are handled using in FORTRAN the statements `READ (*,...)` and `WRITE (*,...)` or in C the library functions `scanf` and `printf`.

Under WINDOWS we emulate the reading of one line of standard input by the two statements

```
CALL WNRDST(STRING, MAXLEN)
READ (STRING, ...) ...
```

Here `STRING` is a character string of length `MAXLEN`. The subroutine `WNRDST` copies the contents of the command line at the bottom of the main window into the character string `STRING` when the `ENTER` key is pressed. The second statement performs conventional formatted reading from a string. In C the corresponding statements are

```
wnrdst(string, &maxlen);
sscanf(string, ...);
```

Note that if the maximal length of the character string `string` you want to read with `wnrdst` is `maxlen` $\leq n$, then you have to declare the length of the character string `string` to be $n + 1$ and set the last character in that string to null, `string[`n`] = 0;`.

Correspondingly, alphanumeric output to the main window is achieved by formatted writing into a string with standard tools and subsequent display of that string in the window by a subroutine. In FORTRAN one needs the two statements

```
WRITE(STRING, ...) ...
CALL WNWRST(STRING, LENGTH)
```

Here `LENGTH` is the length of the string to be displayed. The corresponding C statements are

```
sprintf(string, ...);
wnwrst(string, &length);
```

We conclude this section by the formal description of the programs `WNRDST` and `WNWRST`.

Program G.4: Subprogram `WNRDST` reads a character string from the command line of the main window

CALLING SEQUENCE: `CALL WNRDST(STRING, MAXLEN)`

INPUT ARGUMENT: `MAXLEN` — maximum length of string

OUTPUT ARGUMENT: `STRING` — character string read from the command line

REMARKS: Reading is performed after input on the command line is terminated by pressing the `ENTER` key.

Program G.5: Subprogram `WNWRST` displays a character string in the main window

CALLING SEQUENCE: `CALL WRWRST(STRING, LENGTH)`

INPUT ARGUMENTS: `STRING` — character string
`LENGTH` — number of characters in string

G.5 Program Installation

For **DOS** or **WINDOWS** the installation consists of two steps:

Step 1: Insert the CD-ROM accompanying this book into a suitable drive of your computer. If your computer is not in the DOS mode then go into DOS mode (e.g., by clicking on the DOS prompt symbol under WINDOWS). Enter the command

d:\INSTALL

(Here *d* stands for the letter by which your CD-ROM drive is described.) You will now be taken through a dialog with the computer, at the end of which all files needed will be on your hard disk.

Step 2: Set the *search path* of your computer to contain the directory `\DATAN` and the relevant directories of your compiler. Consult the file `README.DOC` in the directory `\DATAN` for specific instructions.

For **Linux** the installation consists of three steps which you have to perform as "root" user:

Step 1: Insert the CD-ROM accompanying this book into the drive. Mount your CD-ROM drive by a command like

```
mount -o ro /dev/hdd /mnt
```

where `/dev/hdd` is the device special file and `/mnt` stands for the directory which is the mounting point of the CD-ROM.

Step 2: Enter the commands

```
cd /
tar -zvxf /mnt/linux/datan.tgz
```

Once the command is executed all files needed will be on your hard disk.

Step 3: Make sure that a line

```
/usr/local/lib
```

is present in the file `/etc/ld.so.conf.in` (or in `/etc/ld.so.conf`). If it is not, insert that line and give the command

```
ldconfig
```

You have in this way set the path to our libraries. See also the file `README.DOC` for alternative installation possibilities.

G.6 The Directory Structure for DOS and WINDOWS

The installation procedure creates and fills the directory `\DATAN` on your hard disk. It contains a short file `README.DOC` with a summary of the present section, details supplementing Sects. G.8 and G.9, and possibly some additions made after the text of this book was finalized. In addition it contains a graphics initialization file `GRINIT.DAT`, cf. Sect. F.9.1, as well as files which simplify compiling and linking of programs with the compiler you have chosen during the installation procedure. Finally, the directory `\DATAN` contains a total of five subdirectories.

G.6.1 The Subdirectory `\DATAN\DATSRC`

This subdirectory contains the source files of the programs.

G.6.2 The Subdirectory `\DATAN\EXASRC`

This subdirectory contains the source files of the example programs. It also contains an executable file for each example program which can be directly run as demonstration.

G.6.3 The Subdirectory `\DATAN\SOLSRC`

This subdirectory contains the source files and executable files of programs which are solutions to the programming problems.

G.6.4 The Subdirectory `\DATAN\DLIB`. The Data Analysis Library

This subdirectory contains a library file, the *data analysis library*. In this library all object programs which are produced by compiling the source programs of the subdirectory `\DATAN\DATSRC` are combined.

In addition to the library file the subdirectory `\DATAN\DLIB` still contains one or several command files which help you to create a new library after changing one or several of the source programs (cf. Sect. G.8).

G.6.5 The Subdirectory `\DATAN\GLIB`. The Graphics Library

This subdirectory contains a library of the graphics programs (*graphics library*).

G.7 The Directory Structure for Linux

The installation procedure creates and fills the directory `/usr/local/datan` and adds files to the existing directories `/usr/local/lib` and `/usr/local/bin`.

G.7.1 The Directory `/usr/local/datan`

It contains a short file `README.DOC` with a summary of the present section, details supplementing Sects. G.8 and G.9, and possibly with some additions made after the text of this book was finalized. In addition it contains a graphics initialization file `GRINIT.DAT`, cf. Sect. F.9.1. Moreover, it contains five subdirectories.

G.7.1.1 The Subdirectory `/usr/local/datan/datsrc` This subdirectory contains the source files of the programs.

G.7.1.2 The Subdirectory `/usr/local/datan/exasrc` This subdirectory contains the source files of the example programs. It also contains an executable file for each example program which can be directly run as demonstration.

G.7.1.3 The Subdirectory `/usr/local/datan/solsrc` This subdirectory contains the source files and executable files of programs which are solutions to the programming problems.

G.7.1.4 The Subdirectory `/usr/local/datan/include` This subdirectory contains include files needed for programming in C.

G.7.1.5 The Subdirectory `/usr/local/datan/utils` This subdirectory contains shell files and make description files which may be used for compiling the data analysis library or for creating executables as supplement to the shell files under `/usr/local/bin`, see below.

G.7.2 Libraries

The directory `/usr/local/lib` contains three additional library files.

- The file `libdatsrc.a` is a *static library*. It contains the *data analysis library* in which all object programs produced by compiling the source programs of the subdirectory /usr/local/datan/datsrc are combined.

- The file libgrsrc.a is a *static library* of the graphics programs (*graphics library*).
- The file libdatan.so.1.0 combines the functions of the data analysis library and the graphics library. It has the form of a *shared library*.

G.7.3 Shell files

The directory /usr/local/bin contains additional shell files for compiling and linking, cf. Sect. G.9.

G.8 Modifying the Data Analysis Library

If you want to change programs from the data analysis library, e.g., because of the array size limitations mentioned in Sect. G.10, you first have to change the source programs in the subdirectory \DATAN\DATSRC (for Linux: /usr/local/datan/datsrc). Then you have to compile the modified source programs to ensure that the changed programs are used in your application. You must give the names of the resulting object files during linking, cf. Sect. G.9, or include the new object files into the library so that they replace the original ones. Consult the handbook of your compiler and of the corresponding library manager on how to replace parts of a library.

It may be simpler for you to generate a completely new data analysis library after having changed one or several files in the subdirectory \DATAN\DATSRC (for Linux: /usr/local/datan/datsrc). Consult the file README.DOC in the directory \DATAN (for Linux: /usr/local/datan) on how to do that with your compiler.

G.9 Compiling, Linking, and Executing Programs

After the installation (Sect. G.5) has successfully been completed you can write any program, compile it, link it with the data analysis library and – if needed – with the graphics library, and finally execute it. Consult the file README.DOC in the directory \DATAN (for Linux: /usr/local/datan) for details pertinent to your compiler and operating system.

G.10 Array Size Limitations in Some Programs

In rare cases, namely if you want to use very large data arrays, you may have to change some of the programs in the data analysis library. This has the following technical reason. The FORTRAN compiler reserves storage space for an array which is used by both the calling program and a subprogram called by it. This storage is situated within the calling program. The address of that array (and in case of an array with more than one dimension the size of the array) is made available to the called program in its argument list. In this way subprograms can be written independently of such array sizes. In some of the subprograms of the data analysis library, lengthy vector and matrix calculations are performed for which intermediate results have to be stored in arrays. These arrays often are defined only in the subprogram in order to keep the argument lists short. In these cases, however, maximum array sizes have to be assumed. They are defined in parameter statements. Should you want to use larger array sizes, you must change these parameter statements accordingly, recompile the programs, and replace the corresponding object programs in the data analysis library (cf. Sect. G.8). Table G.2 contains a list of all programs with array size limitations, the names and the maximum sizes of arguments as well as the parameters which, if necessary, have to be increased.

Technical hint: Because of the technical particularities of storage administration by the Microsoft FORTRAN compiler, large arrays have been included in `COMMON` blocks. In programs that do not communicate with each other, the same `COMMON` blocks can be used for intermediate results, since those have to be available only during one passage through the subprogram. Table G.2 contains also the names of the `COMMON` blocks used by the particular subprograms.

In the C version of the programs array size limitations are much weaker since dynamic storage allocation is used. For the 32-bit C compilers and the Borland 16-bit C compiler the only limitation is given by the available main memory at run time. For the 16-bit Microsoft compilers the maximum field size is 64 k byte. (Note for programming under Linux: The libraries were created from C sources. Therefore there are no array size limitations.)

Table G.2. Programs with array size limitations in the FORTRAN version.

Program	Limitation	Parameter to be increased in order to lift limitation	Name of COMMON block for intermediate results
LSQPOL	$N \le 1000$	NMAX	DASV04
LSQNON	$N \le 1000$	NMAX	DASV02
LSQMAR	$N \le 1000$	NMAX	DASV02
LSQASN	$N \le 1000$	NMAX	DASV03
LSQASM	$N \le 1000$	NMAX	DASV03
LSQGEN	$N + NR \le 1050$	LMAX	DASV03
LSQASG	$N \le 1000$	NMAX	DASV04
LSQCOG	$N \le 1000$	NMAX	DASV04
	$NR \le 100$	NRMAX	
MINFND	$N \le 50$	MAXN	
MINSIM	$N \le 50$	MAXN	
MINPOW	$N \le 50$	MAXN	
MINMAR	$N \le 50$	MAXN	
MINCOV	$N \le 50$	MAXN	
MINCNT	$N \le 50$	MAXN	
MINASY	$N \le 100$	NMAX	
REGPOL	$N \le 1000$	MAXN	DASV04
TIMSER	$2*K+1 \le 1000$	MAX	DASV03
MTXSVD	$N \le 1000$	NMAX	DASV01
MTXDEC	$N \le 1000$	NMAX	DASV04
MTXSV1	$N \le 1000$	NMAX	DASV00
MTXMAR	$N \le 1000$	NMAX	DASV01
MTXSVM	$N \le 1000$	NMAX	DASV00
MTXLSC	$N \le 1000$	NMAX	DASV02
	$L \le 1050$	LMAX	DASV02

H. Collection of Formulas

Probability

$A, B, \ldots$ are *events*; $\bar{A}$ is the event "*not* A".

$(A+B)$ and (AB) combine events with *logical "or"* or *logical "and"*.

$P(A)$ is the *probability* of the event A.

$P(B|A) = P(AB)/P(A)$ is the probability for B given the condition A (*conditional probability*).

The following rules hold:
For *every* event A

$$P(\bar{A}) = 1 - P(A) \quad ,$$

for *mutually exclusive* events $A, B, \ldots, Z$

$$P(A+B+\cdots+Z) = P(A) + P(B) + \cdots + P(Z) \quad ,$$

for *independent* events $A, B, \ldots, Z$

$$P(AB\cdots Z) = P(A)P(B)\cdots P(Z) \quad .$$

Single Random Variable

Distribution function: $\quad F(x) = P(\mathsf{x} < x)$

Probability density (for $F(x)$ differentiable):
$f(x) = F'(x) = \mathrm{d}f(x)/\mathrm{d}x$

Moments of order ℓ:

(a) about the point c: $\alpha_\ell = E\{(\mathsf{x} - c)^\ell\}$

(b) about the origin (central moments): $\lambda_\ell = E\{\mathsf{x}^\ell\}$

(c) about the mean: $\mu_\ell = E\{(\mathsf{x} - \widehat{x})^\ell\}$

Table H.1. Expectation values for discrete and continuous distributions.

	x discrete	x continuous; $F(x)$ differentiable
Probability density	—	$f(x) = F'(x) = \frac{\mathrm{d}F(x)}{\mathrm{d}x}$, $\int_{-\infty}^{\infty} f(x)\,\mathrm{d}x = 1$
Mean of x (Expectation value)	$\widehat{x} = E(\mathsf{x})$ $= \sum_i x_i P(\mathsf{x} = x_i)$	$\widehat{x} = E(\mathsf{x})$ $= \int_{-\infty}^{\infty} x f(x)\,\mathrm{d}x$
Mean of the function $H(\mathsf{x})$	$E\{H(\mathsf{x})\}$ $= \sum_i H(x_i) P(\mathsf{x} = x_i)$	$E\{H(\mathsf{x})\}$ $= \int_{-\infty}^{\infty} H(x) f(x)\,\mathrm{d}x$

Variance: $\sigma^2(\mathsf{x}) = \mathrm{var}(\mathsf{x}) = \mu_2 = E\{(\mathsf{x} - \widehat{x})^2\}$

Standard deviation or *error* of x: $\Delta\mathsf{x} = \sigma(\mathsf{x}) = +\sqrt{\sigma^2(\mathsf{x})}$

Skewness: $\gamma = \mu_3/\sigma^3$

Reduced variable: $\mathsf{u} = (\mathsf{x} - \widehat{x})/\sigma(\mathsf{x}) \quad ; \qquad E(\mathsf{u}) = 0 \quad , \qquad \sigma^2(\mathsf{u}) = 1$

Mode (most probable value) x_m defined by: $P(\mathsf{x} = x_m) = \max$

Median $x_{0.5}$ defined by: $F(x_{0.5}) = P(\mathsf{x} < x_{0.5}) = 0.5$

Quantile x_q defined by: $F(x_q) = P(\mathsf{x} < x_q) = q \quad ; \qquad 0 \le q \le 1$

Several Random Variables

Distribution function:

$$F(\mathbf{x}) = F(x_1, x_2, \ldots, x_n) = P(\mathsf{x}_1 < x_1,\ \mathsf{x}_2 < x_2,\ \ldots,\ \mathsf{x}_n < x_n)$$

Joint probability density (only for $F(\mathbf{x})$ differentiable with respect to all variables):

$$f(\mathbf{x}) = f(x_1, x_2, \ldots, x_n) = \partial^n F(x_1, x_2, \ldots, x_n)/\partial x_1 \partial x_2 \ldots \partial x_n$$

Marginal probability density of the variable x_i:

$$g_i(x_i) = \int_{-\infty}^{\infty}\int_{-\infty}^{\infty} \cdots \int_{-\infty}^{\infty} f(x_1, x_2, \ldots, x_n)\,\mathrm{d}x_1\,\mathrm{d}x_2 \ldots \mathrm{d}x_{i-1}\,\mathrm{d}x_{i+1} \ldots \mathrm{d}x_n$$

Expectation value of a function $H(\mathbf{x})$:

$$E\{H(\mathbf{x})\} = \int H(\mathbf{x}) f(\mathbf{x})\, \mathrm{d}\mathbf{x}$$

Expectation value of the variables x_i:

$$\widehat{x}_i = E(\mathrm{x}_i) = \int x_i f(\mathbf{x})\, \mathrm{d}\mathbf{x} = \int_{-\infty}^{\infty} x_i g_i(x_i)\, \mathrm{d}x_i$$

The variables $\mathrm{x}_1, \mathrm{x}_2, \ldots, \mathrm{x}_n$ are *independent*, if

$$f(x_1, x_2, \ldots, x_n) = g_1(x_1) g_2(x_2) \ldots g_n(x_n)$$

Moments of order $\ell_1, \ell_2, \ldots, \ell_n$:

(a) about $\mathbf{c} = (c_1, c_2, \ldots, c_n)$:
$\alpha_{\ell_1 \ell_2 \ldots \ell_n} = E\{(\mathrm{x}_1 - c_1)^{\ell_1} (\mathrm{x}_2 - c_2)^{\ell_2} \cdots (\mathrm{x}_n - c_n)^{\ell_n}\}$

(b) about the origin: $\lambda_{\ell_1 \ell_2 \ldots \ell_n} = E\{\mathrm{x}_1^{\ell_1} \mathrm{x}_2^{\ell_2} \cdots \mathrm{x}_n^{\ell_n}\}$

(c) about $\widehat{\mathbf{x}}$: $\mu_{\ell_1 \ell_2 \ldots \ell_n} = E\{(\mathrm{x}_1 - \widehat{x}_1)^{\ell_1} (\mathrm{x}_2 - \widehat{x}_2)^{\ell_2} \cdots (\mathrm{x}_n - \widehat{x}_n)^{\ell_n}\}$

Variance of x_i: $\sigma^2(\mathrm{x}_i) = E\{(\mathrm{x}_i - \widehat{x}_i)^2\} = c_{ii}$

Covariance between x_i and x_j: $\mathrm{cov}(\mathrm{x}_i, \mathrm{x}_j) = E\{(\mathrm{x}_i - \widehat{x}_i)(\mathrm{x}_j - \widehat{x}_j)\} = c_{ij}$
For $\mathrm{x}_i, \mathrm{x}_j$ *independent*: $\mathrm{cov}(\mathrm{x}_i, \mathrm{x}_j) = 0$

Covariance matrix: $C = E\{(\mathbf{x} - \widehat{\mathbf{x}})(\mathbf{x} - \widehat{\mathbf{x}})^{\mathrm{T}}\}$

Correlation coefficient:

$$\rho(\mathrm{x}_1, \mathrm{x}_2) = \mathrm{cov}(\mathrm{x}_1, \mathrm{x}_2)/\sigma(\mathrm{x}_1)\sigma(\mathrm{x}_2) \quad ; \qquad -1 \le \rho \le 1$$

Rules of computation:

$$\begin{aligned} &\sigma^2(c\mathrm{x}_i) = c^2 \sigma^2(\mathrm{x}_i) \quad , \\ &\sigma^2(a\mathrm{x}_i + b\mathrm{x}_j) = a^2 \sigma^2(\mathrm{x}_i) + b^2 \sigma^2(\mathrm{x}_j) + 2ab\, \mathrm{cov}(\mathrm{x}_i, \mathrm{x}_j) \quad ; \\ &a, b, c \quad \text{are constants} \end{aligned}$$

Transformation of Variables

Original variables: $\mathbf{x} = (\mathrm{x}_1, \mathrm{x}_2, \ldots, \mathrm{x}_n)$

Probability density: $f(\mathbf{x})$

Transformed variables: $\mathbf{y} = (\mathrm{y}_1, \mathrm{y}_2, \ldots, \mathrm{y}_n)$

Mapping: $\mathrm{y}_1 = \mathrm{y}_1(\mathbf{x}),\ \mathrm{y}_2 = \mathrm{y}_2(\mathbf{x}),\ \ldots,\ \mathrm{y}_n = \mathrm{y}_n(\mathbf{x})$

Probability density: $g(\mathbf{y}) = |J| f(\mathbf{x})$

with the *Jacobian determinant*

$$J = J\left(\frac{x_1, x_2, \ldots, x_n}{y_1, y_2, \ldots, y_n}\right) = \begin{vmatrix} \dfrac{\partial x_1}{\partial y_1} & \dfrac{\partial x_2}{\partial y_1} & \cdots & \dfrac{\partial x_n}{\partial y_1} \\ \vdots & & & \\ \dfrac{\partial x_1}{\partial y_n} & \dfrac{\partial x_2}{\partial y_n} & \cdots & \dfrac{\partial x_n}{\partial y_n} \end{vmatrix}$$

Error Propagation

The original variables $\mathbf{x}$ have the covariance matrix C_x. The *covariance matrix of the transformed variables* $\mathbf{y}$ is

$$C_y = TC_xT^{\mathrm{T}} \qquad \text{with} \qquad T = \begin{pmatrix} \dfrac{\partial y_1}{\partial x_1} & \dfrac{\partial y_1}{\partial x_2} & \cdots & \dfrac{\partial y_1}{\partial x_n} \\ \dfrac{\partial y_2}{\partial x_1} & \dfrac{\partial y_2}{\partial x_2} & \cdots & \dfrac{\partial y_2}{\partial x_n} \\ \vdots & & & \\ \dfrac{\partial y_m}{\partial x_1} & \dfrac{\partial y_m}{\partial x_2} & \cdots & \dfrac{\partial y_m}{\partial x_n} \end{pmatrix} \quad .$$

The formula is only exact for a linear relation between $\mathbf{y}$ and $\mathbf{x}$, but is a good approximation for small deviations from linearity in a region around $\widehat{\mathbf{x}}$ of the magnitude of the standard deviation. *Only for vanishing covariances* in C_x does one have

$$\sigma(y_i) = \Delta y_i = \sqrt{\sum_{j=1}^{m} \left(\frac{\partial y_i}{\partial x_j}\right)^2 (\Delta x_j)^2} \quad .$$

The Law of Large Numbers

A total of n observations are carried out, which are characterized by the random variable x_i (= 1, if on the ith observation the event A occurs, otherwise = 0). The *frequency* of A is

$$\mathsf{h} = \frac{1}{n}\sum_{i=1}^{n} \mathsf{x}_i \quad .$$

For $n \to \infty$ this frequency is equal to the probability p for the occurrence of A,

$$E(\mathsf{h}) = h = p \quad , \qquad \sigma^2(\mathsf{h}) = \frac{1}{n}p(1-p) \quad .$$

Table H.2. Distributions of Discrete Variables

Distribution	Probability for observing $\mathsf{x} = k$ ($\mathsf{x}_1 = k_1, \ldots, \mathsf{x}_l = k_l$)	Mean	Variance (Elements of covariance matrix)
Binomial	$W_k^n = \binom{n}{k} p^k (1-p)^{n-k}$	$\widehat{x} = np$	$\sigma^2(\mathsf{x}) = np(1-p)$
Multinomial $\sum_{j=1}^{l} p_j = 1$	$W_{k_1,k_2,\ldots,k_l}^n = \frac{n!}{\prod_{j=1}^{l} k_j!} \prod_{j=1}^{l} p_j^{kj}$	$\widehat{x}_j = np_j$	$c_{ij} = np_i(\delta_{ij} - p_j)$
Hypergeometric $L = N - K$, $l = n - k$	$W_k = \binom{K}{k}\binom{L}{\ell} : \binom{N}{n}$	$\widehat{x} = n\frac{K}{N}$	$\sigma^2(\mathsf{x}) = \frac{nKL(N-n)}{N^2(N-1)}$
Poisson	$f(k) = \frac{\lambda^k}{k!} \mathrm{e}^{-\lambda}$	$\widehat{x} = \lambda$	$\sigma^2(\mathsf{x}) = \lambda$

Central Limit Theorem

If x_i are independent variables with mean a and variance σ^2, then $(1/n)\sum_{i=1}^{n} \mathsf{x}_i$ for $n \to \infty$ follows a normal distribution with mean a and variance σ^2/n.

Convolutions of Distributions

The probability density of the sum $\mathsf{u} = \mathsf{x} + \mathsf{y}$ of two independent random variables x and y is

$$f_u(u) = \int_{-\infty}^{\infty} f_x(x) f_y(u - x)\,\mathrm{d}x = \int_{-\infty}^{\infty} f_y(y) f_x(u - y)\,\mathrm{d}y \quad .$$

— The convolution of two Poisson distributions with the parameters λ_1 and λ_2 is a Poisson distribution with the parameter $\lambda_1 + \lambda_2$.

— The convolution of two Gaussian distributions with means a_1, a_2 and variances σ_1^2, σ_2^2 is a Gaussian distribution with mean $a = a_1 + a_2$ and variance $\sigma^2 = \sigma_1^2 + \sigma_2^2$.

Table H.3. Distributions of continuous variables.

Distribution	Probability density	Mean	Variance (Covariance matrix)
Uniform	$\begin{cases} 0; \; x > a, \; x \geq b \\ \frac{1}{b-a}; \; a \leq x < b \end{cases}$	$\frac{1}{2}(b+a)$	$(b-a)^2/12$
Gaussian	$\frac{1}{b\sqrt{2\pi}} \exp\left(-\frac{(x-a)^2}{2b^2}\right)$	a	b^2
Stand. Gaussian	$\frac{1}{\sqrt{2\pi}} \exp(-\frac{1}{2}x^2)$	0	1
Gaussian of several variables	$k \exp\left\{-\frac{(\mathbf{x}-\mathbf{a})^{\mathrm{T}} B (\mathbf{x}-\mathbf{a})}{2}\right\}$	$\mathbf{a}$	$C = B^{-1}$
χ^2	$\frac{1}{\Gamma(\frac{1}{2}f)2^{\frac{1}{2}f}} (\chi^2)^{\frac{1}{2}f-1} \exp(-\frac{1}{2}\chi^2)$	f	$2f$
Fisher's F	$\left(\frac{f_1}{f_2}\right)^{\frac{1}{2}f_1} \frac{\Gamma(\frac{1}{2}(f_1+f_2))}{\Gamma(\frac{1}{2}f_1)\Gamma(\frac{1}{2}f_2)}$ $\times F^{\frac{1}{2}f_1-1}\left(1+\frac{f_1}{f_2}F\right)^{-\frac{1}{2}(f_1+f_2)}$	$\frac{f_2}{f_2-2}$, $f_2 > 2$	$\frac{2f_2^2(f_1+f_2-2)}{f_1(f_2-2)^2(f_2-4)}$, $f_2 > 4$
Student's t	$\frac{\Gamma(\frac{1}{2}(f+1))}{\Gamma(\frac{1}{2}f)\sqrt{\pi}\sqrt{f}} \left(1+\frac{t^2}{f}\right)^{-\frac{1}{2}(f+1)}$	0	$\frac{f}{f-2}$, $f > 2$

— The convolution of two χ^2-distributions with f_1 and f_2 degrees of freedom is a χ^2-distribution with $f = f_1 + f_2$ degrees of freedom.

Samples

Population: An infinite (in some cases finite) set of elements described by a discrete or continuously distributed random variable x.

(Random) sample of *size* N: A selection of N elements $(\mathsf{x}^{(1)}, \mathsf{x}^{(2)}, \ldots, \mathsf{x}^{(N)})$ from the population. (For the requirements for a sample to be random see Sect. 6.1.)

Distribution function of the sample: $W_n(x) = n_x/N$,

Table H.4. Samples from different populations.

	Sample of size n from a continuously distributed population	Sample of size n from a discrete population of size N. Variable of the population is y, Variable of the sample is x.
Population mean	$E(\mathsf{x}) = \widehat{x}$	$\bar{y} = \frac{1}{N}\sum_{j=1}^{N} y_j$
Population variance	$\sigma^2(\mathsf{x})$	$\sigma^2(y) = \frac{1}{N-1}\sum_{j=1}^{N}(y_j - \bar{y})^2$
Sample mean	$\bar{\mathsf{x}} = \frac{1}{n}\sum_{i=1}^{n} \mathsf{x}_i$	$\bar{\mathsf{x}} = \frac{1}{n}\sum_{i=1}^{n} \mathsf{x}_i$
Sample variance mean	$\sigma^2(\bar{\mathsf{x}}) = \frac{1}{n}\sigma^2(\mathsf{x})$	$\sigma^2(\bar{\mathsf{x}}) = \frac{\sigma^2(y)}{n}\left(1 - \frac{n}{N}\right)$
Variance of the sample	$\mathsf{s}^2 = \frac{1}{n-1}\sum_{i=1}^{n}(\mathsf{x}_i - \bar{\mathsf{x}})^2$	$\mathsf{s}^2 = \frac{1}{n-1}\sum_{i=1}^{n}(\mathsf{x}_i - \bar{\mathsf{x}})^2$

where n_x is the number of elements in the sample for which $\mathsf{x} < x$.

Statistic: An arbitrary function of the elements of a sample

$$\mathsf{S} = \mathsf{S}(\mathsf{x}^{(1)}, \mathsf{x}^{(2)}, \ldots, \mathsf{x}^{(N)}) \quad .$$

Estimator: A statistic used to estimate a parameter λ of the population. An estimator is *unbiased*, if $E(\mathsf{S}) = \lambda$ and *consistent*, if

$$\lim_{N\to\infty} \sigma(\mathsf{S}) = 0 \quad .$$

Maximum-Likelihood Method

Consider a population described by the probability density $f(\mathbf{x}, \boldsymbol{\lambda})$, where $\boldsymbol{\lambda} = (\lambda_1, \lambda_2, \ldots, \lambda_p)$ is a set of parameters. If a sample $\mathbf{x}^{(1)}, \mathbf{x}^{(2)}, \ldots, \mathbf{x}^{(N)}$ is obtained, then the *likelihood function* is $L = \prod_{j=1}^{N} f(\mathbf{x}^{(j)}, \boldsymbol{\lambda})$ and the *log-likelihood function* is $\ell = \ln L$. In order to determine the unknown parameters $\boldsymbol{\lambda}$ from the sample the maximum-likelihood method prescribes the value of $\boldsymbol{\lambda}$

for which L (or ℓ) is a maximum. That is, one must solve the *likelihood equation* $\partial\ell/\partial\lambda_i = 0$; $i = 1, 2, \ldots, p$ or (for only one parameter) $\mathrm{d}\ell/\mathrm{d}\lambda = \ell' = 0$.

Information of a sample: $I(\lambda) = E(\ell'^2) = -E(\ell'')$

Information inequality: $\sigma^2(\mathsf{S}) \geq \{1 - B'(\lambda)\}^2/I(\lambda)$
Here S is an *estimator* for λ, and $B(\lambda) = E(\mathsf{S}) - \lambda$ is its *bias*.

An estimator has *minimum variance* if $\ell' = A(\lambda)(\mathsf{S} - E(\mathsf{S}))$, where $A(\lambda)$ does not depend on the sample.

The maximum-likelihood estimator $\widetilde{\lambda}$, i.e., the solution of the likelihood equation, is *unique*, *asymptotically unbiased* (i.e., for $N \to \infty$), and has *minimum variance*.

The *asymptotic form* of the likelihood function for one parameter λ is

$$L = \text{const} \cdot \exp\left(-\frac{(\lambda - \widetilde{\lambda})^2}{2b^2}\right) \quad ,$$

$$b^2 = \sigma^2(\widetilde{\lambda}) = \frac{1}{E(\ell'^2(\widetilde{\lambda}))} = -\frac{1}{E(\ell''(\widetilde{\lambda}))}$$

and for several parameters

$$L = \text{const} \cdot \left\{-\frac{1}{2}(\boldsymbol{\lambda} - \widetilde{\boldsymbol{\lambda}})^{\mathrm{T}} B(\boldsymbol{\lambda} - \widetilde{\boldsymbol{\lambda}})\right\}$$

with the covariance matrix $C = B^{-1}$ and

$$B_{ij} = -E\left(\frac{\partial^2\ell}{\partial\lambda_i\partial\lambda_j}\right)_{\boldsymbol{\lambda}=\widetilde{\boldsymbol{\lambda}}} \quad .$$

Testing Hypotheses

Null hypothesis $H_0(\boldsymbol{\lambda} = \boldsymbol{\lambda}_0)$: Assumption of values for the parameters $\boldsymbol{\lambda}$ that determine the probability distribution $f(\mathbf{x}, \boldsymbol{\lambda})$ of a population.

Alternative hypotheses $H_1(\boldsymbol{\lambda} = \boldsymbol{\lambda}_1)$, $H_2(\boldsymbol{\lambda} = \boldsymbol{\lambda}_2)$, ...: Other possibilities for $\boldsymbol{\lambda}$, against which the null hypothesis is to be tested by consideration of a sample $X = (\mathbf{x}^{(1)}, \mathbf{x}^{(2)}, \ldots, \mathbf{x}^{(N)})$ from the population.

A hypothesis is *simple* if the parameters are completely determined, e.g., $H_0(\lambda_1 = 1,\ \lambda_2 = 5)$, otherwise it is *composite*, e.g., $H_1(\lambda_1 = 2,\ \lambda_2 < 7)$.

Test of a hypothesis H_0 with a *significance level* α or *confidence level* $1 - \alpha$: H_0 is rejected if $X \in S_c$, where S_c is the *critical region* in the sample space and

$$P(X \in S_c | H_0) = \alpha \quad .$$

Error of the first kind: Rejection of H_0, although H_0 is true. The probability of this error is α.

Error of the second kind: H_0 is not rejected although H_1 is true. The probability of this error is $P(X \notin S_c|H_1) = \beta$.

Power function:

$$M(S_c, \boldsymbol{\lambda}) = P(X \in S_c|H) = P(X \in S_c|\boldsymbol{\lambda}) \quad .$$

Operating characteristic function:

$$L(S_c, \boldsymbol{\lambda}) = 1 - M(S_c, \boldsymbol{\lambda}) \quad .$$

Most powerful test of H_0 with respect to H_1 has $M(S_c, \boldsymbol{\lambda}_1) = 1 - \beta = \max$. A *uniformly most powerful test* is a most powerful test with respect to all possible H_1.

An *unbiased test* has $M(S_c, \boldsymbol{\lambda}_1) \geq \alpha$ for all possible H_1.

NEYMAN–PEARSON LEMMA: A test of H_0 with respect to H_1 (both simple hypotheses) with the critical region S_c is a most powerful test if $f(X|H_0)/f(X|H_1) \leq c$ for $X \in S_c$ and $\geq c$ for $X \notin S_c$, where c is a constant only depending on α.

Test statistic $T(X)$: Scalar function of the sample X. By means of a mapping $X \to T(X)$, $S_c(X) \to U$ the question as to whether $X \in S_c$ can be reformulated as $T \in U$.

Likelihood-ratio test: If ω denotes the region in the parameter space corresponding to the null hypothesis and Ω denotes the entire possible parameter region, then the test statistic

$$T = f(x; \widetilde{\boldsymbol{\lambda}}^{(\Omega)})/f(x; \widetilde{\boldsymbol{\lambda}}^{(\omega)})$$

is used. Here $\widetilde{\boldsymbol{\lambda}}^{(\Omega)}$ and $\widetilde{\boldsymbol{\lambda}}^{(\omega)}$ are the maximum-likelihood estimators in the regions Ω and ω. H_0 is rejected if $T > T_{1-\alpha}$ with $P(T > T_{1-\alpha}|H_0) = \int_{T_{1-\alpha}}^{\infty} g(T)\,\mathrm{d}T = \alpha$; $g(T)$ is the conditional probability density of T for a given H_0.

WILKS *theorem* (holds for weak requirements concerning the probability density of the population): If $H_0(\lambda_1 = \lambda_{10}, \ldots, \lambda_r = \lambda_{r0})$ specifies r out of the p parameters, then $-2\ln T$ (where T is the likelihood-ratio test statistic) follows a χ^2-distribution with $f = p - r$ degrees of freedom in the limit $n \to \infty$. For $r = p$ one has $f = 1$.

Table H.5. Frequently used statistical tests for a sample from a normal distribution with mean λ and variance σ^2. (Case 1: σ known; case 2: σ unknown (Student's test); case 3: χ^2-test of the variance; case 4: Student's difference test of two samples of sizes N_1 and N_2 on the significance of s_Δ^2, cf. (8.3.19); case 5: F-test of two samples.)

Case	Test statistic	Null hypothesis	Critical region for test statistic	Number of degrees of freedom
1	$T = \frac{\bar{\mathsf{x}}-\lambda_0}{\sigma/\sqrt{N}}$, $\bar{\mathsf{x}} = \frac{1}{N}\sum_{j=1}^{N} \mathsf{x}^{(j)}$	$\lambda = \lambda_0$ $\lambda \le \lambda_0$ $\lambda \ge \lambda_0$	$\|T\| > \Omega(1-\alpha/2)$ $T > \Omega(1-\alpha)$ $T < \Omega(\alpha)$	—
2	$T = \frac{\bar{\mathsf{x}}-\lambda_0}{\mathsf{s}/\sqrt{N}}$, $\mathsf{s}^2 = \frac{1}{N-1} \times \sum_{j=1}^{N} (\mathsf{x}^{(j)} - \bar{\mathsf{x}})^2$	$\lambda = \lambda_0$ $\lambda \le \lambda_0$ $\lambda \ge \lambda_0$	$\|T\| > t_{1-\alpha/2}$ $T > t_{1-\alpha}$ $T < -t_{1-\alpha} = t_\alpha$	$N-1$
3	$T = (N-1)\frac{\mathsf{s}^2}{\sigma_0^2}$	$\sigma^2 = \sigma_0^2$ $\sigma^2 \le \sigma_0^2$ $\sigma^2 \ge \sigma_0^2$	$\chi^2_{1-\alpha/2} < T < \chi^2_{\alpha/2}$ $T > \chi^2_{1-\alpha}$ $T < \chi^2_\alpha$	$N-1$
4	$T = \frac{\bar{\mathsf{x}}_1-\bar{\mathsf{x}}_2}{\mathsf{s}_\Delta}$	$\lambda_1 = \lambda_2$	$\|T\| > t_{1-\frac{1}{2}\alpha}$	$N_1 + N_2 - 2$
5	$T = \mathsf{s}_1^2/\mathsf{s}_2^2$, $\mathsf{s}_i^2 = \frac{1}{N_i-1} \times \sum_{j=1}^{N_i} (\mathsf{x}_i^{(j)} - \bar{\mathsf{x}}_i)^2$	$\sigma_1^2 = \sigma_2^2$ $\sigma_1^2 \le \sigma_2^2$ $\sigma_1^2 \ge \sigma_2^2$	$F_{1-\alpha/2} < T < F_{\alpha/2}$ $T > F_{1-\alpha}$ $T < F_\alpha$	$f_1 = N_1 - 1$ $f_2 = N_2 - 1$

χ^2-Test for Goodness-of-Fit

Hypothesis: The N measured values y_i with normally distributed errors σ_i are described by given quantities f_i.

Test function: $T = \sum_{i=1}^{N}(y_i - f_i)^2/\sigma_i^2$.

Critical region: $T > \chi^2_{1-\alpha}$.

Number of degrees of freedom: N (or $N - p$, if p parameters are determined from the measurements).

The Method of Least Squares

One considers a set of m *equations* $f_k(\mathbf{x}, \boldsymbol{\eta}) = 0$; $k = 1, \ldots, m$ relating the r-vector of the *unknowns* $\mathbf{x} = (x_1, x_2, \ldots, x_r)$ to the n-vector of *measurable quantities* $\boldsymbol{\eta} = (\eta_1, \eta_2, \ldots, \eta_n)$. Instead of $\boldsymbol{\eta}$, the quantities $\mathbf{y}$ are measured, which deviate from them by the *measurement errors* $\boldsymbol{\varepsilon}$, i.e., $\mathbf{y} = \boldsymbol{\eta} + \boldsymbol{\varepsilon}$. The quantities $\boldsymbol{\varepsilon}$ are assumed to be normally distributed about zero. This is expressed by the covariance matrix $C_y = G_y^{-1}$. In order to obtain the solution $\widetilde{\mathbf{x}}, \widetilde{\boldsymbol{\eta}}$, one expands the f_k for the *first approximations* $\mathbf{x}_0, \boldsymbol{\eta}_0 = \mathbf{y}$. Only the linear terms of the expansion are kept and the second approximation $\mathbf{x}_1 = \mathbf{x}_0 + \boldsymbol{\xi}$, $\boldsymbol{\eta}_1 = \boldsymbol{\eta} + \boldsymbol{\delta}$ is computed. The procedure is repeated iteratively until certain convergence criteria are met, for example, until a scalar function M no longer decreases. If the f_k are linear in $\mathbf{x}$ and $\boldsymbol{\eta}$, then only one step is necessary. The method can be interpreted as a procedure to minimize M. The function M corresponding to the solution depends on the measurement errors. It is a random variable and follows a χ^2-distribution with $f = m - r$ degrees of freedom. It can therefore be used for a χ^2*-test* of the goodness-of-fit or of other assumptions, in particular the assumption $C_y = G_y^{-1}$. If the errors $\boldsymbol{\varepsilon}$ are not normally distribution, but still distributed symmetrically about zero, then the least-squares solution $\widetilde{\mathbf{x}}$ still has the smallest possible variance, and one has $E(M) = m - r$ *(Gauss–Markov theorem).*

Table H.6. Least squares in the general case and in the case of constrained measurements.

	General case	Constrained measurements
Equations	$f_k(\mathbf{x},\boldsymbol{\eta}) = 0,\ k = 1,\ldots,m$	$f_k(\boldsymbol{\eta}) = 0$
First approximations	$\mathbf{x}_0,\ \boldsymbol{\eta}_0 = \mathbf{y}$	$\boldsymbol{\eta}_0 = \mathbf{y}$
Equations expanded	$\mathbf{f} = A\boldsymbol{\xi} + B\boldsymbol{\delta} + \mathbf{c} + \cdots$ $\{A\}_{kl} = (\partial f_k/\partial x_l)_{\mathbf{x}_0,\boldsymbol{\eta}_0}$ $\{B\}_{kl} = (\partial f_k/\partial \eta_l)_{\mathbf{x}_0,\boldsymbol{\eta}_0}$ $\mathbf{c} = \mathbf{f}(\mathbf{x}_0,\ \boldsymbol{\eta}_0)$	$\mathbf{f} = B\boldsymbol{\delta} + \mathbf{c} + \cdots$ — $\{B\}_{kl} = (\partial f_k/\partial \eta_l)_{\boldsymbol{\eta}_0}$ $\mathbf{c} = \mathbf{f}(\boldsymbol{\eta}_0)$
Covariance matrix of the measurements	$C_y = G_y^{-1}$	$C_y = G_y^{-1}$
Corrections	$\widetilde{\boldsymbol{\xi}} = -(A^{\mathrm{T}}G_BA)^{-1}A^{\mathrm{T}}G_B\mathbf{c}$ $\widetilde{\boldsymbol{\delta}} = -G_y^{-1}B^{\mathrm{T}}G_B(A\widetilde{\boldsymbol{\xi}} + \mathbf{c})$ $G_B = (BG_y^{-1}B^{\mathrm{T}})^{-1}$	— $\widetilde{\boldsymbol{\delta}} = -G_y^{-1}B^{\mathrm{T}}G_B\mathbf{c}$ $G_B = (BG_y^{-1}B^{\mathrm{T}})^{-1}$
Next step	$\mathbf{x}_1 = \mathbf{x}_0 + \widetilde{\boldsymbol{\xi}},\ \boldsymbol{\eta}_1 = \boldsymbol{\eta}_0 + \widetilde{\boldsymbol{\delta}},$ new values for A, B, $\mathbf{c}$, $\widetilde{\boldsymbol{\xi}}$, $\widetilde{\boldsymbol{\delta}}$	$\boldsymbol{\eta}_1 = \boldsymbol{\eta}_0 + \widetilde{\boldsymbol{\delta}},$ new values for B, $\mathbf{c}$, $\widetilde{\boldsymbol{\delta}}$
Solution (after s steps)	$\widetilde{\mathbf{x}} = \mathbf{x}_{s-1} + \widetilde{\boldsymbol{\xi}},$ $\widetilde{\boldsymbol{\eta}} = \boldsymbol{\eta}_{s-1} + \widetilde{\boldsymbol{\delta}},$ $\widetilde{\boldsymbol{\varepsilon}} = \mathbf{y} - \widetilde{\boldsymbol{\eta}}$	$\widetilde{\boldsymbol{\eta}} = \boldsymbol{\eta}_{s-1} + \widetilde{\boldsymbol{\delta}},$ $\widetilde{\boldsymbol{\varepsilon}} = \mathbf{y} - \widetilde{\boldsymbol{\eta}}$
Minimum function	$M = (B\widetilde{\varepsilon})^{\mathrm{T}}G_B(B\widetilde{\varepsilon})$	$M = (B\widetilde{\varepsilon})^{\mathrm{T}}G_B(B\widetilde{\varepsilon})$
Covariance matrices	$G_{\widetilde{x}}^{-1} = (A^{\mathrm{T}}G_BA)^{-1}$ $G_{\widetilde{\eta}}^{-1} = G_y^{-1}$ $-G_y^{-1}B^{\mathrm{T}}G_BBG_y^{-1}$ $+\,G_y^{-1}B^{\mathrm{T}}G_BA(A^{\mathrm{T}}G_BA)^{-1}$ $\times\, A^{\mathrm{T}}G_BBG_y^{-1}$	$G_{\widetilde{\eta}}^{-1} = G_y^{-1}$ $-\,G_y^{-1}B^{\mathrm{T}}G_BBG_y^{-1}$

Table H.7. Least squares for indirect and direct measurements.

	Indirect measurements	Direct measurements of different accuracy
Equations	$f_k = \eta_k - g_k(x) = 0$	$f_k = \eta_k - x$
First approximations	$\mathbf{x}_0,\ \boldsymbol{\eta}_0 = \mathbf{y}$	$x_0 = 0,\ \boldsymbol{\eta}_0 = \mathbf{y}$
Equations expanded	$\mathbf{f} = A\widetilde{\boldsymbol{\xi}} - \boldsymbol{\varepsilon} + \mathbf{c} + \cdots$ $\{A\}_{kl} = \left(\frac{\partial f_k}{\partial x_l}\right)_{\mathbf{x}_0}$ $\mathbf{c} = \mathbf{y} - \mathbf{g}(\mathbf{x}_0)$	$\mathbf{f} = A\widetilde{\boldsymbol{\xi}} + \boldsymbol{\varepsilon} + \mathbf{c}$ $A = -\begin{pmatrix} 1 \\ 1 \\ \vdots \\ 1 \end{pmatrix}$ $\mathbf{c} = \mathbf{y}$
Covariance matrix of the measurements	$C_y = G_y^{-1}$	$C_y = G_y^{-1}$ $= \begin{pmatrix} \sigma_1^2 & & 0 \\ & \ddots & \\ 0 & & \sigma_n^2 \end{pmatrix}$
Corrections	$\widetilde{\boldsymbol{\xi}} = -(A^{\mathrm{T}} G_y A)^{-1} A^{\mathrm{T}} G_y \mathbf{c}$	—
Next step	$\mathbf{x}_1 = \mathbf{x}_0 + \widetilde{\boldsymbol{\xi}}$, new values for A, $\mathbf{c}$, $\widetilde{\boldsymbol{\xi}}$	—
Solution (after s steps)	$\widetilde{\mathbf{x}} = \mathbf{x}_{s-1} + \widetilde{\boldsymbol{\xi}}$, $\widetilde{\boldsymbol{\varepsilon}} = A\widetilde{\boldsymbol{\xi}} + \mathbf{c}$	$\widetilde{\xi} = \widetilde{x} = \frac{\sum_k y_k/\sigma_k^2}{\sum_k 1/\sigma_k^2}$, $\widetilde{\varepsilon}_k = y_k - \widetilde{x}$
Minimum function	$M = \widetilde{\varepsilon}^{\mathrm{T}} G_y \widetilde{\varepsilon}$	$M = \widetilde{\varepsilon}^{\mathrm{T}} G_y \widetilde{\varepsilon}$
Covariance matrices	$G_{\widetilde{x}}^{-1} = (A^{\mathrm{T}} G_y A)^{-1}$ $G_{\widetilde{\eta}}^{-1} = A(A^{\mathrm{T}} G_y A) A^{\mathrm{T}}$	$G_{\widetilde{x}}^{-1} = \sigma^2(\widetilde{x})$ $= \left(\sum_k \frac{1}{\sigma_k^2}\right)^{-1}$ —

Analysis of Variance

The influence of external variables on a measured random variable x is investigated. One tries to decide by means of appropriately constructed F-tests whether x is independent of the *external variables*. Various *models* are constructed depending on the number of external variables and the assumptions concerning their influence.

A simple model is the *crossed two-way classification* with multiple observations. Two external variables are used to classify the observations into the classes A_i, B_j ($i = 1, 2, \ldots, I$; $j = 1, 2, \ldots, J$). Each class A_i, B_j contains K observations x_{ijk} ($k = 1, 2, \ldots, K$). One assumes the model

$$x_{ijk} = \mu + a_i + b_j + (ab)_{ij} + \varepsilon_{ijk} \quad ,$$

where the error of the observation ε_{ijk} is assumed to be normally distributed about zero and a_i, b_j, and $(ab)_{ij}$ are the influences of the *classifications* in A, B and their *interactions*. Three *null hypotheses*

$$H_0^{(A)}(a_i = 0,\ i = 1, \ldots, I) \quad , \qquad H_0^{(B)}(b_j = 0,\ j = 1, \ldots, J) \quad ,$$

$$H_0^{(AB)}((ab)_{ij} = 0,\ i = 1, \ldots, I,\ j = 1, \ldots, J)$$

can be tested with the ratios $F^{(A)}$, $F^{(B)}$, $F^{(AB)}$. They are summarized in an analysis of variance (ANOVA) table. For other models see Chapter 11.

Polynomial Regression

Problem: The true values $\eta(t)$, for which one has N *measurements* $y_i(t_i)$ with normally distributed measurement errors σ_i, are to be described by a *polynomial* of order $r - 1$ in the *controlled variable* t. Instead of $\eta(t) = x_1 + x_2 t + \cdots + x_r t^{r-1}$ one writes

$$\eta(t) = x_1 f_1(t) + x_2 f_2(t) + \cdots + x_r f_r(t) \quad .$$

Here the f_j are *orthogonal polynomials* of order $j - 1$,

$$f_j(t) = \sum_{k=1}^{j} b_{jk} t^{k-1} \quad ,$$

whose coefficients b_{jk} are determined by the *orthogonality conditions*

$$\sum_{i=1}^{N} g_i f_j(t_i) f_k(t_i) = \delta_{jk} \quad , \qquad g_i = 1/\sigma_i^2 \quad .$$

The *unknowns* x_j are obtained by least squares from

$$\sum_{i=1}^{N} g_i \left\{ y_i(t_i) - \sum_{j=1}^{r} x_j f_j(t_i) \right\}^2 = \min \quad .$$

The covariance matrix of the x_j is the r-dimensional unit matrix.

Time Series Analysis

One is given a series of *measured values* $y_i(t_i)$, $i = 1, \ldots, n$, which (in an unknown way) depend on a *controlled variable* t (usually time). One treats the y_i as the sum of a *trend* η_i and an *error* ε_i, $y_i = \eta_i + \varepsilon_i$. The measurements are carried out at regular time intervals, i.e., $t_i - t_{i-1} = \text{const}$. In order to minimize the errors ε_i, a *moving average* is constructed for every t_i ($i > k$, $i \leq n - k$), by fitting a polynomial of order ℓ to the $2k + 1$ measurements situated symmetrically about measurement i. The result of the fit at the point t_i is the moving average

$$\tilde{\eta}_0(i) = a_{-k} y_{i-k} + a_{-k+1} y_{i-k+1} + \cdots + a_k y_{i+k} \quad .$$

The coefficients $a_{-k}, \ldots, a_k$ are given in Table 13.1 for low values of k and ℓ. For the beginning and end points t_i ($i < k$, $i > n - k$), the results of the fit can be used with caution also for points other than at the center of the interval of the $2k + 1$ measurements.

I. Statistical Tables

Table I.1. Quantiles $\lambda_P(k)$ of the Poisson distribution.

$$P = \sum_{n=0}^{k-1} \mathrm{e}^{-\lambda_P} \lambda_P^n / n!$$

	P						
k	0.0005	0.0010	0.0050	0.0100	0.0250	0.0500	0.1000
1	7.601	6.908	5.298	4.605	3.689	2.996	2.303
2	9.999	9.233	7.430	6.638	5.572	4.744	3.890
3	12.051	11.229	9.274	8.406	7.225	6.296	5.322
4	13.934	13.062	10.977	10.045	8.767	7.754	6.681
5	15.710	14.794	12.594	11.605	10.242	9.154	7.994
6	17.411	16.455	14.150	13.108	11.668	10.513	9.275
7	19.055	18.062	15.660	14.571	13.059	11.842	10.532
8	20.654	19.626	17.134	16.000	14.423	13.148	11.771
9	22.217	21.156	18.578	17.403	15.763	14.435	12.995
10	23.749	22.657	19.998	18.783	17.085	15.705	14.206
11	25.256	24.134	21.398	20.145	18.390	16.962	15.407
12	26.739	25.589	22.779	21.490	19.682	18.208	16.598
13	28.203	27.026	24.145	22.821	20.962	19.443	17.782
14	29.650	28.446	25.497	24.139	22.230	20.669	18.958
15	31.081	29.852	26.836	25.446	23.490	21.886	20.128
16	32.498	31.244	28.164	26.743	24.740	23.097	21.292
17	33.902	32.624	29.482	28.030	25.983	24.301	22.452
18	35.294	33.993	30.791	29.310	27.219	25.499	23.606
19	36.676	35.351	32.091	30.581	28.448	26.692	24.756
20	38.047	36.701	33.383	31.845	29.671	27.879	25.903
21	39.410	38.042	34.668	33.103	30.888	29.062	27.045
22	40.764	39.375	35.946	34.355	32.101	30.240	28.184
23	42.110	40.700	37.218	35.601	33.308	31.415	29.320
24	43.449	42.019	38.484	36.841	34.511	32.585	30.453
25	44.780	43.330	39.745	38.077	35.710	33.752	31.584

Table I.1. *(continued)*

$$P = \sum_{n=0}^{k-1} \mathrm{e}^{-\lambda_P} \lambda_P^n / n!$$

	P						
k	0.9000	0.9500	0.9750	0.9900	0.9950	0.9990	0.9995
1	0.105	0.051	0.025	0.010	0.005	0.001	0.001
2	0.532	0.355	0.242	0.149	0.103	0.045	0.032
3	1.102	0.818	0.619	0.436	0.338	0.191	0.150
4	1.745	1.366	1.090	0.823	0.672	0.429	0.355
5	2.433	1.970	1.623	1.279	1.078	0.739	0.632
6	3.152	2.613	2.202	1.785	1.537	1.107	0.967
7	3.895	3.285	2.814	2.330	2.037	1.520	1.348
8	4.656	3.981	3.454	2.906	2.571	1.971	1.768
9	5.432	4.695	4.115	3.507	3.132	2.452	2.220
10	6.221	5.425	4.795	4.130	3.717	2.961	2.699
11	7.021	6.169	5.491	4.771	4.321	3.491	3.202
12	7.829	6.924	6.201	5.428	4.943	4.042	3.726
13	8.646	7.690	6.922	6.099	5.580	4.611	4.269
14	9.470	8.464	7.654	6.782	6.231	5.195	4.828
15	10.300	9.246	8.395	7.477	6.893	5.794	5.402
16	11.135	10.036	9.145	8.181	7.567	6.405	5.990
17	11.976	10.832	9.903	8.895	8.251	7.028	6.590
18	12.822	11.634	10.668	9.616	8.943	7.662	7.201
19	13.671	12.442	11.439	10.346	9.644	8.306	7.822
20	14.525	13.255	12.217	11.082	10.353	8.958	8.453
21	15.383	14.072	12.999	11.825	11.069	9.619	9.093
22	16.244	14.894	13.787	12.574	11.792	10.288	9.741
23	17.108	15.719	14.580	13.329	12.521	10.964	10.397
24	17.975	16.549	15.377	14.089	13.255	11.647	11.060
25	18.844	17.382	16.179	14.853	13.995	12.337	11.730

Table I.2. Normal distribution $\psi_0(x)$.

$$P(\mathsf{x} < x) = \psi_0(x) = \frac{1}{\sqrt{2\pi}} \int_{-\infty}^{x} \exp(-x^2/2)\,\mathrm{d}x$$

x	0	1	2	3	4	5	6	7	8	9
-3.0	.001	.001	.001	.001	.001	.001	.001	.001	.001	.001
-2.9	.002	.002	.002	.002	.002	.002	.002	.001	.001	.001
-2.8	.003	.002	.002	.002	.002	.002	.002	.002	.002	.002
-2.7	.003	.003	.003	.003	.003	.003	.003	.003	.003	.003
-2.6	.005	.005	.004	.004	.004	.004	.004	.004	.004	.004
-2.5	.006	.006	.006	.006	.006	.005	.005	.005	.005	.005
-2.4	.008	.008	.008	.008	.007	.007	.007	.007	.007	.006
-2.3	.011	.010	.010	.010	.010	.009	.009	.009	.009	.008
-2.2	.014	.014	.013	.013	.013	.012	.012	.012	.011	.011
-2.1	.018	.017	.017	.017	.016	.016	.015	.015	.015	.014
-2.0	.023	.022	.022	.021	.021	.020	.020	.019	.019	.018
-1.9	.029	.028	.027	.027	.026	.026	.025	.024	.024	.023
-1.8	.036	.035	.034	.034	.033	.032	.031	.031	.030	.029
-1.7	.045	.044	.043	.042	.041	.040	.039	.038	.038	.037
-1.6	.055	.054	.053	.052	.051	.049	.048	.047	.046	.046
-1.5	.067	.066	.064	.063	.062	.061	.059	.058	.057	.056
-1.4	.081	.079	.078	.076	.075	.074	.072	.071	.069	.068
-1.3	.097	.095	.093	.092	.090	.089	.087	.085	.084	.082
-1.2	.115	.113	.111	.109	.107	.106	.104	.102	.100	.099
-1.1	.136	.133	.131	.129	.127	.125	.123	.121	.119	.117
-1.0	.159	.156	.154	.152	.149	.147	.145	.142	.140	.138
-0.9	.184	.181	.179	.176	.174	.171	.169	.166	.164	.161
-0.8	.212	.209	.206	.203	.200	.198	.195	.192	.189	.187
-0.7	.242	.239	.236	.233	.230	.227	.224	.221	.218	.215
-0.6	.274	.271	.268	.264	.261	.258	.255	.251	.248	.245
-0.5	.309	.305	.302	.298	.295	.291	.288	.284	.281	.278
-0.4	.345	.341	.337	.334	.330	.326	.323	.319	.316	.312
-0.3	.382	.378	.374	.371	.367	.363	.359	.356	.352	.348
-0.2	.421	.417	.413	.409	.405	.401	.397	.394	.390	.386
-0.1	.460	.456	.452	.448	.444	.440	.436	.433	.429	.425
0.0	.500	.496	.492	.488	.484	.480	.476	.472	.468	.464

Table I.2. *(continued)*

$$P(\mathsf{x} < x) = \psi_0(x) = \frac{1}{\sqrt{2\pi}} \int_{-\infty}^{x} \exp(-x^2/2)\, \mathrm{d}x$$

x	0	1	2	3	4	5	6	7	8	9
0.0	.500	.504	.508	.512	.516	.520	.524	.528	.532	.536
0.1	.540	.544	.548	.552	.556	.560	.564	.567	.571	.575
0.2	.579	.583	.587	.591	.595	.599	.603	.606	.610	.614
0.3	.618	.622	.626	.629	.633	.637	.641	.644	.648	.652
0.4	.655	.659	.663	.666	.670	.674	.677	.681	.684	.688
0.5	.691	.695	.698	.702	.705	.709	.712	.716	.719	.722
0.6	.726	.729	.732	.736	.739	.742	.745	.749	.752	.755
0.7	.758	.761	.764	.767	.770	.773	.776	.779	.782	.785
0.8	.788	.791	.794	.797	.800	.802	.805	.808	.811	.813
0.9	.816	.819	.821	.824	.826	.829	.831	.834	.836	.839
1.0	.841	.844	.846	.848	.851	.853	.855	.858	.860	.862
1.1	.864	.867	.869	.871	.873	.875	.877	.879	.881	.883
1.2	.885	.887	.889	.891	.893	.894	.896	.898	.900	.901
1.3	.903	.905	.907	.908	.910	.911	.913	.915	.916	.918
1.4	.919	.921	.922	.924	.925	.926	.928	.929	.931	.932
1.5	.933	.934	.936	.937	.938	.939	.941	.942	.943	.944
1.6	.945	.946	.947	.948	.949	.951	.952	.953	.954	.954
1.7	.955	.956	.957	.958	.959	.960	.961	.962	.962	.963
1.8	.964	.965	.966	.966	.967	.968	.969	.969	.970	.971
1.9	.971	.972	.973	.973	.974	.974	.975	.976	.976	.977
2.0	.977	.978	.978	.979	.979	.980	.980	.981	.981	.982
2.1	.982	.983	.983	.983	.984	.984	.985	.985	.985	.986
2.2	.986	.986	.987	.987	.987	.988	.988	.988	.989	.989
2.3	.989	.990	.990	.990	.990	.991	.991	.991	.991	.992
2.4	.992	.992	.992	.992	.993	.993	.993	.993	.993	.994
2.5	.994	.994	.994	.994	.994	.995	.995	.995	.995	.995
2.6	.995	.995	.996	.996	.996	.996	.996	.996	.996	.996
2.7	.997	.997	.997	.997	.997	.997	.997	.997	.997	.997
2.8	.997	.998	.998	.998	.998	.998	.998	.998	.998	.998
2.9	.998	.998	.998	.998	.998	.998	.998	.999	.999	.999
3.0	.999	.999	.999	.999	.999	.999	.999	.999	.999	.999

Table I.3. Normal distribution $2\psi_0(x) - 1$.

$$P(|\mathsf{x}| < x) = 2\psi_0(x) - 1 = \frac{1}{\sqrt{2\pi}} \int_{-x}^{x} \exp(-x^2/2)\, \mathrm{d}x$$

x	0	1	2	3	4	5	6	7	8	9
0.0	.000	.008	.016	.024	.032	.040	.048	.056	.064	.072
0.1	.080	.088	.096	.103	.111	.119	.127	.135	.143	.151
0.2	.159	.166	.174	.182	.190	.197	.205	.213	.221	.228
0.3	.236	.243	.251	.259	.266	.274	.281	.289	.296	.303
0.4	.311	.318	.326	.333	.340	.347	.354	.362	.369	.376
0.5	.383	.390	.397	.404	.411	.418	.425	.431	.438	.445
0.6	.451	.458	.465	.471	.478	.484	.491	.497	.503	.510
0.7	.516	.522	.528	.535	.541	.547	.553	.559	.565	.570
0.8	.576	.582	.588	.593	.599	.605	.610	.616	.621	.627
0.9	.632	.637	.642	.648	.653	.658	.663	.668	.673	.678
1.0	.683	.688	.692	.697	.702	.706	.711	.715	.720	.724
1.1	.729	.733	.737	.742	.746	.750	.754	.758	.762	.766
1.2	.770	.774	.778	.781	.785	.789	.792	.796	.799	.803
1.3	.806	.810	.813	.816	.820	.823	.826	.829	.832	.835
1.4	.838	.841	.844	.847	.850	.853	.856	.858	.861	.864
1.5	.866	.869	.871	.874	.876	.879	.881	.884	.886	.888
1.6	.890	.893	.895	.897	.899	.901	.903	.905	.907	.909
1.7	.911	.913	.915	.916	.918	.920	.922	.923	.925	.927
1.8	.928	.930	.931	.933	.934	.936	.937	.939	.940	.941
1.9	.943	.944	.945	.946	.948	.949	.950	.951	.952	.953
2.0	.954	.956	.957	.958	.959	.960	.961	.962	.962	.963
2.1	.964	.965	.966	.967	.968	.968	.969	.970	.971	.971
2.2	.972	.973	.974	.974	.975	.976	.976	.977	.977	.978
2.3	.979	.979	.980	.980	.981	.981	.982	.982	.983	.983
2.4	.984	.984	.984	.985	.985	.986	.986	.986	.987	.987
2.5	.988	.988	.988	.989	.989	.989	.990	.990	.990	.990
2.6	.991	.991	.991	.991	.992	.992	.992	.992	.993	.993
2.7	.993	.993	.993	.994	.994	.994	.994	.994	.995	.995
2.8	.995	.995	.995	.995	.995	.996	.996	.996	.996	.996
2.9	.996	.996	.996	.997	.997	.997	.997	.997	.997	.997
3.0	.997	.997	.997	.998	.998	.998	.998	.998	.998	.998

Table I.4. Quantiles $x_P = \Omega(P)$ of the Normal distribution.

$$P = \frac{1}{\sqrt{2\pi}} \int_{-\infty}^{x_P} \exp(-x^2/2)\,\mathrm{d}x$$

P	0	1	2	3	4	5	6	7	8	9
0.0	$-\infty$	−2.33	−2.05	−1.88	−1.75	−1.64	−1.55	−1.48	−1.41	−1.34
0.1	−1.28	−1.23	−1.17	−1.13	−1.08	−1.04	−0.99	−0.95	−0.92	−0.88
0.2	−0.84	−0.81	−0.77	−0.74	−0.71	−0.67	−0.64	−0.61	−0.58	−0.55
0.3	−0.52	−0.50	−0.47	−0.44	−0.41	−0.39	−0.36	−0.33	−0.31	−0.28
0.4	−0.25	−0.23	−0.20	−0.18	−0.15	−0.13	−0.10	−0.08	−0.05	−0.03
0.5	0.00	0.03	0.05	0.08	0.10	0.13	0.15	0.18	0.20	0.23
0.6	0.25	0.28	0.31	0.33	0.36	0.39	0.41	0.44	0.47	0.50
0.7	0.52	0.55	0.58	0.61	0.64	0.67	0.71	0.74	0.77	0.81
0.8	0.84	0.88	0.92	0.95	0.99	1.04	1.08	1.13	1.17	1.23
0.9	1.28	1.34	1.41	1.48	1.55	1.64	1.75	1.88	2.05	2.33

Table I.5. Quantiles $x'_P = \Omega'(P)$ of the normal distribution.

$$P = \frac{1}{\sqrt{2\pi}} \int_{-x'_P}^{x'_P} \exp(-x^2/2)\,\mathrm{d}x$$

P	0	1	2	3	4	5	6	7	8	9
0.0	0.000	0.013	0.025	0.038	0.050	0.063	0.075	0.088	0.100	0.113
0.1	0.126	0.138	0.151	0.164	0.176	0.189	0.202	0.215	0.228	0.240
0.2	0.253	0.266	0.279	0.292	0.305	0.319	0.332	0.345	0.358	0.372
0.3	0.385	0.399	0.412	0.426	0.440	0.454	0.468	0.482	0.496	0.510
0.4	0.524	0.539	0.553	0.568	0.583	0.598	0.613	0.628	0.643	0.659
0.5	0.674	0.690	0.706	0.722	0.739	0.755	0.772	0.789	0.806	0.824
0.6	0.842	0.860	0.878	0.896	0.915	0.935	0.954	0.974	0.994	1.015
0.7	1.036	1.058	1.080	1.103	1.126	1.150	1.175	1.200	1.227	1.254
0.8	1.282	1.311	1.341	1.372	1.405	1.440	1.476	1.514	1.555	1.598
0.9	1.645	1.695	1.751	1.812	1.881	1.960	2.054	2.170	2.326	2.576

P	0	1	2	3	4	5	6	7	8	9
0.90	1.645	1.650	1.655	1.660	1.665	1.670	1.675	1.680	1.685	1.690
0.91	1.695	1.701	1.706	1.711	1.717	1.722	1.728	1.734	1.739	1.745
0.92	1.751	1.757	1.762	1.768	1.774	1.780	1.787	1.793	1.799	1.805
0.93	1.812	1.818	1.825	1.832	1.838	1.845	1.852	1.859	1.866	1.873
0.94	1.881	1.888	1.896	1.903	1.911	1.919	1.927	1.935	1.943	1.951
0.95	1.960	1.969	1.977	1.986	1.995	2.005	2.014	2.024	2.034	2.044
0.96	2.054	2.064	2.075	2.086	2.097	2.108	2.120	2.132	2.144	2.157
0.97	2.170	2.183	2.197	2.212	2.226	2.241	2.257	2.273	2.290	2.308
0.98	2.326	2.346	2.366	2.387	2.409	2.432	2.457	2.484	2.512	2.543
0.99	2.576	2.612	2.652	2.697	2.748	2.807	2.878	2.968	3.090	3.291

P	0	1	2	3	4	5	6	7	8	9
0.990	2.576	2.579	2.583	2.586	2.590	2.594	2.597	2.601	2.605	2.608
0.991	2.612	2.616	2.620	2.624	2.628	2.632	2.636	2.640	2.644	2.648
0.992	2.652	2.656	2.661	2.665	2.669	2.674	2.678	2.683	2.687	2.692
0.993	2.697	2.702	2.706	2.711	2.716	2.721	2.727	2.732	2.737	2.742
0.994	2.748	2.753	2.759	2.765	2.770	2.776	2.782	2.788	2.794	2.801
0.995	2.807	2.814	2.820	2.827	2.834	2.841	2.848	2.855	2.863	2.870
0.996	2.878	2.886	2.894	2.903	2.911	2.920	2.929	2.938	2.948	2.958
0.997	2.968	2.978	2.989	3.000	3.011	3.023	3.036	3.048	3.062	3.076
0.998	3.090	3.105	3.121	3.138	3.156	3.175	3.195	3.216	3.239	3.264
0.999	3.291	3.320	3.353	3.390	3.432	3.481	3.540	3.615	3.719	3.891

Table I.6. χ^2-distribution $F(\chi^2)$.

$$F(\chi^2) = \int_0^{\chi^2} f(\chi^2; f)\, d\chi^2$$

χ^2	f = 1	2	3	4	5	6	7	8	9	10
0.1	0.248	0.049	0.008	0.001	0.000	0.000	0.000	0.000	0.000	0.000
0.2	0.345	0.095	0.022	0.005	0.001	0.000	0.000	0.000	0.000	0.000
0.3	0.416	0.139	0.040	0.010	0.002	0.001	0.000	0.000	0.000	0.000
0.4	0.473	0.181	0.060	0.018	0.005	0.001	0.000	0.000	0.000	0.000
0.5	0.520	0.221	0.081	0.026	0.008	0.002	0.001	0.000	0.000	0.000
0.6	0.561	0.259	0.104	0.037	0.012	0.004	0.001	0.000	0.000	0.000
0.7	0.597	0.295	0.127	0.049	0.017	0.006	0.002	0.000	0.000	0.000
0.8	0.629	0.330	0.151	0.062	0.023	0.008	0.003	0.001	0.000	0.000
0.9	0.657	0.362	0.175	0.075	0.030	0.011	0.004	0.001	0.000	0.000
1.0	0.683	0.393	0.199	0.090	0.037	0.014	0.005	0.002	0.001	0.000
2.0	0.843	0.632	0.428	0.264	0.151	0.080	0.040	0.019	0.009	0.004
3.0	0.917	0.777	0.608	0.442	0.300	0.191	0.115	0.066	0.036	0.019
4.0	0.954	0.865	0.739	0.594	0.451	0.323	0.220	0.143	0.089	0.053
5.0	0.975	0.918	0.828	0.713	0.584	0.456	0.340	0.242	0.166	0.109
6.0	0.986	0.950	0.888	0.801	0.694	0.577	0.460	0.353	0.260	0.185
7.0	0.992	0.970	0.928	0.864	0.779	0.679	0.571	0.463	0.363	0.275
8.0	0.995	0.982	0.954	0.908	0.844	0.762	0.667	0.567	0.466	0.371
9.0	0.997	0.989	0.971	0.939	0.891	0.826	0.747	0.658	0.563	0.468
10.0	0.998	0.993	0.981	0.960	0.925	0.875	0.811	0.735	0.650	0.560
11.0	0.999	0.996	0.988	0.973	0.949	0.912	0.861	0.798	0.724	0.642
12.0	0.999	0.998	0.993	0.983	0.965	0.938	0.899	0.849	0.787	0.715
13.0	1.000	0.998	0.995	0.989	0.977	0.957	0.928	0.888	0.837	0.776
14.0	1.000	0.999	0.997	0.993	0.984	0.970	0.949	0.918	0.878	0.827
15.0	1.000	0.999	0.998	0.995	0.990	0.980	0.964	0.941	0.909	0.868
16.0	1.000	1.000	0.999	0.997	0.993	0.986	0.975	0.958	0.933	0.900
17.0	1.000	1.000	0.999	0.998	0.996	0.991	0.983	0.970	0.951	0.926
18.0	1.000	1.000	1.000	0.999	0.997	0.994	0.988	0.979	0.965	0.945
19.0	1.000	1.000	1.000	0.999	0.998	0.996	0.992	0.985	0.975	0.960
20.0	1.000	1.000	1.000	1.000	0.999	0.997	0.994	0.990	0.982	0.971

Table I.7. Quantiles χ_P^2 of the χ^2-distribution.

$$P = \int_0^{\chi_P^2} f(\chi^2; f)\, \mathrm{d}\chi^2$$

	P				
f	0.900	0.950	0.990	0.995	0.999
1	2.706	3.841	6.635	7.879	10.828
2	4.605	5.991	9.210	10.597	13.816
3	6.251	7.815	11.345	12.838	16.266
4	7.779	9.488	13.277	14.860	18.467
5	9.236	11.070	15.086	16.750	20.515
6	10.645	12.592	16.812	18.548	22.458
7	12.017	14.067	18.475	20.278	24.322
8	13.362	15.507	20.090	21.955	26.124
9	14.684	16.919	21.666	23.589	27.877
10	15.987	18.307	23.209	25.188	29.588
11	17.275	19.675	24.725	26.757	31.264
12	18.549	21.026	26.217	28.300	32.909
13	19.812	22.362	27.688	29.819	34.528
14	21.064	23.685	29.141	31.319	36.123
15	22.307	24.996	30.578	32.801	37.697
16	23.542	26.296	32.000	34.267	39.252
17	24.769	27.587	33.409	35.718	40.790
18	25.989	28.869	34.805	37.156	42.312
19	27.204	30.144	36.191	38.582	43.820
20	28.412	31.410	37.566	39.997	45.315
30	40.256	43.773	50.892	53.672	59.703
40	51.805	55.758	63.691	66.766	73.402
50	63.167	67.505	76.154	79.490	86.661
60	74.397	79.082	88.379	91.952	99.607
70	85.527	90.531	100.425	104.215	112.317
80	80.000	101.879	112.329	116.321	124.839
90	107.565	113.145	124.116	128.299	137.208
100	118.498	124.342	135.807	140.169	149.449

Table I.8. Quantiles F_P of the F-distribution.

$$0.900 = P = \int_0^{F_P} f(F; f_1, f_2)\,\mathrm{d}F$$

f_2	f_1 = 1	2	3	4	5	6	7	8	9	10
1	39.86	49.50	53.59	55.83	57.24	58.20	58.91	59.44	59.86	60.19
2	8.526	9.000	9.162	9.243	9.293	9.326	9.349	9.367	9.381	9.392
3	5.538	5.462	5.391	5.343	5.309	5.285	5.266	5.252	5.240	5.230
4	4.545	4.325	4.191	4.107	4.051	4.010	3.979	3.955	3.936	3.920
5	4.060	3.780	3.619	3.520	3.453	3.405	3.368	3.339	3.316	3.297
6	3.776	3.463	3.289	3.181	3.108	3.055	3.014	2.983	2.958	2.937
7	3.589	3.257	3.074	2.961	2.883	2.827	2.785	2.752	2.725	2.703
8	3.458	3.113	2.924	2.806	2.726	2.668	2.624	2.589	2.561	2.538
9	3.360	3.006	2.813	2.693	2.611	2.551	2.505	2.469	2.440	2.416
10	3.285	2.924	2.728	2.605	2.522	2.461	2.414	2.377	2.347	2.323

$$0.950 = P = \int_0^{F_P} f(F; f_1, f_2)\,\mathrm{d}F$$

f_2	f_1 = 1	2	3	4	5	6	7	8	9	10
1	161.4	199.5	215.7	224.6	230.2	234.0	236.8	238.9	240.5	241.9
2	18.51	19.00	19.16	19.25	19.30	19.33	19.35	19.37	19.38	19.40
3	10.13	9.552	9.277	9.117	9.013	8.941	8.887	8.845	8.812	8.786
4	7.709	6.944	6.591	6.388	6.256	6.163	6.094	6.041	5.999	5.964
5	6.608	5.786	5.409	5.192	5.050	4.950	4.876	4.818	4.772	4.735
6	5.987	5.143	4.757	4.534	4.387	4.284	4.207	4.147	4.099	4.060
7	5.591	4.737	4.347	4.120	3.972	3.866	3.787	3.726	3.677	3.637
8	5.318	4.459	4.066	3.838	3.687	3.581	3.500	3.438	3.388	3.347
9	5.117	4.256	3.863	3.633	3.482	3.374	3.293	3.230	3.179	3.137
10	4.965	4.103	3.708	3.478	3.326	3.217	3.135	3.072	3.020	2.978

Table I.8. (*continued*)

$$0.975 = P = \int_0^{F_P} f(F; f_1, f_2)\,\mathrm{d}F$$

	f_1									
f_2	1	2	3	4	5	6	7	8	9	10
1	647.8	799.5	864.2	899.6	921.8	937.1	948.2	956.7	963.3	968.6
2	38.51	39.00	39.17	39.25	39.30	39.33	39.36	39.37	39.39	39.40
3	17.44	16.04	15.44	15.10	14.88	14.73	14.62	14.54	14.47	14.42
4	12.22	10.65	9.979	9.605	9.364	9.197	9.074	8.980	8.905	8.844
5	10.01	8.434	7.764	7.388	7.146	6.978	6.853	6.757	6.681	6.619
6	8.813	7.260	6.599	6.227	5.988	5.820	5.695	5.600	5.523	5.461
7	8.073	6.542	5.890	5.523	5.285	5.119	4.995	4.899	4.823	4.761
8	7.571	6.059	5.416	5.053	4.817	4.652	4.529	4.433	4.357	4.295
9	7.209	5.715	5.078	4.718	4.484	4.320	4.197	4.102	4.026	3.964
10	6.937	5.456	4.826	4.468	4.236	4.072	3.950	3.855	3.779	3.717

$$0.990 = P = \int_0^{F_P} f(F; f_1, f_2)\,\mathrm{d}F$$

	f_1									
f_2	1	2	3	4	5	6	7	8	9	10
1	4052.	5000.	5403.	5625.	5764.	5859.	5928.	5981.	6022.	6056.
2	98.50	99.00	99.17	99.25	99.30	99.33	99.36	99.37	99.39	99.40
3	34.12	30.82	29.46	28.71	28.24	27.91	27.67	27.49	27.35	27.23
4	21.20	18.00	16.69	15.98	15.52	15.21	14.98	14.80	14.66	14.55
5	16.26	13.27	12.06	11.39	10.97	10.67	10.46	10.29	10.16	10.05
6	13.75	10.92	9.780	9.148	8.746	8.466	8.260	8.102	7.976	7.874
7	12.25	9.547	8.451	7.847	7.460	7.191	6.993	6.840	6.719	6.620
8	11.26	8.649	7.591	7.006	6.632	6.371	6.178	6.029	5.911	5.814
9	10.56	8.022	6.992	6.422	6.057	5.802	5.613	5.467	5.351	5.257
10	10.04	7.559	6.552	5.994	5.636	5.386	5.200	5.057	4.942	4.849

Table I.9. Quantiles t_P of Student's distribution.

$$P = \int_{-\infty}^{t_P} f(t; f)\,\mathrm{d}t$$

	P						
f	0.9000	0.9500	0.9750	0.9900	0.9950	0.9990	0.9995
1	3.078	6.314	12.706	31.821	63.657	318.309	636.619
2	1.886	2.920	4.303	6.965	9.925	22.327	31.599
3	1.638	2.353	3.182	4.541	5.841	10.215	12.924
4	1.533	2.132	2.776	3.747	4.604	7.173	8.610
5	1.476	2.015	2.571	3.365	4.032	5.893	6.869
6	1.440	1.943	2.447	3.143	3.707	5.208	5.959
7	1.415	1.895	2.365	2.998	3.499	4.785	5.408
8	1.397	1.860	2.306	2.896	3.355	4.501	5.041
9	1.383	1.833	2.262	2.821	3.250	4.297	4.781
10	1.372	1.812	2.228	2.764	3.169	4.144	4.587
11	1.363	1.796	2.201	2.718	3.106	4.025	4.437
12	1.356	1.782	2.179	2.681	3.055	3.930	4.318
13	1.350	1.771	2.160	2.650	3.012	3.852	4.221
14	1.345	1.761	2.145	2.624	2.977	3.787	4.140
15	1.341	1.753	2.131	2.602	2.947	3.733	4.073
16	1.337	1.746	2.120	2.583	2.921	3.686	4.015
17	1.333	1.740	2.110	2.567	2.898	3.646	3.965
18	1.330	1.734	2.101	2.552	2.878	3.610	3.922
19	1.328	1.729	2.093	2.539	2.861	3.579	3.883
20	1.325	1.725	2.086	2.528	2.845	3.552	3.850
30	1.310	1.697	2.042	2.457	2.750	3.385	3.646
40	1.303	1.684	2.021	2.423	2.704	3.307	3.551
50	1.299	1.676	2.009	2.403	2.678	3.261	3.496
60	1.296	1.671	2.000	2.390	2.660	3.232	3.460
70	1.294	1.667	1.994	2.381	2.648	3.211	3.435
80	1.292	1.664	1.990	2.374	2.639	3.195	3.416
90	1.291	1.662	1.987	2.368	2.632	3.183	3.402
100	1.290	1.660	1.984	2.364	2.626	3.174	3.390
200	1.286	1.653	1.972	2.345	2.601	3.131	3.340
500	1.283	1.648	1.965	2.334	2.586	3.107	3.310
1000	1.282	1.646	1.962	2.330	2.581	3.098	3.300

Literature

Literature Cited in the Text

[1] A. KOLMOGOROV, *Ergebn. Math.* **2** (1933) 3

[2] D.E. KNUTH, *The Art of Computer Programming*, vol. 2, Addison-Wesley, Reading MA 1981

[3] P. L'ECUYER, *Comm. ACM* **31** (1988) 742

[4] B.A. WICHMANN and I.D. HILL, *Appl. Stat.* **31** (1982) 188

[5] G.E.P. BOX and M.E. MULLER, *Ann. Math. Stat.* **29** (1958) 611

[6] L. VON BORTKIEWICZ, *Das Gesetz der kleinen Zahlen*, Teubner, Leipzig 1898

[7] D.J. DE SOLLA PRICE, *Little Science, Big Science*, Columbia University Press, New York 1965

[8] M.G. KENDALL and A. STUART, *The Advanced Theory of Statistics*, vol. 2, Charles Griffin, London 1968

[9] S.S. WILKS, *Ann. Math. Stat.*, **9** (1938) 60

[10] A.H. ROSENFELD, H.A. BARBERO-GALTIERI, W.J. PODOLSKI, L.R. PRICE, P. SÖDING, CH.G. WOHL, M. ROOS, and W.J. WILLIS, *Rev. Mod. Phys.* **33** (1967) 1

[11] PARTICLE DATA GROUP, *Physics Letters* **B204** (1988) 1

[12] W.H. PRESS, B.P. FLANNERY, S.A. TEUKOLSKY and W.T. VETTERLING, *Numerical Recipes*, Cambridge University Press, Cambridge 1986

[13] R.P. BRENT, *Algorithms for Minimization without Derivatives*, Prentice-Hall, Englewood Cliffs NJ 1973

[14] J.A. NELDER and R. MEAD, *Computer Journal* **7** (1965) 308

[15] M.J.D. POWELL, *Computer Journal* **7** (1965) 155

[16] D.W. MARQUARDT, *J. Soc. Ind. Appl. Math.* **11** (1963) 431

[17] C. LANCZOS, *SIAM J. Numerical Analysis* **1** (1964) 86

[18] C.L. LAWSON and R.J. HANSON, *Solving Least Squares Problems*, Prentice-Hall, Englewood Cliffs NJ 1974

[19] G.H. GOLUB and W. KAHN, *SIAM J. Numerical Analysis* **2** (1965) 205

[20] P.A. BUSINGER and G.H. GOLUB, *Comm. ACM* **12** (1969) 564

[21] G.H. GOLUB and C. REINSCH, in *Linear Algebra* (J.H. WILKINSON and C. REINSCH, eds.), p. 134, Springer, Berlin 1971

[22] J.G.F. FRANCIS, *Computer Journal* **4** (1960) 265, 332

[23] F. JAMES and M. ROOS, *Nuclear Physics* **B172** (1980) 475

[24] M.L. SWARTZ, *Nuclear Instruments and Methods* **A294** (1966) 278

[25] V.H. REGENER, *Physical Review* **84** (1951) 161

[26] G. ZECH, *Nuclear Instruments and Methods* **A277** (1989) 608

[27] H. RUTISHAUSER, *Numerische Mathematik* **5** (1963) 48

[28] W. ROMBERG, *Det. Kong. Norske Videnskapers Selskap Forhandlinger* **28** (1955) Nr. 7

[29] K.S. KOELBIG, in *CERN Computer Centre Program Library*, Program D401, CERN, Geneva 1990

Bibliography

The following short list contains several books which can serve to extend the breadth and depth of the present material. It is of course in no way complete, but instead merely indicates a number of books of varied levels of difficulty.

Probability

L. BREIMAN, *Probability*, Addison-Wesley, Reading MA 1968

H. CRAMÈR, *The Elements of Probability Theory*, Wiley, New York 1955

B.V. GNEDENKO, *The Theory of Probability, 4th ed.*, Chelsea, New York 1967

KAI LAI CHUNG, *A Course in Probability Theory*, Harcourt, Brace and World, New York 1968

K. KRICKEBERG, *Wahrscheinlichkeitsrechnung*, Teubner, Stuttgart 1963

Mathematical Statistics

P.R. BEVINGTON, *Data Reduction and Error Analysis for the Physical Sciences*, McGraw-Hill, New York 1969

B.E. COOPER, *Statistics for Experimentalists*, Pergamon, Oxford 1969

H. CRAMÈR, *Mathematical Methods of Statistics*, University Press, Princeton 1946

W.J. DIXON and F.J. MASSEY, *Introduction to Statistical Analysis*, McGraw-Hill, New York 1969

D. DUMAS DE RAULY, *L'Estimation Statistique*, Gauthier-Villars, Paris 1968

W.T. EADIE, D. DRIJARD, F.E. JAMES, M. ROOS and B.SADOULET, *Statistical Methods in Experimental Physics*, North-Holland, Amsterdam 1971

W. FELLER, *An Introduction to Probability Theory and its Allpications*, 2 Vols., Wiley, New York 1968

M. FISZ, *Probability Theory and Mathematical Statistics*, Wiley, New York 1963

D.A.S. FRASER, *Statistics; An Introduction*, Wiley, New York 1958

H. FREEMAN, *Introduction to Statistical Inference*, Addison-Wesley, Reading, MA 1963

A.G. FRODENSEN and O. SKJEGGESTAD, *Probability and Statistics in Particle Physics*, Universitetsforlaget, Bergen 1979

P.G. HOEL, *Introduction to Mathematical Statistics, 4th ed.*, Wiley, New York 1971

M.G. KENDALL and A. STUART *The Advanced Theory of Statistics, 4th ed., 3 vols.*, Charles Griffin, London 1977

L. LYONS, *Statistics for Nuclear and Particle Physicists*, Cambridge University Press, Cambridge 1986

J. MANDEL, *The Statistical Analysis of Experimental Data*, Interscience, New York 1964

S.L. MEYER, *Data Analysis for Scientists and Engineers*, Wiley, New York 1975

L. SACHS, *Statistische Auswertungsmethoden, 5. Aufl.*, Springer, Berlin 1978

E. SVERDRUP, *Laws and Chance Variations*, 2 Bde., North-Holland, Amsterdam 1967

B.L. VAN DER WAERDEN, *Mathematische Statistik, 3. Aufl.*, Springer, Berlin 1971

S.S. WILKS, *Mathematical Statistics*, Wiley, New York 1962

T. YAMANE, *Elementary Sampling Theory*, Prentice-Hall, Englewood Cliffs, NJ 1967

T. YAMANE, *Statistics, An Introductory Analysis*, Harper and Row, New York 1967

Numerical Methods. Matrix Programs

Å. BJÖRK and G. DAHLQUIST, *Numerische Methoden*, R. Oldenbourg Verlag, München 1972

R.P. BRENT, *Algorithms for Minimization without Derivatives*, Prentice-Hall, Englewood Cliffs, NJ 1973

C.T. FIKE, *Computer Evaluation of Mathematical Functions*, Prentice-Hall, Englewood Cliffs, NJ 1968

G.H. GOLUB and C.F. VAN LOAN, *Matrix Computations*, Johns Hopkins University Press, Baltimore 1983

R.W. HAMMING, *Numerical Methods for Engineers and Scientists, 2nd ed.*, McGraw-Hill, New York 1973

D.E. KNUTH, *The Art of Computer Programming*, 3 vols., Addison-Wesley, Reading MA 1968

C.L. LAWSON and R.J. HANSON, *Solving Least Squares Problems*, Prentice-Hall, Englewood Cliffs NJ 1974

W.H. PRESS, B.P. FLANNERY, S.A. TEUKOLSKY, and W.T. VETTERLING, *Numerical Recipes*, Cambridge University Press, Cambridge 1986

J.R. RICE, *Numerical Methods, Software and Analysis*, McGraw-Hill, New York 1983

J. STOER and R. BURLISCH, *Einführung in die Numerische Mathematik*, 2 Bde., Springer, Berlin 1983

J.H. WILKINSON and C. REINSCH, *Linear Algebra*, Springer, Berlin 1971

Collections of Formulas. Statistical Tables

M. ABRAMOWITZ and A. STEGUN, *Handbook of Mathematical Functions*, Dover, New York, 1965

R.A. FISHER and F. YATES, *Statistical Tables for Biological, Agricultural and Medical Research*, Oliver and Boyd, London 1957

U. GRAF and H.J. HENNING, *Formeln und Tabellen zur Mathematischen Statistik*, Springer, Berlin 1953

A. HALD, *Statistical Tables and Formulas*, Wiley, New York 1960

G.A. KORN and T.M. KORN, *Mathematical Handbook for Scientists and Engineers, 2nd ed.*, McGraw-Hill, New York 1968

D.V. LINDLEY and J.C.P. MILLER, *Cambridge Elementary Statistical Tables*, University Press, Cambridge 1961

D.B. OWEN, *Handbook of Statistical Tables*, Addison-Wesley, Reading MA 1962

E.S. PEARSON and H.O. HARTLEY, *Biometrica Tables for Statisticians*, University Press, Cambridge 1958

W. WETZEL, M.D. JÖHNK, and P. NAEVE, *Statistische Tabellen*, Walter de Gruyter, Berlin 1967

List of Computer Programs

Index